Diesel Engine Repair

Diesel Engine Repair

John F. Dagel
Diesel Department Supervisor
Lake Area Vocational Technical Institute
Watertown, South Dakota

John Wiley and Sons, Inc.
New York Chichester Brisbane Toronto Singapore

Cover Illustration by Lee Leung

Library of Congress Cataloging in Publication Data:

Dagel, John F.
  Diesel engine repair.

  Includes index.
  1. Diesel motor—Maintenance and repair.
I. Title.
TJ799.D33   621.43'68   81-615
ISBN 0-471-03542-4   AACR2

Printed in the United States of America

20 19 18

Dedication

To my wife Marlis, who helped me make the initial decision to write this book and for four long years typed, retyped, and organized the manuscript, artwork, and all the other miscellaneous items that go into putting together a book. Without her assistance this book would not have been a reality.

RUDOLF DIESEL, OF MUNICH, GERMANY

Honorary Doctor of Engineering and of the Technical Sciences, University of Munich.
Director, Verein deutscher Ingenieure. Honorary Member, American Society of Mechanical Engineers. Director and Consulting Engineer, Busch-Sulzer Bros.-Diesel Engine Co., Saint Louis.

Preface

We are currently witnessing the dawning of a new era in diesel engine development and application. The boundary lines of diesel engine usage have been expanded tremendously, and we now routinely find diesel engines in all applications from small single cylinder engines to large electric generator sets.

The diesel engine has come of age because of its high efficiency, simplicity, relatively low emissions, and dependability. It has become one of the most widely used and popular power plants in the world.

This book deals with diesel engines that are normally in mobile applications, although many of the engines used in mobile applications are used in stationary situations such as water pumps and generators. It was written by a mechanic and teacher with a sincere effort to "tell it like it is," and to pass on to the beginning mechanic 25 years of experience and knowledge acquired in the trade.

Each chapter is organized in a manner that will help the student comprehend and remember the material. An introduction briefly describes the material to be presented in the chapter along with motivator to stimulate student interest. After the introduction specific objectives are listed that will help the student determine what will be expected of him or her upon completion of the chapter. A general information section follows, leading the student into the subject matter in an easily understood, direct approach. Component parts and nomenclature make up the next section of each chapter, allowing the student to become familiar with terms and parts before attempting to understand how they work. Then theory of operation, repair and overhaul, assembly and adjustment are presented. Procedures are highlighted for easy reference.

Each chapter has, if applicable, a section on troubleshooting that is intended to aid the student in becoming a proficient mechanic. At the end of the chapter are questions to help the student or instructor review the important parts of the chapter.

Although the chapters are presented in a logical order of development, some instructors may want to change the order of presentation or delete chapters to meet the level of the course or students, or to correspond with the course outline being used. All chapters throughout this book were designed to stand alone, so that one subject or component can be covered without reference to another. Since each chapter is a module in itself, this can be easily done.

Every attempt has been made throughout the book to deal with the practical aspect of becoming a diesel mechanic; therefore, every chapter emphazises the "nuts and bolts" approach dealing with "how to do it" in the correct and acceptable manner. As a result, we believe this book is the most logical and comprehensive guide available for diesel mechanics; we hope you will agree.

John F. Dagel

Acknowledgments

I thank the following people for their contribution to this book: Kenneth Dagel, my brother, for his work on the fuel injection chapters and photographs; Jeanette Bergh for proofreading and encouragement; and A. J. Sherrill for her assistance in typing the final manuscript. I also appreciate the encouragement I received from my family, friends, and associates who helped keep me going when the task seemed too large. In addition I acknowledge the contribution of the following technical reviewers: Michael J. Kelly, Bailey Technical School, St. Louis, Missouri; Richard Biby, Okmulgee, Oklahoma; Roger Miller, Salina Vo-Tech School, Salina, Kansas; Lyman Savory, SUNY-Alfred, Wellsville, New York; James P. Brown, Universal Technical Institute, Phoenix, Arizona; Cliff Bierwagen, Casper, Wyoming; Fred Barbarossa, Los Angeles Trade College, Los Angeles, California; Dick Radock, Westmoreland County Community College, Youngwood, Pennsylvania; Don Winton, formerly Los Angeles Trade College, Los Angeles California.

The photographs and illustrations used throughout this book are intended to make the book easily understood and useful as a source of information. Many of the photographs and illustrations were supplied by various engine and/or fuel injection equipment manufacturers. I thank the following companies who assisted in the preparation of this book by supplying photographs and illustrations:

American Bosch Division of United Technologies Corporation, Springfield, Massachusetts
Cummins Engine Company, Inc.
Delco Remy, Division of General Motors Corporation
Detroit Diesel Allison, Division of General Motors Corporation
Gould, Inc., Engine Parts Division
International Harvester Company
J. I. Case Company
John Deere and Company
L. S. Starrett Company
Lucas/CAV Limited
Robert Bosch Corporation
Snap On Tool Corporation
Stanadyne/Hartford Division

Contents

Diesel Engine Repair

1

The Development of the Diesel Engine

A diesel engine is an internal combustion engine. This means that combustion (burning) of fuel occurs within the engine cylinder. Diesel engines are available in many cylinder arrangements and sizes (Figure 1.1). Since 1895, when the first practical diesel engine was developed by Dr. Rudolf Diesel, diesel engines have been a source of reliable, efficient, long-lasting power.

OBJECTIVES

Upon completion of this chapter the student will be able to:

1 Trace the development of the diesel engine from its beginning to the present.

2 Explain why the development of a small lightweight fuel system makes the diesel engine more attractive as a "prime mover" (i.e., a major source of power).

3 List and explain four advantages of a diesel engine.

4 List five applications of the modern diesel engine.

5 List and explain three disadvantages of a diesel engine.

GENERAL INFORMATION

The modern day diesel engine is a direct result of developmental work that started in 1794 when an inventor named Street developed the internal combustion engine. His basic ideas were further developed in 1824 by a young French engineer named Sadi Carnot. Although Carnot did not actually build an engine, he presented ideas that were to be utilized in building the diesel engine. He stated that highly compressed air created by a compression ratio of 15:1 would generate enough heat to ignite dry wood. He also suggested that the air used for combustion be highly compressed before ignition. Engines previous to this time had used air for combustion at atmospheric pressure. He also suggested that the cylinder walls should be cooled since the heat from combustion would make them very hot, affecting the operation of the engine.

In 1876 Dr. Nickolaus Otto constructed the first four stroke cycle internal combustion engine that ran on gasoline using flame ignition. This engine used ideas suggested, but not tried, by other engineers. It was successful and became the model for all succeeding four stroke cycle engines, gas as well as diesel.

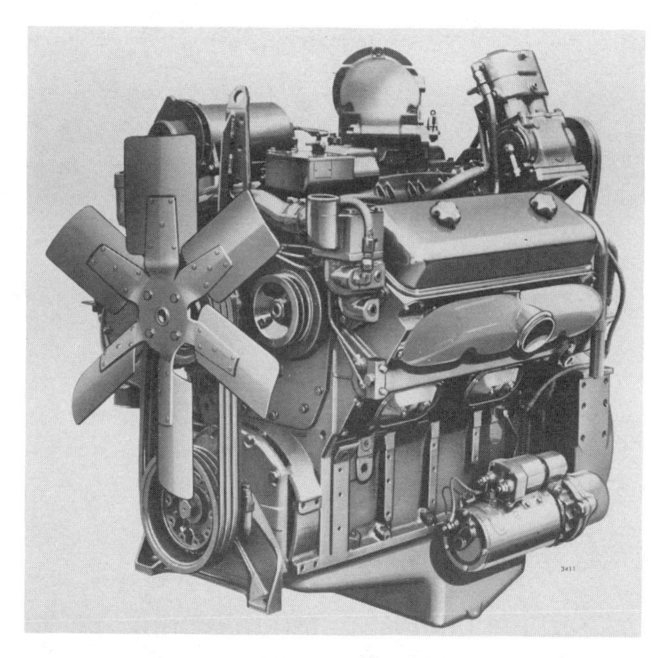

Figure 1.1 Several modern diesel engines.

In 1892 Dr. Rudolf Diesel, a young German engineer, patented a compression ignition engine that attempted to use coal dust as fuel. Coal dust proved to be an unworkable fuel, presenting many problems that caused Diesel to search for another fuel. Lamp petroleum was selected to replace coal dust, and the development of the compression ignition fuel burning engine as we know it today was underway. After much trial and error, Dr. Diesel's third engine was considered a success. It used the compression ignition principle with air fuel injection. Within two years the diesel engine had been adopted as a power source for many applications in Germany.

The catalyst that provided the initiative for continued research and development of the diesel engine was the same in the early 1900s as it is today, that is, the search for a more fuel efficient engine because of the high cost of fuel. In the early 1900s the gas engine or "petrol engine," as it was known, had been king. With the successful operation of Dr. Diesel's compression ignition fuel burning engine a new era dawned. Application of this new engine as prime mover in applications when the petrol engine had been used was not an easy task. The huge size and tremendous weight of the diesel made it impossible to use in any but stationary situations.

The development of a small compact fuel injection system by the German inventor Robert Bosch in 1927 helped to free the diesel engine from one if its major limitations—the fuel system's size. It was compact (barely larger than a carburetor), lightweight and contained a built-in governor. Now all the bulky compressors, reservoirs, lines, and control valves used by the previous air-injection systems could be eliminated. Engine control under varying loads and speed conditions was equal to or better than the carburetor used on petrol engines.

With this major limitation removed, engine manufacturers began seriously considering the diesel engine as a power source. Work began in earnest to further simplify and refine the diesel engine.

As the diesel came to the United States, engines manufactured by Cummins, Caterpillar, General Motors, and others started to impress vehicle owners by their ruggedness and fuel economy.

For many years the engine stayed in the heavy equipment and large truck market, gaining popularity and proving in the face of skepticism that it was a dependable engine.

In the 1950s the development of a smaller rotary lightweight pump by Vernon Roosa paved the way for the diesel's entry into another major field of application, farm tractors. Almost overnight farm tractors around the world became diesel powered. The diesel engine has continued to be developed, entering new markets and experiencing greater usage.

Figure 1.2 A Caterpillar crawler tractor.

A recent example of diesel engine development is the addition of a small lightweight turbocharger (an exhaust-driven air pump that pumps air into the intake manifold). This allowed further reduction of engine weight and a decrease in exhaust smoke, making the engine what it is today, a highly efficient, lightweight, pollution-free engine.

Modern diesels are found in every application and play an important part in our daily lives. Examples of this are:

1 *Construction* (Figure 1.2). Almost all construction equipment is powered by diesel engines. This equipment is used to build roads, buildings, and manufacturing plants.

2 *Power generation* (Figure 1.3). Many millions of kilowatts are generated annually by diesel engines in municipal and privately owned generating

Figure 1.3 Generator set powered by a diesel engine.

Figure 1.4 A typical diesel truck and car.

plants. Important and critical operations, such as hospitals, civil defense, and national defense installations, have standby power through the use of diesels that is instantly available in case of an interruption of the normal power supply.

3 *Transportation* (Figure 1.4). Within the last 20 years almost all trucks and buses have been equipped with diesel engines. Recently the small, high speed, lightweight diesel is being applied to passenger cars and light trucks. This move has been given momentum by the energy crisis and the threat of polluted air created by the gas engine. Many industries depend on fast, efficient diesel transportation, such as the movement of raw material to manufacturing plants and the transportation of food products from the farm to market.

4 *Food production* (Figure 1.5). The United States has the most efficient and dependable agricultural system in the world. This can be attributed in part to the diesel-powered farm tractor. In sizes from 15 to 400 hp the tractor has given the farmer a versatile tool to work the land. Today 90 percent of the farm tractors in the United States are diesel powered. In addition, combines, balers, and many other farm implements are diesel powered.

Figure 1.5 A diesel farm tractor.

5 *Aerospace and national defense*. Many of our country's first line defenses have adopted the engine as a power source to generate electrical power. The space program that put a man on the moon used diesel engines to move the huge rockets into place for takeoff and to provide launch power. Thousands of military vehicles are diesel powered and some use multifuel engines that employ concepts developed for the diesel engine.

6 *Marine*. Almost all small fishing boats and pleasure craft use diesel engines. The diesel engine has proven to be a reliable power source in these applications.

7 *Irrigation and water pumps*. Many thousands of acres of land are now producing food for a hungry world as the result of irrigation. Many of these irrigation systems are powered by diesel engines.

Why has the diesel engine become so popular as a power plant? There are many reasons but here are the major ones:

1 *Efficiency*. Diesel engines burn less fuel to generate a given amount of horsepower. As a result they are 20 to 30 percent more efficient than a comparable gasoline engine.

2 *Engine durability*. Because a diesel engine must be built heavier to withstand the pressure within the engine, it can be expected to run many hours longer than a gas engine.

3 *More lugging power.* Torque rise under load is an inherent characteristic of diesel engines. As diesel engines are pulled under load the resulting increase in torque is brought about by an increase in volumetric efficiency and extra fueling devices in the injection pump. This feature makes the engine a very desirable power source.

4 *Less pollution.* The exhaust from a diesel engine, although visible at times, contains low levels of toxic elements that may be harmful to people. The visible exhaust smoke seen coming from a diesel exhaust in modern engines has been reduced to a light haze and is made up of particles (carbon or soot). In the last 10 years great strides have been made toward eliminating this smoke.

5 *Dependability.* Diesel engines are not affected by wet weather (because they do not use an ignition system) or extreme heat as gas engines might be.

Although many things about diesel engines are positive, there are some disadvantages:

1 *Horsepower-to-weight ratio.* Although much improvement has been made over early engines, the horsepower-to-weight ratio is greater than gas engines because of the heavy parts required in a diesel.

2 *Hard starting in cold weather.* Again, many improvements have been made in cold weather starting, but diesels still do not start as well as gas engines.

3 *High initial purchase price.* Because of the more complex fuel system and heavier parts, diesel engines cost more than gas engines.

4 *Availability of service and repair parts.* Many local repair stations do not have service personnel or parts for diesel engines at present.

In time these disadvantages will be further minimized or eliminated altogether. At present it would

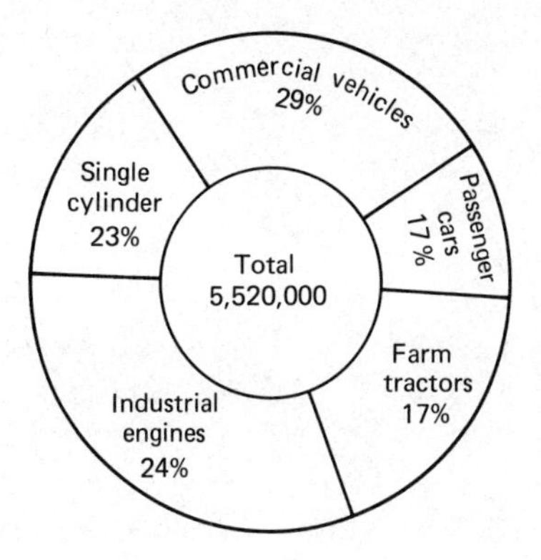

Figure 1.6 World market distribution of diesel engines in 1976 (courtesy Robert Bosch Corporation).

seem that the usage of diesel engines will increase in the foreseeable future. All predictions are that sales of diesel engines will continue to increase rapidly.

Midrange trucks, light duty pickups, and automobiles present the greatest opportunity for growth. Until recently the smaller diesel engine that is used in mid range and smaller vehicles was not sufficiently developed to guarantee successful usage. At the current level of diesel engine development, however, complete diesel saturation is likely in this category.

The present world market for diesel engines is shown in Figure 1.6. In 1976, 5,520,000 diesel engines were produced and sold. Projection into the 1980s in the United States alone indicates a total sale of approximately 700,000 to a million per year. All diesel engine and fuel injection equipment manufacturers are expanding plants or building new ones to keep up with the demand for their products. The world market for diesel engine sales is expected to double by 1985.

In view of the rapid growth rate predicted, the diesel field offers many career opportunities. The need for trained service personnel to diagnose, repair, and rebuild the diesel and its fuel injection equipment is increasing. People who enjoy working with machines and fixing them will find this field challenging and rewarding.

REVIEW QUESTIONS

1 Who invented the first practical diesel engine?

2 Why was the development of the fuel system by Robert Bosch an important factor in the development of diesel engines?

3 List five advantages of diesel engines over other types of engines.

4 List four applications of modern day diesel engines.

5 List four disadvantages of diesel engines.

6 In what market is the greatest predicted growth for diesel engines?

2

Operating Principles

Diesel engines resemble gas engines in many ways. The diesel engine's appearance is nearly the same as a gas engine, and at first glance the engine's internal parts resemble those of a gas engine; however, closer inspection reveals that most internal parts are made stronger and heavier to withstand the greater pressures within the diesel engine. The major difference between the two engines is the ignition and fuel systems. Gas engines employ a carburetor, distributor, and spark plugs, while a diesel engine is a compression ignition engine using a fuel injection pump and injection nozzles.

OBJECTIVES

Upon completion of this chapter the student will be able to:

1. List eight engine component parts and give their function.
2. Explain and point out on a chart the physical differences between a gas and diesel engine.
3. Explain and draw a chart of the valve timing sequence in a four stroke cycle.
4. Explain and draw a chart of the valve timing sequence in a two stroke cycle.
5. Write a brief description of the term "valve overlap."

GENERAL INFORMATION

Diesel engines run by using air, fuel, and ignition just like gasoline engines. Differences between gas and diesel engines are:

1. *Type of fuel.* Diesel fuel is a less volatile fuel than gasoline but possesses a greater number of BTUs per gallon. As a result more total horsepower is obtained from a gallon of diesel fuel than from a gallon of gasoline.
2. *Type of ignition.* The fuel and air mixture in a gas engine cylinder is ignited by a spark plug. In a diesel engine the mixture is ignited by the heat from compression.
3. *Fuel and air mixing.* In gasoline engines the fuel and air are mixed in the carburetor and intake manifold. In a diesel engine the diesel fuel is mixed with the air when the fuel is injected into the cylinder.

NOTE Diesel engines have combustion chambers that are specially designed to aid in the mixing of fuel and air. Since mixing must be done immediately following injection of the fuel, combustion chamber design and manufacture are very important factors in the effective operation of a diesel engine.

I. Engine Component Parts

To fully understand the diesel engine and how it works, the correct nomenclature of all engine parts must be known and understood. Figure 2.1 shows a modern diesel engine and its working parts. A part by part description follows:

A *Cylinder block.* This is considered the "backbone" of an internal combustion engine to which all other engine parts are bolted or connected. The block has many drilled and tapped holes for capscrews, which allow other parts to be connected. Also contained in the block are:

1. Bores or saddles for supporting the crankshaft.
2. Drilled bores for supporting the camshaft.
3. Cylinder holes for the cylinder sleeves.

B *Crankshaft and main bearings.* A crankshaft is a long shaft inserted in the bottom of the block with offset crankpin journals or throws formed onto it. It is used to change up and down motion of the pistons and rods to rotary motion. Contained within the crankshaft are drilled passageways that supply oil to the main and rod bearings.

The main bearings are friction type bearings that support the crankshaft in the block.

C *Cylinder sleeves or liners.* Most diesel engines use a replaceable cylinder sleeve so that if the sleeve becomes worn it can be replaced easily without reboring the cylinder or replacing the block. Cylinder sleeves will be one of two types:

1. *Dry type.* A sleeve fitted into a bored hole in the block with no O rings or other sealing devices on it. It is sometimes called a replaceable cylinder.
2. *Wet type.* A cylinder sleeve that fits into the block and comes in contact with the coolant water. Since water is allowed to circulate around it, the sleeve must be sealed at the top and bottom. Sealing is accomplished on the top end by fitting the sleeve into a counterbore cut into the block. O rings made of oil and water resistant neoprene (a synthetic) are fitted to the bottom of the sleeve and prevent the coolant water from leaking into the crankcase or oil pan.

D *Piston, rings, and connecting rod.* The function of the piston and the rings that are fitted in grooves on the piston is the transmission of pressure from the burning fuel and air to the connecting rod that is connected to the crankshaft. The connecting rod's function is as the name implies: connecting the piston to the crankshaft. Holding the piston and connecting rod together is the piston pin, usually a full floating type (this means the pin floats in both the piston and the rod).

E *Camshaft and timing gears.* The camshaft in a diesel engine operates the intake and exhaust valves and in some engines may drive the oil pump and/or injection pump. The camshaft is timed to the crankshaft by a timing gear or camshaft gear that is meshed into a gear on the front of the crankshaft. This drives the camshaft and insures that the engine valves will stay in time with the crankshaft and pistons.

F *Camfollowers.* The camfollowers (sometimes called lifters) are mounted in drilled holes in the block and ride on the cam lobes. Inserted into the camfollowers are long rods or hollow tubes called push rods, which operate the valves.

G *Cylinder head and valves.* The cylinder head's main function is to provide a cap for the cylinder. In addition, it provides a passageway that allows air into the cylinder and allows exhaust gases to pass out. The ports are opened and closed by poppet type valves that fit into guides in the cylinder head.

H *Rocker arms and push rods.* The rocker arms are mounted on a shaft with one end on the valves and the other end on a push rod. Movement of the push rod causes the arm to rock on its pivot shaft, hence the name rocker arm. Push rods are solid or hollow rods that fit into the camfollowers and transmit the cam action of the camshaft.

I *Oil pan.* A pan-shaped cover that bolts onto the bottom of the block and acts as a reservoir for the engine oil.

J *Lubricating oil pump.* Generally a positive displacement type gear (a pump that delivers a given quantity of oil every revolution). It supplies oil under pressure to the engine.

K *Water pump.* A nonpositive displacement centrifugal pump (a pump that does not deliver a given

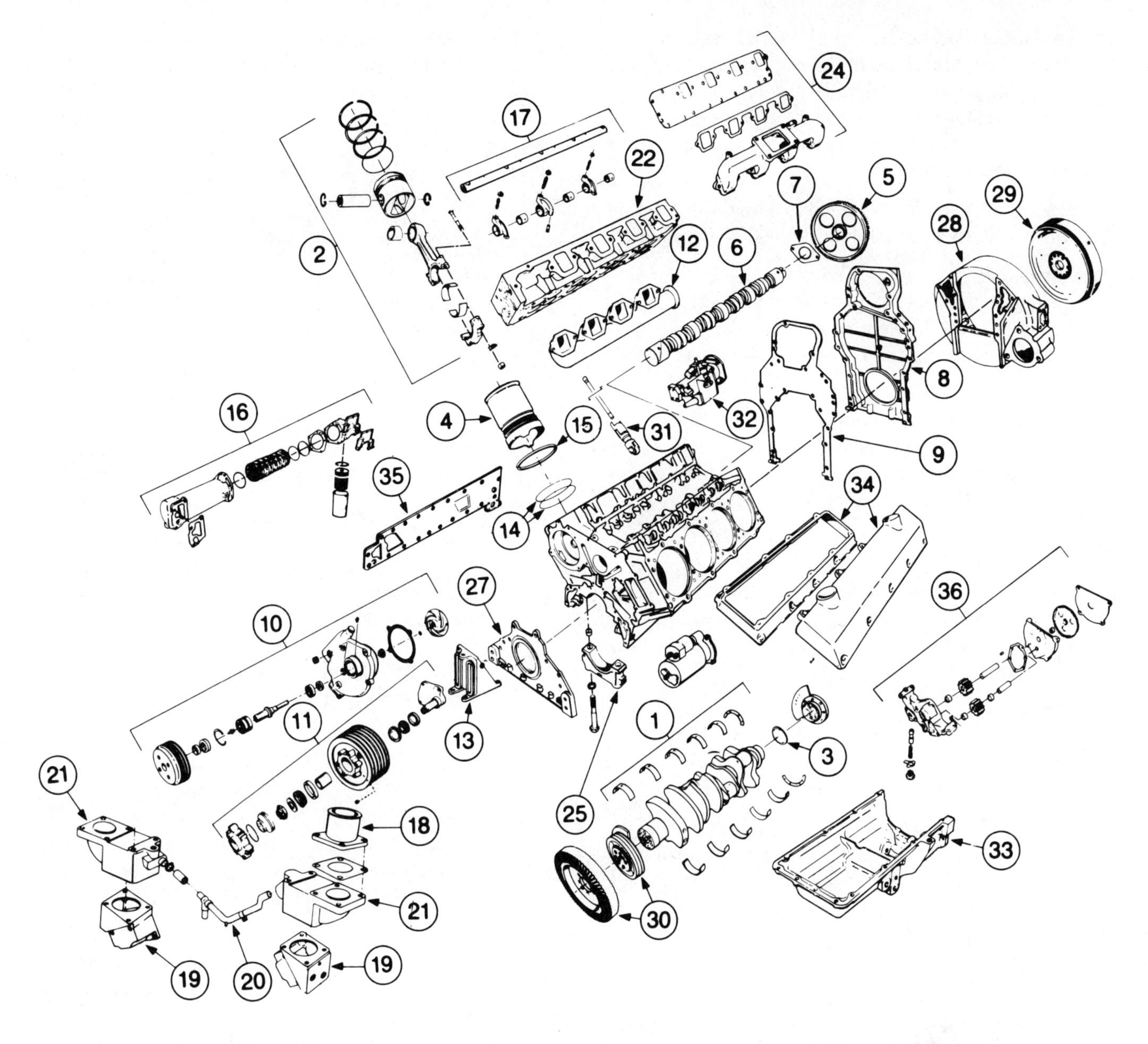

1. Main Bearings
2. Connecting Rod/Piston Assembly
3. Crankshaft Gear And Adapter
4. Cylinder Liner
5. Camshaft Gear
6. Camshaft
7. Camshaft Thrust Plate
8. Gear Housing
9. Gear Housing Plate
10. Water Pump Assembly
11. Fan Hub Assembly
12. Exhaust Manifold
13. Fan Hub Bracket
14. Cylinder Packing Ring
15. Crevice Seal
16. Lube Oil Cooler/Filter
17. Rocker Levers/Shaft
18. Water Outlet
19. Adapter
20. Water By-Pass Tube
21. Thermostat Housing
22. Cylinder Head
23. Cranking Motor
24. Intake Manifold
25. Main Bearing Cap
26. Cylinder Block
27. Front Cover
28. Flywheel Housing
29. Flywheel
30. Vibration Damper/Pulley
31. Tappet/Push-Tube
32. Fuel Pump
33. Oil Pan
34. Cover/Rocker Housing
35. Water Header Cover
36. Lube Oil Pump

Figure 2.1 A modern diesel engine (courtesy Cummins Engine Company).

amount of water every revolution). It aids the flow of coolant water through the engine block and radiator. The pump is generally mounted on the front of the engine block, driven by a V belt from the crankshaft, and connected to the cooling system with rubber hoses.

L *Radiator.* A device designed to allow water to flow through it, thereby cooling the water by radiation.

M *Oil cooler.* A device used to cool the engine oil during engine operation. The construction of the cooler allows coolant water and engine oil to circulate through the cooler simultaneously without being mixed.

N *Flywheel housing.* A round circular housing bolted to the back of the engine that serves as an engine and transmission mount. Enclosed in the flywheel housing is the flywheel.

O *Flywheel.* A heavy metal wheel bolted onto the rear of the crankshaft that provides a place to mount the starter ring gear and the transmission clutch. Engine power impulses are absorbed and stored in the heavy metal wheel that is mounted on the rear end of the crankshaft.

P *Torsional vibration damper.* The main function of the vibration damper is to dampen out torsional twisting vibration that occurs when the engine runs. The damper is mounted on the front of the crankshaft.

Q *Intake manifold.* Bolted to the cylinder head or intake port, the intake manifold provides passageways for the passage of clean air from the air cleaner to the engine.

R *Exhaust manifold.* Connected to the engine exhaust outlets, the exhaust manifold provides a means for collecting the exhaust gases and routing them to the muffler.

S *Fuel system.* The fuel system delivers the correct amount of fuel to the engine cylinders at the correct time, depending on the engine load and speed.

T *Starter.* An electric motor used to start the engine. It is powered by storage batteries.

U *Alternator.* Used to charge the storage batteries and supply current for vehicle lights and other accessories.

V *Turbocharger.* An exhaust-driven air pump that supplies air to the engine under pressure. This pressure boost, called supercharging, increases the engine's efficiency.

II. Engine Operation

Once the engine parts are known by name and their functions understood, the operation of the diesel engine can be studied. Further study of the diesel engine will indicate that most diesel engines are designed utilizing the "four stroke cycle" with a few engines using the "two stroke cycle."

A *Four Stroke Cycle.* The four stroke cycle is a cycle of events that are completed by the engine. This cycle is made up of four strokes of the engine piston that are called:

1 *Intake.* Intake is accomplished in the engine cylinder when the intake valve opens and air is allowed to rush into the cylinder. This air movement is created by the downward movement of the piston that creates a vacuum or void.

2 *Compression.* In a diesel engine only air is compressed during the up-stroke of the piston (compression). Under this intense compression (15:1 to 22:1) the air is heated to as much as 800 to 900° Fahrenheit (427 to 482° Celsius). When the piston reaches the top of its stroke (end of compression) the heated air is ready for fuel to be injected.

3 *Power.* To start the power stroke, fuel is injected by the injector into the cylinder in a highly atomized form. On contact with the heated air the fuel and air ignite spontaneously. As the air and fuel burn, expansion occurs and the piston is pushed down, creating the power stroke.

4 *Exhaust.* As the piston reaches the bottom of its stroke the exhaust valves are opened, allowing the gases to start flowing out. The piston is now on the up-stroke and aids in clearing the exhaust from the cylinder. As the piston reaches the top of its stroke the exhaust valve or valves close and the cycle starts all over again. As the piston begins its downward stroke the intake valve opens, starting the intake stroke.

Figure 2.2 shows valve timing in a four stroke cycle engine, indicating the degrees of crankshaft travel for each stroke. Valve timing, which controls the duration of each stroke in a four stroke cycle engine, is explained below:

1 *Intake valve opens* 10 to 12° before top dead center, as the piston moves down on the intake

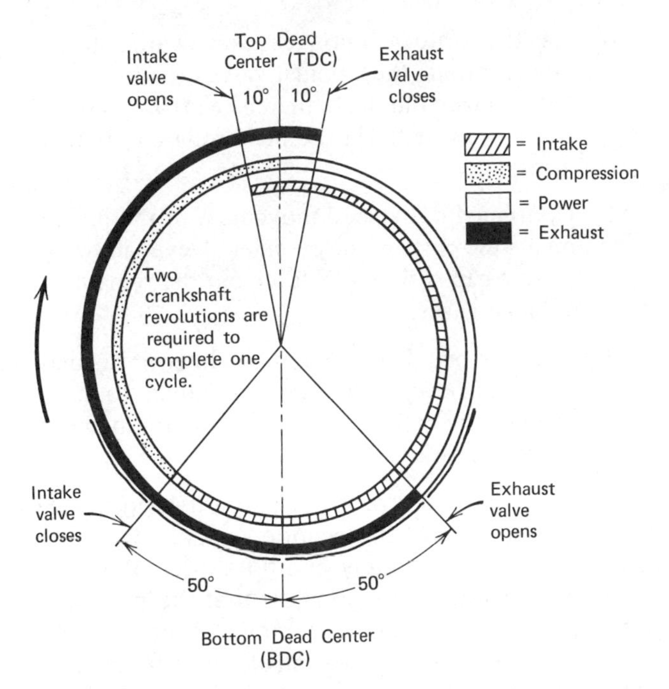

Figure 2.2 A typical four stroke cycle diesel engine valve timing chart.

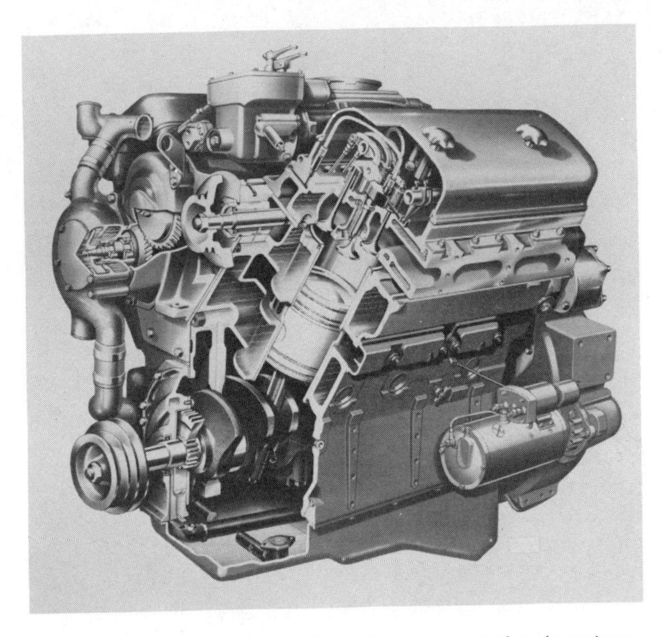

Figure 2.3 A two stroke cycle engine cross section (courtesy Detroit Diesel Allison, Division of General Motors Corporation).

stroke. Intake valve remains open until piston has moved 40 to 50° beyond bottom dead center and started back up.

2 *Intake valve closes* 40 to 50° after bottom dead center, piston moving upward on the compression stroke. Both intake and exhaust valves are closed.

3 *Fuel injection occurs* approximately 25 to 28° before piston top dead center. Both intake and exhaust valves are closed.

4 *After an ignition lag* fuel starts to burn, exerting pressure on the piston.

5 *Power stroke.* As fuel continues to burn and expand, it exerts pressure on the piston, forcing it down. (Both intake and exhaust valves are closed.)

6 *Exhaust valve opens* 40 to 50° before the piston reaches bottom dead center since very little power is lost by opening the exhaust valve early and scavenging of the cylinder gases will be more complete if the exhaust stroke duration is greater than 180°.

7 *Exhaust stroke continues* as piston moves upward. Exhaust valve remains open 10° after piston reaches top dead center, making the total exhaust stroke approximately 200°. At 10° before top dead center the intake valve opens so that the cylinder can start to fill with air for the next succeeding power stroke.

B *Two stroke cycle* (Figure 2.3). Some diesel engines are designed to eliminate the two strokes that pump air and exhaust in a four stroke cycle engine. Two stroke cycle engines have a power impulse occurring every revolution of the crankshaft in contrast to four stroke cycle engines, which have a power stroke every two revolutions of the crankshaft. Theoretically this would mean that the two stroke cycle engine could develop twice as much power as a four stroke cycle engine of the same cubic inches.

In reality this does not happen because a two stroke cycle engine does not breathe very well. In other words, the scavenging or "cleaning out" of exhaust gases from the cylinder after the power stroke is not complete, because of the short period of time available. An explanation of two stroke cycle operation (Figure 2.4) follows:

1 As the piston moves upward, air is forced into the ports in the cylinders – the intake ports are placed about halfway up in the cylinder sleeve (Figure 2.5) from the bottom – and air is forced into them by a gear-driven blower.

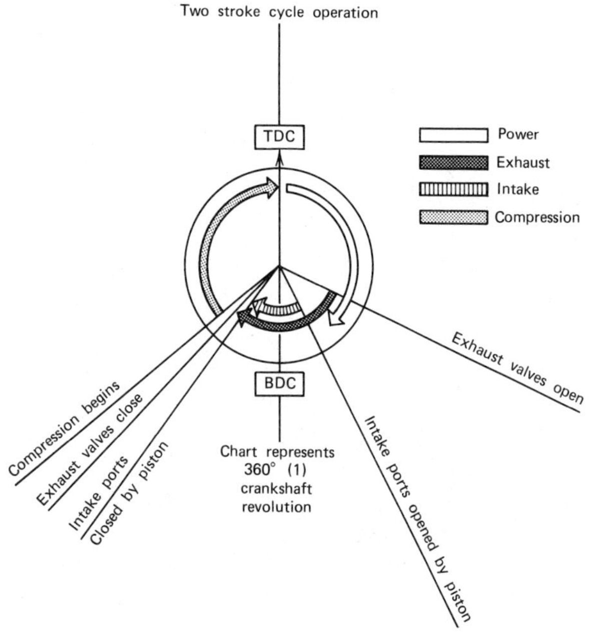

Two stroke cycle operation

TDC

BDC

Power
Exhaust
Intake
Compression

Exhaust valves open
Compression begins
Exhaust valves close
Intake ports
Closed by piston
Chart represents 360° (1) crankshaft revolution
Intake ports opened by piston

Figure 2.4 A typical two stroke cycle diesel engine timing chart.

Figure 2.5 Cylinder sleeve, two stroke cycle engine.

2 As the piston continues to move upward, it closes off the intake ports and compression starts. Continued upward movement of the piston completes compression and heats the air to approximately 800 to 1000° F (427 to 538°C).

3 Fuel is then injected into the heated air and burning starts. The resulting expansion forces the piston downward on its power stroke.

4 As the piston continues downward on its power stroke, the exhaust valves (poppet type valves placed in the cylinder head) are opened by the camshaft via the push rods and rocker arms.

5 Continued downward movement of the piston opens the ports in the cylinder sleeve, allowing fresh air to rush in and force out the remaining exhaust gases.

6 The piston then continues downward and starts back up with air continuing to flow through the cylinder in preparation for the next power stroke.

NOTE The piston's upward stroke, pushing the exhaust gases out of the cylinder, is called scavenging. Unfortunately, the time interval allowed for scavenging is so short that the exhaust gases are never completely removed from the cylinder. As a result, the power stroke that follows is not as efficient because it does not have a full charge of clean air to burn.

C *Valve overlap.* Simply stated, valve overlap is the time when both the intake and exhaust valves are open; this occurs at the end of the exhaust stroke. You will notice that the intake valve was opened 10 to 20° before top dead center and the exhaust valve did not close until 10 to 20° after top dead center. It can be seen, then, that as the piston reaches the top of the exhaust stroke both intake and exhaust valves are open. This period when both valves are open at the same time is known as "valve overlap."

Valve overlap enables the inertia of the exhaust gases to be utilized in creating a vacuum or suction that will start the intake air moving into the cylinder. Filling the cylinder with air after the intake valve opens is of prime importance in the efficient operation of the engine. Most naturally aspirated engines achieve about a 75 to 80 percent cylinder fill. This filling of the cylinder with air (or volumetric efficiency, as it is called) can be increased somewhat by the addition of a turbocharger (an exhaust-driven air pump), which forces air into the cylinder when the intake valve is open.

In addition to utilizing exhaust gas inertia, valve overlap is used to compensate for ignition lag and crankshaft and connecting rod angularity. It also allows the theoretical stroke of the piston to be increased beyond 180°. For example, the intake stroke may be 200° in actual operation of the engine instead of the theoretical 180°.

SUMMARY

Once the combustion cycle is understood for one cylinder, two or four stroke cycle, imagine an engine with many cylinders in various stages of the cycle. Four cylinder engines have a new power stroke starting every 180°, six cylinder engines every 120°, and eight cylinder engines every 90° of crankshaft rotation. The more cylinders an engine has, the closer each succeeding power stroke is to the last. This results in a much smoother operating engine since more than one cylinder is supplying power to the crankshaft at the same time.

The crankshaft functions as the collection point for this power since all cylinder pistons are connected to it via the connecting rods. The total power is then transmitted to the engine flywheel.

Emphasis in this chapter has been on engine operating theory, which must be understood before proceeding into component theory and rebuild. If further questions exist about how the diesel engine runs, consult your instructor.

REVIEW QUESTIONS

1. Although diesel engines resemble gas engines in many ways, list three physical differences between a diesel and gas engine.
2. List and explain each stroke in a four stroke cycle diesel engine.
3. Explain the difference between a four stroke cycle diesel engine and a two stroke cycle engine.
4. List four engine component parts and give their function.
5. Explain the term "valve overlap."
6. Explain why an eight cylinder engine is a smoother running engine than a four cylinder engine.
7. Explain how the intake stroke (in degrees of crankshaft travel) can be lengthened by using valve overlap.

3

Combustion Chamber Design and Operation

When fuel is injected into the combustion chamber of a diesel engine, it must be mixed with air before combustion can occur. The time available for this mixing process is very short; therefore the combustion chamber of a diesel engine must be specially designed. This situation exists only in a diesel engine since gasoline engines start the mixing process (fuel with air) in the carburetor and continue it throughout the intake manifold until it reaches the combustion chamber. Many different types of combustion chambers have been designed and used in diesel engines. This chapter will discuss in detail some of those designs.

OBJECTIVES

Upon completion of this chapter the student will be able to:

1 List and explain two advantages and disadvantages of an open (direct) combustion chamber.

2 List and explain two advantages of a precombustion chamber.

3 Describe in detail why diesel engines have a unique problem with air and fuel mixing.

4 Explain why the mixing of air and fuel in a diesel engine is more important in modern diesel engines.

5 Explain the difference between a turbulence chamber and a precombustion chamber.

GENERAL INFORMATION

Many different types of combustion chambers have been developed over the years for use in diesel engines. As engine speed and horsepower requirements along with increasingly tighter restrictions on diesel engine exhaust emissions evolve, the need for more efficient mixing of the air and fuel at the time of injection becomes evident. Four basic types of combustion chambers are used in today's engine. The operation, inspection and cleaning procedures for these four cells will be discussed in this chapter.

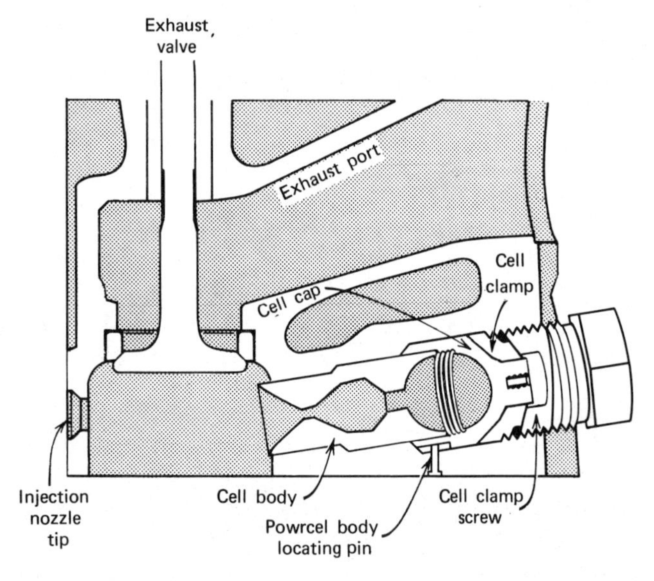

Figure 3.1 Diesel engine energy cell type combustion chamber.

I. Energy Cell

(This cell is sometimes called the Lanova cell after its inventor.) As the name energy cell implies, this chamber is designed to induce a high energy swirl to the air and fuel in the chamber, and generally uses a pintle type injection nozzle. It also allows quieter engine operation. The energy cell will contain about 15 to 20 percent of the total cylinder volume with the piston at top dead center.

A *Component parts and nomenclature* (Figure 3.1).

1. Main combustion chamber.
2. Flat-top piston.
3. A small energy cell (about one-third the size of the main chamber).

NOTE Made up of two parts or halves that can be removed from the cylinder head, the cell is made of much harder material.

4. Passageway connecting the main combustion chamber to the energy cell.

B *Energy cell operation.*

Fuel is injected into the combustion chamber by the injection nozzle, which injects fuel across the chamber at right angles to the piston. Because of the design of the nozzle tip (pintle type) only a small portion of the fuel is injected into the energy cell, which lies directly across the chamber from the injection nozzle. In reality the solid stream of fuel that is contained in the center of the injected fuel is the only part making its way into the energy cell. The remaining fuel is highly atomized and starts to burn almost the instant it hits the heated air in the main combustion chamber.

As the combustion occurs, the piston is forced downward and the air and fuel that were trapped in the energy cell flow back into the main chamber at a very rapid rate. Air and fuel movement create a high degree of turbulence within the combustion chamber, mixing the air and fuel thoroughly for complete combustion.

C *Type of injection nozzle or injector used.*

A pintle type injection nozzle (see Chapter 17 for a complete definition of a pintle nozzle) is used with the energy cell combustion chamber. The nozzle opening pressure is generally in the range of 1800 to 2000 psi (pounds per square inch) (127 to 141 kg/cm^2 kilograms per square centimeter).

D. Procedure for Inspection and Cleaning (Figure 3.2)

The energy cell normally requires very little service but should be inspected on a routine basis when the cylinder head is repaired or the engine overhauled. The energy cell on some engines can be removed and cleaned.

The complete cell must also be removed from the cylinder head when the cylinder head is cracked beyond repair, making it no longer usable. Removal of the cell usually requires a puller of some type.

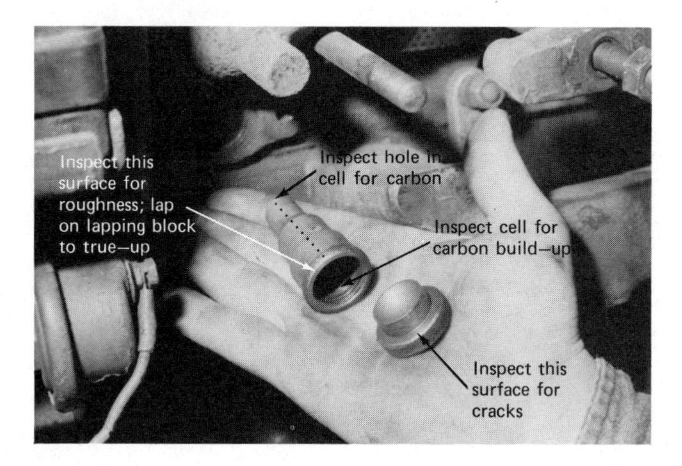

Figure 3.2 Inspection of an energy cell.

1 Inspect the energy cell and its passageway for coking (being plugged with solid carbon).

NOTE A plugged energy cell usually indicates a problem with the engine or injection nozzle, such as excess oil consumption and a worn or misaligned injection nozzle.

2 Inspect the cell closely for cracks or pitting. (Replace if some question about the condition exists.)

II. Direct Injection (Figure 3.3)

As the name implies, a direct injection combustion chamber has a nozzle or injector that injects fuel directly into the combustion chamber. The nozzle is placed in the cylinder head so that the fuel is injected directly on the piston on which some type of swirl-inducing design has been formed. Generally this type chamber uses hole type nozzles of the multihole type.

A *Component parts and nomenclature.*

1 Combustion chamber formed in the piston head.

2 Types. Many different types of piston designs are used in engines that use the open combustion chamber. Two common ones are:

a *Semitoridal,* sometimes called Mexican Hat because of its shape.

b *M.A.N.* (Figure 3.4), a chamber designed in Europe, sometimes called the M chamber. The M.A.N. chamber has a unique bowl-shaped design within the head of the piston. The design is noted for its quietness of operation.

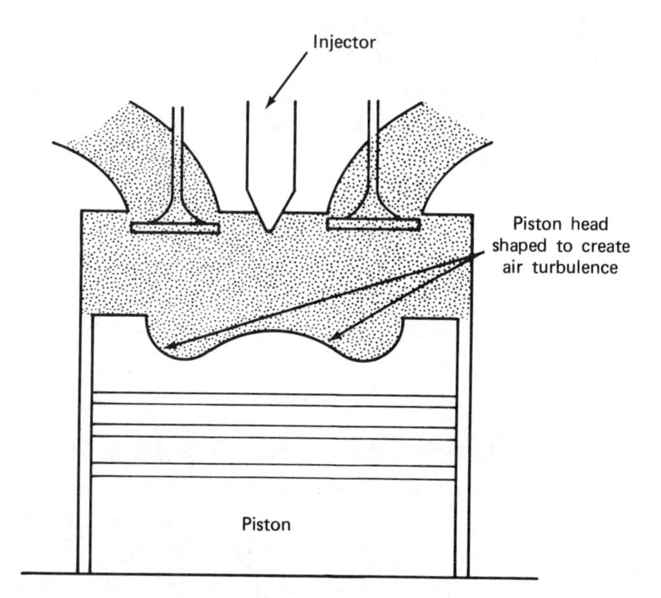

Figure 3.3 Direct injection type combustion chamber.

B *Direct injection operation.*

1 Intake port-induced swirl. As the piston moves downward on the intake stroke, incoming air rushes into the cylinder. Specially designed intake ports cause this incoming air to swirl into the chamber much like a tornado. As the intake valve closes and the piston moves upward, this intake-induced swirl is continued until fuel is injected into the chamber. The incoming highly atomized fuel and the swirling air combine to form a highly volatile mixture that burns cleanly and evenly. Used in many modern engines, the direct injection type engine starts easily and is highly efficient, but it is not as efficient as a precombustion engine in removing exhaust emissions.

2 Piston-induced swirl or squish. As the intake valve closes and piston starts upward on its compression stroke the design on the piston, the Mexican hat, forces the trapped air to rotate or swirl very rapidly by the time the piston reaches the end of its compression stroke. Highly atomized fuel is then injected into the combustion chamber containing the rapidly swirling heated air, and combustion occurs immediately.

C *Type of injection nozzle or injection used.*

A multihole injection nozzle or injector is used with a direct injection combustion chamber design. This multihole arrangement is needed to distribute the fuel throughout the cylinder and to atomize it. Nozzle opening pressures are usually in the range of 2500 to 3000 psi (176 to 211 kg/cm²).

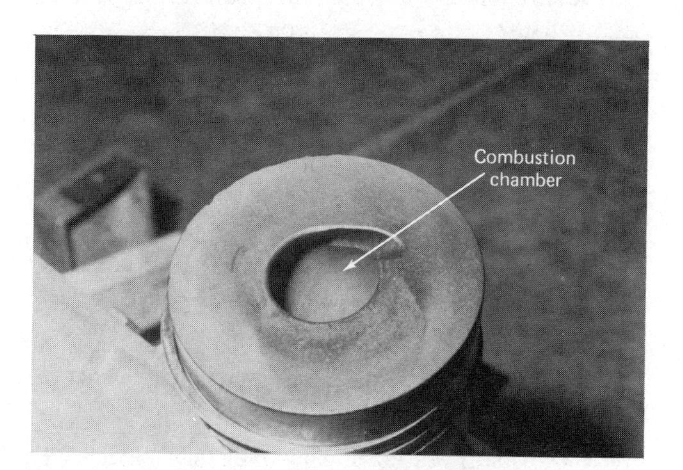

Figure 3.4 M.A.N. type combustion chamber.

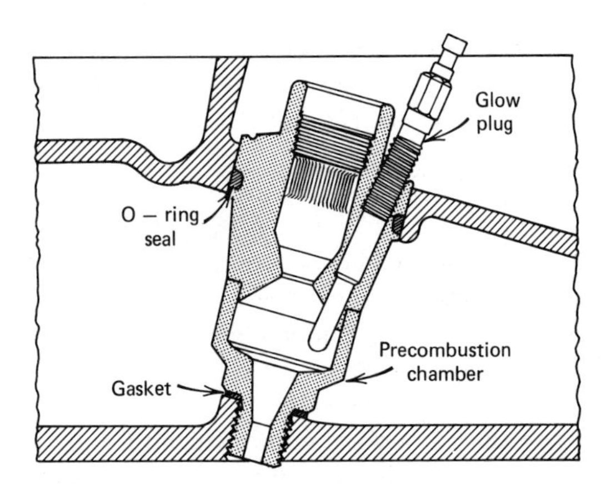

Figure 3.5 A diesel engine precombustion chamber.

D. Procedure for Inspection and Cleaning

The direct injection type combustion chamber requires no special cleaning or inspection during engine overhaul. Normal cleaning of engine parts such as cylinder head and pistons is sufficient.

III. Precombustion Chambers (Figure 3.5)

The precombustion chamber differs from the energy cell in that fuel is injected into the prechamber rather than the main chamber as in the case of the energy cell. The precombustion chamber will contain approximately 20 to 35 percent of the combustion chamber's total top dead center (T.D.C.) volume. Prechambers are connected to the main chamber by a direct passageway.

Precombustion chambers are used on many modern diesel engines and have many advantages such as less exhaust emission and adaptability to various grades of fuel; they also require less atomization of injected fuel. Disadvantages include hard starting and less efficiency. Most prechamber engines are equipped with a cylinder type glow plug for easier starting.

A *Component parts and nomenclature.*

1 A single or two-piece chamber either screwed into the cylinder head or held in place by the injection nozzle.

2 A piston head designed with a concave section in it.

3 In many cases a glow plug (an electrically heated coil or plug that aids in starting) is threaded into the nozzle body or holder and protrudes into the prechamber.

B *Precombustion chamber operation.*

As the piston reaches the top of its compression stroke, heated air is trapped in the main chamber and in the prechamber. At this point fuel is injected into the precombustion chamber. Although the mixture (fuel and air) in the prechamber is excessively rich at the point of injection, burning begins and the rapidly expanding fuel and air rush through the connecting passageway into the main chamber where burning is completed. As can be seen, the fuel and air mixture rushing from the prechamber into the main chamber cause a high degree of turbulence and create a mixture of air and fuel that will burn evenly and cleanly.

C *Type injection nozzle or injector used.*

Precombustion chamber engines use a single or double hole nozzle since atomization requirements are not great. Nozzle opening pressure can be greatly reduced also. Common opening pressures are 1800 to 2000 psi (127 to 141 kg/cm^2) as opposed to 2500 to 3000 psi (176 to 211 kg/cm^2) in direct injection engines.

D. Procedure for Inspection and Cleaning

Prechambers require some special attention when the engine is being serviced since they are not a part of the cylinder head. Many of them are fitted into the water jacket and have O rings and copper gaskets that must be replaced whenever the chamber is removed. Special tools are required to remove and/or replace most prechambers. Check your service manual or with your instructor for the proper procedures.

NOTE Do not overlook the prechamber gaskets or seals when overhauling the cylinder head. Many prechambers are fitted into the water jacket and can cause water leakage into the main combustion chamber.

IV. Turbulence Chambers (Figure 3.6)

A turbulence chamber is very similar to a precombustion chamber in that it is a separate, smaller chamber connected to the main chamber. It differs in that it usually contains approximately 50 to 75 percent of the T.D.C. cylinder volume and is connected to the main

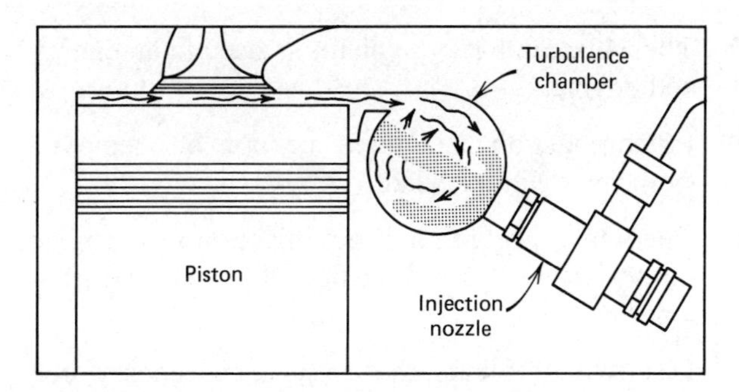

Figure 3.6 Turbulence chamber (courtesy of Deere & Co.).

chamber with a passageway that may run at right angles to the main chamber.

A *Component parts and nomenclature.*

1 Turbulence chambers may be an integral part of the cylinder head or, like the precombustion chamber, may be a separate part that is installed into the cylinder head.

2 They usually have flat-top pistons since the fuel and air mixture does not strike the piston at a right angle when it leaves the chamber. In most cases the passageway is designed so that the fuel and air mixture will enter the chamber parallel to the top of the piston or at a 15 to 20° angle.

3 Engines with this type of turbulence chamber may use a cylinder glow plug for ease in starting.

B *Turbulence chamber operation*

As the piston reaches the top of its compression stroke, air is trapped in the turbulence chamber and the main combustion chamber. Fuel is injected into the turbulence chamber where burning occurs immediately and the resulting expansion forces the air and fuel mixture into the main chamber with considerable force and speed. Because of the design of the passageway connecting the chamber with the main combustion chamber the fuel and air mixture enters the main chamber at an angle and creates a high degree of turbulence in the main chamber. This turbulence aids in mixing the fuel with the air, enabling complete combustion.

C *Type of injection nozzle or injector used.*

A single or double hole nozzle is used in most turbulence chamber engines. This chamber is somewhat similar in operation to the precombustion chamber. A high degree of atomization is not required. Nozzle opening pressure is usually in the range of 1800 to 2000 psi (127 to 141 kg/cm^2).

D. Procedure for Inspection and Cleaning

Inspection and cleaning should be done when the cylinder head is serviced. Turbulence chambers may be fitted into the cylinder head similar to precombustion chambers so that they may need new gaskets and O rings placed on them during cylinder head overhaul. Some turbulence chambers are an integral part of the cylinder head and cannot be removed. They should be inspected and cleaned of all carbon.

SUMMARY

This chapter has attempted to unravel the complexity of prechambers, precombustion chambers, and energy cells, since in actual practice they look much the same and confusion exists in the repair field about them. Remember that the primary reason for any type of prechamber or energy cell is the diesel engine's necessity to mix fuel and air together promptly for efficient engine operation.

REVIEW QUESTIONS

1 State two advantages of a direct injection type combustion chamber.

2 List and explain one disadvantage of a direct injection type combustion chamber.

3 Explain the difference between a precombustion chamber and a turbulence chamber.

4 State two reasons why diesel engines have a special problem in regard to mixing the fuel and air in the combustion chamber.

Questions 5 to 11 are true-false questions.

5 The injection nozzle in a precombustion chamber engine injects fuel directly into the main combustion chamber.

6 The energy cell type chamber contains 50 to 80 percent of the TDC (top dead center) cylinder volume.

7 Direct injection engines usually use flat-top pistons.

8 The Mexican hat type piston is used in an energy cell engine.

9 Precombustion chambers cannot be removed from the cylinder head.

10 The intake ports in a direct injection type engine do not aid in the combustion chamber air turbulence.

11 List two problems that may cause energy cell coking or carboning up.

4

Hand Tools, Test Equipment, and Shop Equipment

One of the most important aspects of learning to become a mechanic is remembering the names and correct use of the hand tools and test equipment that are available. Until you gain experience, tools and testing equipment will be an invaluable aid to you. As you gain experience, situations will become commonplace and some test equipment will not be needed for simple problems. For now, as you start your education as a mechanic, follow the recommendations for use of tools, test equipment, and safety devices to the letter. Some of it may seem repetitious, but repetition builds habits that will stay with you for a lifetime. Work at becoming wrench and bolt "size wise"; for example, can you look at a capscrew or nut and know its size? Once you develop this ability you are well on your way to becoming an accomplished tool handler and mechanic. This chapter will help to familiarize you with hand tools, engine testing equipment, and shop equipment.

OBJECTIVES

Upon completion of this chapter the student will be able to:

1. List and explain several tools that are considered absolutely necessary for a diesel mechanic.
2. List and explain the use of two different sockets.
3. Explain the use of a tubing wrench.
4. List two safety procedures for the use of screwdrivers.
5. List two common types of hammers and their use.
6. List two common uses of a dial indicator.
7. Demonstrate the use of an outside micrometer and explain how to read it.
8. Describe and demonstrate the use of an impact wrench.

9 List and describe two common types of shop equipment.

10 List and identify 15 common hand tools.

11 List and explain two power tools used by a diesel mechanic.

12 List 10 safety procedures that should be followed when using hand tools.

GENERAL INFORMATION

By now you probably have heard the saying, "The mechanic is only as good as his tools." This cliché, although not entirely true, has some truth to it. In most cases, good mechanics have a good selection of tools that fit their style and the equipment they are required to repair. Use caution when buying tools while you are still a newcomer to the trade. Don't buy a large, high-priced assortment just because someone suggested you do so. Start with a small assortment of basic tools and build on it. Many students spend thousands of dollars only to find later that they have bought many tools they do not need. Also guard against becoming a tool collector. Some mechanics buy a tool simply because it looks like something that would be nice to have. Others buy large numbers of tools to cover up a feeling of inadequacy or lack of skill. Tools cost a lot of money and will represent your biggest investment as a mechanic. Buy only what you need. Buy tools *you* like and keep them in good condition.

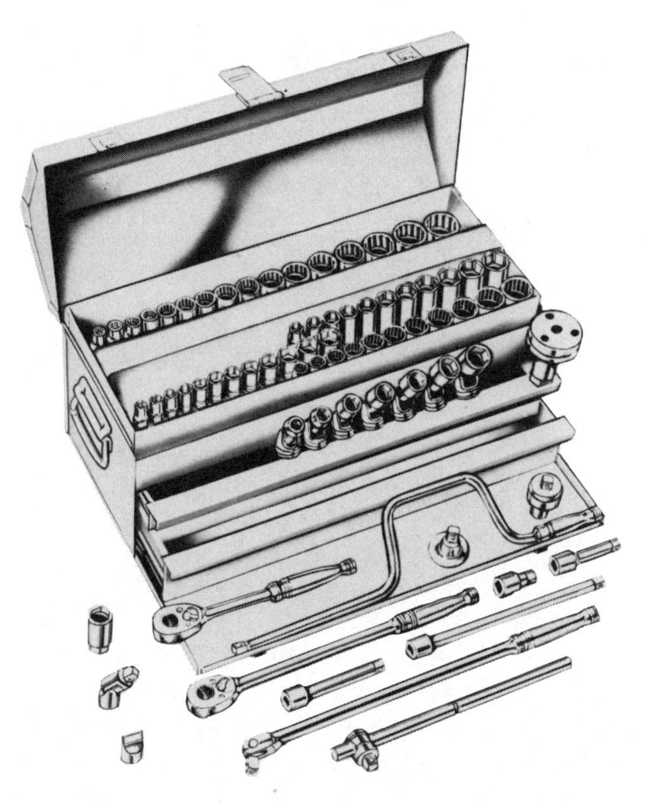

Figure 4.1 Typical socket set with accessories (courtesy Snap-on Tool Corporation).

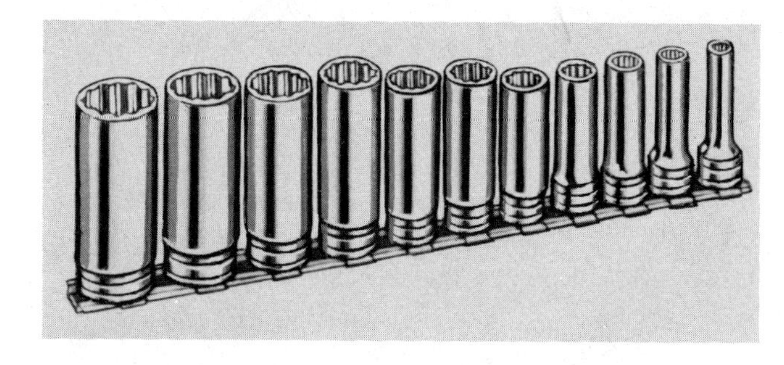

Figure 4.2 Deep socket set (courtesy Snap-on Tool Corporation).

I. Hand Tools

This section deals with the smaller tools that are entirely for use by hand. Hand tools are designed to remove and install all the various fasteners that hold a diesel engine together. These tools make up your basic tool group. Included in this group are the following:

A *Sockets and socket drive handles* (Figure 4.1). A diesel engine mechanic's basic tool set will be a socket set. Sockets are called sockets because they are designed so that a nut or capscrew on which they are used fits into the socket. This socket end may be found in many different configurations, such as twelve point, six point, or square. This socket, when fitted onto the nut or capscrew, must be turned by a handle of some sort since it cannot be turned by hand. Opposite the socket end that fits onto the capscrew or nut is a square hole into which the driving handle is fitted. This square will denote size of the set, for example, $\frac{1}{2}$ in. drive set, $\frac{3}{8}$ in. drive set.

1 Socket sets. These sets will contain many different size sockets, fractional sizes or metric. Socket sets of $\frac{3}{8}$ in. will generally be found in the following drive sizes: $\frac{1}{4}$ in., $\frac{3}{8}$ in., $\frac{1}{2}$ in. (most common), $\frac{3}{4}$ in., 1 in.

2 Special sockets. Two special sockets that are necessary for a mechanic are:

a The deep socket (Figure 4.2). Deeper than a regular socket, the deep socket is for use on nuts that have been turned down on a long bolt or rod.

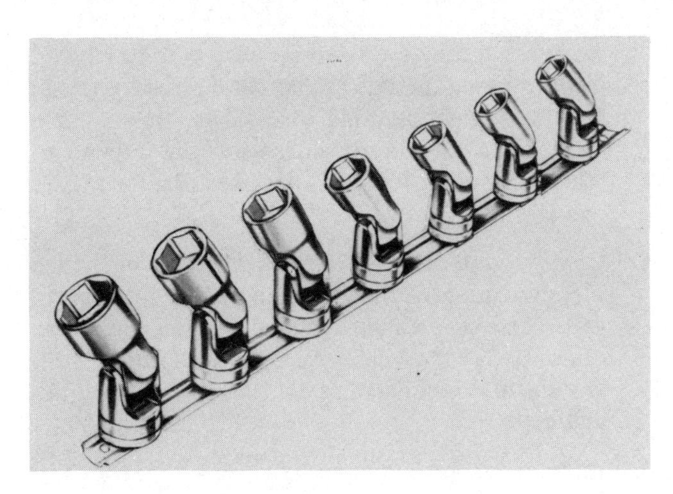

Figure 4.3 Flex sockets ⅜″ drive (courtesy Snap-on Tool Corporation).

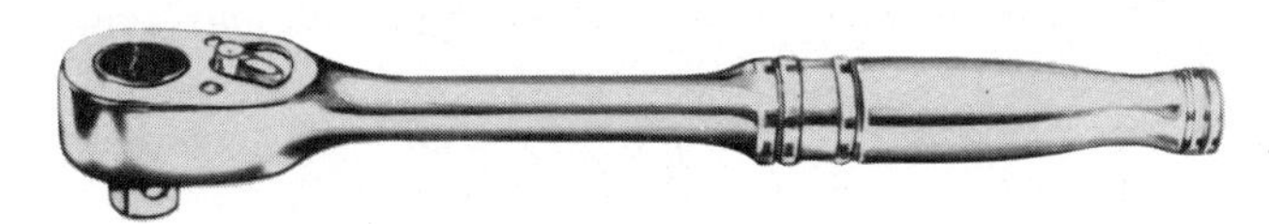

Figure 4.4 A ratchet handle (courtesy Snap-on Tool Corporation).

b Flex sockets (Figure 4.3). The flex socket is a real time-saver, designed like a socket but with a universal joint added to it. When used in conjunction with a speed handle, the flex socket provides speed and flexibility. This set is a must for mechanics and can be bought in sets or individually. Normal sets would include sizes from $\frac{3}{8}$ to $\frac{3}{4}$ in. (9 to 19 mm).

3 Socket set handles and drives. As stated previously, the sockets must be fitted to some type of drive handle so that the nut or capscrew can be turned. Some common socket drive handles are:

a Ratchet (Figure 4.4). A ratchet handle has a spring-loaded pawl arrangement within the head so that as the handle is pulled one way to turn the nut and swing back for a new bite, the handle ratchets. This eliminates the need to remove the ratchet and socket from the nut or capscrew each time you back up with the handle. The ratchet mechanism can be reversed by flipping a small lever on the ratchet head.

b Flex handle or breaker bar (Figure 4.5). A flex handle or breaker bar is a long (approximately 18 in. or 450 mm) rod bar with a swivel head that fits into a socket. It can be used at a 90° angle from the socket or straight up from the socket.

c T handle (Figure 4.6). A T handle is a bar that has a sliding square shank that fits into the socket formed on it. When installed in the socket it forms a T configuration, hence the name. T handles are used on bolts or nuts where torque must be applied with no offside loading of the shaft. Tightening the turbocharger compressor wheel is a good example.

d Speed handle, sometimes called a spinner (Figure 4.7). A speed handle is like a crank with a square drive on one end that fits the socket. The speed handle is very handy in removing bolts or capscrews that are relatively loose.

4 Extensions (Figure 4.8). An extension is a long or short round rod that can be used to extend the square drive end on the handle. This allows the socket to be used in a hole or any other

Figure 4.5 Flex handle (courtesy Snap-on Tool Corporation).

Figure 4.6 A T handle (courtesy Snap-on Tool Corporation).

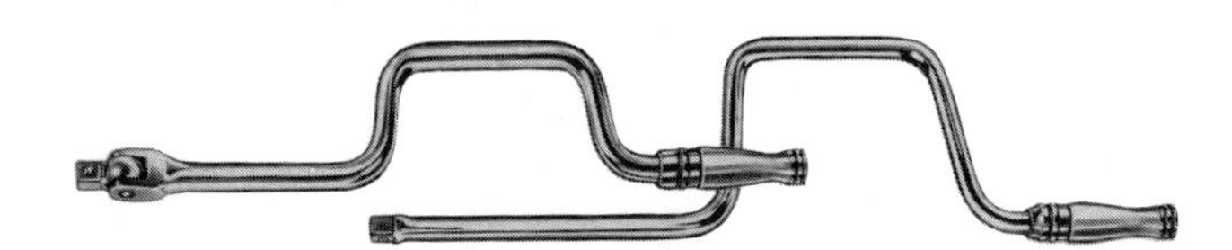

Figure 4.7 Speed handle (courtesy Snap-on Tool Corporation).

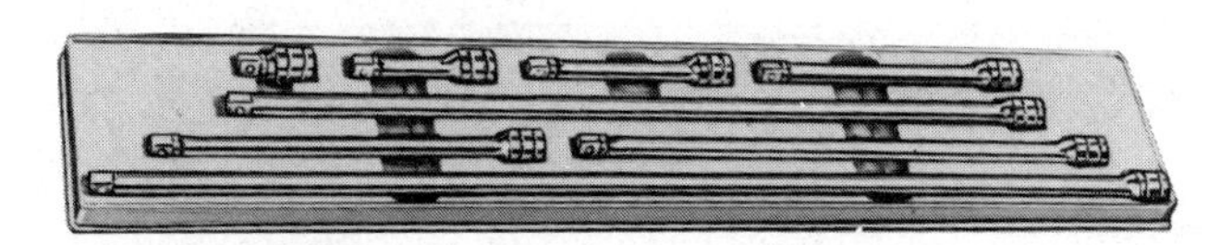

Figure 4.8 Extensions (courtesy Snap-on Tool Corporation).

place where a normal ratchet—socket combination could not reach. Extensions come in many sizes. Some common ones are 3 in. (75 mm), 6 in. (150 mm), 10 in. (250 mm).

B *Box wrenches* (Figure 4.9). The box wrench has a socket head on each end that will fit a nut much like a socket. This wrench is available in either a twelve or six point, and usually the socket ends are of different sizes, for example $\frac{1}{2}$ in. and $\frac{9}{16}$ in. (12 mm and 14 mm). A twelve point box wrench can be used in tight, hard-to-reach places since very little swing (about 15°) is needed to get a new bite on the nut or capscrew. Box wrenches can be purchased individually or in sets. A standard box wrench will be flat with an angle or offset to provide working clearance. Many box wrenches are offset so that the handle is not in line with the head. A variation of the standard box wrench is the "tubing wrench" or "flare nut" (Figure 4.10), so-called because of its intended use for removal of tubing nuts and fittings. The tubing wrench is much like a box wrench except that it has a cut-out portion that allows it to be slipped over a line and onto the nut. This type of wrench allows more nut contact than a straight box wrench would. The tubing wrench is a recommended tool for all mechanics.

C *Open end wrenches* (Figure 4.11). Open end wrenches are designed much like box wrenches except that they are open on each end. Since you have to move the nut 30° or more before you can flop the wrench and get a new bite, this is not a convenient wrench for places that are difficult to reach. This type wrench is not used as often as box end wrenches.

D *Combination wrench* (Figure 4.12). This wrench is a combination of the box and open end wrench, with a box head on one end and an open end on the other. Both ends will be the same size. It is a popular, useful wrench and is available in sets or individually.

E *Allen or hex key wrenches* (Figure 4.13). They are shaped like the letter L and used to remove hex head screws or bolts. Usually allen wrenches are bought in sets with 10 to 13 different sizes. This wrench set is a must for the diesel mechanic.

F *Screwdrivers* (Figure 4.14). Screwdrivers are used to turn screws in or out and have various type heads. Many are made of a round shank of different lengths with handles made of plastic or wood.

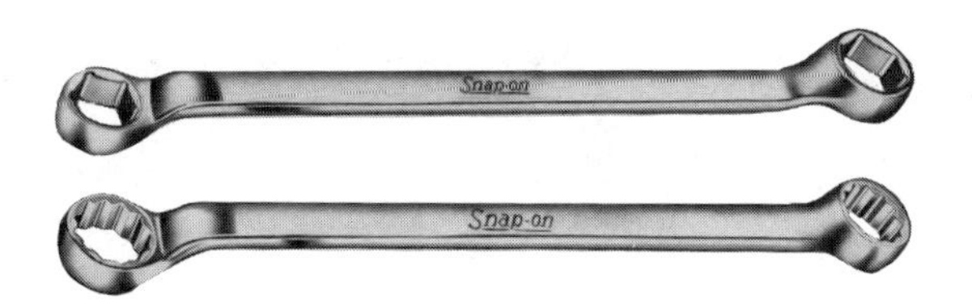

Figure 4.9 Box wrenches (courtesy Snap-on Tool Corporation).

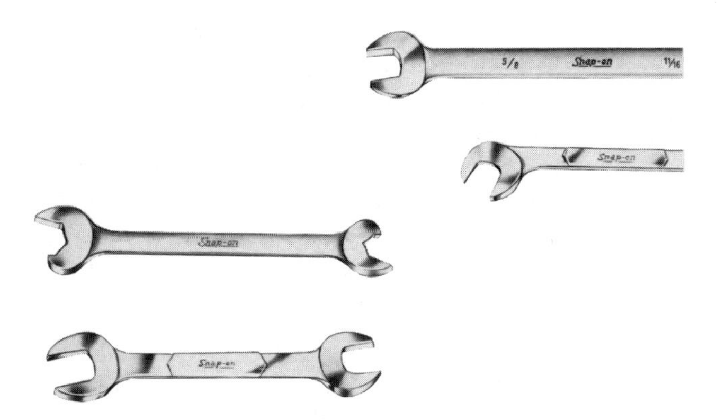

Figure 4.11 Open end wrenches (courtesy Snap-on Tool Corporation).

Figure 4.10 Tubing wrenches (courtesy Snap-on Tool Corporation).

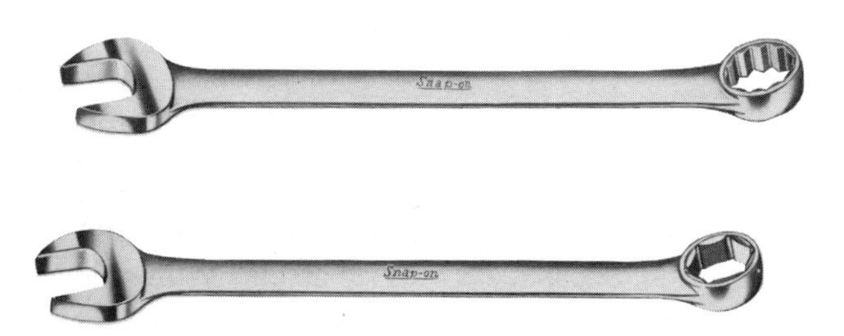

Figure 4.12 Combination wrenches (courtesy Snap-on Tool Corporation).

1. Common. The most often used screwdriver is the slotted head or common screwdriver, which comes in many sizes and lengths.
2. Phillips head. Many screws have a Phillips type recess that requires a Phillips screwdriver. The Phillips screwdriver is available in no. 1, 2, and 3 sizes.
3. Clutch-head (Figure 4.15). The clutch-head screwdriver is not used a great deal in the diesel engine industry but may be encountered occasionally by the mechanic.
4. Other types of screwdrivers. In many situations the common screwdriver will not allow you to reach a screw that is not readily accessible. If this is the case, an offset screwdriver (Figure 4.16) can be used. Another screwdriver, the starting screwdriver (Figure 4.17), can hold the screw on its tip without falling off.

NOTE All types of screwdrivers should be used only on the screws that they are designed to fit.

Some suggestions for screwdriver use are:

1. Do not use a screwdriver as a pry bar.
2. Do not use pliers or vise grip on the shank of a screwdriver to help turn it.
3. Do not drive on screwdriver handles with a hammer.
4. Make sure the slot in the screw head is free from paint or dirt before attempting to remove it.
5. Push downward on the screwdriver and twist at the same time to remove a very tight screw.
6. If you have trouble removing a screw, use a drive pin punch the same size as the screw, place it on the screw head, and give it a sharp tap with a hammer.
7. Keep all straight shank screwdrivers in good condition. The bit must be sharp yet straight to provide a good fit into the screw.

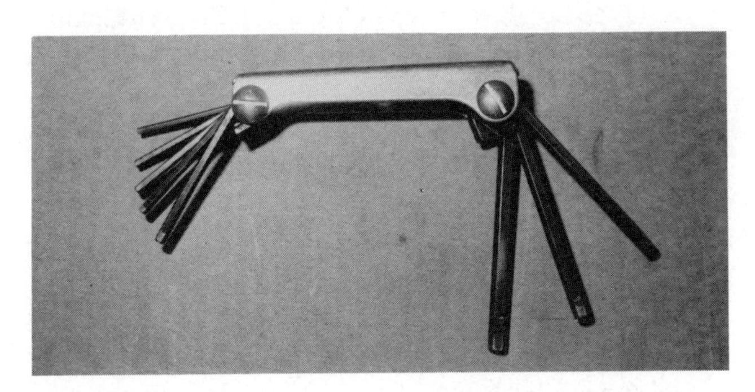

Figure 4.13 Allen or hex key wrenches.

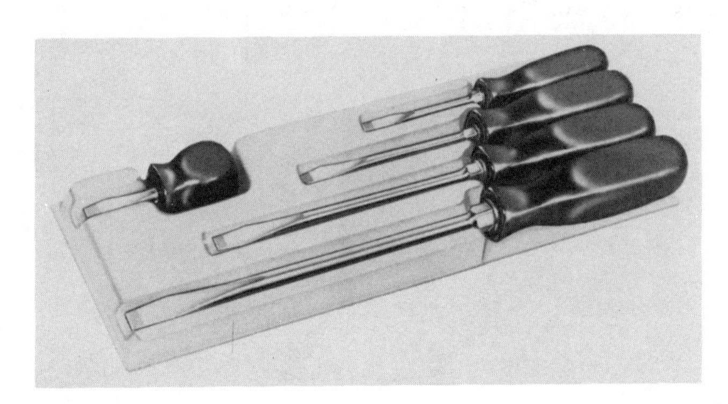

Figure 4.14 Screwdrivers (courtesy Snap-on Tool Corporation).

Figure 4.15 Clutch-head screwdriver (courtesy Snap-on Tool Corporation).

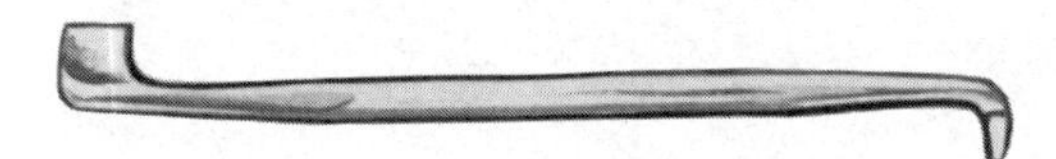

Figure 4.16 Offset screwdriver (courtesy Snap-on Tool Corporation).

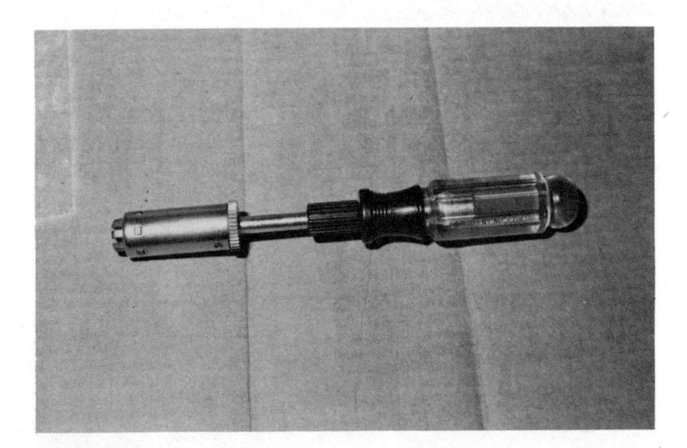

Figure 4.17 Starting screwdriver (courtesy Snap-on Tool Corporation).

8 When placing force on the screwdriver, use caution so that it does not slip from the screw and injure you.

9 Screwdrivers can be dressed or straightened up by using a file or grinding wheel.

CAUTION Use goggles when sharpening screwdrivers on the grinding wheel.

G *Pliers*. One of the common tools you will have in your assortment are the pliers. They come in many different types and sizes and will save you much time. These are some of the pliers you will be required to have:

1 Slip joint pliers (Figure 4.18). These are used for holding and twisting.

2 Channel lock or interlock pliers, sometimes called water pump pliers (Figure 4.19). These are variations of a common slip joint pliers and can be used by mechanics for many odd jobs.

3 Diagonal cutter pliers, sometimes called Dikes (Figure 4.20). These pliers are used for cutting electrical wire, mechanics wire, cotter pins. They also can be used for pulling cotter pins from nuts and shafts, or they may be used for pulling woodruff keys from shafts. They can be purchased in various sizes.

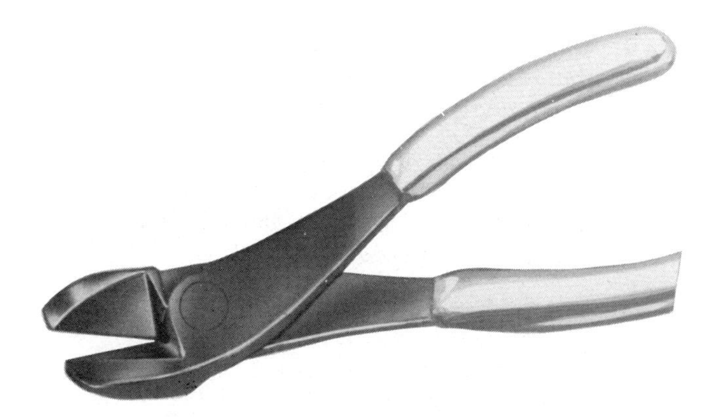

Figure 4.20 Diagonal cutter pliers (courtesy Snap-on Tool Corporation).

4 Needle nose pliers (Figure 4.21). These are used for reaching small objects or parts that cannot be reached by hand. They can be considered an extension of your fingers and are frequently used in electrical work when soldering wires together.

5 Grip lock pliers, sometimes called vise grip after one company that manufactures them (Figure 4.22). They are designed like slip joint pliers but instead of a slip joint they have one adjustable jaw with a snap overcenter mechanism that locks the jaw onto whatever

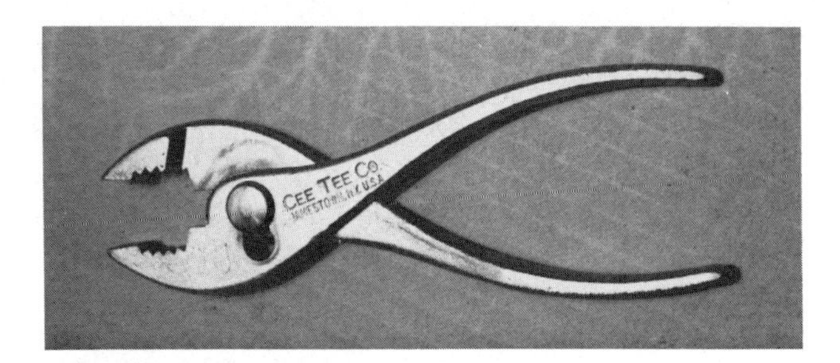

Figure 4.18 Slip joint plier.

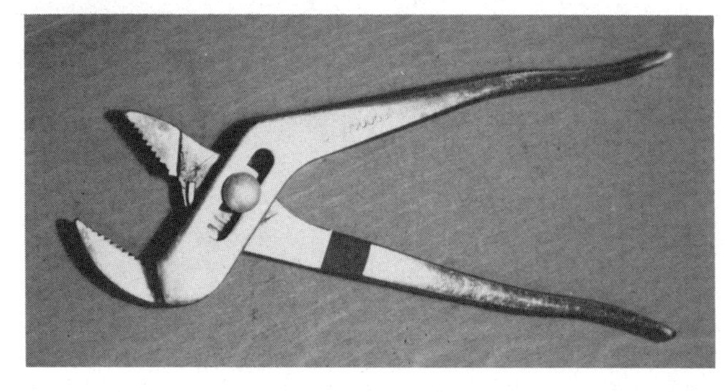

Figure 4.19 Channel lock pliers.

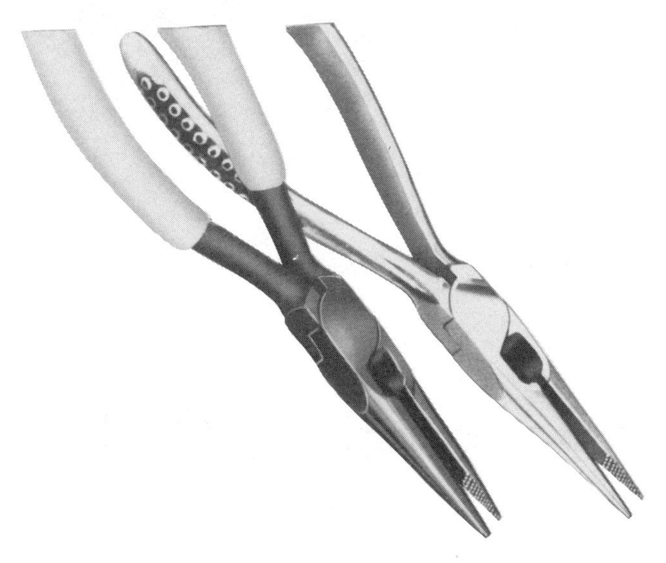

Figure 4.21 Needle nose pliers (courtesy Snap-on Tool Corporation).

you are holding. This tool is excellent for removing ruined or flattened nuts or tubing fittings.

6 Snap ring pliers (Figure 4.23). They are used to remove and install snap rings. There are several different types available for the different types of snap rings that you find used throughout the diesel engine and fuel system.

7 Hose clamp pliers (Figure 4.24). These pliers have specially designed jaws that fit onto the spring type hose clamp. This will save you time and frustration when removing or installing this type of hose clamp.

H *Open end adjustable wrench,* sometimes called Crescent after the manufacturer (Figure 4.25). An open end adjustable wrench is a very handy wrench with a multitude of uses. It has one adjustable sliding jaw that is adjusted by a worm screw arrangement located directly below the jaw opening. The wrench must be used correctly to prevent breaking the movable jaw. It should be placed on the nut or capscrew so that the solid part of the jaw absorbs the force of pull.

CAUTION Make certain the adjustable jaw is adjusted solid against the nut before attempting to loosen the nut, or the wrench will slip.

I *Cold chisels* (Figure 4.26). A chisel is a round or octagonal-shaped rod that has a sharp bit ground on one end. A chisel is used in conjunction with a hammer on any metal that is softer than the chisel. It may be used to cut bolts, split nuts, remove broken-off bolts; and it has many other uses. Cold chisels usually are one of the following types;

1 Flat. A cold chisel may have many different size bits, but they are usually found in $\frac{3}{8}$ in., $\frac{1}{2}$ in., or $\frac{5}{8}$ in. (9 mm, 12 mm, or 15 mm) and are used for multipurpose cutting.

2 Diamond point. This chisel has a three-cornered point.

3 Round nose. The round nose chisel has a rounded bit and can be used in a variety of jobs from removing broken bolts to oil seals.

4 Cape. This chisel can be used to remove keys from a keyway or clean up a damaged keyway.

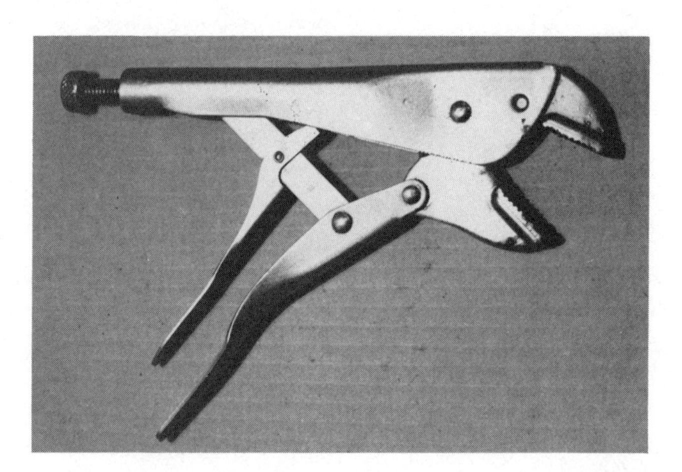

Figure 4.22 Grip lock pliers.

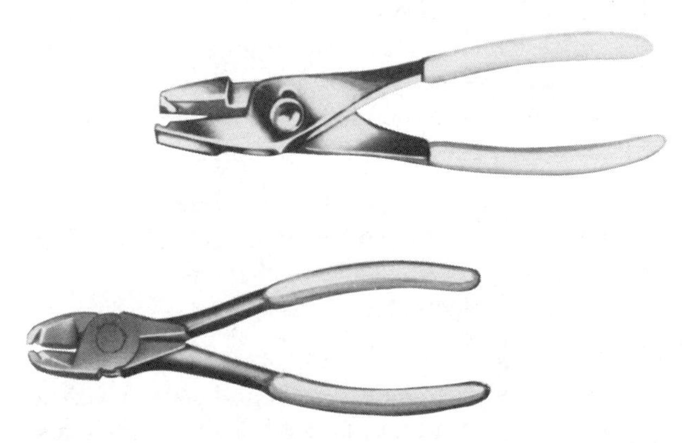

Figure 4.24 Hose clamp pliers (courtesy Snap-on Tool Corporation).

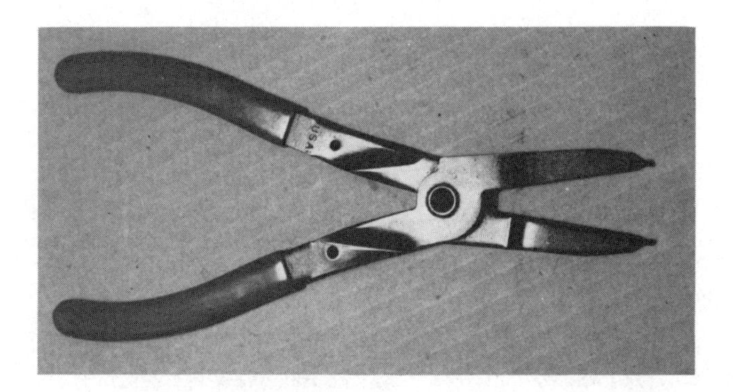

Figure 4.23 Snap ring pliers.

Figure 4.25 Open-end adjustable wrench.

NOTE Chisel heads tend to mushroom after being used. The chisel head along with the cutting edge must be reground at regular intervals so that small metal chips do not break off and injure the person using them.

J *Punches* (Figure 4.27). Punches are used to punch out a bolt or rivet, or center a drill when drilling a hole. The common types of punches are:

1. Pin punch. This punch is shaped like a pin and is available in various sizes from $\frac{1}{16}$ to $\frac{1}{2}$ in (1.5 to 13 mm).
2. Center punch. The center punch is a pointed nose punch used to punch a center on a piece of metal in which you wish to drill a hole.
3. Aligning punch. This is a long, tapered punch used to align two holes in adjacent pieces, such as mounting brackets to engine block and engine motor mounts.

K *Hammers.* A hammer is one tool that is used by a mechanic every day. Hammers are found in many sizes, shapes, and types. Listed below are some of the many types available:

1. Ball peen (Figure 4.28). The ball peen is a universal hammer made for all around use and is available in many sizes and weights. One end of the head is a rounded ball portion that can be used for peening rivets and bolts.
2. Plastic (Figure 4.29). This hammer has a plastic head, used to drive on items that should not be damaged. Most plastic hammers have replaceable plastic ends.
3. No bounce (Figure 4.30). The no bounce hammer, filled with shot pellets and coated with a rubberlike material, is used to drive on items that should not be damaged or scored. Because of the lead filling this hammer does not bounce like an all-rubber hammer. The hammers with wood handles should be checked frequently to make sure the head is securely fastened to the handle. Loose hammer heads may fly off and injure a fellow worker.

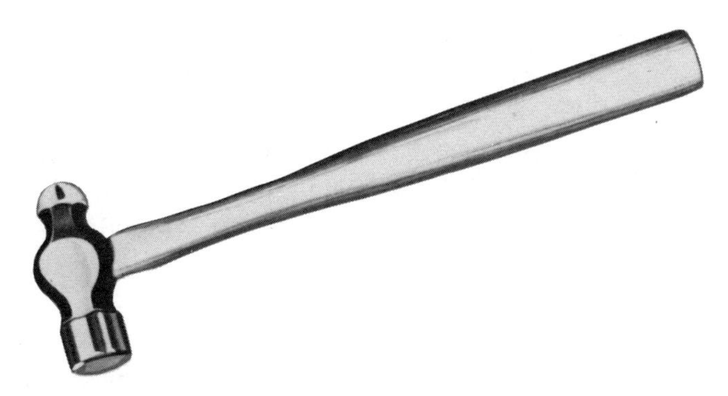

Figure 4.28 Ball peen hammer (courtesy Snap-on Tool Corporation).

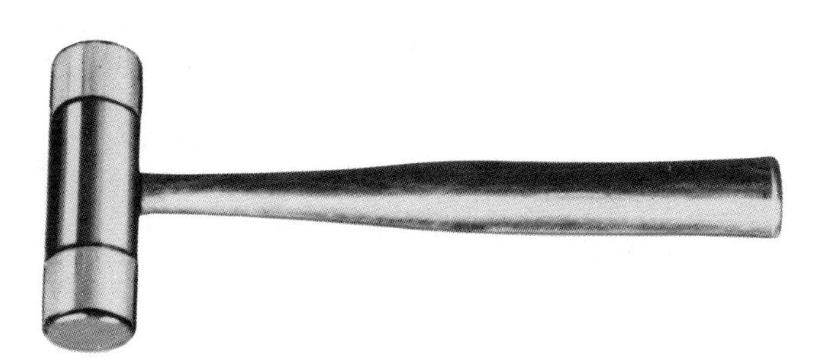

Figure 4.29 Plastic hammer (courtesy Snap-on Tool Corporation).

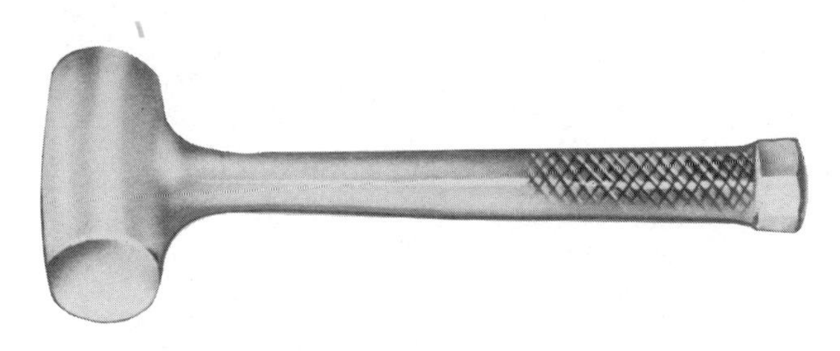

Figure 4.30 No bounce hammer (courtesy Snap-on Tool Corporation).

Figure 4.26 Cold chisel (courtesy Snap-on Tool Corporation).

Figure 4.27 Punch (courtesy Snap-on Tool Corporation).

CAUTION When driving an object that may have sharp, loose metal fragments on it, use safety glasses.

L *Hacksaws* (Figure 4.31). A hacksaw is designed to saw metal. Hacksaws are equipped with a replace-

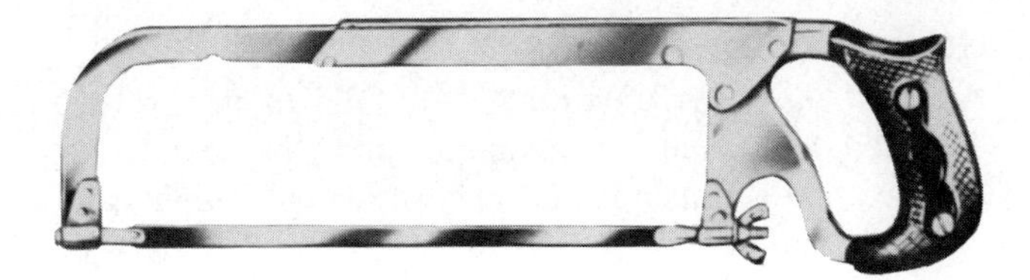

Figure 4.31 Hacksaw (courtesy Snap-on Tool Corporation).

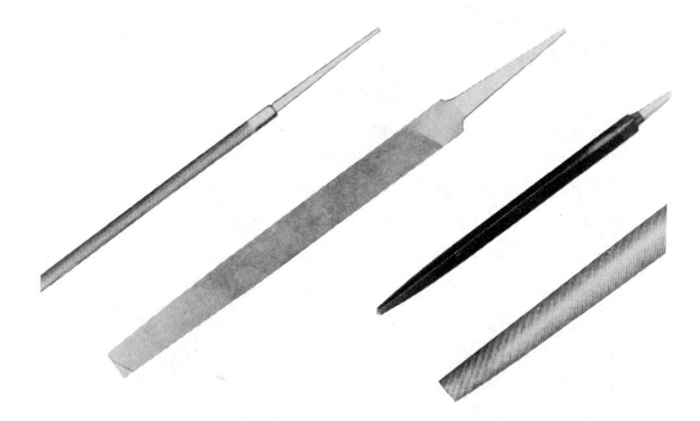

Figure 4.32 Files (courtesy Snap-on Tool Corporation).

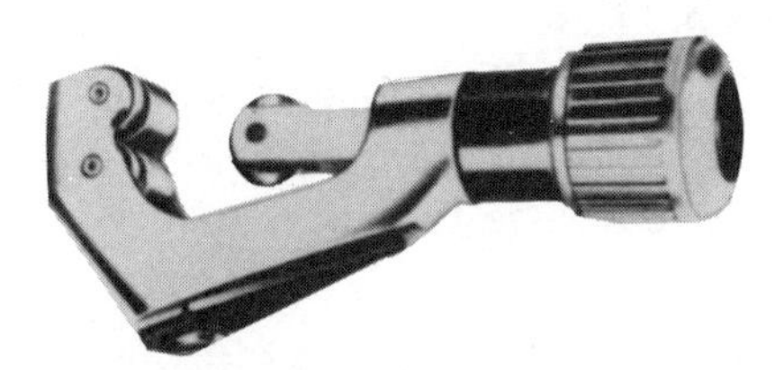

Figure 4.33 Tubing cutter (courtesy Snap-on Tool Corporation).

able blade that can be purchased with different numbers of teeth per inch for cutting and different lengths for cutting various metals. Blades are available with 14, 18, 24, and 32 teeth per inch. Install a new blade with the teeth facing away from you. When sawing with a hacksaw, do not attempt to force it through the metal too quickly. Attempting to do so will ruin the saw teeth and require replacement of the blade.

M *Files* (Figure 4.32). Files are designed for removing small amounts of metal from tools and other metal objects that need straightening. Files are available in many different shapes and tooth design. When using a file, use enough pressure to make the file cut and do not force it. Forcing the file will only leave a rough finish and may ruin the file. Files should be cleaned by tapping them lightly on the handle end or by brushing them with a small wire brush.

N *Tubing cutter* (Figure 4.33). Tubing cutters are used to cut copper or steel tubing. A tubing cutter leaves a smooth, even edge on the tubing when cut. Never use a hacksaw to cut tubing for flaring.

O *Flaring tool* (Figure 4.34). If the end of a copper or steel pipe needs to be flared, a flaring tool must be used.

P *Twist drills* (Figure 4.35). Twist drills are used in a drill to drill holes in metal. Twist drills come in many sizes: fractional, number, and metric. It is common for twist drills to be sold and stored in a drill index (Figure 4.36). When using a twist drill, make sure it is tightened securely in the chuck to prevent damage to the drill. Twist drills must be kept sharp to insure the proper cutting action. Different metals require that the twist drill be ground at different angles. Before doing any extensive drilling it is recommended that you look up the correct drill angles to be used.

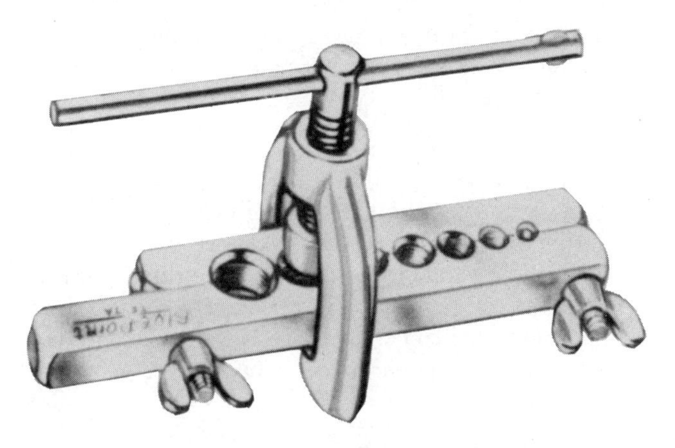

Figure 4.34 Flaring tool (courtesy Snap-on Tool Corporation).

Figure 4.35 Twist drill (courtesy Snap-on Tool Corporation).

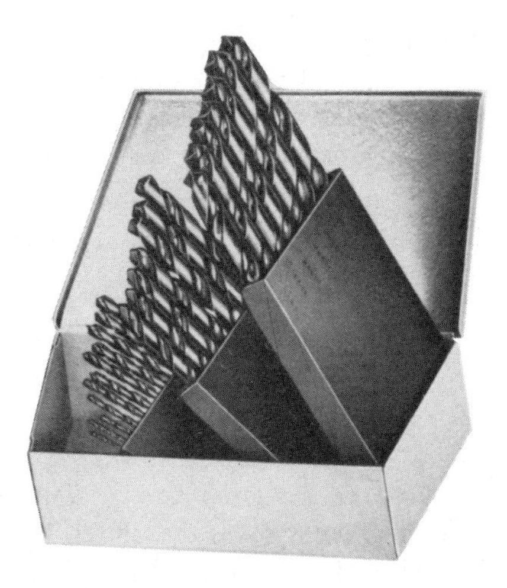

Figure 4.36 Drill index used to store twist drills (courtesy Snap-on Tool Corporation).

NOTE A machinist's handbook will have this information in it.

Q *Pullers* (Figure 4.37). As the illustration of this puller set shows, there are many different types of pullers the mechanic can use. Pullers can be utilized to pull pulleys, gears, and drive hubs from shafts.

R *Feeler gauges* (Figure 4.38). Feeler gauges, thin strips of metal of various sizes, are used to measure the clearance between rocker arms (tappets) and the valve stem. They also may be used to measure the camshaft end clearance and many other clearances throughout the engine.

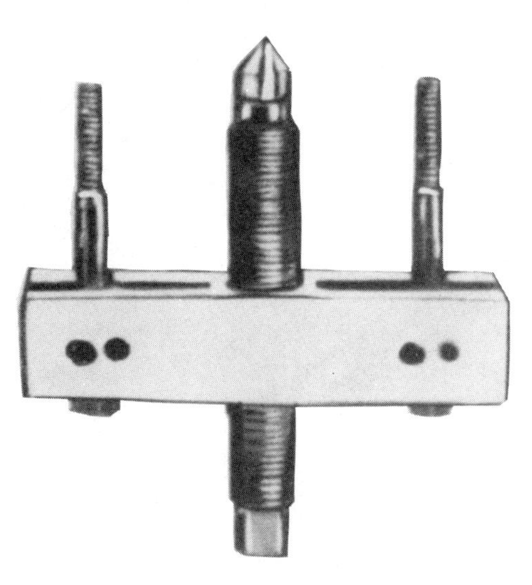

Figure 4.37 Puller (courtesy Snap-on Tool Corporation).

II. Precision Tools

A *Dial indicator* (Figure 4.39). Dial indicators are used to measure movement, such as crankshaft end play, camshaft end play, flywheel runout, and many others.

B *Micrometers*. Micrometers generally are one of three different types: outside, inside, and depth. Outside micrometers (Figure 4.40) are used to measure diameters of shafts, pistons, and the thickness of gasket material or shim stock. Outside micrometers have a measuring range of 1 in. They are available in 0 to 1 in., 1 to 2 in., 2 to 3 in. and so on up to approximately 24 in. The 1 inch measuring range is graduated or marked off in intervals of 0.001 in. (1 thousandth of an inch) on the micrometer thimble, and graduations on the barrel of the micrometer are 0.025 (25 thousandths of an inch). This means that when one complete turn of the thimble is made, the spindle that is attached to the thimble will have moved 0.025. If the thimble is moved less than one complete turn, the distance in thousandths can be determined by counting the marks on the thimble from its original position to its present position.

NOTE Micrometers are also available in metric sizes for use in measuring engine parts and other items that are metric. Because the micrometer

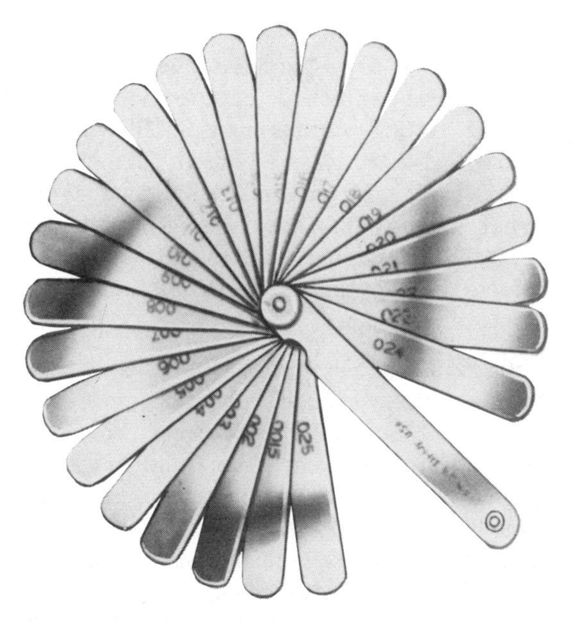

Figure 4.38 Feeler gauge set (courtesy Snap-on Tool Corporation).

Figure 4.39 Typical dial indicator with base (courtesy Snap-on Tool Corporation).

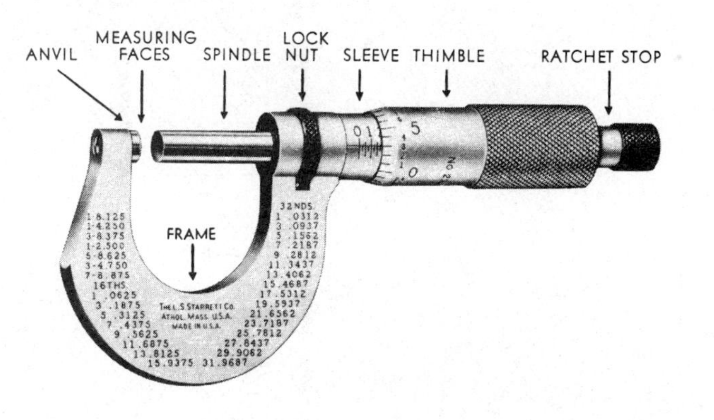

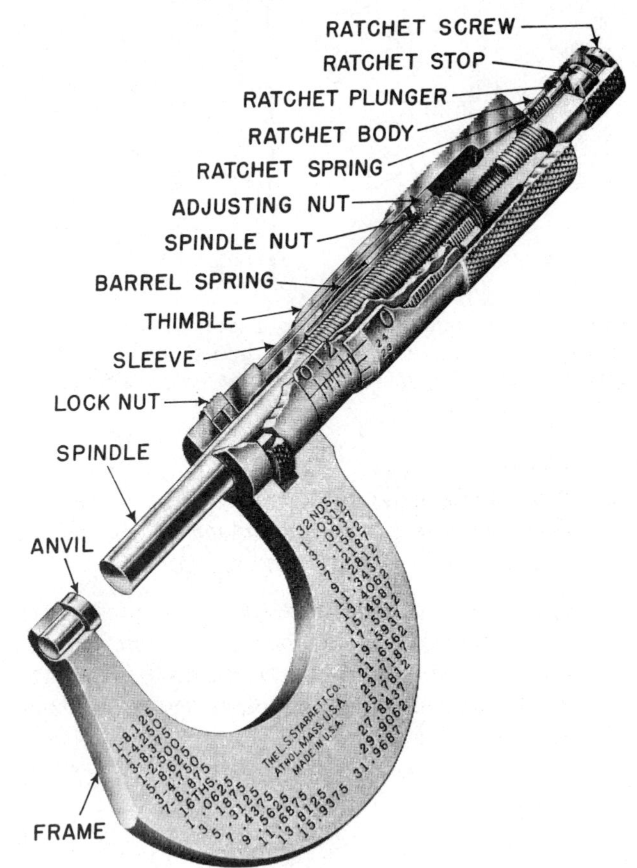

Figure 4.40 Outside micrometer (courtesy L. S. Starrett Company).

that is calibrated in inches is more commonly found, it will be used in the following explanation.

Procedure for Using a Micrometer

1 Select a micrometer of correct size. If you are measuring a shaft larger than 3 in., a 3 to 4 in. micrometer is used.

2 Clean the spindle and anvil face with a clean cloth or rag.

3 Back the spindle away from the anvil face by turning the thimble counterclockwise until the micrometer is considerably larger than the shaft to be measured.

4 Clean the shaft that is to be measured with a rag or cloth.

5 Place the micrometer over the shaft to be measured and turn the thimble until the spindle face contacts the shaft lightly. Move the micrometer back and forth on the shaft to find the tightest spot.

6 Tighten the thimble until a slight drag is felt when you move the micrometer back and forth.

7 If your micrometer is equipped with a ratchet stop, turn the ratchet stop until it clicks. This indicates that the thimble has been tightened against the shaft correctly.

NOTE Since some micrometers are not equipped with a ratchet, you should practice tightening the thimble until you develop a feel for the correct tightness.

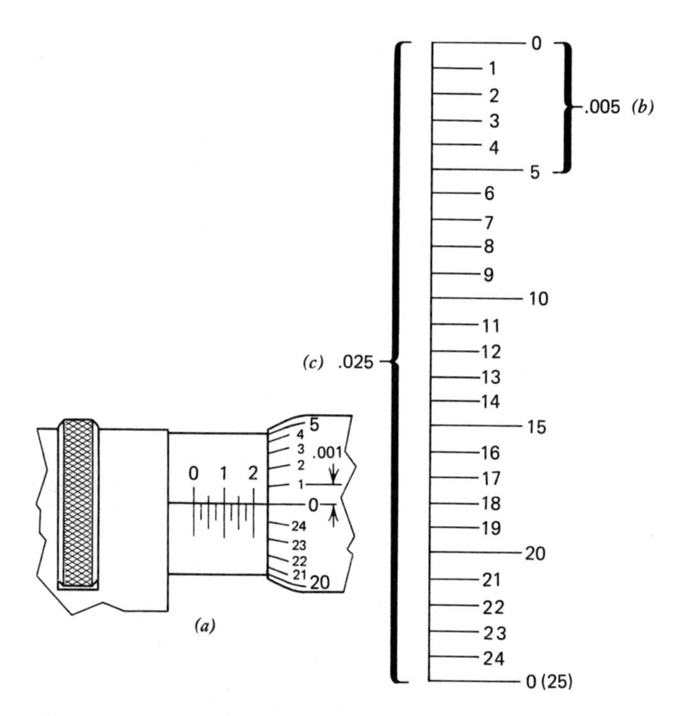

Figure 4.41 Micrometer thimble scale (*a*) Each small mark represents 0.001 of an inch. (*b*) Each large mark represents 0.005 of an inch. (*c*) Total graduations represent 0.025 of an inch (courtesy Gould, Inc.).

8 When the spindle has been tightened against the shaft, the size can be determined by reading the numbers on the thimble and barrel.

9 The thimble of the micrometer is marked off in twenty-five 0.001 of an inch graduations (Figure 4.41). If the graduations were taken from the thimble and illustrated on paper, they would appear as in Figure 4.41.

- **a** Each small mark represents 0.001 of an inch.
- **b** Each large mark represents 0.005 of an inch.
- **c** Total graduations around the thimble represent 0.025 of an inch.

10 The barrel (Figure 4.42) of the micrometer is marked off in ten 0.025 in. graduations for a total of 1000 thousandths. Figure 4.42 illustrates how the micrometer barrel is graduated.

- **a** Each long mark above and below the datum line represents 0.100 (100 thousandths).
- **b** Each of the four divisions between 0 and 1 represents 0.025 (25 thousandths of an inch) for a total of 0.100 (100 thousandths).

11 If the micrometer is in the position seen in Figure 4.43 after it has been adjusted to the shaft diameter, read this position as follows:

- **a** Because the number 2 is the largest number exposed on the barrel, and each number represents 0.100, you have 0.200 (200 thousandths).
- **b** Then note which line on the thimble lines up with the datum line. In this case it is 15. Because each graduation on the thimble equals 0.001, the thimble reading is 0.015 (15 thousandths).
- **c** Add the measurement from the barrel with the measurement from the thimble to determine the total number of thousandths of an inch:

$$\begin{array}{r} 0.200 \\ +\underline{0.015} \\ 0.215 \end{array}$$

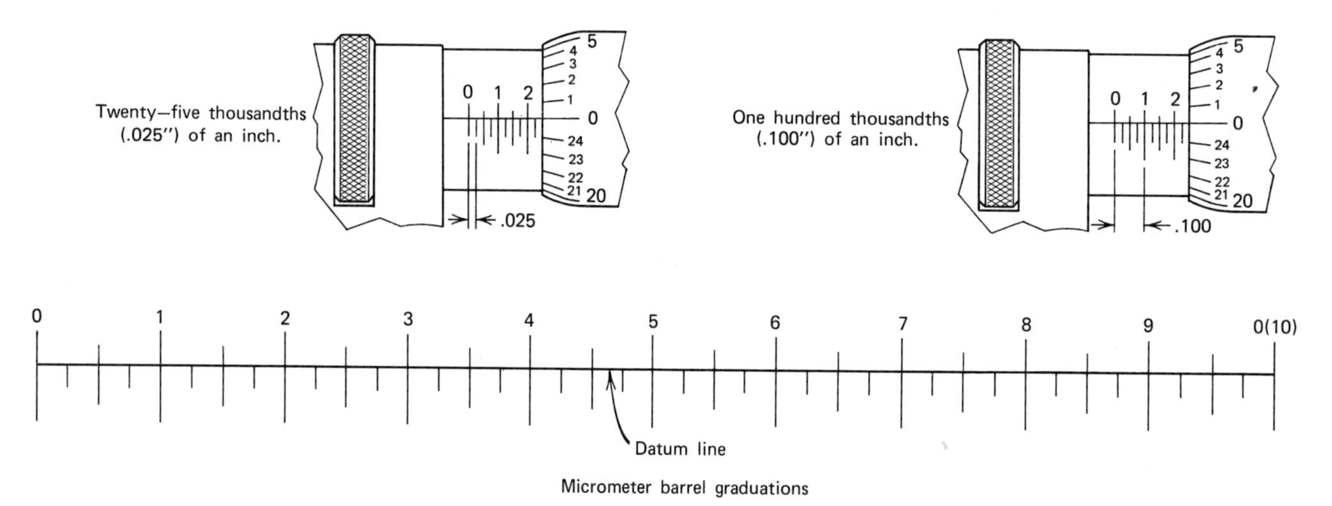

Figure 4.42 Micrometer barrel scale (courtesy Gould, Inc.).

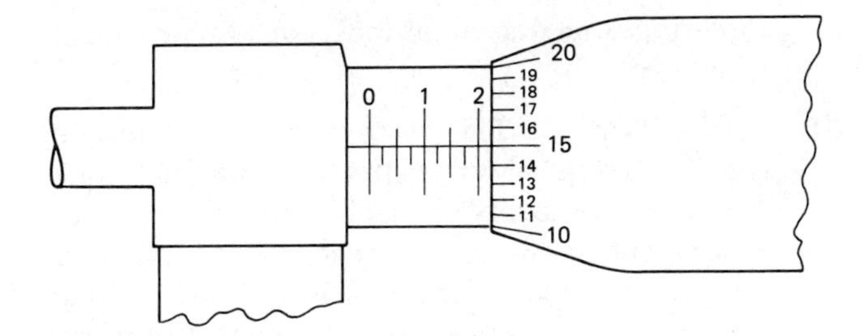

Figure 4.43 A typical micrometer reading.

Figure 4.44 Micrometer vernier scale.

d The micrometer reading then is 0.215 plus the micrometer size in inches. If the micrometer is 0 to 1 in., the reading would be 0.215 thousandths.

e If you use a 4 to 5 in. micrometer, the reading would be expressed as 4.215 (four inches 215 thousandths).

12 If the micrometer thimble stops at a midpoint between a thimble graduation the measurement would be less than 0.001 (thousandths) and should be read in 0.0001 (10 thousandths). To enable you to read this number, an additional scale called a vernier scale (Figure 4.44) has been added to the micrometer barrel. Taken from the barrel and laid out, it would look like the one shown in Figure 4.44. Not all micrometers will have a vernier scale.

a This scale effectively divides each one-thousandth graduation into 10 parts.

b The reading then can be read in 0.0001 (10 thousandths).

13 The micrometer in Figure 4.45 should be read in 0.0001 (10 thousandths).

a Determine the largest number that is visible on the barrel. The largest visible number is 4. Because each number represents 0.100 (100 thousandths), this measurement is 0.400 (400 thousandths).

b Count the number of small marks that are exposed beyond the 4. As previously indicated, each mark represents 0.025 (25 thousandths) of an inch; therefore, the total reading on the barrel is 0.400 + 0.050 = 0.450.

c Read the number on the thimble that is closest to the datum line on the barrel. It can be seen that no line on the thimble will line up exactly with the datum line.

d The reading is very close to 15 but does not match exactly. Each mark on the thimble equals 0.001 (one thousandth); therefore, the reading is more than 0.014 but less than 0.015.

e To determine the exact reading in 0.0001 (one 10 thousandth), the vernier scale must be used. Note the mark on the vernier scale that lines up with any mark on the thimble.

f In this case, the vernier line 6 is the only line that exactly matches any line on the thimble.

g The final result of the measurement is to add together the readings from the barrel, the thimble, and the vernier scale. The resulting micrometer reading is: 0.400 + 0.050 + 0.014 + 0.0006 = 0.4646. This figure is read as four thousand six hundred fifty-six ten thousandths. Reading a micrometer takes practice. Measure different items with the outside mi-

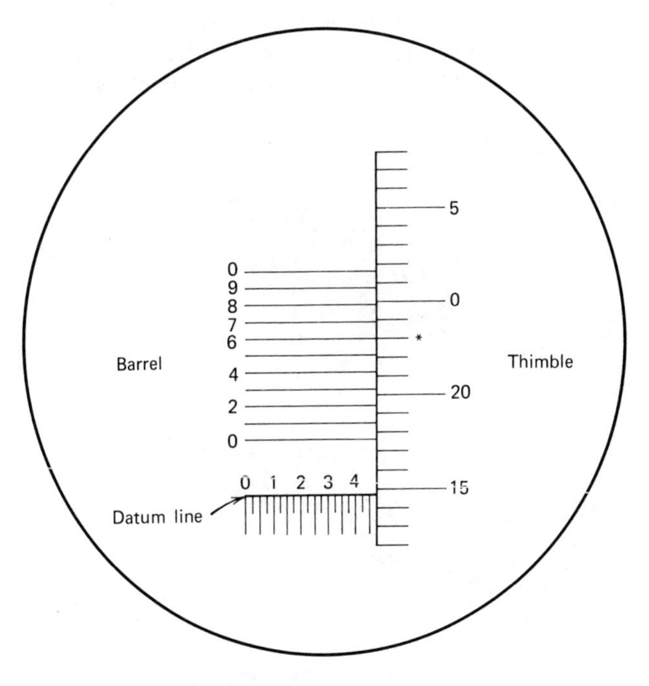

Figure 4.45 Add together all readings.

Figure 4.46 Inside micrometer (courtesy Snap-on Tool Corporation).

crometer until you can read the micrometer correctly.

14 Inside micrometers (Figure 4.46) are used to measure the inside diameter of cylinder sleeves and bushings. The micrometer is expanded until the exact diameter of the cylinder is measured. The scale on the inside micrometer is the same as the scale on the outside micrometer.

15 Depth micrometers (Figure 4.47) are used to measure the depth of holes or counterbores. The micrometer is placed on the object to be measured and turned until the micrometer spindle comes in contact with the bottom of the counterbore or hole. The scale on a depth micrometer is read differently from an outside or inside micrometer. The scale numbering on a depth micrometer is reversed from an outside or inside micrometer. (An outside micrometer scale is numbered 0,1,2,3,4,5,6,7,8,9 while a depth micrometer is numbered 0,9,8,7,6,5,4,3,2,1.) Micrometers of all three types are essential tools for any proficient mechanic.

Figure 4.47 Depth micrometer (courtesy Snap-on Tool Corporation).

16 *Torque wrench.* This wrench measures torque (twisting effort). Many different kinds of torque wrenches are available, such as the micrometer type (Figure 4.48), the dial type (Figure 4.49) and the beam type (Figure 4.50). Most torque wrenches are calibrated in foot pounds (ft lb) and inch pounds (in. lb) in the English system and Newton meter in the metric system. A torque wrench is essential in diesel engine and injection pump repair work.

III. Power Tools

A *Impact wrench* (Figure 4.51). Impact wrenches are handheld, motor-driven wrenches that speed up disassembly and reassembly of engines. The most common sizes are $\frac{3}{8}$ in., $\frac{1}{2}$ in., and $\frac{3}{4}$ in. square drive. Impact wrenches are driven by air or electricity. Air impact wrenches are more popular since they generally are lighter in weight and have more power. Impact sockets (heavy duty sockets) should be used with an impact wrench, since normal sockets are not heavy enough and may be ruined.

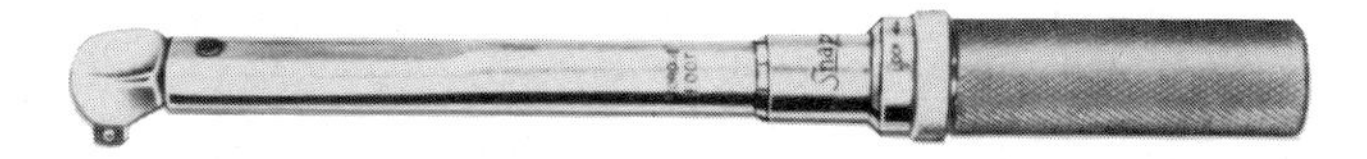

Figure 4.48 Micrometer torque wrench (courtesy Snap-on Tool Corporation).

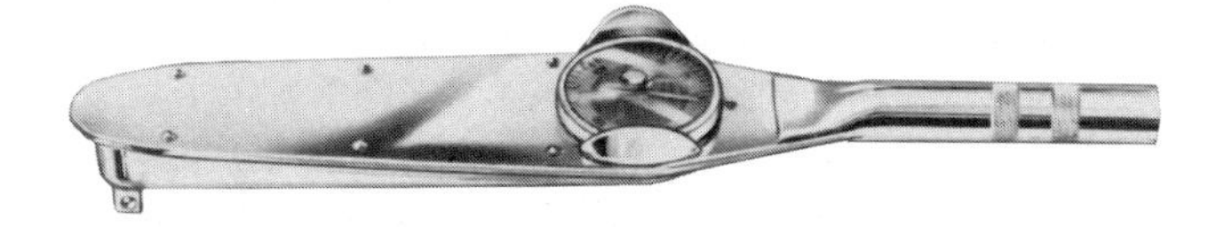

Figure 4.49 Dial type torque wrench (courtesy Snap-on Tool Corporation).

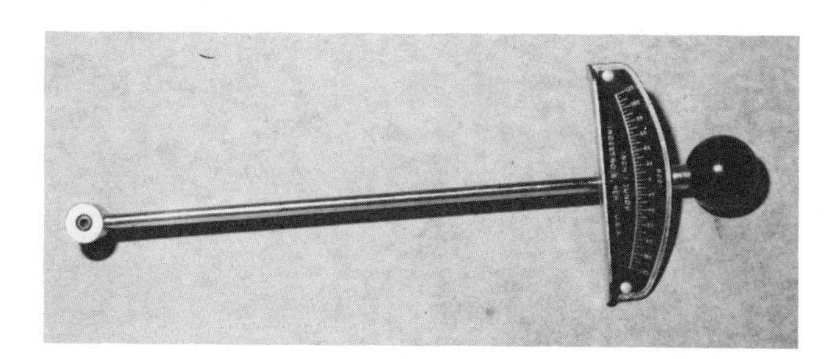

Figure 4.50 Beam type torque wrench.

Figure 4.51 Impact wrench (courtesy Snap-on Tool Corporation).

B *Electric drill.* (Figure 4.52). Electric drills are used to turn drill bits. They are composed of a chuck into which the drill bit is fitted and a small electric motor, and are available in $\frac{1}{4}$ in., $\frac{3}{8}$ in., and $\frac{1}{2}$ in. sizes.

IV. Test Equipment

A *Volt ammeter* (Figure 4.53). Volt ammeters are used to test electrical circuits. In most cases, the voltmeter and ammeter are included in one unit since both may be used to test one circuit.

Voltmeters must always be hooked to the circuit in parallel and ammeters are always connected to the circuit in series.

B *Ohmmeter* (Figure 4.54). Ohmmeters look similar to a volt meter but measure resistance rather than voltage. An ohmmeter is a very useful tester when troubleshooting electrical circuits.

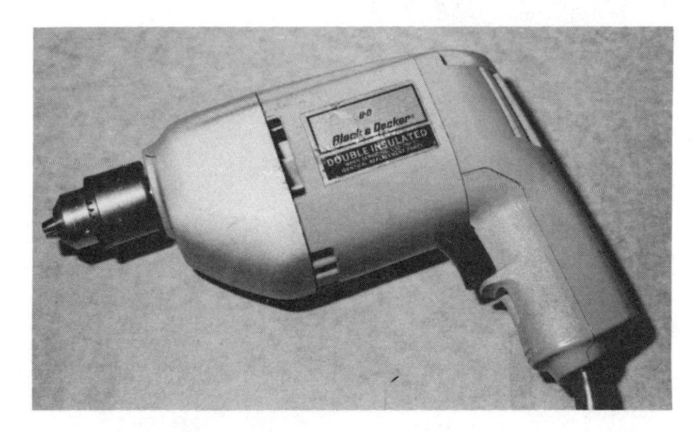

Figure 4.52 Electric drill.

C *Hydrometer* (Figure 4.55). Hydrometers are used to check the specific gravity of battery electrolyte. By checking the electrolyte's specific gravity, you can find out the battery's state of charge.

Figure 4.53 Voltmeter (courtesy Snap-on Tool Corporation).

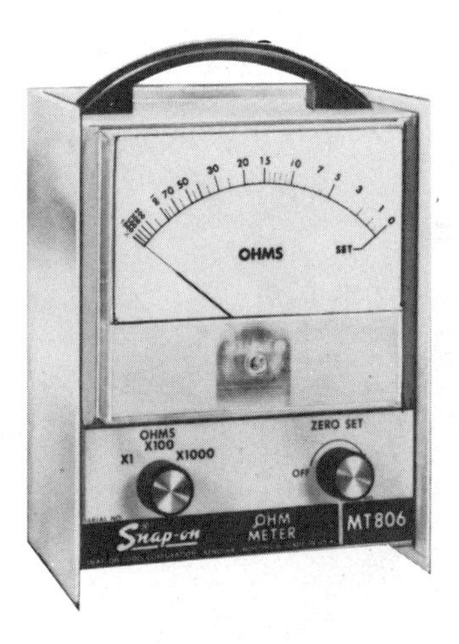

Figure 4.54 Ohmmeter (courtesy Snap-on Tool Corporation).

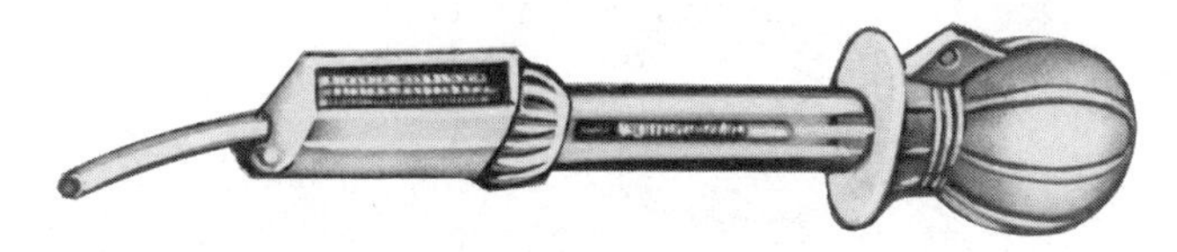

Figure 4.55 Hydrometer (courtesy Snap-on Tool Corporation).

Procedure for Using Hydrometer

To use the hydrometer, squeeze the rubber bulb until all air has been exhausted from it, then continue to grasp it tightly to prevent the bulb from refilling with air. Place the rubber suction tube into the cell to be tested. Release the pressure on the bulb, drawing electrolyte from the cell into the hydrometer. Draw enough electrolyte to cause the float in the hydrometer tube to float. Read the electrolyte's specific gravity by looking at the number on the float that lines up with electrolyte level. A reading of 1250 indicates a fully charged battery. Any reading below that indicates the cell is not fully charged.

D *Armature growler* (Figure 4.56). An armature growler is used to test a generator armature for opens, shorts, and grounds. In addition to the magnet it usually has a test light and voltage meter.

E *Generator-alternator test bench* (Figure 4.57). This test machine is used to test alternators, generators, and regulators. The machine has a universal mounting bracket so that any generator or alternator can be mounted on it. Instruments on the machine include voltmeter, ammeter, and tachometer.

The machine is provided with switches and controls that allow the operator to control the generator output, direction of rotation, speed, and load.

F *Injector nozzle tester* (Figure 4.58). This injection nozzle tester is a hand pump with a reservoir that

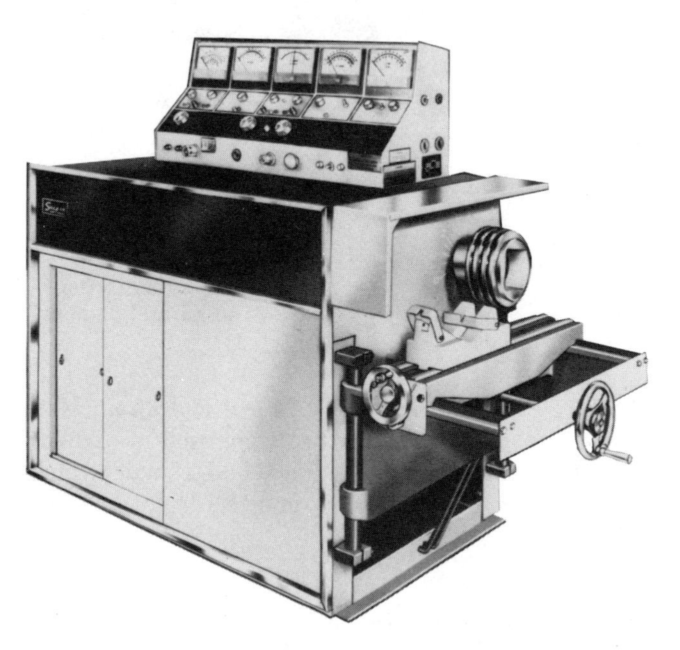

Figure 4.57 Generator, alternator test bench (courtesy Snap-on Tool Corporation).

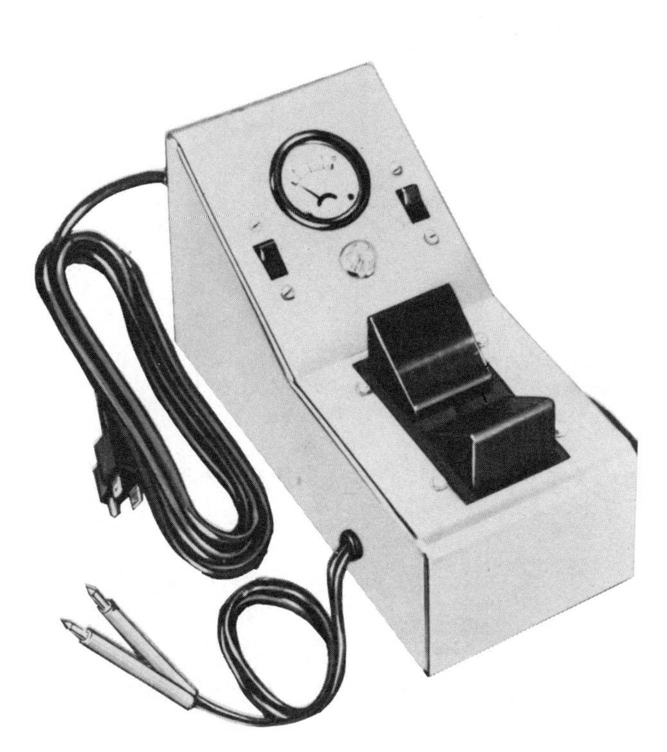

Figure 4.56 Armature growler (courtesy Snap-on Tool Corporation).

Figure 4.58 Injection nozzle tester.

Figure 4.59 Diesel injection pump test stand.

Figure 4.60 Injector comparator.

Figure 4.61 Hydraulic jack.

is filled with calibrating fluid. The pump is fitted with a high pressure gauge so that accurate tests of nozzle opening pressure can be made.

G *Diesel injection pump test stand* (Figure 4.59). A test stand is used to test diesel pumps after they have been repaired or rebuilt. The stand is driven by an electric motor through a variable speed device so that the injection pump can be tested at various speeds. It is equipped with pressure gauges, tachometers, and beakers with which to measure pump pressure, speed, and fuel flow. Test stands are designed to simulate engine operation and are necessary if diesel pumps are to be rebuilt.

H *Injector comparator* (Figure 4.60). An injector comparator is a machine used to test unit injector type injectors. It is designed so that an injector can be installed quickly and tested for delivery. The machine is equipped with pressure gauge, beaker, and counter so that the injector can be checked accuratcly.

V. Shop Equipment

This term generally refers to the equipment in the shop, other than test equipment and hand tools, that the mechanic needs to perform the work in the shop. Shop equipment may include such items as jacks, hoists, and cleaners.

Figure 4.62 Chain hoist.

A *Hydraulic jacks* (Figure 4.61). Hydraulic jacks are used to lift tractors, trucks, and other heavy equipment so that a part or parts can be removed. The jack is a small hydraulic system that contains a reservoir, hand pump, and control valve.

B *Chain hoists, manual and electric* (Figure 4.62). They are used for lifting heavy objects, manual hoists are operated by a chain through a gear reduction. Hoists are also available with an electrically driven gearbox that does the lifting and can be controlled by the operator.

C *Hydraulic floor hoists* (Figure 4.63). Hydraulic floor hoists are sometimes called cherry pickers because of their design. These hoists are mounted on casters that allow them to be rolled from job to job. They have a hand-operated hydraulic pump that raises and lowers the boom. The most common hoists have a rated capacity of two tons.

Figure 6.63 Hydraulic floor hoist.

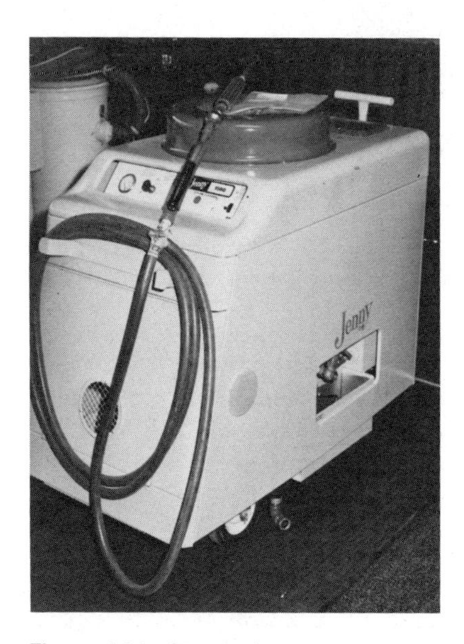

Figure 4.64 Steam cleaner.

D *Steam cleaner* (Figure 4.64). This machine uses high pressure steam to clean engines and equipment. The steam cleaner usually has a boiler arrangement that is heated by a propane or fuel oil fired burner. The hot steam is pumped by a pump at high pressure through a nozzle on the end of the rubber hose.

E *Pressure washer* (Figure 4.65). The high pressure washing device is similar to the steam cleaner but does not use high pressure steam to clean parts and equipment. Instead, a high pressure pump is used to pump a strong soap and water solution through a nozzle at very high pressure. The pressure washer is very popular in many shops since it does not require as much maintenance as a steam cleaner.

F *Jack stands* (Figure 4.66). These stands are made of steel and are used to support a piece of equipment, such as an axle or frame, when the equipment is being worked on. They are used in place of hydraulic jacks for safety reasons.

Figure 4.65 Pressure washer.

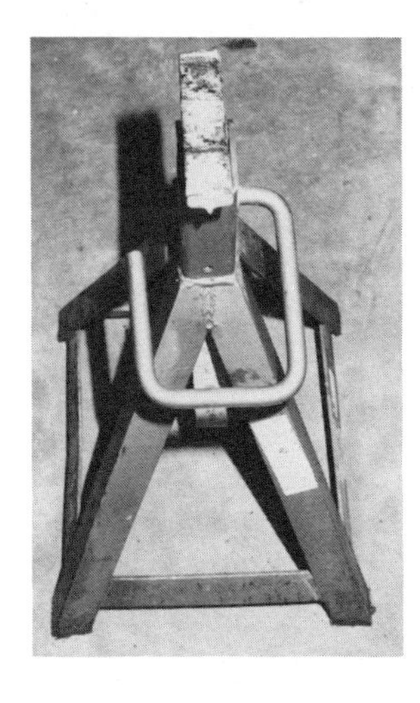

Figure 4.66 Jack stand.

G *Dynamometer* (Figure 4.67). A dynamometer is used to apply a load to an engine for testing purposes. Two common types are found in repair shops: one connnects to the P.T.O. (power take-off) shaft of a farm tractor and another bolts to the flywheel of an engine.

H *Battery charger* (Figure 4.68). A battery charger is a rectifying device that takes line voltage of 110 volts A.C. (alternating current) and supplies 6 or 12 volts D.C. (direct current) for use in charging a wet cell battery. It is normally equipped with a timer, rate of charge meter, rate of charge regulator, and a voltage selection switch.

Most heavy duty battery chargers have a built-in boost circuit of 300 to 400 A (amperes) so that the charge may be used to boost an engine while cranking.

I *Engine repair stands.* They are used to mount engines during repair and rebuild.

Figure 4.67 P.T.O. dynamometer.

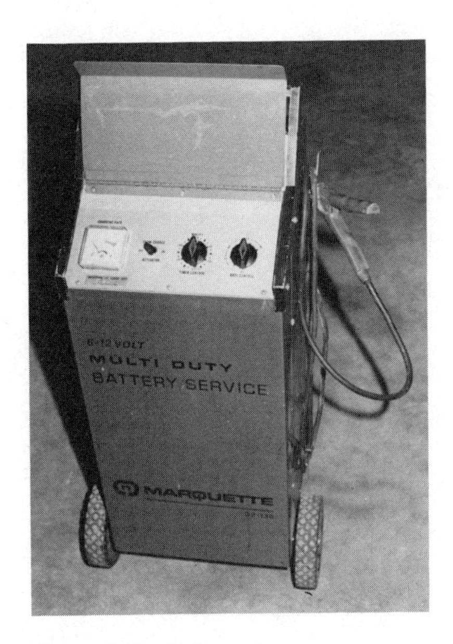

Figure 4.68 Battery charger.

SUMMARY

This chapter has explained some of the tools and equipment you will be using as a mechanic. The tools listed are general in nature and will be supplemented by special tools that may be required to repair the engine you are working on. As you become experienced, the correct procedures for using the seemingly endless lists of tools and special test equipment will become second nature. If you have any questions about the correct usage of a tool or piece of test equipment, ask your instructor.

REVIEW QUESTIONS

1 Name several shop tools that you consider necessary for a diesel mechanic.

2 Which two types of sockets will most mechanics have?

3 A tubing wrench, which is used to remove the tubing nuts, has an advantage over an ordinary open end wrench when used to remove the tubing nuts. Explain this advantage.

4 Plastic and no bounce hammers have many uses. List two uses for each tool.

5 Dial indicators are used by a diesel mechanic in many situations. Explain two situations.

6 Outside micrometers have many uses. List two common uses and where they must be used.

7 Impact wrenches should be used with impact sockets. Why?

8 Many types of shop equipment are used by a diesel mechanic. List two types and explain their use.

5

Engine Disassembly and Diagnosis

Engine disassembly is a very important part of being a proficient diesel mechanic; teardown should be accomplished rapidly but not haphazardly.

Much can and should be learned about the engine during teardown, such as: Did it fail prematurely? Was failure operator or maintenance oriented? Also by the time the mechanic has completed teardown he should have a good idea of what parts will be needed for repair or rebuilding the engine. It can be seen, therefore, that engine teardown or disassembly is one of the most important parts of engine overhaul.

OBJECTIVES

Upon completion of this chapter the student will be able to:

1. List four safety precautions that should be observed before engine disassembly.
2. Correctly disassemble an engine using the service manual.
3. List and explain four items that should be checked before a decision is made to overhaul the engine.
4. List five checks that should be made visually during engine disassembly.
5. List two advantages and two disadvantages of an in-the-vehicle overhaul.
6. List two advantages and two disadvantages of engine removal overhaul.

GENERAL INFORMATION

The decision to disassemble an engine for overhaul should be based on fact, not assumptions, and must be made by the mechanic before any disassembly takes place. To assist in making this decision the engine should be run or operated in some manner, preferably with a dynamometer.

NOTE In the case of a truck or agricultural tractor, most shops will have a dynamometer with which the engine can be tested. If no dynamometer is available, make an effort to run the engine under load by operating it.

I. Engine Diagnosis and Inspection before Disassembly

A *Discuss the engine operation with the operator.* Is the engine being overhauled as a matter of routine because of mileage or hours, or is it being overhauled because of a particular problem such as oil consumption or engine noise? In discussing the engine operation with the operator you may discover that the engine does not need an overhaul; it is possible that an incorrect assumption has been made by the owner or operator. An example of this is excessive engine oil consumption, which may be caused by many things besides worn piston rings.

A thorough check of the following items should be made before the engine is overhauled:

1. Engine valve seals (if used). Seals may be broken, worn out, or improperly installed.
2. Engine front and rear main seals. Check for leakage during operation.
3. Air systems, and air compressor (if used). Check air tank for oil accumulation.
4. Engine turbocharger.
   - a Remove the pipe or hose that connects the turbocharger to the intake manifold. Oil accumulation in this pipe indicates a turbo seal leak.
   - b Oil dripping out the exhaust side of the turbocharger indicates a turbo seal leak.
5. Engine blower.
   - a Remove air inlet pipe to blower (Detroit Diesel engines).
   - b Blower rotors should not be wet with oil; if they are, oil seal leak is indicated.

NOTE The above examples should point out the possibility of a mistake in assuming an engine needs an overhaul just because it may be using an excessive amount of oil. Many times engines are overhauled needlessly because someone did not first check the engine closely and thoroughly diagnose the problem.

After discussing the engine to be overhauled with the owner/operator, the mechanic should make a test run to determine if there are any unusual engine conditions that will require special attention during overhaul. The engine should be checked for:

B *Engine noises.* Noises such as rod-bearing noise or piston slap are generally removed during a complete overhaul. Other noises that come from timing gears and piston pin bushings should be noted so that they are completely checked during engine overhaul.

C *Engine oil pressure.* Engine oil pressure must be considered one of the vital signs of engine condition. For example, if engine oil pressure is low, particular attention must be given to the following items during engine rebuild:

1. Oil pump.
2. Crankshaft journal size and condition.
3. Pressure relief valves.
4. Oil filter bypass valves.
5. Oil cooler bypass valves.
6. Camshaft journals and camshaft bearings.

NOTE Although normal rebuild procedures should return all of these engine parts to like-new condition, in some cases, if a part such as a relief valve spring looks acceptable, it might be put back into the engine as long as no low oil pressure problems had been encountered. If it had been known in advance that a low oil pressure problem existed, the mechanic would have replaced the spring and valve to insure correct operation.

D *Engine temperature.* If the engine temperature is abnormal (higher or lower) during operation, the following items should be given a close check during engine overhaul:

1. Radiator flow and condition.
2. Water pump condition.
3. Thermostats and shutters (if used).
4. Thermostat seals.

E *Engine operation.* Check engine operation for:

1. Excessive smoke.
   - a This indicates that fuel system may be improperly calibrated.
   - b Injectors or injection nozzles are clogged or incorrectly adjusted.
   - c Restricted air cleaner.
2. Low power, no smoke.
   - a Indicates fuel starvation.
   - b Improperly calibrated pump.
   - c Dirty or clogged fuel filter.

II. Types of Engine Overhaul

Engine overhaul usually falls into one of two categories:

A *Overhaul with the engine in the vehicle.* Very often engines are overhauled (in the frame major) with the engine left in the vehicle.

1 Advantages of in-the-vehicle overhaul:

a Time is saved by not having to remove engine.

b The vehicle serves as a place to mount the engine so that it can be worked on without additional stands or brackets.

c Cost to the customer is reduced. As much as 16 hours (two days working time) may be saved by not removing the engine.

2 Disadvantages of in-the-vehicle overhaul:

a All engine seals and gaskets are not replaced, such as front and rear main seals.

NOTE There are exceptions to this, of course, but as a general rule the crankshaft seals are not disturbed.

b The block is not thoroughly cleaned as it would be if it were removed from the vehicle and cleaned in a chemical tank. In particular, the water jacket would be cleaned much better if it were cleaned in a chemical tank.

c The mechanic may have to climb up on a crawler tractor (for instance), causing considerable inconvenience and awkward work access.

B *Overhaul with engine removed from the vehicle.* Most *complete* overhauls are done with the engine removed from the vehicle.

1 Advantages of engine-removed overhaul:

a Engine can be completely disassembled and all gaskets and seals replaced.

b The engine can be mounted on an engine stand, which provides easy access to it.

2 Disadvantages of engine-removed overhaul:

a It takes more time than in-the-vehicle overhaul.

b Heavy lifting stands and brackets to remove engine from vehicle are required.

NOTE The decision to remove the engine from the vehicle is often based on the condition of the crankshaft. A worn crankshaft must be removed from the engine to be reconditioned. This cannot be done with the engine in the vehicle. Common practice in this case is to drop the engine pan and inspect the shaft before making a decision.

III. Procedure for Engine Removal

These procedures are intended to supplement the service manual.

A Remove hood, side panels, or tilt the cab if engine is in a cab over truck.

CAUTION Before tilting the cab of a truck check the inside of the cab for any loose articles that may fall through the windshield as the cab tilts. Many cabs have hydraulic lifts so that one person can easily lift the cab; others have no lifts and require two people to tilt them. Before tilting or raising any cab, read the instructions on the cab near the lift or ask the owner/operator.

NOTE After the cab has been raised make sure the safety catch is in place before you start to work on the engine.

B Visually inspect the engine for oil and water leaks. This may help you in making a repair decision later.

C Steam clean or pressure wash the engine and vehicle in the engine area.

D Drain the coolant from the radiator and engine block.

E Remove the radiator and all connecting hoses if required.

F Disconnect any oil lines that lead to oil filters or gauges.

NOTE If the engine to be removed is in a farm tractor, splitting stands of some type are required. Check with your instructor or service manual for proper stand to use.

G Disconnect all air lines that lead to the engine.

H Disconnect transmission or remove as required. (Refer to your instructor for further instructions.)

I Remove intake and exhaust pipes.

J Disconnect all electrical connections from the vehicle to the engine. Most mechanics identify the electrical connections in some manner so that after the engine is reinstalled there is no question about where to hook them up. This can be done with various colors of spray paint, masking tape or tags. Any method that you have available will save considerable time later.

K Remove any other items that in your opinion may get in the way of engine removal.

L Attach lifting chain or bracket to the engine.

M Move hoist over engine and connect chain to hoist.

NOTE Make sure hoist and chain have sufficient capacity to lift the engine being removed. Diesel engines are HEAVY! A common diesel engine may weigh 2500 lbs (1100 kg). BE CAREFUL!

N Lift engine from vehicle.

CAUTION Lift engine high enough to clear the frame only and make sure no one is standing under it.

O Place engine on the floor with blocks to level it or on an engine stand in preparation for disassembly.

IV. Procedure for Engine Disassembly

Since engine disassembly with the engine removed from the vehicle is the most complete disassembly procedure, it will be discussed in detail in this section. If the engine is to be disassembled in the vehicle, the procedures need only be altered to omit the engine components that are not going to be removed, such as the crankshaft. If the engine has not been cleaned by steaming or high pressure washer previous to removal, it should be cleaned at this time. Place the engine on a suitable stand or cart for disassembly.

NOTE The following disassembly procedure is general in nature and should be used with the engine service manual. Keep in mind that much of the engine disassembly procedure is determined by the mechanic doing the job. In most cases no set procedure must be followed. As you gain experience you will acquire or develop your own procedure for disassembly. You will soon see that in diagnosing engine problems engine disassembly is the one area that requires the most knowledge. Remember, the faster you tear down the engine, the more time you gain on the flat rate, enabling you to meet or exceed the recommended time.

Figure 5.1 Removing rocker arm covers.

CAUTION Do not disassemble the engine in such a manner that damage to component parts occurs.

To aid you in becoming a professional mechanic, many visual checks that you should make as a matter of practice have been included in the following disassembly procedures.

A *Rocker arms covers* (Figure 5.1).

1. Remove bolts from cover and remove cover; place bolts back into cover. Note the condition of the oil clinging to the underside of the rocker cover. A white-colored scum or film of oil clinging to the cover may indicate water leakage into the engine by any of the following:
   - **a** cracked block
   - **b** cracked head (other than combustion chamber area)
   - **c** leaking sleeve seal rings (if wet type sleeve engine)
   - **d** leaking oil cooler

B *Intake manifold.*

NOTE On some engines the fuel injection lines may have to be removed before the intake manifold can be removed.

1. Remove capscrews or bolts that hold intake manifold on, and remove intake manifold (Figure 5.2).

Figure 5.2 Removing intake manifold.

NOTE At this time decide what is to be done with capscrews and bolts as they are removed. Experienced mechanics usually place them in a basket for cleaning or storage; however, beginners often have trouble finding the correct bolts when reassembling the engine. A time-saving suggestion for the beginner would be to place the bolts with the part or parts taken off.

2 Inspect manifold for accumulation of dust or oil.

a Dust or dirt in the manifold would indicate a faulty air cleaner or inlet pipe. (Check it closely before engine reassembly!)

b A wet, oily film in the intake manifold would indicate leaking turbocharger or blower seals. If engine is not equipped with a turbo or blower the air cleaner could be over full (oil bath type of air cleaner).

3 Check manifold for cracks (visual).

C *Rocker arms or rocker box assemblies and push rods* (Figure 5.3).

NOTE Before removing rocker arm assemblies or rocker boxes, the injector (if used) and valve adjusting screws and lock nuts should be loosened and backed out. This is done to prevent damage to valves or push rods when the rocker arm assemblies are replaced during reassembly.

1 Inspect rocker arm assemblies for worn bushings.

a This can be done by grasping the rocker arm with vise-grip or plier and attempting to tip it sideways. Excessive rocking indicates a worn bushing.

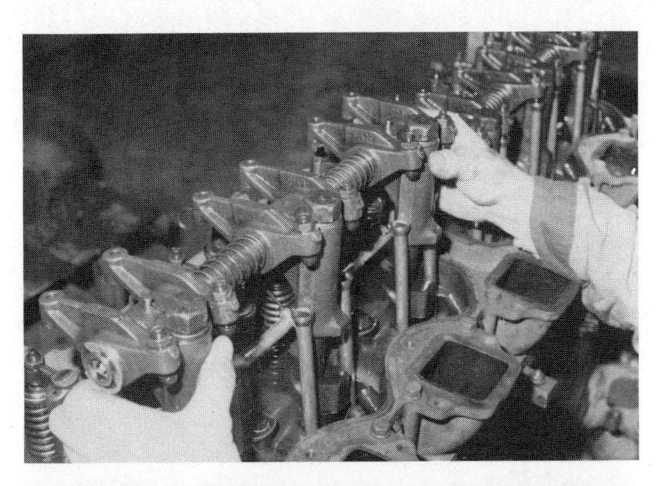

Figure 5.3 Removal of rocker arm assembly from engine.

b Another check on rocker arm conditions can be performed by disassembling the rocker arm assembly and visually inspecting the bushing and shaft.

2 Inspect push rods for straightness.

D *Water manifold and thermostat housing (if used)* (Figure 5.4).

Figure 5.4 Water manifold and thermostat housing removal.

Figure 5.5 Exhaust manifold removal.

Figure 5.6 Fuel line removal.

1 Remove the bolts that attach it to the engine head or block.

2 Inspect water manifold for cracks or rusted spots that may cause water leakage.

E *Turbocharger (if used).*

1 Remove the turbocharger hold-down bolts and any other support brackets.

2 Remove oil inlet and return lines.

3 Inspect turbo outlet for traces of oil film. (This may indicate that the turbo needs an overhaul.)

F *Exhaust manifold* (Figure 5.5).

1 On some engines lock plates (plates that hold the bolts in place) will have to be straightened before the manifold-retaining bolts can be removed.

2 Check manifold for cracks.

G *Injection nozzles or injectors and fuel lines* (Figure 5.6).

1 On engines using injection nozzles, remove the fuel lines from the nozzles and pump. Place plastic caps or plugs on all openings to prevent the entry of dirt.

2 Inspect fuel lines for worn spots that may cause leakage.

3 Loosen hold-down capscrew or screws.

4 Grasp the nozzle and attempt to turn it back and forth, pulling up at the same time. This will remove many nozzles.

5 If stud bolts are used to hold the nozzle in, it cannot be turned; in this case, use a pry bar or similar tool. Wedge it under the nozzle to move it upward out of the cylinder head. Visually inspect nozzle for damage to the tip.

CAUTION Care must be used in prying the nozzle upward, as damage to nozzle could result. (Pencil nozzles manufactured by Roosa Master are easily bent and extreme care must be taken.)

a Since nozzles are often stuck in the cylinder head, a puller or slide hammer must be used to pull them.

6 Removal of injectors (Cummins–Detroit).

a Loosen and remove hold-down capscrew or nut.

b Using a rolling head pry bar or special removal tool. Remove injector by prying it upward.

c Visually inspect injector tip for damage. Make sure injector openings are all capped or plugged, and store injectors in a place where they will not be damaged.

H *Water pump* (Figure 5.7).

1 Remove capscrews that hold the water pump to block or cylinder head and remove water pump.

2 Visually check the water pump impeller for erosion.

3 Check the drive belts for cracks.

4 If fan was bolted to the water pump drive pulley, check it closely for cracks and bent blades.

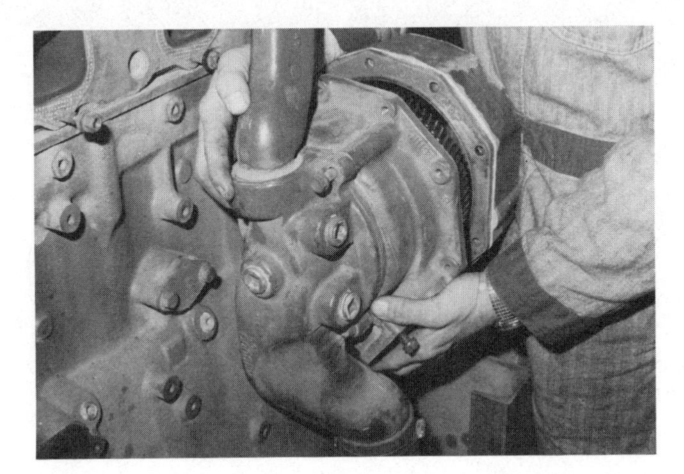

Figure 5.7 Water pump removal.

I *All accessories.*

1 If not previously done, remove all accessories, such as fuel filter housings, fuel hoses, oil filters, water filters, and any other bracket or accessory that may be attached to the engine.

J *Injection pump* (Figure 5.8).

1 Remove all bolts that hold the injection pump to the engine and remove pump.

2 For future reference visually inspect the pump for broken mounting flange and stripped or cross-threaded fittings.

K *Cylinder head or heads* (Figure 5.9).

1 Before you remove the cylinder head, open and/or remove drain plugs to make sure all coolant has been drained from the block. Although the cooling system radiator may have been drained before the engine was removed from the vehicle, some coolant may have remained in the block.

2 Loosen all cylinder head hold-down bolts.

3 Lift out cylinder head bolts, checking each one for erosion and rust. Eroded or rusted bolts should not be used over.

4 Lift cylinder head from the block:

- **a** Using lifting bracket and hoist.
- **b** Manually. (Some heads, such as six cylinder Cummins, Case, or any other engine that has individual heads for every two cylinders, can be easily lifted off by hand.)

5 Inspect the cylinder head combustion chamber surface closely for:

Figure 5.8 Injection pump removal.

- **a** Cracks.
- **b** Pitting.
- **c** Signs of gasket leakage.

NOTE If head is badly cracked or pitted, now is the time to make a decision about replacement or repair. Items such as cylinder heads may not be stocked on a routine basis and must be specially ordered.

6 Inspect cylinder gasket closely. Areas that are blackened or burned out may indicate a warped head or block.

7 Inspect the top of piston for the injection nozzle tracks.

Figure 5.9 Cylinder head removal.

NOTE Fuel injected into the cylinder generally leaves light carbon or soot tracks on the pistons. This "track" or pattern indicates how well the injection nozzle or injector tip is aligned and if any plugged holes exist.

8 Place the cylinder head on blocks or cardboard to protect it from damage.

9 If you wish to disassemble the cylinder head at this time, before further engine disassembly, refer to Chapter 7 on cylinder heads.

L *Clutch and flywheel* (Figure 5.10).

1 Before removing the clutch pressure plate and clutch disc, mark the pressure plate by placing "match marks" on the pressure plate and flywheel. Match marking is usually done with a center punch or chisel and a hammer.

2 Remove the bolts that secure the pressure plate to the flywheel.

CAUTION Some flywheels are recessed and the clutch pressure plate will stay on the flywheel after the bolts are removed. Others are not recessed and, as the last bolt is removed, the flywheel will have to be held by hand to prevent it from dropping to the floor.

3 Lift pressure plate and clutch disc from the flywheel.

4 Inspect pressure plate for:

a Cracks, distortion, or warpage with a straight edge.

b Wear on release finger tips and loose release finger pivot pins.

5 Inspect clutch disc for worn lining.

a Lining is excessively worn if the rivets that hold it on the disc are flush with the lining surface. On new clutch plates the rivets will be recessed about $\frac{1}{16}$ inch (1.6 mm).

6 Match mark flywheel (if not marked by manufacturer). Match marking of the flywheel is done by marking the flywheel and crankshaft if accessible. If the crankshaft cannot be reached with the flywheel installed, sometimes a punch mark can be put on the flywheel and flywheel dowel pin. Many flywheels are marked or so designed at the factory for correct assembly in one of the following ways:

a Off set bolt holes.

b Off set dowel pin holes.

c Match marks or timing marks. In any event, match marking the flywheel insures that you will reinstall it the same way it was installed previously. If there is any doubt about flywheel timing when reinstalling the flywheel, double check the engine service manual!

7 Loosen flywheel attaching bolts and remove flywheel (Figure 5.11).

CAUTION If flywheel does not have dowel pins, use caution when removing the last bolt since the flywheel may fall on the floor and injure you or a fellow worker.

Figure 5.10 Typical diesel engine clutch and flywheel.

Figure 5.11 Flywheel removal.

Figure 5.12 Engine oil pan removal.

8 Check flywheel for the following:

a Cracks.

b Warpage or distortion; use a straight edge.

c Pilot bearing fit (pilot bearing should be a hammer tap fit into the flywheel).

d Bolt holes for oblong-shaped and missing or stripped threads (pressure plate bolt holes).

M *Oil pan* (Figure 5.12).

1 Remove bolts that secure the engine oil pan to block.

2 Remove oil pan.

NOTE The oil pan gasket may cause the oil pan to stick to the engine block, requiring you to wedge a small screwdriver or putty knife between the block and the oil pan to break it loose. Use caution when prying on the pan to prevent damage.

N *Oil pump and pickup screen* (Figure 5.13).

1 Unlock the oil pump bolts that hold the oil pump to the engine block.

2 Remove the oil pump hold-down bolts and remove pump.

NOTE Some engines such as Cummins have an externally mounted oil pump. This type of oil pump can be removed without removing the pan.

Figure 5.13 Oil pump removal.

At this time the author prefers to turn the engine over to allow further disassembly to take place. If the engine is mounted on an engine stand, this is no problem; simply rotate the engine by turning the crank on the engine stand. If the engine is situated on the floor, lift the engine with a hoist to tip it over or lay it down.

CAUTION When lifting or moving the engine with a hoist, get someone to help you. *Make it safe.*

O *Vibration damper* (Figure 5.14). The vibration dampers on most engines require a special puller for removal.

1 Remove bolt or bolts that secure crankshaft damper to crankshaft.

2 Select the correct puller for removal or as indicated by your instructor.

Figure 5.14 Vibration damper removal.

CAUTION Most dampers have puller holes in them to allow them to be removed. Connect the puller only at this point or serious damage to the damper may result.

3 After removal of the damper, check it visually for:

   a Worn areas where engine front seal rides.

   b Nicks or marks on the flywheel part of damper.

4 For further information and checks to be made on the damper, refer to Chapter 8 on crankshaft, main bearings, vibration damper, and flywheels.

P *Timing gear cover* (Figure 5.15).

1 Remove bolts that hold timing gear cover to the engine block.

2 Remove timing cover by tapping it with a plastic hammer.

3 If the cover cannot be removed by tapping with a plastic hammer, a screwdriver may be wedged between the cover and block to "break" it loose.

CAUTION Care must be exercised when wedging or driving a screwdriver between cover and block as damage to the cover may result.

Q *Flywheel housing (bell housing)* (Figure 5.16).

1 Remove the bolts from the flywheel housing.

2 Remove the housing and inspect it for cracks.

NOTE Most flywheel housings are aligned to the block with dowel pins and may require the use of a plastic hammer to jar them loose. If the hammer does not loosen the housing, you may have to use a bar or a screwdriver to pry it off.

Figure 5.15 Timing gear cover removal.

CAUTION Care must be used when prying the housing off or damage to the housing may result.

R *Pistons and connecting rods.*

NOTE If the engine block is not on an engine stand enabling you to rotate the engine, have someone help you tip it over or use a hoist. The engine block should be in the horizontal position when removing the pistons. If the engine block is mounted on an engine stand, it can be rotated easily so that pistons can be removed.

1 Before attempting to remove the pistons, the carbon and/or ridge should be removed from the top of the cylinder bore. If only carbon is at the top of the bore, it can be removed easily with emery paper or a carbon scraper (Figure 5.17). If a ridge is worn at the top of the cylinder, a ridge reamer must be used to remove it.

NOTE Most diesel engines using sleeves have very little or no ridge regardless of the time on the engine: this is a result of the lubricating quality of diesel fuel. Ridge removal, then, is generally necessary on engines that do not have sleeves.

Figure 5.16 Flywheel housing removal (courtesy of Detroit Diesel Allison, Division of General Motors Corporation).

Figure 5.17 Removing carbon from top of sleeve.

Figure 5.18 Removing rod caps.

2 Check rod bearing caps and rods for "match marks." If rods have not been factory marked, mark them with a punch or number marking set to insure that the rod cap and rod are placed together during inspection and reassembly.

3 Remove rod cap bolts and remove rod caps (Figure 5.18).

4 Push piston and rod assembly out with a wooden driver or plastic hammer handle (Figure 5.19).

CAUTION Do not attempt to drive connecting rod and piston assemblies out with a metal driver. Serious damage to the connecting rod may occur.

S *Main bearing caps and crankshaft.*

1 Remove main bearing bolts.

2 Check main bearing caps for match marks or

Figure 5.19 Removing piston and connecting rod from engine.

Figure 5.20 Removing main bearing caps.

numbers. If caps are not marked, use a punch or number marking set and mark the caps in relationship to the block to insure that the caps are reinstalled in the same manner.

3 After "match marking" or checking the factory marks, remove main bearing caps (Figure 5.20).

NOTE In many cases the main bearing caps have an interference fit with the block and will require a slight tap with a plastic hammer to remove. If this does not remove the cap, insert a main bearing bolt into the cap partway and tip sideways on the bolt. This will cause the cap to tip. Continue working the cap from side to side in this manner to allow you to remove it easily.

4 After the main bearing caps have been removed, inspect the main bearings in an effort to detect any unusual wear patterns that may indicate problems with the block or crankshaft.

Figure 5.21 Removing crankshaft.

For a detailed explanation of bearing failures see Chapter 8.

5 Remove crankshaft using a lifting hook and hoist (Figure 5.21).

NOTE In cases where a small in-line engine or V-8 engine is being worked on, the crankshaft can be easily removed by hand. Have someone help you and lift the shaft straight up and out.

6 Lay the crankshaft on the floor to support it or stand it on end and secure it to a workbench or other solid structure.

NOTE If the crankshaft is to be laid on the floor, it should be placed on a clean piece of cardboard and in an area where no damage can occur.

7 Visually inspect the crankshaft for ridging and roughness.

An evaluation should be made at this time regarding the condition of the crankshaft since it may have to be sent out for inspection and grinding, which takes a considerable amount of time. This decision should be made now to prevent holding up the reassembly of the engine at a later date. See Chapter 8 for more detailed information on how to check and evaluate the condition of the crankshaft.

T *Camshaft and timing gears.*

1 Remove bolts that secure the camshaft to the block and remove the camshaft (Figure 5.22).

Figure 5.22 Camshaft removal.

Figure 5.23 Cam follower removal.

2 Visually inspect the camshaft for worn lobes or bearing journals.

3 Inspect the timing gear or gears for wear.

U *Cam followers or cam follower boxes.*

1 Remove cam followers by lifting them from the bores in the block (Figure 5.23).

NOTE Most engines (except Cummins and Detroit Diesel) have cam followers that ride in bored holes in the block and can be removed by simply lifting them out of the bore. Cummins engines have the cam followers anchored to a plate called a box, which is bolted to the side of the engine. To remove this box requires the removal of the six bolts that hold it in place.

2 Visually inspect cam followers for wear, pitting, and flaking. For more detailed information on checking cam followers see Chapter 10, "Camshaft, Cam Followers, Push Rods, and Timing Gear Trains."

V *Other brackets, pulley, and miscellaneous items.*

To prepare the block for cleaning and inspection all other brackets and soft plugs and cylinder liner sleeves (for cylinder sleeve removal instructions, see Chapter 6) should be removed before placing the block in the chemical tank. Complete cleaning, inspection and repair of the cylinder block is covered in Chapter 6.

SUMMARY

This completes the general disassembly of the engine. If the recommended visual inspections were made as the engine was disassembled, you should know the general condition of the engine so that you have some idea of what parts will be needed to repair it. At this time a further component inspection and repair check should take place so that a complete parts listing can be compiled. Components such as cylinder head, oil pump, and fuel injection pumps are covered separately in other chapters of this book.

REVIEW QUESTIONS

1 List three items that should be considered before making a decision to overhaul an engine.

2 List two advantages of an in-the-vehicle overhaul.

3 List two disadvantages of an in-the-vehicle overhaul.

4 What should be checked before removal of the connecting rod caps?

5 Give two reasons why the visual inspection of engine components during engine disassembly is important.

6 List two safety precautions that should be followed when removing an engine from the vehicle.

7 Why should the clutch pressure plate and flywheel be "match marked" before the clutch pressure plate is removed?

8 In what way will placing the bolts back into the components as they are removed help you?

9 List two items you would check closely during engine disassembly if the engine had been overheating.

10 Why should the valve adjusting screws and lock nuts be backed off before removal of the rocker arms?

11 If a cracked cylinder head is found during disassembly, why should the decision about replacement or repair be made immediately?

6

The Cylinder Block

Most modern diesel engines use a cylinder block similar in construction to the one pictured in Figure 6.1. The cylinder block may be described as the largest single part and the main structure or "backbone" of the diesel engine. All other engine parts are bolted or connected to the cylinder block in some way.

OBJECTIVES

Upon completion of this chapter the student should be able to:

1 List five component parts of a cylinder block.

2 Identify three types of cylinder blocks in use today.

3 Explain the procedures used in inspection, overhaul, and testing of a cylinder block.

4 Demonstrate the ability to remove wet and dry type sleeves.

5 Demonstrate the proper procedures for installing a sleeve, wet or dry.

6 Demonstrate the ability to install a cam bushing correctly with reference to oil holes.

7 List five causes for sleeve wear or damage other than normal wear.

8 Demonstrate the ability to install an expansion plug of the cup type, indicating the type of sealer that should be used.

GENERAL INFORMATION

Contained within the cylinder block are the following:

1 Coolant passages and water jacket.

2 Holes or bores for the piston and sleeve assembly.

3 Bores or supports for the cam bushings and camshaft.

4 Main bearing bores that hold the main bearings and support the crankshaft.

5 Drilled or cored passageways for the engine lubrication system.

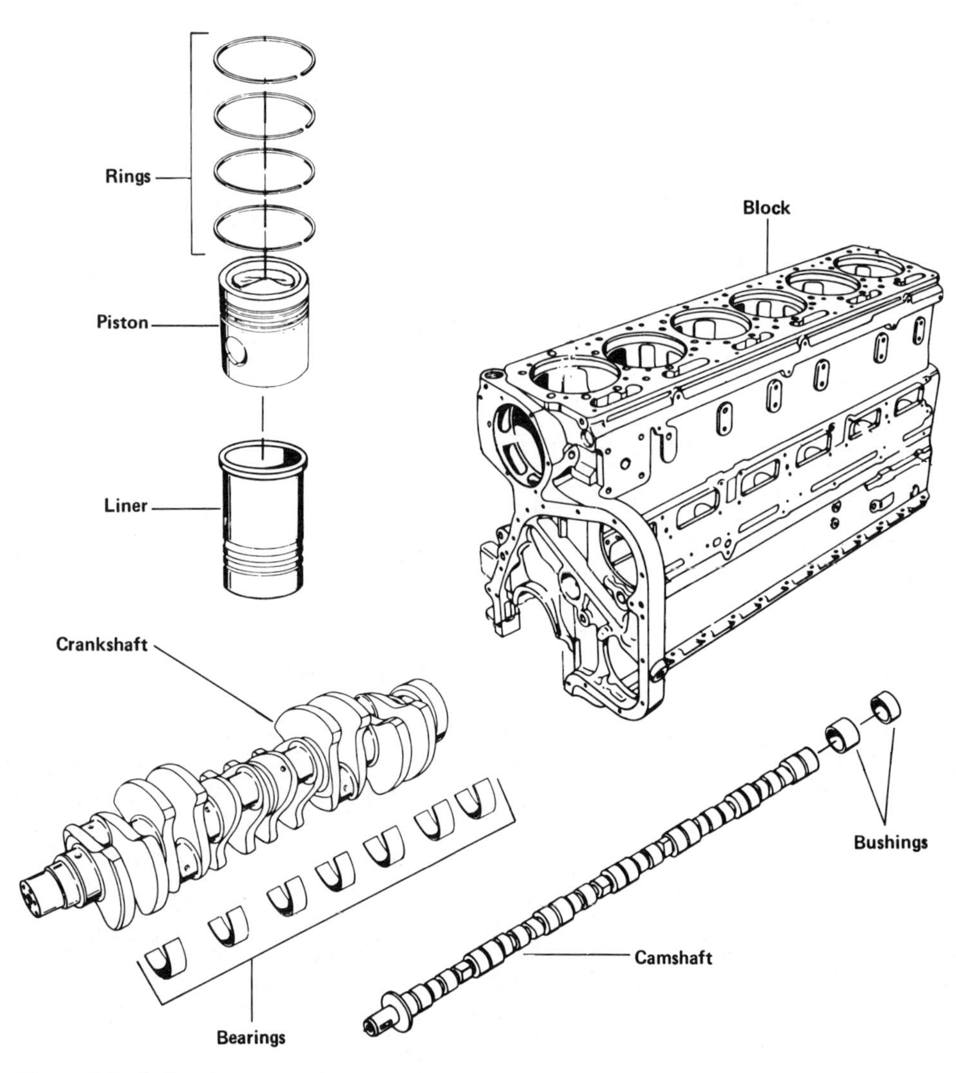

Figure 6.1 A diesel engine cylinder block and related parts (courtesy Cummins Engine Company).

6 Holes or bores in the water jacket that allow insertion of the freeze or expansion plugs.

7 Many drilled and tapped holes utilizing various types of threads that allow the cylinder head or heads and other engine parts to be bolted or connected to it with some type of fastener or bolt.

I. Diesel Engine Cylinder Blocks

Diesel engine cylinder blocks may be one of four types: wet sleeve, dry sleeve, bored without a sleeve, or air cooled.

A *The wet sleeve type* (Figure 6.2) is designed with a number of large holes in which the cylinder sleeves are inserted. These holes are designed so that the

Figure 6.2 Diesel engine cylinder block with wet type sleeves.

coolant will be circulated around the cylinder sleeve or liner. The coolant is prevented from leaking into the crankcase of the engine by O ring seals at the bottom of the liner. At the top of the block a counterbore is cut into the block for the lip or flange of the sleeve to fit onto and prevent coolant leakage. The uppermost part of this lip may be slightly larger than the lower part. This larger diameter provides an interference fit with the block when the sleeve is installed.

1 Advantages:

a The major advantage is the contact of coolant directly with the sleeve, enabling rapid and positive heat transfer from the combustion chamber to the coolant.

b Sleeves are easily removed and installed during engine rebuild to bring the cylinder block back to like-new condition.

c Cylinder sleeves may be replaced individually if they become worn or damaged prematurely.

2 Disadvantages:

a The major disadvantages are the problems encountered in maintaining a coolant seal between the bottom of the sleeve and the block. The seals used (O ring and crevice seals) sometimes do not have the same longevity as might be expected from the engine.

b This seal leakage generally occurs at the bottom of the sleeve and contaminates the lube oil.

B *The dry sleeve type* (Figure 6.3) is designed with a bored or honed hole in the block that allows no coolant contact with the cylinder sleeve. The sleeve is inserted into the bored hole. A counterbore is bored into the block to accommodate the sleeve lip and to help position the sleeve as in the wet sleeve type. The sleeve is held in place by the cylinder head gasket and cylinder head bolted onto the block.

1 Advantages:

a The dry sleeve type does not have coolant in contact with the cylinder sleeve since the sleeve is fitted into a bored hole in the block. This is a major advantage in that sealing the sleeve at the bottom is not required.

Figure 6.3 Diesel engine cylinder block with dry type sleeves.

b There is no lube oil contamination as a result of the leaking of coolant by the sleeve seals.

c The block can be brought back to like-new condition easily by the installation of new sleeves.

d Cylinder sleeves may be replaced individually if they become worn prematurely or damaged.

2 Disadvantages:

a Since the coolant is not in direct contact with the sleeve, heat transfer from the combustion chamber to the coolant water is not as rapid as it should be.

b This slow heat transfer may result in short engine life and cylinder damage.

C *The bored or no sleeve type* (Figure 6.4) has holes bored for the cylinder with the pistons and piston rings inserted directly into this hole. No provision in the block for wet or dry type sleeves is made.

1 Advantages:

a The major advantage is the initial cost of construction in that the machining and fitting of sleeves is not required.

b No provision has to be made for O ring grooves and no contact area is needed.

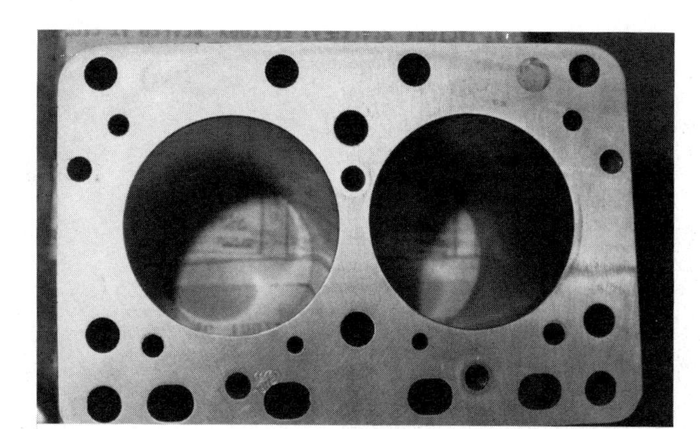

Figure 6.4 Diesel engine cylinder block with cylinders bored directly into the block.

- **c** The block can be made lighter because of thin cylinder wall construction.

2 Disadvantages:

- **a** A major disadvantage of this type block is that during rebuild or repair of the engine a worn cylinder must be rebored or honed.
- **b** A premature failure of an individual cylinder sleeve cannot be repaired as easily as a sleeve type block. It must be rebored. Reboring requires special equipment and the engine must be completely disassembled.

D *The air cooled type* is similar to a bored type block in that it does not have cylinder sleeves but bored holes for the piston. It has no coolant passageways or water jackets; fins have been added to the cylinder block to dissipate heat. Cooling then is accomplished by the passage of air around the fins.

1 Advantages:

- **a** It is much lighter in weight because the water jackets have been eliminated.
- **b** No coolant is required, which in itself eliminates problems that go with liquid cooled engines, such as leakage, freezing, rust formation, and inadequate cooling.
- **c** This eliminates the need for a radiator, water pump, and thermostat.

2 Disadvantages:

- **a** It generally does not have sleeves, making replacement or reboring necessary if one cylinder becomes worn or damaged.
- **b** It has no coolant with which to operate hot water heaters used in trucks, and tractors with cabs.
- **c** It needs some type of cooling fan, usually belt driven.
- **d** Cooling fins around cylinders can become clogged by dirt and engine oil, creating an overheated cylinder.

II. Procedure for Disassembly, Inspection, and Cleaning of Cylinder Block

At this point all major components and accessories should have been removed from the block. If not, refer to Chapter 5 on engine disassembly. Further disassembly should include removal of the following:

A *Oil galley or passageway plugs.*

B *All cover plates (oil and water).*

C *Expansion plugs (soft plugs).*

Removal of expansion plugs can be accomplished quickly and easily by:

1. Driving a sharp punch or chisel through them (Figure 6.5).
2. Twisting or turning them sideways.
3. Prying them from block with a bar, using caution not to damage the block, which will prevent a new plug from sealing.
4. Inspect the expansion plugs after removal in an attempt to determine if the engine coolant was being properly maintained.

NOTE If expansion plugs show signs of high corrosive action within the cooling system, a check should be made of the water filter or conditioner.

D *Oil pressure relief valves.*

1. Remove valve and spring.
2. Make sure valve moves freely in its bore and spring is not broken.

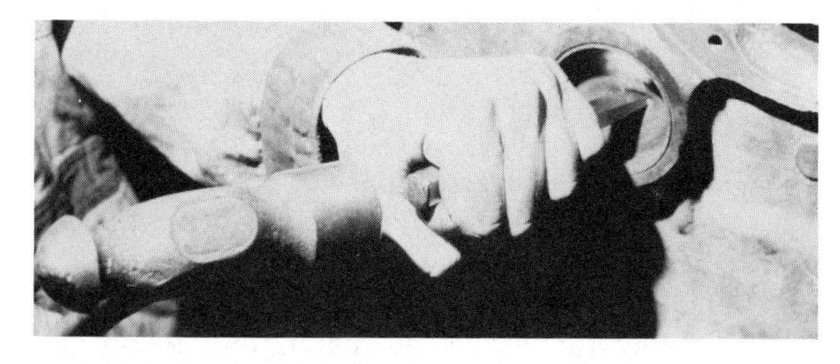

Figure 6.5 Expansion plug removal.

E. Procedure for Cam Bushing Removal

Removal of cam bushings or bearings should not be attempted without special bushing drivers or damage to the cam bearing bore in the block may result. Two types of bushing installation and removal tools are common in most shops: the solid nonadjustable type (Figure 6.6) made to fit one size bushing or the adjustable type (Figure 6.7) which can be adjusted to fit any size bushing within a given range.

Before removal of the cam bushings, inspect them to determine if normal wear has occurred or if some malfunction or lubrication problem exists. Also check all oil supply holes before removal of the bushings so that no question exists about the proper alignment when installing the bushing. The cam bushings should be removed as follows:

1 Select the correct size bushing driver.

2 Place the bushing driver into the bushing to be removed.

3 Place the driver guide cone on the driving bar and insert assembly into the block and bushing driver.

4 Make sure guide cone is held securely into another bushing or bushing bore so that no misalignment of driving bar can occur.

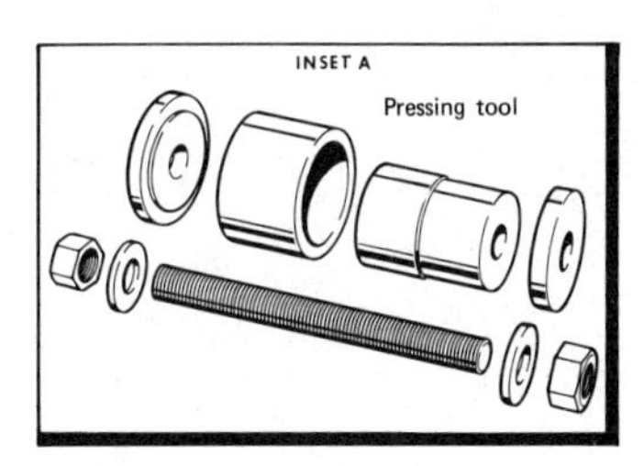

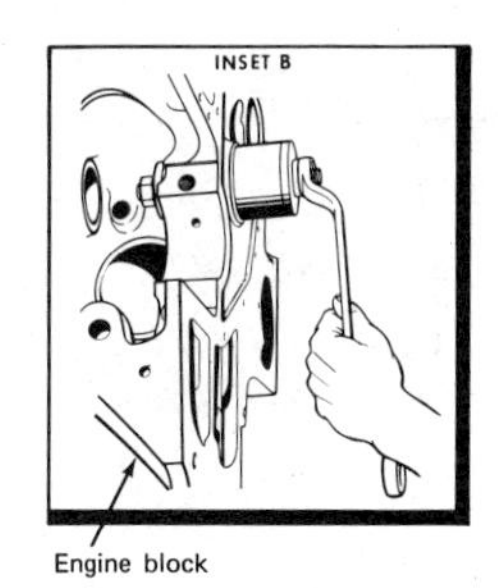

Figure 6.6 Nonadjustable type cam bushing installation tool (courtesy J. I. Case Company).

Figure 6.7 Adjustable type cam bushing installation tool.

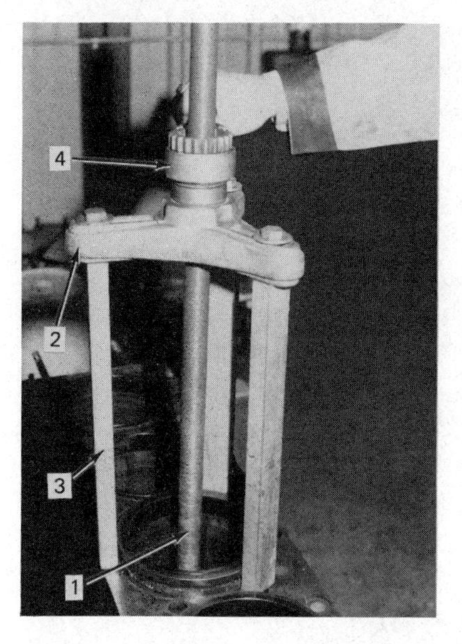

Figure 6.8 Sleeve puller.

5 Holding the bar with one hand, strike it on the driving end with solid, firm hits with a large hammer.

6 Drive bushing until it clears block bore.

7 Follow the above procedure and remove all of the cam bushings.

F *Cylinder sleeves.*

Cylinder sleeves must be pulled or pushed out of the block with a sleeve puller. The most common type of sleeve puller in use is similar to the one shown in Figure 6.8. The numbered parts are keyed to the illustration: (1) long threaded through bolt or rod, (2) support bracket, (3) supports or legs, (4) through bolt nut and rachet, (5) adapter plates to fit different size sleeves, not shown.

1. Procedure for Pulling Wet Type Sleeves from Block

a Select the correct adapter plate that will fit the sleeve.

b Make sure the plate fits snugly in the sleeve (to prevent cocking) and that the outside diameter of the puller plate is not larger than the sleeve outside diameter. (An adapter plate larger than the sleeve may damage the block.)

c Attach the adapter plate to the through bolt.

Figure 6.9 Sleeve puller adapter plate being installed in cylinder.

NOTE If the adapter plate is the type that can be installed from the top of the sleeve, it will have a cutaway or milled area on each side (Figure 6.9). This, along with the swivel on the bottom of the through bolt, allows the plate to be tipped slightly and inserted from the top, eliminating the need to install the adapter plate in the bottom of the sleeve and then insert through bolt and attach nut.

d After the adapter plate and through bolt have been installed in the sleeve, hold the through bolt and adapter plate firmly in the sleeve with one hand and install the support bracket with the other hand.

e Screw the through bolt nut down on the through bolt until it contacts support bracket. This will hold the adapter plate and through bolt snugly in place.

CAUTION Before tightening the sleeve puller nut, make sure the sleeve puller supports or legs are positioned on a solid part of the block. Tighten nut with ratchet; the sleeve should start to move upwards. If it does not and puller nut becomes hard to turn, stop and recheck your puller installation before proceeding.

f On wet type sleeves, after the sleeve has been pulled from the block far enough to clear the O

Figure 6.10 Sleeve removal (courtesy Deere and Company).

rings (Figure 6.10), then tip or swivel the sleeve puller adapter plate and remove the sleeve puller assembly.

g The sleeve can now be lifted out by hand.

NOTE Engines with tight-fitting dry type sleeves may require a special hydraulic puller.

2. Procedure To be Used When Pulling Dry Type Sleeves with a Hydraulic Puller

a Select adapter plate to fit sleeve. (See note under step 1a.)

b Assemble through bolt plate and hydraulic ram.

c Adjust through bolt nut so that adapter plate fits snugly in sleeve and hydraulic ram sits firmly on supports or legs.

CAUTION Make sure legs or supports are positioned on solid part of block to prevent cracking of block.

d Pump hydraulic hand pump that is connected to hydraulic ram.

NOTE If sleeve does not move upward after considerable hydraulic pressure has been applied, it may be necessary to tap sleeve adapter plate from the bottom, using a bar and hammer to break it loose.

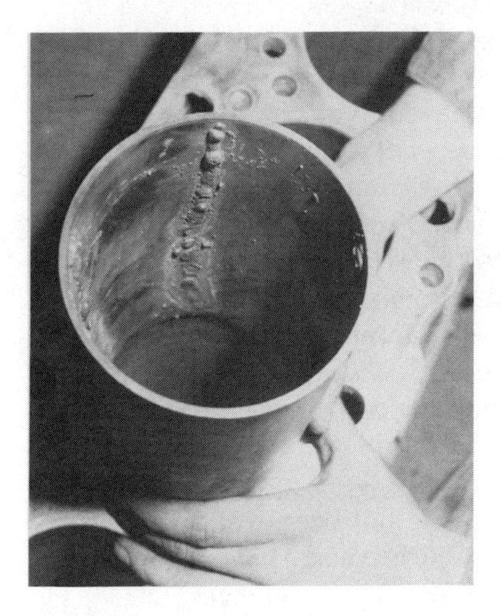

Figure 6.11 Welds in dry type sleeve.

CAUTION Under no circumstances should a hydraulic ram or hand pump be overloaded by using an extension handle on the hand pump. This may cause a hydraulic hose to burst, causing serious injury to the operator from high pressure hydraulic fluid escaping.

e Pump hydraulic hand pump until the puller cylinder has moved its full length.

f Open pump control valve to allow cylinder to retract.

g Adjust through bolt so that adapter plate fits snugly into the sleeve as in step 2c.

h Repeat steps e to g until sleeve can be lifted from the block.

i If sleeve cannot be moved using this procedure, another procedure used by some mechanics is to use an electric welder and weld several beads vertically along the full length of the sleeve from top to bottom (Figure 6.11). This heating and cooling of the sleeve may shrink it enough to allow removal.

j In some cases, press-fit dry type sleeves cannot be successfully removed using any one or all of the procedures outlined above. If this is the case, the cylinder block must be taken to an automotive machine shop and the sleeve bored out.

After all sleeves have been removed, a preliminary visual inspection should be made at this time to determine if the block can be repaired and reused or if it requires replacement. Items to check at this point are:

1. Visual cracks in water jacket internally and externally.
2. Check cored or drilled passageways for cracks.
3. Check main bearing and cylinder head bolt holes for cracked or broken threads.
4. Block top surface for excessive erosion or cracks.
5. Main bearing caps for cracks.

After determining that the block will be reusable, all gasket material and heavy accumulation of grease or oil should be scraped or wiped off. It is common practice at this time to soak the block in a hot or cold tank of cleaning solution, which should remove all carbon, grease, scale, and lime deposits. After the block is removed from the tank, it can be cleaned with a steam cleaner, high pressure washer, or water hose. During steaming of all passageways, oil galleys, and water jackets, use a stiff bristle brush or other suitable device to dislodge all foreign material that may be lodged or caked in the block. Also at this time the block should be visually checked again for cracks that may have opened up in the hot tank.

NOTE Special attention should be given to the removal of scale or sludge accumulations within the water jackets because they will act like insulation and prevent heat from traveling into the coolant water. Poor heat transfer may cause scuffing or scoring of the cylinder and rings and excessive oil consumption. Consideration of what caused the sludge formation should be given at this point. Is it a normal accumulation or has the cooling system maintenance been neglected?

It is recommended after the block has been steamed that it be sprayed with a light coat of preservative or rust preventive oil or solution.

III. Final Inspection, Testing, Reconditioning, and Assembly

At this time, it is suggested that the block be placed on a suitable engine stand so that it can be rotated and tilted to allow access to all areas. If an engine stand is not available, a clean workbench will be sufficient.

NOTE It is also suggested that the block be checked with an electric crack detector if it is available. Every attempt should be made at this point to insure that the block is not cracked, since in the following steps the block is being readied for reassembly.

A. Procedure for Checking and/or Resurfacing Block Top Surface

1 The cylinder block top must be checked for straightness throughout its length and for erosion around water outlets, using the following procedures.

NOTE Erosion can be checked only after block has been thoroughly cleaned.

a Clean top surface of block by hand with sandpaper or with an electric or air driven sander (Figure 6.12).

b To determine the extent of erosion damage use a new head gasket. Lay gasket on the block. Visually check to see that erosion will or will not interfere with the gasket sealing.

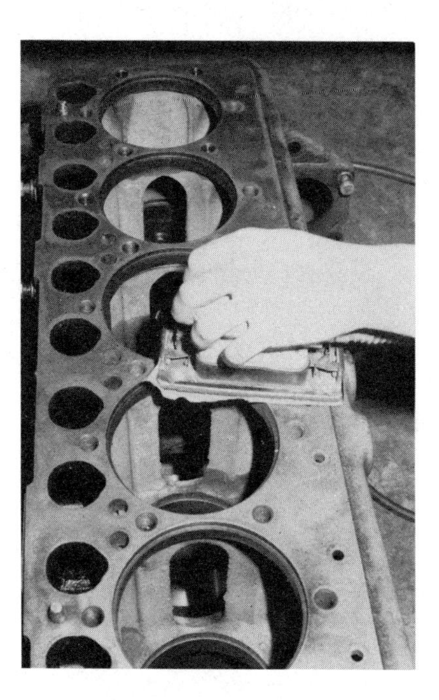

Figure 6.12 Sanding cylinder block to remove rust and corrosion.

c If the erosion around water holes is excessive, the block top surface must be resurfaced or the water holes sleeved. No further checks can or should be made until resurfacing of top surface has been done. (Resurfacing or machining of block top surface requires special equipment and should not be attempted in a general repair shop.) Most automotive machine shops have equipment to perform the resurfacing operation. If the water holes are to be sleeved, refer to the engine service manual for the correct procedure.

d If top surface has not been resurfaced and is considered usable because of lack of erosion, it should be checked for straightness.

e Using an accurate straight edge, check block by setting straight edge on the top of the block (Figure 6.13).

f Hold straight edge with one hand. Using a 0.0015 to 0.002 in. (0.04 to 0.05 mm) feeler gauge, try to insert feeler gauge between the block and straight edge. Most engine manufacturers recommend that if the block is warped 0.004 in. (10 mm) or more it should be remachined.

B. Procedure for Checking Main Bearing Bore Size and Alignment

Checking main bearing bore is a very important step in engine or block rebuild. The main bearing bore should be checked for correct diameter and out-of-roundness, as shown in Figure 6.14. In addition to this check, many manufacturers recommend the use of a master bar (Figure 6.15) to check main bearing bore alignment.

NOTE Main bearing bore alignment checks are not made by some mechanics on engines that have a tendency to have a problem with bore alignment. The main bearing bore is simply redone whenever the engine is rebuilt, as a matter of routine. With this in mind, check with your instructor or someone who has had experience with engine rebuilding. If in doubt about main bearing bore alignment, send the block to a shop that has the capability to check and/or bore the main bearing bores.

Figure 6.13 Checking cylinder block with straight edge.

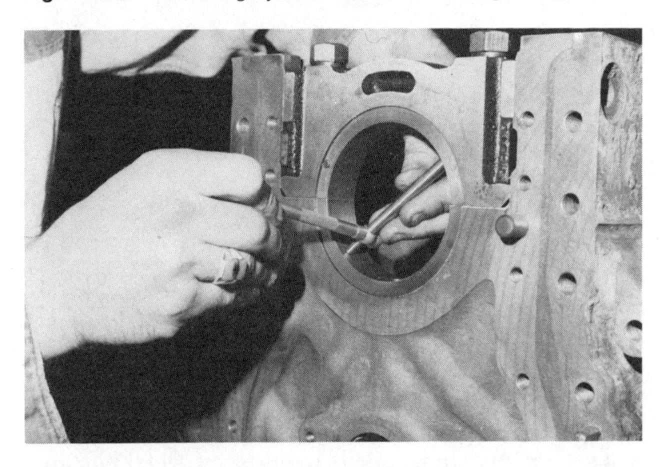

Figure 6.14 Checking main bearing bore for out-of-roundness.

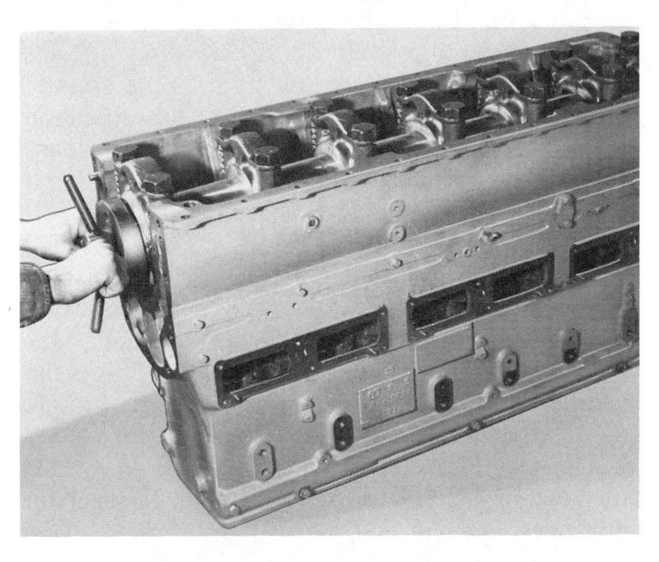

Figure 6.15 Using master bar to check main bearing bore alignment (courtesy Cummins Engine Company).

C. Procedure for Checking and/or Reconditioning Cylinder Sleeve Counterbore

1 Block counterbore and packing ring area must be completely cleaned of all rust, scale, and grease and should not have any rough, eroded areas that might cut or ruin a sleeve, O ring or crevice seal.

  a Check the sleeve counterbore closely for cracks; if cracks are found, the block can be salvaged by resleeving of the counterbore. Resleeving of the counterbore should be attempted only by experienced mechanics using the correct equipment.

2 Cleaning of block packing ring area and counterbore lip can be done by hand with a small piece of crocus cloth, wet-dry sandpaper, or emery paper of 100 to 120 grit (Figure 6.16).

3 Block counterbore top depth must be measured to insure that sleeve protrusion will be correct after sleeve is installed. Counterbore should also be uniform in depth around the circumference of the bore.

4 Measurement of block counterbore should be done with a depth micrometer or dial indicator mounted on a fixture (Figure 6.17). If a depth micrometer is used, make sure that the micrometer is held firmly on the block surface when making measurement.

5 Counterbore depth should be checked in at least four positions around the circumference of the counterbore to determine if depth is within specifi-

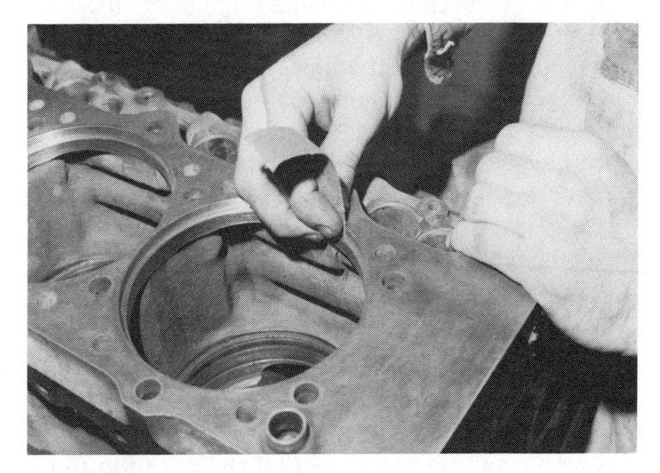

Figure 6.16 Cleaning block counterbore before sleeve installation.

Figure 6.17 Measuring block sleeve counterbore with depth micrometer.

cations. The counterbore depth should not vary more than 0.001 in. (0.025 mm) at all four positions.

6 After measuring counterbore depth, sleeve lip or flange should be measured with an outside micrometer.

NOTE Use a new sleeve for this measurement or check service manual.

The block counterbore depth can then be subtracted from this figure to obtain an estimated sleeve protrusion of 0.001 to 0.005 in. (0.025 to 0.127 mm).

7 If counterbore does not meet manufacturer's specifications, it should be reworked. On many engines, reworking the counterbore is a simple operation that can be accomplished easily if the correct tools are available. The counterbore tool is designed to fit into the block and recut the counterbore to a uniform depth (Figure 6.18). If the block counterbore cannot be reworked within your shop, many automotive machine shops can perform this type of work. Since correct sleeve protrusion determined by block counterbore depth and condition is vitally important to correct head gasket sealing, the block counterbore must be correct before engine reassembly can continue.

NOTE After recutting, the counterbore depth has been increased in depth. As a result, the sleeve protrusion will not be correct. This can be remedied by placing shims of the proper thickness on the sleeve to make up for the metal that has been removed from the block. Not all sleeves fit in the block with protrusion. Some engine sleeves when installed are below the top surface of the block.

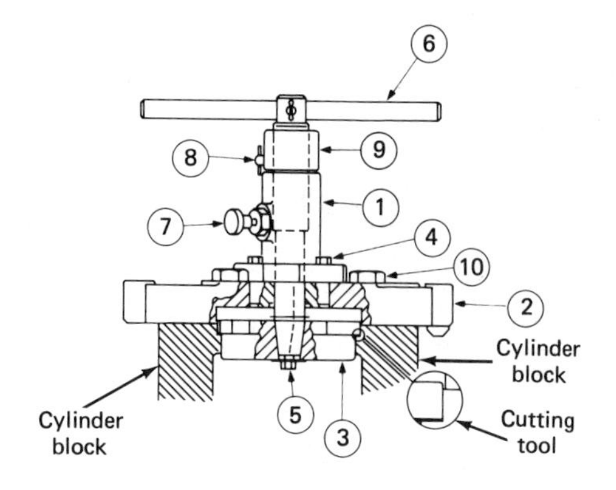

1. Driver
2. Adapter
3. Tool holder
4. Capscrew
5. Capscrew
6. Handle
7. Plunger
8. Locking screw
9. Adjusting nut
10. Capscrew and washer

Figure 6.18 Counterbore cutting tool (courtesy Cummins Engine Company).

Check sleeve position according to manufacturer specifications.

After you have completed the above outlined checks, the block is ready for reassembly.

D. Procedure for Cam Bushing Installation

Installation of cam bushings requires a special bushing driver or drivers, as described in the removal section.

CAUTION Particular attention must be given to bushing alignment during installation to insure lubrication to various parts of the engine. Many engines pump oil to the cam bushings and then to the rocker arms via a drilled passageway. Alignment of the bushing oil feed hole with the passageway is critical.

Before bushing installation, check all cam bushing bores in the block for nicks, scratches, and rust. Most bushing bores are tapered slightly on one or both sides to make bushing installation easier. Make sure the taper has no nicks or burrs that may damage the new bushing. Select bushings and determine their proper location in the block. Cam bushings may be of different widths and of internal and external diameters in any one given engine.

1 Place the bushing on the driver without the driver bar.

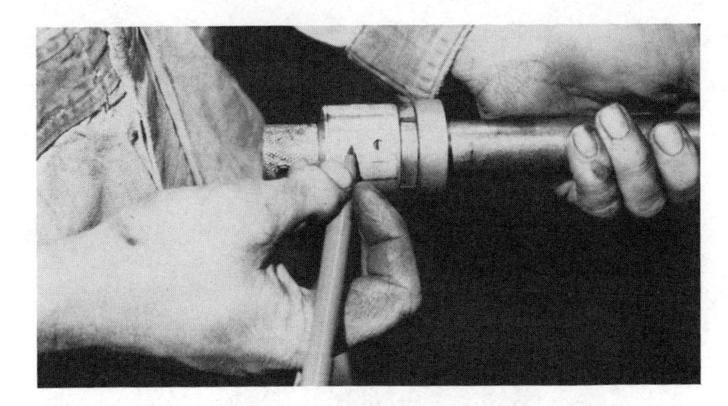

Figure 6.19 Marking cam bushing with felt tip pen.

2 Place the bushing and driver in front of the hole or bore that the bushing is supposed to be driven into.

3 Make sure the bushing is aligned with the block oil holes.

4 Mark the bushing driver in line with lube hole in the bushing (Figure 6.19).

5 Mark block in line with lube hole in bushing bore, using a magic marker pen or similar device.

6 Insert driving bar with driving cone into bushing and driver.

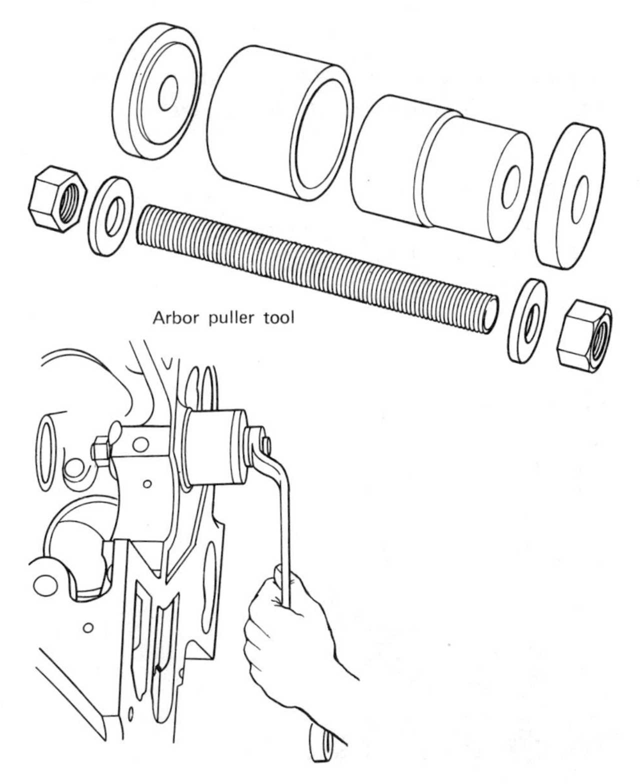

Figure 6.20 Cam bushing puller type installing tool (courtesy J. I. Case Company).

7 Tap driving bar lightly to start bushing into the bore, recheck alignment, and then drive the bushing into place with firm, solid hits with the hammer.

CAUTION When driving cam bushings into block, use care to prevent bushing from tipping sideways; this would ruin the bushing.

Cam bushings can also be installed using a puller type tool (Figure 6.20). This tool is very similar to the one mentioned previously; the main difference is that the bushings are not driven in with a hammer but pulled in. The driving rod or through bolt has been threaded and the bushings can be pulled or pushed in place by tightening a nut screw onto the threaded through bolt. This particular type of puller has an advantage because very little or no damage is done to the cam bushing, which sometimes happens when a driving type installer is used.

E. Procedure for Installation of Galley Plugs, Expansion Plugs, Cover Plates, and Oil Pressure Relief Valves

NOTE Select a suitable sealer such as Permatex, pipe joint sealer, or 3-M compound.

1 Apply sealer to galley plugs in small amounts and install and tighten plugs securely.

2 Apply sealer to expansion plugs and, using a driver, drive into block with a hammer.

a Cup type plugs can be driven in with a bushing or seal driver that just fits into the plug (Figure 6.21).

Figure 6.21 Cup type expansion plug driver.

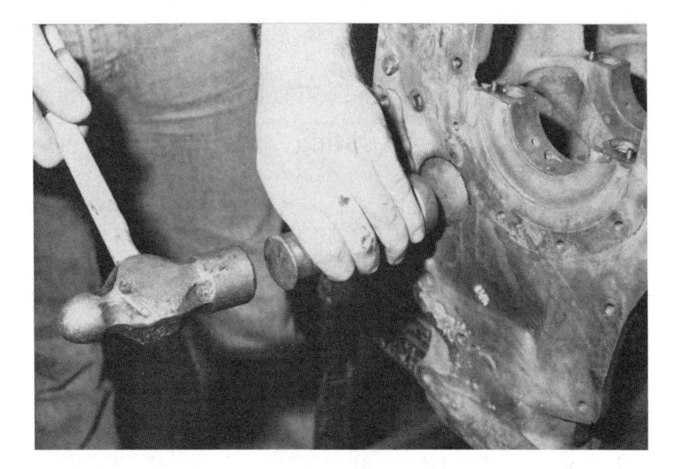

Figure 6.22 Installing convex type plug with ball peen hammer and driver.

b The convex plug can be expanded when in place by striking with a ball peen hammer and driver (Figure 6.22).

F. Procedure for Sleeve Installation (Wet Type)

After block counterbore depth and protrusion have been checked and are considered acceptable, the old sleeves should be checked to determine if they are reusable. The sleeve or sleeves should be cleaned using a cleaner in which they can be completely immersed for removal of all grease, carbon, and rust. Another very good method of cleaning is the glass bead blasting machine. This method is preferred over the immersion method if it is available. After cleaning, the sleeve should be inspected closely and, if failures are evident, the sleeve should be discarded (Figures 6.23 and 6.24).

NOTE Although used sleeves sometimes can be reconditioned and reused, it is rarely done in a major engine rebuild, as every attempt must be made to bring the engine back to a like-new condition and make it reliable for many more hours. Therefore, it is recommended that new sleeves be installed whenever possible.

An attempt should be made at this time to determine what caused the failure if it is not normal wear. Correction of the condition that brought about the sleeve failure is a very important step in any engine rebuild. In addition to failures that can be detected visually, the inside diameter of the sleeve should be checked for wear. A normal wear pattern would involve greatest wear at the top of the ring travel. The

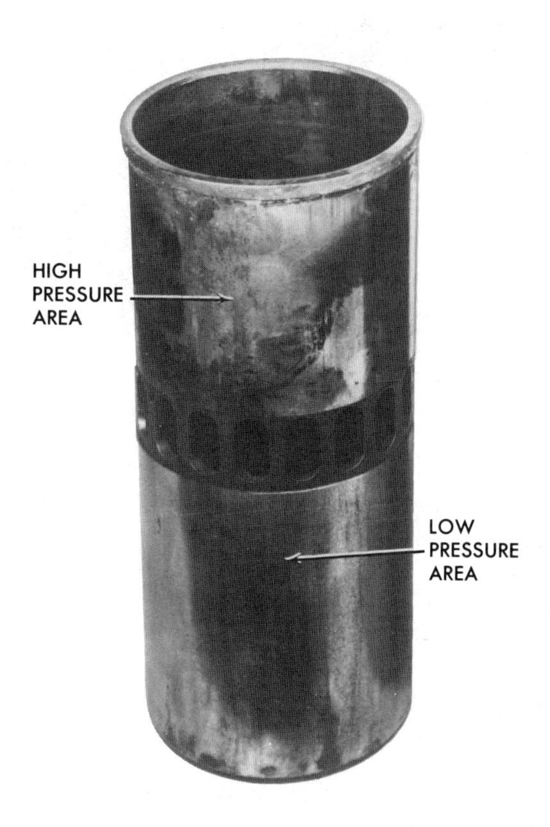

Figure 6.23 Sleeve failure (courtesy Detroit Diesel Allison, Division of General Motors Corporation).

sleeve should be checked for taper with one of the following: an inside micrometer, snap gauge, or cylinder gauge.

1 Check taper by measuring sleeve at top of ring travel and at bottom of ring travel. The difference is the cylinder taper.

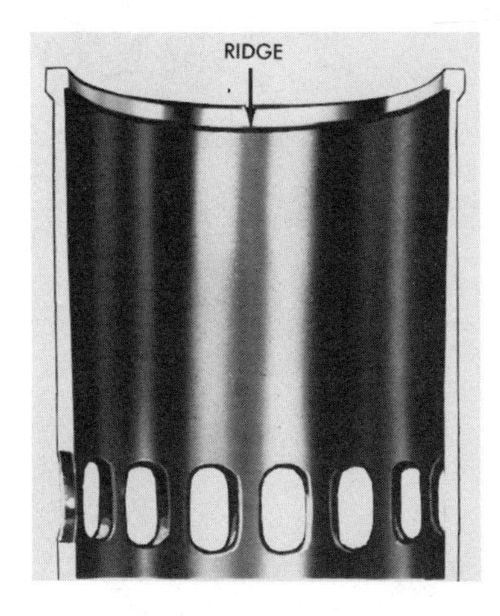

Figure 6.24 Sleeve failure (courtesy Detroit Diesel Allison, Division of General Motors Corporation).

2 Measure the sleeve for out-of-roundness, measuring the sleeve diameter at one point and then moving the micrometer 180° and measuring the diameter again.

If the sleeve taper is not excessive and the sleeve is considered reusable, some engine manufacturers suggest that the sleeve be honed or deglazed. If the sleeve is to be deglazed, a hone similar to the flex-hone type should be used (Figure 6.25). This deglazing tool plateau hones the cylinder and is recommended by most engine manufacturers.

NOTE Always wash the cylinder with soap and water after honing to remove all traces of abrasive material. After washing, dry and oil cylinder walls.

If the sleeve does not have an interference lip on it, it can be installed without the O rings as a check to see if there is any obstruction that will prevent the sleeve from fitting in the block as it should. Sleeve protrusion can be checked at this time and used as a reference later after the sleeve has been installed with O rings.

NOTE Refer to the section on checking sleeve protrusion for correct procedure.

CAUTION Make sure sleeve is removed and O ring installed before final engine assembly is attempted.

Install O rings and/or crevice seal on sleeve, making sure that O rings are not twisted when in place.

NOTE A small pick or screwdriver can be inserted under the O ring and drawn around the entire sleeve to allow the O ring to straighten or align itself in the groove on the sleeve. Consult instructions fŭrnished with sleeve or manufacturer's service manual for correct O ring placement. Some engines use two or three different types of O rings and they must be in the correct location on the sleeves or in the block (Figure 6.26).

Figure 6.25 The flex-hone tyle cylinder hone.

Figure 6.26 O ring placement on a cylinder sleeve.

Lubricate the O ring and the lower packing ring area of the block with a lubricant recommended by the engine manufacturer. Many different types are used, such as engine oil, liquid soap, and silicone.

Insert sleeve into the block carefully. Do not damage O rings on sharp ridges on top of the block. Allow sleeve to set into block on its own weight and rotate back and forth with a slight circular motion to insure that O rings are setting on the lower packing ring chamfer.

In most cases sleeve can now be pushed into place by hand or with a specially designed driver (Figure 6.27).

CAUTION Do not use excessive force when pushing in sleeve. If sleeve does not slip into place easily or bounces back up after pressure is removed, O rings may be twisted or cocked.

If this condition exists the sleeve must be removed and the O rings replaced or straightened before reinstalling the sleeve. A sleeve with an interference lip must be driven into place approximately the last $\frac{1}{8}$ inch (3 mm).

After sleeve is firmly seated, it is advisable to hold it in place with suitable bolts and washers or spacers before the final checks on sleeve protrusion are made.

Figure 6.27 Driving sleeve in place with driver (courtesy Cummins Engine Company).

Figure 6.28 Checking sleeve protrusion with a dial indicator (courtesy Cummins Engine Company).

Procedure for Checking Sleeve Protrusion

Some common used tools are:

1 Straight edge and feeler gauge.

2 Depth micrometer.

3 Dial indicator and some type of holding fixture or bracket (Figure 6.28).

Once the sleeves in a block or bank have been installed and protrusion checked, a comparison measurement should be made to determine that all sleeves are within approximately 0.001 in. (0.025 mm) of the adjacent sleeve in protrusion. This can be made with a straight edge and feeler gauge by placing a straight edge on top of the installed sleeves lengthwise with the block and using a feeler gauge of 0.0015 to 0.002 in. (0.04 to 0.05 mm) thick between each sleeve and straight edge. If 0.0015 or 0.002 in. (0.04 or 0.05 mm) can be inserted between the sleeve and straight edge, the sleeve is too low or the adjacent one is too high. Recheck protrusion and correct sleeve that is not within specification, using procedure stated earlier.

The inside of the sleeve should be checked for out-of-roundness in the packing ring area with a dial indicator, snap gauge, or inside micrometer. If the sleeve is out-of-round in excess of 0.002 to 0.003 in. (0.05 to 0.08 mm), this indicates that an O ring or crevice seal has been cocked or doubled during installation, and the sleeve must be removed and O rings checked for damage.

Before attempting to reinstall the sleeve, recheck the block packing ring area and O ring to determine reason why the sleeve could not be installed properly.

G. Procedure for Installation of Dry Type Sleeve

The installation of dry type sleeves is a somewhat different procedure since dry type sleeves do not require O rings. Particular attention must be given to the condition of the block bore and sleeve fit in the block bore. Depending on the engine, a dry type sleeve may be fit into the block bore with a 0.001 in . (0.025 mm) interference fit or with a 0.0015 in. (0.04 mm) clearance. Consult engine service manual for specifications.

Dry type sleeves require good contact with the block bore to enable combustion heat to be transferred from the sleeve to the block. As a result, the block bore must be checked for straightness and carbon or rust formation.

The block cylinder bore should be thoroughly cleaned with a rigid hone or flex-hone. It should then be checked with a cylinder gauge or inside micrometer for out-of-roundness, taper, and high or low spots.

CAUTION Care must be used in cleaning the cylinder with a hone so that no metal is removed, as this may enlarge the cylinder bore. A rigid hone, if used properly, will allow the mechanic to tell at a glance if the block bore is distorted because it will polish or ride on the high spots and not touch anywhere else.

If the block bore is distorted it should be bored or honed to accommodate an oversize sleeve. New oversize sleeves are supplied by most engine manufacturers in sizes of 0.002, 0.005, and 0.010 in. oversize. If the block is to be rebored and good quality reboring equipment with an experienced operator is not available, it should be taken to an automotive machine shop specializing in this type of work.

If it is decided to hone the block to accommodate the next size oversize sleeve, it is recommended that a rigid, adjustable type of hone similar to the Sunnen Model AN-100 be used (Figure 6.29).

NOTE Do not attempt to use this procedure unless you have honed cylinder bores before. If you need help in making this decision, ask your instructor.

This hone has two adjustable stones and two adjustable wiping pads that contact the cylinder wall. It should be driven by a $\frac{1}{2}$ in. electric drill that will run at 350 to 375 rpm. Select stones of the proper grit (usually a 100 series stone for sleeve fitting and 200 series for deglazing or cylinder resizing). Install them in or onto the hone and then insert the hone into the cylinder or sleeve to be honed. Adjust the hone so that the stone and wipers are firmly seated against the cylinder wall.

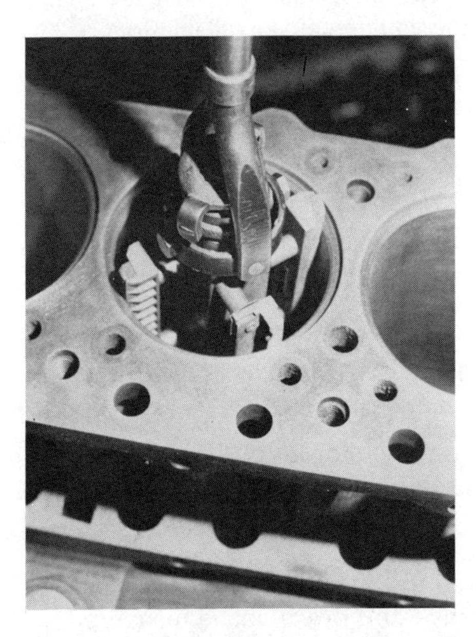

Figure 6.29 The Sunnen Model AN-100 ridgid cylinder hone.

NOTE Most hones of this type work better if used dry. Do not use oil or solvent.

CAUTION If a sleeve or cylinder is to be honed with the crankshaft and other engine parts installed, a rag should be placed over the crankshaft to prevent the hone material from getting into crankshaft passageways.

Start the drill, moving hone up and down in the cylinder, making sure the hone is moved the full length of the cylinder and clears or comes out of the sleeve by about $\frac{1}{4}$ in. (7 mm) at each end.

When honing a cylinder block bore to accommodate an oversize sleeve, hone for a short period of time, remove hone, and clean the cylinder wall with a rag and lightweight engine oil or until all abrasive material from hone has been removed. Take the sleeve that is to be used and insert into block, and check for correct fit. Continue the honing and fitting procedure until the sleeve will fit into block correctly.

CAUTION Care should be used to prevent removing too much material from the block bore through honing, making it necessary to use the next larger sleeve.

When honing a cylinder or sleeve in which a piston and rings will be directly installed, attention should be directed to the crosshatch pattern that is being established by the up and down movement of the hone. Generally a 200 to 250 grit stone is used, and the hone should be moved rapidly enough up and down to produce a crosshatch pattern 15 to 35 microinches (Figure 6.30).

NOTE When honing a used cylinder or sleeve, the cylinder will generally be tapered or smaller on the bottom of the ring travel. An attempt should be made to hone the cylinder straight by honing the smaller inside diameter more than the larger, keeping the sleeve inside diameter within specifications.

Figure 6.30 Crosshatch pattern left by hone (courtesy Deere and Company).

CAUTION Do not attempt to hone a cylinder sleeve out of the engine block by using a vise as a holding fixture because damage to the sleeve may result. Use a sleeve-holding tool or insert sleeve into block.

After sleeve, block bore, and counterbore are cleaned and checked, sleeves should be inserted into block cylinder bore by hand (Figure 6.31). The sleeve should start in the bore easily and slide down into the block bore within one inch or so of being all the way in. On a block with "loose fitting sleeves," the sleeves then should have to be driven or pulled the rest of the way into the block. It can be driven into the block with a specially made driving tool.

Some blocks use "tight fitting sleeves," which will probably just start into the block by hand and have to be pulled or pushed in the rest of the way. A method sometimes used on this type sleeve is to cool the sleeve with dry ice or in a deep freeze and then insert it into the block. After cooling, the sleeve will be smaller and will be much easier to install.

These methods are offered only as suggestions. It would be advisable to follow engine service manuals very closely in this critical area.

If, when checking the sleeve fit in the block, it is found that the sleeve fit is too loose, the block must be rebored to accommodate an oversize outside diameter sleeve.

After the sleeve has been installed and sleeve protrusion checked as outlined under wet type sleeve installation, the bore of the sleeve should be checked with a cylinder gauge or dial indicator.

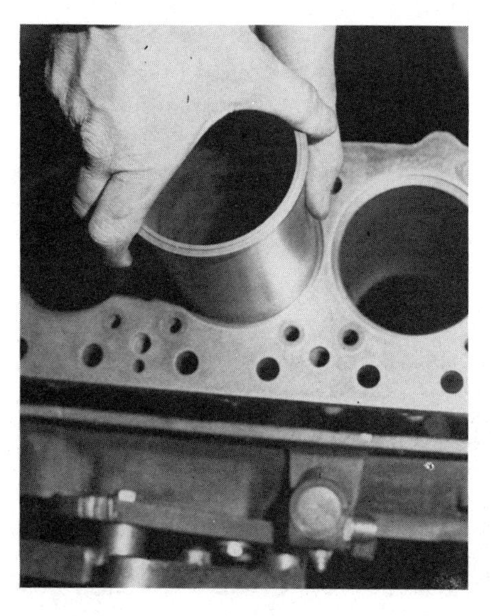

Figure 6.31 Inserting dry type sleeve in cylinder block.

H. Procedure for Checking and/or Reconditioning Bored Block

A bored type of block must be checked for out-of-roundness, taper, and scoring as outlined in the sleeve section. This type of cylinder is reconditioned by honing or reboring to an oversize. Of course, the installation of an oversize piston and rings is required.

I. Procedure for Final Testing of Block for Water Leaks after Sleeves (Wet Type) Have Been Installed

Testing the cylinder block for water leaks before final assembly is recommended by some manufacturers.

NOTE This procedure is rarely used in the field, since, if the block is in good condition and the sleeves are installed correctly, water leaks rarely occur.

Many methods of checking blocks for water leaks are used in the field, dictated primarily by the equipment available. Two methods are listed here. Select the method that is recommended by the engine manufacturer or that the equipment is available for.

1. Fill the block with antifreeze after sleeves and expansion plugs have been installed. If equipment is available, pressurize with compressed air and check for leaks.
2. Pressurize the block water jacket with air and immerse the block in a tank of hot water. Check for bubbles. Any area leaking air should be checked.

The block should now be ready for further assembly of engine parts and complete engine assembly.

NOTE If block is to be stored for a time before further engine assembly, it should be protected from rusting by first covering any openings and then painting it. A heavy coat of oil or grease should be applied to the inside of cylinder liners as a rust preventive and preservative measure.

SUMMARY

The procedures outlined in this section for block overhaul and repair are the normal procedures that would be followed in a major engine rebuild. If less than a major rebuild is to be done, such as replacing one sleeve or replacing O rings on one sleeve, those procedures that apply to that area, such as sleeve protrusion and counterbore measurement, can and should be used.

REVIEW QUESTIONS

1. List and explain the function of five component parts of a cylinder block.
2. List two common types of cylinder blocks used in modern diesel engines.
3. List four critical areas that should be checked on the cylinder block during engine overhaul.
4. List the two most common types of sleeves used in a diesel engine.
5. Wet type sleeves are sealed on the bottom by O rings. If these seals leak, what engine damage may occur?
6. Explain the procedure for installing cam bushings into a cylinder block.
7. Give four reasons for sleeve replacement.
8. Describe a situation where engine sleeves may be successfully reused.
9. List two types of expansion plugs.
10. Describe a procedure for checking "main bearing bore alignment."

7

The Cylinder Head and Components

Diesel engine cylinder heads are similar in structure to the one shown in Figure 7.1. The cylinder head's main function is to provide a head or cap to the engine cylinder. Cylinder heads may be found in different configurations depending on the number of cylinders they cover. Different configurations are designed to deal with such factors such as weight, warpage, and ease of handling. It is common to find cylinder heads that are designed for one, two, three, four, and six cylinders.

OBJECTIVES

Upon completion of this chapter the student should be able to:

1. List four component parts of a diesel engine cylinder head.
2. Explain the procedures used in inspection, overhaul, and testing of a cylinder head.
3. Demonstrate the ability to grind a valve seat correctly.
4. Demonstrate the ability to narrow a valve seat using either a grinding stone or cutter.
5. Demonstrate the ability to install a valve guide.
6. List two types of valve seals.
7. Demonstrate the proper procedure for torquing a cylinder head.
8. Demonstrate the ability to adjust valve clearance (set tappets).

GENERAL INFORMATION

The diesel engine cylinder head (see Figure 7.1) consists of:

1. A single piece casting (1) that may cover one or more cylinders.
2. Valve guides (3) that guide the valves as they are opened and closed.
3. Passageways or ports (11) that are opened and closed by the valves.
4. Drilled bores or seats for the injector or injection nozzles, water passageways, ports for intake air, and exhaust gases.

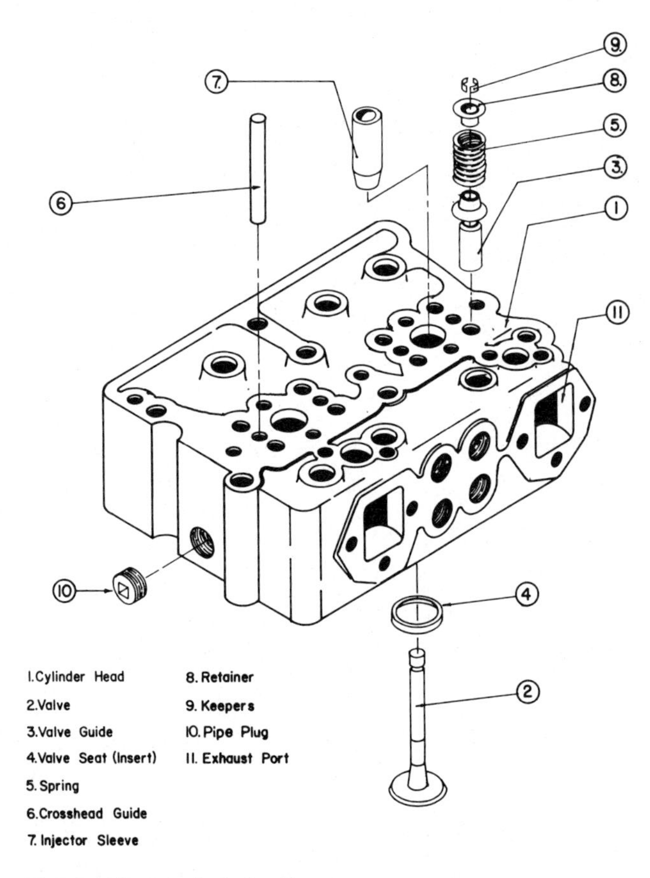

Figure 7.1 A typical diesel engine cylinder head.

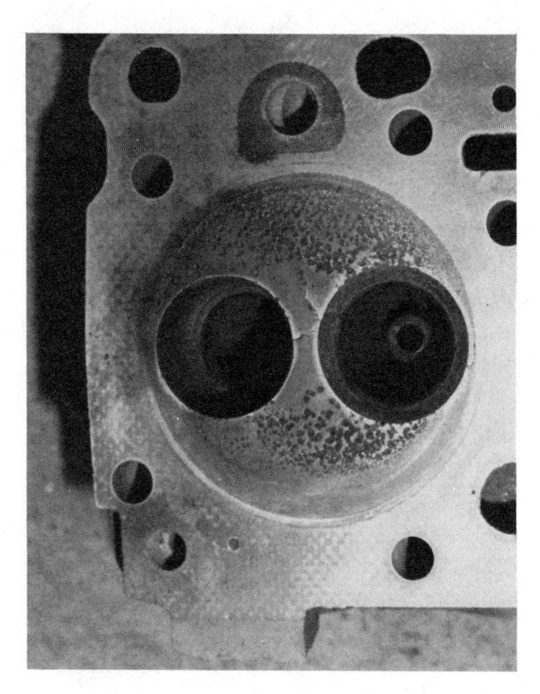

Figure 7.2 A cracked diesel engine cylinder head showing the effects of water in the combustion chamber.

5 Special steel alloy inserts or seats (4) for the valves.

6 Precombustion chambers (some heads).

In modern diesel engines, cylinder head service must be an important part of major engine overhaul. Cylinder head service is sometimes performed hastily and with little consideration of the important functions the cylinder head must perform. Along with the cylinder and rings the cylinder head aids in the development of compression and oil control. It is recommended then that the cylinder head service be performed with care and accuracy to provide long hours of trouble-free engine operation. If the cylinder head is being repaired, if a routine major overhaul is being done, or if the head has experienced a premature valve failure, all the following service recommendations should be performed.

I. Procedure for Disassembly of Cylinder Head

Assume that the cylinder head or heads have been removed from the engine and are ready to be disassembled and reconditioned. If not, refer to chapter on dissassembly. Before disassembly of cylinder head, all loose grease and dirt should be removed either by using a steam cleaner or high pressure washer.

CAUTION All injectors or injection nozzles should be removed before the head is steamed or washed.

A visual inspection should be made to determine if the head is usable. Any visible cracks in the cylinder head will make the head unfit for further use.

NOTE There are some situations where small cracks in the cylinder are not damaging or detrimental to the operation of the engine. Each engine design may have some peculiarity in this regard. If the mechanic does not have experience with a particular model of cylinder head, it must be taken to a shop or repair station that has been rebuilding or servicing cylinder heads of that type. Figure 7.2 shows a cracked cylinder head that is no longer usable.

Many times cracked cylinder heads are repaired by welding or pinning, and in many cases they have proven to be dependable. A firm that has considerable experience should be selected if the cylinder head is to be repaired, since a rebuilt head that does not stand up

Figure 7.3 Using C clamp type valve spring compressor.

in service may ruin the rest of the engine by allowing coolant to leak into the engine lube oil. If a cracked head is discovered during a major rebuild, it is my personal opinion that the head should be replaced with a new one. The increased cost will be offset in the long run by increased engine life and dependability. If there is some question about rebuilding or replacing a cylinder head, check with your instructor.

The valve springs, keepers, and retainer should be removed with a valve spring compressor. Many types of compressors can be used, but the most common one in the field is the C clamp type (Figure 7.3).

A With the valve spring compressor in the open position, place it on the valve and valve spring.

B Adjust the spring end jaws with the adjusting screw so that the jaws clamp snugly on the valve spring retainer.

C Compress the spring by closing the valve spring compressor.

Figure 7.4 Tapping valve spring retainer with hammer.

NOTE It may be impossible to break the retainer loose from the retaining clips and compress the spring with the force exerted by the valve spring compressor. A slight tap with a hammer on the retainer while attempting to compress the spring will aid in loosening the retainer (Figure 7.4).

D When the valve spring retainer is loosened, compress the spring far enough to remove keepers and retainer.

E Loosen compressor and remove spring and valve from cylinder head.

F Visually inspect valves and valve seats for signs of damage and wear.

G Remove any other part or parts from the cylinder, such as water plates, thermostat housings, and brackets.

II. Procedure for Cleaning Cylinder Head

A Immerse the cylinder head into a dip or soak type cleaning tank and allow it to soak until all baked-on grease, oil, and paint have been removed. In most cases two hours will be sufficient.

B Remove the cylinder head from the cleaning tank and steam clean or rinse with hot water.

C Blow out all passageways with compressed air.

D Place the head on some type of suitable stand or bench and sand the cylinder head gasket surface with a sander (Figure 7.5).

Figure 7.5 Using an orbital sander to clean the cylinder head surface.

III. Procedure for Testing and Checking Cylinder Heads for Cracks

At this time the cylinder head must be checked for cracks that may not have been detected during visual inspection. The three most common methods available to perform this check are:

A *Electromagnetic crack detector.* An electromagnetic crack detector is a U-shaped device that is set on the surface of the cylinder head and energized with an electrical power source (Figure 7.6). Metal filings are sprinkled around the detector and, if the head has a crack, the crack will attract the metal filings, making the crack visible.

B *Pressure testing.* Plates or plugs to cover all water inlets and outlets must be available to pressure test a cylinder head. After plates have been bolted on, connect air pressure to the head with appropriate fittings and immerse in a tank of water. Any cracks will be pinpointed by the air bubbles escaping from them. Identify the source of the bubbles to make sure they are not coming from the plates or the plugs being used to seal the head. Particular attention must be given to the area around the valve seats and injector sleeves. Mark the leaking sleeves for replacement.

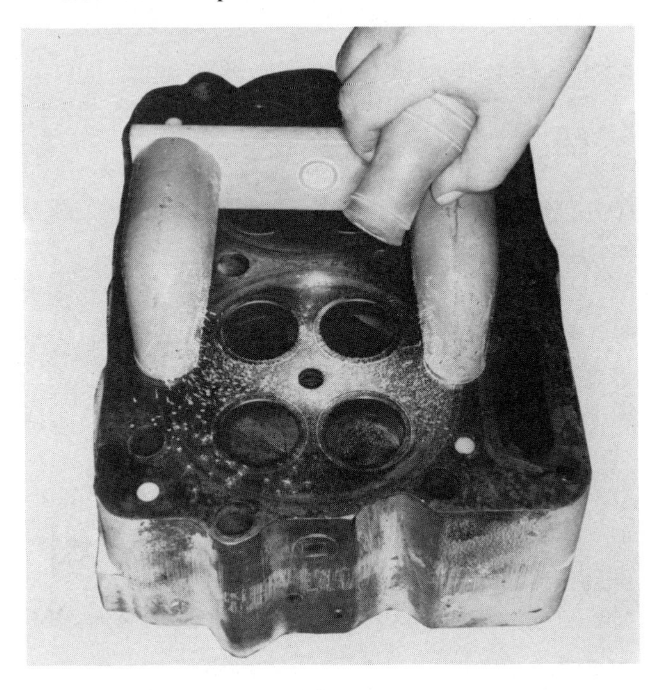

Figure 7.6 Checking a cylinder head for cracks using electromagnetic tester (courtesy Cummins Engine Company).

NOTE When testing Cummins diesel cylinder heads using air pressure, it is recommended that the sleeve be held in place using a scrap injector or a sleeve-holding tool. Check the engine service manual to determine if the cylinder head requires a sleeve hold-down device.

C *Dye penetrant.* Dye penetrant is a crack-detecting method that requires no special equipment with the exception of a can of spray type penetrant and a can of spray developer. When using the dye penetrant, spray the area to be checked and wipe off or remove all excess dye. Spray on the developer. It will draw the dye penetrant from the crack, making it visible.

Of the three types of crack detection listed here, the electromagnetic and dye penetrant would be used in areas where they can be seen. The pressure testing method should be used where there is a possibility of a crack in an area that cannot be seen, such as valve ports, combustion chambers, and all other areas not visible.

IV. Procedure for Testing Cylinder Head for Warpage

Check the cylinder head for warpage using a straight edge and a 0.004 (0.10 mm) feeler gauge (Figure 7.7). The feeler gauge must not pass between the straight edge and cylinder head at any point. If it does, the cylinder head must be resurfaced.

NOTE Engine manufacturers may vary in their recommendations as to what point during the cyl-

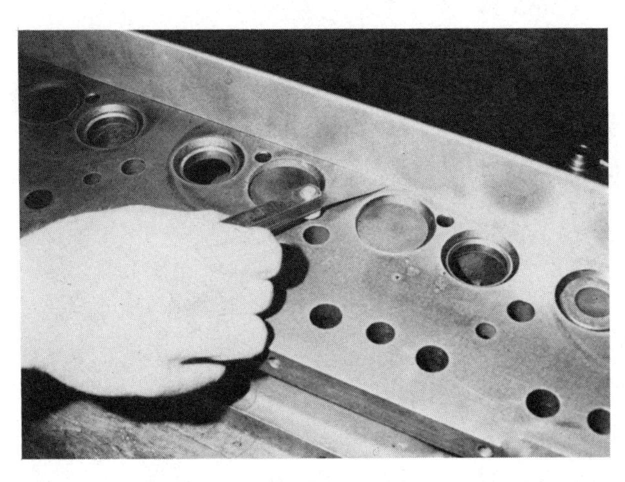

Figure 7.7 Checking a cylinder head for warpage by using a feeler gauge and a straight edge.

inder head rebuild the head should be resurfaced. Check your engine service manual for instructions at this point. Most automotive machine shops have the necessary equipment to resurface the cylinder head.

V. Procedure for Testing and Replacement of Injector Sleeves

Most diesel engine cylinder heads have a copper sleeve into which the injector is installed. This copper sleeve is installed directly into the water jacket and must be sealed at top and bottom. If leakage occurs at the copper sleeve, water may leak into the combustion chamber. As a result, the sleeve seal must be checked carefully during cylinder head repair for water leakage. Leakage testing of the injector sleeves will involve pressurizing the coolant passageways as outlined under pressure testing of the cylinder head.

If it is found that the injector sleeves leak and they are to be replaced, the following procedure should be followed:

A Remove the injector sleeve following the manufacturer's instructions. Most engine manufacturers provide specially designed tools to install and remove the injector sleeves (Figure 7.8).

B Before sleeve installation is attempted, the bore in the cylinder head that the sleeve fits into must be thoroughly cleaned, using compressed air or by sanding with emery paper.

NOTE Many cylinder heads use O rings in the sleeve bore to help seal the sleeve to the cylinder head. During cleaning of the sleeve bore make sure the O ring grooves are cleaned.

C Install the new sleeve on the installation tool and insert sleeve and installation tool in the cylinder head. Some sleeves are simply driven in place (Figure 7.9), while others must be rolled over on the combustion chamber end and reamed (Figure 7.10) after installation.

Figure 7.8 A typical injector sleeve installation tool (courtesy Detroit Diesel Allison, Division of General Motors Corporation).

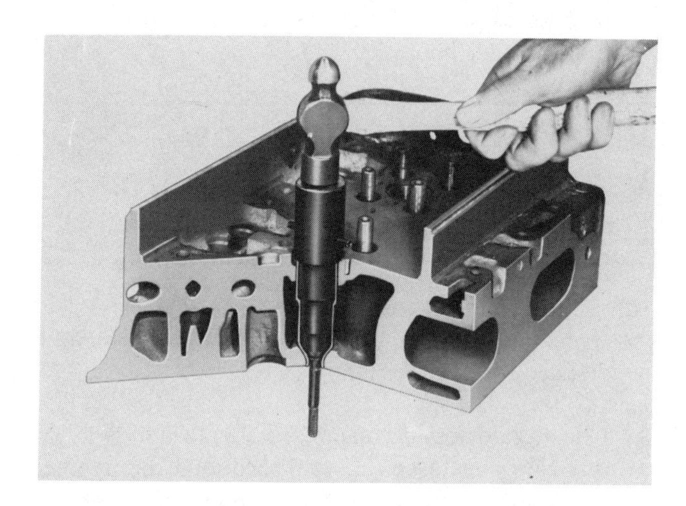

Figure 7.9 Installing an injector copper sleeve (courtesy Detroit Diesel Allison, Division of General Motors Corporation).

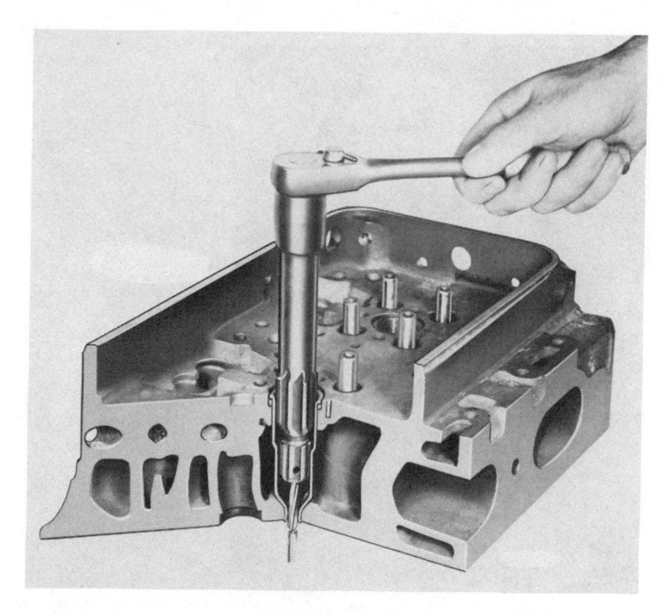

Figure 7.10 Reaming a copper sleeve after installation (courtesy Detroit Diesel Allison, Division of General Motors Corporation).

D After injector sleeves have been installed it is recommended that the cylinder head be pressure checked to insure that a leak tight seal has been established to prevent water leakage around the sleeves.

E After a new injector sleeve has been installed, injector tip protrusion should be checked and compared to manufacturer's specifications. Injector tip protrusion is the distance the injector tip protrudes below the surface of the cylinder head gasket surface. Too much or too little protrusion may cause the injector spray to strike the piston in the wrong place or strike the cylinder wall. This incorrect positioning of the injector spray can cause incorrect cylinder operation.

VI. Procedure for Valve Guide Checking and/or Replacement

After the cylinder head has been checked or resurfaced and is considered usable, the valve guides should be checked for wear as follows:

A Check guide inside diameter with a snap, ball, or dial gauge in three different locations throughout the length of the guide (Figure 7.11).

NOTE Experienced mechanics usually can determine if the valve guide is worn excessively by inserting a new or unworn valve into the guide to within approximately ¼ in. (7 mm) of the cylinder head or valve seat and moving it from side to side. The method of measurement used will be determined by the mechanic's experience and the degree of accuracy desired.

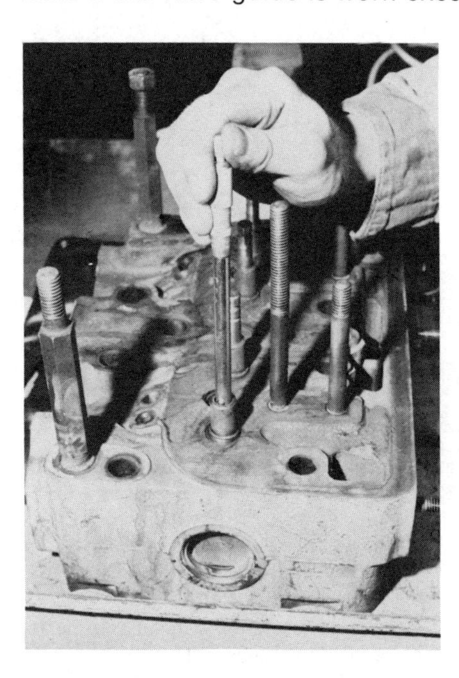

Figure 7.11 Measuring a valve guide with a ball gauge.

B. Procedure for Valve Guide Replacement

If guides are worn excessively, they should be replaced if they are the replaceable type. In most cylinder heads the valve guides can be removed by using a driver. Select a driver that fits the valve guide. After selection of a driver the guide should be removed as follows:

1. Support the cylinder head on a block or stand to allow the guide to be driven out.
2. If guide is to be pressed out, place the cylinder head in the press and align the driver with the press ram.
3. Note the position of the guide in relationship to the cylinder head for reference when installing a new guide.
4. Drive or press the guide out of the cylinder head, making sure that the driver is driven or pressed straight (Figure 7.12). If it is not pressed straight,

Figure 7.12 Removing a valve guide from cylinder head with a press.

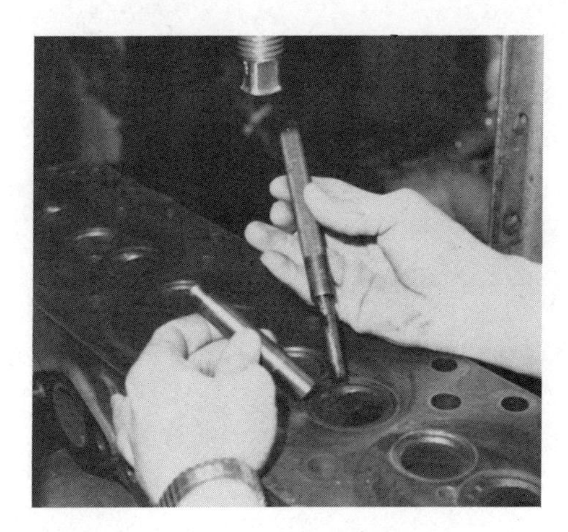

Figure 7.13 Valve guide driver and valve guide.

damage to the valve guide or cylinder head may result.

NOTE Integral guides (guides machined into the head) that cannot be replaced should be reconditioned by knurling or by drilling the guide out and inserting a guide sleeve.

5 After the guide has been removed, check the guide bore for scoring. A badly scored guide bore may have to be reamed out to accommodate the next larger size guide.

6 After guide bore is checked, select the correct guide (intake or exhaust) and insert it in guide bore (Figure 7.13).

Figure 7.14 Inserting guide and guide driver in valve guide bore.

NOTE Guide insertion can be made easier by the use of a press fit lubricant such as supplied by Sunnen Manufacturing Company.

7 Insert guide driver and drive or press guide into cylinder head until correct guide position is reached (Figures 7.14 and 7.15). If manufacturer's specifications are not available, position guide in the same position that the old guide was in.

8 Many manufacturers recommend that a new guide be hand reamed after installation to insure that the guide inside diameter did not change during the installation process.

NOTE Some engine manufacturers have guides that are manufactured by a special process to make them wear longer; these guides should not be reamed. Check engine manufacturer's recommendation closely in this area.

VII. Procedure for Checking Valve Crossheads (Bridges) and Guides

Valve crossheads are used on some types of diesel engines that use "four valve heads." Four valve heads used on four stroke cycle engines have two intake and two exhaust valves per cylinder. Two stroke cycle

Figure 7.15 Pressing guide into cylinder head.

Figure 7.16 Detroit Diesel valve crosshead.

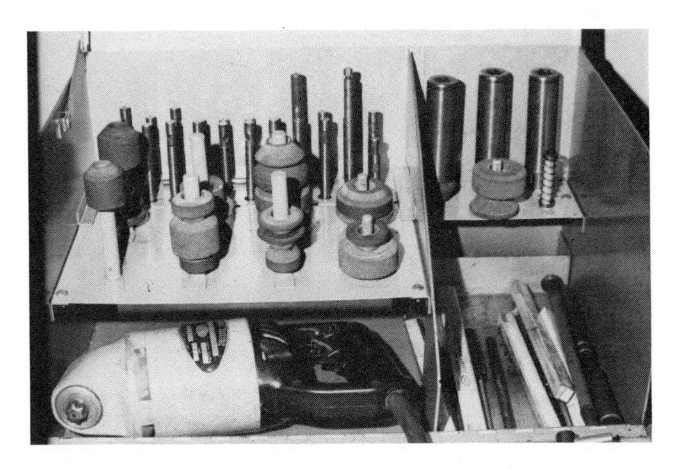

Figure 7.17 Valve seat grinding tool set.

engines such as Detroit Diesel use four exhaust valves in each cylinder. The crosshead is a bracket or bridge-like device that allows a single rocker arm to open two valves at the same time (Figure 7.16).

The crossheads (bridge) must be checked for wear as follows:

A Check crosshead for cracks visually and with magnetic crack detector if available.

B Check crosshead inside diameter for out-of-roundness and excessive diameter.

C Visually check for wear at the point of contact between rocker lever and crosshead.

D Check adjusting screw threads for broken or worn threads.

E Check crosshead guide pin for diameter with micrometer.

F Check crosshead guide pin to insure that it is at right angles to head-milled surface.

G If guide pin requires replacement, check engine service manual for correct procedure.

VIII. Procedure for Valve Seat Checking and Reconditioning

After guides have been replaced or reconditioned, valve seat checking and reconditioning should be done. Valve seats must be checked for looseness by tapping the seat lightly with the peen end of a ball peen hammer. A loose seat will produce a sound different from the sound produced while tapping on the cylinder head. In some cases a loose seat can be seen to move while tapping on it. If the seat is solid, check it for cracks and excessive width. Normal valve seat width is 0.0625 to 0.125 in. (1.59 to 3.175 mm).

If the seat passes all checks it should be reconditioned as outlined, using a specially designed valve seat grinder. Most valve seat grinders are similar to the one shown in Figure 7.17.

A. Procedure for Valve Seat Reconditioning or Grinding

1 Select the proper mandrel pilot.

NOTE Selection of the mandrel pilot is done by measuring the valve stem with a micrometer or caliper or by referring to manufacturer's specifications. After experience has been gained in this area, selection of the pilot is easily done by a visual check.

2 Pilots are usually one of two types, expandable or tapered. An expandable pilot (Figure 7.18) is inserted into the guide and then expanded by tightening the expanding screw on the pilot, which expands the pilot. Expandable pilots are not considered as accurate as tapered pilots and should be used only when a tapered pilot is not available. Some mechanics prefer expandable pilots because they compensate for guide wear better than a tapered pilot; by expanding into the guide the pilot tightens and adjusts to the guide size or wear. A tapered pilot (Figure 7.19) does not have an expanding screw and relies on the taper of the pilot to tighten it into the guide.

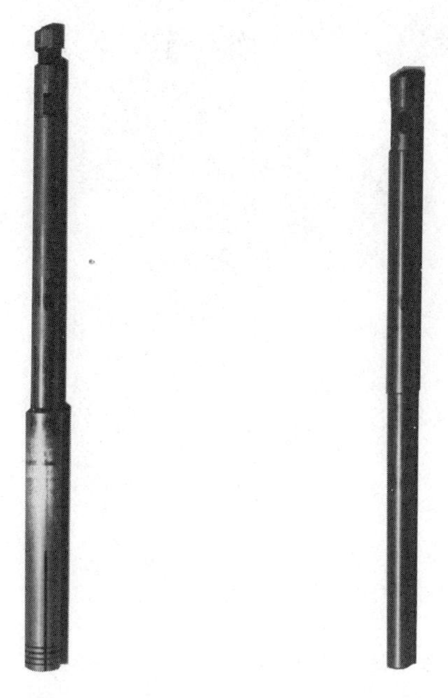

Figure 7.18 Expandable pilot. **Figure 7.19** Tapered pilot.

B After the pilot has been selected and inserted into the guide (Figure 7.20), the grinding wheel or stone must be selected. Selection of the grinding wheel is made by determining the valve seat angle, diameter, and the seat material.

NOTE If seat angle, diameter, and material cannot be determined by visual inspection, the manufacturer's specifications should be consulted.

Stellite-hardened seats usually will require a roughing stone for fast cutting and a finishing wheel to finish. Normal cast iron or steel seats generally can be ground with a finishing wheel.

Figure 7.20 Tapered pilot inserted into the valve guide.

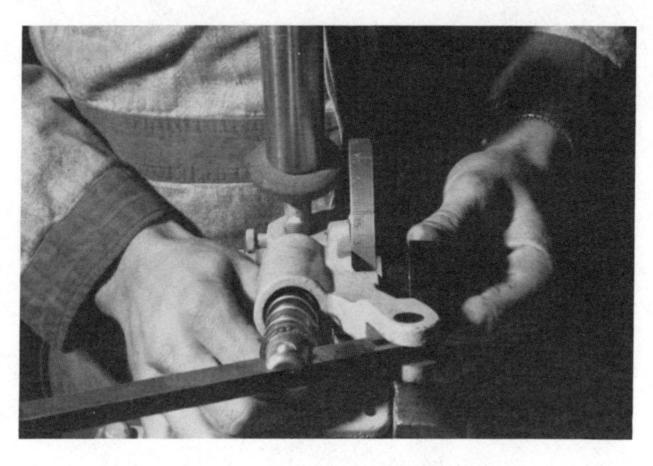

Figure 7.21 Adjusting valve seat grinding stone cutter to correct angle.

C After selection, the grinding wheel should be dressed to the correct angle and diameter of the valve seat to be ground or reconditioned on a dressing tool.

1. Place grinding wheel and holder on dressing tool pilot.
2. Adjust holder to the correct angle (Figure 7.21) so that the dressing tool diamond will contact the stone when the diamond is moved up and down across the face of the stone.
3. Place driver in grinding wheel holder.
4. Start driver and hold with one hand, using the free hand to move the diamond across the face of the stone.
5. Grinding wheel should be dressed at the same angle as the valve seat unless an interference angle is desired or recommended by the manufacturer (Figure 7.22).

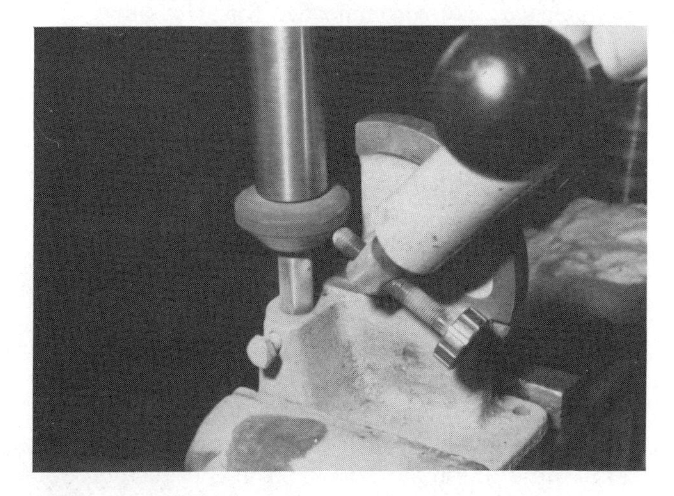

Figure 7.22 Dressing valve seat grinding stone.

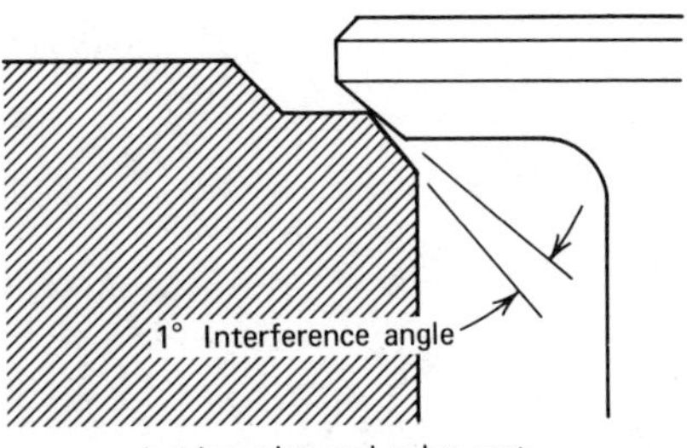

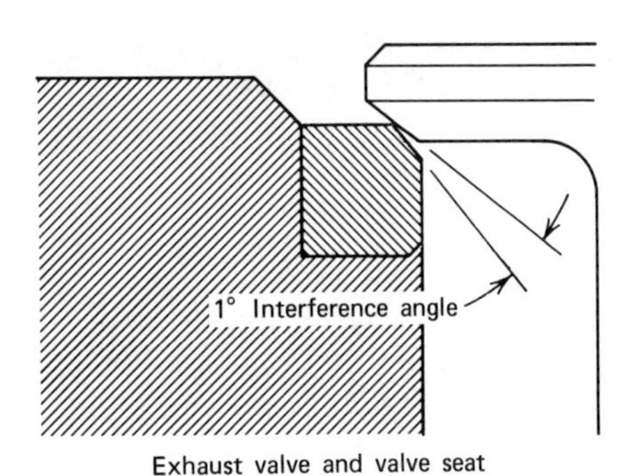

Figure 7.23 Interference angle between valve and valve seat (courtesy J. I. Case Company).

NOTE To establish an interference angle between valve and valve seat, valve and valve seat must be ground at a different angle to provide a sharp or narrow seat to valve contact. For example, seat ground at 45° with valve ground at 44° (Figure 7.23).

This interference angle with its narrow contact area aids in seating the valve during starting by helping to cut through the carbon particles that may accumulate on the valve seat. Interference angles are not recommended for valves that have rotators, as the rotating valve will remove any carbon that may accumulate on the seat or valve face.

D When the grinding wheel has been dressed to the proper angle, place the grinding wheel and holder on the pilot that has been inserted into the valve guide.

1. Place driver into grinding wheel holder (Figure 7.24) and operate driver until seat has been ground to the proper finish.

NOTE The driver can be tipped or swayed from side to side to improve the grinding wheel's cutting characteristics. In addition, on hard valve seats the stone may require several dressings before the valve seat is refinished correctly.

Figure 7.24 Inserting the driver into the stone mandrel.

2. Visually check seat for cracks and seat width (Figure 7.25). (Follow manufacturer's recommendation on seat width.) General specs. $\frac{1}{16}$ to $\frac{1}{8}$ in. (1.59 to 3.175 mm). Valve seat concentricity must be checked at this point with a concentricity indicator (Figure 7.26).

NOTE Valve seat concentricity can best be explained by saying that the valve seat must be the same distance from the center on all sides, as shown in Figure 7.27.

3. If valve seat is cracked it must be replaced. If cylinder head did not have valve seats as original equipment, it will have to be machined or counterbored to accommodate a replacement seat. If originally equipped with the valve seats, the seat must be replaced.

NOTE Special tools and equipment are required to replace and/or install valve seats; see

Figure 7.25 Measuring valve seat width with a small ruler.

Figure 7.26 Checking valve seat concentricity.

Figure 7.28 Narrowing a valve seat using a grinding stone.

Section E. If this equipment is not available, the cylinder head must be taken to an automotive machine shop for repair.

4 If the valve seat is too wide, as measured from top to bottom, it should be narrowed. Seat narrowing is accomplished by removing metal from the top side of the seat, using a grinding stone (Figure 7.28) of a lesser angle or a cutting tool (Figure 7.29) designed especially for this purpose.

NOTE When resurfacing or grinding valves and seats, do not remove any more material than is absolutely necessary to refinish the valve or valve seat. Removal of excess metal may cause the installed valve to be too low, which in turn increases top dead center volume and could result in a cylinder misfire during engine operation or starting.

5 If, after grinding the valve and valve seat, the valve head height is not within specifications, a new seat and occasionally a new valve must be installed to bring the valve protrusion back to specifications. Valve protrusion or head height can be checked using a straight edge and feeler gauge (Figure 7.30).

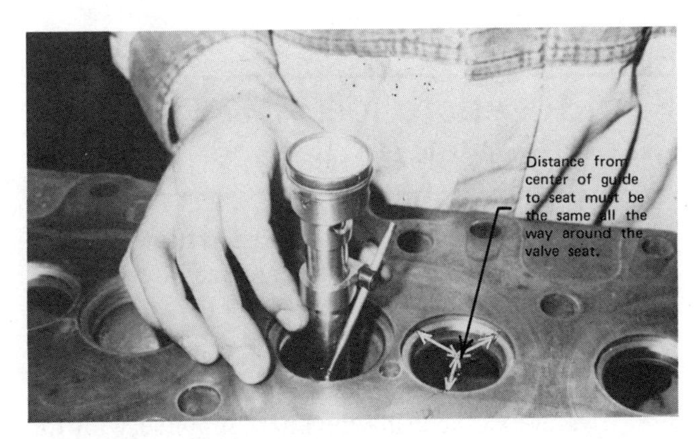

Figure 7.27 Valve seat concentricity.

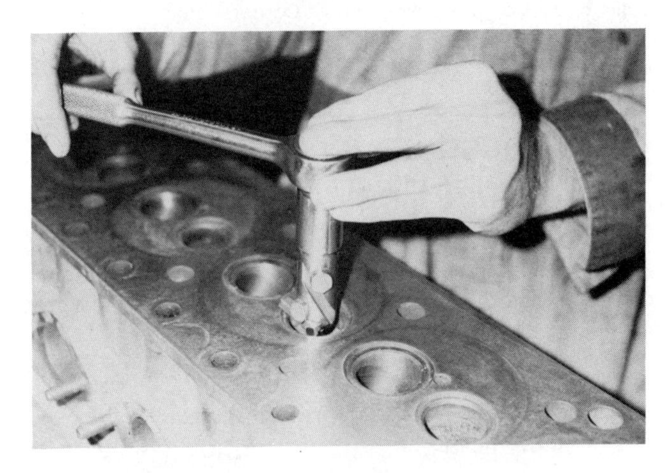

Figure 7.29 Narrowing a valve seat using a cutting tool.

Figure 7.30 Checking valve head height with feeler gauge and straight edge.

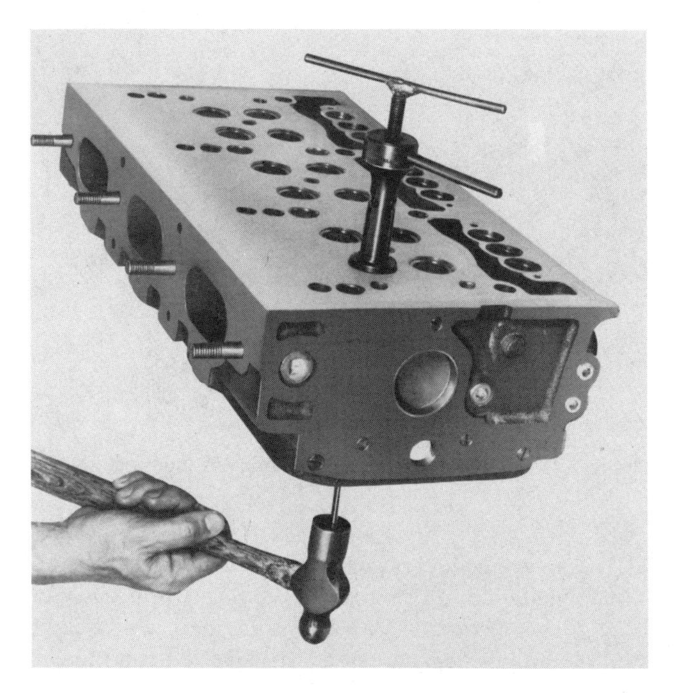

Figure 7.31 Valve seat removing tool (courtesy Detroit Diesel Allison, Division of General Motors Corporation).

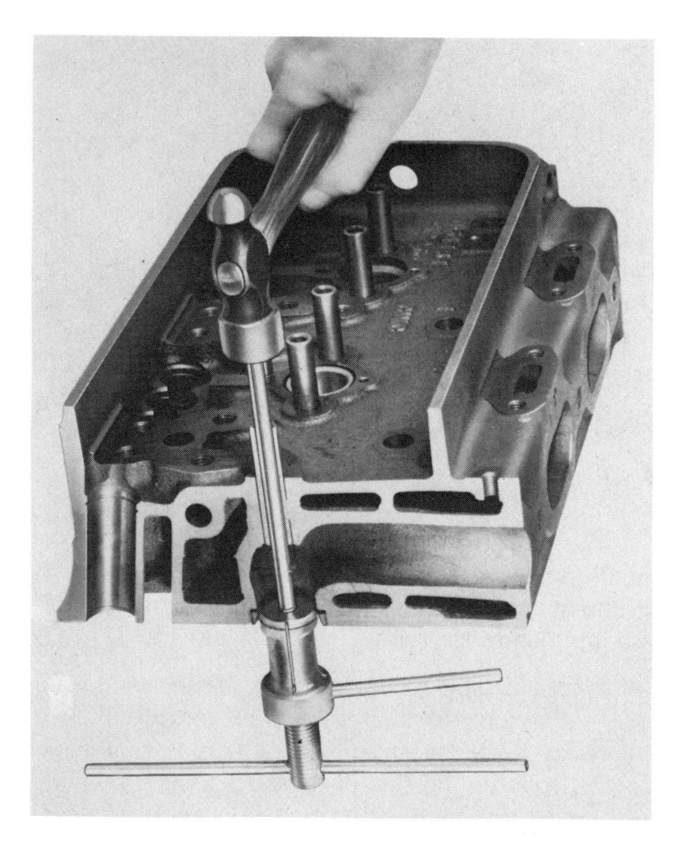

E. Procedure for Valve Seat Replacement

If it is determined that the valve seat must be replaced because of cracks or excessive width, the following procedure should be used:

1 Using a removing tool (Figure 7.31), remove the seat carefully to prevent damage to the cylinder head.

NOTE Use extreme caution when removing the valve seat. Damage to the cylinder head in the valve seat area may render the head unfit for further use.

2 Clean valve port and seat area with a carbon brush and compressed air.

NOTE It is recommended that after removal of a valve seat the seat counterbore should be enlarged to allow installation of an oversize seat. Although this practice is used most often, with additional experience the mechanic will be able to determine the type of cylinder heads that will allow the successful replacement of valve seats without enlargement of the counterbore. A slightly oversize valve seat (0.005 to 0.010 in., 0.013 to 0.25 mm) is generally available and may be used to insure a good seat fit in the counterbore if the counterbore is not recut.

CAUTION Valve guides must be in good condition or replaced before any attempt to cut valve seat counterbore or replace valve seats.

3 If it is decided to enlarge the valve seat counterbore or to cut a counterbore in a head not originally equipped with valve seats, a cutting tool or reseater similar to the one shown should be used (Figure 7.32). Many types of counterbore cutting tools are available. Follow operating instructions of cutter being used to eliminate damage to the cylinder head.

F. Procedure for Installing Valve Seats

1 After valve seat counterbore has been enlarged or considered usable, select a mandrel pilot for the insert that will fit snugly in the valve guide.

2 Obtain a new valve seat insert that will fit the counterbore using the engine parts manual as reference or using the chart supplied with the valve seat insert cutting tool set.

Figure 7.32 Valve seat counterbore cutter.

3 Visually inspect the counterbore, making sure that it is free from metal particles and rough edges. Select a driver that has an O.D. (outside diameter) slightly smaller than the seat.

4 Place ring insert over driver pilot onto the cylinder head counterbore.

5 Place driver onto pilot, and with a hammer drive the valve seat into the counterbore using sharp, hard blows.

NOTE Alternate methods of valve seat installations are: 1. Shrinking valve seat by cooling and then driving them in. 2. Warming the cylinder head in hot water and then installing the seat.

CAUTION Safety glasses should be worn during this operation as valve seat inserts are very brittle and may shatter, causing eye damage.

6 After driving the seat in place, it is recommended that the seat be staked or knurled in place (Figure 7.33). If a knurling or staking tool is not available, a $\frac{1}{4}$ in. (7 mm) round end punch may be used to stake insert around its outer circumference.

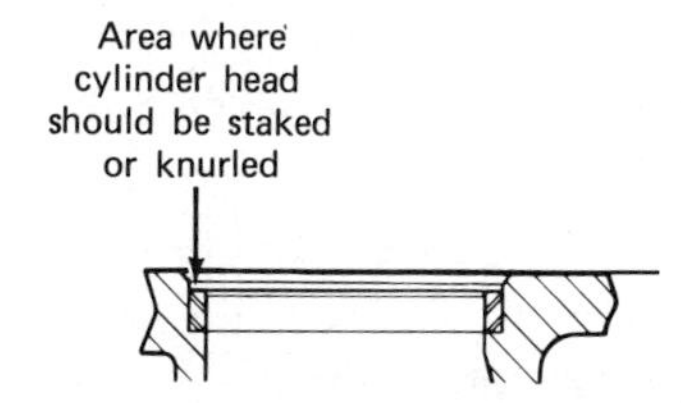

Figure 7.33 Area of cylinder head where cylinder head should be staked or knurled to hold valve seat in.

NOTE If the seat is cast iron, no knurling or staking is necessary as the seat has the same coefficient of expansion as the cylinder head. Seats that are made of steel alloy require staking since their expansion rate does not match that of cast iron. As a result, they may fall out during engine warm-up.

IX. Valve Inspection, Cleaning, and Refacing

A. Procedure for Inspection

A decision must be made at this time to replace or reface the valves. It is common procedure to replace all valves in an engine that has been run several thousand hours. To determine if the valves are reusable they should be inspected for:

1 Carbon buildup on underside of the head. A buildup of carbon indicates that oil has been leaking into the combustion chamber between the valve stem and the valve guide.

2 Stretched stem or cupped head. Valves that are badly cupped (Figure 7.34) or stretched (Figure 7.35) should not be reused, as they could break and ruin the engine. Cupped or stretched valves are usually caused by excessive heat, excessive tappet clearance, engine overspeeding, or weak valve springs.

3 Nicks or marks in the head. Valves with nicks or marks in the head indicate the valve was in a cylinder that had metal particles in it. Metal particles in a cylinder usually come from broken piston rings, broken pistons, or broken valves. Replace all valves that show any sign of damage.

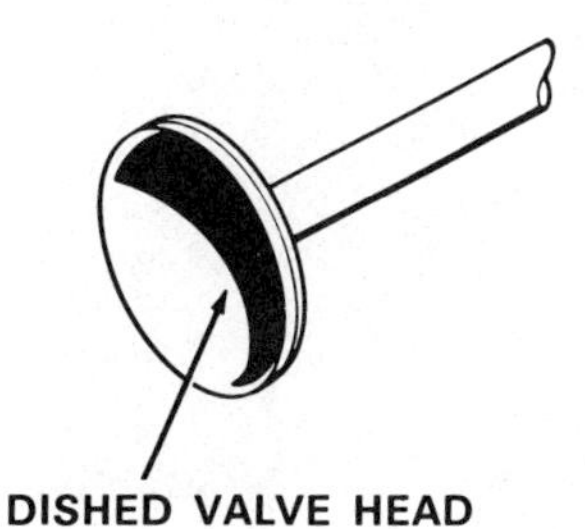

Figure 7.34 Cupped valve.

Figure 7.35 Stretched valve (courtesy Deere and Company).

4 Burned or pitted area in face (Figure 7.36). Burning or pitting is caused by tight tappet adjustment, dirty inlet air, or engine overfueling.

5 Worn keeper (collet) grooves (recesses).

6 Scored or worn stem (Figure 7.37). (Stem diameter should be checked with micrometer.)

7 Margin width.

8 Worn stem end.

B *Cleaning.*

1 If the valve passes all the checks listed above, it must be thoroughly cleaned using a wire buffing wheel.

CAUTION Do not press valve against wire wheel too hard as damage to the valve may result. Safety glasses must be worn during valve buffing.

Figure 7.36 Valve with burned or pitted face.

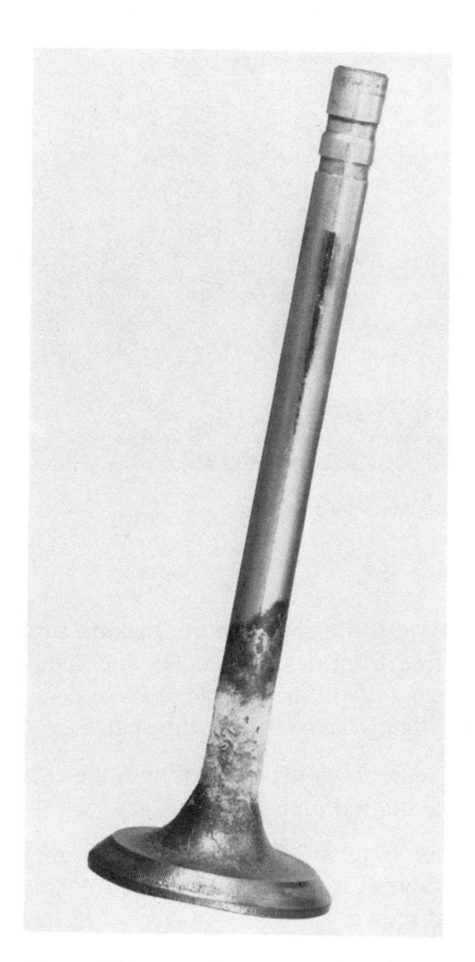

Figure 7.37 Worn or scored valve stem (courtesy Deere and Company).

A much preferred method of valve cleaning is the glass bead blaster if it is available.

2 After cleaning, the valve must be checked for warpage. Experienced mechanics usually check valves for warpage by inserting them into the valve-refacing machine. If the valve is warped it can easily be seen when the valve is moved up to the grinding wheel.

C *Valve refacing.*

Valve refacing is done on a valve-refacing machine similar to the one shown in Figure 7.38.

1 Determine at what angle the valve face is ground by visual inspection or by checking the manufacturer's specifications.

2 Adjust valve chuck head to correspond with the valve face angle (Figure 7.39). If an interference angle is recommended by the engine manufacturer, set the valve chuck head at that angle at this time. Although the valve is gener-

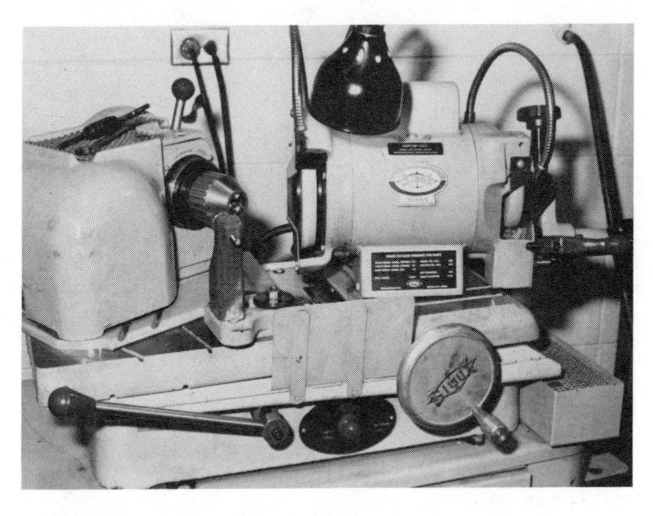

Figure 7.38 Valve refacing machine.

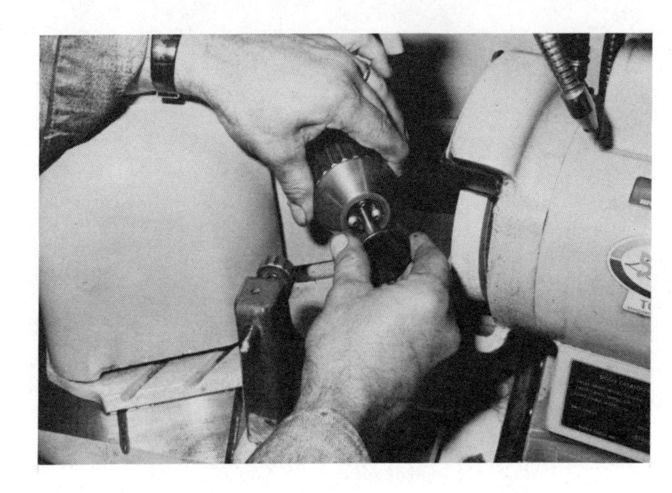

Figure 7.41 Inserting valve into chuck.

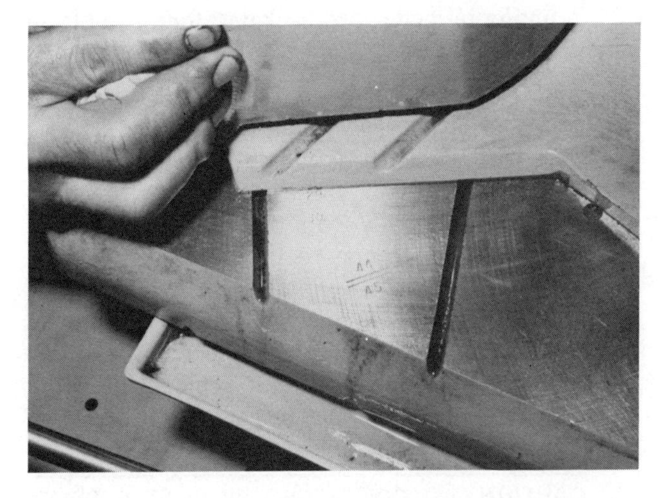

Figure 7.39 Adjusting the valve chuck head to the correct angle.

ally ground with an interference angle, some manufacturers recommend grinding the valve seat to establish the interference angle.

3 With valve grinding machine stopped, install and adjust the grinding wheel dresser so that it will just touch the grinding wheel (Figure 7.40). Start the machine and move the diamond dresser back and forth across the face of the grinding wheel until the wheel surface is smooth and flat all the way across. Remove the dressing attachment.

CAUTION Safety glasses must be worn during valve refacing.

4 Place the valve in the machine chuck (Figure 7.41) and adjust the valve stop (Figure 7.42) so that the valve will be positioned in the chuck

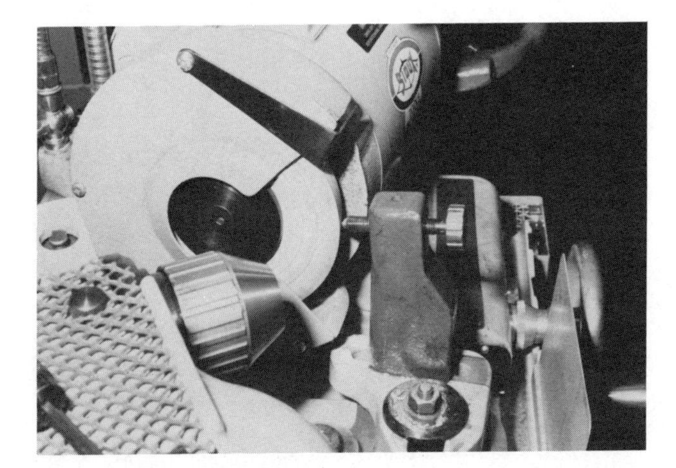

Figure 7.40 Dressing valve refacing stone.

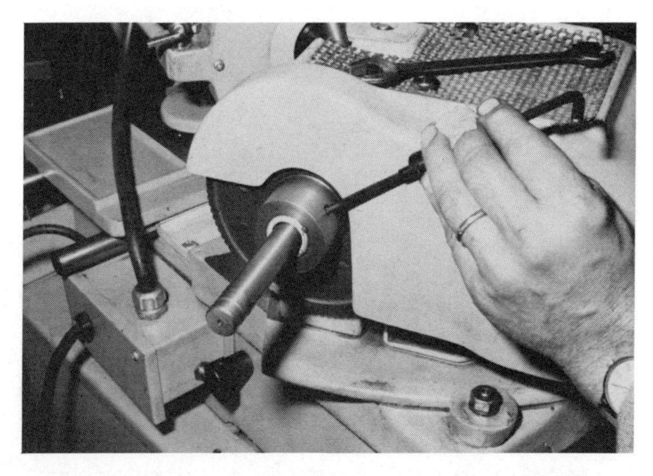

Figure 7.42 Adjusting valve stop.

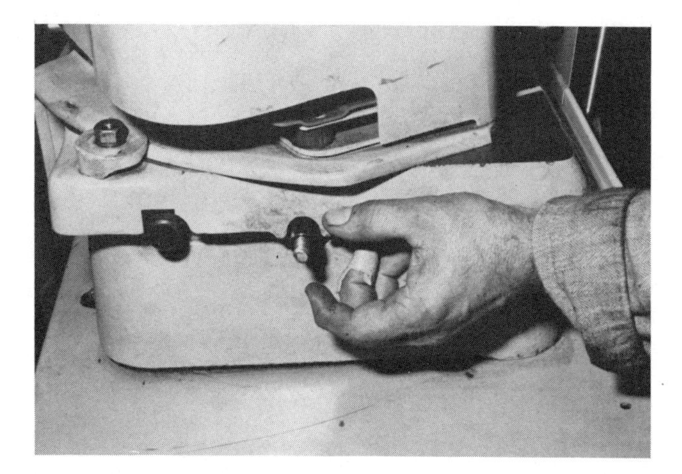

Figure 7.43 Location of valve table stop.

on the uppermost portion of the machined area on the stem.

5 Tighten the chuck on the valve stem.

6 With the machine stopped and grinding stone backed slightly away from valve face, adjust the valve table stop nut (Figure 7.43) so that the stone does not touch the valve stem (Figure 7.44).

7 Start machine; adjust cooling oil flow with adjusting valve on coolant hose or pump so that an adequate flow of oil is established. Move grinding wheel toward the valve until it just touches. Move valve back and forth across the face of the stone with table control lever, moving the stone closer to the valve with the stone feed control as the valve is ground (Figure 7.45).

Figure 7.44 Valve stem touching grinding stone; table stop adjusted incorrectly.

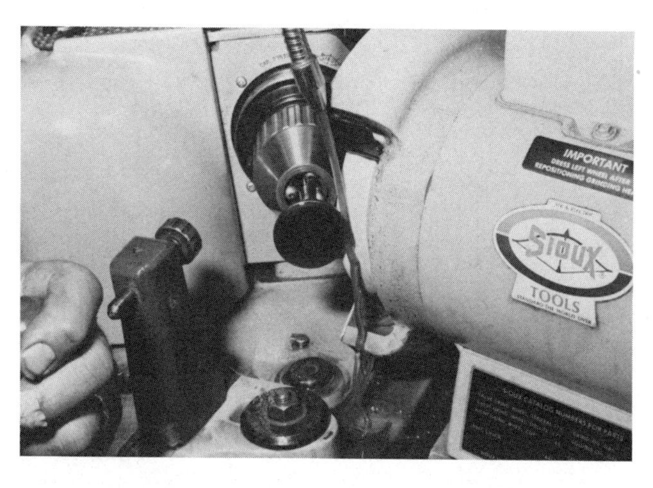

Figure 7.45 Grinding valve face.

8 When the valve appears to be ground or refaced so that it is smooth and free from pits and/or burned spots, back the stone away from the valve and move the valve table and valve clear of grinding stone, using previously mentioned controls.

CAUTION Do not move the valve table away from the grinding stone until the stone has been backed away from the valve, as damage to the valve face may result.

9 Visually check the valve carefully for pits and face condition. If the valve face still has pits and wear marks, continue grinding until the valve face is completely smooth and free of burned spots and pits.

CAUTION Do not remove valve from chuck until the grinding or refacing is completed.

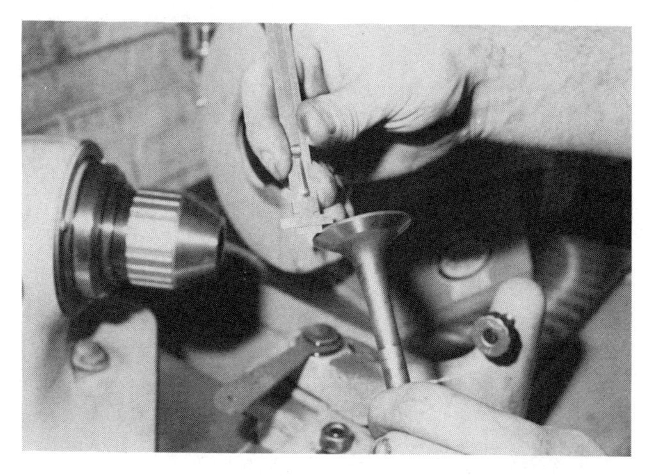

Figure 7.46 Measuring valve margin with a small ruler.

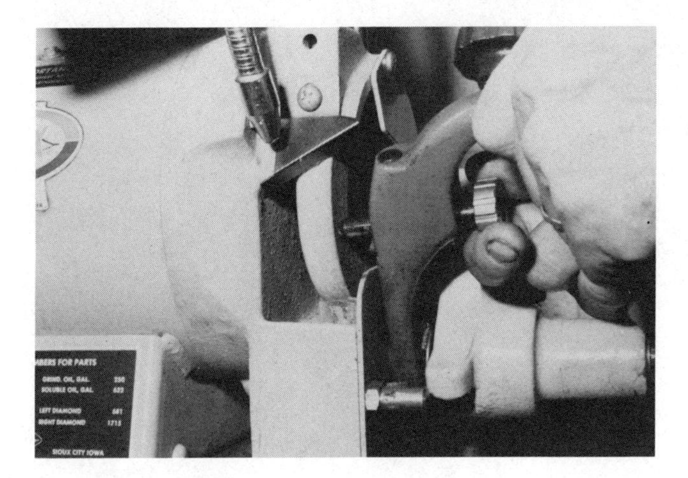

Figure 7.47 Dressing stem end grinding stone.

Figure 7.48 Grinding valve stem end.

Once the valve is removed from the chuck, it is impossible to install it in the same position in the chuck. As a result, the valve will require additional grinding, which would have been unnecessary if it had remained in the chuck. Following removal of the valve from the machine chuck, check the valve margin (Figure 7.46).

NOTE Valve margin is the distance from the valve head or top to the valve face. This margin must be held within manufacturer's specifications to prevent premature burning and subsequent failure. If it is not within manufacturer's specifications, the valve should be replaced.

10. Procedure for Refacing Valve Stem End

After the valve has been refaced, the stem end of the valve should be ground to insure that it is flat. Use the following procedure:

a Start machine; using the dressing diamond, dress the stone (Figure 7.47).

b Then clamp the valve in the holding bracket.

c With the valve-grinding machine operating, move the valve across the stone, removing only enough metal to "true" up or flatten the end of the valve (Figure 7.48).

d Remove valve from the holder and install taper or camfer tool in holder. Start the machine and grind taper (Figure 7.49).

NOTE Taper does not have to be very large. $\frac{1}{32}$ to $\frac{1}{16}$ in. (.79 to 1.59 mm) is considered sufficient.

X. Procedure for Checking Valve Springs

Valve springs are very important to the life of the valve as well as to efficient engine operation. They must be checked before reassembling the cylinder head. The valve spring should be checked for:

A *Straightness.*

Valve springs must be checked for straightness, tension, and breaks by using the following tools and methods:

1 Straightness. Use a T square or similar device (Figure 7.50).

2 Tension and free length. Insert the valve spring in a spring tension gauge to test unloaded or free length and tension at loaded length (Figure 7.51).

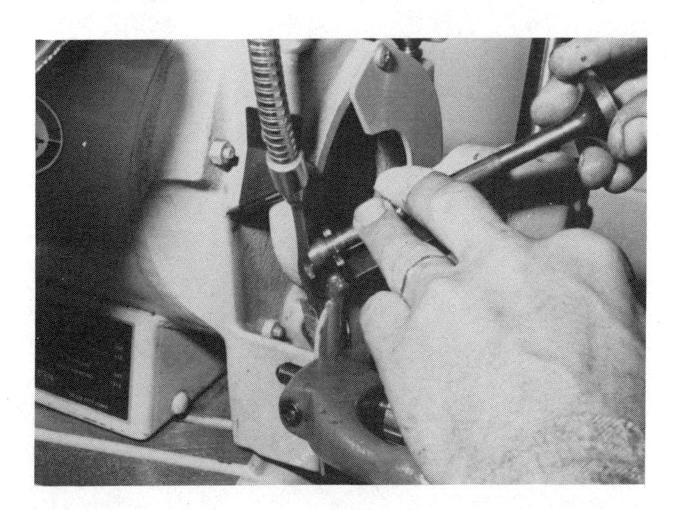

Figure 7.49 Grinding taper on valve stem end.

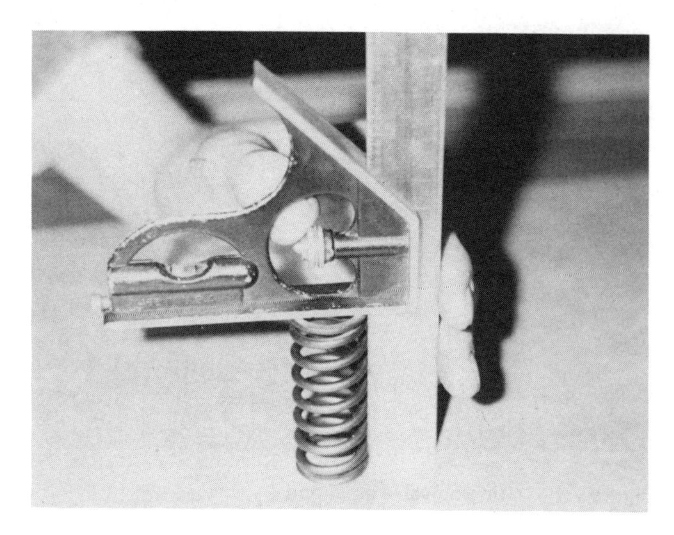

Figure 7.50 Checking a valve spring with a T square.

3 Breaks. By visual inspection check the valve spring carefully after it is cleaned for cracked or broken coils. If any evidence of breaks or cracks is indicated, the valve spring must be replaced.

XI. Valve Rotators and Keepers

Valve rotators can be one of two types, either the free release type or the mechanical, positive type. Valve rotators are attached to the valves to make them rotate during engine operation. This rotation insures that no carbon will collect on the valve face or seat and cause valve burning.

Figure 7.51 Checking valve springs with a spring tester.

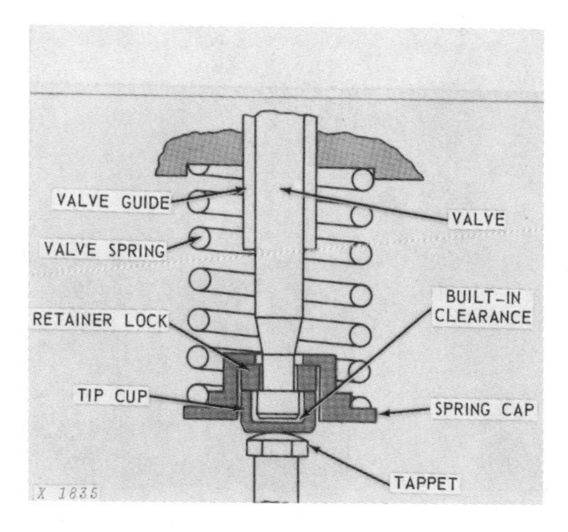

Figure 7.52 A free valve rotator.

A The free valve or release type rotator is designed so that every time the valve is opened and closed the valve has no spring tension on it (Figure 7.52). This release of spring tension allows the valve to be rotated by the outgoing exhaust gases or engine vibration. The free type of valve rotator must be visually checked closely for wear during reassembly and all worn parts replaced.

B The positive type rotator is a mechanical device that mechanically rotates the valve every time it is opened and closed by the rocker arm. The positive type rotator can be checked in the following manner:

1 Tapping with a plastic hammer after the valve spring rotator and keepers have been installed. (Tapping with a hammer simulates engine operation.)

Figure 7.53 Applying Prussian blue to valve face.

NOTE Although all valve rotators may pass the tests or checks and be considered usable, it is good practice during a major engine overhaul to replace all valve rotators to insure a long period of trouble-free operation.

C The valve keepers must be checked closely for wear and replaced if any wear is evident.

XII. Preparation for Cylinder Head Assembly

After the cylinder head has been given a final cleaning (rinsed with cleaning fluid and blown off with compressed air) it should be placed in or on a suitable stand for final assembly.

A. Procedure for Checking Valve Seat to Valve Contact

Install the valves in the head one at a time and check seat-to-valve contact with one of the following methods:

1 Prussian blue. When Prussian blue is being used to check valve seating, apply the blue to the valve face (Figure 7.53). Insert valve into the valve guide and snap it lightly against the seat. Remove valve and inspect face. The bluing should have an even seating mark all the way around the valve face. If not, valve seat is not concentric and must be reground.

2 A lead pencil or felt-tip marking pen. When a lead pencil or felt-tip marker is being used to check the valve to seat contact, place pencil marks about $\frac{1}{8}$ in. (3 mm) apart all the way around valve face (Figure 7.54). After the valve face is marked, place the valve in the guide and snap it against the valve seat. Remove valve and inspect marks made on the face. If the valve-to-seat contact is good, all marks will be broken. If all marks are not broken, valve seat is not concentric and it must be reground. After checking the valve-to-seat contact make sure lead is wiped from valve.

NOTE Lapping or seating of the valve with a lapping compound is not required if the valve and seat have been ground properly. Lapping is not recommended by most engine manufacturers.

Figure 7.54 Pencil marks on valve face.

B. Procedure for Checking Valve Head Height

If the valve head height (distance the valve protrudes above or below the machined surface of the cylinder) has not been checked previously, it should be checked at this time. It should be checked as follows:

1 Place a straight edge across the cylinder head and use a feeler gauge to measure distance between the valve head and straight edge. Compare this reading to manufacturer's specifications.

2. As indicated, some valves may protrude above the surface of the cylinder head; if this is the case, place the straight edge across the valve and use a feeler gauge to measure the distance from the valve head to a cylinder head machine surface.

C *Valve seals.*

After all valves and seats have been checked, the next step in the head assembly is to determine what type of valve seals (if any) are to be used.

NOTE Some engines do not use valve seals because the valve guides have been tapered to prevent oil loss at this point.

Valve seals fit around the valve and prevent oil from running down the valve stem into the combustion chamber, causing oil consumption. Valve seals come in many types and configurations (Figure 7.55). If the cylinder head was originally equipped with valve seals, the engine overhaul gasket set will generally contain the valve seals. If the cylinder head was not originally equipped with

Figure 7.55 Two types of valve seals.

Figure 7.56 Positive type valve seals installed on guides.

Figure 7.57 Valve guide machining tool.

valve seals, select either the positive or umbrella type of valve seal for intake and exhaust valves.

NOTE Although some cylinder heads may have been equipped with valve seals, it may be desirable to select a more modern or positive type of seal for installation on the cylinder head.

The most common types of seals are the rubber umbrella type or the teflon insert type that clamps onto the valve stem (Figure 7.56). Umbrella type oil deflectors generally will not require guide top machining. Positive type seals will require machining if the cylinder head was not originally equipped with them. A valve guide machining tool must be used to machine the guide (Figure 7.57).

D. Procedure for Valve Guide Machining and Seal Installation

1 Select the correct cutting tool by referring to the seal manufacturer's application data.

2 Install the cutting tool in a $\frac{1}{2}$ in. electric drill and machine the top of the guide, using a firm pressure on the drill to prevent cutting a wavy or uneven top on the guide (Figure 7.58).

NOTE The guide should be machined to the height specified in the instruction sheet provided by the seal manufacturer.

After all guides are machined, all metal chips must be thoroughly cleaned from the cylinder head.

3 Blow out all intake and exhaust ports with compressed air to insure that no chips or particles are left in the cylinder.

4 Install valves and valve seals (Figure 7.59).

XIII. Procedure for Final Cylinder Head Assembly

Complete the head assembly using the following procedures:

A Install spring over valve and valve seal.

B Install keeper retainer on valve spring.

C Using the valve spring compressor, compress the spring just far enough to install keepers (Figure 7.60).

Figure 7.58 Machining valve guide.

Figure 7.59 Installing valve guide seal.

D If the cylinder head is to be stored for some time before it is installed, the intake and exhaust ports must be plugged or covered with tape to prevent anything from getting into them until final engine assembly.

XIV. Procedure for Assembly of Cylinder Head onto Engine

The assembly of the cylinder head or heads onto the engine block will involve many different procedures that are peculiar to a given engine. The following procedures are general in nature and are offered to supplement the manufacturer's service manual.

A Before attempting to install the cylinder head, make sure the cylinder head and block surface are free from all rust, dirt, old gasket materials, and grease or oil.

NOTE Before placing the head gasket on the block make sure all bolt holes in the block are free of oil and dirt by blowing them out with compressed air.

B Select the correct head gasket and place it on the cylinder block (Figure 7.61), checking it closely for an "up" or "top" mark that some head gaskets may have on them.

NOTE In most cases head gaskets will be installed dry with no sealer, although in some situations an engine manufacturer may recommend applying sealer to the gasket before cylinder head installation. Take particular note of the recommendations in the service manual or ask your instructor.

Figure 7.60 Compressing valve spring with C clamp compressor.

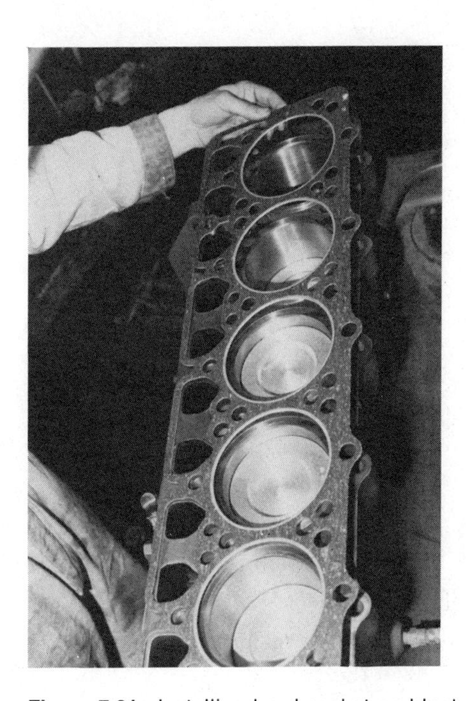

Figure 7.61 Installing head gasket on block.

C After placing the head gasket on the block, place water and oil O rings (if used) in the correct positions.

Some head gaskets will be a one-piece solid composition type, while others will be made of steel and composed of several sections or pieces (Figure 7.62). Detroit Diesel, for example, uses a round circular tin ring that fits on top of the cylinder sleeve to seal the compression. In addition to this sleeve seal are numerous O rings that seal the

Figure 7.62 Two typical engine head gaskets.

coolant and lubricating oil, making up the head gasket.

NOTE The installation of some head gaskets requires the use of dowels (threaded rods) that are screwed in the head bolt holes to hold the head gasket in place during head installation. If dowels are recommended they can be made from bolts by sawing off the heads and grinding a taper on the end. Most modern engines will have dowels or locating pins in the block to aid in holding the cylinder head gasket in place during cylinder head assembly.

D After gasket and all O rings are in place, check the cylinders to make sure no foreign objects have been left in them, such as O rings and bolts.

E Place head or heads on the cylinder block carefully to avoid damage to the head gasket.

F On in-line engines with three separate heads that do not use dowel pins it may be necessary to line up heads by placing a straight edge across the intake or exhaust manifold surfaces.

G Clean and inspect all head bolts or capscrews for erosion or pitting. To clean bolts that are very rusty and dirty use a wire wheel.

CAUTION Do not force or push the threaded part of the bolt into the wire wheel as damage to the threads may result.

NOTE Any bolt that shows any signs of stretching or pitting should be replaced. A broken head bolt can ruin a good overhaul job.

H Coat bolt threads with light oil or diesel fuel and place the bolts or capscrews into the bolt holes in head and block.

CAUTION Before placing bolts into the engine be sure each bolt has a washer (if used). Some engines do not use washers under the head bolts. CHECK! Be sure you have the correct bolt and/or washer combination.

I Using a speed handle wrench and the appropriate size socket, start at the center of the cylinder head and turn the bolts down snug, which is until the bolt touches the head and increased torque is required to turn it.

J When all bolts have been tightened this far, continue to tighten each bolt one-quarter to one-half turn at a time with a torque wrench until the recommended torque is reached (Figure 7.63).

NOTE Bolt tightening must be done according to the sequence supplied by the engine manufacturer when available (Figure 7.64). If sequence is not available, tighten the head starting with the center bolts, working outward toward the ends in a circular sequence.

XV. Procedure for Push Rod (Tube) Checking and Installation

After the cylinder heads have been torqued to the correct amount, the push rods or tubes and rocker arm assemblies may be installed.

Figure 7.63 Tightening head bolts with a torque wrench.

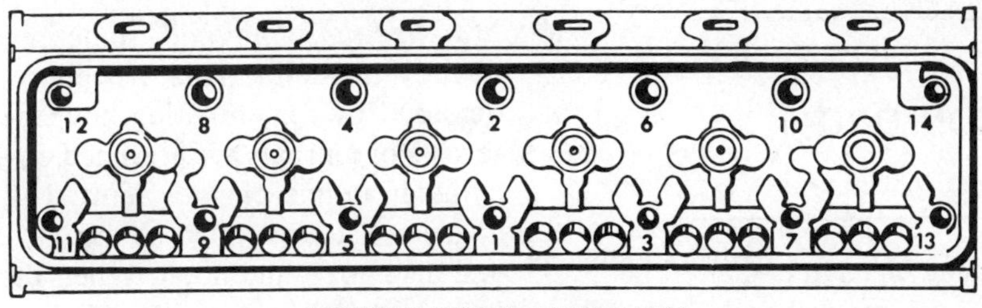

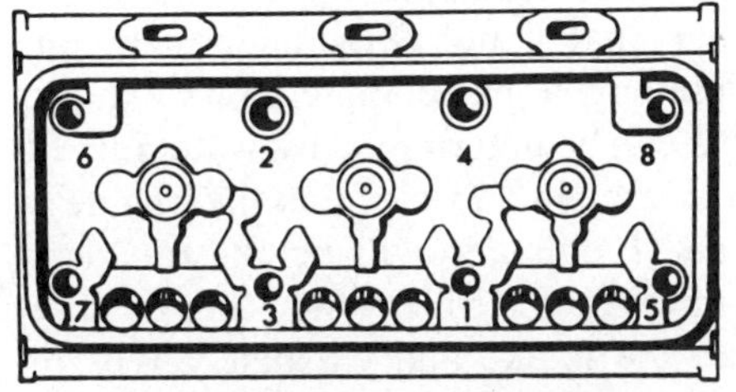

4-CYLINDER ENGINE CYLINDER HEAD

Figure 7.64 A typical cylinder head tightening sequence (courtesy Detroit Diesel Allison, Division of General Motors Corporation).

A Check all push rods and tubes for straightness by rolling them on a flat surface.

NOTE Straightening of push tubes is discouraged. Bent ones should be replaced with new. Push tubes must be checked for breaks or cracks where the ball socket on either end has been fitted into the tube.

B Place rod or tubes in the engine, making sure that they sit into the cam followers or tappets.

XVI. Rocker Arm Checking and Installation

The rocker arm assembly or rocker arm is one component part of the engine that is occasionally overlooked during a diesel engine overhaul. Some mechanics have the mistaken assumption that rocker arms wear very little or not at all. This is not true and rocker arm wear may account for increased engine oil consumption. Oil consumption occurs because the increased clearance allows an excessive amount of oil to splash or leak on the valve stem. This oil will run down the valve and end up in the combustion chamber.

A. Procedure for Rocker Arm Checks

The rocker arms and rocker arm assembly should be visually checked for the following:

1 Rocker arm bushing wear.

2 Rocker arm shaft wear.

Although some manufacturers may give a dimension for the shaft and bushing, the decision to replace should not rest entirely on dimension. The appearance of the shaft and bushing is a factor in determining if a replacement should take place. Look for:

a Scoring.

b Pitting.
It must also be kept in mind that a bushing may be acceptable now but not after an additional 2000 hours of operation.

c Magnetic inspection. Some manufacturers recommend that the rocker arms be checked for cracks using a magnetic type tester. If equipment is available, it is recommended that this check be made.

B. Procedure for Rocker Arm Installation and Adjustment

Place rocker arm assembly or housing on engine, making sure that rocker arm sockets engage the push rods.

NOTE Rocker arm assemblies on some engines may be built into a separate housing called rocker boxes. These box assemblies will require the installation of gaskets between them and the cylinder head.

> **CAUTION** If rocker arm or tappet adjusting screw were not loosened during engine disassembly, it should be done at this time. This insures that no damage will result to the valves or valve train when the rocker arm assembly is pulled in place with the hold-down bolts.

C Install rocker arm hold-down bolts and torque them to specifications.

NOTE Some rocker arms may be held in place with bolts that serve as head bolts in addition to holding the rocker arms. These bolts will be tightened to the same torque as the head bolts.

D After rocker arm assemblies have been installed, adjust the rocker arms or tappets (Figure 7.65).

E Check specifications for tappet clearance and adjustment sequence.

Listed below are several proven methods of setting tappets.

NOTE It should be remembered that they are offered only as guidelines or recommendations and, if the engine service manual outlines a specific procedure, it should be used.

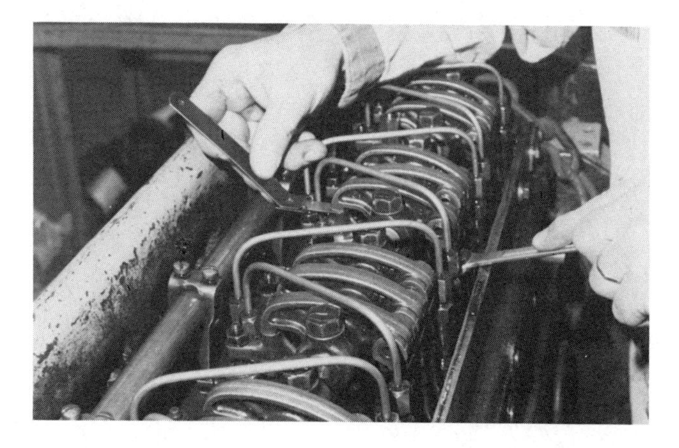

Figure 7.65 Adjusting tappets on a two stroke diesel engine.

1 *Matched throw method,* sometimes called the "buddy method." This procedure is best understood by remembering how the crankshaft throws or journals are arranged on a normal six cylinder in-line engine. Journal throws no. 1 and no. 6 are matched as are no. 2 and no. 5 and also no. 3 and no. 4. When the crankshaft is installed in the engine and the pistons connected to it, pistons no. 1 and no. 6 will always be matched together. Using this information, it can be seen that if no. 6 cylinder in a four stroke cycle engine has completed its exhaust stroke (this can be determined by watching the exhaust valve open and close as the engine is rotated) no. 1 cylinder will be on top dead center compression stroke. When this position is reached, the intake and exhaust valves on no. 1 cylinder can be adjusted to the clearance recommended by the engine manufacturer.

NOTE Very few, if any, engine manufacturers recommend the adjustment of valves or tappets while the engine is running. One reason for this is that on some engines a very small amount of clearance exists between the piston and valve when the piston is at top dead center position and, if a feeler gauge is inserted between the rocker arm and the valve with an improperly adjusted rocker arm, it may cause the valve to hit the piston if the engine is running.

To adjust the remainder of the valves, the engine should be rotated or turned while watching the exhaust valve on the no. 2 cylinder open and close. After it has closed, the valve on the no. 5 cylinder can be adjusted. The mechanic can proceed through the firing order by watching the exhaust valve on the matched throw cylinder of the valves to be adjusted as outlined in the following chart:

To set valves on the no. 1 cylinder watch exhaust valve open and close on no. 6.

To set valves on the no. 5 cylinder watch exhaust valve open and close on no. 2.

To set valves on the no. 3 cylinder watch exhaust valve open and close on no. 4.

To set valves on the no. 6 cylinder watch exhaust valve open and close on no. 1.

To set valves on the no. 2 cylinder watch exhaust valve open and close on no. 5.

To set valves on the no. 4 cylinder watch exhaust valve open and close on no. 3.

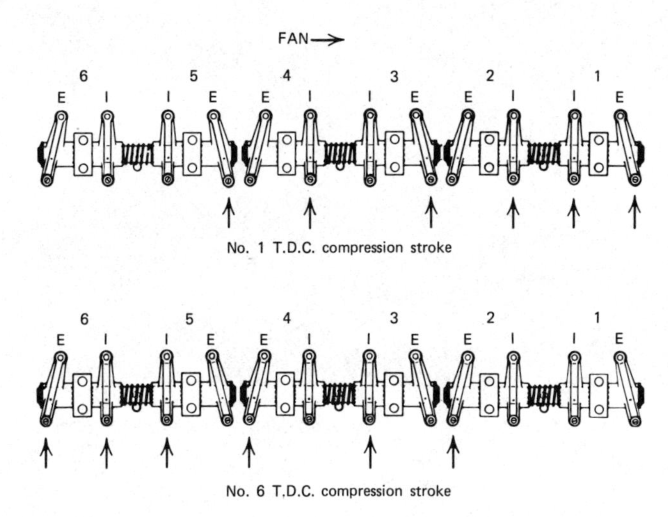

Figure 7.66 Sequence method of valve setting (courtesy of J. I. Case Company).

2 *Sequence or procedure method.* This method is given by many manufacturers in the service manual and can be used only on the engine for which it is listed (Figure 7.66). With the engine positioned at top dead center, no. 1 compression stroke position, valves no. 1,2,3,5,7,9 can be set. To set the remaining valves, turn the engine one complete revolution to the top dead center to the no. 6 position.

3 *The degree method.* The degree method is used by many mechanics and can best be understood by remembering that theoretically a cylinder fires every 120° in a six cylinder inline engine. With this in mind, we can see that when an engine with a firing order of 1,5,3,6,2,4 is in the no. 1 top dead center position, the next cylinder in the firing order, no. 5, will be on top dead center compression 120° of crank travel later. If the tappets are set on the no. 1 cylinder with the engine at top dead center, no. 1 compression position, it then is evident that by turning the engine 120° the tappets on the no. 5 cylinder can be adjusted, no. 3 120° later, and so on throughout the firing order.

This procedure allows adjustment of all the cylinders in two rotations of the engine.

F After tappets have been adjusted, install the rocker arm covers. New gaskets should be used on the cover or covers and glued to the cover with a gasket adhesive.

G Install the intake manifold, exhuast manifold, generator or alternator, thermostat housing, and any other accessory that could not be installed before the heads were installed. After this final assembly, recheck all hose connections and electrical connections to insure that all connections are completed and tight.

SUMMARY

If the engine can be run at this time, check all fluid levels (oil and water) and run the engine until operating temperature is reached. Many engine manufacturers recommend that the cylinder head or heads be retorqued at this time. After retorquing cylinder heads, the tappets should be readjusted. Replace rocker arm covers and give the engine a final visual inspection.

REVIEW QUESTIONS

1 List and explain the function of four component parts of a cylinder head.

2 List two ways a cylinder head may be cleaned.

3 List two ways crack testing of the cylinder head can be done.

4 Explain why and how a straight edge is used to check a cylinder head.

5 If valve guides are worn in excess of specifications, they should be replaced. Explain the proper procedure for valve guide replacement.

6 Describe one method of checking valve seats for looseness.

7 Explain the proper procedures for valve seat regrinding.

8 Explain valve seat concentricity and how it is checked.

9 List one way a valve seat may be narrowed.

10 List four checks that should be made on the valve.

11 Why are valve seals used on the valve stems?

12 Explain two methods that may be used to adjust the valve tappets after the cylinder head has been installed.

Crankshaft, Main Bearings, Vibration Damper, and Flywheel

The crankshaft in a diesel engine is used to change the up-and-down motion of the pistons and connecting rods to usable rotary motion at the flywheel. It is called a crankshaft because it is made with cranks or throws (an offset portion of the shaft), with a rod journal (that connects rod bearing surfaces) machined or manufactured on the end. Different designs and different throw arrangements are used, determined by the number of engine cylinders and engine configurations, such as in-line or V design. On one end, generally the rear of the shaft, a flywheel (a heavy metal wheel) will be attached. Attached to the opposite end will be the vibration damper. This assembly (Figure 8.1) is mounted on the bottom of the engine block by the main bearings.

OBJECTIVES

Upon completion of this chapter the student will be able to:

1. Demonstrate the ability to measure a crankshaft with an outside micrometer.
2. Explain and point out the area on a crankshaft that must be fillet (radius at journal area) ground.
3. Demonstrate and explain the reasons for lifting a crankshaft correctly.
4. List two reasons why a crankshaft should be reground.
5. Explain and demonstrate how a crankshaft vibration damper (rubber element type) should be removed, checked, and reinstalled.
6. List three checks that should be made on an engine flywheel.
7. List three types of premature bearing failures and explain the corrective action.
8. Demonstrate the ability to measure main bearing clearance with plastigage.

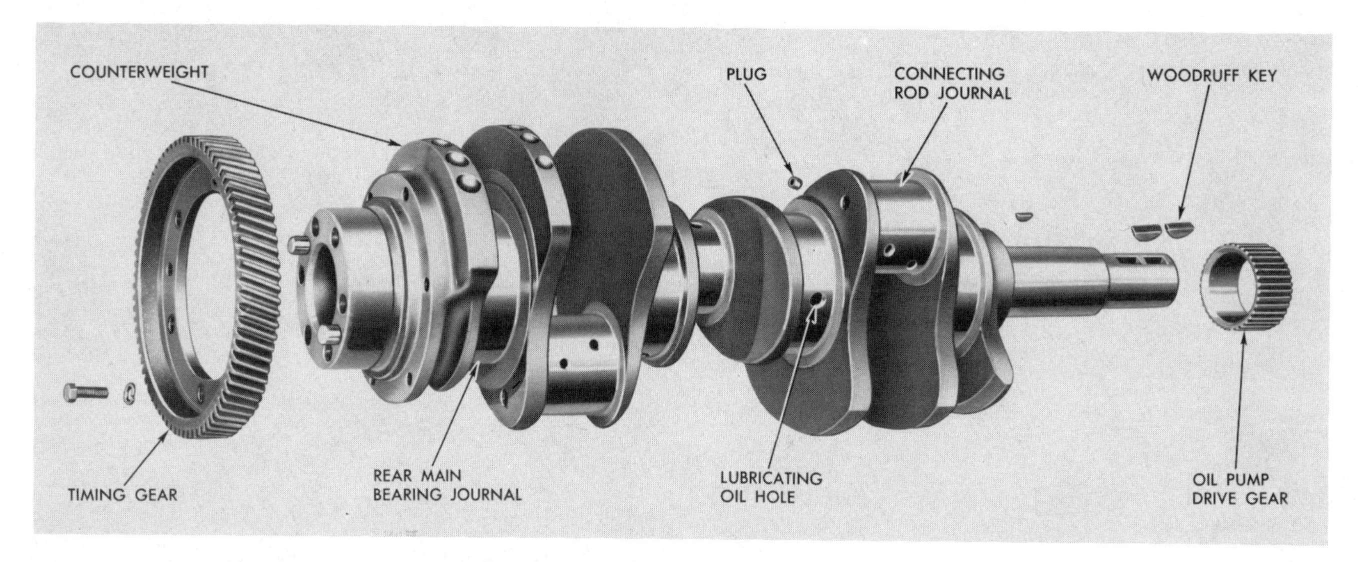

Figure 8.1 Typical diesel engine crankshaft (courtesy Detroit Diesel Allison, Division of General Motors Corporation).

9 Demonstrate the ability to correctly remove and reinstall a crankshaft seal wear sleeve.

10 Demonstrate the ability to correctly install a rear main bearing seal.

11 Demonstrate the ability to correctly remove and install a flywheel ring gear.

GENERAL INFORMATION

Within the diesel engine the pressure developed during operation by the burning fuel and air is trapped in the cylinder by the pistons and rings that are connected to the crankshaft by the connecting rods. The crankshaft then transmits this pressure or power to the flywheel for use outside the engine. To increase this power and produce torque, the crankshaft has been designed with the addition of cranks or throws. These throws extend from the center line of the shaft outward. The distance that they extend outward is determined by the engine manufacturer and is called the stroke of the crankshaft. The crankshaft will have one throw for every cylinder in an in-line engine and one throw for every two cylinders in a V design engine. Throw arrangement or spacing plays a very important part in helping to balance the engine. Figure 8.2 shows typical throw arrangements found in engines used today.

Since the crankshaft must rotate at different speeds over a wide speed range, it must be balanced precisely to avoid vibration. In addition, counterweights must be added to offset the inertia forces generated by the up-and-down movement of the piston-and-rod assembly. Most crankshafts will be constructed with counterweights on them while others may be bolted on.

The crankshaft must be solidly supported in the block to absorb the power from the engine cylinders. This is done by the main bearings, friction type bearings (made from steel and babbitt) that fit into machined bores or saddles in the block. Main bearings are constructed in a way similar to rod bearings and are of the same material, with babbitt the most commonly used material. For a more detailed description of bearing material construction see Chapter 9.

Since the main and rod bearings are friction type bearings, adequate lubrication must be maintained at all times.

Lubrication for the crankshaft and main bearings is provided by engine oil supplied by the oil pump to the oil galleries that are connected to the main bearings. After reaching the bearings it flows through drillings in the crankshaft to the rod bearing journals. It then provides lubrication for the rod bearings and is allowed to drip off into the oil pan.

Mounted on the rear of the crankshaft is the flywheel. This flywheel helps to smooth out the power impulses developed within the engine and provides a place for the attached transmission clutch (a transmission connecting and disconnecting device).

Since the crankshaft now has a heavy flywheel mounted on the back, the free or front end must have a torsional (twisting) vibration damper to prevent twisting of the crankshaft by power impulses as they occur in the engine. This damper is smaller in size than the

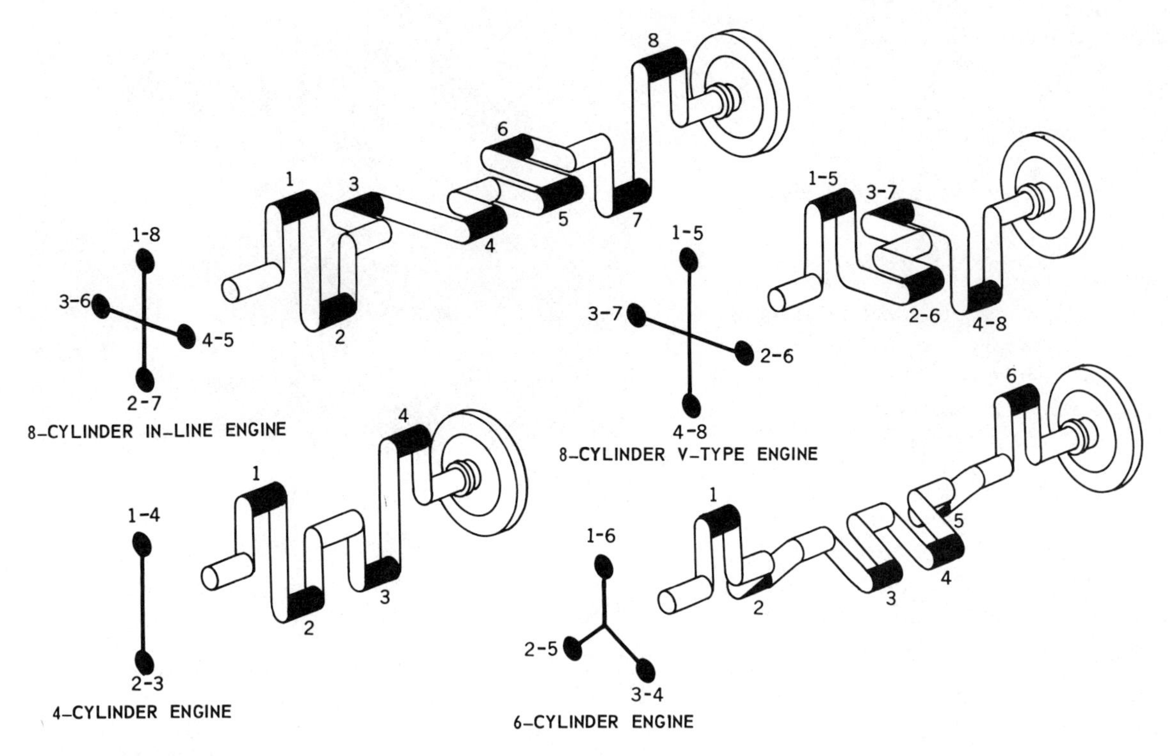

Figure 8.2 Typical crankshaft throw arrangement in a diesel engine (courtesy Deere and Company).

flywheel and is especially designed to prevent crankshaft breakage that may result from torsional vibrations created in the engine during operation.

The flywheel, crankshaft, main bearings, and vibration damper make up the team that transmits the power developed within the engine to the load. During engine overhaul these components require careful, detailed inspection and reconditioning if they are to give many hours of trouble-free service.

Unfortunately, components such as the vibration damper and flywheel sometimes receive at best only a casual inspection during a major rebuild and, as a result, bring about premature engine failure. It is recommended that all the components be checked and reconditioned as described in this chapter.

I. Crankshaft

Before any measurements are made on the crankshaft it should be thoroughly cleaned.

A. Procedure for Cleaning

The best cleaning method is the hot chemical cleaning tank. Before placing the crankshaft in the hot tank you should:

1. Remove all oil passageway plugs.
2. Remove all seal wear sleeves if present.
3. Remove transmission pilot bearing or bushing (if mounted in crankshaft).

After removal from hot tank the crankshaft should be cleaned further by:

1. Using a stiff bristle brush to clean oil passageways and drillings.
2. Using a steam cleaner or high pressure washer to clean the entire shaft.
3. Using compressed air to blow out all oil passageways and blow dry entire crankshaft.

B. Procedure for Visual Inspection

Crankshaft should be visually inspected at this time for the following:

1. Check for cracked or worn front hub key slots.
2. Check rod and main bearing journal visually for excessive scoring and bluing.

3 Check crankshaft dowel pin holes for:

a Cracks.

b Size (oversize or oblong).

4 Check dowel pins for wear or damage and snug fit into crankshaft.

5 Check around all oil supply holes for cracks.

6 Check area on shaft where oil seals ride (front and rear). If a wear sleeve is used it should be replaced. If the wear sleeve is not used and the shaft has a deep groove in it, the groove should be smoothed out with emery paper.

NOTE If at this point you find that the crankshaft is unfit for further use or needs reconditioning, an effort should be made to determine what caused the crankshaft wear or damage so that the problem can be remedied before a new or reground shaft is installed.

The following steps should be used in determining what may have caused the damage to the crankshaft and main bearings:

7 Inspect old main and connecting rod bearings, using the illustrations later in this chapter as a guide.

8 Check bearings and shaft for evidence of insufficient lubrication.

9 Check bearings and shaft for evidence of improper assembly.

10 Check block line bore as outlined in the block section.

11 If crankshaft was broken, check vibration damper.

NOTE If the crankshaft does not appear to be worn or damaged beyond repair, the following checks should be made to accurately determine if the shaft needs to be reconditioned.

C. Procedure for Inspection by Measurement

Accurate measurement of crankshaft rod and main bearing journals must be made with a micrometer in the following manner:

1 *Out-of-roundness.* Check by measuring in at least two different places around the journal diameter with a micrometer as shown in Figure 8.3. If journal diameter is 0.001 to 0.002 in. (0.025 to 0.050 mm) smaller (less) than manufacturer's specifications, the crankshaft should be reground.

Figure 8.3 Measuring connecting rod journal.

NOTE Manufacturer's specifications should be checked for allowable crankshaft wear.

2 *Journal taper.* Check by measuring for connecting rod and main bearing journal diameter with a micrometer. Measure near one edge of the journal next to the crank cheek and then move across the journal, checking the diameter in the middle and opposite edge. If the diameters are different in excess of 0.0015 in (0.381 mm), the crankshaft must be reground.

3 *Crankshaft thrust surfaces.* The crankshaft thrust surfaces should be checked for:

a Scoring (visually).

b Measurement with an inside micrometer (Figure 8.4).

NOTE Thrust surfaces may be reconditioned or reground if they are scored or rough. Most shops that are equipped to regrind crankshafts can perform this repair. It must be remembered that after the thrust surface has been reground an oversize thrust bearing will be required.

If it is to be reconditioned, it should be taken to an automotive machine shop specializing in this type of work. It is recommended that a shop be selected that can grind the fillets (area between the crank cheek and journal) in addition to the crankshaft journals and that can make some type of magnetic or electrical check for cracks.

CAUTION It is a must that all diesel engine crankshafts be fillet ground to prevent breakage of the crankshaft (Figure 8.5).

Figure 8.4 Checking crankshaft thrust width with an inside micrometer (courtesy Cummins Engine Company).

D. Procedure for Crack Detection

One of several methods may be employed by repair shops to check for cracks in a crankshaft. In most cases, crack detecting will be done by the shop doing the grinding. Explanation of two popular methods used is given at this point in case a repair shop does not have the equipment and capability to perform the checks.

1 *Magnetic particle method.* In this method the crankshaft is magnetized using some type of electrical magnet to magnetize a small section of the whole crankshaft at a time. A fine metallic powder is then sprayed on the crankshaft. If the crankshaft is cracked, a small magnetic field forms at the crack and the metal particles are concentrated or gathered at this point.

2 *Spray penetrant method.* This method uses a spray penetrant dye which is sprayed on the crankshaft and the excess is wiped off. The shaft is then sprayed with a developer that draws the penetrant out of the cracks, making them visible.

Figure 8.5 Fillet area of a diesel engine crankshaft (courtesy Detroit Diesel Allison, Division of General Motors Corporation).

II. Main Bearings

Main bearings are generally replaced with new ones during a major engine overhaul or rebuild. As indicated earlier in this chapter, the bearings should be inspected closely for wear and damage to determine if some abnormal condition such as low oil pressure or main bore misalignment exists within the engine that must be corrected before the engine is reassembled.

The following information and illustrations should be used when inspecting main bearings to determine if the bearing wear is normal or if conditions exist within the engine that may cause premature bearing failure. Premature bearing failures are caused by:

1 Dirt – 44.9%

2 Misassembly – 13.4%

3 Misalignment – 12.%

4 Insufficient lubrication – 10.8%

5 Overloading – 9.5%

6 Corrosion – 4.2%

7 Other – 4.5%

A *Surface fatigue* (Figure 8.6).

1 *Damaging act.* Heavy pulsating loads imposed upon the bearing by reciprocating engine cause the bearing surface to crack due to metal fatigue, as illustrated in Figure 8.7

Fatigue cracks widen and deepen perpendicular to the bond line. Close to the bond line, fatigue cracks turn and run parallel to the bond line, eventually joining and causing pieces of the surface to flake out.

2 *Appearance.* Small irregular areas of surface material missing from the bearing lining.

3 *Possible causes.* Bearing failure due to surface fatigue is usually the result of the normal life span of the bearing being exceeded.

4 *Corrective Action.*

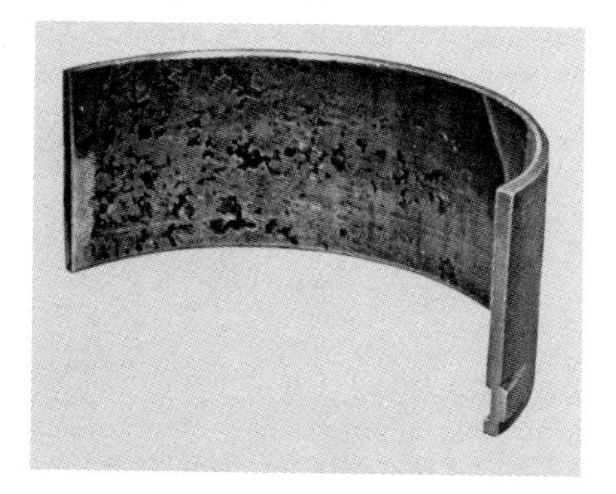

Figure 8.6 Bearing surface showing effects of surface fatigue (courtesy Gould, Inc., Engine Parts Division).

Figure 8.8 Foreign particles embedded in bearing lining (courtesy Gould, Inc., Engine Parts Division).

a If the service life for the old bearing was adequate, replace with the same type of bearing to obtain a similar service life.

b If the service life of the old bearing was too short, replace with a heavier duty bearing to obtain a longer life.

c Replace all other bearings (main connecting rod and cam shaft) as their remaining service life may be short.

d Recommend that the operator avoid "hot rodding" and lugging as these tend to shorten bearing life.

B *Foreign particles in lining* (Figure 8.8).

1 *Appearance.* Foreign particles embedded in the bearing. Scrape marks may also be visible on the bearing surface.

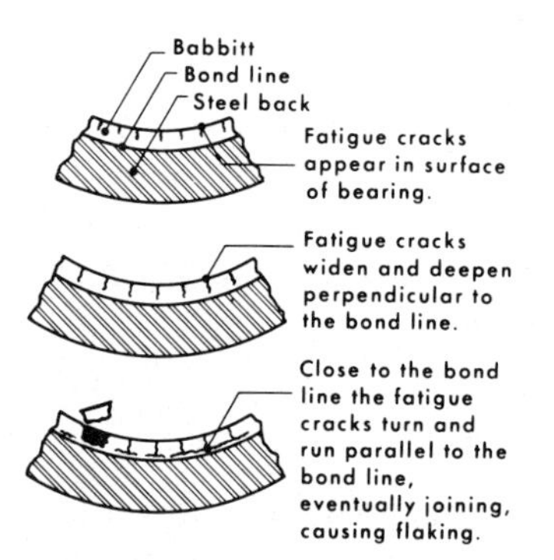

Figure 8.7 Bearing cross section showing internal damage caused by fatigue (courtesy Gould, Inc., Engine Parts Division).

2 *Damaging action.* Dust, dirt, abrasives and/or metallic particles present in the oil supply embed in the soft babbitt bearing lining, displacing metal and creating a high spot (Figure 8.9).

A high spot may be large enough to make contact with the journal causing a rubbing action that can be lead to the eventual breakdown and rupture of the bearing lining. Foreign particles may embed only partially and the protruding portion may come in contact with the journal and cause a grinding wheel action.

3 *Possible causes.* Three factors can lead to bearing failure due to foreign particles.

a Improper cleaning of the engine and parts prior to assembly.

b Road dirt and sand entering the engine through the air-intake manifold.

c Wear of other engine parts, resulting in small fragments of these parts entering the engine's oil supply.

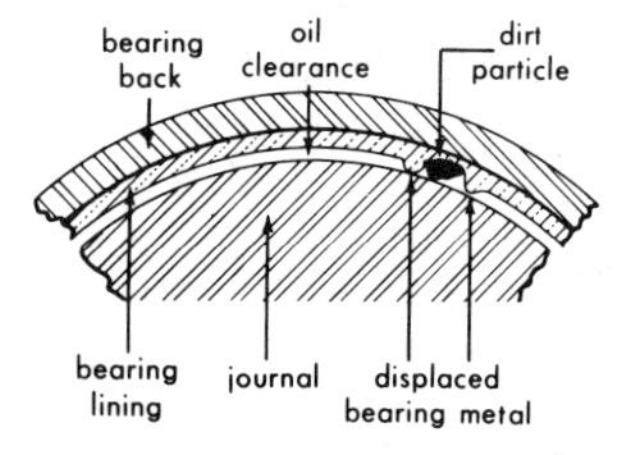

Figure 8.9 Bearing cross section showing how dirt particles are embedded in bearing material (courtesy Gould, Inc., Engine Parts Division).

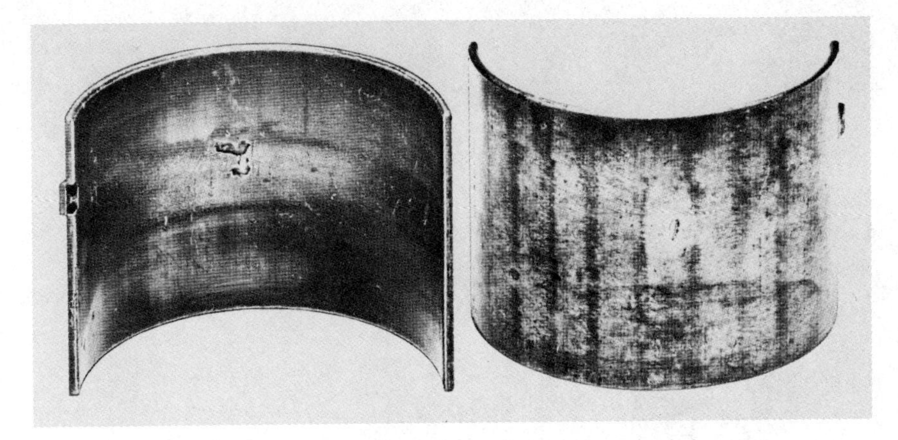

Figure 8.10 Results of bearing being installed with dirt on backside (courtesy Gould, Inc., Engine Parts Division).

4 *Corrective action.*

a Install new bearings, being careful to follow proper cleaning procedures.

b Grind journal surfaces if necessary.

c Recommend that the operator have the oil changed at proper intervals and have air filter, oil filter, and crankcase breather-filter cleaned as recommended by the manufacturer.

C *Foreign particles on bearing back* (Figure 8.10).

1 *Appearance.* A localized area of wear can be seen on the bearing surface. Also, evidence of foreign particle(s) may be visible on the bearing back or bearing seat directly behind the area of surface wear.

2 *Damaging action.* Foreign particles between the bearing and its housing prevent the entire area of the bearing back from being in contact with the housing base (Figure 8.11). As a result, the transfer of heat away from the bearing surface is not uniform causing localized heating of the bearing surface, which reduces the life of the bearing.

Also, an uneven distribution of the load causes an abnormally high pressure area on the bearing surface, increasing localized wear on this material.

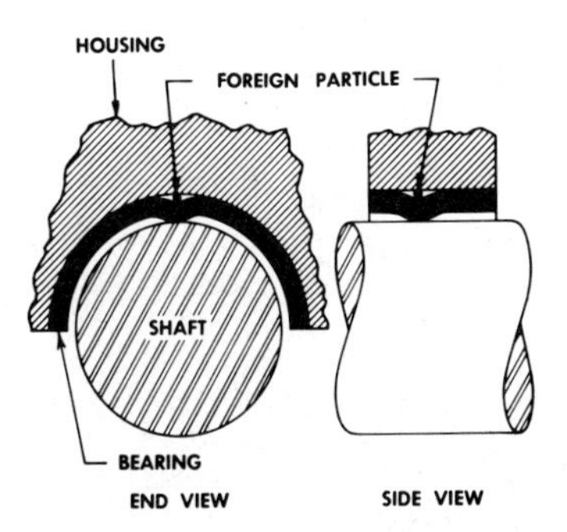

Figure 8.11 Cross section showing dirt under bearing (courtesy Gould, Inc., Engine Parts Division).

3 *Possible causes.* Dirt, dust, abrasives, and/or metallic particles either present in the engine at the time of assembly or created by a burr removal operation can become lodged between the bearing back and bearing seat during engine operation.

4 *Corrective action.*

a Install new bearings following proper cleaning and burr removal procedures for all surfaces.

b Check journal surfaces and if excessive wear is discovered, regrind.

D *Insufficient crush* (Figure 8.12).

1 *Appearance.* Highly polished areas are visible on the bearing back and/or on the edge of the parting line.

2 *Damaging action.* When a bearing with insufficient crush is assembled in an engine, it is loose and therefore free to work back and forth within its housing. Because of the loss of radial pressure, there is inadequate contact with the bearing seat, thus impeding heat transfer away

Figure 8.12 Bearing showing effects of insufficient crush (courtesy Gould, Inc., Engine Parts Division).

from the bearing. As a result, the bearing overheats, causing deterioration of the bearing surface.

3 *Possible causes.* There are five possible causes of insufficient crush:

   a Bearing parting faces were filed down in a mistaken attempt to achieve a better fit, thus removing the crush.

   b Bearing caps were held open by dirt or burrs on the contact surface.

   c Insufficient torquing during installation (be certain bolt doesn't bottom in a blind hole).

   d The housing bore was oversize or the bearing cap was stretched, thus minimizing the crush.

4 *Corrective action.*

   a Install new bearings using correct installation procedures (never file parting faces).

   b Clean mating surfaces of bearing caps prior to assembly.

   c Check journal surfaces for excessive wear and regrind if necessary.

   d Check the size and condition of the housing bore and recondition if necessary.

   e Correct shim thickness (if applicable).

E *Shifted bearing cap* (Figure 8.13).

1 *Appearance.* Excessive wear areas can be seen near the parting lines on opposite sides of the upper and lower bearing shells.

2 *Damaging action.* The bearing cap has been shifted, causing one side of each bearing-half to be pushed against the journal at the parting line.

The resulting metal-to-metal contact and excessive pressure cause deterioration of the bearing surface and above-normal wear areas.

3 *Possible causes.* These are five factors that can cause a shifted bearing cap:

   a Using too large a socket to tighten the bearing cap. In this case, the socket crowds against the cap, causing it to shift.

   b Reversing the position of the bearing cap.

   c Inadequate dowel pins between bearing shell and housing (if used), allowing the shell to break away and shift.

   d Improper torquing of cap bolts, resulting in a "loose" cap that can shift positions during engine operation.

   e Enlarged cap bolt holes or stretched cap bolts, permitting greater than normal play in the bolt holes.

4 *Corrective action.*

   a Check journal surfaces for excessive wear and regrind if necessary.

   b Install the new bearing being careful to use

Figure 8.13 Damage caused to bearing by a shifted or misaligned bearing cap (courtesy Gould, Inc., Engine Parts Division).

the correct size socket to tighten the cap and the correct size dowel pins (if required).

c Alternate torquing from side to side to assure proper seating of the cap.

d Check the bearing cap and make sure it's in its proper position.

e Use new bolts to assure against overplay within the bolt holes.

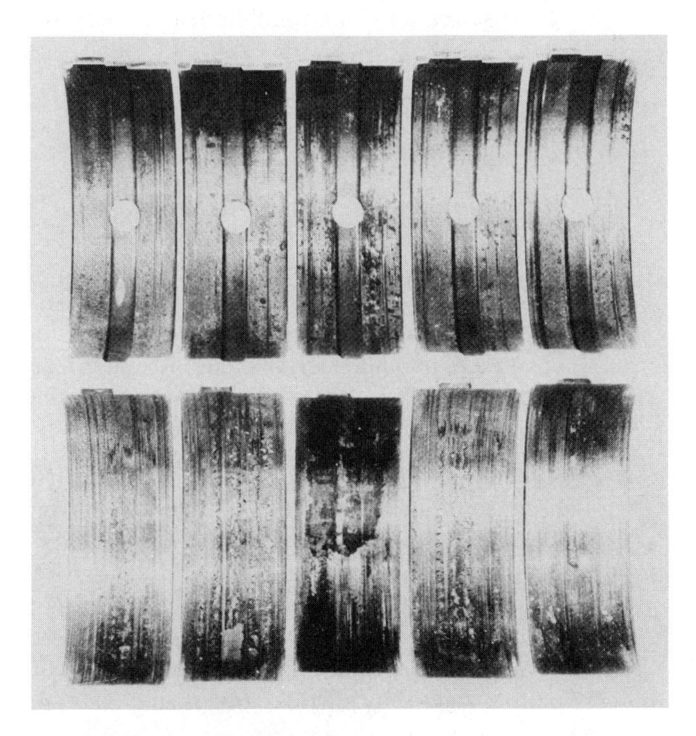

Figure 8.14 Bearings damaged by a distorted crankcase (courtesy, Gould, Inc., Engine Parts Division).

F *Distorted crankcase* (Figure 8.14).

1 *Appearance.* A wear pattern is visible on the upper or lower halves of the complete set of main bearings. The degree of wear varies from bearing to bearing depending upon the nature of the distortion. The center bearing usually shows the greatest wear.

2 *Damaging action.* A distorted crankcase imposes excessive loads on the bearing with the point of greatest load being at the point of greatest distortion. These excessive bearing loads cause excessive bearing wear. Also, oil clearance is reduced and metal-to-metal contact is possible at the point of greatest distortion.

3 *Possible causes.* Alternating periods of engine heating and cooling during operation is a prime cause of crankcase distortion. As the engine heats the crankcase expands, and as it cools the crankcase contracts. This repetitive expanding and contracting causes the crankcase to distort in time.

Distortion may also be caused by:

Extreme operating conditions (for example "hot-rodding" and "lugging").

Improper torquing procedure for cylinder head bolts.

4 *Corrective action.*

a Determine if distortion exists by use of prussian blue or visual methods.

b Align bore the housing (if applicable).

c Install new bearings.

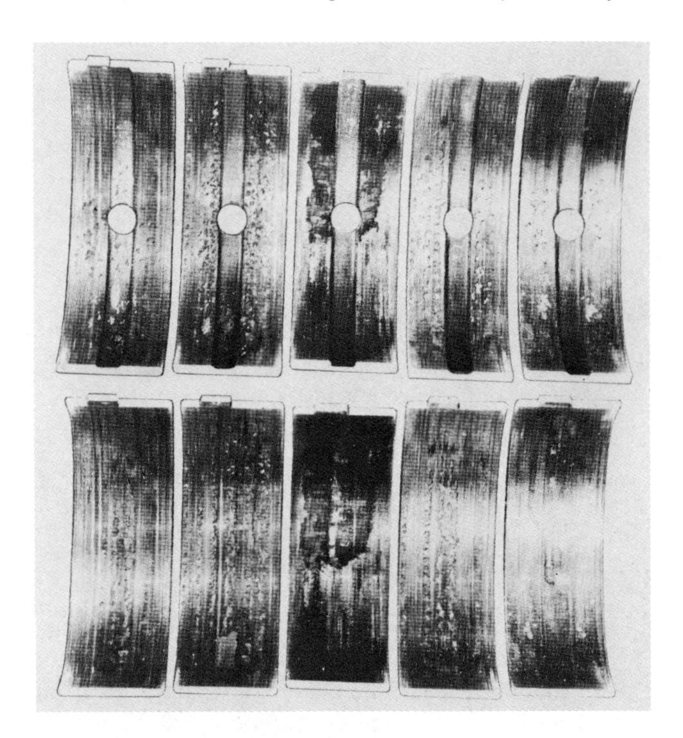

Figure 8.15 Bearings damaged by a bent crankshaft (courtesy Gould, Inc., Engine Parts Division).

G *Bent crankshaft* (Figure 8.15).

1 *Damaging action.* A distorted crankshaft subjects the main bearings to excessive loads, with the greatest load being at the point of greatest distortion (Figure 8.16). The result is excessive bearing wear. Also, the oil clearance spaces between journals and bearings are reduced, making it possible for metal-to-metal contact to occur at the point of greatest distortion.

2 *Appearance.* A wear pattern is visible on the upper and lower halves of the complete set of main bearings. The degree of wear varies from bearing to bearing depending upon the nature of the distortion. The center bearing usually shows the greatest wear.

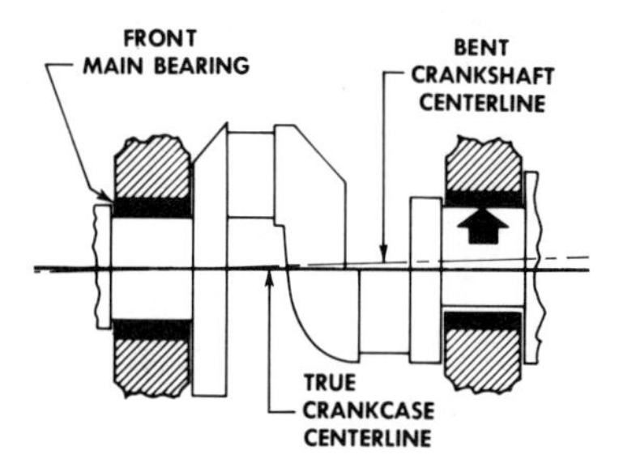

Figure 8.16 Cross section showing how bent crankshaft fits into block (courtesy Gould, Inc., Engine Parts Division)

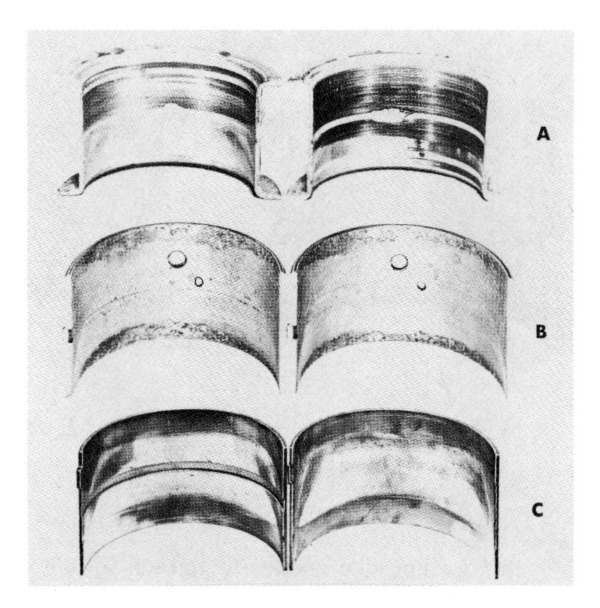

Figure 8.17 Damage to bearings caused by out-of-shape journal (courtesy Gould, Inc., Engine Parts Division).

3 *Possible causes.* A crankshaft is usually distorted due to extreme operating conditions, such as "hot-rodding" and "lugging."

4 *Corrective action.*

a Determine if distortion exists by means of prussian blue or visual methods.

b Install a new or reconditioned crankshaft.

c Install new bearings.

H *Out-of-shape journal* (Figure 8.17).

1 *Damaging action.* An out-of-shape journal imposes an uneven distribution of the load on the bearing surface, increasing heat generated and thus accelerating bearing wear. An out-of-shape journal also affects the bearing's oil clearance, making it insufficient in some areas and excessive in others, thereby upsetting the proper functioning of the lubrication system.

2 *Appearance.* In general, if a bearing has failed because of an out-of-shape journal, an uneven wear pattern is visible on the bearing surface. Specifically, however, these wear areas can be in any one of three patterns: In Figure 8.17 photo A above shows the wear pattern caused by a **tapered** journal. Photo B shows the wear pattern caused by an **hour-glass** shaped journal. Photo C shows the pattern of a **barrel**-shaped journal. See also Figure 8.18.

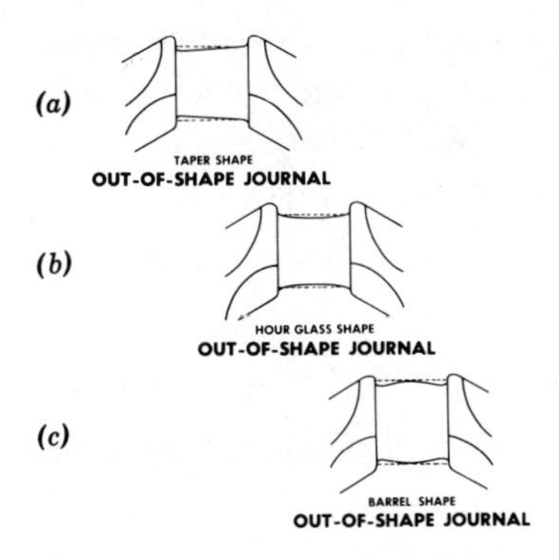

Figure 8.18 Various journal shapes caused by wear (courtesy Gould, Inc., Engine Parts Division).

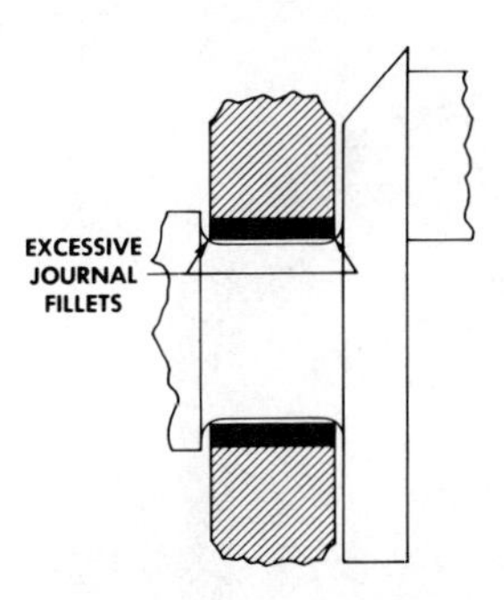

Figure 8.20 Cross section of crankshaft and bearing showing fillet ride (courtesy Gould, Inc., Engine Parts Division).

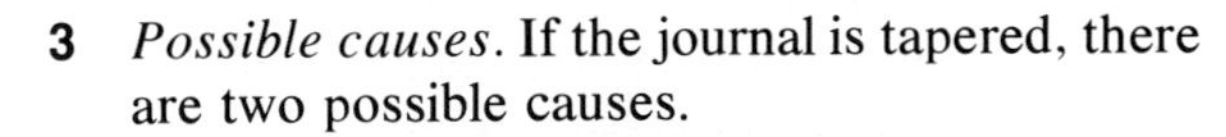

3 *Possible causes.* If the journal is tapered, there are two possible causes.

a Uneven wear at the journal during operation (misaligned rod).

b Improper machining of the journal at some previous time.

If the journal is **hour-glass** or **barrel**-shaped, this is always the result of improper machining.

4 *Corrective action.* Regrinding the crankshaft can best remedy out-of-shape-journal problems. Then install new bearings in accordance with proper installation procedures.

I *Fillet ride* (Figure 8.19).

1 *Appearance.* When fillet ride has caused a bearing to fail, areas of excessive wear are visible on the extreme edges of the bearing surface (Figure 8.20).

2 *Damaging action.* If the radius of the fillet at the corner where the journal blends into the crank is larger than required, it is possible for the edge of the engine bearing to make a metal-to-metal contact and ride on this oversize fillet.

This metal-to-metal contact between the bearing and fillet causes excessive wear, leading to premature bearing fatigue.

3 *Possible causes.* Fillet ride results if excessive fillets are left at the edges of the journal at the time of crankshaft machining.

4 *Corrective action.*

a Regrind the crankshaft paying particular attention to allowable fillet radii.

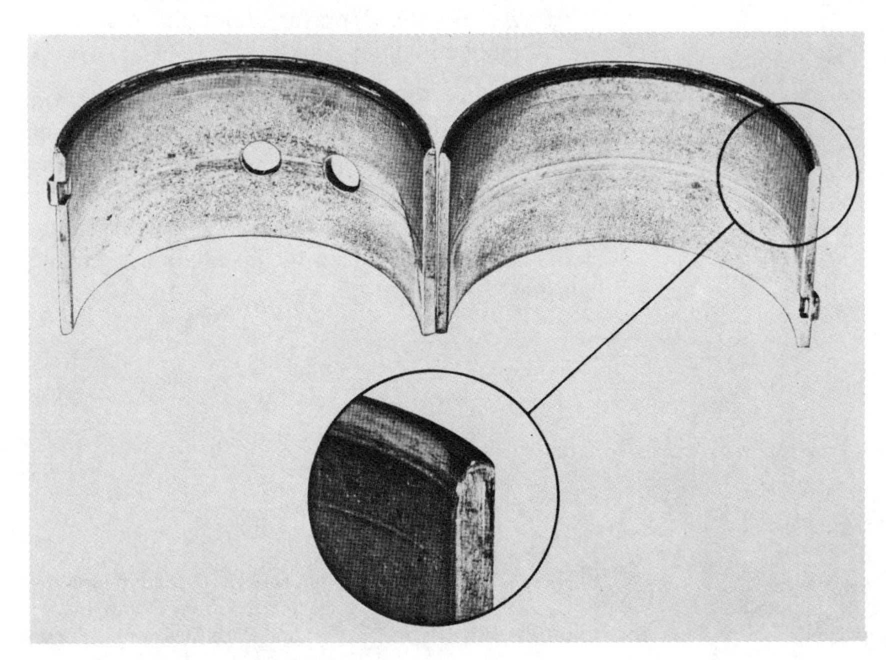

Figure 8.19 Bearing damage caused by fillet wear (courtesy Gould, Inc., Engine Parts Division).

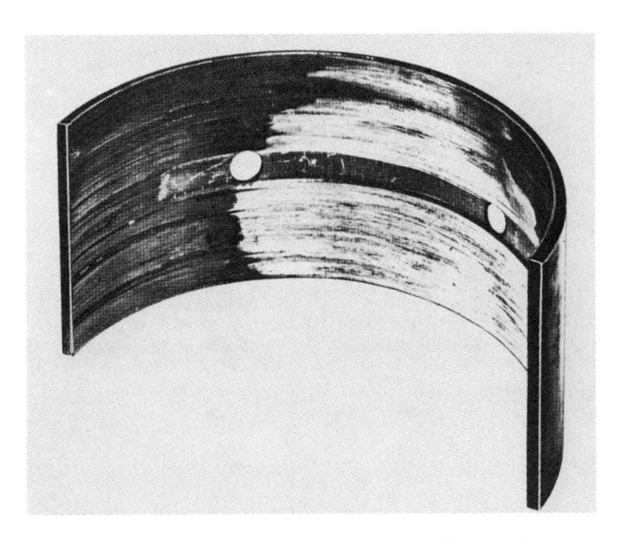

Figure 8.21 Bearing damage caused by oil starvation (courtesy Gould, Inc., Engine Parts Division).

NOTE Be careful not to reduce fillet radius too much, since this can weaken the crankshaft at its most critical point.

b Install new bearings.

J *Oil starvation.*

1 *Appearance.* When a bearing has failed due to oil starvation, its surface is usually very shiny. In addition, there may be excessive wear of the bearing surface due to the wiping action of the journal (Figure 8.21).

2 *Damaging action.* The absence of a sufficient oil film between the bearing and the journal permits metal-to-metal contact. The resulting wiping action causes premature bearing fatigue (Figure 8.22).

3 *Possible causes.* Any one of the following conditions could cause oil starvation:

a Insufficient oil clearance—usually the result of utilizing a replacement bearing that has too great a wall thickness. In some cases, the journal may be oversize.

b Broken or plugged oil passages, prohibiting proper oil flow.

c A blocked oil suction screen or oil filter.

d A malfunctioning oil pump or pressure relief valve.

e Misassembling main bearings metering off an oil hole.

4 *Corrective action.*

a Double check all measurements taken during the bearing selection procedure to catch any errors in calculation.

b Check to be sure that the replacement bearing you are about to install is the correct one for the application (that it has the correct part number).

c Check the journals for damage and regrind if necessary.

d Check engine for possible blockage of oil passages, oil suction screen, and oil filter.

e Check the operation of the oil pump and pressure relief valve.

f Be sure that the oil holes are properly indexed when installing the replacement bearings.

g Advise the operator about the results of engine lugging.

K *Misassembly.* Engine bearings will not function properly if they are installed wrong. In many cases misassembly will result in premature failure of the bearing.

Figures 8.23 to 8.25 show typical assembly errors most often made in the installation of engine bearings.

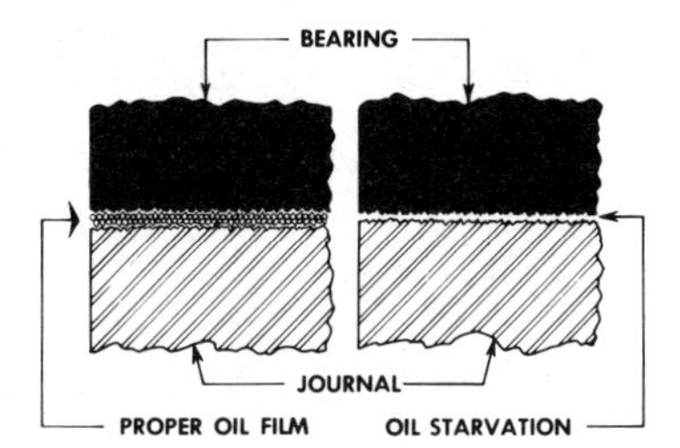

Figure 8.22 How oil starvation would damage engine during operation (courtesy Gould, Inc., Engine Parts Division).

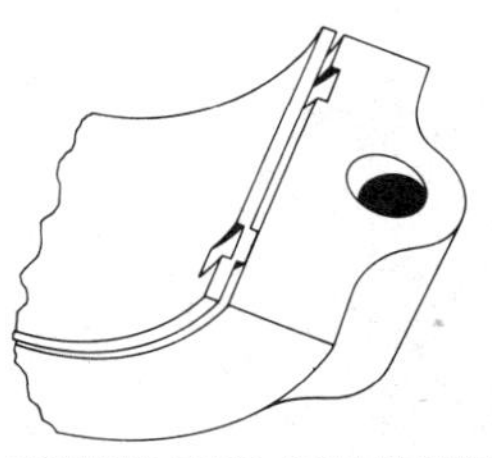

Figure 8.23 Locating lugs not nested (courtesy Gould, Inc., Engine Parts Division).

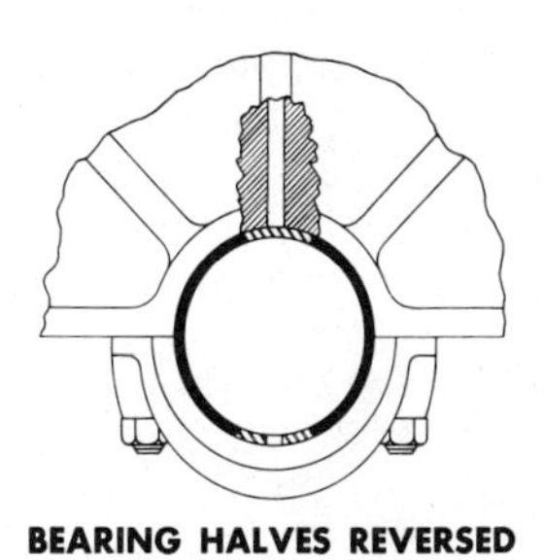

Figure 8.24 Bearing halves reversed, oil hole in wrong place (courtesy Gould, Inc., Engine Parts Division).

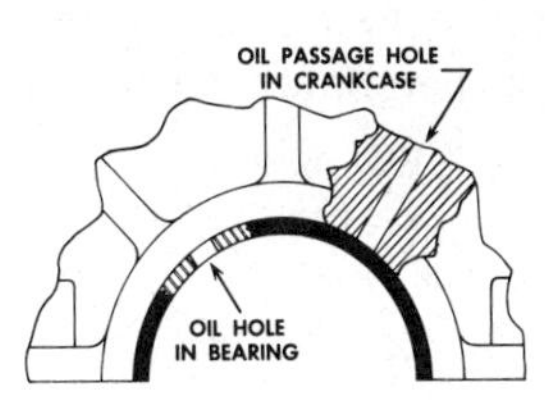

Figure 8.25 Bearing oil hole not aligned with oil passage hole (courtesy Gould, Inc., Engine Parts Division).

L Main bearings for many engines are supplied in standard 0.001, 0.010, 0.020, 0.030, and 0.040 inch sizes. Other engine manufacturers do not recommend that the crankshaft be ground; as a result, no undersize bearings are supplied. After the crankshaft has been reconditioned, the correct size main bearings must be selected to give the recommended running or oil clearance. For example, a crankshaft may be ground to a 0.020 undersize (the correct main bearing then is a 0.020 undersize).

If specifications are not available for the crankshaft on which you are working the following general specifications may be referred to when you are measuring the crankshaft.

General specifications for main bearings and crankshaft tolerances.

1. Crankshaft finish – 20 micro-inches or more.
2. Diameter tolerance – 0.0005 in. for journals up to $1\frac{1}{2}$ in. in diameter.
   – 0.001 in. for journals $1\frac{1}{2}$ to 10 in. in diameter.
3. Out of round – 0.002 in. maximum. (Never use a medium out-of-round journal with a maximum out-of-round bore.)
4. Taper should not exceed
   0.0002 in. for journals up to 1 in. wide
   0.0004 in. for journals from 1 to 2 in. wide
   0.0005 in. for journals 2 in. and wider
5. Hour-glass or barrel-shape condition – use same specifications.
6. Oil holes must be well blended into journal surface and have no sharp edges.

III. Procedure for Main Bearing and Crankshaft Installation

It is assumed that the cylinder block has been checked, cleaned, and reconditioned. If not, refer to block reconditioning in Chapter 6 before attempting to install main bearings or crankshaft.

A Put cylinder block on a clean workbench or engine stand in the inverted position.

B Install main bearing top half (shells) of proper size carefully in cylinder block, making sure that bearing locating lug is correctly aligned with matching slot in block or cap (Figure 8.26).

CAUTION Make sure all main bearing feed holes are lined up with holes in the main bearings.

C Install rear main bearing seal into the block if a split type seal is used. Most rear main seals in modern engines are made of neoprene (an oil-resistant, rubberlike substance) and should be coated with grease before the crankshaft is installed to prevent damage to the seal on initial engine start-up.

D Blow out all oil passageways and remove any protective grease or preservative from crankshaft.

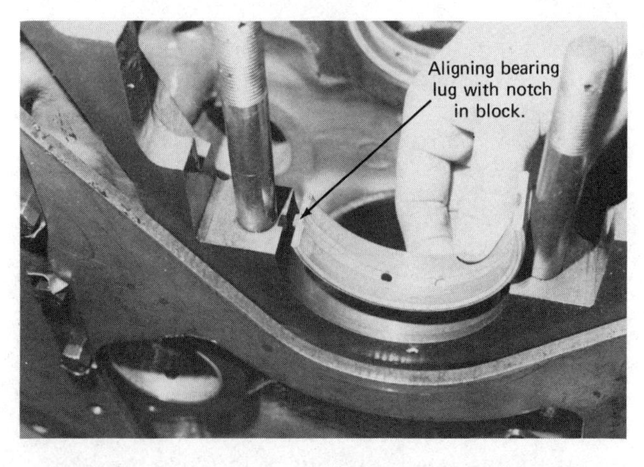

Figure 8.26 Aligning bearing lug with slot in engine block or connecting rod.

Figure 8.27 Lifting crankshaft with a sling.

E Install crankshaft using a lifting sling or bracket as in Figure 8.27.

F If timing gears and camshaft are installed in block, index the timing mark on crankshaft gear with appropriate mark.

G Main bearing clearance should be checked at this time using plastigage.

NOTE Plastigage is the plastic thread that can be broken into the correct length and placed on the rod journal or in the rod cap on the bearing. When the rod cap is installed and torqued, the plastic thread is flattened out to the clearance between the rod journal and rod bearing. The cap is then removed and the width of the plastigage compared to various widths pictured on the package that contained the plastigage. By this comparison the rod bearing clearance can be determined. Carefully clean the plastigage from the rod journal and bearing.

H Break off a piece of plastigage the same width as the crankshaft journal and place it on the crankshaft journal of one main bearing.

I Carefully place the appropriate main bearing and cap on the journal (check number) and tighten down evenly with a torque wrench.

CAUTION Do not drive main bearing caps in place with a hammer since this may dislodge the bearing shell and distort the bearing when the cap is drawn into place.

NOTE Do not turn crankshaft after main bearing cap is torqued with plastigage in place because damage to plastigage will result.

J Loosen main bearing cap bolts completely and remove caps.

K Measure plastigage by comparing it to the pictures on the envelope that it came in (Figure 8.28).

NOTE The envelope is marked off in various widths that represent various thicknesses, such as 0.001, 0.002, 0.003 in. (and also 0.025 mm, 0.050 mm, 0.075 mm etc.)

L If clearance is as recommended by manufacturer, clean plastigage from the bearing and journal. If clearance is not as recommended, check the bearing for correct size with respect to the crankshaft. If bearing clearance is incorrect, remove bearing and check for dirt or metal particles that may have been lodged under it.

M All main bearings should be checked as outlined, then lubricated with engine oil or acceptable lubricant and the caps installed and torqued.

NOTE In most cases the main bearing cap bolts are torqued using a torque wrench to a specified torque, such as 100 ft lb (14 N.m). Some engines, such as Cummins Diesel, use the template or torque turn method to tighten main bolts. (Consult your engine service manual.)

N Turn crankshaft and check for binding. If crankshaft does not turn freely, loosen all caps and

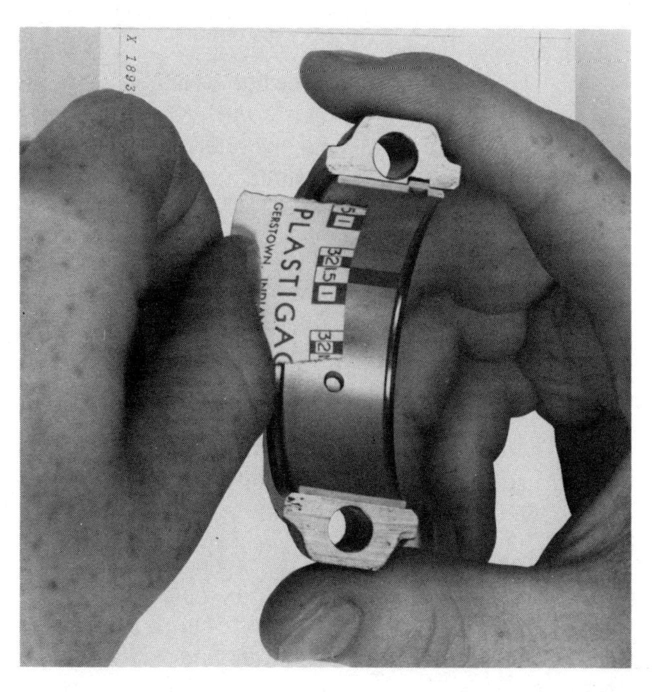

Figure 8.28 Comparing plastigage against envelope (courtesy Deere and Company).

tighten them individually to determine which one is causing the binding.

IV. Procedure for Installation of the Rear Main Seal

If the rear main seal is not of the split, two-piece type and has not been installed in the block as outlined in Section C below, it should be installed at this time. Most modern engines use a single-piece, lip type, rear main seal (Figure 8.29).

A A main bearing seal like the one shown in Figure 8.30 will be mounted into a seal housing that is bolted to the block and holds the seal in place.

NOTE The rear main seal housing that is bolted to the engine block must be centered (sometimes called run out) carefully on the crankshaft seal flange or wear sleeve. Many engine manufacturers use dowel pins to center the housing;

Figure 8.29 A typical rear main seal (courtesy Detroit Diesel Allison, Division of General Motors Corporation).

Figure 8.30 Seal and seal housing in place on cylinder block.

others have no dowel pins and the housing must be centered using a dial indicator or other centering device.

B If wear sleeve has not been replaced, replace it at this time using the following peocedure.

1. Remove old wear sleeve, if not previously done, by cutting it with a sharp chisel.

CAUTION Use extreme caution when cutting old sleeve to prevent damage to the crankshaft.

2. Clean crankshaft flange.
3. Install wear sleeve using a special driver or puller.

C If rear seal housing uses dowel pins for centering, the seal can be installed into the seal housing before the housing is installed onto the engine block. Install the seal using driver to prevent damage to the seal.

D Install housing onto block, tapping it on the dowel pins with a plastic hammer.

NOTE Special attention should be paid to the seal if it is installed in the housing to insure that the seal lip is installed on the crankshaft flange without damage.

E If rear seal housing does not have dowel pins for centering:

1. Install housing with a new gasket onto the block and install all bolts finger tight.

Figure 8.31 Centering main bearing seal housing with a dial indicator (courtesy Cummins Engine Company).

2 Mount a dial indicator on crankshaft flywheel flange with the plunger end of the indicator riding on the inside of the seal housing (Figure 8.31).

3 If no dial indicator is available or another method is desired, a piece of round stock (a round piece of metal similar to a bolt) may be used to center the housing.

NOTE To determine what size round stock should be used, measure the crankshaft seal flange to obtain diameter. Next measure the inside diameter of the seal housing and subtract the flange diameter from it. Divide the answer by two to determine the stock size.

4 With seal housing bolted to block finger tight, insert stock between crankshaft seal flange and seal housing. Roll the stock around the entire circumference of the crankshaft. This will center the seal housing.

5 Carefully tighten the seal housing bolts to recommended torque.

6 Using a seal driver or installer, install seal into the housing.

CAUTION No attempt should be made to install the seal into the housing without a seal driver or installer, as serious damage to the seal may result.

V. Vibration Dampers

Vibration dampers used on diesel engines are designed to help dampen the torsional vibrations created within the crankshaft when the engine is running. The vibration damper is usually connected or mounted onto the free end of the crankshaft opposite the flywheel. It may be made up of two metal units shaped in the form of a circle, bonded together by a rubber element or bonded with a viscous silicone fluid (Figure 8.32).

A *Bonded type.* The inner hub unit of the damper construction is hublike to fit the crankshaft while the outer unit or ring is designed to fit over the hub with the rubber element in between. The entire assembly is held together by the molded rubber.

1 Inspection of bonded damper.

Although vibration dampers appear to be a solid unit that requires little if any inspection, they must be checked before used on a rebuilt engine.

a Some rubber element dampers have index marks that should be checked for mark alignment. If the marks do not line up, the damper should be replaced.

b It should also be checked for wobble (lateral run out) after it has been mounted on the crankshaft. (A dial indicator is used on the inner surface.) If wobble exceeds manufacturer's specifications, a new damper should be installed.

B *Viscous type.* The viscous damper is of a two piece design with a fluid chamber in between. The fluid in the chamber provides the resistance so that the unit absorbs the twisting motion of the crankshaft.

1 Inspection of viscous damper.

Viscous dampers should be checked for nicks, cracks, or bulges. Bulges or cracks may

Figure 8.32 Typical vibration damper (courtesy Detroit Diesel Allison, Division of General Motors Corporation).

indicate that the fluid has ignited and expanded the damper case.

NOTE Many engine manufacturers recommend that the viscous dampers be replaced during a major engine overhaul or rebuild. The manufacturer's suggestions and recommendations concerning replacement should be followed closely in this area to prevent crankshaft breakage.

VI. Flywheels

The engine flywheel is a component that, like the vibration damper, sometimes does not get a close inspection during an engine rebuild. The flywheel has several important functions to perform. Some of these are to:

1 Provide a true machined surface on which to mount the clutch.

2 Provide a place to mount the clutch pilot bearing.

3 Provide a place to mount the starter ring gear (a steel ring gear into which the starter motor pinion gear meshes).

4 Help balance the engine.

5 Provide momentum to keep the engine running under heavy load.

VII. Procedure for Inspection of Flywheel and Starter Ring Gear

The flywheel should be closely inspected during an engine rebuild for the following:

A Check contact surface (face) for:

1 Straightness with a straight edge.

NOTE If flywheel contact surface is not straight or true it can be machined and reused.

2 Heat cracks or checks.

NOTE Small heat cracks in the face of the flywheel are common and are not a reason for discarding the flywheel.

B Other flywheel and ring gear checks.

1 Dowel pin holes in flywheel and bolt holes.

2 Condition of flywheel ring gear. Ring gear teeth should not be worn or chipped. If so, replace.

NOTE The flywheel ring gear is a replaceable type gear that fits onto the outside of the flywheel and can be replaced by supporting the flywheel on a wood block with the face or clutch side up. The blocks used to support the flywheel must be smaller than the inside diameter of the ring gear to allow the ring gear to pass over them.

C. Procedure for Ring Gear Removal and Replacement

1 Heat the ring gear to approximately 300° F (150° C) evenly around the entire circumference with an acetylene torch. (Figure 8.33).

NOTE Some manufacturers do not recommend heating the ring for removal, but it has been my experience that slight heating of the flywheel ring gear before removal will make the gear come off much easier.

2 Using a drift and a hammer, drive the ring gear from the flywheel.

NOTE The position or direction of the flywheel gear chamfer should be noted during removal for reference when reinstalling the new gear.

3 Turn flywheel over (ring gear side up) and support it on a flat surface.

4 Heat new ring gear evenly around its entire circumference.

CAUTION Ring gear should not be heated in excess of 300 to 400° F. Temperature can be checked with tempilstick or soft solder. Simply

Figure 8.33 Heat flywheel ring gear with torch.

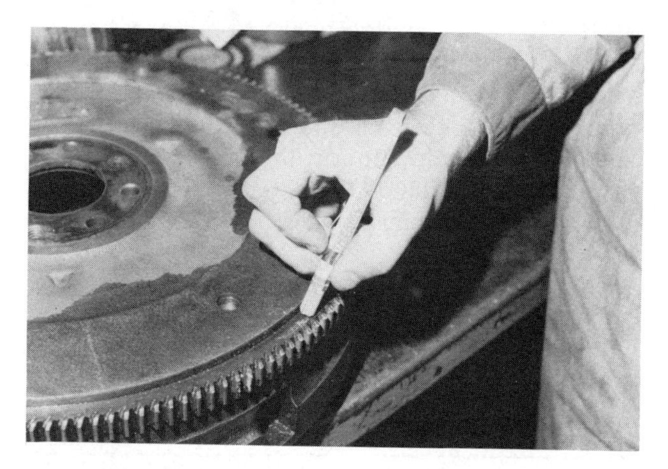

Figure 8.34 Checking temperature of flywheel ring gear with tempilstick.

touch the tempilstick or solder to the heated gear; if they melt, the gear is at about 300 to 400° F or 150 to 200° C (Figure 8.34).

5 Grasp the ring gear with pliers and place on the flywheel recess with chamfered side of teeth facing the same way as the old ring gear was.

6 Tap ring gear into place with drift or large punch and hammer.

VIII. Procedure for Flywheel Installation

Before attempting to install the flywheel, a method of lifting it into place should be selected. In many cases the flywheel is lifted by hand; in other cases the mechanic may select a hoist of some type to lift it into place. Before attempting to install the flywheel, you should:

A Place guide studs or dowels in crankshaft flange. (Guide studs can be easily made from bolts by sawing off the heads.)

B Determine if the flywheel can be installed in more than one position relative to the crankshaft. This is called flywheel timing.

NOTE Flywheel timing is very important since the flywheel may have marks on it that will be used later on to set valves, injectors, and/or time injection pump.

C Install flywheel to crankshaft cap screws and torque to the recommended torque.

NOTE If guide studs were used to guide the flywheel in place, they should be removed and replaced with bolts at this time.

D If a bolt lockplate is utilized, it should be locked or bent at this time.

Installation of the crankshaft, main bearings, vibration damper and flywheel should be complete at this time.

SUMMARY

As noted throughout the chapter, the components mentioned in the chapter, such as crankshaft, vibration damper, and flywheel, are given a very quick and casual inspection by some mechanics. It cannot be emphasized enough that a complete, extensive check be made on these components, because it will prevent costly and embarrassing premature engine failure. Unfortunately, we cannot always count on engine parts failing because the normal life span has been exceeded. Thus a mechanic must not only be a "replacer of parts" but, like a doctor, must be capable of diagnosing the engine to determine why the part failed. Study the failure analysis section of this chapter so that you become proficient in recognizing the telltale signs of engine bearing problems; this will enable you to correct them before they can cause engine damage or failure.

REVIEW QUESTIONS

1 What is the main function of the crankshaft in a diesel engine?

2 Why are counterweights added to the engine crankshaft?

3 What device on the diesel engine has as its sole function the elimination of torsional vibration?

4 List four checks that must be made on the crankshaft.

5 Why should the old main bearings be inspected closely, even if they are to be discarded?

6 When installing main bearing top half shells in engine block, list two items that you must be sure are correct.

7 What can be used to check crankshaft-to-bearing clearance?

8 Why is it not recommended to drive the main bearing caps in place with a hammer?

9 List two ways a rear main seal housing can be centered in relation to the crankshaft seal flange.

10 List two types of vibration dampers used on diesel engines.

11 List four checks to be made on the flywheel during engine overhaul.

12 Why is proper flywheel timing important in a diesel engine?

13 Explain how you would remove and replace a flywheel ring gear.

14 What area of crankshaft inspection do you consider to be the most important?

9

Pistons, Piston Rings, and Connecting Rod Assembly

The piston rings, piston pin, and connecting rod assembly make up what are considered the major rotating parts in a diesel engine (Figure 9.1). The piston and ring assembly provide the plug or seal for the pressure developed by the burning fuel and air within the cylinder. The piston pin attaches the piston and ring assembly to the connecting rod, which in turn is connected to the crankshaft. Power developed within the cylinder is transferred to the crankshaft by this assembly.

OBJECTIVES

Upon completion of this chapter the student should be able to:

1. Demonstrate the correct procedure for removal and installation of a piston on a connecting rod.
2. Demonstrate the proper placement of rings on the piston.
3. Demonstrate the procedure used in checking a rod big end bore for out-of-roundness.
4. Select from a given list four reasons for ring replacement.
5. Identify from a given assortment the following types of rings: compression, scraper, and oil control.
6. List five conditions that may cause scored pistons.
7. List four conditions that may cause stuck or damaged rings.

GENERAL INFORMATION

The piston, piston rings, and connecting rod assembly is one of the most unique and important assemblies within the diesel engine. Probably no other part of the diesel engine is subjected to the extreme heat, pressure, and force that are encountered by the piston, which is the central or main part of this assembly. During normal engine operation the piston is subjected to temperatures of 1200 to 1300° F (650 to 700° C), while

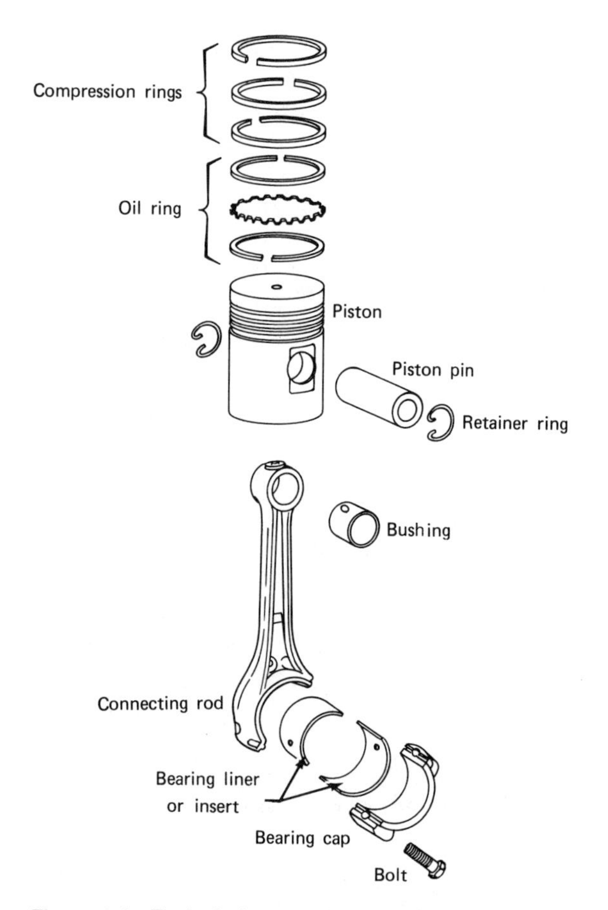

Figure 9.1 Typical piston and connecting rod assembly.

during shutdown it may be at ambient temperature (temperature surrounding engine, or atmospheric temperature).

To accept this extreme temperature change many times through its normal lifetime without failing, the piston must be made from a very durable material. Most diesel engine pistons are made from aluminum alloy, a mixture of copper, silicon, magnesium, manganese, iron, and lead. An exception to this would be Detroit Diesel pistons, which are constructed from iron and plated with tin.

NOTE Aluminum is most commonly used in four stroke cycle diesel engines because of its ability to transfer heat quickly, thereby allowing the piston to run cooler than one made from cast iron. In addition, aluminum is much lighter, making the total reciprocating weight within the engine less; this decreases the inertia (the tendency of a body in motion to stay in motion). Inertia is generated at the top and bottom of the piston stroke by rapid start and stop. Less inertia helps create a better balanced and smoother running engine.

Diesel engine pistons are usually of the trunk type (Figure 9.2), although some diesel engines (usually larger bore engines) do use the crosshead type of piston. The design of the crosshead piston (Figure 9.3) is such that it almost eliminates any side or thrust load on the piston, thereby decreasing wear and increasing ring life.

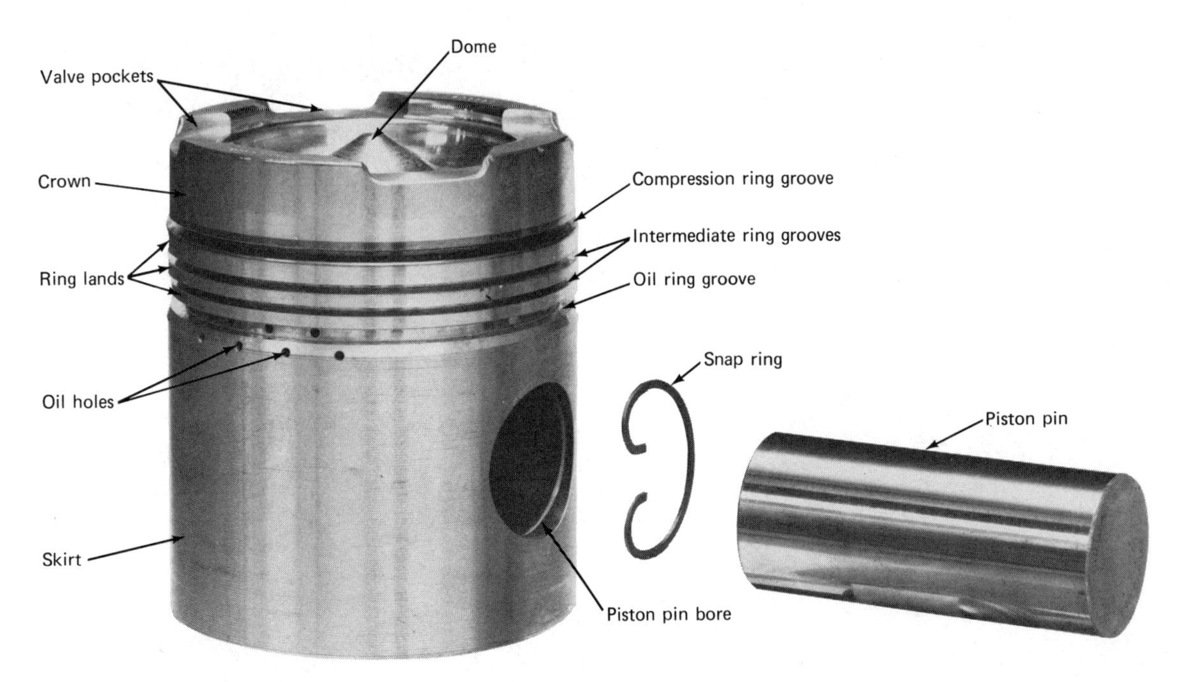

Figure 9.2 Typical trunk type piston (identified) (courtesy Cummins Engine Company).

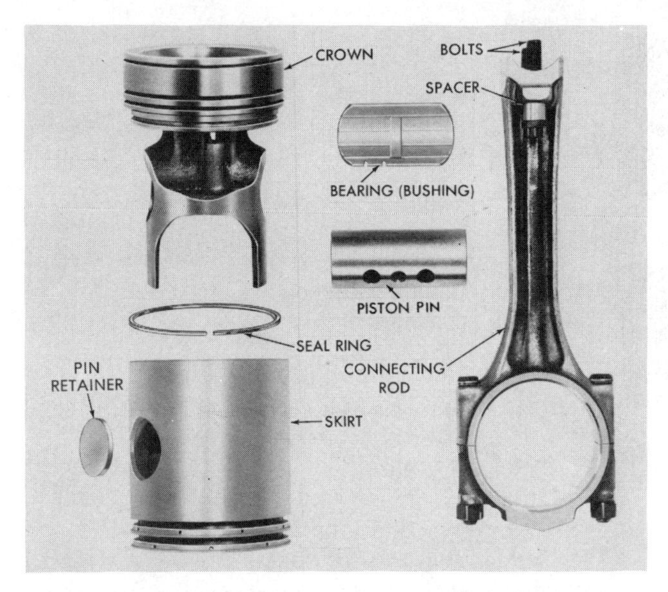

Figure 9.3 Crosshead type diesel engine piston (courtesy Detroit Diesel Allison, Division of General Motors Corporation).

The trunk type of piston is made up of piston head, piston pin boss (bearing area), ring grooves, groove lands, and piston skirt. Since trunk type aluminum pistons are constructed with more metal in the piston pin boss area to support the piston pin, provision must be made for the uneven expansion that results from this design. To insure that the piston will be round after expansion from the heat of combustion, the pistons are cam ground or egg-shaped (elliptical) (Figure 9.4).

Careful consideration must be given to the clearance of the piston within the cylinder. Excessive clearance will allow the piston to pound or knock against the cylinder wall after engine start-up until the piston

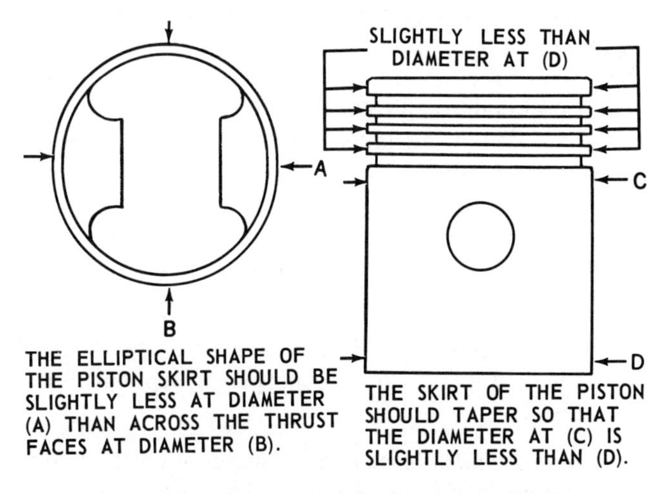

Figure 9.4 Cam ground piston (courtesy Deere and Company).

Figure 9.5 Diesel engine piston head design.

becomes hot. Too little clearance can cause scoring of the piston when it is hot because no clearance remains for lubricating oil between the piston and cylinder wall. It can be seen then that a clearance that gives little or no piston noise during cold operation and provides lubrication clearance during hot operation is the clearance desired. This clearance is generally built into the piston by the manufacturer. For example, a 4 in. (101.60 mm) diameter cylinder would be fitted with a piston 0.001 to 0.002 in. (0.025 to 0.05 mm) smaller in diameter than the cylinder. Clearance between the piston and cylinder depends a great deal on the diameter and size of the piston, as a large piston must have more room for expansion when it becomes hot during engine operation.

The head of the piston may contain (depending on engine design) the cylinder combustion chamber (Figure 9.5). This combustion chamber is designed to aid in mixing the fuel and air together so that complete combustion (burning of the fuel) can occur.

NOTE Complete combustion in a diesel engine is the ultimate goal of all engine manufacturers. How well it is achieved depends on factors such as the design of the combustion chamber, injection nozzle opening pressure, injection nozzle hole size, and compression ratio. This subject is discussed in detail in the chapter on combustion chambers.

The piston pin boss (bearing area) is the part of the piston that provides the support for the piston pin that connects the piston to the connecting rod. The piston boss is made as part of the piston and supported also by ribs or bars on the inside of the piston.

Guiding and supporting the piston within the cylinder are the piston skirts (side walls of the piston below the ring area). When combustion occurs and

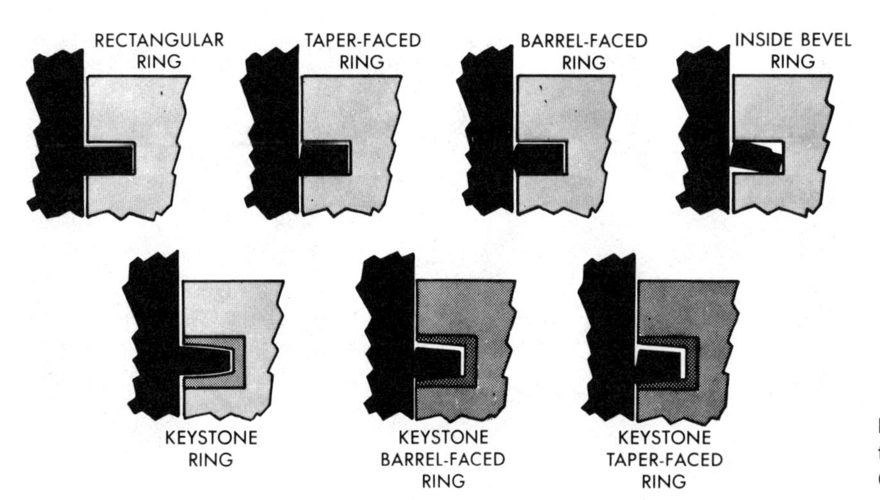

Figure 9.6 Several types of piston rings (courtesy Deere and Company).

force is exerted on the piston, it is held straight in the cylinder by the piston skirts in contact with the cylinder wall.

NOTE In reality piston skirts do not make contact with the cylinder wall since a film of lubricating oil is maintained between wall and piston at all times during engine operation.

Cut in the piston immediately below the head are the ring grooves. These ring grooves are designed or shaped the same as the rings that are fitted into them. Many aluminum pistons have an iron or Ni-Resist insert in the top ring groove. The Ni-Resist area of the piston, in which the ring groove is cut, will be made from a harder metal such as iron to increase the wear qualities of the ring groove.

Installed into the ring grooves to aid the piston in reducing power loss due to blow-by are the piston rings (circular, springlike steel devices) (Figure 9.6). Between each ring, supporting them, are the ring lands.

Piston rings are designed with an uninstalled diameter larger than the cylinder bore so that when the ring is installed radial pressure is applied to the cylinder wall.

The piston will normally have several different types of rings on it. Here are three examples:

1 *Compression ring* (top position). The top or compression ring seals the compression and pressure from combustion in the combustion chamber.

2 *Combination compression and oil scraper ring* (second groove). This second ring is generally a combination compression and oil scraper ring, aiding in controlling combustion loss and oil control.

3 *Oil control ring* (third or fourth groove depending on how many rings are on the piston). The oil control ring is designed to control the flow of oil onto the cylinder wall on the up-stroke of the piston for lubrication and scrape the oil back off on the downstroke.

Not all pistons will have three rings. The number of rings is determined by the engine manufacturer, taking into consideration factors such as bore size, engine speed, and engine configuration (in-line or V). Shown are several different pistons with their respective ring combinations (Figure 9.7).

As stated earlier in this chapter, the pistons and ring combination are connected to the connecting rod by the piston pin, which is held in place by the retainer rings. The piston pin bushing is supported in the end of the connecting rod by a bushing made from brass, bronze, steel, or aluminum. The connecting rod is composed of very strong steel alloy shaped like an I beam with a hole in one end for the piston pin (Figure 9.8). The other end of the rod has a larger hole or bore with a removable cap so that the rod may be connected to the rod journal. Installed in this hole will be a sleeve type friction bearing (Figure 9.9) comprised of two halves, one half in the connecting rod and the other half in the rod cap. Connecting rod bearings are specially designed to meet the following requirements imposed upon them during engine operation:

1 *Fatigue resistance.* The bearing must be able to withstand intermittent loading to which it may be subjected.

2 *Conformability.* The bearing material must be able to creep or flow slightly to compensate for any un-

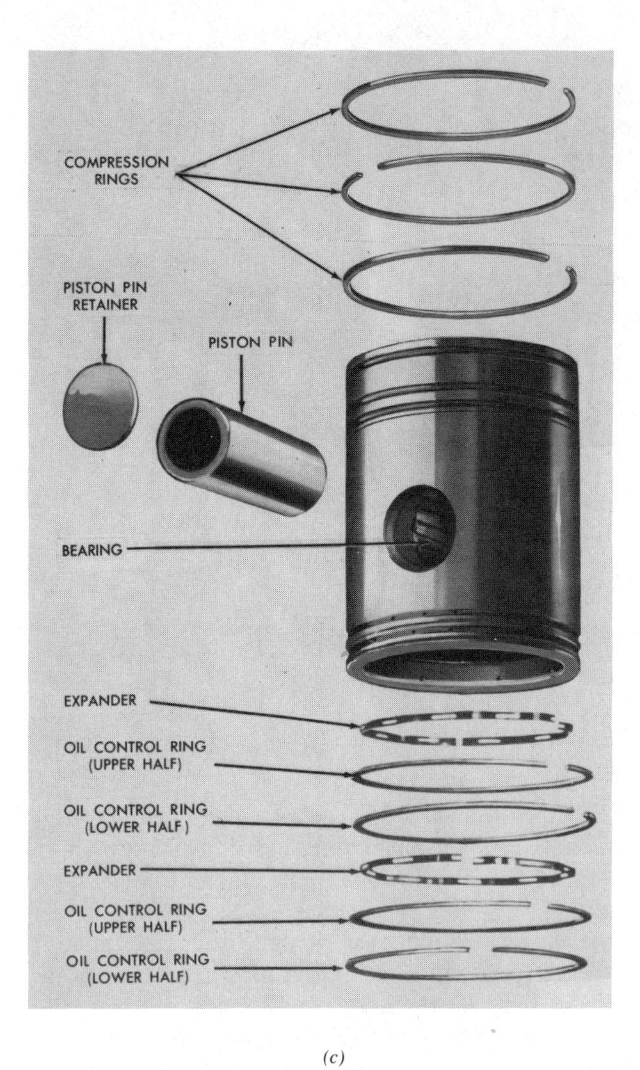

(c)

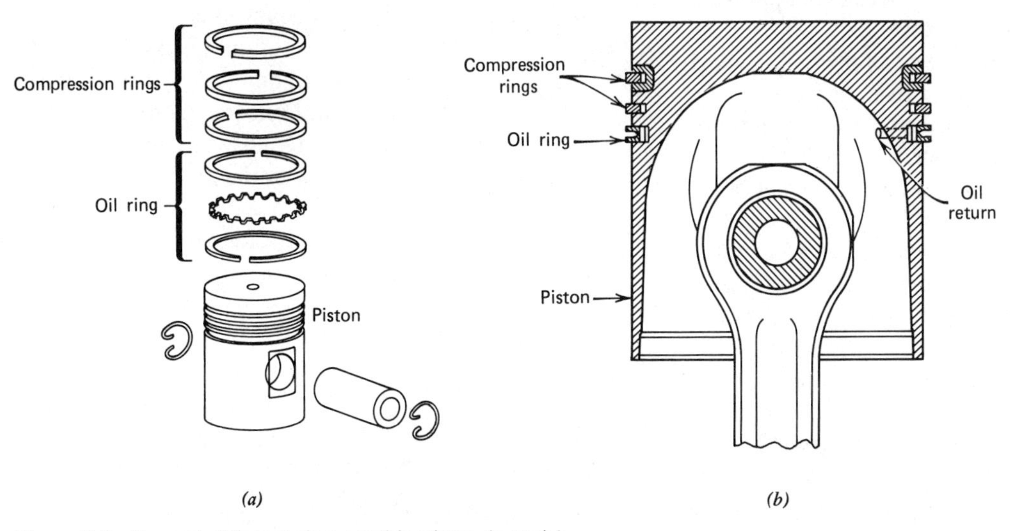

(a) (b)

Figure 9.7 Several different ring combinations. (*a* and *b* courtesy J. I. Case, *c* courtesy Detroit Diesel Allison).

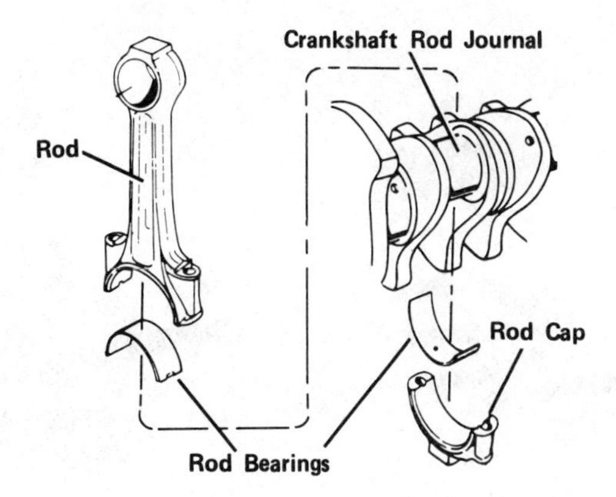

Figure 9.8 Diesel engine connecting rod (courtesy Cummins Engine Company).

avoidable misalignment between the shaft and bearing.

3 *Embeddability.* The ability of the bearing material to absorb foreign abrasive particles that might otherwise scratch the shaft that the bearing is supporting.

4 *Surface action.* The ability of a bearing to resist seizure if the bearing and shaft make contact during engine operation. This situation may occur when an extreme load squeezes the oil film out of the clearance space between the shaft and bearing.

5 *Corrosion resistance.* A bearing characteristic that resists chemical corrosion caused by acids that are the by-product of combustion.

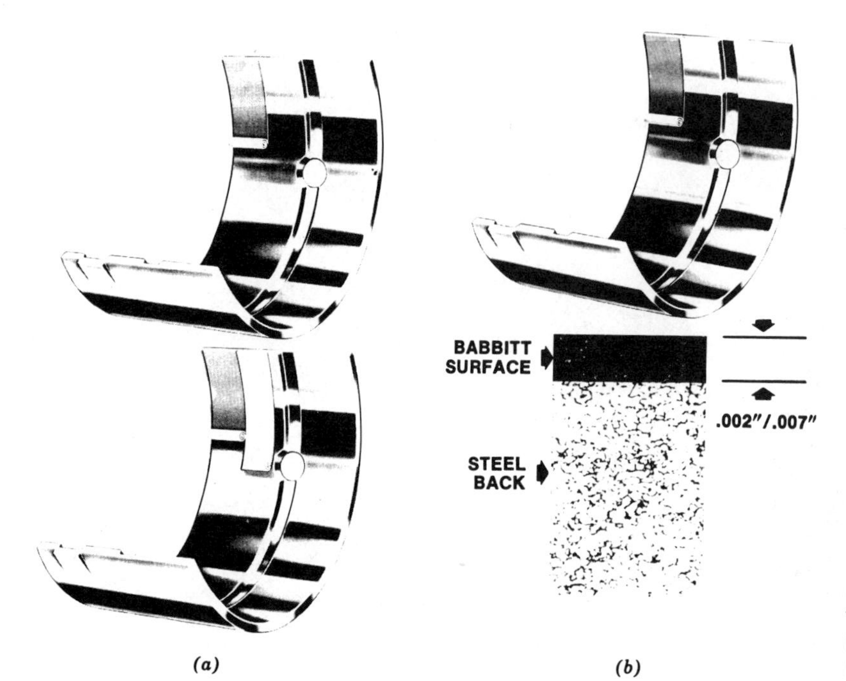

Figure 9.9 (*a*) Sleeve type friction bearing. (*b*) Steel back bearing crosssection (courtesy Gould, Inc.)

6. *Temperature strength.* How well the bearing will carry its load at engine-operating temperature without flowing out of shape or breaking up.
7 *Thermal conductivity.* The ability of a bearing material to absorb heat and transfer it from the bearing surface to the housing. An important factor in bearing longevity.

To meet all of the requirements engine bearings are designed with a steel backing (Figure 9.9) and a liner of bearing material. The bearing surface is the part of a journal bearing that performs the basic antifriction function and thus is considered to be of primary importance. The most common metal used is babbitt, an alloy composed of 83 percent lead, 15 percent antimony, 1 percent tin, and 1 percent arsenic.

Babbitt is considered an almost ideal journal-bearing surface because it provides the slipperiness to overcome friction, the softness required to permit a reasonable amount of foreign particles to embed themselves, and the "flow" necessary so that shaft and bearing will conform to each other.

The four most common materials used in modern engine bearings are: *Babbitt* is divided into two categories, conventional and micro or thin babbitt. It may be in a tin or lead base material. Conventional babbitt bearings differ from micro bearings by the amount of babbitt laminated on the steel back. *Sintered copper lead* is made by sintering metal powders on a steel strip. *Cast copper lead,* a copper lead alloy cast on a steel strip, is available with or without an overlay. *Aluminum* is a widely available and corrosion-resistant material obtainable in solid, bimetal, and trimetal construction.

The rod bearing is lubricated by engine oil supplied under pressure through a drilling in the crankshaft journal. Since there is clearance between the connecting rod bearing and the crankshaft journal, the oil used for lubrication is allowed to leak off into the oil pan or crankcase area of the engine.

One of the most critical wear areas in the engine is the piston rings and pistons because they are subjected to the tremendous heat of combustion and possible dirt-laden air supplied to the cylinder. To insure a long, trouble-free period of operation, particular attention must be given to regular oil, oil filter, and air filter changes. In addition, it is very important during engine overhaul or rebuild that strict attention be paid to detail and manufacturer's recommendations to insure that a quality job can be done.

I. Procedure for Inspection, Cleaning, and Removal of Pistons

A Visual inspection of piston.

Place the piston and rod assembly in a vise as shown in Figure 9.10 and clamp it securely.

Figure 9.10 Correct procedure for clamping rod and piston in a vise.

NOTE It is recommended that a vise with brass jaw protectors or a rag be used to protect the rod when it is clamped into the vise.

Make a visual inspection of the piston rings, land, and skirt area to determine if piston is reusable. The piston should be checked for:

1 Scored skirt area.

2 Cracked skirt.

3 Uneven wear (skirt area).

4 Broken ring lands.

5 Stuck or broken rings.

6 Worn piston pin bores.

7 Burned or eroded areas in head.

After inspecting the piston as mentioned above, it must be determined what caused the piston damage (if any). The condition that caused the piston damage must be corrected before the engine is reassembled.

Piston skirt scoring (Figure 9.11) can be caused by any of the following: engine overheating, excessive fuel settings, improper piston clearance, insufficient lubrication, or improper injection nozzle or injector.

Piston cracking or ring land breakage can be caused by excessive use of starting fluid, excessive piston clearance, or foreign objects in the cylinder.

Piston skirt wear can be caused by normal engine operation, dirty lubricating oil, too little piston clearance, or dirty intake air.

Piston burning or erosion can be caused by plugged nozzle or injection orifices, excessive engine load during cold operation, and water leakage into the cylinder.

Piston pin bearing bore wear can be caused by normal engine operation, dirty engine oil, or insufficient lubrication.

Pistons with stuck rings may be the result of overheating, insufficient lubrication, or excessive fuel settings.

If the piston fails this inspection it should be removed from the rod and discarded (see section on rod removal). If the decision is made to use the piston again, it should be left on the rod and rings removed.

NOTE Leaving the piston on the rod will provide a means for holding the piston during the cleanup and inspection to follow.

Rings can be easily removed by using a ring installation removal tool (Figure 9.12).

Normally pistons and rings are discarded and replaced with new ones during a major overhaul. Situ-

Figure 9.11 Scored piston (courtesy Detroit Diesel Allison, Division of General Motors Corporation).

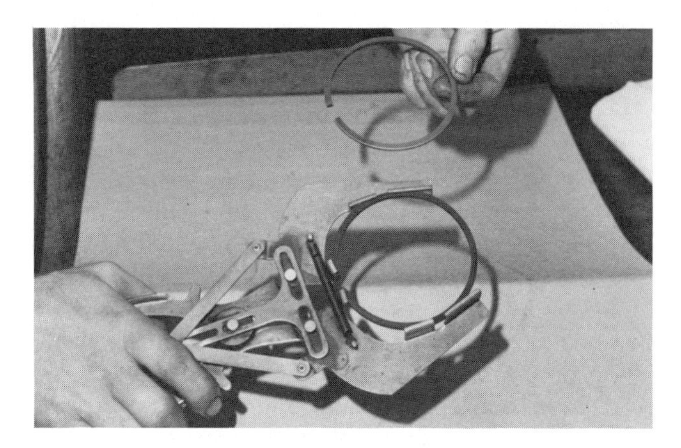

Figure 9.12 Ring installation and removal tool.

ations may develop where exceptions to this recommendation may occur. For example, a new engine with very few hours on it may be disassembled because of excessive oil consumption. Normally engine manufacturers recommend replacing the rings only, not the pistons. Pistons may also be used again in an older engine that, after overhaul, is to be used only occasionally or for a short number of hours. Given these circumstances, rings must be replaced and the piston or pistons may be used over again if they pass inspection.

B. Procedure for Cleaning of Ring Grooves

1 A ring groove cleaner (Figure 9.13) should be used to clean all the carbon from the ring grooves so they may be checked for wear.

2 Select the tool bit that fits the ring groove and install the cleaner on the piston in the ring groove.

Figure 9.13 Piston ring groove cleaner.

3 Operate the groove cleaner by twisting or turning it around the piston.

4 Clean grooves until all carbon has been removed.

CAUTION Care must be exercised when using the groove cleaner to prevent any metal from being removed from the bottom of the ring grooves or piston surface by continuing to turn the cleaner after all the carbon has been cleaned away.

NOTE If a ring groove cleaner is not available, a top compression ring may be broken in half, the end filed square and used as a ring groove cleaner.

C. Procedure for Ring Groove Measurement

The ring grooves can now be checked for wear to determine if the pistons will be reusable. Use the following procedure:

1 Check top and second ring grooves with ring groove gauge if available.

2 If a ring groove gauge is not available, a new ring and a feeler gauge may be used. If the ring and groove are straight, the ring need not be installed on the piston. If the ring and groove are of the keystone type, the ring must be installed on the piston and the ring pushed flush with the piston ring land. Using a 0.006 in. (0.015 mm) feeler gauge, try to insert the gauge between the ring and piston ring land. If the feeler gauge can be inserted and removed easily, the ring groove is worn excessively and the piston must be replaced.

D. Procedure for Piston Removal

Once it has been determined that the piston ring grooves are in usable condition, the piston can now be removed from the connecting rod for further checking. The piston may be removed as follows:

1 Remove piston pin-retaining rings using a plier (Figure 9.14).

NOTE When removing Detroit Diesel piston pin retainers (a thin steel-like cap or plug that holds the piston pin in), a hole should be made in the retainer with a chisel or punch; then insert a bar

Figure 9.14 Removing piston pin retaining rings.

and pry the pin retainer from the piston. Pin can now be easily pushed from the piston and rod by hand.

CAUTION Piston pins must never be driven from an aluminum piston without first heating the piston to approximately 200° F (93° C) in hot water. If this procedure is not followed, serious damage to the piston may result, After the piston is heated, the piston pin can be tapped out very easily using a driver and a hammer. Detroit Diesel iron pistons will not require heating and the piston pin can be pushed easily from the piston after removal of the piston pin retainers.

2 After piston has been removed from the connecting rod and if the piston is to be used again, it should be given a final cleaning in a commercial parts cleaner or a strong detergent and water. Many automotive machine shops are now using glass bead cleaning machines to clean pistons. This method is preferable if it is available.

3 After final cleaning, the piston should be given a critical final inspection for cracks.

4 Store pistons in a clean, dry area where they will be protected from dirt and possible damage until reassembly.

II. Inspection of Connecting Rods

The connecting rod bearings should be removed from the rod and rod cap in preparation for inspection. After removal of bearings the rod cap should be installed on the rod and cap bolts torqued to specifications. After torquing the cap, the rod should be checked as follows:

A. Procedure for Measuring Rod Small End Bore for Out-of-Roundness with a Snap Gauge and Outside Micrometer

1 Place snap or telescoping gauge into the rod bore and determine the bushing diameter (Figure 9.15).

2 Remove snap gauge and use the outside micrometer to accurately measure the snap gauge size.

3 This measurement is the bushing size expressed in thousandths of an inch.

NOTE Although measurement of the piston pin bushing is recommended by most manufacturers to determine if it is worn beyond reusable replacement limits, experience has proven that piston pin bushings should be replaced as a matter of practice during a major overhaul on a high time engine. If piston bushings are not replaced during a major overhaul when a new piston, sleeve, and rings are installed, the load on the piston pin bushing will be increased and this may cause an old pin bushing to break up or wear excessively during engine break-in. In addition, if the pin bushing is not replaced it may fail before the engine is due for another major overhaul.

B The rod big end bore should be measured with a snap gauge and inside micrometer in the same manner that the small end bore was measured. If

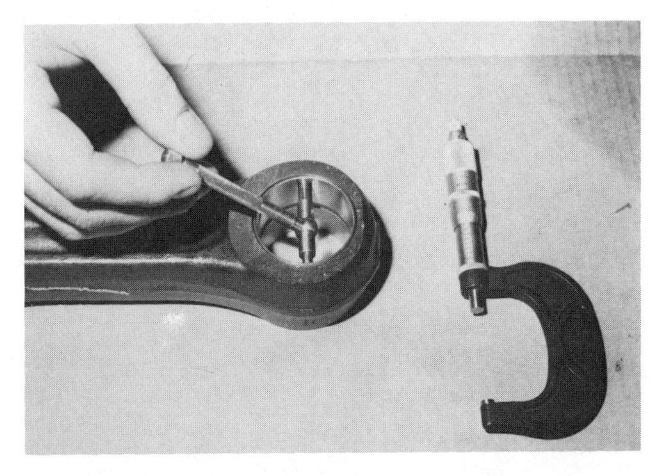

Figure 9.15 Measuring piston pin bushing in rod.

Figure 9.16 Checking connecting rod for straightness.

rod big end bore does not meet specifications, it should be reconditioned or replaced.

NOTE Major rod reconditioning, such as honing the big end bore or straightening, is generally not attempted in a general repair shop since this procedure requires special equipment.

C If rod big end bore passes the checks outlined above, the rod can then be checked for straightness using a rod alignment device (Figure 9.16).

D If rod is not straight, it should be replaced or reconditioned.

NOTE Again, if the correct equipment is not available to straighten and check rod for cracks using a magna-flux machine, the rod should be taken to a shop that specializes in this type of repair.

III. Procedure for Piston Pin Inspection

The piston pin should be measured with an outside micrometer at both ends and in the middle (Figure 9.17). The measurement should be in agreement with manufacturer's specifications. If not, the pin should be replaced.

NOTE This check may be omitted if new sleeves and pistons are being installed since most sleeve and piston kits contain a new piston pin.

IV. Procedure for Final Assembly of Pistons, Piston Rings, and Connecting Rods

A Install the piston on the rod by inserting the piston pin through the rod and piston. Then install the piston pin retainer rings.

NOTE The piston pin can be inserted through an aluminum piston very easily by hand if the piston has been preheated to 200° F (93° C) using hot water (Figure 9.18).

B Using a rag or some other suitable protector for rod, clamp the piston and connecting rod assembly into a vise.

C Allow the piston to rest on vise jaws, in preparation for piston ring installation onto the piston.

D Before installing piston rings on the piston, it is a good practice to check the rings in the cylinder for correct end gap.

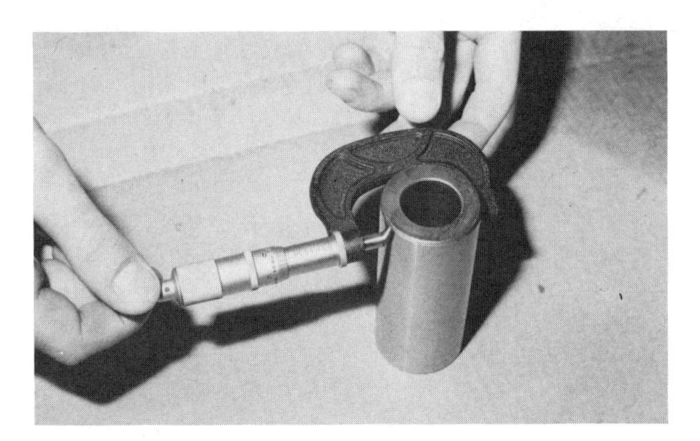

Figure 9.17 Piston pin measurement.

Figure 9.18 Inserting piston pin into piston.

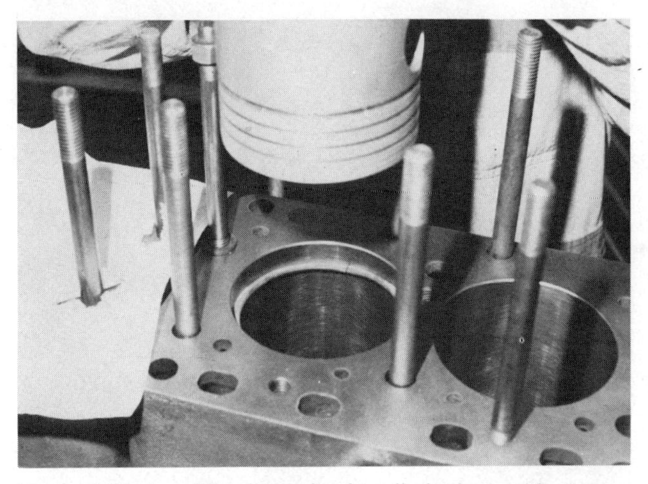

Figure 9.19 Leveling piston ring in cylinder bore with piston.

NOTE Insufficient ring end gap will not allow the ring to expand when heated and may cause ring scuffing and scoring, resulting in compression loss, excessive blow-by, and oil consumption.

E Insert rings vertically one at a time into the cylinder with end gap up. Tip the ring into the horizontal position and place a piston without rings head first into cylinder bore, pushing it down onto the ring, leveling it (Figure 9.19).

F With a feeler gauge measure the gap between the ends of the ring (Figure 9.20). Ring gap should be within specifications provided by the manufacturer. A general specification for ring end gap is 0.004 in. for every 0.001 in. of cylinder diameter (0.01 mm for every 0.025 mm). If ring gap does not meet specifications, check ring set to insure that correct set is being used. All rings in the set should be checked as indicated above.

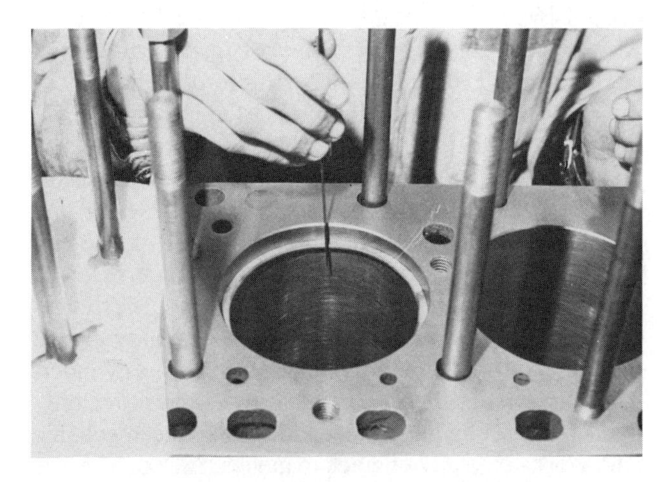

Figure 9.20 Measuring piston ring end gap with feeler gauge.

Figure 9.21 Installing ring expander.

G Carefully read instructions included with the piston rings before attempting to install them.

NOTE Installation of piston rings on the piston is a very important step and allows no room for error. Follow instructions to the letter in this critical area.

H The following instructions are general but are very similar to the instructions included with most ring sets.

1. With piston and rod assembly clamped in a vise, place oil ring expander (if used) (Figure 9.21) in the piston groove where the oil ring will be installed.

CAUTION Do not allow expander (a spring device that fits under the ring, holding it out against the cylinder wall) ends to overlap. This could cause broken rings or excessive oil consumption.

NOTE Since the oil control ring or rings is the ring nearest the bottom of the piston, it should be installed first. If the top rings were installed first, it would be impossible to install the lower rings unless they were installed from the bottom up. Although this can be done, most mechanics prefer to install the rings on the piston starting with the lowest ring and working upward, installing the top ring last.

2. Select the oil ring that fits into the lowest groove on the piston and carefully inspect it to determine which side goes up. This can be easily determined if the ring is marked with a dot or "top" (Figure 9.22). If ring is not marked,

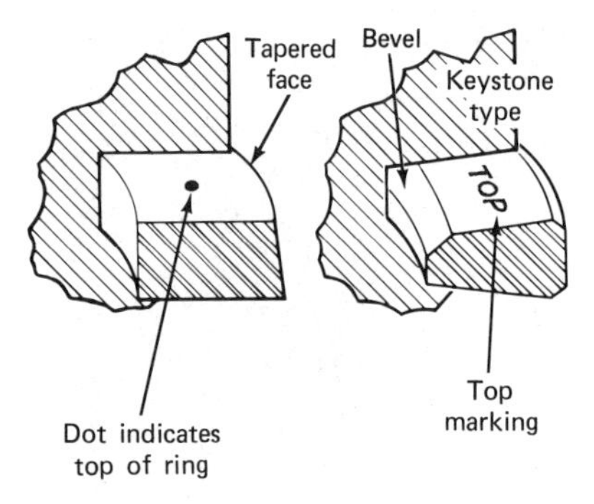

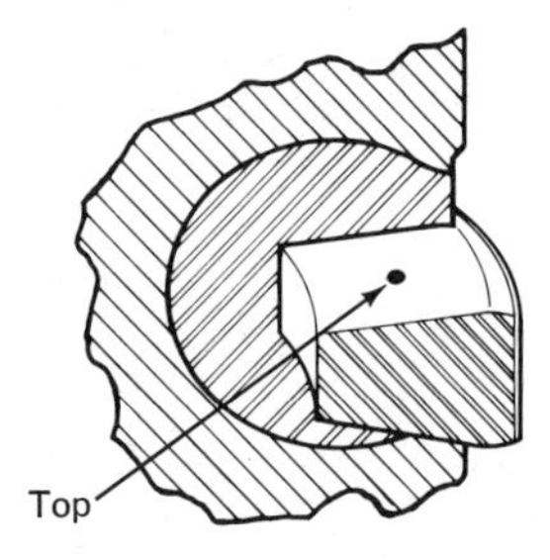

Figure 9.22 Typical ring markings.

Figure 9.23 Universal type ring compressor.

refer to installation instructions. Some oil rings may be tapered, with the taper installed to the top. Others may not have any taper or mark and can be installed either way.

3 Place ring in a ring installation tool. Expand ring so that it will slide down over the piston easily and install it over ring expander. Place ring end gap at a 90° angle away from the expander butt joint.

CAUTION Do not expand the ring any more than is absolutely needed to slide it over the piston, as this may permanently warp or damage the ring.

---

4 After installation of the oil ring, select and install the ring immediately above it, paying close attention to the "top" mark.

5 Continue installing the remaining rings, using the installation instructions as a reference, until all rings are installed.

This completes the assembly of the piston, rings, and rod assembly. The assembly is now ready to be installed in the engine. (See Section V "Installation of Piston and Connecting Rod Assembly" below.) If assembly is not to be installed in the engine immediately, it should be placed in a rack or in some suitable place so that no damage to rings or piston will result.

V. Procedure for Installation of Piston and Connecting Rod Assembly

Installation of the piston and rod assembly into the sleeve or cylinder bore requires a special tool called a ring compressor.

The ring compressor is a device that fits around the piston and compresses the rings so that they may be inserted into the cylinder or sleeve without breakage or damage. Ring compressors are usually of the compression (Figure 9.23) or tapered sleeve type (Figure 9.24).

A sleeve type ring compressor resembles an engine sleeve but has a taper cut into one end. When the piston and rod assembly is inserted into the sleeve, the rings contact the taper and are compressed into position as the piston is pushed into the compressor. This type of a ring compressor can be made for any engine by obtaining and machining a taper on one end of an old sleeve. This type of ring compressor is preferred by many mechanics since it is easier to install on the piston and eliminates the possibility of ring breakage during piston installation into the cylinder. The primary disadvantage is that it will work for only one engine or cylinder size, requiring the mechanic who works on many engines to have a ring compressor for each one.

Figure 9.24 Tapered sleeve type ring compressor.

Figure 9.26 Installing universal type ring compressor.

After selection of a ring compressor, the piston and rod assembly may be installed in the following steps:

A With the piston and rod assembly clamped in the vise, remove the rod bolts and rod cap from the connecting rod.

B Determine what size rod bearing must be used by measuring the crankshaft rod journal with a micrometer. Select and insert the bearing top half into the connecting rod, paying particular attention to the bearing locating lug and the slot in the connecting rod (Figure 9.25).

CAUTION Make sure a final check is made of the rod bearing insert to insure that it is the correct size. Size markings are found on the back of the rod bearing insert. Standard size bearings may or may not be marked indicating their size: 0.010 in., 0.020 in. (0.025 mm, 0.050 mm).

C Lubricate the rod bearing with engine oil or light grease.

D Lubricate rings and pistons with engine oil or light grease.

NOTE Grease is sometimes used when the engine is going to be sitting for a long period of time before start-up. Engine oil will not stay on the rings and pistons for a very long time; consequently, if oil is used, the piston rings may be dry when the engine is started, causing scuffing and scoring of the rings.

E Position rings around piston so that the gaps do not line up. A common recommendation is to stagger ring gaps 90° to 120° apart around the piston.

NOTE Positioning the rings in this manner will prevent excessive blow-by during the initial start-up that would result if all the ring gaps were in line.

F If a clamp or band type of ring compressor is to be used, expand it and place on the piston (Figure 9.26).

G If a tapered sleeve type of ring compressor is being used, the piston and rod assembly must be removed from the vise and inserted into the sleeve compressor.

H After installation of the ring compressor on the piston and rod assembly, it can now be inserted

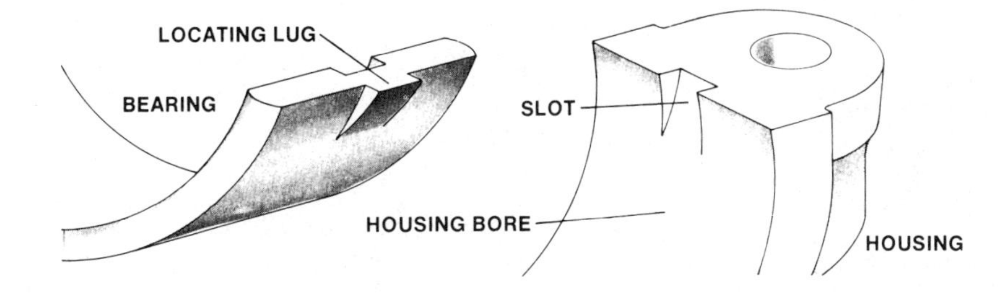

Figure 9.25 Aligning bearing lug with slot in engine block or connecting rod (courtesy Gould, Inc.).

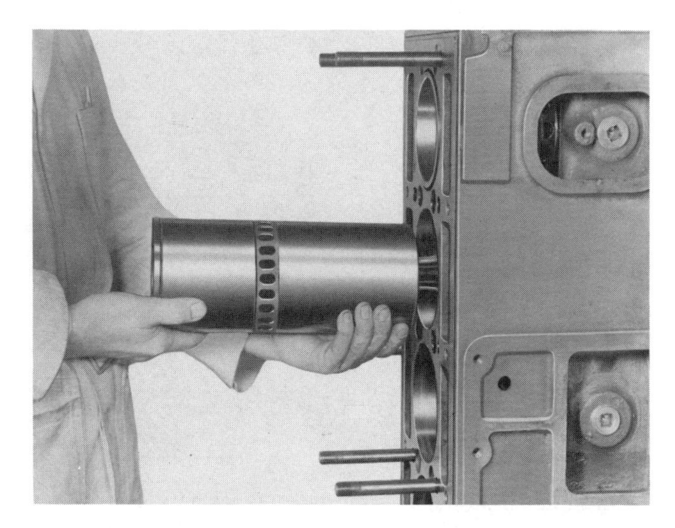

Figure 9.27 Piston and sleeve assembly ready for installation (courtesy Detroit Diesel Allison, Division of General Motors Corporation).

into the cylinder sleeve or cylinder bore; the rod number should face the camshaft on six cylinder engines and the outside of blocks on V-8 engines. The rods are numbered to indicate which cylinder they fit into.

CAUTION Before inserting the piston and rod assembly in the cylinder bore, the crankshaft should be positioned so that the rod journal of the pistons being installed is in the BDC position.

NOTE Some engine manufacturers recommend installing the piston and rod assembly into the sleeve before the sleeve is installed into the block (Figure 9.27).

Figure 9.28 Pushing piston into cylinder on V8 diesel engine.

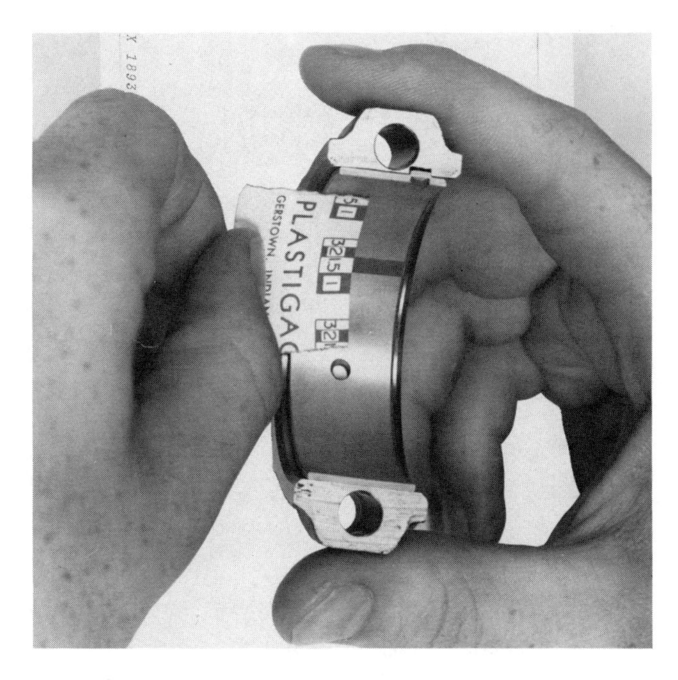

Figure 9.29 Plastigage (courtesy Deere and Company).

I Using a hammer handle, tap or push down on the piston, inserting it into the cylinder (Figure 9.28).

CAUTION When pushing the piston into the cylinder, make sure that the rod is lined up with the rod journal. Failure to do this may result in damage to the rod journal by the rod. If the rod bolts are in the rod, it is a good practice to put a plastic cap or piece of rubber hose on each bolt to protect the rod journal.

J After the piston and rod assembly is in place with the rod and bearing firmly seated on the rod journal, the rod bearing clearance should be checked using plastigage (Figure 9.29).

NOTE Plastigage is thin plastic thread that can be broken into the correct length and placed on the rod journal or in the rod cap on the bearing. When the rod cap is installed and torqued, the plastic thread is flattened out to the clearance between the rod journal and rod bearing. The cap is then removed, and the width of the plastigage compared to various widths pictured on the package that contained the plastigage. By this comparison the rod bearing clearance can be determined. Carefully clean plastigage from rod journal and bearing.

K Lubricate the rod bearing with lubricant recommended by manufacturer and install the rod cap,

Figure 9.30 Rod markings (match marks).

Figure 9.31 Torquing rod cap bolts.

making sure the number on rod and cap match and are on the same side (Figure 9.30).

L Torque rod cap bolts to specifications and, if used, lock the lock plates (Figure 9.31). Then with a feeler gauge check for correct rod side clearance between the connecting rod and the crank journal flange. Turn the crankshaft after each rod and piston assembly has been installed to make sure it moves freely. If the crankshaft does not turn after torquing of the rod cap, recheck the rod for alignment-bearing clearance and rod side clearance to determine the problem. Install all rods and rod caps in the same manner, as described above. This completes the installation of piston and rod assemblies.

After completion of piston and rod installation, it is good practice to recheck all rod torques and numbers, making sure that rods and pistons are in the correct cylinders and rod caps are matched to the correct rod. After this final check the engine is ready for further assembly.

SUMMARY

This chapter has covered the correct procedures for removing, checking, cleaning, and reassembling pistons, connecting rods and piston rings. If you have any further questions concerning the piston and connecting rod assembly, consult the engine service manual or your instructor.

REVIEW QUESTIONS

1. Identify and list four parts of a diesel piston, using correct nomenclature.
2. Why are most pistons in diesel engines made from aluminum?
3. Which pistons are most commonly used in a diesel engine, the trunk type or the crosshead type?
4. Why is piston skirt clearance very important?
5. List the three piston rings normally used in a diesel engine.
6. What type of bearing is the connecting rod big end bore bearing?
7. List four things that a used piston must be checked for during an engine overhaul.
8. Normally pistons are replaced during a major engine rebuild. List one situation where it would be considered normal to reuse a piston.
9. Explain how Detroit Diesel piston pin retainers differ from most other engine piston pin retainers and what the correct removal procedure would be.

10 What special precaution should be taken when piston pins are inserted into an aluminum piston?

11 Why should piston pins be replaced during a major engine overhaul?

12 Why is it extremely important to check piston ring end gap before installing the piston and piston rings into the cylinder?

13 Why should a ring installation tool be used when installing rings on a piston?

14 Explain how connecting rod clearance should be checked using plastigage.

15 Why are the match marks or numbers on the connecting rod and cap important?

10

Camshaft, Cam Followers, Push Rods, Rocker Arms, and Timing Gear Train

During engine operation the camshaft (a long shaft with cams on it), cam followers, push rods, and timing gears (Figure 10.1) work together to open and close the intake and exhaust valves. As the valves open and close, intake air is admitted into the cylinder on the intake stroke and exhaust gases are allowed to move out of the cylinder on the exhaust stroke, allowing the engine to breathe. In addition to providing the mechanism to operate the valves, the camshaft and timing gear train may be utilized to operate the valves. The camshaft and timing gear train may also be utilized to operate the fuel transfer pump (sometimes called a lift pump), injection pump, and the engine oil pump. The camshaft fits into the engine block in bores (drilled holes) fitted with sleeve type bearings or is mounted in bearing supports on top of the cylinder head (overhead camshaft engines). Lubrication is provided by splash in some engines while in others it is provided to the cam bushings or bearings under pressure from the oil pump.

OBJECTIVES

Upon completion of this chapter the student will be able to:

1. Demonstrate the correct procedure for installing a camshaft gear on a camshaft.
2. Demonstrate the correct procedure for lining (timing) the timing marks in the timing gear train.
3. Demonstrate the correct procedure for measuring a camshaft lobe with a micrometer.
4. Describe and explain the term "valve overlap."
5. List four checks that should be made on a camshaft gear during overhaul.
6. List two things a cam follower

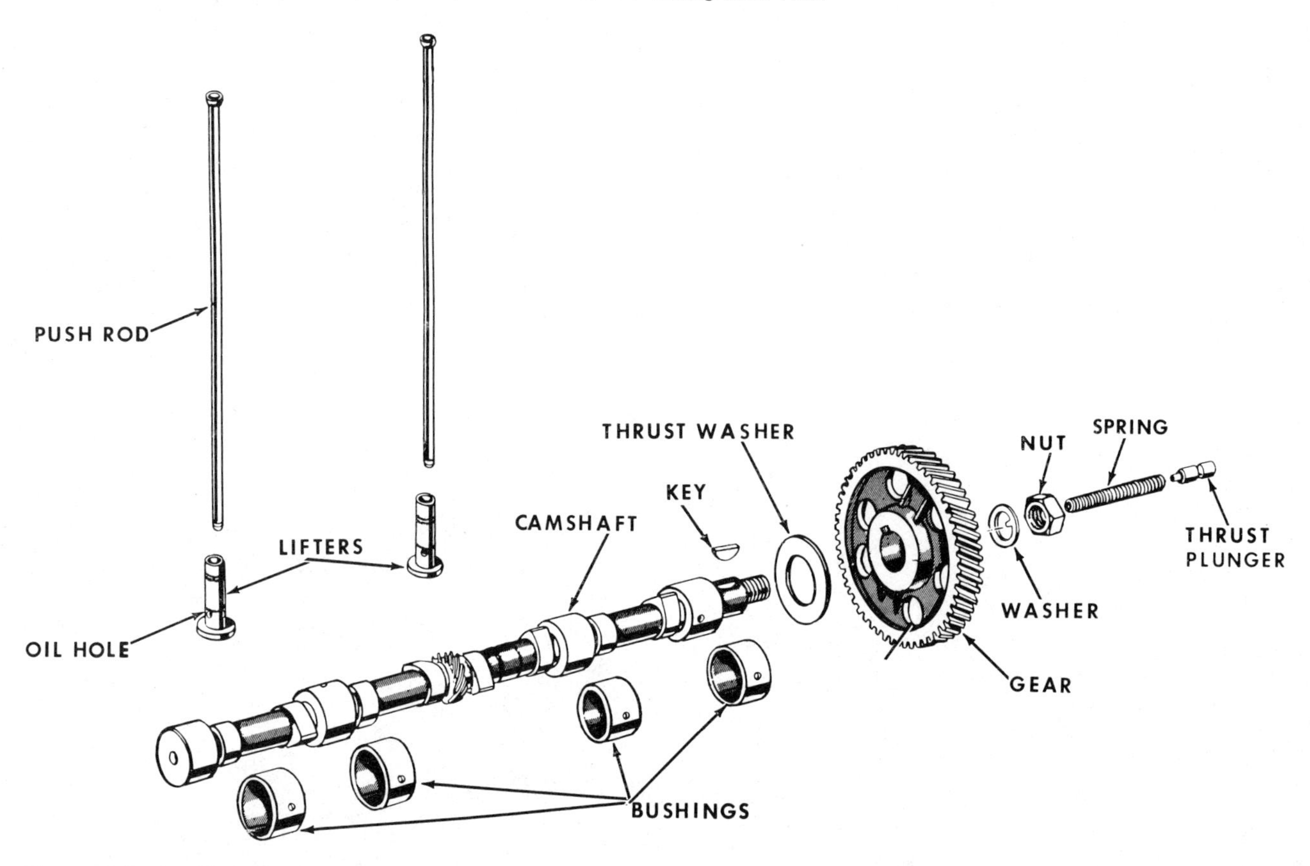

Figure 10.1 A typical diesel engine camshaft with cam followers, push rods and timing gear (courtesy J. I. Case).

should be checked for before it can be used again.

7 Demonstrate the correct procedure for checking a camshaft lobe lift with the camshaft installed in the engine.

8 Demonstrate the proper procedure for disassembly of a roller type cam follower.

9 Explain why an offset woodruff key would be used in a camshaft gear combination.

10 Demonstrate how timing gear backlash can be checked.

GENERAL INFORMATION

The camshaft in a diesel engine is used to operate the intake and exhaust valves. In most diesel engines this camshaft will have two lobes per cylinder to operate the intake and exhaust valves.

NOTE Some engines, such as Detroit, Caterpillar, and Cummins Diesel, will have an additional lobe or cam on the camshaft that is used to operate the injector. In these engines the camshaft will have three lobes per cylinder.

Working with the camshaft to operate the valves are:

1 *Cam followers.* Sleevelike plungers that fit into bored holes in the block and ride on the camshaft. They are also called lifters. (Figure 10.1).

2 *Push rods.* Long, hollow or solid rods that fit into the cam followers on one end with the other end fitting into a ball socket arrangement on the rocker arm (Figure 10.1).

3 *Rocker arms and rocker arm shaft* (Figure 10.1). The rocker arms provide the pivot point between the push rod and valve or injector. Generally mounted on a shaft that is supported by brackets

bolted to the cylinder head, the rocker arms are pushed up on one end by the push rod and rock on the shaft much like a lever. The opposite end pushes the valve down against the spring pressure, allowing air into the cylinder or exhaust gases out. Most rocker arms have an adjusting screw and lock nut that is used to adjust the clearance from rocker arm to valve (tappet).

The power required to turn the camshaft and operate the valves is supplied by the crankshaft through the timing gear train.

The time when the valves open and close is called "valve timing" and is controlled by the timing gears (the cam lobes and their placement on the camshaft). This "valve timing" becomes a very critical part of the engine design, since engine fuel efficiency, power, and smooth operation are dependent upon it.

Since valve timing does play a very important part in the operation of the engine, all component parts related to it must be checked as follows during an engine overhaul.

I. Procedure for Camshaft Cleaning and Inspection

It is assumed at this time that the camshaft has been removed. If not, refer to the chapter on disassembly. Since the camshaft is an internal engine part that constantly runs lubricating oil into the engine, it does not require much cleanup. Generally it can be cleaned by rinsing it in a cleaning solvent of the type used in a shop cleaning tank or by steaming with a steam cleaner.

NOTE If the camshaft has oil galleries or oil passageways, clean them with a wire brush and compressed air.

After cleaning, blow dry with compressed air. Then visually inspect the shaft, cam lobes, and bearing surfaces. This visual inspection must include the following:

1 Inspection of the cam lobes for pitting, scoring, or wear.

2 Inspection of cam bearing journals for scoring, bluing, or wear.

3 Inspection of cam drive gear keyway for cracks or distortion.

NOTE On engines such as the Detroit Diesel the camshaft bearings will have to be removed from the camshaft before inspection can take place. The Detroit Diesel uses a unique camshaft-bearing (bushing) arrangement (Figure 10.2). It is made up of two bearing halves that, when fitted together around the camshaft journal, make up the camshaft bearing. Holding the bearing together until it and camshaft are installed into the block is a spring ring. After installation in the block, the bearing is secured in place with a set screw. See engine repair manual or consult your instructor for correct removal procedures.

If the camshaft does not pass this visual inspection, it must be discarded and replaced with a new one. If camshaft does pass visual inspection it must be inspected further, using a micrometer to measure the cam lobes and bearing journals. Measure the cam-bearing journals as shown in Figure 10.3, using a micrometer and comparing the measurement to specifications supplied by the engine manufacturer. Since engine manufacturers recommend many different ways to check camshaft lobes, the service manual must be consulted before making the cam lobe measurement. In addition, to perform an accurate check, you must have a thorough understanding of cam lobe design. Study the cam lobe shown in Figure 10.4 before making any checks.

Listed below are several procedures that can be used to check a cam lobe:

A Measure with a micrometer from heel to toe with

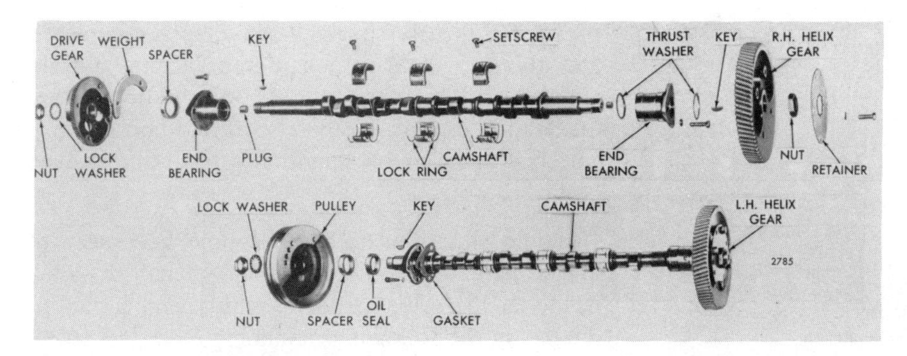

Figure 10.2 Detroit Diesel camshaft bearing arrangement.

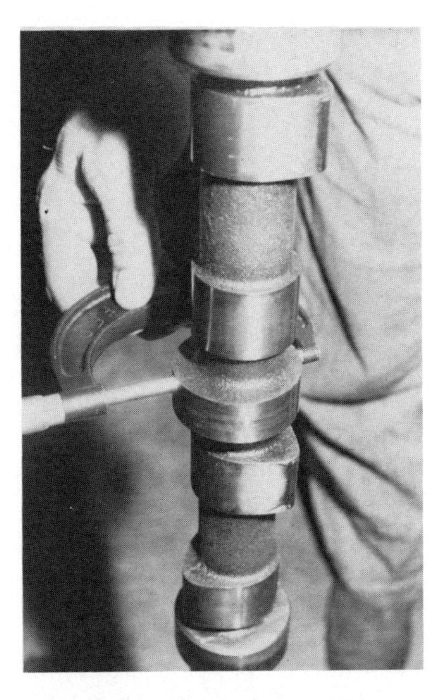

Figure 10.3 Measuring cam bearing journals using an outside micrometer.

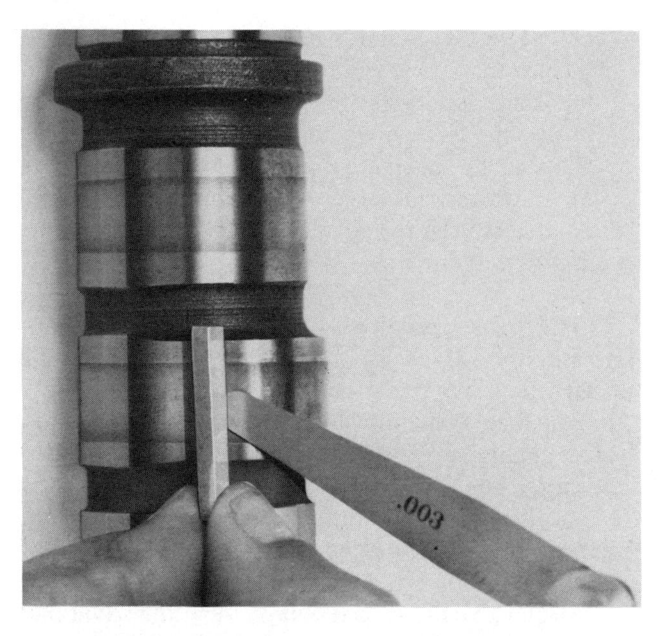

Figure 10.5 Checking cam lobes using a feeler gauge and square stock (courtesy Detroit Diesel Allison, Division of General Motors Corporation).

the outside micrometer to determine if the cam is worn sufficiently to affect lift. (Refer to specifications.)

B Measure with a micrometer at the base circle, then heel to toe, and subtract the base circle from the heel–toe measurement. This answer is called the cam lift.

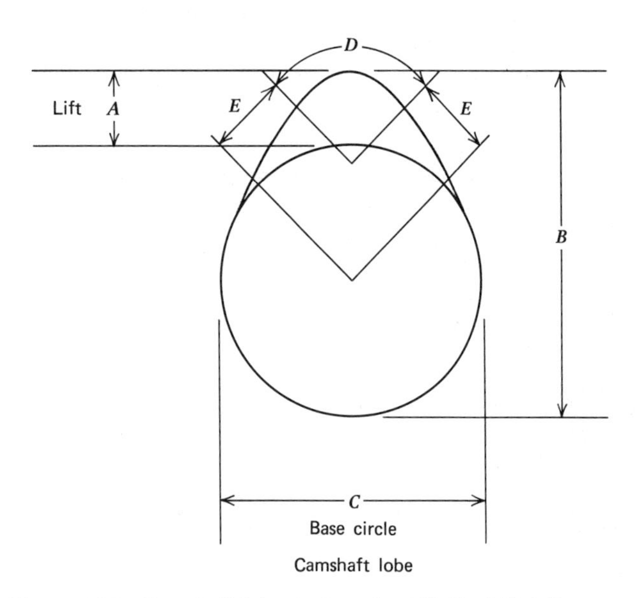

Figure 10.4 Camshaft lobe design. A = Lift B = Lift + Base circle dimension C = Base Circle D = Nose E = Flank.

C Measure cam lobe for wear using a feeler gauge and piece of hard square stock slightly longer than the lobe width. Lay stock across the cam lobe and attempt to insert feeler gauge as shown (Figure 10.5). The cam lobe should not be worn in excess of 0.003 in. (0.076 mm).

D Measure the cam lobe from heel to toe and compare with specifications.

E Some engine manufacturers recommend placing the camshaft in V blocks and using a dial indicator to check runout.

NOTE If the equipment is not available to do step E, a close check on the camshaft runout can be made by inserting the camshaft into its bore in the block. If the camshaft turns freely without binding, it can be assumed that it is straight.

When the above checks have been completed and the camshaft is considered usable, closely check each lobe for roughness or burrs on the cam lobe surface. These small surface imperfections can be removed by sanding with a fine crocus cloth.

F In some situations it may become necessary to check the cam lobe lift with the camshaft installed in the engine. This can be done by using a dial indicator in the following manner:

Figure 10.6 Measuring cam lobe lift with dial indicator (courtesy International Harvester).

1 Remove rocker arm assembly and place dial indicator on the engine as shown in Figure 10.6 with the plunger of indicator on the push rod.

2 Turn the engine until the push rod has bottomed (dial indicator will stop moving). This indicates that the cam follower is on the base circle or heel of the cam lobe. Zero indicator at this point.

3 Turn engine until the dial indicator needle stops moving in one direction (clockwise, for example) and begins to move in the opposite direction. This indicates that the cam follower has moved up on the cam to the point of highest travel and is starting to drop off (or recede).

NOTE It may be necessary to back the engine up for a quarter turn and then turn it back in the direction of rotation to reestablish the point of maximum rise.

4 Record the dial indicator travel from zero to maximum rise. This represents the cam lobe lift and should be compared to the manufacturer's specifications.

5 If cam lobe lift is less than manufacturer's specifications, the camshaft must be replaced.

II. Procedure for Cam Follower Inspection

As previously explained, the cam followers are round, sleeve-shaped devices that ride on the camshaft lobes in a bored guide hole in the block. The followers guide and support the push rods.

NOTE Some cam followers have a roller attached to the end of the sleeve. This roller rolls on the camshaft lobes and aids in the reduction of friction (Figure 10.7).

Sleeve type cam followers must be checked on the thrust face for pitting and scoring.

Roller type cam followers must be disassembled so that a check of the roller, roller bushing, and roller pin can be performed. Consult the engine repair manual before attempting to disassemble roller and pin from a cam follower. Most manufacturers recommend special procedures and fixtures for removal of the pin and roller. If this procedure is not followed, damage to the cam follower may result.

If roller bushing and pin are worn excessively or beyond manufacturer's specifications, they should be replaced with new ones.

NOTE The above procedures apply in general to overhead valve engines (those which have push rods and cam followers). Some diesel engines may have overhead camshaft arrangements.

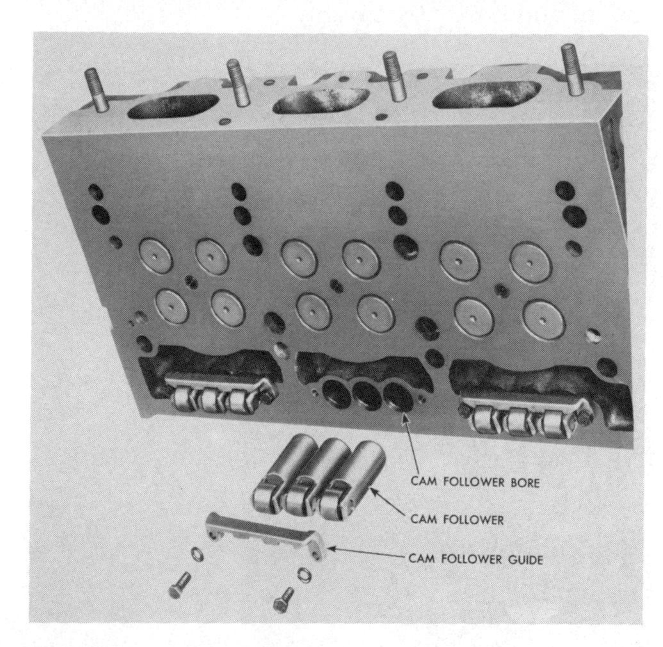

Figure 10.7 Cam followers equipped with rollers (courtesy Detroit Diesel Allison, Division of General Motors Corporation).

Engines with this type of arrangement will have different cam followers that require different checking procedures. The service manual must be consulted when working on these engines.

III. Procedure for Push Rod Inspection

The long metal rod or tube that rides in the cam follower and extends upward to the rocker arm is a very important part of the valve operation mechanism. In normal service very few problems are associated with the push rod. Push rod damage generally occurs when the rocker arms are incorrectly adjusted or the engine is overspeeded.

When the push rods are removed as a result of engine service, they should be checked for the following:

A Straightness by rolling on a flat surface or with a straight edge.

B Ball and socket wear, usually with a micrometer or as recommend by the manufacturer.

NOTE Push tubes such as used by Cummins should be checked to determine if they are full of oil by tapping them lightly on a hard surface. A hollow sound should result. If not, the tube is full of oil and indicates that the ball pressed into the tube is loose. Replace the push tube if this condition is found.

Figure 10.8 Timing gear train.

IV. Procedure for Rocker Arm Inspection

Rocker arms should be inspected and checked as outlined in Chapter 7, Section XVI.

V. Inspection, Replacement, and Assembly of Timing Gear Train

The timing gear train will include all gears that drive the camshaft and generally will be the crankshaft gear, idler gear, and camshaft gear (Figure 10.8). All the gears in the timing gear train should be checked when the camshaft or cam gear is serviced since any wear on an associated gear in the train may have an adverse effect on the camshaft gear and/or engine timing. In addition, worn gears may cause a knocking noise in the engine or fail after a few hours of operation.

A. Procedure for Inspection of Timing Gears

The timing gears should be visually inspected for the following:

1 Chipped teeth.

2 Pitted teeth.

3 Burred teeth.

NOTE Often a close, visual inspection of the timing gears will reveal a slight roll or lip on each gear tooth caused by wear. A gear worn with this type of wear should be replaced.

B. Procedure for Camshaft Gear Removal

If the camshaft gear is to be replaced it must be removed from the camshaft using either a press or a puller. If a press is to be used, some general rules to follow are:

1 Make sure the gear is supported on the center hub next to the shaft to prevent cracking or breaking.

2 Place a shaft protector on the end of the camshaft to protect the shaft.

3 Press shaft from gear (Figure 10.9).

CAUTION Safety glasses should be worn when the press is being used, and engine service

Figure 10.9 Pressing camshaft gear from camshaft (courtesy Detroit Diesel Allison, Division of General Motors Corporation).

manual should be checked to determine if any special pullers, supports, or procedures are to be used.

NOTE If failure of the timing gear train has occurred, it is very important at this point to determine what may have caused it, for example, lack of lubrication, gear misalignment or overloading, wrong gear for the application, or normal wear. Make sure you know what caused the failure before you reassemble the gear train.

C. Procedure for Replacement of Camshaft Gear

Replacement of the camshaft gear on the camshaft requires strict attention to detail since damage to the gear and camshaft may result if proper procedures are not followed. After consulting the engine service manual, the following general procedures must be followed when assembling the camshaft gear to the camshaft:

1 Install a new key in the camshaft keyway.

NOTE Some engines may require offset keys to advance the camshaft timing. (Check your service manual carefully.)

2 Check and install thrust plate or wear washer if used. Some engines will not require a wear washer since the back of the camshaft gear rides directly against the block.

NOTE If not previously done, thrust plate or wear washer should be checked for wear visually and with a micrometer.

3 If gear is to be pressed on shaft, apply grease to the end of shaft.

4 Support camshaft in arbor press in a way to prevent damage to the camshaft. This can be done by using two flat pieces of metal on the press table. Insert the camshaft between them until the shoulder of the camshaft rests firmly onto them.

5 Set gear on shaft, paying close attention to alignment with woodruff key.

6 Select a sleeve or piece of pipe that will fit over the shaft and place it on the gear.

NOTE This sleeve or pipe must have an inside diameter at least $\frac{1}{16}$ in. (1.58 mm) larger than the camshaft.

7 Press the gear onto the camshaft until the gear contacts the shoulder of the camshaft.

8 If gear is to be heated and then installed on the camshaft, the following procedure is to be followed:

NOTE Heating of the gear is recommended by some engine manufacturers. Consult your service manual.

a Heat camshaft gear to 300 to 400° F (150 to 200° C), using an oven or a heating torch.

NOTE A 400° F tempilstick may be used to check the gear temperature. If a tempilstick is not available, a piece of soft solder of 50 percent lead and 50 percent tin melts at approximately 350 to 400° F (175 to 200° C) and can be used to check gear temperature. Touch gear with tempilstick or solder to check temperature.

b With a plier or tongs, place gear on camshaft. Tap slightly with a hammer to insure that it is installed all the way onto the camshaft.

c Install retaining nut if used.

CAUTION Gear should not be overheated and should be allowed to cool normally. *DO NOT* use cold water to cool the gear as it may shrink the gear and cause it to break.

9 When camshaft and gear are assembled, the camshaft can be inserted into the block and timing

Figure 10.10 Timing gear train and timing marks.

marks indexed or lined up (Figure 10.10).

NOTE If camshaft bushings have not been replaced and are worn, refer to Chapter 6, "Cam Bushing Installation."

10 Check timing gear backlash to insure that the gears have the correct clearance between the gear teeth. Too little clearance will cause a whining noise, while too much clearance may cause a knocking noise.

NOTE Two methods of checking gear backlash are acceptable: the dial indicator method and the feeler gauge method.

11 If the dial indicator method is to be used, attach the dial indicator to the engine block by clamping or with magnetic base.

12 Position indicator plunger or gear tooth and zero indicator.

Figure 10.11 Gear backlash.

13 Rock gear forward and back by hand, observing dial indicator movement (Figure 10.11).

NOTE Gear backlash is measured between mating gears only, so that one gear must remain stationary during checking.

14 If the feeler gauge method is to be used in checking, select feeler ribbons of the correct thickness as indicated in the engine repair manual. If, after insertion of correct thickness feeler gauge, there is still gear backlash, the gears are worn excessively and must be replaced.

15 After the camshaft and timing gears have been installed and all timing marks checked, recheck all retaining bolts for proper torque; lock all lock plates if used.

16 If an oil slinger is used, install it on the crankshaft.

NOTE It is a good practice at this point to review and check the oil supply system for the timing gears. Most engines lubricate the timing gear train with splash oil; others may employ pressure lubrication through a nozzle or oil passage. If a complete engine overhaul is being performed, all passageways should have been checked during the block cleaning and inspection. DOUBLE CHECK IT NOW! before the front cover is installed and it is forgotten.

D. Procedure for Front Timing Cover Installation

When a final check has been made of the timing mark alignment, lock plates, bolt torques, and oil slinger installation, the engine is ready for installation of the front cover.

1 If not previously done, clean the cover thoroughly and remove the old crankshaft oil seal and cover gasket.

2 Install a new seal into front cover using a seal driver (Figure 10.12).

3 Glue new cover gasket onto cover with a good gasket sealer.

NOTE On some engines the camshaft will have a retaining device of some type other than a plate that is bolted to the block to hold the camshaft in place. It may be a thrust plate in the cover or a spring-loaded plunger in the end of the camshaft. This (hold-in device) must be checked carefully

Figure 10.12 Installing front seal in front cover.

before installation of the front cover for the proper adjustment or installation procedures.

4 Install front cover, starting it onto the dowel pins (if used) and tapping in place. Install and tighten bolts.

5 Install vibration damper or front pulley in this manner:

a Install woodruff key in crankshaft slot if used.

b Start damper or pulley on shaft, making sure the slot in pulley lines up with woodruff key in the crankshaft.

c Install retaining bolt with washer into end of crankshaft and tighten, pulling damper in place.

NOTE Some dampers may not slide onto the shaft far enough so that the retaining bolt can be started into the crankshaft. If this condition exists, a long bolt may be used to pull the damper in place.

CAUTION Do not drive on the damper with a hammer as damage may result that would cause it to have excess runout or be out of balance.

---

6 Install all other components, such as water pump, alternator, and fan brackets.

SUMMARY

This chapter has presented a detailed explanation of cam lobe nomenclature and valve overlap to assist you in understanding the function of the camshaft and timing gear train. In addition, information concerning camshaft, cam followers, push rods, rocker arms, and timing gear train inspection and assembly has been provided to assist you in repair or replacement of the timing gears and camshaft components. If questions still exist concerning the procedures and recommendations, consult your instructor.

REVIEW QUESTIONS

1 List four functions of the camshaft in a diesel engine.

2 What part of the diesel engine keeps the camshaft in time with the crankshaft?

3 In what way do Detroit Diesel camshaft bearings differ from other diesel engines?

4 Name three parts of a diesel engine camshaft lobe.

5 Name two types of cam followers found in diesel engines.

6 What condition might exist in a diesel engine if the timing gears have excessive backlash?

7 Crankshaft damper installation can seriously affect engine operation if done incorrectly. Why?

11

Lubrication Systems and Lube Oil

Most diesel engine lubrication systems are similar to the one shown in Figure 11.1. The system is composed of oil galleries, oil cooler, oil filter, oil pump, and oil pan or oil sump.

The oil pan or sump will be filled with engine oil. This oil is supplied by the lubrication system throughout the engine to all points of lubrication during engine operation. Without this supply of oil the engine would be quickly destroyed.

OBJECTIVES

After completion of this chapter the student will be able to:

1. Name and describe five parts of a lubrication system.
2. List and explain the differences between the three most common types of lube systems.
3. List and explain five lube system components.
4. List five lube system requirements.
5. List and explain three different types of pumps used in a lube system.
6. List and explain two types of lube oil filters used on diesel engines.
7. List two reasons why a regulating valve must be used in an engine lube system.
8. Explain the operation of one type of oil pressure gauge.
9. Give two reasons for engine oil coolers.
10. Select four engine oil classifications and explain each.
11. Using a typical diesel engine as an example, explain what type of oil should be used in it and why.
12. List four functions of lube oils.
13. Demonstrate the ability to troubleshoot and diagnose a lube system with low oil pressure.
14. Demonstrate the ability to overhaul gear and vane types of oil pumps.

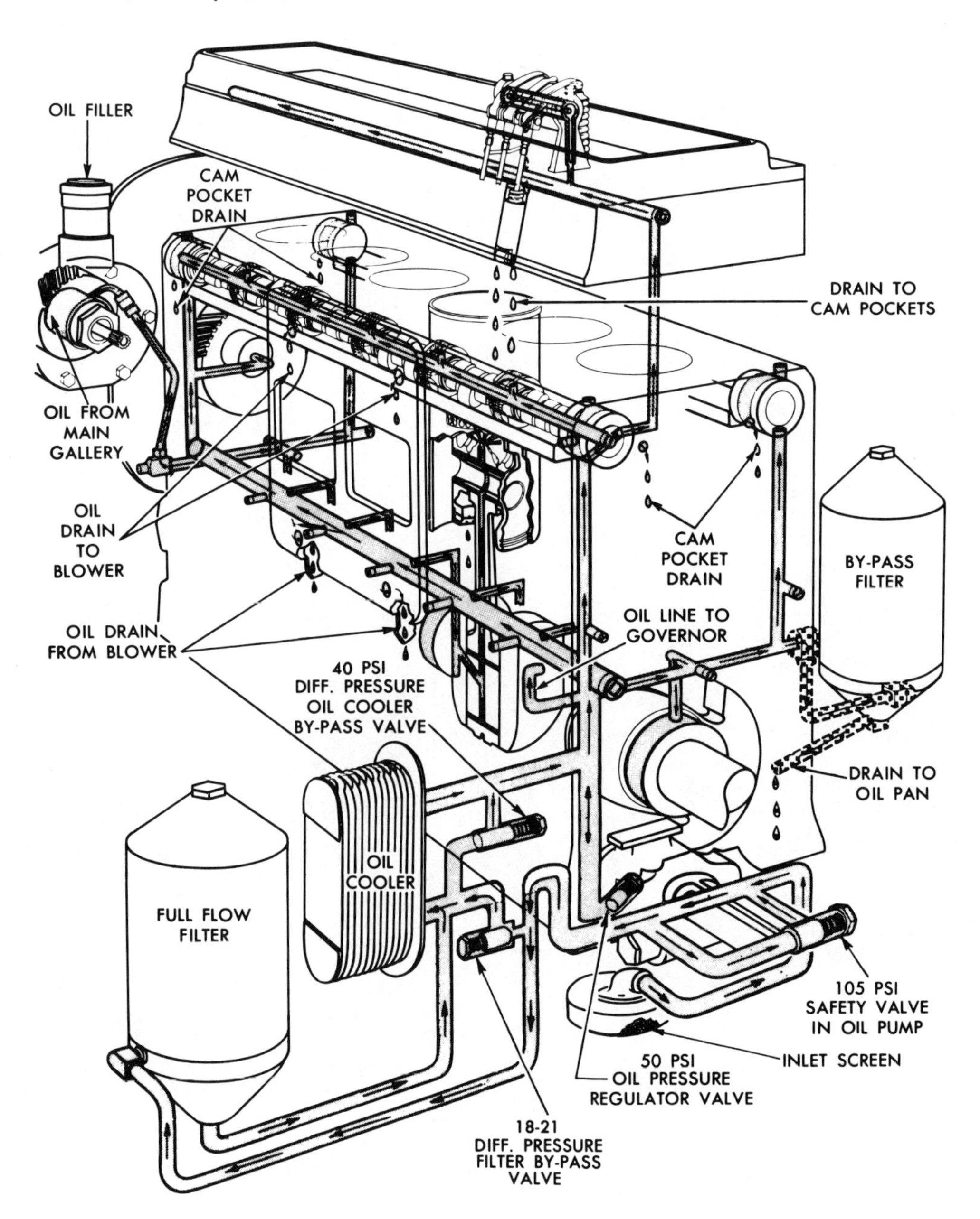

Figure 11.1 A typical diesel engine lubrication system (courtesy Detroit Diesel Allison, Division of General Motors Corporation).

15 Demonstrate the ability to test an oil cooler for internal leaks (oil to water).

GENERAL INFORMATION

The diesel engine lubrication system provides lubrication throughout the engine during operation to prevent friction that would cause the engine to be ruined in a very short period of time. Other functions of the lube system along with the lube oil are:

1. *Dissipate (get rid of) engine heat.* The heat generated by friction within the engine must be controlled to avoid engine damage.
2. *Clean; prevent rust and corrosion.* By-products of combustion (water, acids) must be removed from

the engine by the lube system and lube oil so that parts such as pistons, rings, and bearings remain clean throughout the engine life. In addition, the lube oil must prevent rust and corrosion from occurring, especially during long periods of engine shutdown.

3 *Provide a seal between the piston rings and cylinder wall or cylinder sleeve.* The piston and ring assembly would not be capable of providing a gas-tight seal without the aid of the lube oil provided by the lube system. Excessive blow-by would result if this seal did not exist, resulting in the loss of compression and a poor running engine.

4 *Absorb thrust or slack loads.* As the engine operates, many parts throughout the engine are subjected to shock or thrust type loads that must be absorbed or reduced to prevent engine noise and damage. An example of this would be the force exerted on the connecting rod journal by the connecting rod and piston assembly during the engine power stroke. At full load this force may be as much as 5000 psi (350 kg/cm^2). Without the cushioning effect of lube oil, the rod bearings would be destroyed quickly.

5 *Reduce friction.* The lube system reduces friction by providing and maintaining an oil film between all moving parts. The oil film between two sliding surfaces (like the rod bearing and the crankshaft journal) has two characteristics that if understood make us realize how very important the lubrication of engine parts is. These two characteristics are:

a Oil molecules shaped like very small ball bearings slide over one another freely.

b The oil molecules adhere to the bearing and crankshaft surfaces more readily than to each other.

Figure 11.2 shows the resulting effect. The top layer of oil molecules clings to the surface of the moving metal and moves with it. In so doing, it slides over the second layer of oil molecules to some degree, but does exert some drag that causes the second layer to move, but at a much slower rate. In like manner, the second layer slides over and drags the third layer at a slower speed. This continues through all the layers of oil molecules until the bottom layer is reached. The bottom layer clings to the stationary piece of metal and remains stationary. This action by the lubricant greatly reduces friction and increases bearing life.

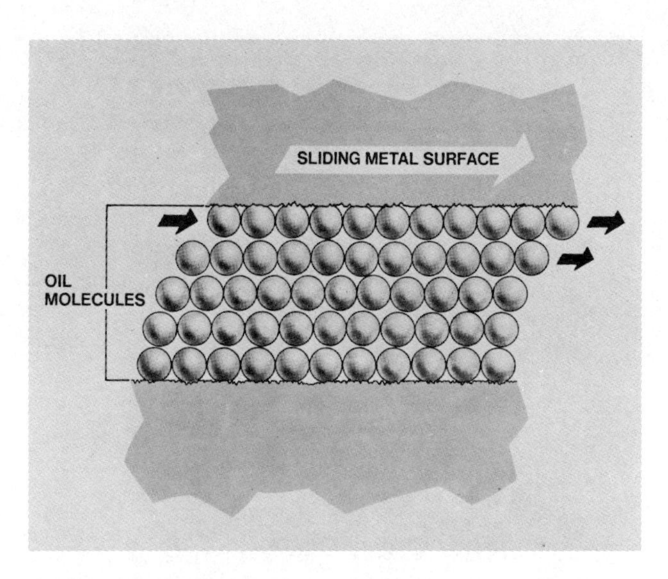

Figure 11.2 Layers of oil molecules between two metal surfaces (courtesy Gould, Inc.).

I. System Design

The system must be designed to fulfill all of the engine requirements for many hours of operation without failure. Three basic systems have been used in internal combustion engines over the years. They are:

A *Splash lubrication system,* a system that utilizes the movement of engine parts to supply splash oil to all moving parts during engine operation. This type of system has many drawbacks since lubrication is provided to those engine parts within the engine crankcase only. No pressure or flow is developed by which oil could be supplied to areas of the engine such as rocker arms or external engine accessories.

B *Force feed lubrication system.* This system supplies oil to most of the engine rotating or rocking parts under pressure supplied by an oil pump. This oil pump is located within the engine crankcase on some engines, while on others it may be located externally. All pumps are driven by the engine timing gear train or camshaft. This type of system is used on many diesel engines.

C *Full force feed lubrication system.* The full force feed lube system is very similar in design to the force feed system with the exception that it supplies oil through a rifle drilling in the connecting rod to the piston pin and underside of the piston. This oil, sprayed on the underside of the piston head, aids in cooling, especially if the engine is turbocharged.

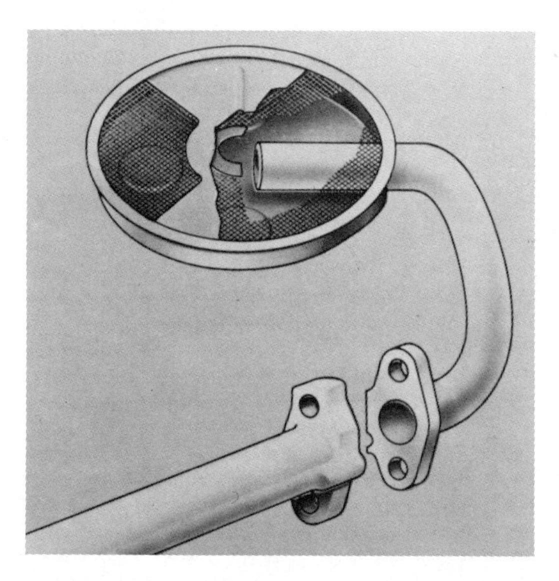

Figure 11.3 Suction strainer (courtesy Gould, Inc.)

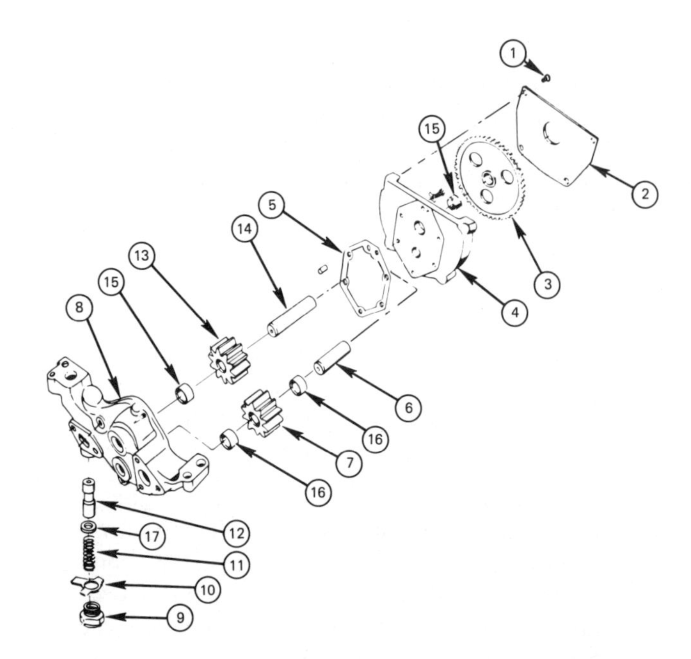

1. Cap Screw
2. Cover Plate
3. Drive Gear
4. Pump Cover
5. Gasket
6. Idler Shaft
7. Idler Gear
8. Pump Body
9. Regulator Cap
10. Lockplate
11. Spring
12. Regulator Plunger
13. Driven Gear
14. Drive Shaft
15. Bushing – Body
16. Bushing – Gear
17. Spacer

Figure 11.4 An external type gear pump.

II. System Components

Lubrication system components vary greatly in design, but little difference exists between component function from engine to engine. A typical full force feed system will have the following components:

A *Reservoir* (oil pan), generally located at the bottom of the engine to hold and collect engine oil. Oil is distributed to the various lubrication points from this pan by the oil pump.

B *Suction strainer or filter* (Figure 11.3), located on the inlet side of the pump to prevent any large dirt particles or other foreign material from entering the pump.

This strainer in many engines is nothing more than fine metal screen; in others it may be a cloth or fabric strainer.

C *Oil pump,* considered the heart of the lubrication system. A sufficient flow of oil to maintain oil pressure to all lubrication points must be supplied by this positive displacement pump. Engine lube oil pumps for high speed diesel engines are generally of two types:

1 *External gear pumps* (Figure 11.4) are most commonly used. This pump consists of two meshed gears, one driving the other, a body or housing in which they are enclosed and an inlet and outlet. As the gears are turned, oil is drawn in the inlet side and is carried in the space between the teeth and the pump housing. As the gears continue to turn, the oil is carried around to the outlet port of the pump and forced out by the meshing of the gear teeth.

2 *Internal gear or crescent pumps* (Figure 11.5) are designed with one gear rotating inside of another gear. The smaller gear drives the larger diameter gear and is offset from the center of the large gear so that the gear teeth are in

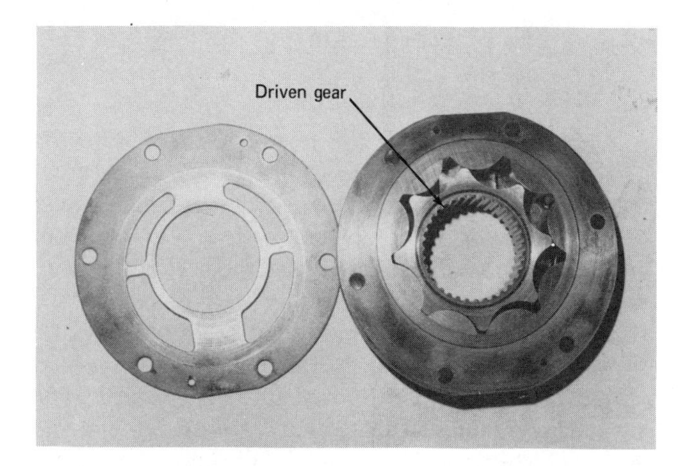

Figure 11.5 Internal gear pump.

Figure 11.6 Partial flow oil filter (courtesy Detroit Diesel Allison, Division of General Motors Corporation).

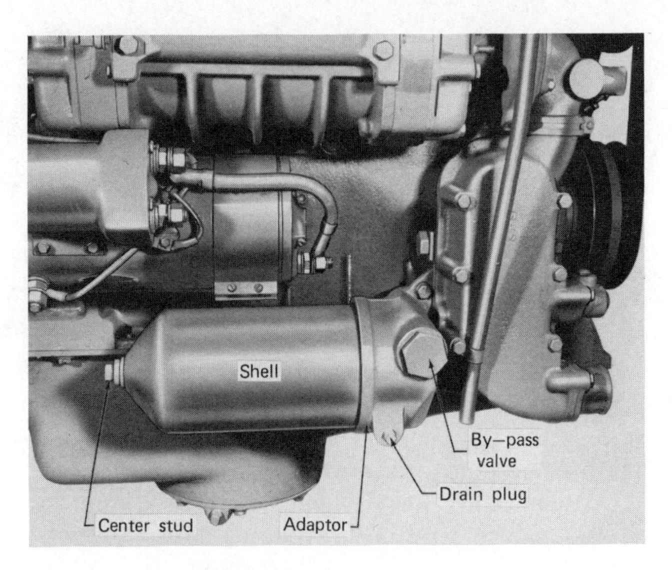

Figure 11.7 Full flow oil filter (courtesy Detroit Diesel Allison, Division of General Motors Corporation).

mesh. As the gears turn, oil is picked up and carried by the space between the large gear teeth. When the gears come into mesh, the oil is forced out the pump outlet by the intermeshing of gears.

NOTE As the impeller turns, oil is trapped in the space between the vane and pump body. Continued turning forces the oil into a progressively smaller area and finally discharges it out of the pump outlet.

D *Filters.* Oil filters generally are one of two common types, partial (Figure 11.6) or full flow (Figure 11.7).

1. *Partial or percentage.* The partial flow filter, as the name implies, filters only a certain percentage of the oil (approximately 30 to 40 percent). It is connected to the engine oil gallery in such a manner that oil does not have to pass through the filter before entering the engine. In fact, oil flows through the filter and then back to the oil pump or oil pan. Most engine manufacturers no longer use this type filter because of its low cleaning efficiency.
2. *Full flow.* Most modern diesel engines use the full flow type filter. In this filter all oil pumped by the oil pump must pass through the oil filter before it enters the engine oil gallery. Since 100 percent of the oil is filtered before it is used for lubrication, this filter's efficiency in dirt removal is much greater, resulting in its widespread use.

Either type of filter, partial or full flow, may be an element type (Figure 11.8) with a replaceable element or a replaceable type of metal can (Figure 11.9) that has the element sealed into it.

The replaceable cartridge for the element type may be constructed in one of two different ways:

1. Cottonwaste (an absorbent or depth type of filter), which absorbs impurities in the oil.
2. Treated paper (more common). The treated paper filter element is considered an adsorbent (or surface) type. This means that it filters from the oil passing through it all dirt particles larger than the holes in the paper.

NOTE As the filter continues to catch dirt, it becomes more efficient since the collection of dirt particles on its surface allows smaller and smaller particles to pass through it. Eventually, however, the filter becomes completely plugged and must be replaced with a new cartridge.

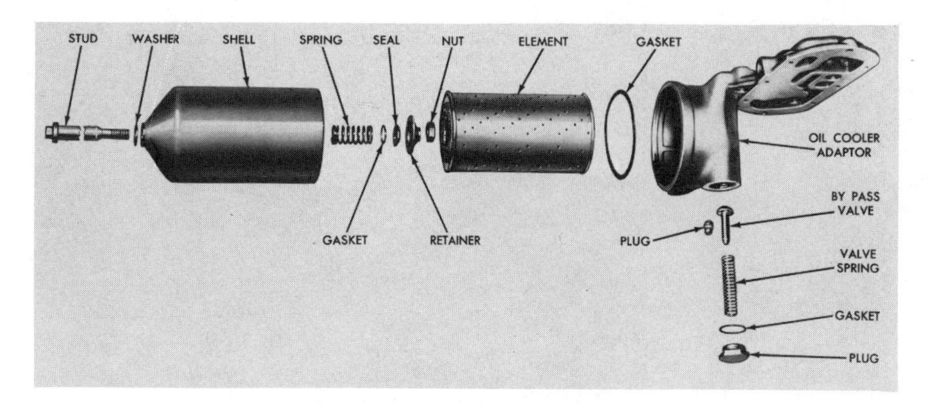

Figure 11.8 Element type filter (courtesy Detroit Diesel Allison, Division of General Motors Corporation).

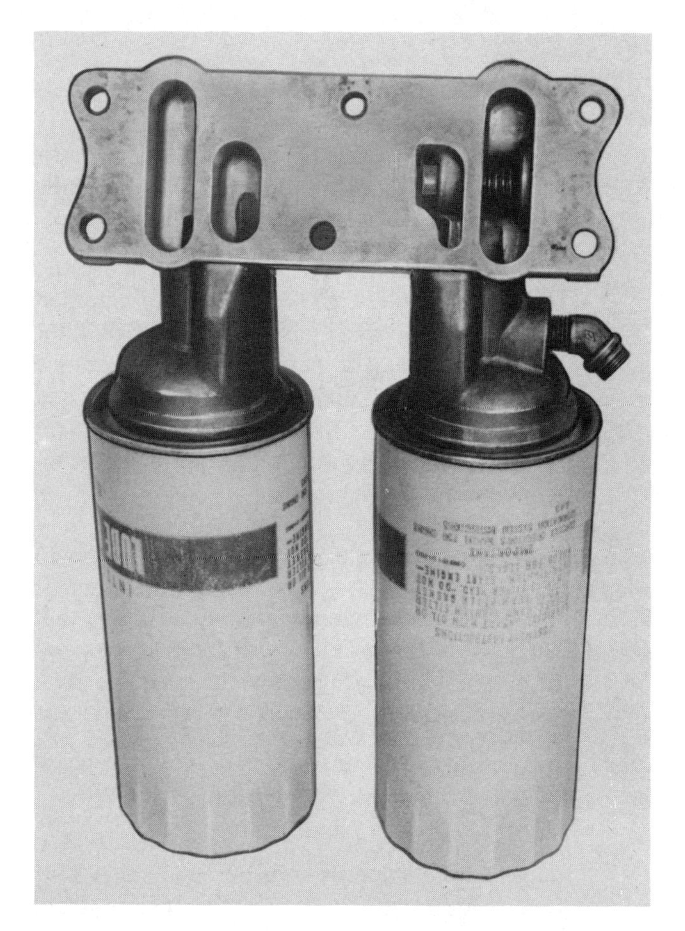

Figure 11.9 Replaceable metal can filter (courtesy International Harvester).

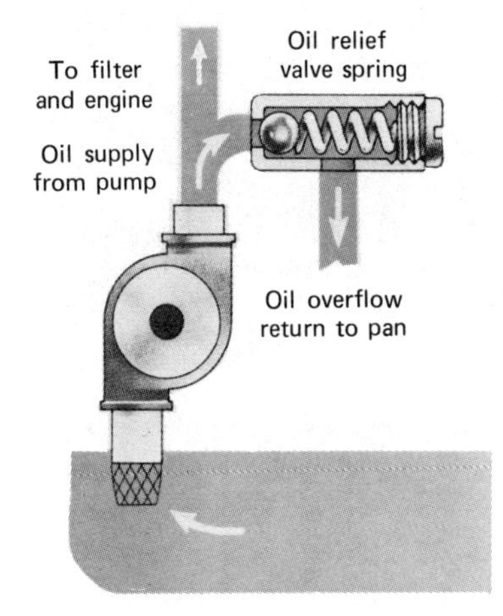

Figure 11.10 Ball type regulating valve (courtesy Gould, Inc.).

E *Pressure regulator and bypass valves.* Pressure within the lubrication system must be regulated as the viscosity of the oil changes. Also, because of temperature changes, filters such as the full flow filter must have an automatic bypass system to prevent engine damage in the event that the filter becomes clogged.

Both these requirements are fulfilled by either a pressure regulator valve or a bypass valve. Design of these valves may be one of two common types:

1 Ball (spring loaded) (Figure 11.10).

2 Plunger (spring loaded) (Figure 11.11).

The main job of the pressure-regulating valve is to prevent excessive oil pressure within the system; in a sense, it becomes a safety valve or limiting valve. Oil pumps are designed to maintain sufficient flow at normal engine temperature and speed. As a result, when oil temperature is low and viscosity is high, excess pressure will be devel-

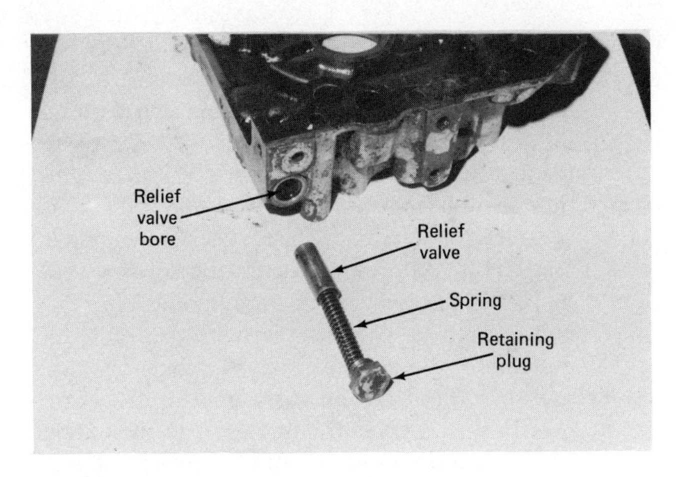

Figure 11.11 Plunger type regulator valve.

oped. A regulating valve is then used to dump off the excess flow to maintain correct pressure.

Since most modern diesel engines have a full flow filter, a bypass valve is designed into the filter or filter base so that the engine receives a sufficient oil supply at all times in the event that a filter cartridge or element becomes plugged. This "safety device" is in the normally closed position during engine operation unless the filter becomes clogged.

F *Oil coolers* (warmers). Oil coolers in many diesel engines are of the oil to water type. Coolers of this type resemble a small radiator enclosed in a housing. Water is pumped through the copper core or element and oil is circulated around it (Figure 11.12).

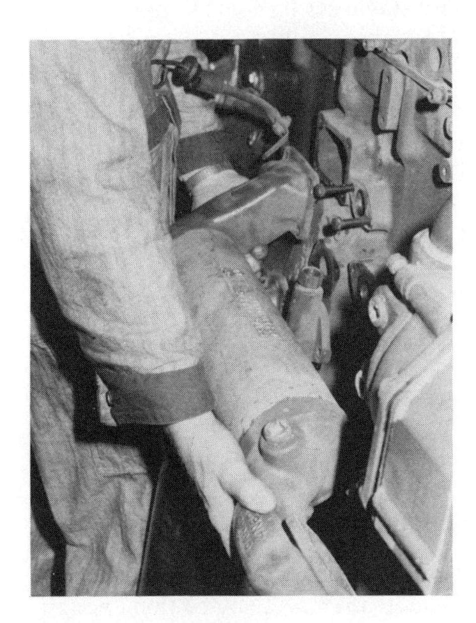

Figure 11.12 Typical oil cooler.

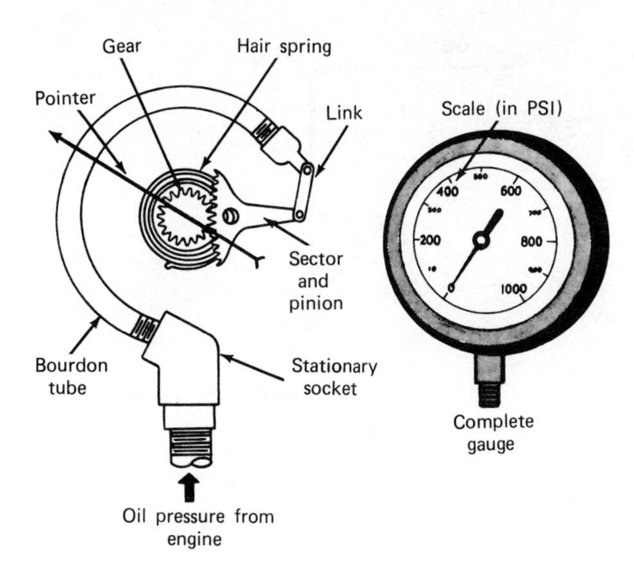

Figure 11.13 Bourdon type oil pressure gauge (courtesy of Deere and Company).

In most situations during engine operation, the oil is hotter than the coolant water, resulting in heat transfer from the oil to water, keeping the oil at a safe operating temperature. Normal oil temperature in most engines will be about 220 to 230° F (104 to 110° C).

Cooling is not the main job of the oil cooler at all times. During operation following a cold start, engine water will reach operating temperature much sooner than engine lube oil. In this situation the oil cooler warms the oil rather than cools it.

G *Supplementary component parts.*

1 Pressure gauge or indicator light.

The oil pressure gauge is calibrated in pounds per square inch (psi). It is used by the operator to determine if the engine oil pressure is correct. Pressure gauges can be one of two types:

a Mechanical. The mechanical type (sometimes called the Bourdon type) (Figure 11.13) is connected to an oil pressure gallery by a tube that transmits the pressure to the gauge mounted on the operator's instrument panel.

b Electrical. The electrical type (Figure 11.14) is made up of an indicating gauge (similar to the one used on the Bourdon tube), which is a wire and pressure sending

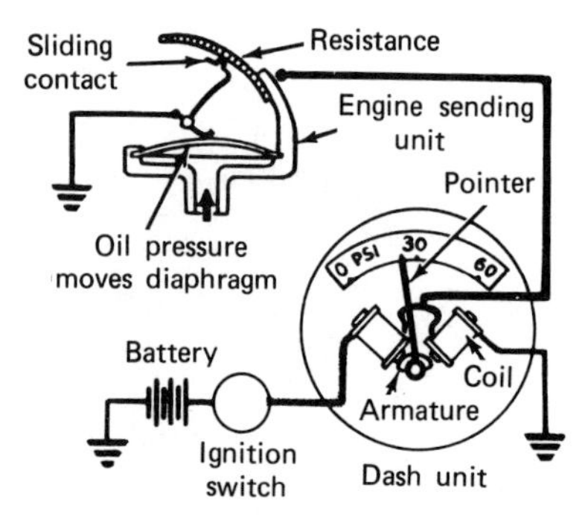

Figure 11.14 Electric type oil pressure gauge (courtesy of Deere and Company).

unit that is screwed into the engine oil gallery at a convenient take-off port.

Engine oil pressure pushing against the diaphragm in the sending unit moves a sliding wiper arm across a resistor (Figure 11.15), changing the resistance value of the sending unit since the sending unit provides the ground for the gauge circuit. The amount of current passing through the gauge, which causes needle movement, will be determined by the amount of ground that the sending unit is providing.

Many modern engines employ an indicating light system in place of a pressure gauge. Making up the system are an indicator light, electrical wire, and sending unit. In this system

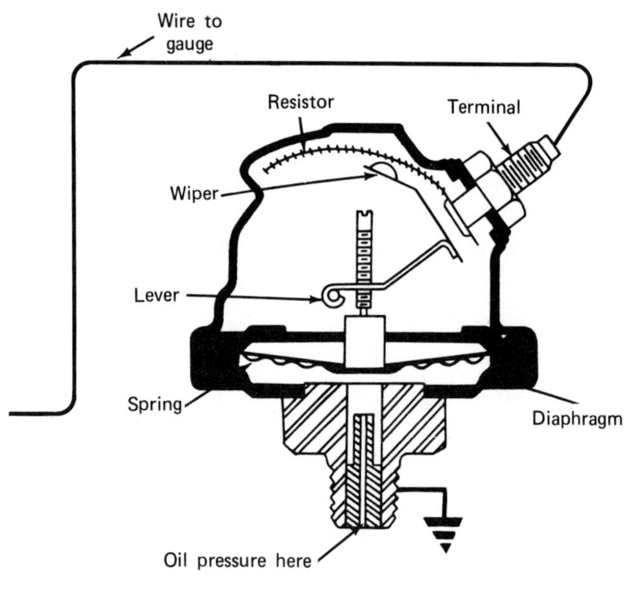

Figure 11.15 Electrical sending unit (courtesy of Deere and Company).

the sending unit inserted into the oil gallery is a pressure-operated switch that operates the lamp circuit. During engine shutdown with no oil pressure the sending unit is closed, providing a ground for the lamp circuit. With the ignition switch in the "on" position, power is supplied to the lamp to light. After the engine is started and oil pressure opens the sending unit by pushing against the diaphragm, the lamp circuit ground is lost and the light goes out, indicating to the operator that the engine has oil pressure. This system has a built-in disadvantage in that it gives the operator no indication of the amount of oil pressure being developed.

2 Dipstick.

The most simple and most widely used method of checking the oil in the crankcase is the dipstick. The dipstick is constructed of a long piece of flat steel that fits into a tube inserted into the engine oil pan from the top, allowing easy operator access.

III. Engine Lube Oil

The lubrication system is not complete without lube oil. Lube oil is a petroleum product made up of carbon and hydrogen along with other additives to make a lubricant that can meet the specifications supplied by engine manufacturers. In general, diesel engine lube oil should meet the following requirements. It must:

1. Be viscous enough at all engine temperatures to keep two highly loaded surfaces apart.
2. Remain relatively stable at all engine temperatures.
3. Act as a coolant and cleaner.
4. Prevent rust and corrosion.

In determining if an oil will meet the engine requirements, it is subjected to the following tests by the company that refines it:

1. *Viscosity*. Viscosity may be defined as the resistance to flow of a liquid. The molecules of a more viscous oil have greater cohesion (stick together more firmly) than a less viscous oil. The higher the number given to the oil, the greater its viscosity (resistance to flow) will be. Temperatures also greatly affect the viscosity of an oil. Hotter oil will flow more rapidly than colder oil.
2. *Pour point*. The pour point of an oil is the lowest temperature at which an oil will still be thin enough to pour.

3 *Flash point*. The flash point is the temperature at which the oil will be sufficiently vaporized to ignite.

In selecting an engine oil, consult the recommendations supplied by the engine manufacturer to insure that the oil selected will meet engine requirements.

To enable the user of engine oil to determine what type oil is being purchased, a classification system has been jointly developed by the American Petroleum Institute, the Society of Automotive Engineers, and the American Society for Testing and Materials. The different types of oil have been classified and divided into two groups:

1 The *S* classes are for passenger cars and light trucks. They are:
   SA for utility engine operation under mild conditions that don't require compounded oils.

   SB for minimum duty gasoline engines that require compounding only for antiscuff capability and resistance to oil oxidation and bearing corrosion.

   SC for gasoline engine operation outlined in manufacturers' warranties from 1964 through 1967. These oils control high and low temperature deposits, wear, rust and corrosion in gasoline engines.

   SD for gasoline engines operating under 1968 or later warranties that provide for even greater protection against temperature deposits, wear, rust and corrosion.

   SE oil meeting the 1972 requirements of the automobile manufacturers. Intended primarily for use in passenger cars. Provides high temperature antioxidation, low temperature antisludge, and antirust performance.

   SF designed for 1980 passenger cars and trucks that use unleaded gas; it has a greater resistance to oxidation and the different types of deposits created by unleaded fuels.

2 The four *C* classes for heavier vehicles such as heavy duty trucks, farm tractors, and industrial equipment:
   CA for light duty diesel engines. These oils protect against bearing corrosion and high temperature deposits in engines burning high quality fuels.

   CB for moderate duty diesel engines burning lower quality fuels that require more protection.

   CC for lightly supercharged diesel engines and heavy duty gasoline engines.

   CD for diesel engines operating under severe conditions. These oils are designed for supercharged diesels that run at high speeds on a wide range of fuels to turn out high horsepower under heavy load.

As can be seen from the preceding information, the lube system and the lube oil in it deserve more than a casual consideration, especially in oil selection. Engine oil must be changed at regular intervals to keep the internal engine parts clean, since in most cases dirt is the engine's primary enemy. Keep the engine clean and it won't wear out.

In addition, the lube system must be closely inspected during an engine overhaul if it is to function correctly for many hours of operation. All too often the components of the system are taken for granted and overlooked during engine overhaul. The result in many cases is a newly overhauled engine with less oil pressure than it should have. It is recommended that all lubrication system components be checked thoroughly during engine overhaul.

IV. Inspection and Overhaul of Components

It is assumed that the oil pump has been removed from the engine at this time. If not, refer to the chapter on "Engine Disassembly."

A. Procedure for Oil Pump Disassembly

1 Remove pump cover from pump body (external gear type pump).

2 Inspect cover for wear as shown (Figure 11.16).

NOTE If the cover is worn, it should be machined or replaced with a new cover.

3 Remove the idler gear and driven gear from the pump housing.

NOTE In most cases the idler gear can be lifted from its supporting shaft (idler shaft) and removed. The driven gear will be keyed to the drive shaft with a woodruff key (half moon) or roll pin (spring steel pin) and may have to be pressed from the shaft (Figure 11.17). If the driven gear must be pressed from the drive shaft, the drive shaft should be removed from the pump body before attempting to remove the gear. In many cases the pump drive shaft will have the oil pump drive gear pressed onto it. This must be removed before the drive shaft and driven gear can be removed from the pump housing. (Check your engine service manual or with your instructor.)

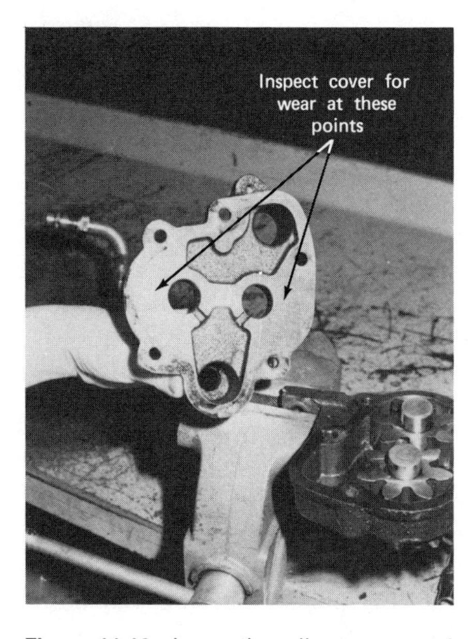

Figure 11.16 Inspecting oil pump cover for wear.

Figure 11.18 Areas where oil pump gears should be checked for wear.

4 Check idler and driven gear closely for pitted and worn teeth. Teeth should be checked closely at the points as shown in Figure 11.18.

5 Check gear width (parallel with center hole) with a micrometer.

6 Check drive shaft diameter at bushing contact with a micrometer. This diameter should meet manufacturer's specifications.

7 Check the pump housing internally for wear as shown in Figure 11.19.

8 Check drive shaft bushing (if used) with snap gauge. Compare with specifications.

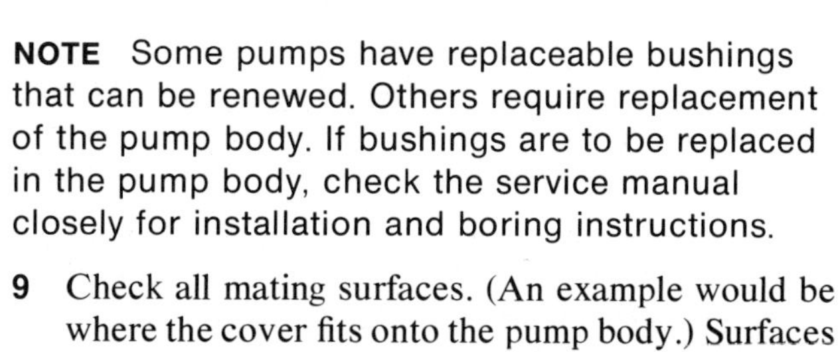

NOTE Some pumps have replaceable bushings that can be renewed. Others require replacement of the pump body. If bushings are to be replaced in the pump body, check the service manual closely for installation and boring instructions.

9 Check all mating surfaces. (An example would be where the cover fits onto the pump body.) Surfaces should not be nicked or burred. If there are nicks or burrs, remove them with a small, flat file.

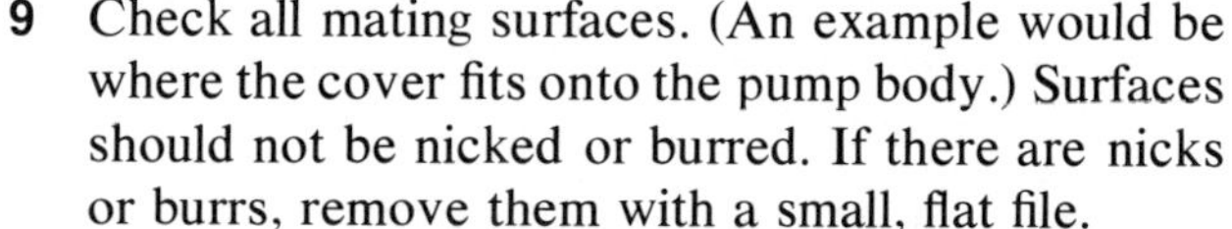

CAUTION Care must be exercised when filing on a pump body, as it can be easily ruined. If body is warped excessively, replace it with a new one.

Figure 11.17 Pressing oil pump gear from shaft.

Figure 11.19 Areas to check when checking oil pump housing for wear.

B. Procedure for Oil Pump Reassembly (general)

After all parts have been inspected, repaired, or replaced, the pump can be reassembled.

1 Insert the oil pump drive shaft with the driven gear installed on it into the pump body.

NOTE To install the pump drive gear, the pump drive shaft must be supported on the opposite end while the pump drive gear is installed. You must do this at this time before the pump cover is installed.

2 Support drive shaft on driven gear end and press the pump drive gear in place (Figure 11.20).

CAUTION Make sure woodruff key or roll pin is installed securing the gear to the shaft, to prevent gear from turning on shaft during operation.

3 Install the idler gear on the idler shaft.

4 Turn the drive shaft, checking for binding or interference as the shaft is turned.

NOTE Any binding usually is caused by nicks on gears. Remove gears and file or stone off nicks.

5 When pump drive shaft and gears have been rotated without binding, pump-cover-to-gear clearance should be checked with plastigage as follows:

a Place plastigage across the face of the gears.

NOTE Plastigage is a thin plastic thread that can be broken into the correct length and used to check the clearance between two closely fitted parts.

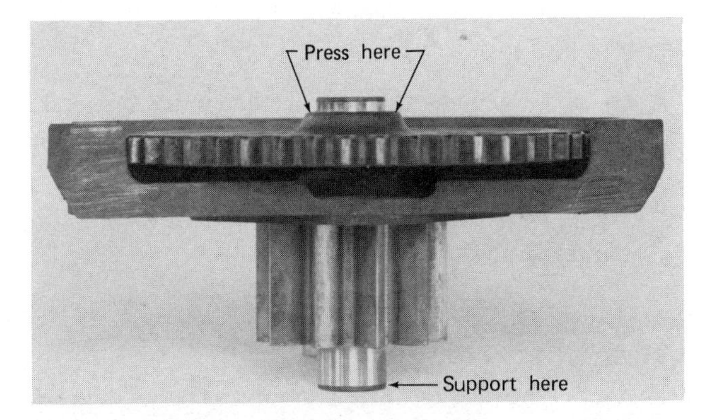

Figure 11.20 Pressing drive shaft and drive gear together (courtesy Cummins Engine Company).

b Install cover and tighten to specifications.

CAUTION Do not turn the pump drive gear with plastigage in it.

c Remove cover and check width of plastigage against plastigage envelope to determine clearance between cover and gears.

d If clearance is excessive or does not meet specifications, it must be corrected by replacing the gears or pump as needed. [General specifications for oil pump gear-to-cover clearance are 0.003 to 0.005 in. (p. 076 to 0.127 mm)].

NOTE Many oil pumps will contain a bypass or regulating valve that should be checked during pump overhaul as outlined in the following section.

C. Procedure for Inspection and Repair of Oil Pressure Regulation and Bypass Valves

Because of different systems' designs and requirements, the regulator or bypass valves may be located anywhere in the engine or oil pump. Regardless of where they are located or what their function is, *all* valves must be disassembled and inspected as follows:

1 Remove the valve spring cap retaining nut.

2 Remove spring, inspect for worn or twisted coils, check free length and spring pressure with spring tester.

3 If the valve is the plunger type, check plunger and plunger seat for scoring and wear. Replace all valves that show excessive wear.

4 Ball type valves should be checked for pitting and wear and replaced if found defective.

5 Check valve seat visually for pitting and uneven wear.

NOTE Some valve seats are replaceable and can be replaced without replacing the valve assembly. Other types of valves will require replacement of the entire valve body or pump assembly, which includes the valve seat.

D. Procedure for Checking Oil Filter Housing or Mounting

Regardless of the type of filter used, element or cartridge type, the base or filter mount must be checked carefully during engine overhaul as follows:

1. If a bypass valve is incorporated into the housing, it should be removed and checked as outlined in the regulator valve section.
2. Check the housing for cracks.
3. Check all gasket surfaces for straightness and nicks.
4. If using an element type filter, which uses a can or shell with a center bolt, make sure the center bolt threads are usable.
5. Make sure springs, gaskets, and spacers (as outlined in service manual or parts book) are in place in element type filters.
6. Check can for signs of collapse on the closed end.
7. Check for cracks around bolt hole, etc.
8. Check housing passageways, making sure all are open and free of obstructions.

E. Procedure for Oil Cooler Testing and Repair

Since oil and water both flow through the oil cooler simultaneously, it becomes very important that mixing of oil and water at this point does not occur due to a leak or crack in the cooler. As indicated in the previous discussion under "General Information," two types of coolers are found on modern diesel engines, plate type and tube type. Many variations of these basic types will be found.

NOTE Complete servicing of the oil cooler should be done on a routine basis during a major engine overhaul. In addition, the cooler may occasionally fail between engine overhaul periods. Indications of cooler failure are oil in the water or water in the oil pan. In either case the following service instructions will apply.

Servicing of the oil cooler assembly should include complete disassembly and cleaning of the core section with a solution recommended by the engine manufacturer. In most cases the cooler core can be cleaned using an oakite type solution or muriatic acid.

CAUTION Some cores contain aluminum, a nonferrous metal that cannot be cleaned as outlined above. In this type of situation, use a normal parts cleaning solution. After cleaning, inspect all parts as follows.

1. Check core visually for:
   - a. Cracks or breaks in welded joints.
   - b. Bulges or bent tubes.
2. If core does not pass visual inspection, replace it with a new one.
3. If core passes visual checks, it should be checked by pressurizing with air in the following manner:
   - a. If a test fixture is available, mount the core in the fixture in preparation for testing.

NOTE If a test fixture is not available, a plate or plates can be made to test the cooler core. In general, the plate should be designed so that air pressure can be supplied to one side (oil or water) of the cooler only.

   - b. After plates have been attached to core, pressurize it with 35 to 40 psi (2.5 to 2.8 kg/cm^2) and immerse it in water.
   - c. Inspect carefully for air leaks.
4. If core passes all checks outlined above, clean housing and install new O rings or gaskets and reassemble core to cooler.

NOTE Generally, cooler assembly will be a simple matter of installing the core in the housing with the cooler housing, using new gaskets or O rings. An exception to this is the Cummins tube type cooler that requires indexing (timing) in the housing during assembly. Figure 11.21 is an example of index marking.

5. Before cooler assembly is installed onto engine, the bypass valve should be checked as outlined under the regulator valve section.
6. Using new gaskets, mount cooler on engine and tighten all bolts.

V. System Testing

Testing of the lube oil system involves testing the pressure at various engine speeds and temperatures. In general, testing is done at the following times:

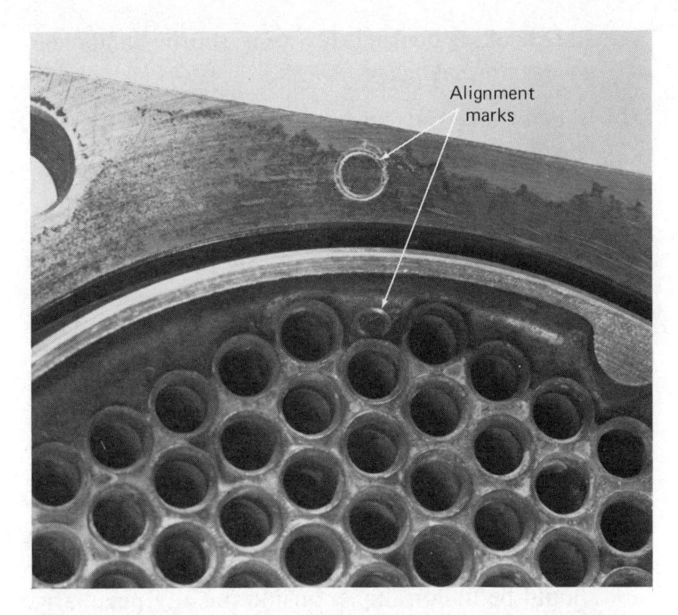

Figure 11.21 Oil cooler index markings (courtesy Cummins Engine Company, Division of General Motors Corporation).

A *After an engine overhaul.* The lube system must be thoroughly checked and monitored during engine start-up and rebuild. Proper oil pressure is vital to long engine life.

B *Lube system problems between engine overhauls.* Many times low lube oil pressure will occur long before the engine is due for a major rebuild. The cause of the problem can only be found by testing and checking the lube system.

C *Before an overhaul.* Many mechanics like to check lube oil pressure before the engine is disassembled for rebuild as an indication of what to be especially alert for during rebuild. (For example, if lube pressure is low, the clearances throughout the engine should be carefully checked and held within recommendations.)

VI. Procedure for Testing and Troubleshooting the Lube System

To accurately test and determine the condition of a lubrication system, the following procedure should be followed:

A Test oil level and check oil for correct weight, viscosity, and possible dilution. If any doubt exists about the condition of the lube oil, it should be replaced before proceeding with further tests.

B Change lube oil filter and inspect old filter for metal particles.

NOTE If a can type spin-on filter is used it may be necessary to cut it apart to check for metal particles. If the filter is filled with many particles of bearing material, it is a good possibility that a worn or damaged bearing is causing the low oil pressure. In most cases the engine bearings should be checked before proceeding with further testing of the lube system.

C Warm engine thoroughly by driving the vehicle or by loading engine on a dynamometer.

D Install a master pressure gauge somewhere in the system where it will indicate system pressure.

CAUTION Make sure gauge reads as high or higher than the highest expected oil pressure or damage to gauge may result.

E Discuss complaint with customer if complaint is customer oriented. Make a mental note of causes that may be creating problems.

F Determine from the engine service manual what correct oil pressure should be and at what speed testing should be done. (Many engines require a minumum of 40 psi (2.8 kg/cm^2) at rated load or high idle, for example.)

G Run engine and check the oil pressure.

H If engine oil pressure meets specifications, no further checks need be made. If not, proceed with step I.

I If oil pressure was too low or too high, an attempt should be made to determine what the cause might be.

NOTE Some engines will have a pressure adjustment for oil pressure. If your engine is so equipped, make the adjustment to bring the oil pressure into the specified range.

J If oil pressure cannot be adjusted and does not meet specifications, the system pressure regulating valve should be checked.

NOTE The order and the type of checks that will be made are largely dependent on the mechanic and the condition of the engine. For example, a newly rebuilt engine would be tested knowing that all clearances and regulator valves were checked during assembly. A key item to look for in rebuilt engines is improperly installed or incorrect type gaskets at places like the oil filter base, oil pump mounting, blower mounting, and other places where pressure oil could be incorrectly routed. In contrast to this would be an engine that had been performing correctly with good oil pressure for many hours and then develops low oil pressure. Troubleshooting of this engine must be approached with the idea that it was correct at one time, but wear or malfunction of some part has created a low pressure situation.

K If the pressure regulator valve and spring are in good condition, check all bypass valves used on oil filters and oil coolers.

L Check oil pump. In most cases the engine oil pan will have to be removed to gain access to the oil pump. (Refer to Section IV-A for pump overhaul.)

M If no problem exists with the oil pump, main and rod bearings should be checked by removing caps and checking bearings for wear, scoring, and clearance with plastigage.

N If main and rod bearing clearances are all right, check the camshaft bushings.

NOTE Since checking of the camshaft bushings requires an extensive amount of engine teardown, a decision should be made at this time about the engine condition. It might be that a complete engine overhaul should be done at this time. This decision would be dependent on the overall general condition of the engine, number of hours on it, and the owner's wishes.

O At this time most of the points within the engine that contribute to low oil pressure have been indicated. If the problem has not been remedied, a thorough study of the engine lubrication system should be undertaken, considering any peculiarity the engine being worked on may have.

SUMMARY

This chapter has explained the lube oil and system requirements in a diesel engine. The function, operation, testing, and overhaul or replacement of each component are covered in sufficient detail to allow you to understand, test, and repair the system. When troubleshooting, it is very important to view the lube oil, lube system, and its components as a complete unit, using common sense and a systematic simple-to-complex method of checking to isolate and repair problems.

REVIEW QUESTIONS

1. Name five component parts of a lubrication system.
2. Name five lube system requirements.
3. What three types of oil pumps are usually found in a diesel engine lubrication system?
4. Why are regulating valves used in lube systems?
5. What function does the oil pressure gauge perform?
6. Why do most diesel engines use oil coolers?
7. Why are engine oils blended with different characteristics?
8. Name four functions of a good lubricating oil.
9. When working on a diesel engine with a low oil pressure complaint, list three checks you would make first (not necessarily in correct order).
10. List four checks that should be made when overhauling or checking a gear type pump.
11. For what reason would an oil cooler be checked and how should it be done?

12

Cooling Systems and Controls

When a diesel engine runs, heat is created by the burning of fuel and by the friction of engine parts rubbing together. Much of the heat created by the burning fuel is transferred into usable power by the engine. Since all of the engine parts are subjected to heat during operation, many engine parts would become overheated and ruined if the engine temperature were not controlled. The cooling system (Figure 12.1) maintains and controls the engine temperature over a wide range of ambient (surrounding air temperature) temperatures.

OBJECTIVES

Upon completion of this chapter the student will be able to:

1. Describe four cooling system components and their functions.
2. Explain the difference between the bypass and blocking type thermostat.
3. Demonstrate the ability to test an entire cooling system.
4. List three reasons why an engine would overheat.
5. Explain why cooling system filters and conditioners are used.
6. Demonstrate the ability to overhaul a water pump.
7. Demonstrate the ability to test a pressure cap.

GENERAL INFORMATION

Diesel engines are called upon to operate in extreme cold and heat, with the cooling system protecting the engine at all times. Specifically, the cooling system must:

1. Provide quick engine warm-up.
2. Maintain a predetermined temperature during all phases of engine operation. (For example, engine temperature must be maintained under light load as well as full load.)

Although the liquid cooling system is most commonly used on diesel engines, many engines will employ direct air-cooling systems. Engines with these systems are called air-cooled engines. Since most

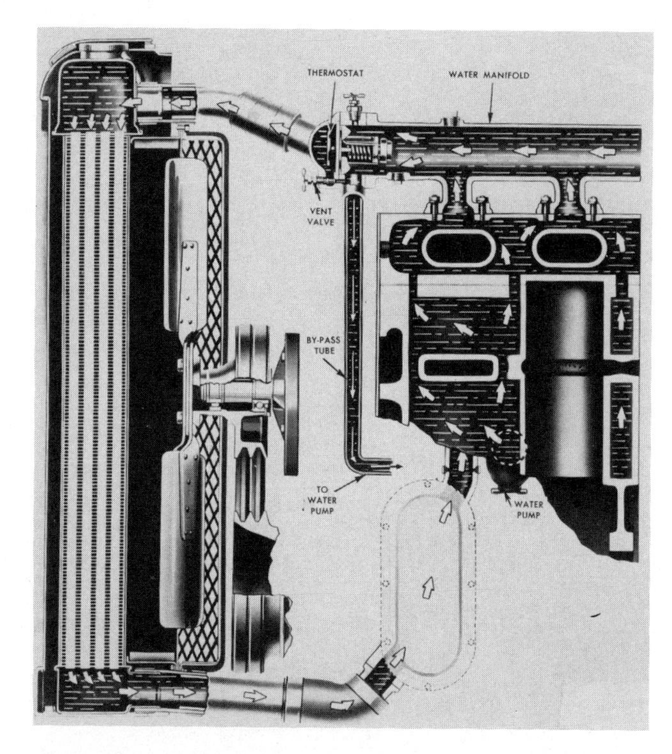

Figure 12.1 Diesel engine cooling system and controls (courtesy Detroit Diesel Allison, Division of General Motors Corporation).

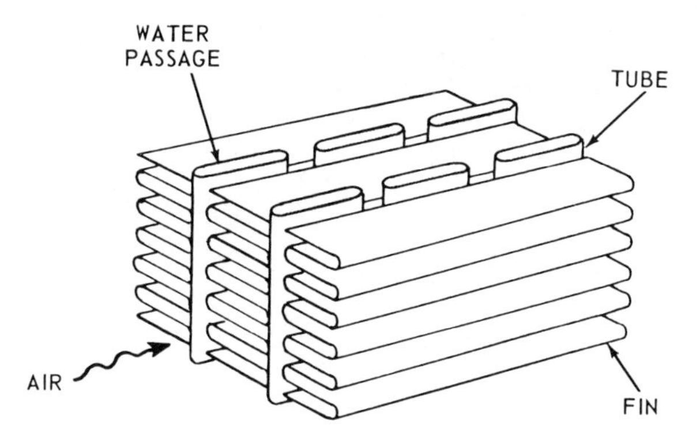

Figure 12.2 Tube and fin type radiator (courtesy Deere and Company).

diesel engines do have liquid cooling systems, most of the following material will be devoted to the inspection and overhaul of the liquid cooling system and its components.

I. Component Description, Operation, and Function

All liquid type cooling systems have basically the same components except for the temperature controls, such as fan clutches, shutters, and thermostats. A typical liquid cooling system will have the following components:

A *Radiator.* A device that performs two important functions:

  1 Provides a storage tank for the engine coolant.

  2 Provides a surface where engine heat can be dissipated to the surrounding air.

    a Radiator cores (the radiator surface that dissipates the heat) are generally tube and fin type (Figure 12.2).

B *Water jackets.* Water jackets surround the engine block and provide a storage area for the coolant. They also provide a place for the coolant to circulate through the block, and pick up the excess engine heat.

C *Water pump.* A centrifugal nonpositive displacement pump used to pump the water through the block (water jacket) and radiator. It may be driven by a gear or pulley and belt arrangement.

D *Thermostat.* A temperature-controlled valve that regulates the flow of coolant through the engine and radiator. This flow regulation maintains engine temperature. The thermostat can be one of two types:

  1 *Pellet type.* A thermostat that uses a wax pellet to operate the flow valve. The pellet senses the temperature and regulates the valve of the thermostat accordingly (Figure 12.3).

  2 *Bellows type.* The bellows type thermostat does not contain a temperature-sensitive pellet but is operated by metal bellows that are filled with an alcohol liquid having a low boiling point. As the water warms up, it warms the liquid, causing it and the bellows to expand, opening the valve (Figure 12.4).

E *Fan.* The fan is mounted on the water pump drive pulley hub or on a separate fan mount. It provides air movement across the radiator so that heat can be dissipated. The fan can be thermostatically or electrically controlled. Depending on the application and air flow requirements, fans may be suction or blower type.

F *Temperature gauge.* A gauge that tells the operator what the engine coolant temperature is. It can be one of two types:

  1 Electric (Figure 12.5).

  An electric temperature gauge has a send-

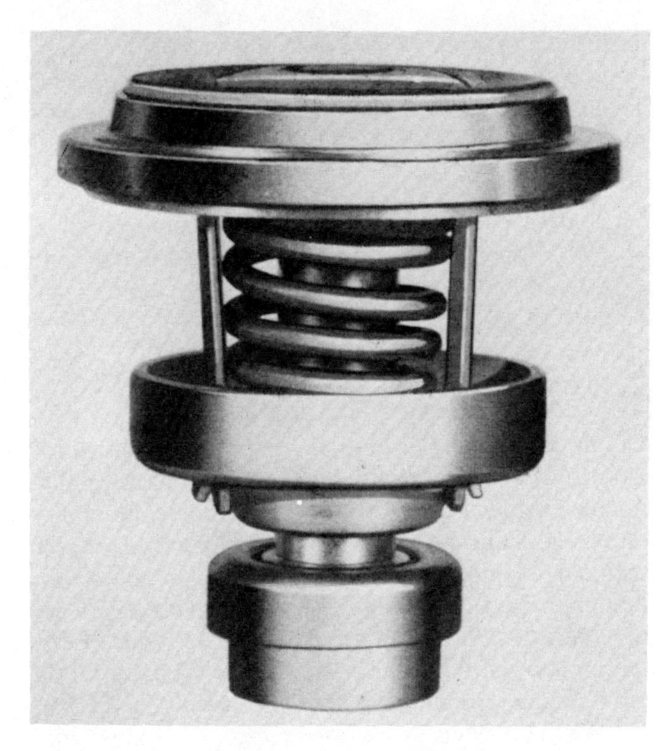

Figure 12.3 Pellet type thermostat (courtesy Detroit Diesel Allison, Division of General Motors Corporation).

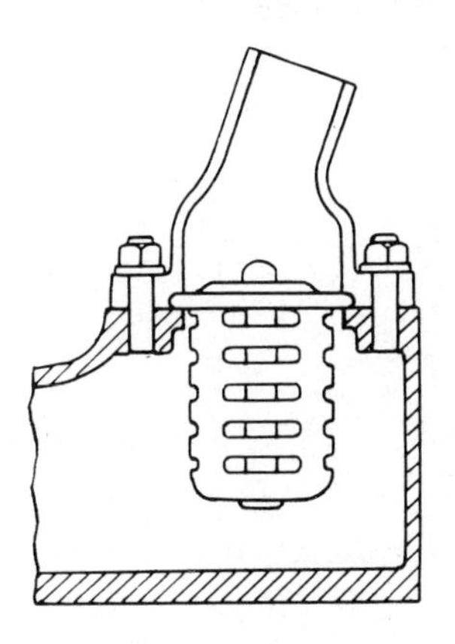

Figure 12.4 Bellows type thermostat (courtesy Deere and Company).

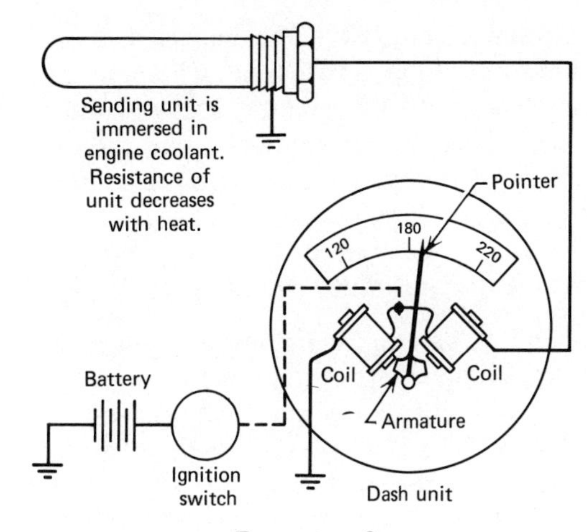

Figure 12.5 Electric type temperature gauge (courtesy Deere and Company).

Figure 12.6 Expansion type temperature gauge.

ing unit threaded into the water jacket or manifold to sense engine temperature. This sending unit provides the ground for the gauge circuit that includes the indicator gauge and circuit supplied to the circuit from the battery via the ignition switch. When the coolant is cold, the sensing unit provides no ground for the circuit; when it heats up, a ground is provided, causing current to flow in the gauge circuit and to indicate a reading on the gauge.

2 Expansion (Figure 12.6).

The Bourdon expansion type gauge unit is made up of a gauge, a long copper or steel tube with a protective cover, and a sensing unit or bulb that fits in the block or water manifold. The tube and gauge expansion unit is filled with a liquid that expands rapidly when heated. When the coolant warms up, the sensing unit and the liquid in it warm up and the gauge mechanism is operated, showing coolant temperature.

G *Shutters.* Shutters are louverlike panels that are mounted in front of the radiator and, when closed, prevent air flow across the radiator. This restricted air flow decreases warm-up time after a cold engine start and provides a means for regulating water temperature during engine operation. Most shutters are air operated while some may be operated by a thermostat through direct linkage.

H *Radiator cap.* A cap that maintains a given pressure within the cooling system. This pressure allows coolant temperatures to run hotter without boiling. (Every pound of pressure exerted on the coolant increases the boiling point by $3\frac{1}{2}°$ F.) Included in the cap is a vacuum valve that allows air to enter the system when the coolant cools and contracts. If the cap did not have a vacuum valve, pressure inside the radiator may fall so low that outside air pressure may cause the radiator to collapse.

I *Water conditioner and filter.* A filter containing an element that conditions the coolant and prevents it from becoming too acidic. An over-acid coolant can cause cavitation that can erode the sleeves and cylinder block.

J *Coolant.* The coolant is generally a water and antifreeze mixture. Even in climates where freezing is not a problem, antifreeze and water are the most popular coolant because of the antifreeze's rust inhibitor capabilities. Antifreeze and water are generally installed in the cooling system in a 50–50 ratio.

K *Shutterstat.* The shutterstat is an air control valve that is operated by coolant temperature and controls the air supply to the shutter operating cylinder.

II. Cooling System Operation

A typical diesel engine cooling system has to warm up the engine quickly to prevent premature wear, maintain the engine temperature when the ambient temperature is high and when load on the engine is light, and maintain engine temperature when the ambient air temperature is high and the engine load is high. To better understand how the cooling system meets these requirements, the system coolant flow (Figure 12.7) and operating principles will be discussed.

When the engine runs, the heat generated by combustion and friction is absorbed by the coolant within the engine water jacket. This heated water rises, aided by the water pump, to the top of the engine and eventually works its way to the top water outlet. Depending on the type of thermostat the system uses, the

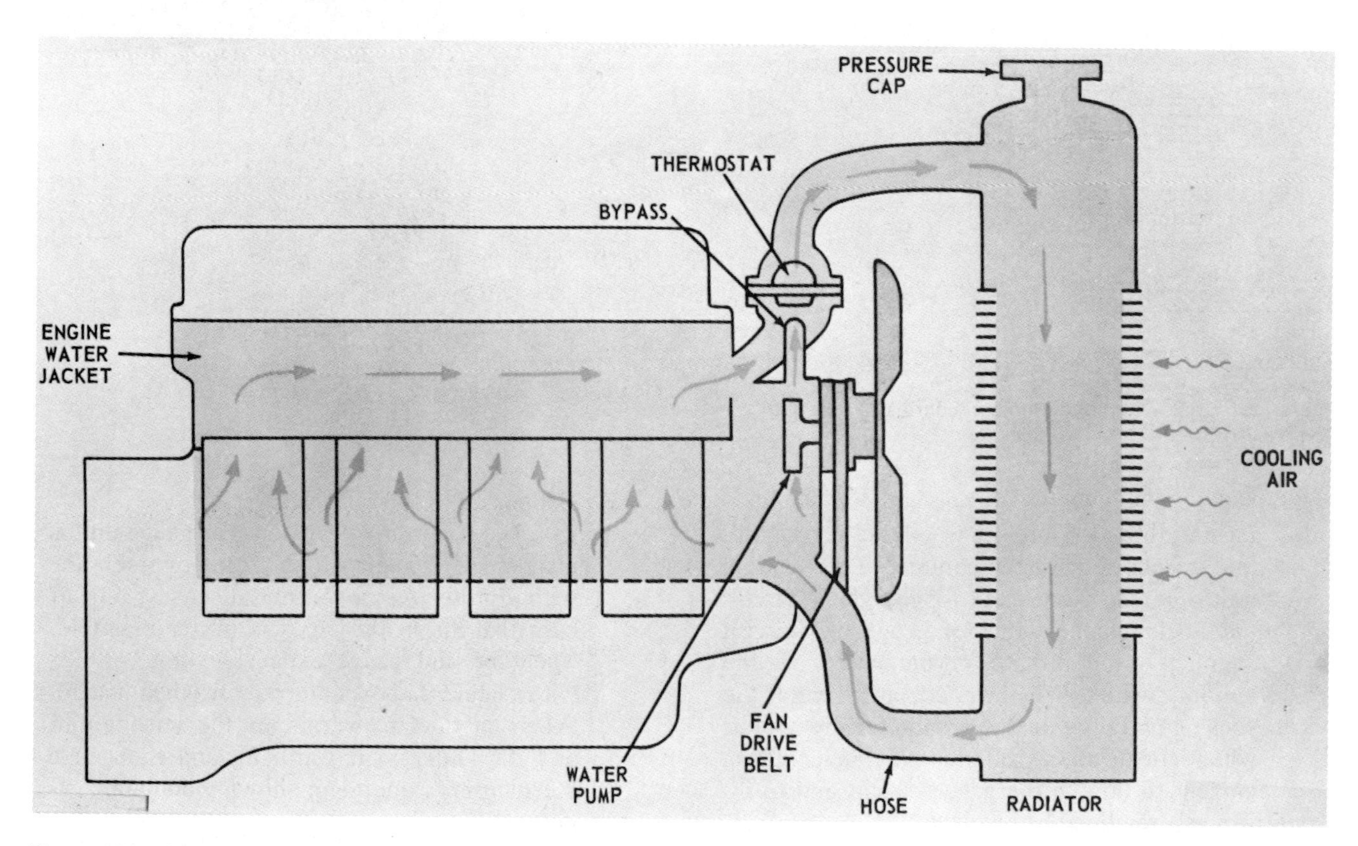

Figure 12.7 System coolant flow (Courtesy Deere and Company).

heated water may be allowed to circulate back to the coolant pump via a bypass pipe (bypass type thermostat), or it may be trapped at this point until it becomes hot enough to open the thermostat (full restriction type thermostat).

In the bypass thermostat system, the thermostat in its closed position doesn't allow water to circulate through the engine for cooling. Instead, it is diverted to a bypass pipe that directs it back to the coolant pump. The coolant is circulated in this manner until its temperature reaches engine operating temperature (180 to 195° F, 82 to 91° C). At this point the thermostat blocks off the bypass pipe and allows the coolant to circulate through the radiator. For example, if the coolant would be cooled to 170° F (76° C), the thermostat would close again, causing water to be directed back to the pump via the bypass pipe.

This thermostat system allows for quick engine warm-up while allowing water to be circulated through the block, preventing hot spots and block or head warpage. This is a commonly used system on most diesel engines.

The blocking type thermostat system differs in that the water is not allowed to circulate in the block via a bypass pipe and is not allowed to circulate through the radiator until the coolant has reached operating temperature (180 to 195° F, 82 to 91° C).

NOTE Although the thermostat is closed, blocking most of the flow from the block to the radiator, a small vent hole in the thermostat valve does allow a very minimal amount of circulation. This circulation is so slight that it does nothing in the way of cooling.

With the thermostat open, the coolant circulates through the radiator to dissipate engine heat. If coolant temperature drops below 180° F (82° C), the thermostat closes and circulation stops until the coolant has again reached 180° F (82° C), when the thermostat then starts to open. This system is not used on many diesel engines because the circulation within the block during warm-up or the time when the thermostat is closed is minimal.

Regardless of what type of thermostat is used, the function of the radiator remains the same. As the heated water reaches the radiator, it is accumulated in the top tank and then circulated downward through the cooling tubes. These tubes have metal (copper or aluminum) fins attached to them, increasing their total surface area. As the coolant flows through the tubes, air is forced across them by the fan. Heat then moves rapidly from the radiator tubes to the air moving across them. Unwanted or excess engine heat then is dissipated to the surrounding air, keeping the engine at a safe operating temperature.

III. Inspection and Overhaul of Components

Individual components of the cooling system may need testing, overhaul, and/or replacement. The following information must be used in checking the system components during a major rebuild or engine repair.

A. Procedure for Radiator Inspection and Testing

In most cases the radiator can be checked without removing it from the vehicle. The most common problem with the radiator is leakage. Checking the radiator for leaks should be done in the following manner:

1 Visually check the radiator for coolant leaks. If severe leaks are noted, the radiator should be removed and taken to a radiator repair shop for repair. If no leaks are visible, proceed with test 2.

2 Remove the radiator cap and install a cooling system pressurizer as shown in Figure 12.8.

3 Operate the tester to build pressure in the system.

NOTE When pressurizing the system, do not exceed the pressure stamped on the radiator cap by more than two or three pounds. To do so may cause the radiator or heater core to burst at the solder seams.

4 Carefully inspect radiator for leakage. If leaks are

Figure 12.8 Cooling system pressurizer installed on radiator.

found, the radiator should be removed and taken to a radiator repair shop.

5 If an overheating problem exists and there is some question about the flow capability of the radiator and all other possible causes have been eliminated, remove the radiator and take it to a radiator shop for cleaning.

NOTE Although reverse flushing of a radiator or cooling system may be done without removing the radiator from vehicle, it is rarely done in most repair shops since this has proven to be an ineffective way of cleaning a clogged radiator.

B. Procedure for Water Pump Overhaul

It is assumed that the water pump has been removed from the engine. If not, refer to Chapter 5 on engine disassembly.

In most cases, if a major overhaul is being performed on the engine, the water pump should be overhauled or replaced. The following steps must be followed when overhauling a water pump:

NOTE Make sure you check the manufacturer's service manual since each water pump may have to be disassembled in a special way. The following instructions are general and do not apply to all pumps.

1 *Disassembly.*

- a Remove pump drive hub or pulley by pulling it from the pump drive shaft with a puller.
- b Remove snap rings or retaining rings that may hold the pump drive shaft and bearing in place.
- c Remove impeller by supporting pump housing and pressing out shaft and bearing assembly.

2 *Inspection.*
After disassembly, check the following pump parts:

- a Pump housing for cracks.
- b Interference fit between impeller seal and housing.
- c Interference fit between impeller and pump shaft.
- d Face on impeller seal area. (Replace impeller seal seat if rough or nicked.) It should be carefully inspected.
- e Rough or worn bearings.
- f Drive pulley grooves for wear and interference fit between pulley and pump shaft.
- g Inspect weep hole in housing, making sure it is not clogged with grease or dirt.

NOTE The weep hole is cut into the housing so that any coolant escaping by the impeller seal can leak to the outside of the water pump rather than into the bearing.

3 *Reassembly.*
After determining what parts need replacement, proceed with pump reassembly as follows:

NOTE Pump repair kits are supplied by most engine manufacturers and include the following items: bearings, impeller seal, shaft, and needed gaskets.

- a Install shaft and bearing assembly into pump housing and install retaining clip or snap ring.
- b Support shaft on pulley or hub end in a press and start impeller on shaft as shown in Figure 12.9.
- c Obtain impeller-to-housing clearance from manufacturer's specifications.
- d Press impeller onto shaft until specified impeller-to-housing clearance is obtained (Figure 12.10).

NOTE Impeller to housing clearance is usually measured one of two ways: from back side of impeller-to-housing or from impeller-to-housing cover plate surface. This is a very important clearance; make sure you have it correct.

- e Turn pump over and support pump shaft on impeller end.

CAUTION Make sure the pump shaft is resting on the support, not the impeller. Serious damage to the impeller can result if the pump is not supported properly.

- f Install pulley or hub on pump shaft and press into place.

NOTE Make sure pulley is positioned correctly with the pump housing and according to manufacturer's specifications.

C. Procedure for Checking the Fan Belt

Fan belts must be checked for cracks and wear every time they are removed. During a major engine rebuild

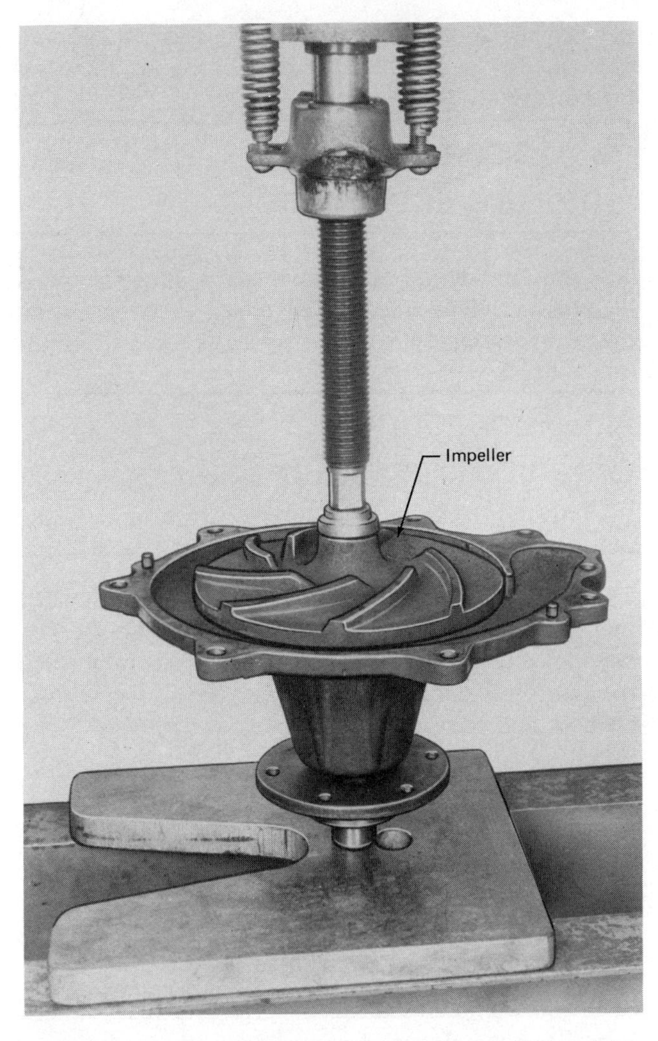

Figure 12.9 Pressing impeller on water pump shaft (courtesy International Harvester Company).

it is a good practice to replace all belts unless they are new. Belt replacement at this time may prevent serious damage to the engine.

D. Procedure for Checking the Thermostat

Thermostats should be checked during engine overhaul and replaced if any question of their accuracy exists.

NOTE Most repair shops do not check thermostats; they replace them. Check with your instructor for further instructions on replacement or checking thermostats.

If the thermostat is to be checked, it can be checked using a testing device similar to the one

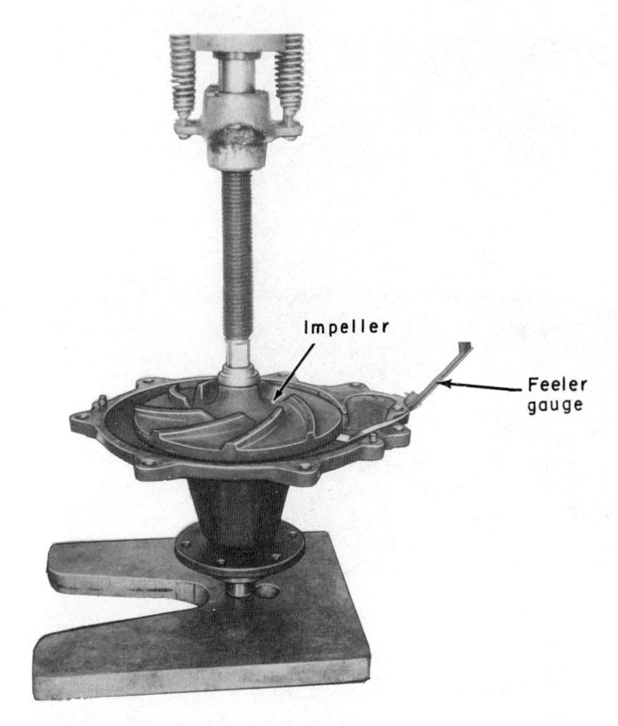

Figure 12.10 Impeller-to-housing clearance (courtesy International Harvester Company).

shown in Figure 12.11. To test thermostats, proceed as follows:

1 Fill tester can with water.

2 Turn on or plug in can heater.

3 Using your fingers, pry open the thermostat slightly and insert a short piece (six inches) of string into the opening.

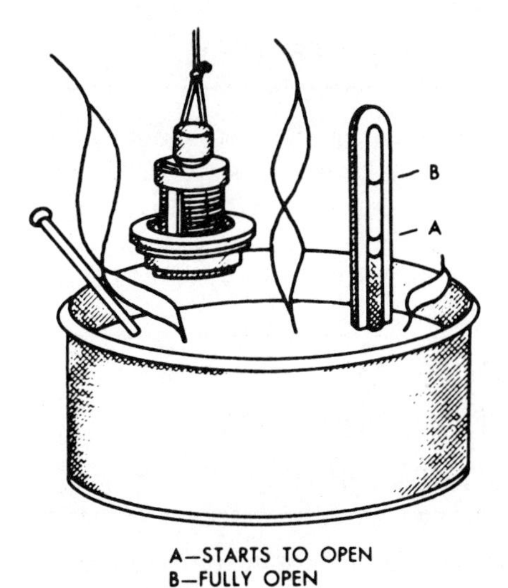

Figure 12.11 Thermostat testing device (courtesy Detroit Diesel Allison, Division of General Motors Corporation).

Figure 12.12 Testing radiator pressure cap.

4 Allow thermostat to close on the string. (This will allow the thermostat to be suspended in the water can.)

5 Suspend thermostat from the can bracket so that it is immersed in the water.

6 Set thermometer in the can.

7 Watch thermostat closely. As the water temperature rises, the thermostat will open, causing it to drop off the string.

8 Read the temperature on the thermometer. This is the temperature at which the thermostat will open.

E. Procedure for Testing the Pressure Cap

The radiator pressure cap rubber seal should be visually inspected for cracks or for signs of deterioration. If the seal is not damaged in any way, the cap should be tested for operating pressure using a testing device similar to the one shown in Figure 12.12.

F. Procedure for Checking the Shutterstat (shutter control valve)

The shutterstat is best checked for operation when it is installed in the system. It should not leak any air from either the closed or open position and must open and close at the correct temperature. If shutterstat cannot be adjusted to maintain the correct engine temperature, it should be replaced.

G. Procedure for Checking the Shutters

The shutters should open completely when the air is bled off and close when air is applied to them. Check the air operating cylinder for air leaks when the shutters are closed.

H. Procedure for Cleaning the Water Filter or Conditioner

The water filter usually requires no maintenance other than periodic changing and cleaning. When the filter element is changed, the can must be cleaned and flushed to remove all sludge and sediment. Install a new filter element and cover gasket when reassembling.

NOTE Some water filters require a special element if water is to be used as a coolant. If antifreeze solution is to be used, a different number element must be used. Other filters (depending on brand) have a common element that can be used with both water and antifreeze. Make sure the replacement filter you install is the correct one.

I. Procedure for Testing, Inspection, Repair, and Replacement of Temperature Gauges

As described under component parts, the temperature gauge can be of two different types, expansion (Bourdon tube type) and electric. Since the expansion type gauge cannot be repaired, inspect it for worn or broken spots in the tube. If any are found that have allowed the gas to escape, the gauge assembly must be replaced. If doubt exists about the gauge, test it by inserting the bulb end in the thermostat testing device to see if it indicates the correct temperature.

NOTE Accuracy can be checked at this time by comparing gauge reading to the temperature gauge in the thermostat tester.

Testing and repair of the electric gauge must be done in the following manner:

1 If gauge is not working, remove the sending unit wire and turn the vehicle key switch to the "on" position.

2 Ground wire to engine block or vehicle frame. Gauge should show a hot indication at this time.

   a If it does, the sending unit should be replaced with a new one.

   b If it does not show hot, check the power supply at the gauge with the key switch on.

   c If power is being received at the gauge and gauge does not show an indication with the sending unit wire grounded, replace gauge.

J. Procedure for Coolant Testing and Inspection

The coolant used in the system should be checked for cleanliness on a regular basis. This can be done visually either by removing the radiator cap and inspecting the coolant or by draining a small amount into a bucket and inspecting it. In climates where freezing temperatures are encountered, the coolant should be a mixture of antifreeze and water to prevent the system from freezing. A 50–50 mixture is considered adequate for most temperatures encountered. If some doubt exists about the strength of the antifreeze and water solution, it can be checked using a hydrometer designed especially for that purpose (Figure 12.13).

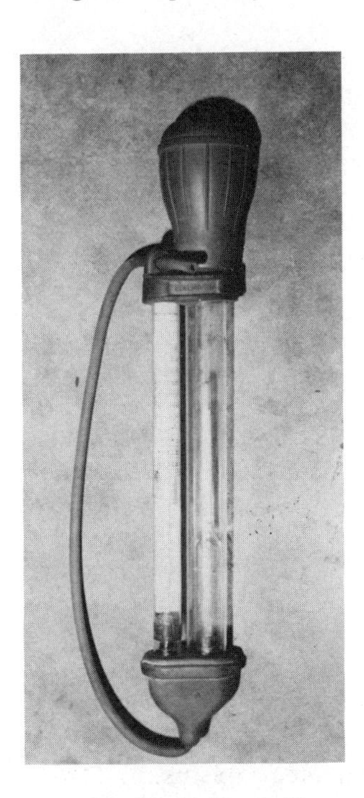

Figure 12.13 Antifreeze tester.

The following steps should be followed when checking the coolant solution:

1 Run engine to bring operating temperature up to normal.

2 Remove radiator cap and insert hose from hydrometer.

3 Depress bulb, then allow it to expand and draw coolant into the glass chamber.

4 Read letter at the top of fluid level on float and note temperature on temperature gauge.

5 Look up letter on chart printed on unit.

6 Match letter and temperature on chart. Immediately read the strength of antifreeze given next to letter and temperature.

IV. Complete System Testing, Inspection, and Troubleshooting

At various times problems will occur with the cooling system, such as overheating and coolant loss. When these problems occur, the system must be checked as a unit in the following manner:

A Inspect all hose clamps for tightness. Tighten or replace as needed.

B Check hoses for cracks, softness, or mushy feeling and replace if needed.

C Visually inspect all components, such as radiator, water pump, coolant filter, head gaskets, and water manifold, for any indication of leaks.

NOTE Usually leaking antifreeze can be detected by the red or green color residue that accumulates at the leak.

D Check fan and water pump drive belts. Belts should not show any signs of cracks or fraying.

NOTE To further test belts for condition, remove belt and bend belt backwards with your fingers and check for cracks. A deteriorated or old belt will crack easily when checked in this manner.

E If no visual leaks are evident, pressurize the system with a pressurizer and watch hoses, radiator tank joints, head gaskets, and other possible leakage points in the system closely.

F If no external leaks can be found:

   1 Test lube oil for evidence of coolant.

NOTE A simple way to test lube oil for coolant is

to allow the engine to cool (preferably overnight) and carefully remove pan drain plug, allowing a small amount of oil to drain out. If coolant is leaking into the oil, water will start to drip from the pan plug before the oil.

2 Operate engine under load to see if coolant is being blown out of the radiator overflow. If a head gasket is leaking pressure into the combustion chamber water, the cooling system will be filled with excessive pressure, which will force water by the pressure cap.

NOTE Many times when loading an engine to check for cooling system pressurization, the system and engine will function normally until the engine load is removed and the engine is allowed to idle. In some cases, if a problem exists it is at this time that the coolant will be forced by the radiator cap and out the overflow pipe. If this occurs, it is a good indication that pressure is being allowed into the cooling system by a leaking head gasket or cracked cylinder head.

G If system is not maintaining proper temperature (overheating, underheating):

1 Operate the engine under load. Check operation of shutters if used. Adjust shutterstat (shutter control) to obtain normal engine operating temperature.

2 Remove and test or replace thermostat. Replace with a thermostat of the proper opening temperature.

3 Make sure radiator fins are not clogged with foreign material, such as leaves, dirt, grease, or other obstructions.

4 Observe water in radiator top tank after system operating temperature has been reached. If no flow is apparent, a check of the water pump and/or radiator must be made.

The only sure way to test a radiator for restriction is to remove and take it to a radiator repair shop that can flow check it. Although it is considered an unusual situation, there are times when the water pump (because of mechanical failure, such as impeller broken or impeller loose on drive shaft) will not pump water. If this occurs, the water pump must be repaired or replaced.

If all the above outlined checks have been made and malfunctions corrected, the cooling system should operate normally.

SUMMARY

This chapter has covered material essential to good cooling system overhaul and maintenance. Remember, the cooling system and its related components should be closely inspected during a diesel engine overhaul to prevent an engine failure at a later date.

REVIEW QUESTIONS

1 What is the engine cooling system's main function?

2 List four cooling system components and explain their functions.

3 Why must a Bourdon expansion gauge be discarded when a hole is worn in the tube that connects the sensing unit and the gauge?

4 What may happen if the vacuum valve in a radiator pressure cap is stuck in the closed position?

5 What condition within the engine is a bypass type cooling system on a diesel engine used to prevent?

6 When overhauling a water pump, why is the impeller-to-housing clearance important?

7 Although thermostats are generally replaced during engine overhaul or cooling system troubleshooting, why should they be checked?

8 Coolant solution strength can best be checked by using what instrument?

9 Explain one check that can be made to determine if water is leaking into the fuel oil.

10 Pressurization of the cooling system may be caused by a number of engine problems. List two.

13

Air Intake Systems

The air intake system of a diesel engine (Figure 13.1) is a very important part of the total engine. Clean, cool intake air is vital to the operation of the engine. Without a supply of clean air, the internal engine parts would be quickly ruined by air-borne dust particles. The air intake system must prevent this from happening, but also must be correct in size and shape, easy to service, and efficient in dirt removal.

OBJECTIVES

After completion of this chapter the student will be able to:

1. Give two reasons for having an air intake system.
2. List and demonstrate five preventive maintenance checks that should be performed on a complete air intake system.
3. Demonstrate the ability to remove and clean or replace a dry type air cleaner cartridge.
4. Demonstrate the ability to clean and refill a wet type air cleaner with oil.
5. Explain the operation of an oil bath air cleaner.
6. Explain the operation of a dry type air cleaner.
7. List three checks that should be made on a turbocharger during a routine maintenance inspection.
8. Demonstrate the ability to remove, repair, and replace a turbocharger.
9. Demonstrate the ability to test, remove, and replace a roots type blower.
10. List and explain four checks that should be made on a complete air intake system during engine overhaul.

GENERAL INFORMATION

Diesel engine air intake systems must supply clean air to the engine at all times, regardless of the operating conditions. Diesel engines are called upon to operate

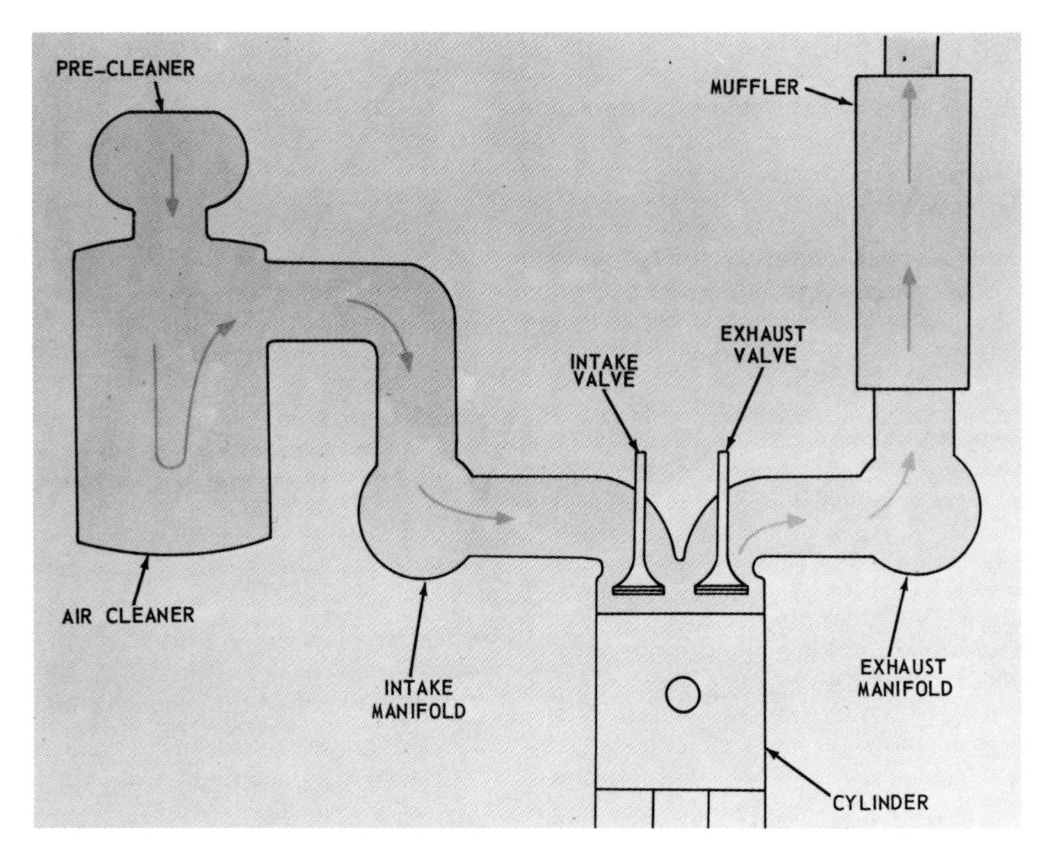

Figure 13.1 Air intake system schematic (courtesy Deere and Company).

efficiently and over long periods of time in hot, dusty, cold, and wet conditions.

If the system is not efficient and allows dirt to enter the engine, it will be ruined quickly. Air-borne dirt and dust particles that find their way into the engine combustion chamber are turned into glasslike particles by the heat from combustion. These small particles of very hard material account for engine wear. Impossible as it may seem, early diesel engines used no air cleaners or else cleaners that were highly inefficient. As engine wear was analyzed, it became apparent that dirt was the engine's number one enemy, and air cleaners as we know them today were developed.

In recent years pollution laws have placed increasingly stringent demands on the smoke emitted from the diesel engine under load. This fact, along with a desire to build smaller and higher horsepower engines, has given emphasis to the turbocharger, an exhaust-driven air pump. Many modern diesel engines utilize turbochargers in the intake air system to increase the air supplied to the engine cylinder. As the volumetric efficiency (ability to breathe) is increased by the turbocharger, the engine combustion process is better, making a more powerful and a cleaner running engine.

I. System Components and Function

A *Air cleaners.*

1. *Centrifugal.* Centrifugal air cleaners are often called precleaners and, as the name implies, are designed to be used in conjunction with either an oil bath or dry type air cleaner. Designed to use centrifugal force to clean the air, they are intended to remove the larger particles of dirt from the air before they reach the main air cleaner. Precleaners are usually located at the top of the pipe that leads to the main air cleaner and are extensively used on farm tractors and heavy equipment diesel engines. They are generally not found on truck engines.

2. *Oil bath* (Figure 13.2). Designed to be used in conjunction with the centrifugal precleaners or as a single unit, the oil bath type air cleaner uses oil as the cleaning agent to remove the dirt particles from the air. It is designed not only to

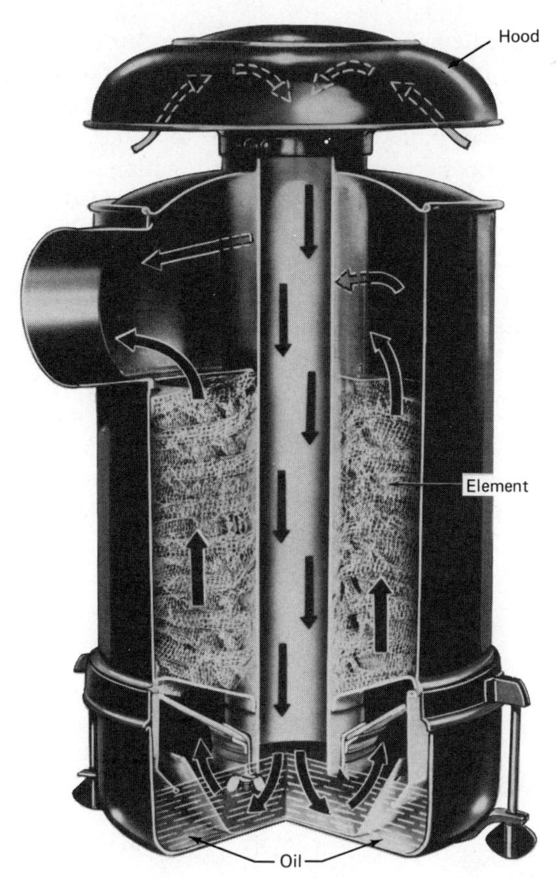

Figure 13.2 Typical oil bath air cleaner (courtesy Detroit Diesel Allison, Division of General Motors Corporation).

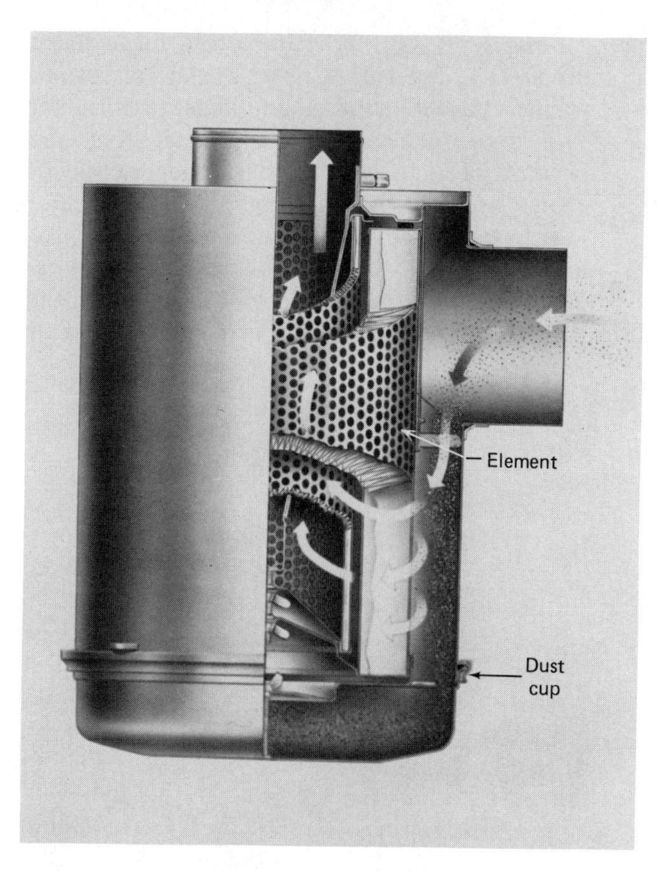

Figure 13.3 Diesel engine dry type air cleaner (courtesy Cummins Engine Company).

clean the air but to silence the engine air intake also. The oil bath air cleaner is not efficient over the entire engine operating range, although it can be 97 percent effective at rated engine speed.

NOTE The reason the oil bath air cleaner is not efficient at speeds less than engine rated speed is that air cleaning depends on air velocity, and at slow engine speeds air velocity is reduced.

3 *Dry type* (Figure 13.3). This is an air cleaner that contains no oil and uses a dry type filter cartridge made of resin impregnated paper to clean the air. The dry type air cleaner is very efficient (99.6 to 99.9 percent) at all engine speeds and is used almost exclusively on modern day diesel engines.

B *Piping, tubing, and hoses.* The pipe, tubes, and hoses are used to transport the air from the air cleaner to the intake manifold or turbocharger. They must be airtight, rugged, yet somewhat flexible, and designed in such a way so that no restriction to air flow exists.

C *Supercharger.* A supercharger, which is a gear-driven air pump powered by the engine, forces air into the engine cylinder. This air pump increases volumetric efficiency from 75 to 80 percent (common for nonsupercharged engines) to almost 100 percent.

D *Intercooler or aftercooler.* This is a radiator-like device installed in the intake manifold on many engines through which the air must pass after leaving the turbocharger or supercharger. Intercoolers usually are air to water, meaning that the air, as it passes through the core of the intercooler, is cooled by the coolant.

NOTE Some engine manufacturers call the air-cooling device an aftercooler since it is located after the turbocharger. This terminology doesn't alter the function of the cooler, regardless of what it's called. Its job is to cool the air between the turbocharger and the intake manifold.

E *Turbocharger.* A turbocharger is an exhaust-driven centrifugal air pump that pumps air into the engine cylinder, improving volumetric efficiency. It is used on many modern day diesel engines because of its flexibility in adapting to specific engine requirements. The turbocharger requires very few special design changes in the engine since it is driven by exhaust gases. Since it is not gear-driven, it is not limited to engine rotational speed. As a result, engine fuel can be increased, and the increased engine exhaust causes the turbocharger to turn faster, thereby creating more intake air to the engine under load. The turbocharger rotational speed may be as high as 60,000 to 100,000 rpm. The turbocharger has one other function: exhaust silencing. On some engines it is the sole device used to silence engine exhaust.

F *Intake manifold.* The intake manifold is generally a cast aluminum housing that is bolted to the

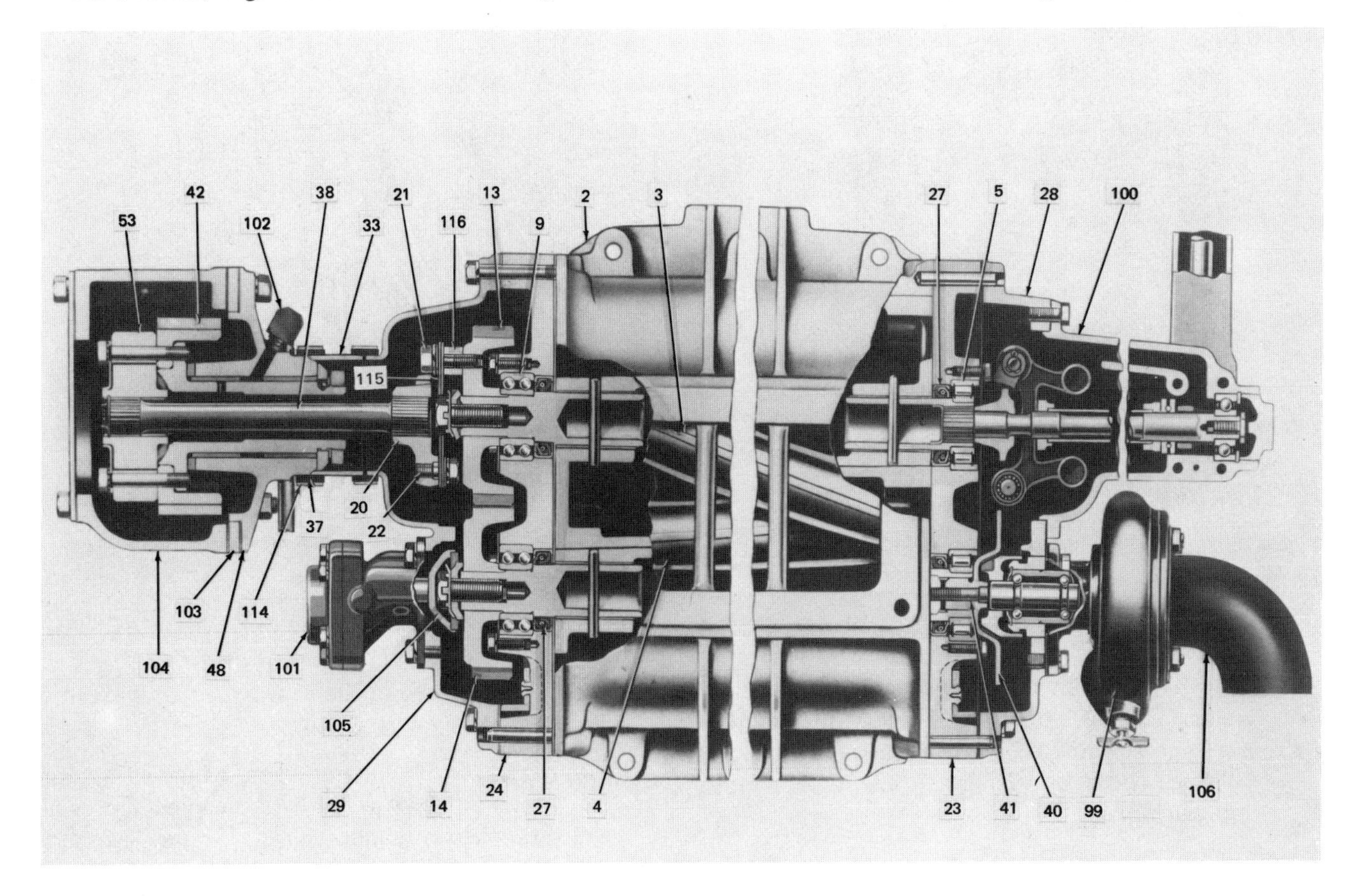

2. Housing — — blower
3. Rotor–blower–upper R.H. helix
4. Rotor–blower–lower L.H. helix
5. Bearing (roller) — front
9. Bearing (ball) — rear — doble row thrust
13. Gear–rotor–lower R.H. helix
14. Gear–rotor–lower L.H. helix
20. Hub–rotor drive gear
21. Bolt–plate to gear
22. Bolt–plate to hub
23. End plate–front
24. End plate–rear
27. Oil seal–end plate
28. Cover–end plate–front
29. Cover–end plate–rear
33. Cover–blower drive shaft
37. Seal–drive shaft cover
38. Shaft–blower drive
40. Coupling assembly, water pump drive
42. Gear–blower drive
48. Support–blower drive gear hub
53. Coupling assembly, blower drive
99. Pump–fresh water
100. Governor
101. Pump — fuel
102. Elbow (90°) — oil line to blower drive
103. End plate — cylinder block—rear
104. Housing — flywheel
105. Fork — fuel pump drive
106. Cover–water pump inlet
114. Clamp–drive cover seal
115. Plate–blower rotor drive hub
116. Spacer–plate to gear

Figure 13.4 Roots type blower (courtesy Detroit Diesel Allison, Division of General Motors Corporation).

engine cylinder head air inlets. It must be leakproof to prevent the entry of dirt and be light in weight.

G *Restriction indicator.* A gauge indicating to the operator if the air cleaner has excessive restriction due to dirt. It is operated by vacuum, created as a result of the restriction in the air cleaner, much like a vacuum gauge. It can be electrically operated by a sensing unit placed in the intake manifold and connected to an indicator light on the vehicle instrument panel. This gauge can only be used with a dry type air cleaner.

II. Types of Systems and System Operation

Many different systems have been designed, dependent on the requirements established by the engine manufacturer.

A *Naturally aspirated system.* This system uses no air pump of any type, and relies solely on the difference in pressure between the engine cylinder and the atmosphere to move the air into the engine. It is one of the most common systems in use on older four stroke cycle diesel engines. It generally employs an oil bath or dry type air cleaner for cleaning of the air.

B *Roots type blower system.* The roots type blower is used primarily on two stroke cycle engines, although it can be found on four stroke cycle engines. The roots type blower (Figure 13.4) is used to force air into the engine cylinder and exhaust gases out (called scavenging). It is used on two stroke cycle engines since a two stroke cycle engine does not have an intake stroke or an exhaust stroke (the Detroit Diesel, for example).

C *Supercharger system.* Many four stroke sycle engines use a supercharger (a gear-driven blower) to improve efficiency and performance. A supercharger forces air into the cylinder under slight pressure, supercharging the engine cylinder with air.

D *Turbocharger system.* Turbochargers may be used on four stroke cycle or two stroke cycle engines to improve engine power and efficiency. The turbocharger forces air into the engine cylinder as does a supercharger, but differs from the supercharger in that it is exhaust-driven. Being lightweight and small in size, the turbocharger is by far the most popular intake air boost system in use today.

E *Turbocharger and intercooler system.* Air compressed by the turbocharger during engine operation is heated to approximately 300° F (150° C). Since cool, dense air is best for engine efficiency, the heated air coming from the turbocharger should be cooled before entering the intake manifold.

Cooling the air between the turbo and intake manifold is the intercooler. Air passing through the intercooler may be cooled by as much as 100° F 23° C).

III. System Maintenance and Inspection

Since clean air is vital to long engine life, maintenance of the air cleaner and the intake air system is a very important part of daily care. Although daily maintenance is usually the engine operator's responsibility, all mechanics should be familiar with the servicing, inspection, and repair of system components.

A. Procedure for Visual Inspection of System

A visual inspection should be made of the air cleaner, hose connections, and all possible points where the system could leak. Make sure all rubber hoses (if used) are in good condition. Replace hoses if in doubt. When replacing hoses, always use a sealer or cement to insure an airtight seal. The precleaner should be inspected for dirt accumulation and cleaned if needed.

B. Procedure for Air Cleaner Cleaning and/or Replacement

1 The wet type (oil bath) air cleaner cup or bowl should be removed (Figure 13.5) and cleaned, then refilled with clean oil the same weight as the oil being used in the engine crankcase.

a If the filter has a removable wire mesh filter, remove and clean it.

CAUTION Do not clean air filter elements in kerosene, gasoline, or any flammable solvents. Engine runaway could result upon starting.

b Reinstall the wire mesh filters and oil cup, making sure the cup is seated onto the air cleaner housing correctly. Then tighten the cup clamp securely by hand.

Figure 13.5 Removing oil bath air cleaner bowl.

2 Dry type air cleaners. Remove the cartridge (paper element cartridge) from the air cleaner housing and clean or replace it. Some manufacturers recommend that the paper element cartridge be cleaned in one of the following ways:

a Blowing the dirt from the cartridge with compressed air.

NOTE If compressed air is used to blow dirt from the filter cartridge, use an air supply of 40 psi (2.8 kg/cm²). Hold the nozzle approximately two to four inches from the filter cartridge. Holding the nozzle directly against the filter cartridge may rupture the cartridge, making it unfit for further use.

b Washing the element in detergent and water.

CAUTION After the element has been washed, it should not be placed back into the air cleaner for use until it is completely dry. Also, carefully check the rubber seal or lip on the cartridge. Make sure it is not damaged before reusing the cartridge.

C. Procedure for Checking System for Air Restriction

An easy way to check for air system restriction is with a manometer (a very sensitive pressure vacuum-measuring device). Connect the manometer to a fitting in the air inlet pipe approximately 2 to 3 in. (50 to 75 mm) from the manifold inlet fitting. If the engine is equipped with a turbocharger, connect the manometer just in front of the turbo air inlet side.

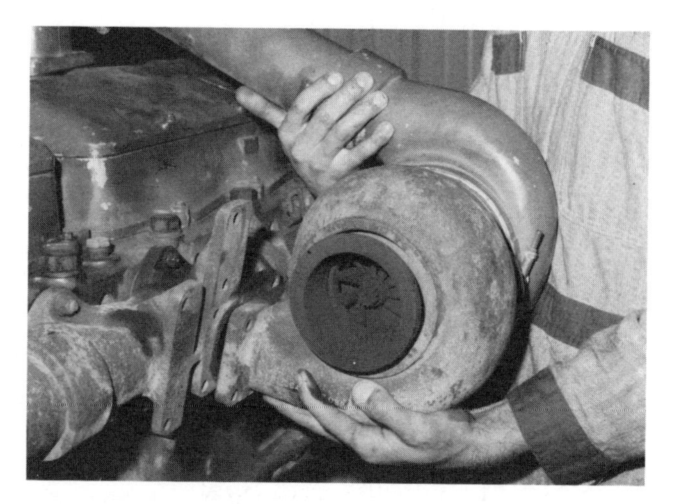

Figure 13.6 Turbocharger.

After connecting the manometer, run the engine at a specific speed and record the manometer reading. Next disconnect the air tubing and air cleaner. Run the engine at the same speed as with the air cleaner and tubing connected, compare readings, and check manufacturer's specifications.

NOTE Each system will have different readings; for example, Detroit V-71 engine with a clean air cleaner and a precleaner and no turbo will have a reading of 15 to 16 in. (380 to 405 mm) of water at 2100 rpm. Check specifications and know what reading the system you are working on should have.

D. Procedure for Turbocharger Inspection

It is assumed that the turbocharger (Figure 13.6) has been removed from the engine exhaust manifold. If not, refer to the engine disassembly chapter.

The diesel engine turbocharger is a precision unit that will require inspection during an engine rebuild. The turbocharger should be inspected visually for:

1 External oil leaks.

2 Oil in the air pipe that connects the turbocharger to the intake manifold or the turbocharger compressor housing.

NOTE A wet film of oil in this pipe indicates that the turbocharger is leaking oil into the engine intake and must be repaired.

3 Visually inspect turbine wheels and housings for rubbing. For example, the intake turbine wheel

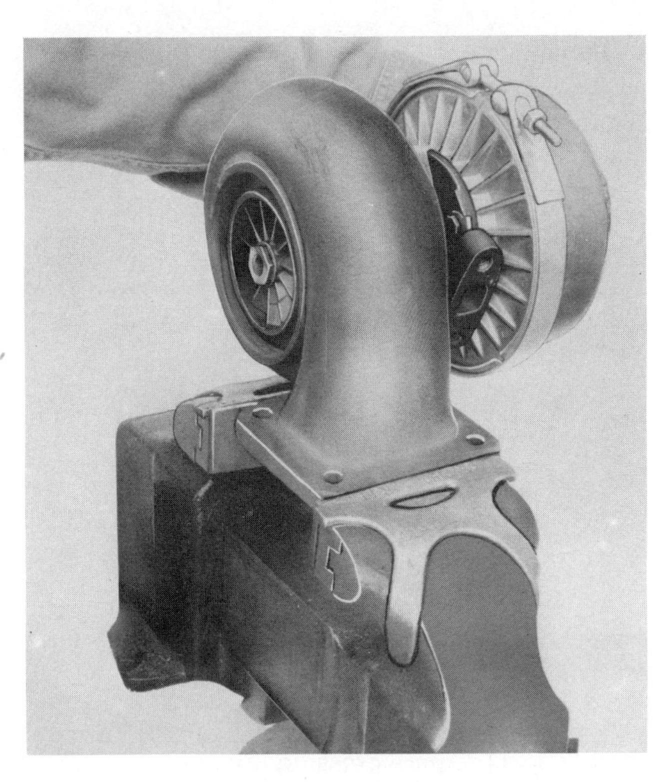

Figure 13.7 Turbocharger mounted in vise (courtesy International Harvester Company).

Figure 13.8 Checking turbocharger end clearance (courtesy International Harvester Company).

should not rub on the housing. Rubbing indicates that the bearings and shaft are worn excessively and must be replaced.

NOTE If it is found the turbocharger intake wheel has been rubbing on the housing, steps 4 and 5 can be eliminated. Proceed to Section E, "Turbocharger Disassembly and Inspection."

4 Mount the turbo in vise (Figure 13.7) equipped with brass jaws by clamping onto the lip of the exhaust housing and tightening the vise.

CAUTION Do not tighten the vise excessively as damage to the housing may result.

After turbocharger is mounted securely, spin the turbo wheel by hand. It should spin freely. If it doesn't a problem exists within the turbocharger and it must be repaired.

5 Bearing clearance checks.

a Shaft bearing radial clearance check. With the turbo mounted in a vise (Figure 13.7), use a dial indicator with a magnetic base. Mount the dial indicator with the plunger inserted into the turbocharger drain hole.

NOTE In many cases, a plate with a fitting in it that covers the drain outlet will have to be removed before the dial indicator can be inserted.

b Position the plunger of the dial indicator directly on the turbine shaft.

c Set the dial indicator on zero.

d Grasp the intake and exhaust turbine wheel. Apply pressure to both wheels evenly and attempt to move the shaft upward toward the dial indicator. Note the reading on the dial indicator. This is the shaft-to-bearing clearance.

e Move the shaft back and forth several times to insure that maximum bearing-to-shaft clearance is obtained.

f If the total radial clearance is less than 0.003 in. (0.076 mm) or greater than 0.006 in. (0.153 mm), the shaft-to-bearing clearance is incorrect and the turbo must be disassembled and repaired (refer to Section E).

g Position dial indicator so that its plunger is contacting either end of the turbine as shown in Figure 13.8.

h Manually force the turbine wheel and shaft as far one way as it will go.

i Zero dial indicator.

j Manually force the turbine wheel and shaft toward the dial indicator. Note indicator movement. This is end clearance.

k If end clearance is less than 0.001 in. (0.025 mm) or more than 0.004 in. (0.1 mm), the turbocharger must be disassembled and overhauled.

E. Procedure for Turbocharger Disassembly and Inspection

If the turbo has failed the clearance checks as outlined previously and is to be repaired, the following procedures must be followed:

NOTE Many repair shops fix turbochargers on a routine basis while others may replace a worn turbo with a rebuilt unit. Since it is highly likely that a diesel mechanic will at some time or another be called upon to service a turbocharger, the following information on teardown and rebuild is presented.

1 If the turbocharger was not cleaned previously, give it a thorough external cleaning at this time.

2 Before loosening the clamp band or any of the bolts that hold the turbocharger together, make reference or match marks on the turbine housing, compressor housing, and the center housing (Figure 13.9).

NOTE Reference or match marks can be made with a punch and hammer or electric etcher, whichever is convenient. These marks will be used during the reassembly of the turbocharger. Lining up the marks insures proper alignment of the compressor housing outlet, oil inlet, and drain lines when the turbo is installed on the engine.

3 Bend down the lock plates that secure the compressor housing bolts. Remove bolts and compressor housing.

4 Bend down the lock plates that secure the turbine housing bolts and remove the bolts and turbine housing.

NOTE In many cases the turbine housing will be rusted onto the center section (bearing housing) and will require pressing the bearing housing from the turbine housing. Support turbo in press as shown in Figure 13.10 and press the center section downward.

Figure 13.9 Match marks on turbocharger (courtesy International Harvester Company).

CAUTION Make sure the center section is not allowed to drop on the floor as it is pressed free.

5 Select a socket that will fit the hex head on the exhaust turbine wheel and fit it to a flex handle. Clamp the flex handle and socket in a vise with the socket facing upward. Set the turbine wheel and center housing onto the socket with the compressor wheel facing upward.

Figure 13.10 Pressing center section from exhaust housing.

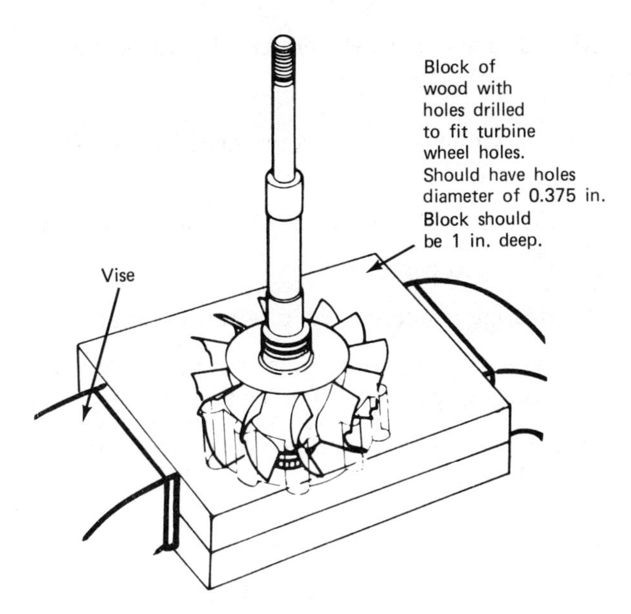

Figure 13.11 Compressor wheel holding fixture.

NOTE If this procedure for holding the compressor wheel is not satisfactory, a holding fixture similar to the one shown in Figure 13.11 can be made up.

6 Loosen and remove compressor wheel lock nut.

7 Lift the compressor wheel from the shaft.

NOTE If compressor wheel is tight on the shaft, removal may require a slight tap on the turbo shaft with a plastic hammer. Support the turbo in a vise and tap downward on the turbine shaft. Be careful to catch the turbine wheel and shaft as it drops from the housing (Figure 13.12).

Figure 13.12 Taking turbine shaft and wheel from housing.

8 If compressor wheel lifted off easily and turbine shaft and wheel are still in the center section, pull it from the housing.

9 To unlock the four lock plates on the cap screws that hold the center housing to the back plate, bend them down.

10 Remove the four cap screws.

11 Separate the back plate from the center housing.

12 Lift the thrust bearing and thrust washer from the center housing.

13 Remove the retaining rings that hold the turbo bearings in place. Remove only the outside snap rings. Bearings can then be removed without removing the inner snap rings.

14 Place all parts in a cleaning solution such as carburetor cleaner.

15 After parts have soaked in the cleaner for at least an hour, remove and clean all parts in a solvent tank or with hot water.

16 Blow all parts dry with compressed air and a nozzle.

17 Inspect the bearing housing for the following:

- a Wear in the bearing bores. Bearing bores should be smooth and show no signs of scoring or scuffing.

NOTE It is a good practice to fit a new bearing into the housing to determine if the bearing housing bores are worn. A new bearing should fit snugly but be free to turn when installed. Replace bearing housing if bearing bores are scored or bearings fit loose.

- b Further checks of the bearing bore may be made by measuring the bore with a snap gauge. The bearing bore size should be 0.6228 in. (15.96 mm).

18 Inspect turbine wheel and shaft for wear as follows:

- a Check the turbine for bent or nicked blades. Since the blades cannot be straightened, if either condition is found, replace turbine wheel and shaft.
- b Rub marks on outer diameter of turbine blades.

c Inspect shaft bearing journals for wear in the area where the bearings ride on the shaft. Bearing journals should be checked with a micrometer and their diameter should not be less than 0.3994 in. (10.15 mm).

d Inspect the threads on the opposite end of the shaft from the turbine wheel. The threads must be free of nicks or burrs so that proper torque is obtained during reassembly.

19 Inspect the back plate for wear as follows:

a Closely inspect the back plate bore for wear caused by the sealing ring. (Any indication of scratches or wear at this bore will make it necessary to replace the back plate housing.) Bore diameter should not be greater than 0.501 in. (12.72 mm).

b Rusted or cracked thrust spring. Replace the thrust spring if it shows any signs of rusting.

20 Inspection of compressor wheel and housing. Inspect the compressor wheel and housing for the following:

a Worn or nicked blades on the compressor wheel. Replace the wheel if any worn or broken blades are found.

b Check compressor housing for cracks or damage of any kind.

21 Thrust collar and bearing. Check thrust collar and bearing for wear. If any wear exists on the collar thrust face, it should be replaced.

NOTE In most cases the thrust collar and bearing assembly is generally replaced as a matter of practice during a turbocharger overhaul. This, of course, is dependent on the situation. If a relatively new turbo is disassembled, it may not be necessary to replace the bearing and thrust washer.

F. Procedure for Turbo Failure Analysis

After parts inspection, if damage to the turbo parts is found, a determination of what caused the failure should be made. Most turbo failures are caused by lack of lubricant, foreign objects in the compressor or turbine section, and contamination of the lubricating oil. In analyzing turbo failure each one of the causes mentioned above can be identified by visual inspection.

1 Lack of lubricant can be determined by the following:

a Discoloration of turbine shaft (bluish in color).

b Damaged bearings.

c Bearing material welded to turbine shaft.

d Outer diameter of turbine or compressor wheel blades worn, as result of bearing failure.

2 Ingestion of foreign material can be determined by the following:

a Turbine wheel tip damage.

b Damaged compressor wheel blades.

c Worn compressor blades, caused by loose inlet air hose.

3 Contaminated lubricant can be determined by the following:

a Polished and grooved bearing journals.

b Worn and scored thrust bearing.

c Inner and outer bearing surfaces scored.

d Contaminates embedded in bearings (aluminum).

e Channel or groove wear on outer bearing diameter.

G. Procedure for Turbocharger Reassembly

After inspection of all turbo parts, obtain new parts for the ones that are worn excessively.

NOTE A turbo repair kit is available for turbocharger overhaul. Two different kits are available. One kit has all the sealing rings plus gaskets and O rings. The other kit has all bearings, back plate, all gaskets, and sealing rings.

1 Give the center bearing housing one last visual inspection for dirt and carbon, then install new bearings and their lock rings.

CAUTION Turbo assembly should be done in a clean area so that no dirt can enter during the assembly.

2 Lubricate the bearings with engine oil and turn each bearing to spread the lubricant. Double-check bearing retainers to make sure they are seated in the grooves.

3 The center housing is now ready for installation of the turbine wheel and shaft. Install a new seal ring in the groove on the turbine shaft.

CAUTION Two types of housings are used in the TO-4 and TO-4B turbochargers. One has a stepped seal bore and the other has a straight seal bore. Make sure you have the correct seal ring for the housing being used. The straight seal bore has a diameter of 0.703 in. (17.86 mm) and the stepped seal bore has a diameter of 0.713 in. (18.11 mm). Excessive oil leakage will be the result if the wrong ring is installed.

4 Install the turbine shroud over the exhaust side of the center housing.

5 Lubricate the turbine shaft seal ring with engine oil.

6 Install the turbine wheel and shaft into the center housing and the bearing shroud.

NOTE As the turbine shaft is inserted, caution must be exercised when the seal ring comes in contact with the center housing bore. Forcing the turbine wheel and shaft at this point may break the ring unless caution is used. Rocking the turbine wheel from side to side as you push it on will make installation easier (Figure 13.13).

7 Install a new seal on the thrust bearing and attach it to the thrust collar.

8 With the turbine wheel and center housing resting on the turbine wheel side, place the thrust collar and bearing assembly onto the turbine wheel shaft.

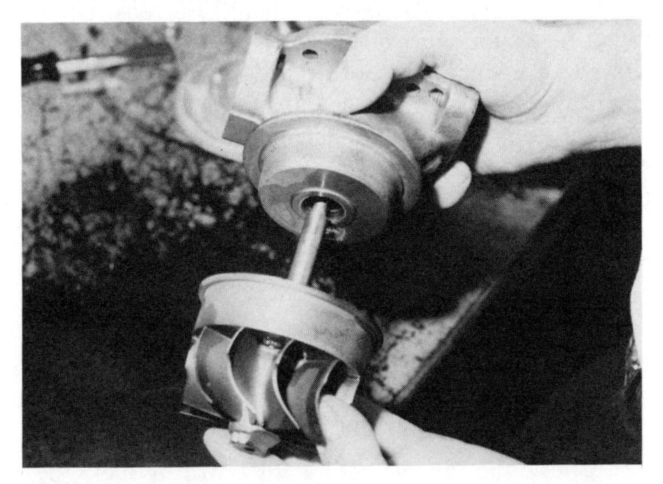

Figure 13.13 Installation of turbine wheel and shaft.

Figure 13.14 Installing thrust bearing and washer.

9 Slide the bearing assembly down onto the center housing, making sure that the bearing thrust washer engages the two alignment pins in the housing (Figure 13.14).

10 Make sure that a new seal ring is installed in the center housing and then place the back plate assembly over the turbine wheel shaft and slide it down until it contacts the bearing and seal ring. Rock it gently so that the ring will slide into the back plate bore.

11 Line up match marks that were previously placed on the center housing and the back plate; then, using new lock plates, install lock plates and the four capscrews that hold the back plate onto the center housing. Torque capscrews to 75–90 in. lb (8.47–10.16 N m). Bend lock plates against bolts to prevent them from loosening.

12 Place the compressor wheel on the turbine wheel shaft and install the retaining nut.

13 Secure the turbine wheel by fitting a socket to the serrated end or by placing it in the holding fixture and clamping it in a vise.

14 Tighten compressor wheel nut to 18–20 in. lb (2.03–2.25 N m) using a T handle and socket to prevent imposing any side loads on the turbine wheel shaft.

15 After torquing the compressor wheel nut to 18–20 in. lb (2.03–2.25 N m) advance it an additional 90° or $\frac{1}{4}$ turn. This is the final torque.

16 After final tightening of the compressor wheel nut, remove the turbo from the vise and check it by spinning it to check for freedom of movement.

17 After this assembly of the center section has been completed, it is ready for installation into the exhaust turbine housing.

NOTE Before installing the assembled center section into the exhaust turbine housing, carefully inspect the recess that it fits into for carbon and rust. All carbon and rust must be removed before assembly.

18 Install the center section into the exhaust housing, rotating it back and forth to insure that it is seated into the housing (Figure 13.15).

19 Align center section using the match marks made during disassembly.

20 Using new lock plates, install the capscrews and clamps. Torque capscrews to 100–130 in. lb (11.29–14.68 N m).

21 Assemble the diffuser if used into the compressor housing and install the center section and exhaust housing, using the match marks made during disassembly as a guide to align the center section and compressor housing.

22 If lock plates and capscrews are used to secure the compressor housing to center section, torque them to 100-130 in. lb (11.29–14.68 N m). If a clamp is used to secure the compressor housing to the center section, torque the clamp nut to 40–80 in. lb (4.51–9.03 N m).

23 This completes the assembly of the turbo. A final check should be made for turbine and shaft freedom by spinning the turbo. As the turbo is being spun, lubricate it through the oil inlet hole with clean engine oil.

Figure 13.15 Installing center section in exhaust housing.

24 If the turbo is not to be installed on the engine, cap all openings to prevent entrance of foreign materials.

25 If the turbo is to be installed on the engine, clean the exhaust manifold turbo mounting flange and install a new gasket.

26 Install the turbo and tighten the capscrews or nuts that hold it to the exhaust manifold.

27 Install the oil return line and pour about a pint of clean engine oil into the turbo oil inlet.

28 Install inlet oil line.

29 Start the engine and run it at a slow idle for approximately one minute.

30 Run engine at 900 to 1000 rpm and check for oil leaks at the turbo inlet and outlet.

This completes the testing and/or rebuild of the turbocharger. As stated previously, the information presented applies to Airesearch TO-4 and TO-4B turbochargers. If a different make and model of turbo is to be overhauled, this information can be used as a general guideline, but specifications and torques must be obtained for the particular turbo being worked on.

H. Procedure for Inspection of Roots Type Blower (Detroit Diesel)

The roots type blower used on Detroit Diesel engines and some older four stroke cycle engines must be inspected during a major engine rebuild to determine if it needs to be repaired. External inspection will give a good indication if the blower should be repaired or replaced. Major overhaul of the blower will not be discussed here since it is not as widely used on Diesel engines as turbochargers are. It is suggested that if a blower needs to be overhauled, the service manual should be used as reference or the blower should be taken to a qualified repair station.

The blower can be inspected while it is on as well as off the engine. Use the following procedures when inspecting the blower:

1 Remove the air shutdown housing by removing the bolts that secure it to the blower housing and disconnecting the control cable (Figure 13.16).

2 Check the blower rotors for burrs or nicks by turning the blower by hand if it is not on the

Figure 13.16 Removing air shutdown housing.

engine, or if on the engine by turning the engine slowly with a bar.

3 Badly scored rotors should be replaced. Rotors that are scored a small amount may be dressed down with a file and used again. In either case the blower must be removed and disassembled.

4 Check blower rotors for oil accumulation. Oil on the blower rotors indicates that the blower seals are leaking. To check this condition further, run the engine at idle and observe the end plate through the rotor compartment opening. If a seal is leaking, a thin film of oil will accumulate on the end plate.

NOTE When inspecting the end plates, a flashlight may be needed to see into the rotor compartment.

CAUTION Care must be exercised when working around the blower air intake with the engine running. Do not allow clothing, hair, or fingers to get caught in the blower rotors.

5 Inspect the blower drive by attempting to rotate the blower rotors back and forth. Normal rotor movement is $\frac{3}{8}$ to $\frac{5}{8}$ in. (9 mm to 16 mm) on a V-71 series engine blower.

6 When testing this movement, a springlike resistance will be felt. When released, the rotors should spring back.

7 If the above inspection indicates that the rotors don't move or do not spring back when released, it can be assumed that the drive coupling is worn or damaged and should be replaced.

8 Inspect rotors for rubbing between the rotors. If rubbing is evident, the rotor bearings probably have failed. The blower must be disassembled and the bearings replaced if this condition exists.

9 Inspect the rotors for rubbing along their entire length. If rubbing is evident in this area, it indicates the blower gears are worn and should be replaced.

10 Check the lubricating oil connection that supplies oil to the blower from the blower drive support. Leakage at this point indicates that the O ring seal is damaged and should be replaced.

I. Procedure for Installation of Blower

If the blower passes the above inspection or has been repaired, it can now be installed onto the engine.

NOTE The following instructions are general and should be used in conjunction with the service manual.

1 Clean the block blower mounting, making sure that no old gaskets, sealer, or dirt are left on the block.

2 Cement a new blower-to-block gasket to either the block or the blower.

3 Place the blower on the engine and start bolts.

NOTE On an in-line engine, which has the blower mounted on the side, it is recommended that guide studs or dowels be used to guide it in place.

4 Tighten bolts to the recommended torque.

5 Connect the blower oil supply tube used on six cylinder engines.

6 Connect fuel lines to the fuel pump.

7 Install the air shutdown housing and connect the control cable.

CAUTION Make sure the proper gaskets along with the screen are installed on the air shutdown housing. Improper assembly may prevent the shutdown plate from shutting off the engine air when the shutdown cable is pulled.

8 Install the blower drive shaft and the retaining ring. Install the blower drive shaft housing cover.

9 Install any other brackets or control cables that may have been removed.

10 Run the engine and check for oil leaks.

11 With engine running at idle, check for correct operation of air shutdown assembly by pulling the emergency stop control.

CAUTION Make sure that this test is run at *idle only!* Pulling the emergency stop at high idle may cause damage to the blower oil seals.

SUMMARY

This chapter should give you a better understanding of the maintenance, inspection, and repair of a diesel engine air intake system. Turbochargers, the key to modern diesel engine efficiency, have been covered in detail to assist you in their installation, inspection, and repair. As can be seen, a properly maintained air intake system holds the key to long and trouble-free engine life. As a diesel mechanic, you will maintain the air intake system maintenance and advise operators on the correct procedures to use when servicing their equipment.

REVIEW QUESTIONS

1 Give two reasons diesel engines require air cleaners.

2 Which air cleaner, dry type or wet type, is more efficient over all engine speeds?

3 Why are precleaner type air cleaners used on diesel engines?

4 Why must the oil be changed on a wet (oil bath) type air cleaner?

5 Why are turbochargers used on diesel engines?

6 How do superchargers differ from turbochargers?

7 Prepare a list of items that should be checked during an intake air system check.

8 Why are intercoolers used sometimes with a turbocharged diesel engine?

9 What one factor is the key item in determining if a diesel engine turbocharger needs overhaul?

10 In what way does a roots type blower (as used on the Detroit Diesel) differ from a turbocharger?

14

Exhaust Systems

The diesel engine exhaust system is a very important part of the engine. It must carry the exhaust gases away from the engine, vehicle, and operator. It must reduce the noise that is created within the engine by combustion. Without the exhaust system the engine would be very noisy and objectionable to operate. This system must be designed so that it does not interfere or get in the way of other engine or vehicle components.

OBJECTIVES

Upon completion of this chapter the student will be able to:

1 Explain the purpose of an exhaust system used on a diesel engine.

2 List four component parts of a typical exhaust system.

3 Explain why exhaust systems need periodic maintenance.

4 List and explain two ways to find exhaust system leaks.

GENERAL INFORMATION

Sometimes taken for granted, the diesel engine exhaust system lends much to the overall performance and acceptance of a diesel engine. Without it, noise levels from the engine would far exceed the threshold of pain that the human ear can withstand. In addition, the operator would not be protected from smelly and toxic gases that would be allowed to enter the cab or operator's area. Excess noise would not be tolerated by the general public who may be in the area where the engine is being operated.

As you become familiar with exhaust systems, you will recognize many different styles and designs used on various types of equipment.

I. System Components

The exhaust system will be made up of:

A *Manifold.* The manifold collects and directs exhaust from all the cylinder outlets into a single pipe that is connected to the muffler.

B *Exhaust pipe.* A pipe designed to carry the exhaust gases from the manifold to the muffler.

C *Muffler.* A device that is designed to decrease the sound of the engine exhaust as the gases pass through it. The muffler must reduce noise but not create any back pressure on the flow of exhaust gases. Mufflers may be of two types:

1 *Straight through.* On straight-through mufflers the exhaust gases flow through the muffler with no change in direction.

2 *Reverse flow.* Reverse-flow mufflers are designed so that the exhaust gases must change direction when flowing through the muffler.

D *Muffler extension or pipe.* A pipe fitted to the top or back end of the muffler that directs the exhaust gases further up or away from the vehicle.

E *Turbocharger.* An exhaust-driven air blower that utilizes some of the energy in the exhaust gases that would normally be wasted.

F *Clamps and gaskets.* Clamps are used to hold the pipe to the manifold and muffler and to attach the complete assembly to the vehicle frame.

G *Rain cap.* This prevents rain from entering the exhaust system.

II. Types of Systems

Many different exhaust systems are used on diesel engines. The specific type used is generally dependent on the engine application.

A Farm tractor and industrial engine type.

Farm tractors and industrial engines generally have a very simple exhaust system. This system may be:

1 Exhaust manifold, turbocharger, exhaust pipe, and rain cap (Figure 14.1).

NOTE Many engines that use turbochargers do not use a muffler since the turbocharger acts as the muffler.

2 Exhaust manifold, muffler, exhaust pipe, and rain cap (Figure 14.2).

NOTE Very few mounting brackets are needed on a farm tractor exhaust system since it usually is mounted directly to the exhaust manifold and requires no further support.

B Motor truck and bus engine types.

Diesel engines installed in trucks and cars require a more complex system to prevent excessive noise and provide a means of routing the exhaust away from the cab or inside of the vehicle. These systems may be single or double.

III. System Maintenance and Testing

Exhaust system maintenance usually amounts to repair and/or replacement of parts in the system, such as pipes, clamps, and mufflers.

A To check the system for leaks, look for signs of leakage at all of the connection points in the system.

NOTE Leaking connections are usually indicated by a dark accumulation of soot that is blown out from the connections.

1 Repair of the leak may only require tightening of the clamp in many cases. Other cases may

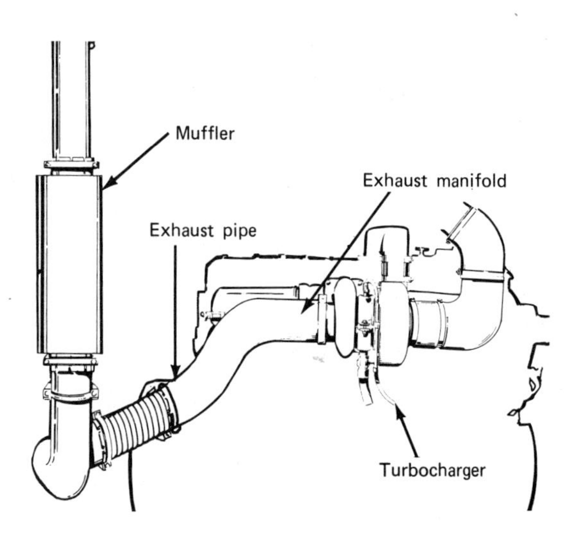

Figure 14.1 A typical diesel engine exhaust system with turbocharger.

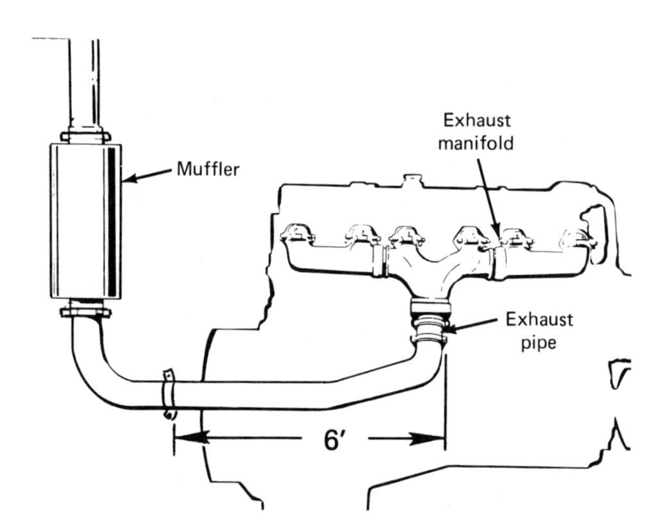

Figure 14.2 A typical diesel engine exhaust system with a muffler.

require that the exhaust pipe or muffler to which it is connected must be replaced along with a new clamp.

2 Many leaks can be located by holding your hand around the exhaust pipe at the connection, while you accelerate and deaccelerate the engine quickly.

3 Another way a hard-to-find leak can be found is to run the engine at idle and place a board or flat piece of metal over the muffler pipe. This creates enough back pressure so that a leaking connection can be found.

B Maintenance.

1 System maintenance usually involves visually checking clamps, pipes, rusted pipes, and mufflers.

2 Other checks include checking the turbocharger for oil leaks and operation.

NOTE For complete checks on the turbocharger, see Chapter 13 on air intake systems.

3 Checking system back pressure. The exhaust system should create very little or no back pressure on the engine during operation.

NOTE A small amount of back pressure can cause a drop in engine horsepower.

SUMMARY

Exhaust systems, although not too complicated to repair, must be kept in good shape to prevent undesirable exhaust gas leaks or noise.

REVIEW QUESTIONS

1 What is the function of the exhaust system?

2 List two components that are included in an exhaust system.

3 Describe one way to check an exhaust system for leaks.

15

Fuel Injection Systems

The diesel engine fuel injection system is considered the control center or heart of the diesel engine. Many different types of fuel injection systems are used on diesel engines today. Regardless of the design or type, the major function of all of them is the injection of fuel into the engine in correct amounts and at the correct time.

This chapter will deal with the basic requirements of a diesel fuel system and their relationship to engine performance.

OBJECTIVES

Upon completion of this chapter the student will be able to:

1 List and explain four requirements of a diesel fuel system.

2 Explain two major differences between the carburetor and the fuel injection system.

3 List and explain two differences between a mechanically operated injector and an injection nozzle.

4 List four fuel contaminants that are detrimental to the operation of a fuel injection system.

5 List and explain two types of fuel systems used on modern diesel engines.

GENERAL INFORMATION

Fuel injection systems as we know them today are the result of many years of development and research by their manufacturers.

Early fuel injection systems were very large and unreliable, while modern injection systems are light in weight, small in size, and one of the most reliable units on the diesel engine.

As your study of fuel injection systems continues, names like Robert Bosch, American Bosch, Stanadyne injection systems, Detroit Diesel, Cummins, Caterpillar, CAV, and many others will become commonplace. These are some of the companies that have pioneered and developed the fuel injection systems in use throughout the world today.

I. Fuel System Requirements

If the engine is to develop full power and operate efficiently, its fuel system must do the following:

A *Meter (measure).* The fuel injection system must measure the fuel supplied to the engine very accurately since fuel requirements vary greatly from low to high engine speed. Fuel is measured within the injection pump or injector by measuring it as it fills the pumping chamber (inlet metering) or as it leaves the pumping element (outlet metering). Although many variations of these two concepts exist, the basic principles have changed very little.

B *Time.* The timing of fuel injected into the cylinder is very important during engine starting, full load, and high speed operation. Diesel engines start best when fuel is injected at or very close to top dead center (TDC), since it is at this point that air in the chamber is the hottest. After the engine is started and running at high speed, the injection timing may have to be advanced to compensate for injection lag, ignition lag, and other factors that influence combustion within the engine cylinder.

Many modern injection pumps have an automatic timing device built into them that automatically changes the timing as the engine speed changes. These devices have been given names by their manufacturers, such as automatic advance, speed advance, intravance, and many others. Their major purpose is the varying of fuel injection timing to produce a powerful yet efficient engine.

C *Pressurize.* The fuel system must pressurize the fuel to open the injection nozzle (a spring-loaded valve) or the injector tip. In addition to the pressure required to open the nozzle, some pressure is required to inject fuel into the combustion chamber to offset the pressure of compression, which may be 350 to 450 psi (25 to 32 kg/cm^2).

The pressure setting of the injection nozzle or injector tip is directly related to the degree of atomization required. As the fuel is pumped through the holes in the tip (multihole type nozzle) or around the pintle (pintle type nozzle) at high pressure, 1500 to 4000 psi (105 to 280 kg/cm^2), atomization occurs. This atomization can be compared to the atomization that occurs when you attach a spray nozzle to the end of a garden hose.

D *Atomize* (the breaking up of fuel into small particles). The fuel must be atomized when it is injected into the combustion chamber since unatomized fuel will not burn easily. The degree of atomization required will vary from engine to engine depending on the combustion chamber design. Consider the following examples:

1. A precombustion engine will require very little atomization since the fuel is injected into the prechamber first and then into the main combustion chamber. In most prechamber engines, the fuel is heated in the prechamber to start burning; as it burns, the resulting expansion forces it through a passageway into the main chamber. Turbulence created by the rapidly burning and expanding fuel help mix air and fuel for complete combustion. Therefore, a high degree of atomization is not required during initial injection.

2. A direct injection engine relies solely on atomization of fuel during injection and piston crown design to mix air and fuel for combustion. This engine design will obviously require a higher degree of atomization if complete combustion is to occur. A multihole type nozzle tip is generally used with this combustion chamber design.

E *Distribute.* Closely related to timing, the distribution of fuel must be accurate and according to the engine's firing order. Distributor pumps deliver fuel to each pump outlet in succession and the lines are hooked to the cylinders in the correct firing order, much like a distributor used on a gas engine. In-line pumps have the camshaft designed to permit the pump outlets to fire in the required engine cylinder firing order. Along with distributing the fuel to the various cylinders, the fuel system must distribute the fuel within the combustion chamber during initial injection.

The fuel must be injected throughout the chamber so that all of the air within the chamber is utilized. This requirement is fulfilled by the injection nozzle or injector, its hole size, and angle.

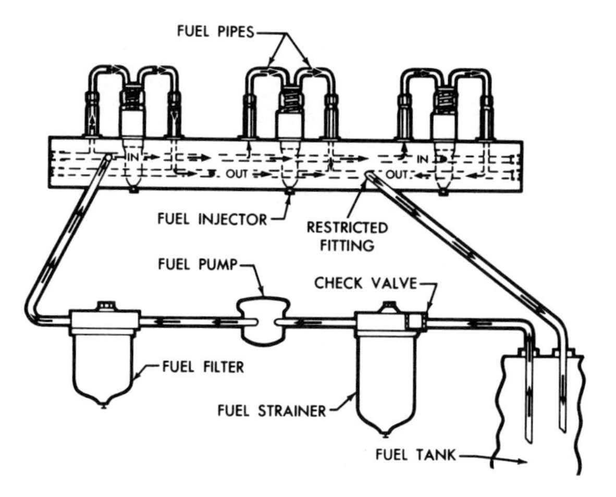

Figure 15.1 Typical unit injector system (courtesy Detroit Diesel).

F *Control start and stop injection.* Injection of fuel must start quickly and end quickly. Any delay in beginning will alter the pump-to-engine timing, causing hard starting and poor running engines.

Any delay in injection ending can cause a smoky exhaust and irregular exhaust sound. The end of injection should be instantaneous with no dribbling or secondary injections. In many systems this is accomplished by a valve called a delivery valve or retraction valve. Other pumps have a camshaft designed with a sharp drop on the cam lobe that stops injection very rapidly.

II. Types of Fuel Systems

Many types of fuel systems are used on modern diesel engines. Some of the more commonly used types are given below.

A *Unit injector* (Figure 15.1) used by Detroit Diesel, manufactured by them for use on their engines. The system is composed of:

1. *A low pressure gear pump* that moves fuel from the tank to the fuel filter and to the camshaft-operated injector.
2. *Injector.* The injector meters, times, and pressurizes the fuel. It is camshaft-operated through a push rod, rocker arm arrangement, and one injector is utilized for each cylinder.
3. *Filters.* Several fuel filters are employed throughout the system to protect the highly machined parts from dirt and water.
4. *Governor.* A separate governor is connected to the fuel control rack that controls the position of the injector plunger. This governor may be hydraulic or mechanical.

B *Pressure time system* (common rail) (Figure 15.2). This system is manufactured by Cummins diesel and is used exclusively on Cummins diesel engines. The system is sometimes called a unit injector system, but in reality it resembles a common rail system since not all fuel system functions are performed within the injector. Major component parts of the system and their functions are:

1. *Gear pump.* Moves fuel from tank through the system to the injector; and also supplies fuel to lubricate the governor and other working parts in the main pump housing.

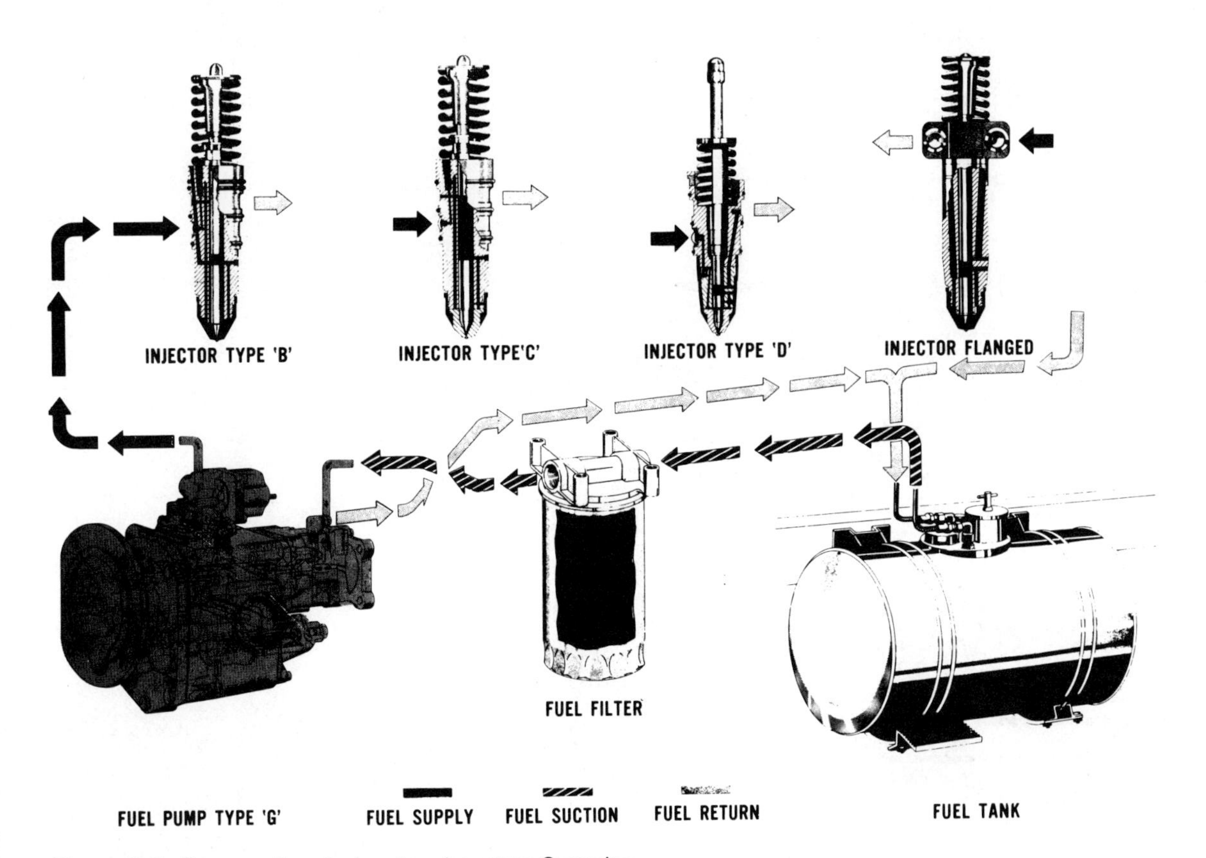

Figure 15.2 Pressure time fuel system (courtesy Cummins Engine Company).

2 *Governor.* Flyweight type governor controls maximum fuel pressure to prevent engine overfueling, controls engine idle, and also prevents excessive top engine speed by controlling fuel supply. Contained within the main pump housing.

3 *Throttle.* The throttle is controlled by the operator and it controls the fuel flow and pressure to the injector. The throttle is simply a shaft with a hole in it, fitted into the pump housing. As the shaft is turned, alignment of the shaft hole with fuel passageways within the housing changes, thereby regulating fuel flow and pressure.

4 *Fuel injector.* The injector is operated by the engine camshaft through a push rod and rocker arm. Fuel is supplied to the injector from the fuel passageways within the cylinder head. To enter the injector, fuel must pass through a small orifice much like the jet in a carburetor. The orifice, which is replaceable, can be selected to allow specific amounts of fuel so that the injector can be used in many different Cummins engines. Fuel flows into the injector and fills the cup; then, as the injector plunger is forced downward by the action of the camshaft, fuel is injected into the engine. One injector per cylinder is utilized.

C *In-line pump and injection nozzle system* (Figure 15.3). This common type of system is manufactured by several companies (American Bosch, Robert Bosch, and CAV–Simms) and is widely used on all types and makes of engines. Main components and their function are:

1 *Engine driven injection pump* that meters, times, pressurizes, and controls the fuel being delivered to the injection nozzle. It is composed of single pumping elements for each cylinder, fitted into a common housing, that are operated by the pump camshaft.

2 *Governor.* The governor, generally of the mechanical flyweight type, may be mounted in the pump housing or in a separate housing mounted on the main injection pump housing. The governor controls fuel delivery to regulate fuel control at all engine speeds (variable speed governor) or controls high idle and low idle only (limiting speed type).

3 *High pressure steel lines.* These lines deliver fuel from the pump to the injection nozzle.

4 *Injection nozzle.* The injection nozzle is a spring-loaded, hydraulically operated valve that is inserted into the combustion chamber. This nozzle may be of several different types, but all serve the same basic function, that is, atomization of the injected fuel. (For a more detailed description of injection nozzles see Chapter 17.)

5 *Fuel filters.* Filters and water traps are em-

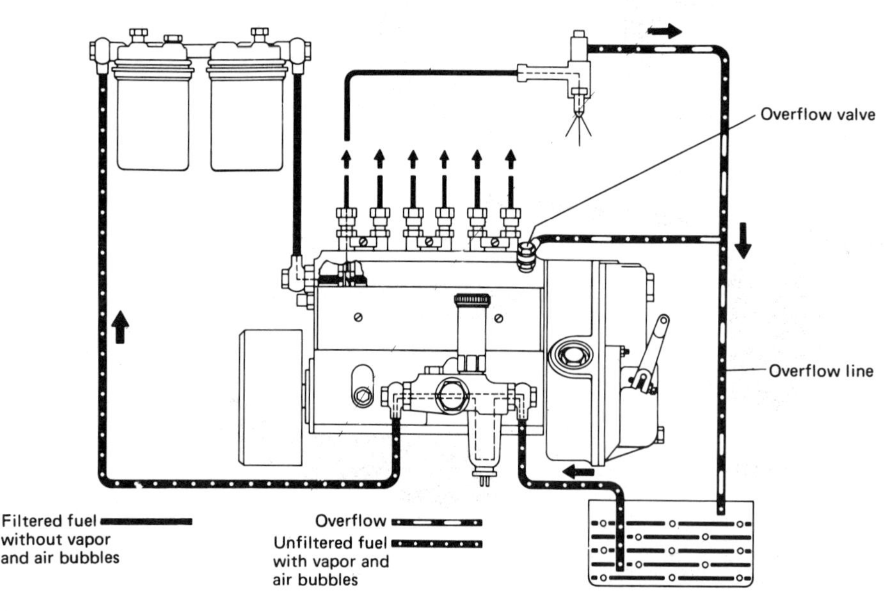

Figure 15.3 In-line pump and injection nozzle system (courtesy Robert Bosch Corporation).

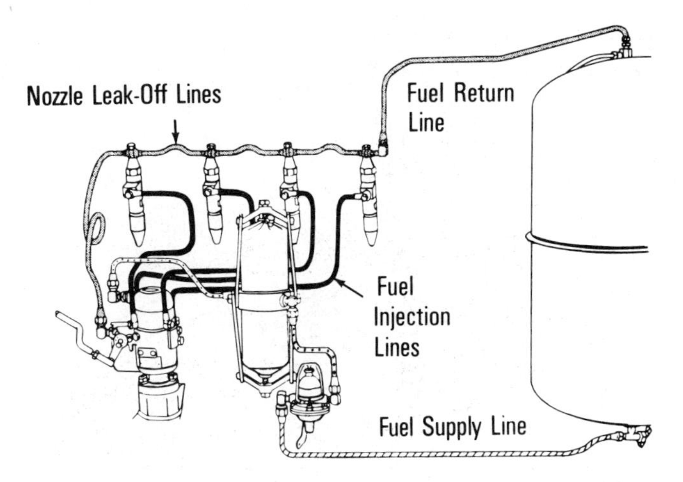

Figure 15.4 Distributor type pump and injection nozzle system (courtesy of Deere and Company).

ployed throughout the system to prevent damage to the system by dirt and water.

D *Distributor type pump and injection nozzle system* (Figure 15.4). A relatively small compact system designed for small and medium horsepower engines, this is manufactured by many different pump manufacturers such as Robert Bosch, American Bosch, CAV–Simms, and Stanadyne Injection Systems. It is used on many makes and models of engines. Main components and their functions are:

1 *Engine drive rotary type injection pump.* This pump usually employs one or two pumping plungers that supply fuel to a distributor that then distributes the fuel to the various cylinders much as a distributor on a gas engine distributes spark. Fuel metering and injection timing are controlled by the pump.

2 *Governor.* The governor may be of the mechanical flyweight type or hydraulic depending on engine application. Available as limiting speed or variable speed type.

3 The following components are identical to the in-line pump system.

a High pressure steel lines.

b Injection nozzle.

c Fuel filters.

III. Fuel Metering (Measurement)

As you study each of the following chapters of fuel systems, you will see many different types of fuel metering devices employed. The key to a good fuel system is the method by which the fuel is controlled. Some common methods are the port and helix, inlet metering, and sleeve control types.

A *Port and helix* (Figures 15.5). This is probably one of the most common types of fuel control systems

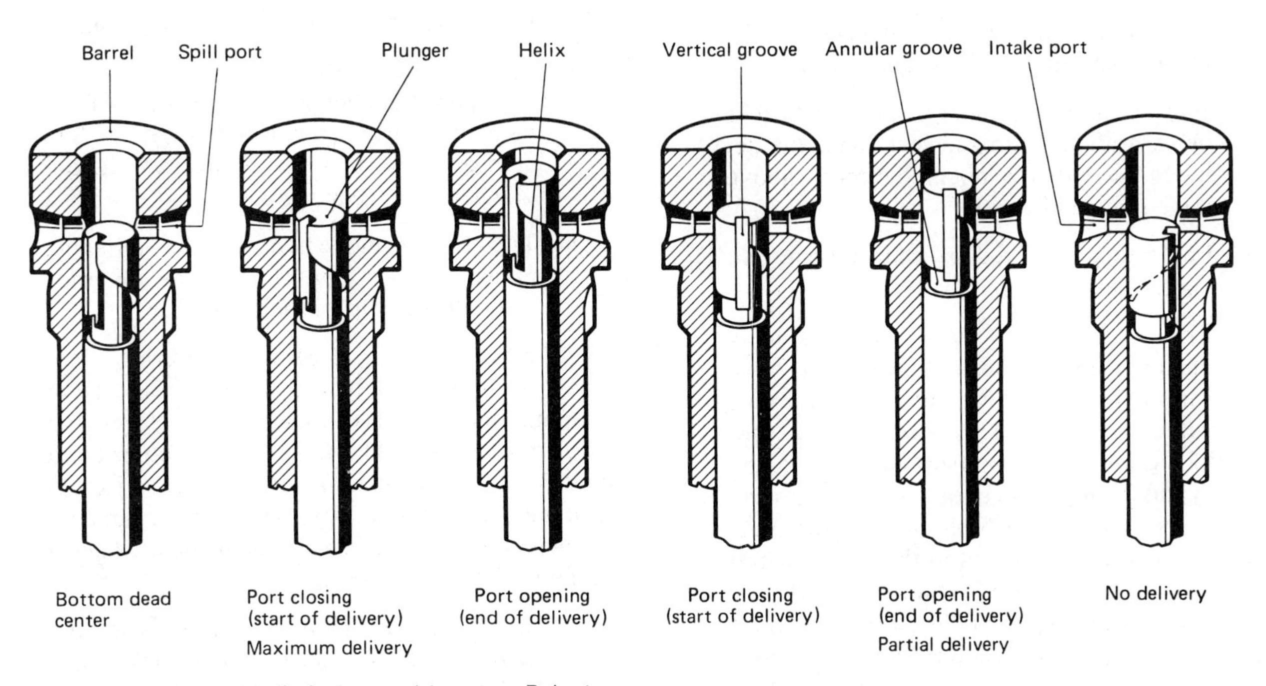

Figure 15.5 Port and helix fuel control (courtesy Robert Bosch Corporation).

in use today. It is called "spill port" metering because it controls the amount of fuel pumped by opening a port and by spilling off high pressure fuel.

1 *Component parts and nomenclature*. The port and helix type pumping unit is composed of:

   a Barrel and plunger unit fitted or lapped together with a very small clearance between them to allow enough fuel to enter between the mating parts for lubrication. This clearance is approximately 0.0002 in. (0.005 mm) to prevent excess leakage of high pressure fuel around the plunger.

   b Helix and vertical groove. If the pumping plunger unit did not have a helix or control groove machined on it, the pumping element would pump the same amount of fuel at all times, giving the operator no control over the engine.

2 *Fuel flow and operation*.

   a With the helix and vertical groove, the pump output can be easily varied by turning the pumping plunger in relation to the barrel.

   b As the pumping plunger is forced upward and covers the inlet and outlet ports in the barrel, fuel is trapped above the pumping plunger.

   c The chamber and the vertical groove in the plunger are filled with pressurized fuel.

   d As the pumping plunger moves further upward, the pressurized fuel opens the delivery valve that is mounted directly above the pumping element.

   e Fuel is then delivered to the injection nozzle via the fuel injection line.

   f End of delivery occurs when the helix uncovers an inlet port, allowing high pressure fuel to rush down the vertical groove cut in the plunger. This lowers the pressure in the pumping chamber. Delivery to the cylinder stops since the injection nozzle and delivery valve both close as the pressure drops. The output of the pumping unit can be controlled by rotating the pumping plunger, causing the helix to open or close the spill port for a longer period of time. Some pumping plungers may have the helix designed to control the ending of fuel injection, while others may control the beginning of fuel injection. A specially designed type of plunger used in some injection systems uses two helixes, one that controls the beginning and the other the end of injection.

   g In a pump application, the position of the plunger or helix is controlled by the fuel control rack that is connected to the plunger via a gear segment.

   h The rotary movement of the pumping plunger is then controlled by the governor through the control rack and gear.

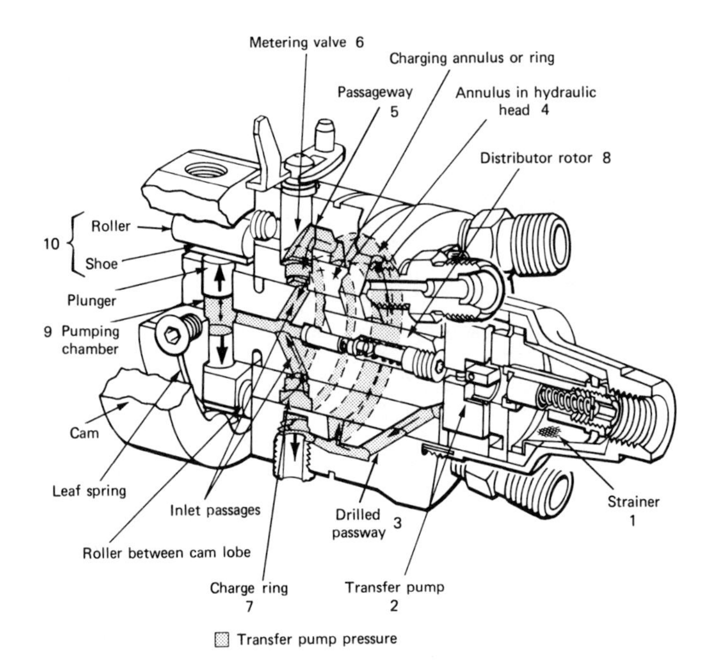

Figure 15.6 Inlet metering principle. (Courtesy Stanadyne/Hartford)

B *Inlet metering* (Figure 15.6). Inlet metering (measurement of the fuel before it reaches the main pumping element) was pioneered and developed by Vernon Roosa. This concept was considered unworkable by the manufacturers of fuel injection equipment at that time. After much testing and development, its use became accepted and it changed the fuel injection pump industry.

1 *Component parts and nomenclature*. The inlet metering system is made up of a metering valve and a drilled bore into which the valve is fitted. This bore is drilled into a hydraulic head that

makes up the main part of the injection pump. Intersecting this bore at a right angle is a drilling or passageway through which fuel is supplied to the metering valve bore. The metering valve, in the shape of a round rod, is approximately $\frac{1}{4}$ in. (6 mm) in diameter and is 1 in. (25 mm) long. It is designed with an angular groove cut on one side to regulate fuel delivery and a slot on top for attachment of the governor control linkage.

2 *Fuel flow and operation.* Fuel is supplied to the metering valve under low pressure, 60 to 80 psi (4.2 to 5.6 kg/cm²), from a vane type transfer pump through a drilled hole in the hydraulic head.

a The delivery valve and its bore are connected via a drilling to the ring or annulus area of the hydraulic head.

b As seen in Figure 15.6, when the rotor has turned, allowing the charge port in the rotor to align with a charge port in the hydraulic head, fuel is supplied to the pumping plungers.

c The amount of fuel delivered is dependent on the position of the metering valve and its angular groove.

d If the groove has a large part of the supply drilling from the transfer pump covered, a small amount of fuel will be delivered to the pumping plungers during charging.

e If the groove covers very little of the supply drilling, a large amount of fuel will be delivered to the pumping plungers during charging.

For a more detailed explanation of this principle, see Chapter 21 on Roosa Master fuel systems.

C *Sleeve control* (Figure 15.7). This type fuel control has been used by American Bosch for many years, first in the P.S.B. pump and recently in the 100 Series pump. It is called sleeve control because a sliding sleeve is used to control fuel bleed-off or port opening.

1 *Component parts and nomenclature.* The sleeve control system is composed of a movable sleeve and pumping plunger. The sleeve and the plunger are lapped together to make a matched set. Controlling the sleeve position is a control unit or control lever that in turn is connected to the governor.

2 *Fuel flow and operation.* Fuel is supplied to the cavity immediately above the pumping plunger under low pressure from the transfer pump. As the pumping plunger moves upward, it closes off the inlet ports (Figure 15.7,A). Fuel is then trapped in the chamber above the plunger (Figure 15.7,B) and forced down through the vertical drilling in the plunger. This vertical drilling (Figure 15.7,C) is connected to a horizontal drilling that is covered by the control sleeve (Figure 15.7,D). As plunger continues to move upward, the trapped fuel above it is forced out through the delivery valve and distributed to the correct cylinder. This delivery will take place as long as the horizontal hole in the

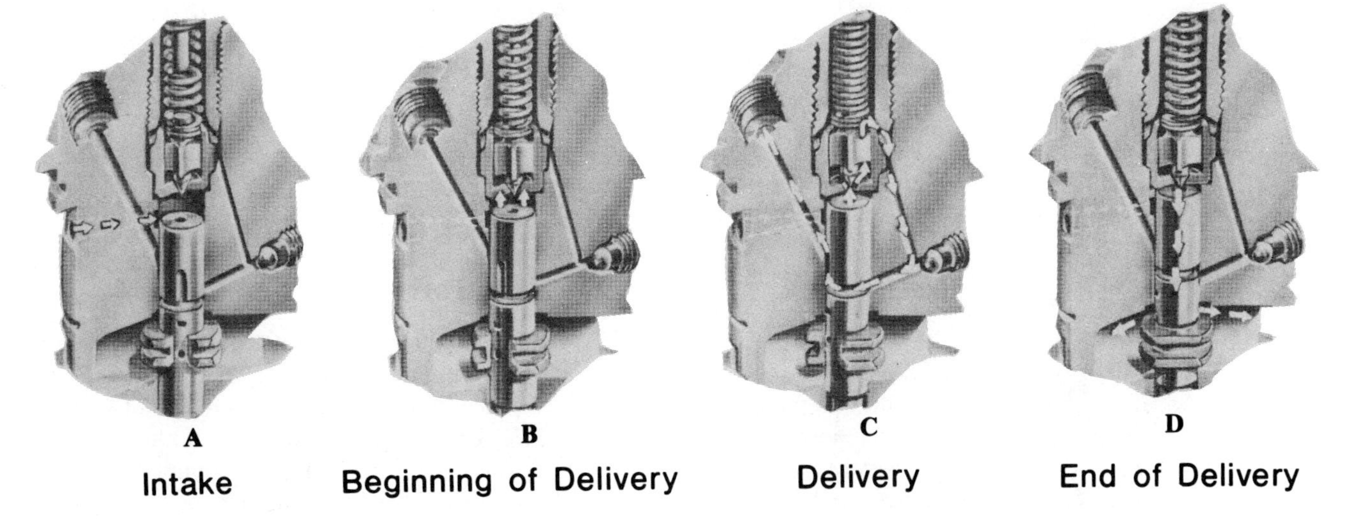

Figure 15.7 Sleeve control as used by American Bosch.

plunger is covered by the control sleeve. When the plunger has risen far enough to move the horizontal hole out of the control sleeve, the trapped fuel above the plunger rushes down the vertical drilling in the plunger and exits into the low pressure area of the hydraulic head. It can be seen then that movement of the control sleeve upward or downward by the control unit will control port opening and consequently fuel delivery.

For a more detailed description of this method of fuel control see Chapter 18 on American Bosch fuel systems.

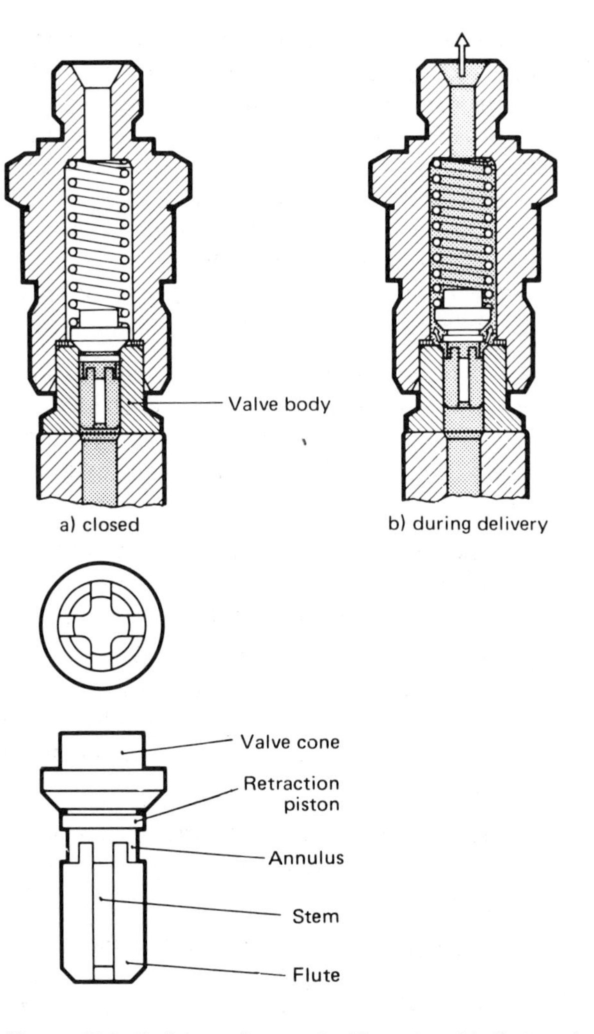

Figure 15.8 Delivery valve used with port and helix type fuel system (courtesy Robert Bosch Corporation).

IV. Delivery (Retraction) Valves

Another concept that must be understood during the study of fuel injection pumps is injection line retraction, which simply means retracting (or pulling back) a small amount of fuel in the injection line, which lowers the line pressure and allows the injection nozzle to close quickly, bringing injection to a rapid ending. As stated earlier, this is a fuel system requirement that must be met if the engine is to develop full power and run with clean exhaust.

A A typical delivery valve (Figure 15.8) used with a port and helix type of fuel system.

1 *Component parts and nomenclature*. The delivery valve assembly is composed of a valve, spring, and fitting (holder) into which the valve fits. This assembly is mounted in the fuel pump housing directly above the pumping plunger and barrel assembly.

2 *Operation*. As fuel is trapped above the plunger and the delivery valve is forced off its seat, fuel is allowed to flow around it, then through the delivery pipe to the nozzle.

a The valve will remain open until fuel flow stops, which occurs when the pumping element stops pumping fuel.

b The pumping element stops pumping fuel when the port is uncovered by the helix (as outlined in item III, A).

c As the fuel flow from the element is stopped by opening the port, the pressure that had been holding the valve open is released. The fuel above the valve and the spring pressure seat the valve.

d As the delivery valve seats, it moves downward and displaces a given amount of area directly above it.

e This area is immediately filled with fuel from within the line and line pressure is lowered.

f The lowered line pressure allows the injection nozzle valve to seat and stop fuel delivery to the cylinder.

g The delivery valve once seated maintains a certain amount of line pressure, so that the succeeding injection of fuel doesn't have to fill the fuel line with fuel. If the fuel line were not full of fuel, the timing would be retarded, causing the engine to operate poorly.

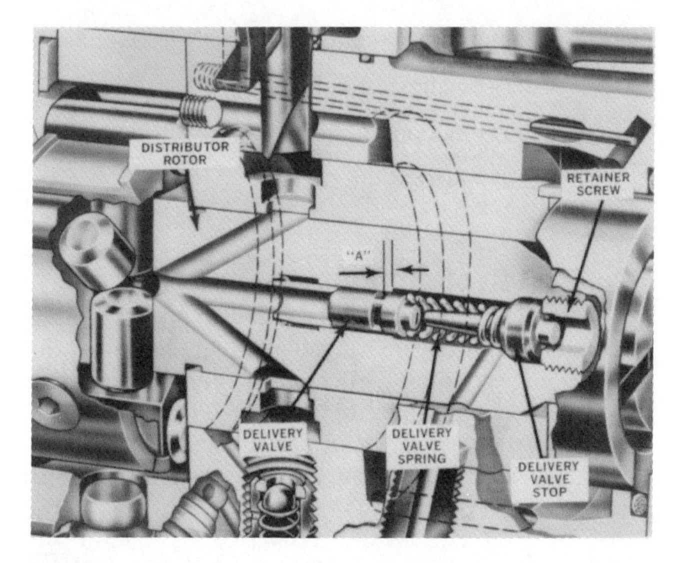

Figure 15.9 Sleeve type delivery valve. (Courtesy Stanadyne/Hartford)

h Delivery and retraction all happen very quickly during engine operation, taking place in approximately two milliseconds.

B A sleeve type of delivery valve (Figure 15.9) is used with some inlet metering distributor type pumps.

1 *Component parts and nomenclature.* This type of delivery valve is composed of:

a Piston type valve with holes cut in it (to close delivery passageway).

b Spring (to aid in closing the valve).

c Stop (to prevent damage to spring).

2 *Operation.* The delivery valve is positioned in the delivery passageway in the center of the rotor so that fuel leaving the pumping plungers must move it outward and then pass through it.

a As fuel continues to pass up the delivery passage, the valve is held in open position.

b As the rollers move off the cam lobes and fuel flow stops, the valve is returned to its seated position in the delivery passage by the fuel pressure and spring pressure above it.

c As the valve returns to its seat in the bore, fuel pressure in the injection line is reduced, allowing the nozzle valve to seat quickly.

The two delivery valve or retraction systems mentioned are commonly used in many modern diesel pumps and are very similar to other versions being used.

V. Diesel Fuel Properties and Selection

Diesel fuel is the source of energy that runs the engine in addition to lubricating the fuel injection system. Diesel fuel is made from crude oil and is made up of hydrocarbons. Diesel fuel is generally supplied in three grades, 1D, 2D, and 4D. Numbers 1D and 2D are used in high speed diesel engines, while 4D is used in large, low speed engines.

A *Diesel fuel properties.*

The properties or characteristics that constitute a good diesel fuel are as follows:

1. *Volatility.* Volatility is the property that affects burning. A very volatile fuel will make the engine start better but cause an inefficient engine, since a very volatile fuel has a low heat value. The volatility must be a compromise between starting, burning, and total heat energy.

2 *Pour point.* The lowest temperature at which the fuel will flow is known as the pour point.

3 *Cloud point.* The lowest temperature to which the fuel can be subjected before it begins to cloud or form paraffin crystals. This is closely related to the pour point and becomes very important if you are attempting to operate the engine during cold weather. The paraffin or wax in the fuel will clog the fuel filters and cause an engine shutdown.

4 *Viscosity.* Generally known as the resistance to flow. All parts of the diesel fuel system are designed to operate correctly with a certain viscosity. If viscosity is too low or too high, the injection system will not operate correctly.

5 *Flash point.* The flash point of diesel fuel is the temperature to which the fuel must be heated before it will give off sufficient vapor to ignite.

6 *Sulfur.* All diesel fuels contain a certain amount of sulfur, but sulfur in excessive amounts may be detrimental to the engine. Sulfur does not burn except at extremely high temperature, so in many cases it simply accumulates in the combustion chamber. High sulfur content fuel tends to increase engine wear and contaminates the engine lube oil.

7 *Carbon residue*. The carbon residue of a fuel oil to a degree determines how clean the engine and lube oil stay during operation. Carbon residue, as the name implies, is the residue left over after combustion occurs.

8 *Gravity*. Gravity is used to determine the heat value of the fuel. Generally heavier fuels contain more heat per gallon. Gravity has very little to do with engine performance.

9 *Water and other impurities*. All diesel fuels will contain impurities to some degree. No matter how clean you keep the fuel, a certain percentage of rust, dirt, metal, neoprene, and paint will work their way into the fuel during handling.

B *Fuel selection*.

Fuel selection must be based on the requirements provided by the manufacturer with particular emphasis on grades and the cetane number.

1 *Grades*. As stated earlier, the common grades for high speed diesel engines are 1D and 2D.

a 1D is usually selected for operation in cold weather because its viscosity and pour point are lower than that of 2D.

b 2D is usually used during warm weather in most diesels because it has more BTUs per gallon. In addition, its greater viscosity provides better lubrication for the fuel injection system.

2 *Cetane number*. The cetane number is the ignition quality (how easily the fuel will ignite) of diesel fuel and is generally in the 33 to 64 cetane rating range. Cetane and octane, which is used to rate gasoline, are not alike, since octane represents the ability of gasoline to resist rapid burning.

a When selecting a fuel for a particular engine, check the manufacturer's recommendation on cetane number.

Selection of fuel for a diesel engine is usually done by the operator or owner, sometimes, with the mechanic's advise. Burning the fuel to provide energy is only half the battle. It must then be pumped by an expensive injection system that requires lubrication or it will be ruined. Efficiency is of prime importance also. A fuel that meets all the requirements but has a low BTU content per gallon should not be used. The selection process then is a matter of the:

1 Engine requirements.

2 Fuel specifications.

3 Operating conditions.

4 Availability of the fuel.

VI. Fuel System Maintenance (General)

Many of the component parts within an injection pump are finely lapped (fitted) and require clean fuel for operation. As stated earlier, most fuel systems have several filters and water traps included in the system to prevent dirt, water, or other contaminants from harming the system. Maintenance of the system then must include:

1 Using clean fuel and making sure no dirt or dust is allowed to enter the system during filling.

2 Changing filters and draining water traps at regular intervals.

3 Periodically checking all lines, steel and rubber, for worn or frayed spots. (Broken lines would allow leakage or suction leaks, causing air to enter the system.)

4 Repairing any leaks immediately to prevent dirt and dust accumulation or eventual system malfunction or failure.

Generally speaking, all diesel fuel systems are easily maintained until complete failure (engine stoppage or improper engine operation) occurs. When this happens, the fuel pump or injection nozzles should be removed from the engine for further service.

VII. Future Systems

Present fuel systems, although considered by many to be a technological marvel, have shown only the beginning. Systems of tomorrow will have to incorporate many changes to meet the demands of an ever-changing diesel engine. The engine changes will be brought about by the demand for cleaner running (pollution-free) and more efficient engines. This means high nozzle opening pressures, 4000 to 10,000 psi (280 to 700 kg/cm^2), better governors with much more emphasis on electronic governors for better speed sensing, and more and better automatic advance devices that will tailor timing to the engine's operating conditions. All these improvements and design changes are on the drawing boards now for use in the foreseeable future.

SUMMARY

The information provided in this chapter was intended to give you an overview of fuel injection system types. In addition to knowing the difference between the common fuel systems in use today, such as in-line pump systems or the P.T. system as used by Cummins, you should have a better understanding of how fuel systems measure and distribute fuel and the major components in a common fuel system.

In addition to the fuel system, a detailed description of diesel fuel requirements should give you enough background to aid engine operators in fuel selection. Included also is a quick look at what you are going to be working on in the future if you choose to work on fuel injection systems. A thorough understanding of the concepts and principles will enable you to grasp much more easily the information in the chapters on specific fuel systems.

REVIEW QUESTIONS

1. List and explain four requirements of a diesel fuel system.
2. What type of injection nozzles do direct injection fuel systems usually use?
3. Explain the operation of a unit injector type system.
4. Explain the operation of a pump and nozzle system.
5. List the two most common types of fuel control used in today's fuel systems.
6. What is the function of a delivery valve?
7. The cetane number of a diesel fuel is:

   a Similar to the octane rating of gasoline.

   b Ease with which the fuel will ignite.

   c Indicates the number of BTUs in a gallon of fuel.

   d None of the above.
8. Sulfur in the diesel fuel will:

   a Cause excessive piston and ring wear.

   b Cause no harm because it is burned and passed out the exhaust pipe.

   c Act as ignition improver.

   d Collect on valves, piston, and combustion chambers but cause no damage.
9. The specific gravity of diesel fuel is important because it:

   a Indicates how much water and sediment are in the fuel.

   b Indicates the heat value of the fuel oil.

   c Has a great effect on the diesel fuel's performance in the engine.

   d None of the above.
10. Which of the following is not used in the selection of diesel fuel?

    a The engine requirements.

    b The fuel specifications.

    c The brand of fuel used.

    d The operating conditions.

16

Governors

Since the speed of the engine is directly related to its power, speed must be maintained during operation. This is the job of the governor, which is considered the brain of the engine. The diesel engine governor controls the engine speed under various load conditions by changing the amount of fuel delivered to the engine cylinders. Governors, like engines, may be of many types and designs, but all will be designed to accomplish engine speed control under low idle, high speed, and full load conditions.

OBJECTIVES

Upon completion of this chapter the student will be able to:

1 List two different types of governors.

2 List and explain two different classifications of governors.

3 List two engine operating problems that can be caused by worn governors.

4 Explain the governor's function on a diesel engine.

5 List four reasons why a governor is needed on a diesel engine.

6 List four items that must be carefully checked during governor overhaul.

7 List two items that may cause excessive governor hunting.

8 Give one application of a variable speed governor and why it is used in that application.

GENERAL INFORMATION

Probably one of the least understood components of the diesel engine is the governor. The governor may be an integral part of the injection pump or may be a separate unit mounted by itself. The governor's main function is engine control—under all conditions at all times. If, for example, a truck engine did not have a governor, the operator would have to control the engine speed at idle manually, since the engine would not idle unattended. On the other end of the speed range, the top speed of the engine would have to be limited by the operator or the engine would overspeed and could cause engine damage. It is obvious that a governor on a truck engine is a much needed component. Without it, the operator would have difficulty in controlling the engine properly.

Consider another example: A farm tractor or heavy equipment operator would have considerable difficulty in maintaining the correct engine speed and power under all the different load conditions that he would encounter in the operation of the machine. Because most farm and industrial equipment engines are equipped with a variable speed governor, the engine speed may be adjusted by the operator to fit the conditions. This speed may be anywhere in the speed range from idle to high speed; then as the machine is operated, it may encounter a change in load many times a minute, causing the governor to change the fuel delivery accordingly. This fuel delivery change, in turn, maintains steady engine speed with sufficient power to pull the load. The operator could not possibly anticipate the rapid load change encountered by the engine so that he could maintain a steady engine speed as well as sufficient power to pull the load.

To be sensitive to load and speed changes encountered by the engine, the governor must be connected to the engine either as an integral part of the injection pump or a separate unit driven by a belt or gear arrangement.

Since the governor is being run or turned by the engine, any change in speed is immediately sensed, causing the governor to react. This is the vital link or sensing line that provides the governor with information needed to control the fuel delivery quantity.

Obviously the governor must be kept in top operating condition if the engine is going to develop its full power. The governor then becomes a component that must be given a careful inspection during fuel injection pump overhaul.

I. Basic Governor Components

Regardless of governor type, most governors operate with many of the same basic components. These components should be understood before further governor study can take place.

The basic mechanical governor (Figure 16.1) is a speed-sensing device that employs centrifugal force and spring tension to govern the engine speed. It is made up of the following parts:

A *Governor drive shaft.* The governor drive shaft provides a place to mount other component parts and transmits power from the engine to the governor.

B *Flyweights.* Flyweights are usually made from metal in rectangular, round, or square configurations. These weights may be mounted to the shaft in a carrier (Figure 16.2) or by a pin-and-shoulder arrangement on the governor drive shaft. As speed or governor spring tension changes, the weights pivot outward or inward.

C *Governor spring.* The governor spring provides tension on the governor weights via the control sleeve or thrust sleeve. In variable speed governors this spring is adjusted (tightened or loosened) when the operator moves the speed control lever. In constant speed governors the governor spring tension is changed by adding or subtracting shims placed over the governor spring pack.

D *Thrust sleeve.* The governor thrust sleeve is the component that transfers the outward or inward movement of the flyweights to the fuel control rod or linkage.

E *Drive gear.* Many diesel engine governors use a drive gear to drive the governor; others need no special drive gear because it is an integral part of the fuel injection pump drive shaft.

F *Bearings and bushings.* Throughout the governor there will be bearings and bushings that are utilized to eliminate friction and promote easy, prompt movement to any speed change by the governor.

II. Governor Terms

The following terms should be understood before further study of the governor takes place.

A *Percent of regulation.* The amount of speed change expressed in a percentage from high idle no load to high idle full load. Expressed another way, it would be the percentage of high idle speed lost when engine is loaded at its rated load.

B *Speed droop.* The amount of speed droop that must occur before the governor will sense a speed change and cause a fuel change setting. Speed droop can be calculated in the following manner:

$$\text{Speed droop in percent} = \frac{\text{high idle} - \text{rated load speed}}{\text{rated load speed}} \times 100$$

C *Hunting.* Governor hunting is caused by the governor overcorrecting; for example, if the engine

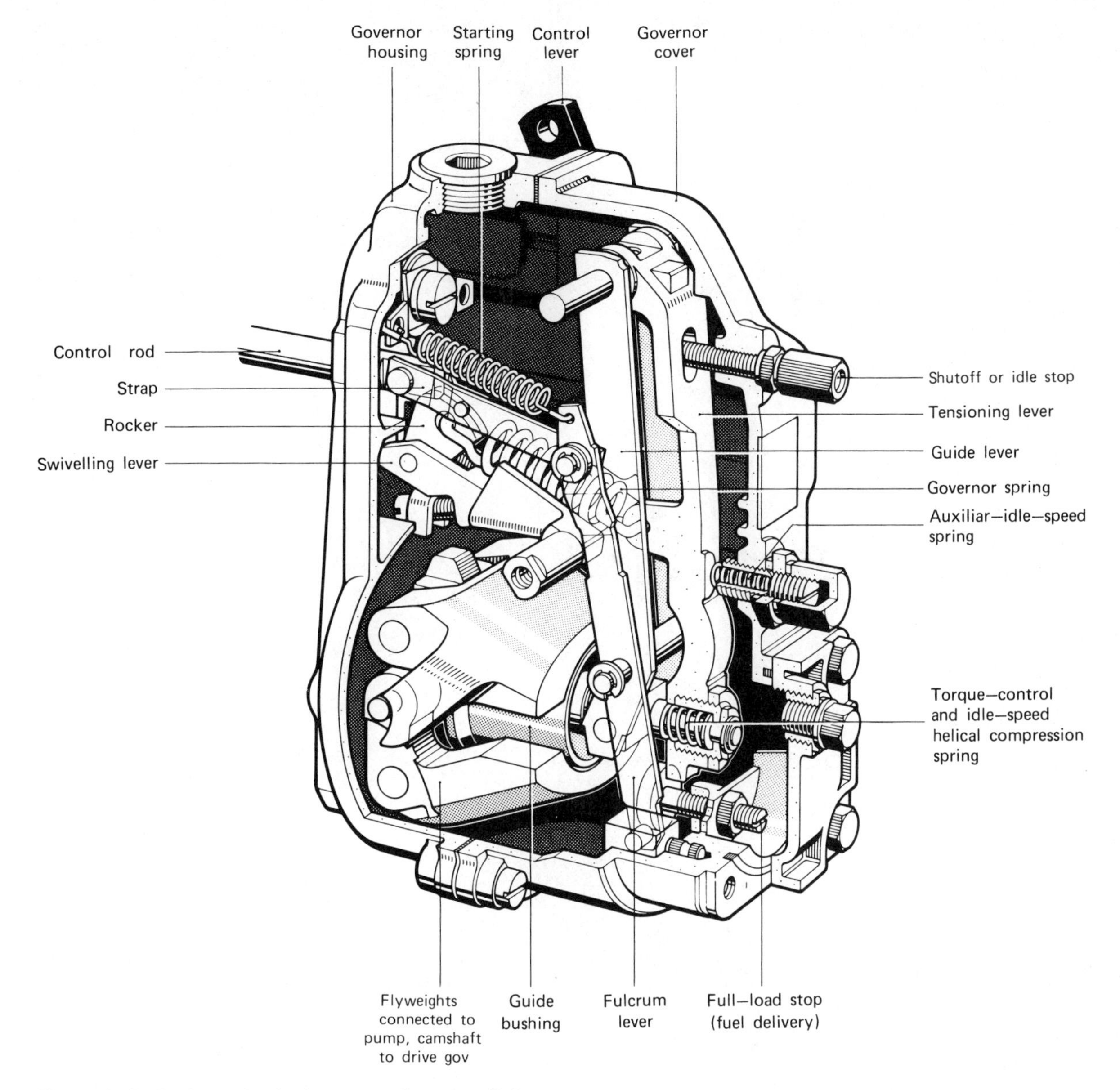

Figure 16.1 Basic mechanical governor (courtesy Robert Bosch Corporation).

speed attempts to exceed the high idle setting and the governor corrects by decreasing the fuel in excess of what was needed to slow down the engine, the engine will then run too slow. The governor again attempts to correct the engine speed by adding more fuel than is required to speed it up. It then overspeeds, causing the governor to once again decrease fuel. This cycle may continue indefinitely, causing a very unstable engine.

NOTE This condition generally occurs only at the times when the engine is under no load.

D *Low idle.* The governor or throttle setting that allows the engine to run at idle (usually about 600 to 700 rpm).

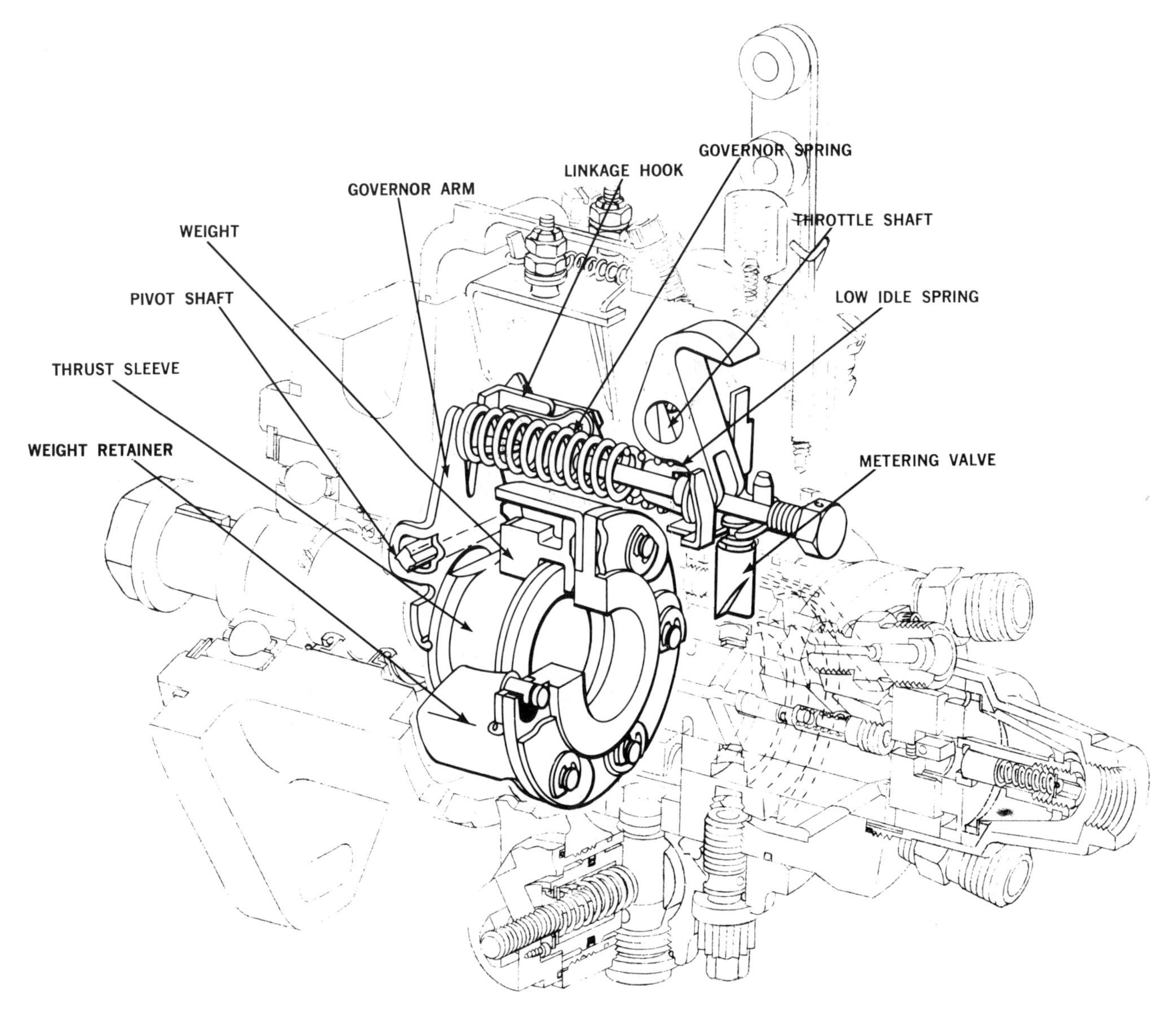

Figure 16.2 Weights mounted to the shaft in a carrier (courtesy Stanadyne/Hartford).

E *High idle* (W.O.T.—wide open throttle). The throttle setting that allows the engine to run at its recommended high speed. (Modern engines run high idle anywhere from 2000 to 3500 rpm). This speed should not be exceeded or engine damage may result.

F *Stability*. The ability of the governor to maintain a given or set engine speed without fluctuations (changes in engine rpm). Hunting will result if a governor does not have stability.

G *Sensitivity*. A good governor must be sensitive to load and/or speed changes, but must not overcorrect, which may cause hunting.

H *Throttle or governor control*. A device with which the speed of the engine is changed by changing the governor spring tension.

I *Under run*. This term refers to the governor's inability to maintain low idle speed as the engine speed is dropped quickly from a high speed. The governor allows the engine to fall below idle speed, and some cases allow it to stop.

J *Over run*. This term refers to the inability of the governor to maintain high idle speed as the engine is accelerated from idle. Engine speed may be exceeded by as much as 200 to 500 rpm.

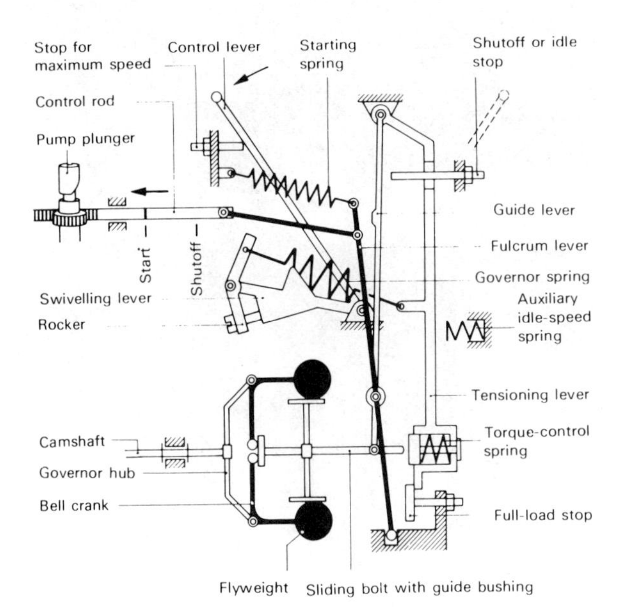

Figure 16.3 Governor weights in full fuel position (courtesy Robert Bosch Corporation).

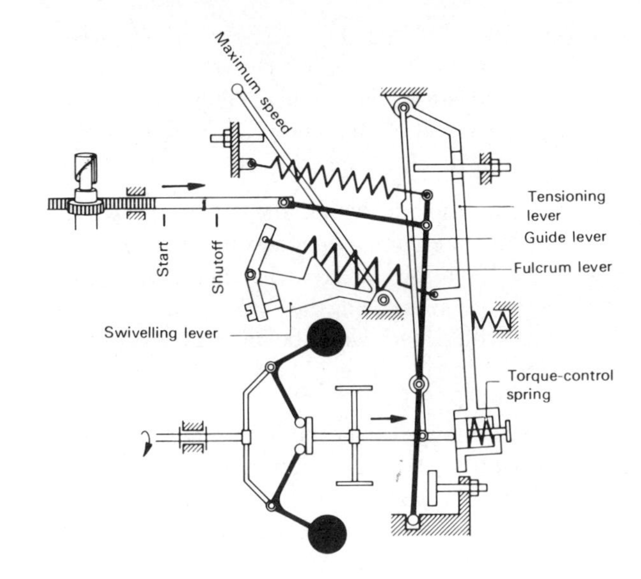

Figure 16.4 Governor weights tipping outward (courtesy Robert Bosch Corporation).

III. Basic Governor Operation (Variable Speed Governor)

Since governor overhaul and adjustment are an important part of diesel engine and/or fuel injection pump overhaul, the mechanic must have a good understanding of how a governor functions. This understanding will help him make the correct decisions during governor overhaul and adjustment.

The following governor operation pertains to a variable speed, mechanical governor.

A With the governor at rest, that is, flyweights all the way in (Figure 16.3), the fuel control rod or linkage will be in the full fuel position. It is held in this position by the governor spring. This is the position of the governor during engine start.

B As the engine is started, the rotating shaft carries the weights with it, creating centrifugal force that causes the weights to tip outward (Figure 16.4) against the tension of the governor spring.

C This movement is transmitted through the thrust sleeve and linkage to the fuel control rack or metering valve. As the flyweights move outward, the fuel delivery quantity is decreased from the overfuel or start position. This outward movement will continue until the centrifugal force generated by the rotating weights is equal to the governor spring tension.

D At this point of balance or equilibrium, the movement of the fuel control rod will stop with fuel delivery and engine speed stabilized.

E The engine speed and governor flyweight position will stay the same until a load is applied to the engine, slowing it down. This speed change alters the governor weight centrifugal force, which allows the governor spring to force the governor weights inward. This causes the thrust sleeve and control rack or metering valve to move, changing the amount of fuel delivered to the engine.

F This movement of the fuel control supplies more fuel to the engine, allowing it to develop enough power to maintain speed. Since the governor spring has moved the governor weights inward, it has lost some of its tension and once again an equilibrium is reached between the governor weight force and the governor spring. It should be noted that this speed is lower than the original high idle. This difference in speed is the result of governor droop (the difference between governor steady speed no load and steady speed rated load).

G Engine speed will be stabilized at the speed indicated in step E until the load or governor spring tension changes.

H If the governor spring tension is changed by the operator, as in the case of a variable speed governor, the rotating flyweight assembly would move further outward until a balance between the weight assembly and spring is reached. This movement would decrease the fuel being delivered to the engine with a resulting speed decrease.

I Another condition in which the governor would react to prevent engine damage would be when the load is quickly removed from the engine. If a governor were not employed, the fuel required to run the engine with a load would be far in excess of the fuel required to run the engine without load.

All governors work on the principles just explained, although hydraulic, electric, and pneumatic governors may not use flyweights for sensing the engine speed. Each variation or type will be explained in more detail later in this chapter.

IV. Governor Types

A *Mechanical.* A mechanical governor is one in which the power supplied by the governor to move the fuel control is provided by flyweights. Generally mounted on a gear, flyweights can pivot in or out, depending on the centrifugal force. Connected or pushing against the weights will be a thrust sleeve or washer. This sleeve is connected to the fuel control mechanism by linkage. The weights sense engine speed change as centrifugal force increases or decreases.

B *Hydraulic (servo-type).* A hydraulic governor (Figure 16.5) is called hydraulic because the force used to move the fuel control mechanism is supplied by a hydraulic serve piston instead of flyweights. Flyweights are generally used in this governor to move the control valve, which, in turn, directs oil under pressure to move the servo piston.

C *Hydraulic (nonservo).* Hydraulic governors that do not use servos (Figure 16.6) usually control fuel flow, either to the pumping element, such as in the DPA pump by CAV, or control the amount of pressurized fuel delivered to the nozzles by allowing a given amount to be spilled off through a helix and port arrangement, such as used in the EP/VA Robert Bosch rotary pump.

D *Pneumatic.* Pneumatic governors (Figure 16.7) use a vacuum (a difference in air pressure) to control engine speed. The governor has two main parts, a venturi in the intake manifold and a diaphragm connected to the injection pump control rack. Mounted in the intake manifold inlet, the

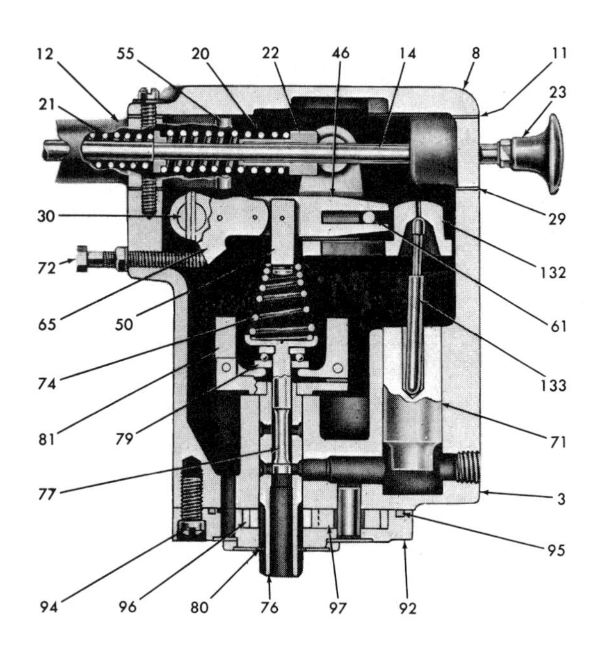

3. Housing--Governor
8. Cover--Governor
11. Gasket--Cover
12. Subcap
14. Rod--Fuel
20. Spring--Shut-Down
21. Spring--Fuel Rod
22. Collar--Fuel Rod
23. Knob--Fuel Rod
29. Gasket--Housing to Subcap
30. Shaft--Speed Adjusting
46. Lever--Floating
50. Fork--Spring
55. Pin--Stop
61. Bracket--Droop Adjusting
65. Lever--Speed Adjusting
71. Piston--Power
72. Screw--Maximum Speed Adjusting
74. Spring--Speeder
76. Ball Head Assy.
77. Plunger--Pilot Valve
79. Bearing--Plunger
80. Lock Ring
81. Flyweight
92. Base--Governor
94. Screw--Base to Housing
95. Ring--Housing to Base Seal
96. Gear--Oil Pump Drive
97. Gear--Oil Pump Driven
132. Lever--Terminal
133. Pin--Terminal Lever to Piston

Figure 16.5 Hydraulic governor (servo type) (courtesy Detroit Diesel Allison, Division of General Motors Corporation).

venturi contains a butterfly connected to the throttle. As the position of the butterfly is changed by the operator, the vacuum in the intake manifold is changed. A line is connected from the intake manifold to the diaphragm unit to supply it with vacuum. Retarding the diaphragm unit's movement is a strong spring, which will position the fuel control rack in the full fuel position when no vacuum exists. Depending then on the butterfly position and the speed of the engine, varying amounts of vacuum will be supplied to the diaphragm that will control the rack position and, subsequently,

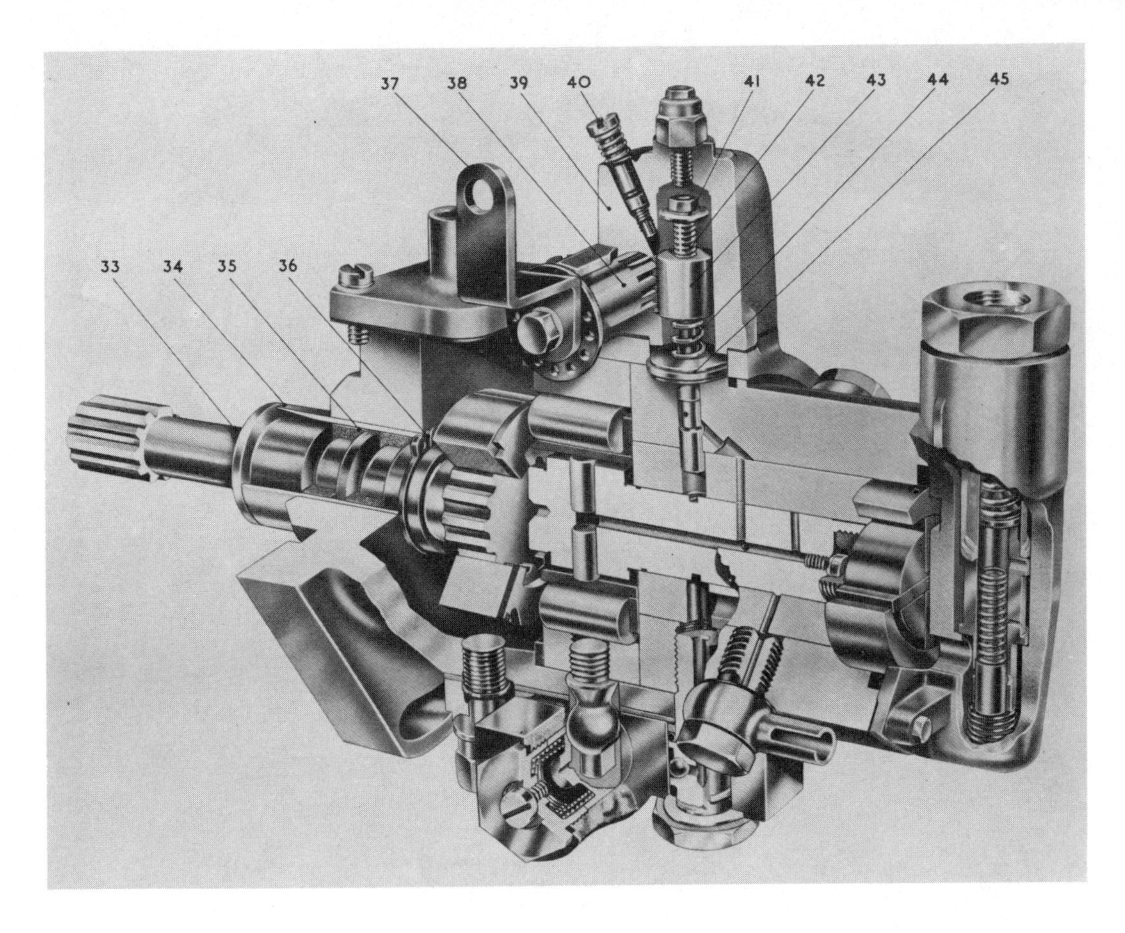

33 — Splined drive shaft
34 — Pilot tube
35 — Seals
36 — Snap ring
37 — Control lever
38 — Pinion shaft
39 — Housing
40 — Idiling stop screw
41 — Washer
42 — Idiling spring
43 — Rack
44 — Governor spring
45 — Damping washers

Figure 16.6 Hydraulic governor (nonservo type) (courtesy C.A.V. Simms).

engine speed. If, because of a load, the engine speed changes, air flow through the venturi slows down, decreasing the amount of vacuum supplied to the diaphragm. The diaphragm spring then takes over and moves the control rack toward the full fuel position.

E *Electric.* Electric governors are not generally found on diesel engines used in mobile equipment, since constant engine speed is not a requirement. Used on generator sets, electric governors maintain engine speed, within a few rpm's of a set speed. This speed must be maintained in most cases to maintain the frequency output of the generator that the engine is driving. An electric governor generally uses some form of electronic sensing, such as a frequency measurement, to sense the speed change. As speed change information is sensed, it is relayed to a servo motor or solenoid that changes the throttle position. Electric governors have very fast response with almost no droop. Generally no speed change is noticed from load to no load.

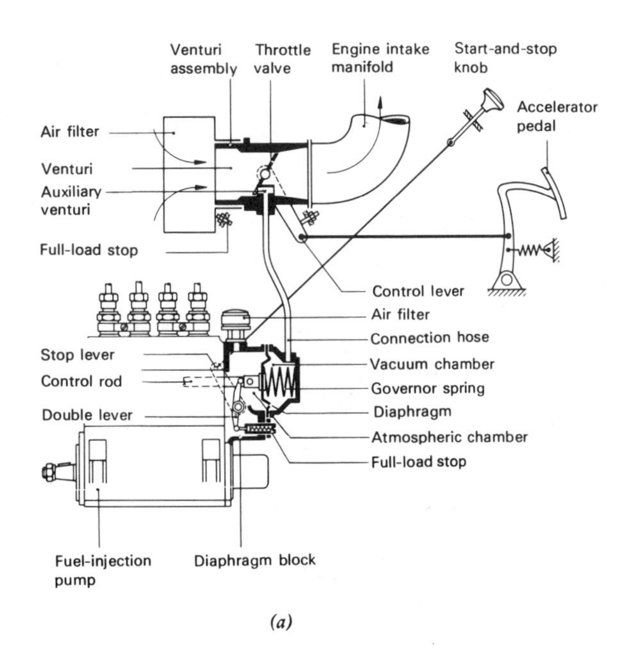

(a)

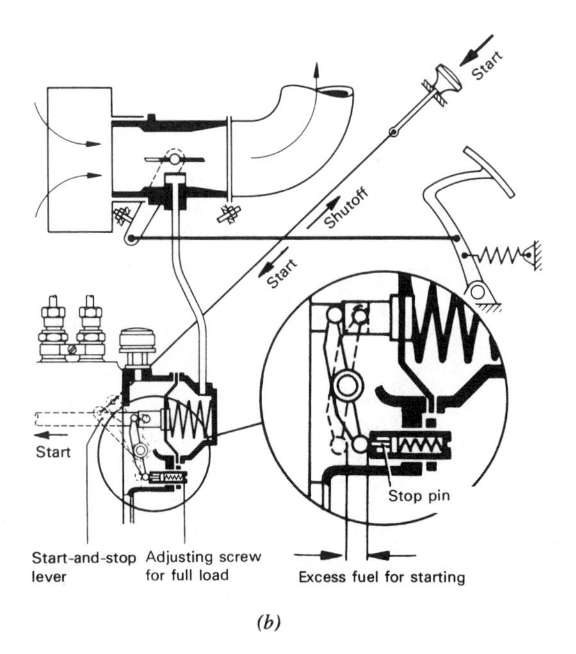

(b)

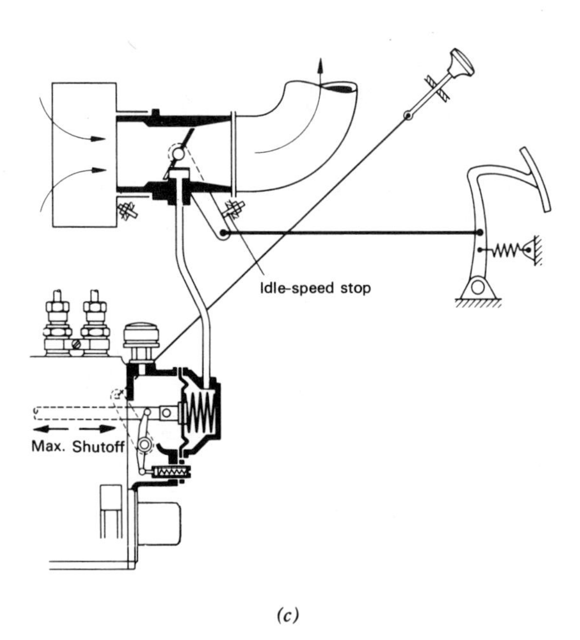

(c)

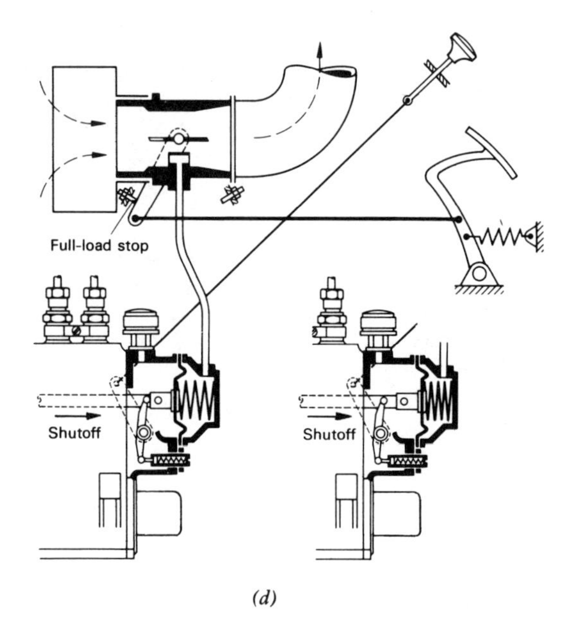

(d)

Figure 16.7 Pneumatic governor (courtesy Robert Bosch Corporation). (*a*) Schematic of pneumatic governor. (*b*) Diaphragm block in pneumatic governor with built-in-control rod stops for full load and for the excess fuel for starting. (*c*) Idle position. (*d*) Full load and high idle.

V. Governor Classification

Regardless of the type of governor used by the engine manufacturer, mechanical, hydraulic, or pneumatic, the governor will fall into one of the following categories:

A *Variable speed.* A variable speed governor is designed so that it can be adjusted by the engine operator to control engine speed within a given range at speeds from just above idle to full speed (high idle). Generally, variable speed governors are found on industrial engines and farm tractors. The major advantage for these applications is the ability to control engine speed at any speed selected by the operator between low idle and high idle.

B *Constant speed.* Constant speed governors are used on engines used to drive generators. Constant speed governors maintain the engine speed at a constant rpm.

C *Limiting speed.* Limiting speed governors are generally used on diesel truck and passenger car engines; they control engine low idle and high speed (high idle). The in-between speeds on this engine are controlled safely by the operator. The main function of this governor is to prevent the engine from stalling at low idle (speed) and from exceeding a safe top high idle.

All kinds of governors are classified according to regulation of speed. An example would be: A mechanical governor of the variable speed type could be used either on an industrial engine or a farm tractor. It would control engine speed at any speed between low idle and high idle, depending on the position of the speed control.

VI. Governor Troubleshooting

Governor troubleshooting, in most cases, must be done with the governor installed on the engine. Even though the governor is installed in the injection pump and the governor can be tested on a test stand, engine operating conditions cannot be duplicated exactly so that the governor cannot be fully tested on the test stand. All injection pump manufacturers give governor settings for the test stand, although they suggest that engine speeds be checked after the pump and governor are installed.

NOTE *Do not* attempt to adjust high or low idle or correct any other governor malfunctions unless the fuel filter or filters have been changed. Fuel starvation, resulting from a clogged or restricted filter, may cause problems similar to governor problems.

Some common problems that may be encountered when working on a diesel engine governor are:

A *Engine under run at low idle.* This may be caused by:

1. Worn governor parts (weights, pivot pins, linkage).
2. Sticking metering valve (inlet metering pump).
3. Incorrect idle speed adjustment.
4. Binding governor linkage.
5. Incorrect linkage adjustment.
6. Sticking or binding rack (in-line pumps).

B *Engine high idle not correct.* In most cases engine high idle can be changed by adjustment of the high idle adjusting screw. If not, the problem might be:

1. Defective governor spring.
2. Worn governor parts (weights, pivot pins, linkage).
3. Incorrect governor linkage adjustment.

C *Engine does not develop full power under load.* This problem is one that mechanics are confronted with very often. Many times repairs and adjustments are made to the injection pump or other engine components when in reality the problem was located in the governor. Engines are continually being overfueled by adjusting the injection pump to correct for a low horsepower problem when the governor is at fault. Admittedly, this condition requires some special skill in troubleshooting and testing that a beginning mechanic may not have. The following outline is intended as a guide in checking an engine for a low horsepower complaint:

1. The governor should be suspected when all other components of the engine have been checked and have been found to be in good working order.
2. The best way to test a governor for a low power complaint is with a dynamometer.
3. Make sure the high idle setting is correct before loading the engine.
4. Load the engine until rated rpm is reached. Engine horsepower should be close to specifications.
5. If horsepower is low and the engine's condition is good, the governor may be at fault.
6. Apply a load to the engine quickly; the governor should respond with no lag or hesitation. If not, the governor is probably faulty.
7. Replace the governor spring if it is easily accessible. (In many cases the spring loses its stretchability.)
8. If, after replacement of the governor spring, the governor does not work properly, remove the governor and/or injection pump for repair.

D *Erratic speed changing, hunting and no stability.* This may be caused by:

1. Worn governor parts (weights, thrust bearings, and linkage).
2. Worn governor spring.
3. Binding linkage.
4. Binding control rack or metering valve.

Many times governor troubleshooting will involve testing and checking a governor with the above procedures. As you gain experience, it will become easier to determine where the problem exists.

VII. Governor Repair and Overhaul

As stated earlier in this chapter, governor repair is usually done when the injection pump is overhauled. If the governor is repaired without the injection pump, the following governor parts should be checked as outlined in the manufacturer's service manual:

A *Governor weights.* Governor weights should be checked closely at the pivot points, such as bushings.

B *Linkage.* All linkage contact points should be checked for wear, such as abrasion, burrs, and flat spots. All parts that show wear should be replaced or repaired.

C *Thrust washers, bearings, and sleeves.* All governor thrust parts should be closely checked for pitting and ridging.

D *Governor springs.* Governor springs are generally replaced during governor overhaul as a matter of routine. The governor springs are very important to the correct operation of the governor.

E *Bearings and bushings.* All bearings and bushings should be replaced during governor overhaul.

After all parts have been repaired or replaced, the governor should be reassembled and adjusted according to the manufacturer's service manual.

Run the engine and check for correct high idle and low idle.

SUMMARY

The material presented in this chapter will help you understand the operation of the basic mechanical governor and to recognize and do some preliminary troubleshooting. The governor is one of the most important components of the diesel engine when troubleshooting a diesel engine for a low horsepower complaint.

To effectively troubleshoot you must be familiar with the operating principles, terms, and types that have been explained in this chapter. Remember when overhauling the engine it is a good practice to overhaul and adjust the governor.

REVIEW QUESTIONS

1. List and explain the function of five governor component parts.
2. Name two types of governors that are used on modern day diesel engines.
3. Regardless of type, diesel engine governors will be classified according to use. List and explain two classifications.
4. Why will worn governor weights and pins cause a drop in engine horsepower?
5. Explain why a diesel engine requires a governor.
6. Name four items that must be carefully checked during governor overhaul. How should these be repaired?
7. Speed droop as applied to governors is:
   - **a** Only found in variable speed governors.
   - **b** Needed to insure proper governor response.
   - **c** The change in governor speed from high idle to rated load speed.
   - **d** Of very little concern since it has little effect on engine performance.
8. Variable speed governors are:
   - **a** Used only on engines with power generator sets.

b Used on engines in applications where engine speed must be adjusted to different operating speeds by the operator.

c Capable of holding the engine speed within one percent of high idle speed when load is applied.

d Always of the hydraulic type.

9 Which of the following is not a symptom of a worn governor?

a Low engine power.

b High idle over run.

c Low idle under run.

d A rough running engine.

10 A weak governor spring may cause all of the following except:

a Low engine power.

b Slow governor response.

c Excess engine power.

d Low engine top speed.

11 Erratic engine speed changes and severe hunting may be caused by all of the following except:

a Worn governor parts.

b Binding linkage.

c Worn metering valve.

d Improper high idle adjustment.

17

Injection Nozzles

Many diesel engines use injection nozzles similar to the ones shown in Figure 17.1. Because of the different sizes and shapes of engines, the nozzles must be made in many different configurations that allow them to be fitted within a very small area in the cylinder head. The injection nozzle has two purposes, injection and atomization, that occur simultaneously during engine operation when fuel is pumped by the injection pump through the nozzle.

OBJECTIVES

Upon completion of this chapter the student will be able to:

1 Demonstrate the procedure for testing an injection nozzle on a nozzle tester.

2 Demonstrate the procedure to be used in locating (isolating) a misfiring nozzle.

3 List in detail the proper steps to be used in cleaning and servicing an injection nozzle.

4 List two different types of nozzle valve assembly variations.

5 Select from a given list three engine problems that a clogged (plugged) nozzle will create.

6 Give two reasons why injection nozzles should be cleaned on a regular basis.

GENERAL INFORMATION

Unlike the gasoline engine, which utilizes an electrical spark to ignite a mixture of fuel and air, the diesel engine must rely solely upon the intense heat of compression, high turbulence in the combustion chamber, and extremely fine atomization (breaking up) of the fuel to obtain ignition.

The type of nozzle to be selected depends to a great degree on the type of combustion chamber used in the engine. As a general rule, indirect (precombustion) combustion chamber engines use pintle nozzles and direct injection engines use orifice or hole type nozzles. The selection of injection nozzles is critical as injection nozzles greatly affect the proper burning of the fuel. Engine manufacturers have placed added emphasis on nozzle design as reduced exhaust emissions (air pollution) become increasingly important to our society.

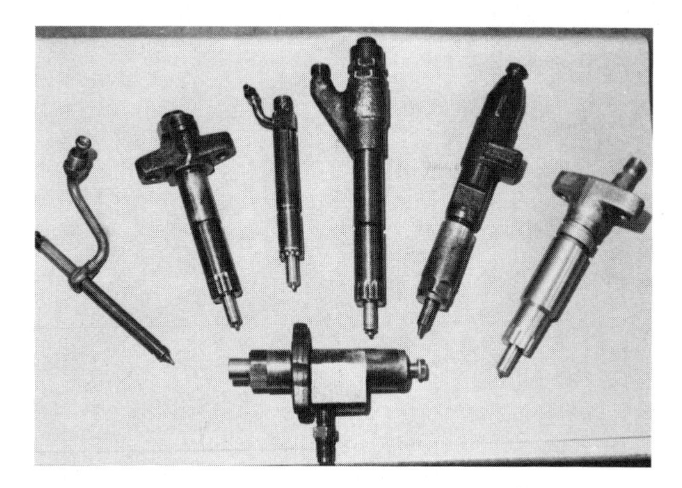

Figure 17.1 Several different sizes and types of injection nozzles.

Injection nozzles (Figure 17.2) are simple hydraulic valves operated by fuel pressure. Fuel flow generated by the injection pump enters the nozzle holder at the fuel inlet and proceeds down the feed channel and into the nozzle sac chamber. When the pressure of the fuel against the annular area of the needle valve exceeds the preset pressure of the pressure spring, the needle valve is raised from its seat. Then a metered amount of fuel is injected through the orifices on a hole type nozzle or by the pintle on a pintle type nozzle and into the combustion chamber.

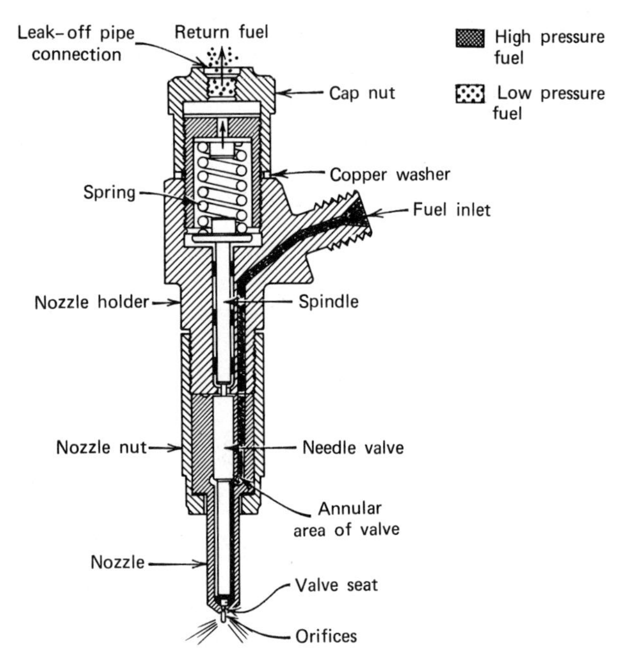

Figure 17.2 Operational schematic (courtesy C.A.V. Simms).

During operation a small amount of fuel will leak through the needle valve to help lubricate and cool the valve. This fuel accumulates in the pressure spring area and is returned to the supply tank by a fuel return line.

Early diesel engine injection nozzles commonly employed opening pressures of 800 to 1800 psi (6 to 127 kg/cm²) when used with precombustion chambers or energy cells. Modern diesel engines using a precombustion chamber utilize a nozzle opening pressure of 1800 to 2200 psi (127 to 155 kg/cm²). Early direct injection engines used injection nozzles with opening pressure in the 2400 to 2600 psi (169 to 182 kg/cm²) range, while modern direct injection engines use injection nozzles with opening pressures between 2600 and 4000 psi (182 and 280 kg/cm²). Predictions by fuel injection manufacturers are that within five years injection pressures in some engines may be as high as 10,000 psi (700 kg/cm²).

All of the pressure changes and nozzle variations have been made in an effort to make the diesel engine a clean running engine. This effort has proven to be effective. The exhaust from a diesel engine today contains less smoke and other harmful emissions than ever before.

I. Parts Identification and Function

A *Nozzle holder.* The nozzle holder (Figure 17.3) is the main structural part of the injection nozzle. It provides a means of holding the nozzle to the engine cylinder head; it routes fuel from the injection pump to the nozzle; and it sometimes contains passageways for leakoff fuel coming from the nozzle and going back to the fuel tank or injection pump. Excluding occasional breakage or thread damage due to poor handling, the nozzle holder is very reliable. Information listed directly on the holder includes:

1. Holder type number (varies with engine application).
2. Holder part number (manufacturer's part number).
3. Application part number (on some types).
4. Nozzle opening pressure (on some types).

B *Pressure spring.* The pressure spring determines the opening pressure of the nozzle valve. Tension of the pressure spring can be adjusted in most cases by an adjusting screw located above it, or by a shim pack.

C *Cap nut.* The cap nut provides a dust seal for the

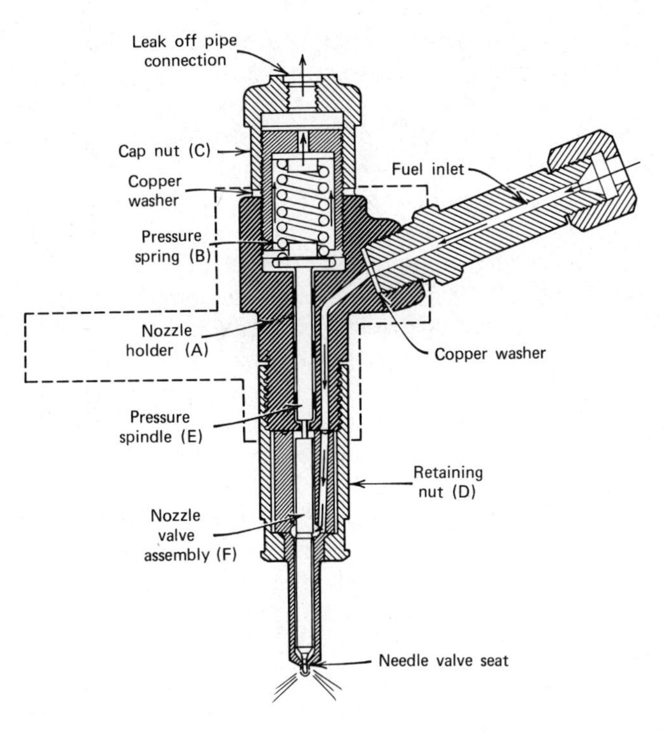

Figure 17.3 Nozzle component parts (courtesy C.A.V. Simms).

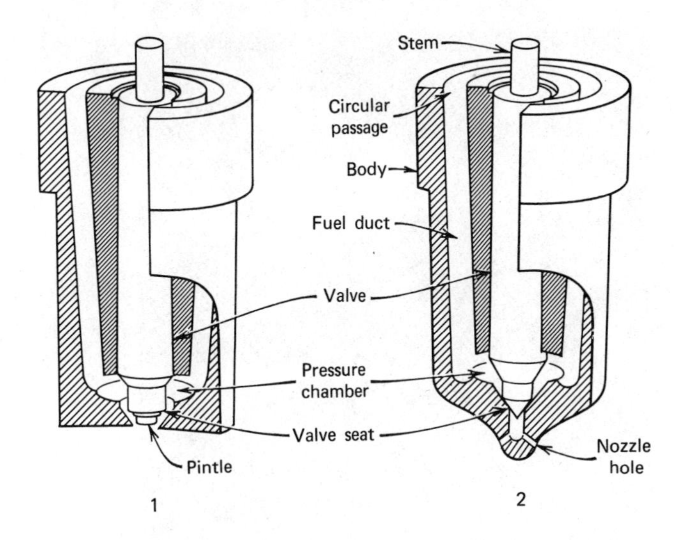

Figure 17.4 Sectional view of pintle and orifice type nozzles (courtesy C.A.V. Simms).

nozzle holder and usually incorporates a connection for leakoff fuel. Some nozzles using a shim pack to set nozzle opening pressure do not require a cap nut.

D *Retaining nut.* The retaining nut connects the nozzle body to the nozzle holder and also serves as a compression seal in the cylinder head.

E *Pressure spindle.* The pressure spindle is a metal rod that transfers the force of the pressure spring to the nozzle valve.

F *Nozzle valve assembly.* The nozzle is the heart of the injection nozzle assembly. The valve and body of the nozzle are lapped together and are not interchangeable. The valve has a special tapered seat that effectively seals off nozzle fuel pressure and does not allow any fuel to dribble into the combustion chamber. Two basic types of nozzles are employed (Figure 17.4). These are (1) pintle nozzles and (2) orifice nozzles or hole type.

1 *Pintle nozzles.*

The standard pintle nozzle is used with indirect combustion chamber engines. Operating pressures range from 1500 to 2100 psi (105 to 147 kg/cm²). Pintle nozzles have a conical-shaped spray, the angle of which is determined by the angle of the pintle that moves in and out of the nozzle orifice. A popular variation of the standard pintle nozzle is the throttling or delay nozzle (Figure 17.5), which uses a modified pintle to produce an initial pilot spray designed to enter the energy cell or precombustion chamber, initiating the combustion process. A variation of the throttling pintle nozzle is shown in Figure 17.6. This nozzle was devel-

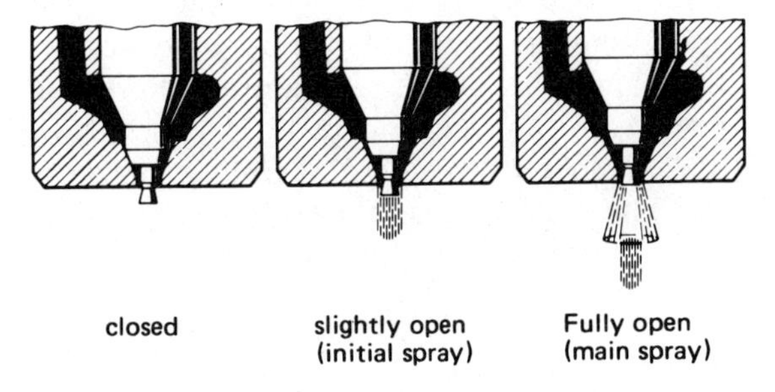

Figure 17.5 Throttling pintle nozzle (courtesy Robert Bosch Corporation).

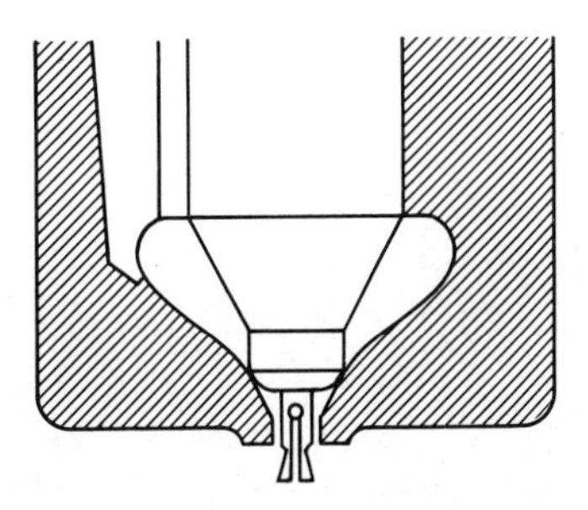

Figure 17.6 Robert Bosch nozzle for Mercedes Benz (courtesy Robert Bosch Corporation).

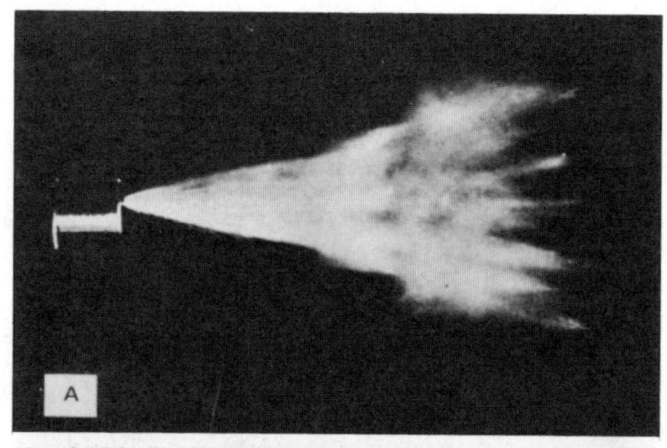

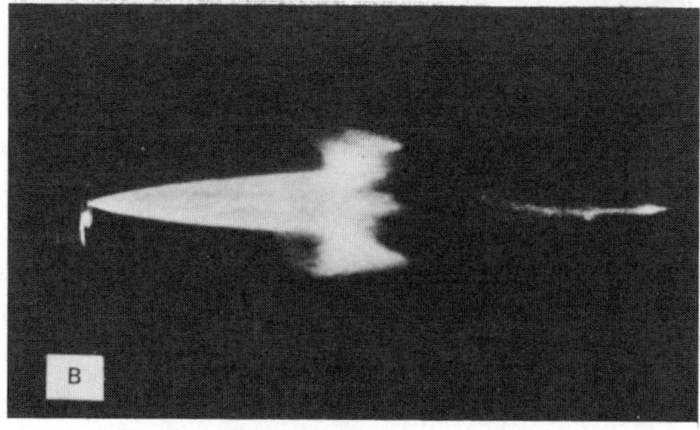

Figure 17.7 Spray patterns of (*a*) standard and (*b*) throttling pintle nozzles (courtesy Robert Bosch Corporation).

oped by Robert Bosch for use in the diesel passenger cars to reduce combustion noise at low idle. The small hole allows fuel to be injected earlier in the cycle, which gives a longer burning period and eliminates knock at low idle. Following the pilot spray is a conical-shaped spray similar to the standard pintle nozzle. Spray patterns of the standard and throttling nozzle are shown in Figure 17.7.

A second variation of the pintle nozzle is the Pintaux CAV nozzle (Figure 17.8). The Pintaux nozzle is identical in operation to the throttling pintle nozzle except for the auxiliary spray hole. This extra hole assists in the cold starting of the engine. At starting speed the pintle doesn't completely clear the pintle hole, and fuel is sprayed from the auxiliary hole to a localized hot spot within the cylinder where it ignites more quickly than normal. With the engine running, the pintle now completely clears the pintle hole and a major portion of the fuel will be discharged through it.

A less widely used variation of the pintle nozzle is the outward opening type (Figure 17.9). These nozzles, although actuated hydraulically like the inward opening types, differ in the operation of the pintle. As the name implies, the nozzle pintle is forced down and out of the nozzle and fuel is then sprayed around the angled face of the pintle. These nozzles, which are of the capsule type, are usually not serviceable. Pressure is preset and the capsule is replaced as a unit. Opening pressures range from 500 to 1800 psi (35 to 126 kg/cm^2).

2 *Orifice or hole type nozzles.*

Orifice or hole type nozzles (Figure 17.10) differ from pintle nozzles because the tip or pintle of the needle does not protrude through the nozzle body. The body has a number of

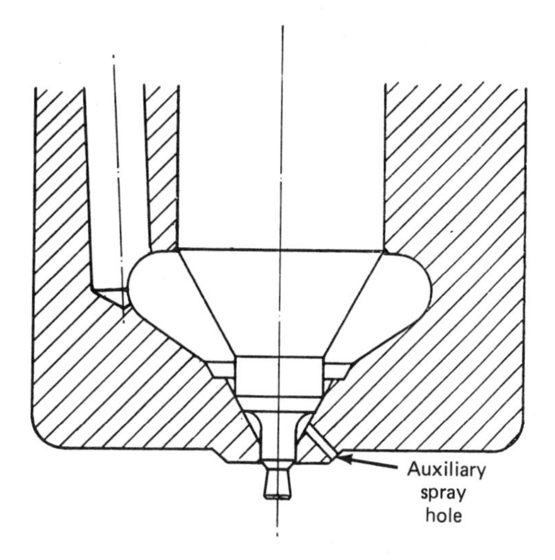

Figure 17.8 The Pintaux variation.

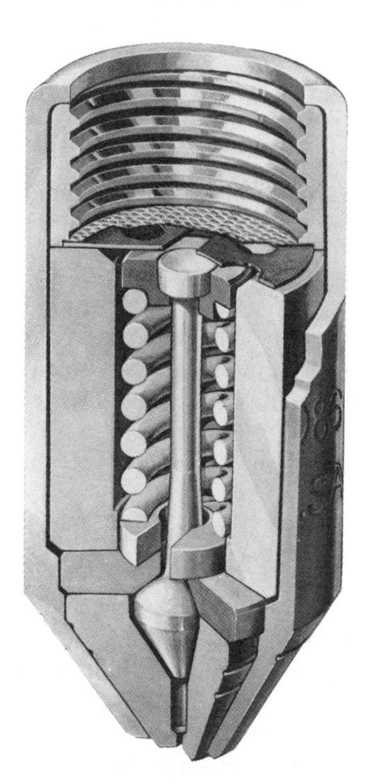

Figure 17.9 Caterpillar outward opening nozzle.

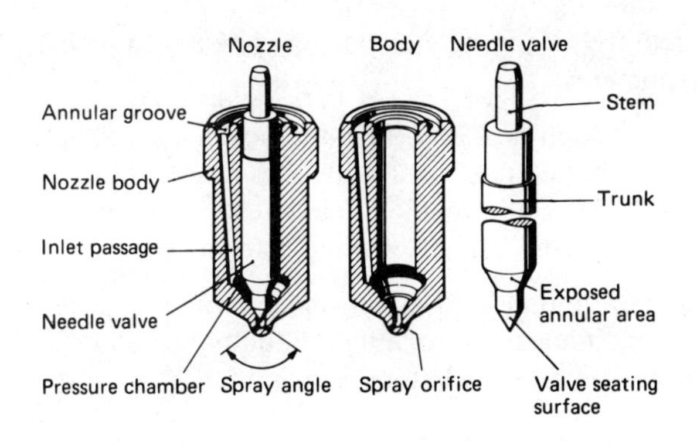

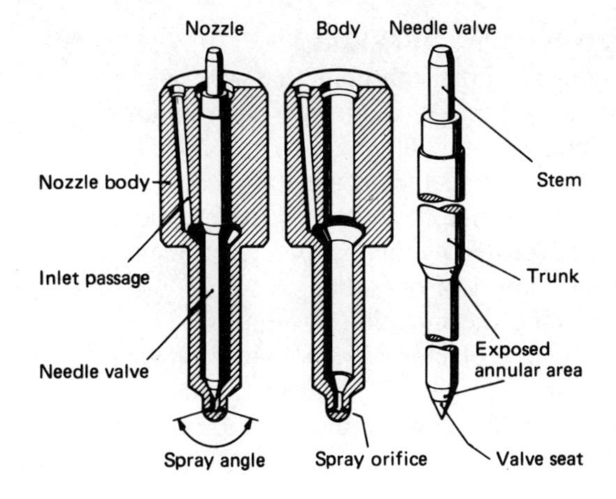

Figure 17.10 Standard orifice nozzle (left) and long stem variation (right) (courtesy Robert Bosch Corporation).

small orifices through which the fuel is sprayed and thereby atomized.

Because of the fine atomization brought about by the small holes and a higher opening pressure, 2000 to 4000 psi (140 to 280 kg/cm²), orifice nozzles are most often used on engines with direct injection. Anywhere from 2 to 12 holes may be used, varying in size from 0.006 to 0.60 in. (0.15 to 1.5 mm).

A nozzle code number, printed on the nozzle body, gives full information concerning the nozzle. For example, the nozzle may have the code DLL 145 S 318 stamped near the top.

D = The D identifies this item as an injection nozzle from Robert Bosch.

NOTE Other major nozzle manufacturers include American Bosch (AD), CAV (BD), and Nippondenso (ND).

LL = The two L's signify a hole type nozzle with long shank and long collar (ring groove marking).

145 = Spray angle in degrees of spray cone.

S = Size of nozzle body collar in millimeters.

318 = Variation number. This number varies with nozzle application. It should be used when ordering replacement nozzles.

The chart (Figure 17.11) explains each number and letter of all Robert Bosch nozzles.

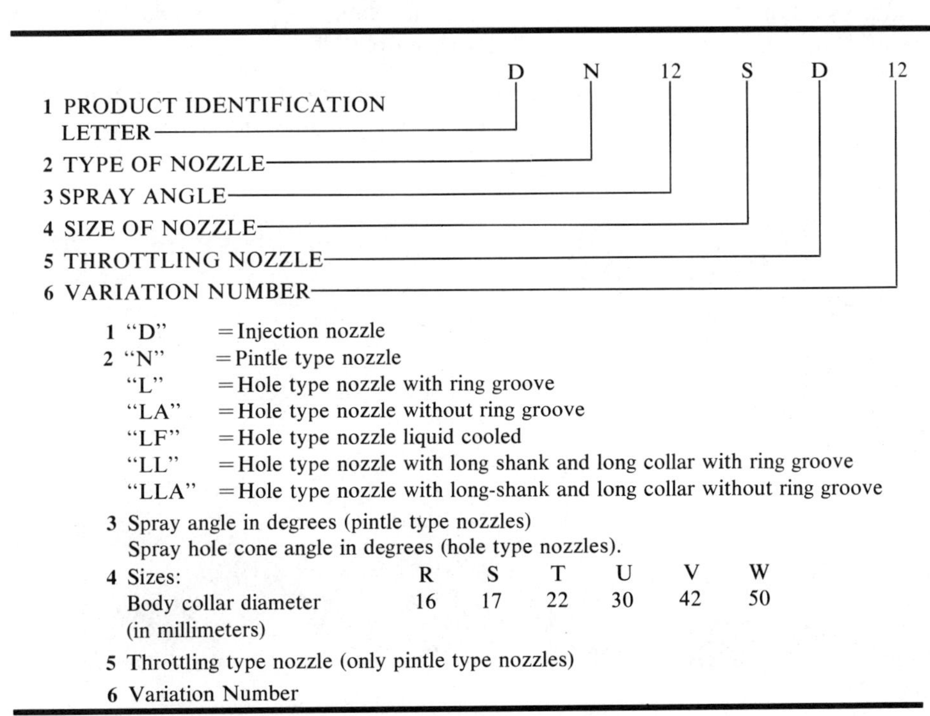
D N 12 S D 12

1 PRODUCT IDENTIFICATION LETTER
2 TYPE OF NOZZLE
3 SPRAY ANGLE
4 SIZE OF NOZZLE
5 THROTTLING NOZZLE
6 VARIATION NUMBER

1 "D" = Injection nozzle
2 "N" = Pintle type nozzle
"L" = Hole type nozzle with ring groove
"LA" = Hole type nozzle without ring groove
"LF" = Hole type nozzle liquid cooled
"LL" = Hole type nozzle with long shank and long collar with ring groove
"LLA" = Hole type nozzle with long-shank and long collar without ring groove

3 Spray angle in degrees (pintle type nozzles)
Spray hole cone angle in degrees (hole type nozzles).

4 Sizes:	R	S	T	U	V	W
Body collar diameter (in millimeters)	16	17	22	30	42	50

5 Throttling type nozzle (only pintle type nozzles)
6 Variation Number

Figure 17.11 Code for identifying Robert Bosch injection nozzles. The code is used with the nozzle application list (courtesy Robert Bosch Corporation).

II. Injection Nozzle Servicing

Injection nozzles are precision units that will operate for extremely long periods if given proper maintenance, that is, changing fuel filters at the prescribed intervals and using clean fuel from the start.

General maintenance includes a complete cleaning at intervals suggested by the manufacturer, usually after 600 to 1000 hours of operation.

When servicing the injection system of a diesel engine that is giving problems, always follow a logical, step-by-step method to give the quickest, most economical results. Listed below are the steps to follow in the proper order:

1 Run engine to check operation. (Run it on a dynamometer if possible.)
2 Check fuel supply up to fuel filter(s).
3 Change fuel filter(s).
4 Check air supply.
5 Check injection pump timing. (Correct if necessary.)
6 Check for proper engine speeds. (Correct if necessary.)
7 Check for proper injection advance (if applicable).
8 Check and/or clean injection nozzles.
9 Repair or adjust injection pump.

If, after performing the first seven steps, the problem still exists, then the nozzles should be checked. In some instances it is helpful to isolate the faulty nozzle(s) to aid in repair. Isolate the nozzle as follows:

1 Run engine up to operating temperature.
2 Accurately determine engine low idle speed with a tachometer.
3 Cut out each nozzle in turn by loosening the high pressure fuel line fitting at the nozzle (Figure 17.12). This allows the fuel to escape before it reaches the nozzle, preventing it from operating.
4 The nozzle that least affects engine low idle speed and smoothness is probably the faulty one and can be marked as such before removal.

Procedure for Removal of Injection Nozzles

Clean the engine thoroughly by steam cleaning or washing with water. All connections and lines must be absolutely clean.

CAUTION Never steam clean or spray cold water directly on an injection pump while it is rotating. Seizure of pump parts may result (damage to pump internal parts, brought about by insufficient lubrication or clearance).

1 Disconnect all high pressure lines leading to nozzles (Figure 17.13).
2 Disconnect all fuel return lines leading to the nozzles (Figure 17.14).
3 Remove nozzle clamping nuts, studs, or special gland nuts (Figure 17.15).
4 Remove nozzle from cylinder head carefully (Figure 17.16). A pry bar may be necessary in some cases.

Figure 17.12 Isolating a faulty nozzle.

Figure 17.13 Disconnecting high pressure lines.

Figure 17.14 Disconnecting return fuel lines.

CAUTION Be extremely careful not to bend the nozzle holder.

NOTE Make sure nozzle sealing washers come out with the nozzle. If not, remove them with a tapered, serrated tool.

At this time all connections and openings should be covered with plastic caps or aluminum foil. Do not use tape or rags because of the danger of lint or gummy residue getting into the lines.

Place the nozzles in an area where they will not be damaged until they are worked on or take them immediately to a shop specializing in this work.

III. Cleaning Fuel Injection Nozzles

The cleaning of injection nozzles should be done in an area that is absolutely clean. Dirt and dust in the air, filings on benches, and greasy rags will contribute to faulty nozzle operation and early failure. Tools and equipment necessary for the cleaning of nozzles are:

1 Parts cleaner (solvent or ultrasonic type).

2 Clean pans.

3 Lint-free towels.

4 Nozzle cleaning kit.

5 Nozzle holder.

6 Hand tools.

7 Clean diesel fuel.

Shown in Figure 17.17 are the items included in most nozzle cleaning kits.

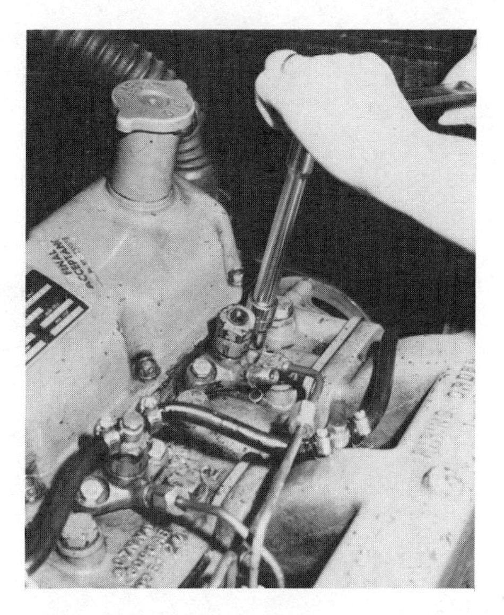

Figure 17.15 Removing nozzle hold down cap screw.

Figure 17.16 Removing nozzle from head.

Procedure for Cleaning

After nozzles are received for cleaning, clean the exterior with solvent to remove loose dirt and grease. Loosen the cap nut and nozzle retaining nut (Figure 17.18). Place nozzles in a suitable parts cleaner to loosen carbon and remove varnish.

After soaking for a minimum of one-half hour, the nozzles should be rinsed in solvent.

By disassembling and overhauling only one nozzle at a time, the chance of a parts mix-up is greatly reduced.

A Mount the nozzle on a holder or clamp in vise (with brass jaws or rags) if unable to use holder.

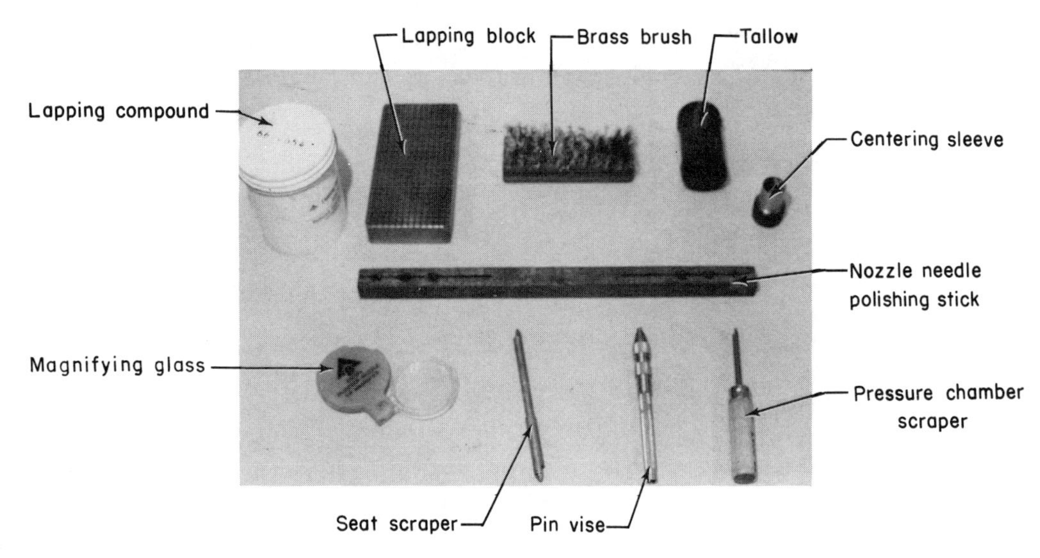

Figure 17.17 Parts of a nozzle cleaning kit.

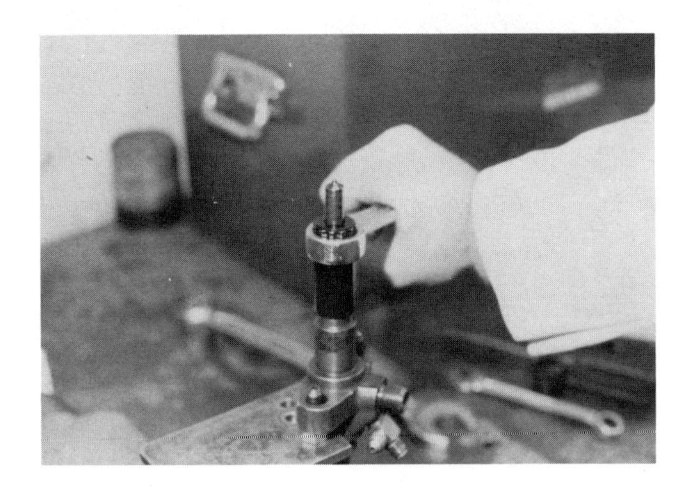

Figure 17.18 Loosening nozzle retaining nut.

Figure 17.19 Loosening pressure adjusting screw.

Release pressure on nozzle spring by removing cap nut and loosening pressure-adjusting screw (Figure 17.19).

CAUTION Failure to remove spring pressure may result in dowel pin breakage when the retaining nut is loosened.

B Invert nozzle in holder and remove nozzle retaining nut and nozzle assembly. Be careful not to drop the nozzle needle!

C Remove nozzle needle from the nozzle. Usually after a thorough soaking the needle will remove easily. If it is stuck, a hydraulic nozzle extractor, which attaches to the nozzle test stand, can be used to free the needle. If it remains stuck, discard the nozzle, as it is probably scored beyond repair.

NOTE Be certain the nozzle needle is kept with the nozzle body it was removed from, because nozzle needles are a selective fit in the nozzle body and cannot be interchanged from one nozzle body to another.

D Examine the needle carefully for scoring, blue spots, excessive wear, and corrosion. If any are found, discard the nozzle.

E Clean needle completely with brass wire brush (Figure 17.20) to remove all carbon and varnish.

CAUTION Never use steel wire bristle brushes on precision nozzle parts. Always use brass wire brushes.

Figure 17.20 Cleaning nozzle needle with brass brush.

Figure 17.21 Polishing the end of the needle.

F Using the pintle cleaning block, polish the tapered end of the needle with mutton tallow (Figure 17.21). Place tallow on needle, insert needle in cleaning block, and rotate gently to polish the needle seat. Rinse off excess tallow in clean diesel fuel or calibrating oil.

CAUTION Never use abrasives such as lapping compound, crocus cloth, or jewelers rouge to polish the needle. Always use tallow.

G Using the brass brush, clean the nozzle body to remove loose carbon deposits (Figure 17.22).

H If cleaning orifice type nozzles, clean the holes with the proper size cleaning wire.

NOTE Most nozzle valves will have the hole size stamped or etched on them. If the hole size is not stamped on the nozzle valve, refer to the manufacturer's specifications.

The cleaning wire should be fitted in a pin vise as shown in Figure 17.23, letting the wire protrude approximately $\frac{1}{16}$ in. (1.5 mm). This lessens the danger of breaking wires off in the holes, since they are extremely hard to remove when broken. Most popular-sized wires are contained in the nozzle cleaning kits.

I Using the special pressure chamber scraper shown in Figure 17.24, clean the chamber by rotating and exerting an upward pressure on the tool. Five or six turns are usually sufficient.

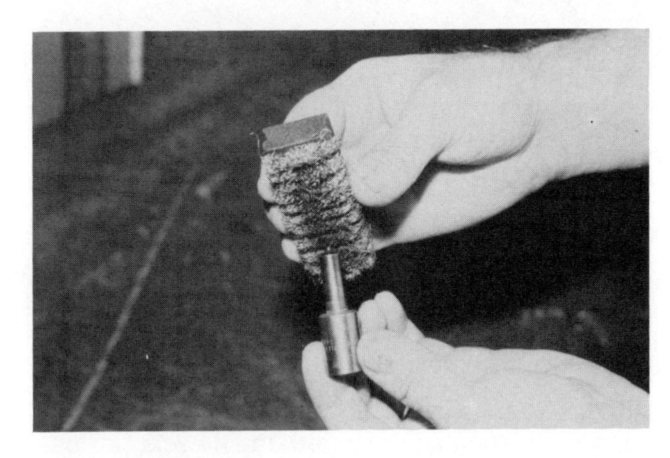

Figure 17.22 Removing carbon from valve body.

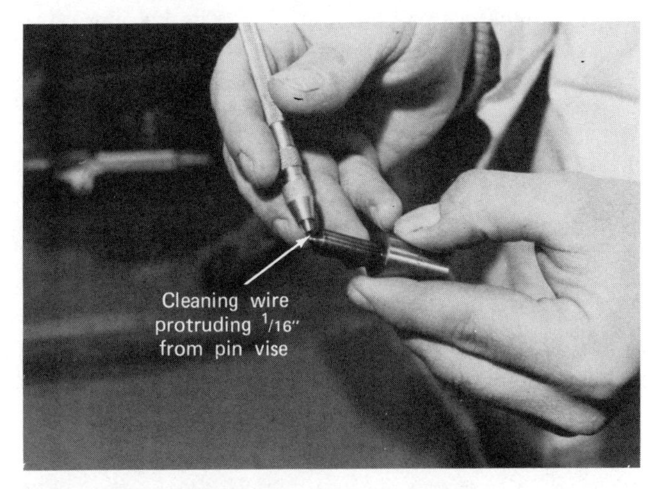

Figure 17.23 Cleaning orifices.

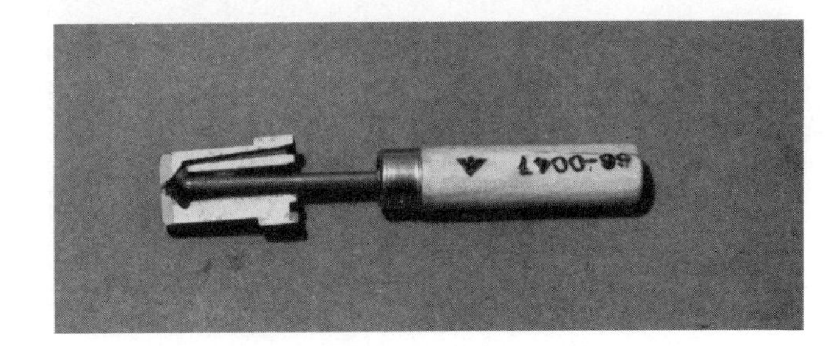

Figure 17.24 Sectional view showing how the pressure chamber is cleaned.

J The nozzle valve seat scraper (Figure 17.25) is used to clean carbon from the valve seat. Two sizes are contained on the same tool for varying nozzle sizes. Rotate the tool to clean the seat.

K Apply a small amount of tallow to a polishing stick and thoroughly clean and polish the valve seat in the nozzle body (Figure 17.26).

L The surface of the nozzle body that contacts the nozzle holder as well as the holder surface must be lapped before reassembly.

NOTE Nozzles using dowel pins are not always lapped. If lapping is required, the dowels can be removed using a diagonal pliers.

1. Place a small amount of nozzle lapping compound on the lapping plate. Hold the nozzle so that pressure will be exerted evenly on the entire surface. Move the nozzle smoothly and steadily in a figure 8 motion (Figure 17.27).

CAUTION Do not rock the nozzle from side to side.

2. Lap only until the nozzle mating surfaces are clean and flat. Rinse the nozzle completely in clean diesel fuel to remove all traces of lapping compound.
3. Lap the nozzle holder in the same manner (Figure 17.28). Steady the holder near the lower end to prevent it from rocking.

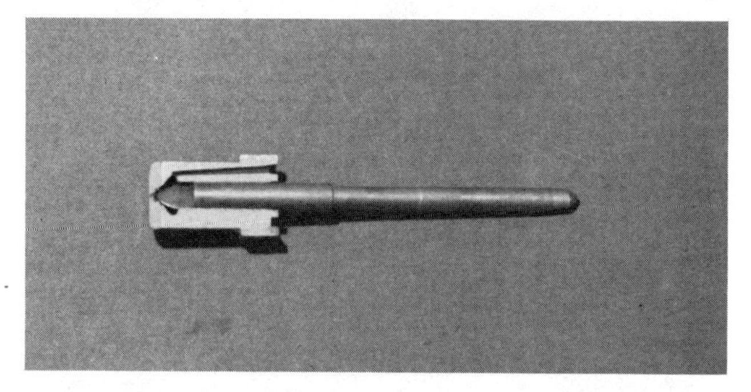

Figure 17.25 Cleaning the valve seat.

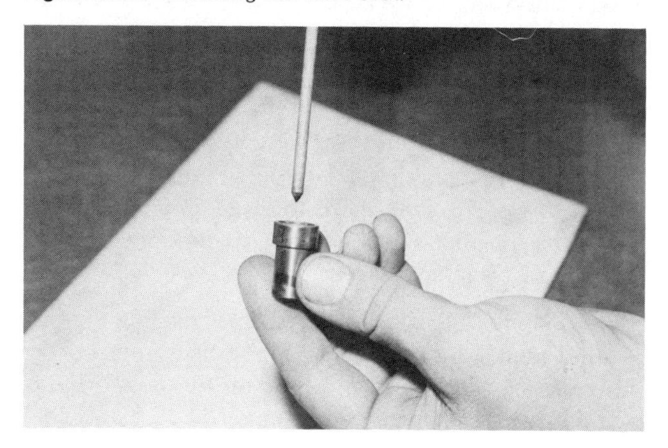

Figure 17.26 Polishing the nozzle valve seat.

M Using a small screwdriver, scrape all loose carbon from the nozzle retaining nut and check for cracks and damaged threads. The sealing surface for the nozzle retaining nut may be cleaned up by rubbing it on the emery cloth as shown in Figure 17.29.

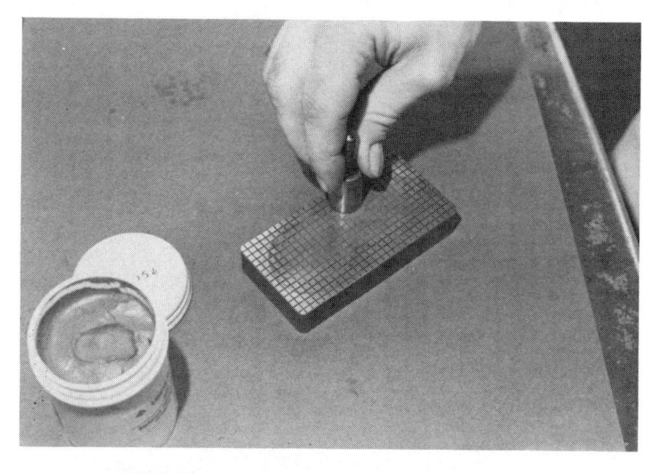

Figure 17.27 Lapping the nozzle mating surface.

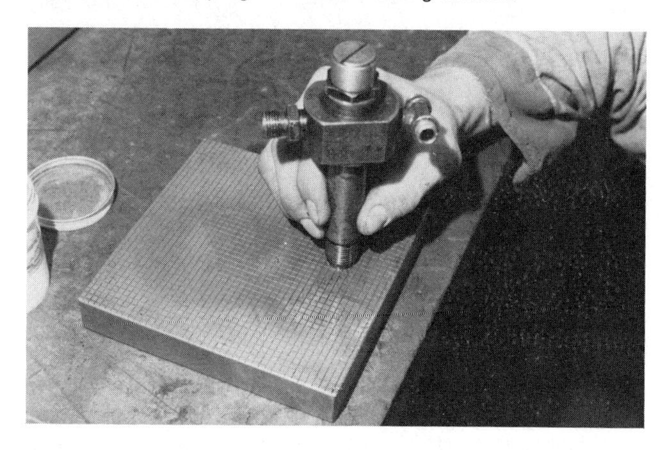

Figure 17.28 Lapping the nozzle holder.

Figure 17.29 Cleaning the sealing surface of the retaining nut.

IV. Procedure for Injection Nozzle Reassembly

A Start reassembly by rinsing nozzle needle and body in clean diesel fuel and checking the valve fit (Figure 17.30). This can be done by holding nozzle at a 45° angle and pulling the needle one-third of the way up. It should fall freely back to its seat. If it does not, remove the needle, rinse the parts, and try again.

B Rinse the sealing surfaces of the nozzle holder and nozzle body in diesel fuel and assemble.

CAUTION Since no sealing rings of any type are used at this point, the mating surfaces must be absolutely clean. Do not use compressed air to clean the surfaces as lint and dust will remain.

Make certain when assembling that locating dowels (if used) are in alignment with holes in the holder (Figure 17.31). On some nozzle types the spray tip is separate from the nozzle body and must be aligned by means of timing lines (Figure 17.32). Hold the tip with a small wrench while snugging up the retaining nut.

C On pintle nozzles, before final torquing of the retaining nut, the nozzle must be centered in the nut to insure proper operation.

NOTE Do not center nozzles used with Robert Bosch holders and retaining nuts as they are self-centering.

To center the nozzle, a special sleeve (Figure 17.33) is used (supplied with nozzle cleaning kits). Carefully fit the centering sleeve over the nozzle body. The tapered end of the sleeve centers the nozzle within the retaining nut bore and on the holder. With the sleeve in place, tighten the nut finger tight. Make sure the sleeve turns freely.

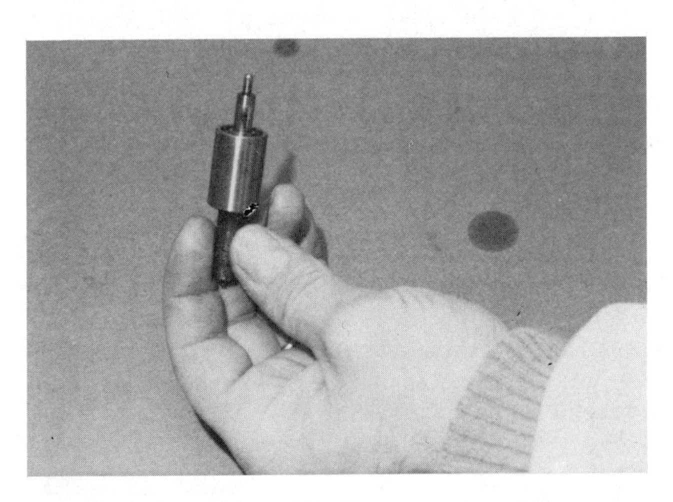

Figure 17.30 Checking nozzle valve for free operation.

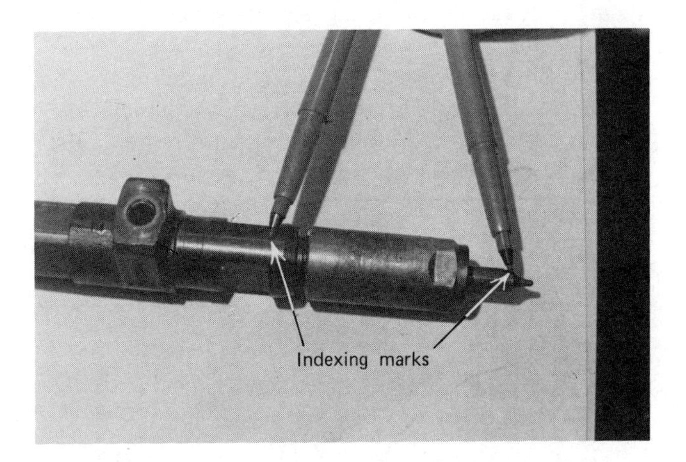

Figure 17.32 Using alignment marks to properly position spray holes.

Figure 17.31 Align the nozzle valve with the dowel pins if used.

Figure 17.33 Using centering sleeve.

Torque the nut to manufacturer's specifications using a deep well socket.

NOTE On orifice type nozzle valves, centering is not required. Simply torque the nut to specifications.

V. Nozzle Testing

After the nozzle has been completely cleaned and reassembled, it must be adjusted on an injection nozzle test stand (Figure 17.34).

Suitable testers are available from the major nozzle manufacturers. Clean calibrating oil should be used in the tester.

Procedure

Basically, there are five tests to be performed with the nozzle tester: (1) opening pressure, (2) chatter test, (3) seat leakage, (4) spray pattern, and (5) back leakage. All tests should be done according to manufacturer's testing procedures and using correct specifications.

A *Opening pressure.* Attach nozzle to the test stand and flush thoroughly by operating the handle. Adjust the opening pressure with the pressure adjusting screw or shims as required (Figure 17.35).

CAUTION Close gauge isolating valve before operating tester handle to prevent damage to the pressure gauge.

Set to correct opening pressure as listed in manufacturer's specifications.

CAUTION Do not allow high pressure fuel to come in contact with hands or arms as it may puncture the skin and cause infection or blood poisoning.

For cleaned nozzles, set all nozzles to within 50 psi (3.5 kg/cm²) of each other. When installing new nozzles or new springs, set pressure 100 psi (7.0 kg/cm²) above the listed pressure lock adjusting screw and install cap nut with new gasket.

B *Chatter test.* When performing nozzle chatter test, isolate tester gauge to avoid damaging gauge. Operate tester handle at 60 strokes per minute and listen for the rapid opening and closing of the nozzle needle, which is called chatter. Although it isn't imperative that a nozzle chatter, it does indicate that the valve is free.

C *Seat leakage.* To prevent nozzle dribble and excess smoke from the engine, the nozzle needle must seat perfectly. This is checked on the test stand by raising nozzle pressure to 200 psi (14 kg/cm²) below opening pressure and observing the nozzle tip for any signs of leakage. No dripping should occur. If dripping is evident, replace the nozzle.

D *Spray pattern.* Isolate the gauge. Operate tester handle at 60 strokes per minute and observe the spray pattern (Figure 17.36). It should be a crisply defined pattern with no offshoot streams of fuel and it should be well atomized. On pintle nozzles it should not shoot off to one side but rather straight out, which will enable the fuel to enter the energy cell or precombustion chamber.

Figure 17.34 Nozzle test stand.

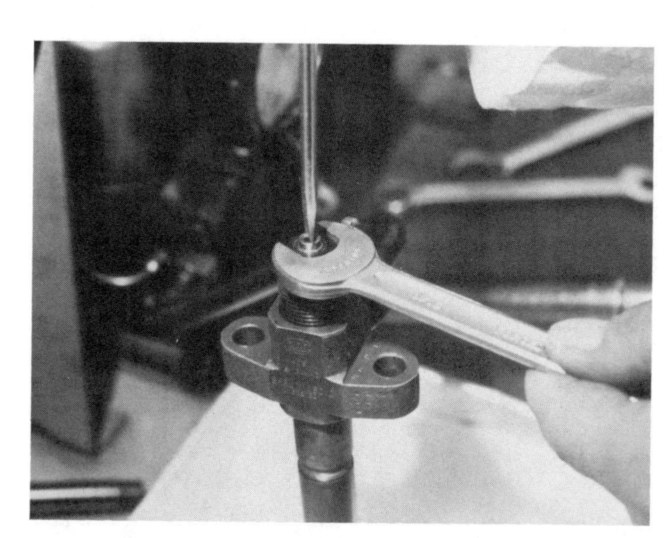

Figure 17.35 Adjusting nozzle opening pressure.

Figure 17.36 Checking nozzle spray pattern.

E *Back leakage.* Back leakage of nozzles is the fuel that leaks past the nozzle needle for lubrication and returns to the tank. Excessive leakage will reduce the efficiency of the nozzle, necessitating replacement. The amount of back leakage allowed is usually not listed in service manuals, but if the nozzle drips excessively from the return it should be replaced.

If the nozzle passes all of the above tests, it may be placed back in service.

VI. Installation of Cleaned Nozzles to Engine

Procedure

The nozzle recess in the cylinder head must be thoroughly cleaned. This cannot be overemphasized. Lack of a good, clean contact surface can result in combustion gas blow-by or cocking of the nozzle, which can cause sticking. Be certain the old copper sealing washer has been removed and the seat wiped free of any loose carbon. Any hardened carbon particles may require the use of the special tool shown in Figure 17.37.

Always install a new sealing washer on the nozzle. Since there are various types of washers in use, make certain that the correct washer is being used.

Insert the nozzle, using no lubricant of any kind. Lubricants will turn to carbon under the intense heat in the cylinder head and could cause removal problems.

Install nozzle clamp nuts finger tight. Some nozzles use gland nuts as shown in Figure 17.38.

Torque the clamp nuts evenly to the torque specifications listed by the manufacturers. After nozzle has been torqued, inspect and clean the end of each injection line before connecting it to the nozzle. The line nut should be left loose to allow for bleeding (air purging). Systems should be bled by placing the fuel control in the full fuel position and cranking the engine until fuel is visible at all nozzle connections.

CAUTION Do not crank the engine with the starting motor over 30 seconds at a time without a cool-down period or serious damage to the starting motor may result.

After all nozzle lines have been bled, tighten connecting nuts and start the engine. While engine is running, check all connections for evidence of leakage.

NOTE If a line leak develops, it may help to loosen the connecting nut slightly with engine running and allow a small amount of fuel to leak from the connection, removing any foreign material that may have been lodged in the connection. Tighten the connecting nut and recheck for leakage.

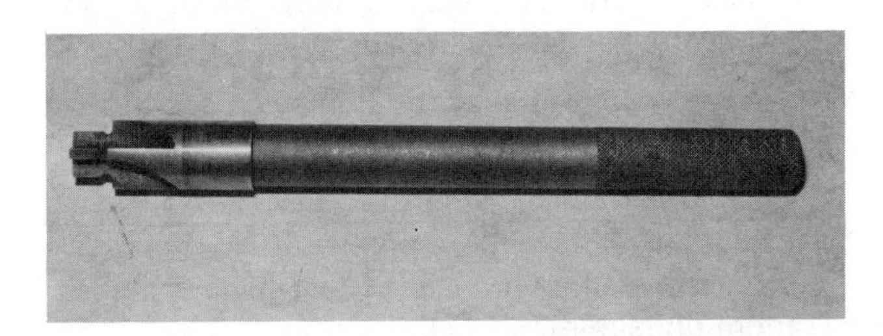

Figure 17.37 Tool used to clean nozzle bore and cylinder head.

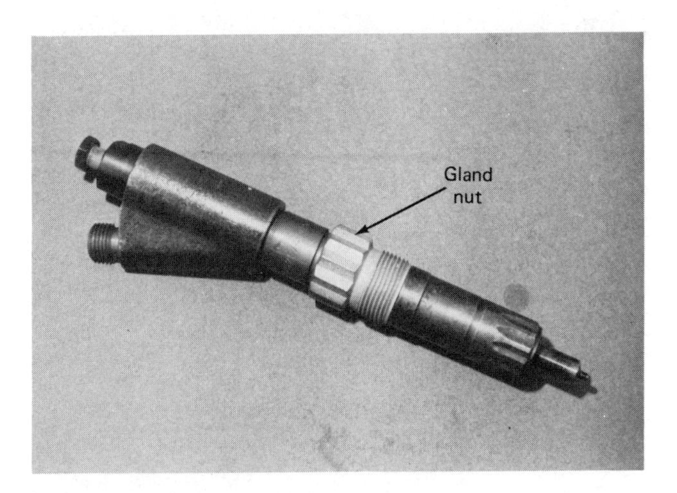

Figure 17.38 Nozzle using a gland type nut.

VII. Servicing Roosa Master Pencil Nozzles

The Roosa Master pencil nozzles represent a completely new idea in injection nozzle design. As can be seen in Figure 17.39, the nozzle is much smaller in size than the nozzles that have been discussed so far. As a direct result of the different design, servicing procedures for the Roosa Master nozzle are considerably different from other nozzles. This procedure must be followed closely when servicing any Roosa Master pencil nozzle.

A. Procedure for Removing Pencil Nozzles from Engine

1 Completely clean the exterior of the engine with a steam cleaner or high pressure washer.

2 Remove return line boots from nozzles and remove return line.

3 Disconnect high pressure fuel lines.

4 Remove nozzle by pulling upward with a slight twisting motion (Figure 17.40).

NOTE Some nozzles may not come out easily and a slide hammer type puller may be required to remove them.

CAUTION Do not use a pry bar to remove pencil nozzles; they are easily bent (which makes them unusable).

5 Plug the holes to prevent entry of contamination.

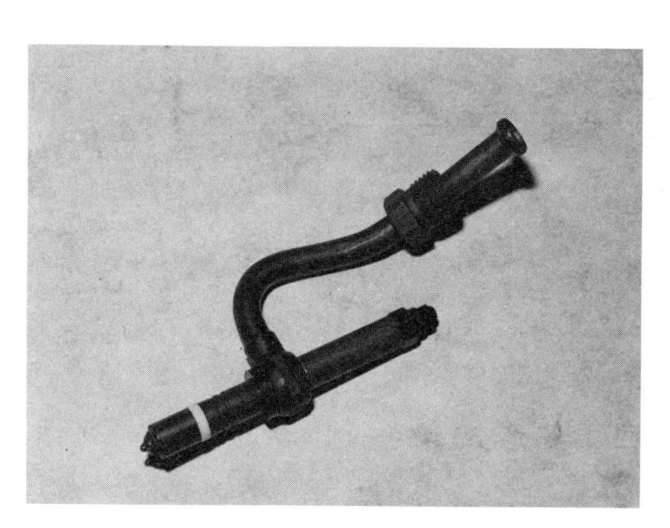

Figure 17.39 Roosa Master pencil nozzle as used on GM engines.

B. Procedure for Cleaning the Pencil Nozzle

It is not recommended that the pencil nozzle be disassembled and cleaned as a general practice. The proper method is to first clean the nozzle exterior with a solvent and remove the carbon dam and upper dust seal. Place the nozzles in parts cleaner so that only the tips are submerged and soak for one-half hour.

CAUTION Do not submerge the nozzle completely because the thin film of teflon on the exterior would be removed, causing rusting.

After soaking, remove the nozzles and check all orifices for erosion or clogging. Replace any nozzles that have eroded holes. If holes are clogged, they may be cleaned using the proper sized cleaning wire.

Thoroughly brush nozzle tip and seal groove with brass wire brush.

C. Procedure for Testing and Adjusting the Pencil Nozzle

1 Using the special test pump adapter (Figure 17.41), attach the nozzle to the nozzle tester.

CAUTION Do not attempt to use the standard Ermeto fitting on the pencil nozzles, as damage to the flange will result.

2 Loosen locknut and back out lift-adjusting screw approximately five turns (Figure 17.42).

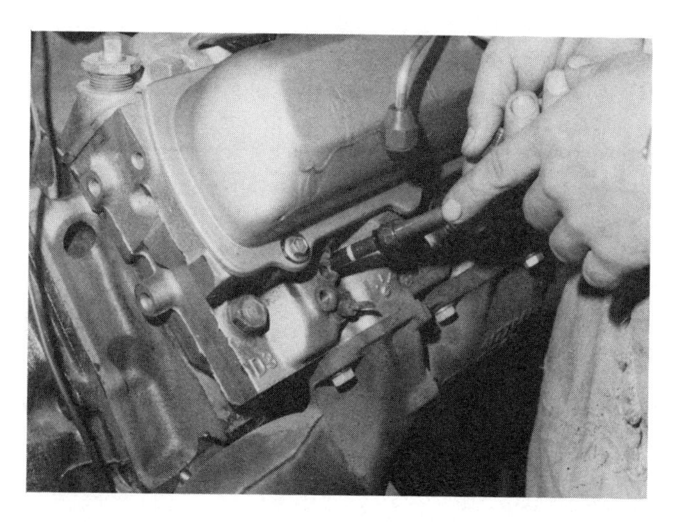

Figure 17.40 Removal of pencil nozzle.

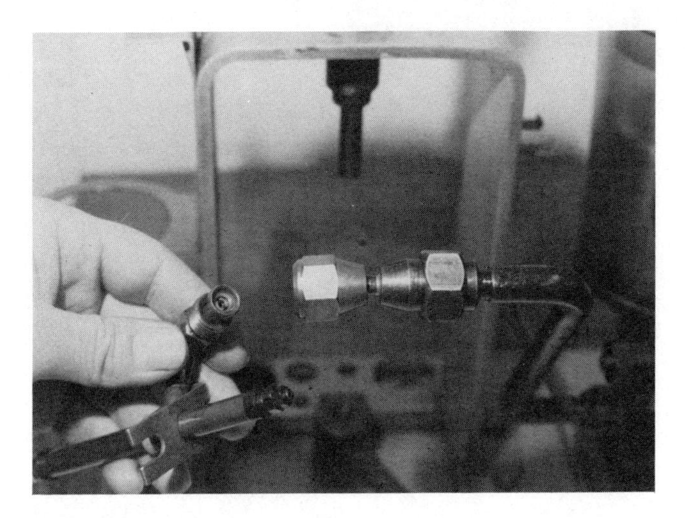

Figure 17.41 A special adapter is used to attach pencil nozzles to the tester.

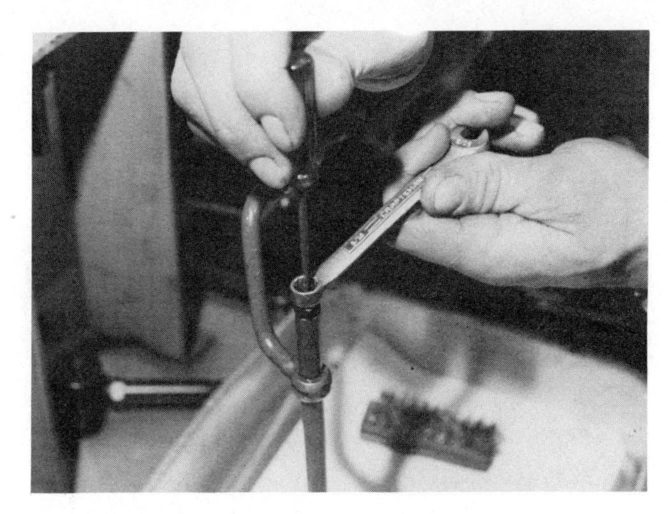

Figure 17.42 Backing out the lift-adjusting screw.

3 Flush nozzle with fuel. Set opening pressures with pressure-adjusting screw to pressure recommended (Figure 17.43).

NOTE On earlier nozzles using one locknut, leave the locknut loose at this time. With later nozzles two locknuts are used, one for the pressure-adjusting screw and one for the lift-adjusting screw. With the latter type, lock the pressure screw after pressure has been set.

4 Set the valve lift by first bottoming the screw gently. Check for bottoming of the valve by raising the pressure 500 psi (35 kg/cm²) over the nozzle opening pressure. Fuel should not dribble from the tip. From this position, back the valve out the specified amount, usually $\frac{1}{2}$ to $\frac{3}{4}$ turn. Secure the locknut and recheck opening pressure.

NOTE Always set lift before checking the chatter and spray pattern, as the amount of lift will affect the chatter.

5 Torque the locknut with adapter shown in Figure 17.44. Tighten the nut to 70 to 75 in. lb (7.90-8.47 N.m) on later nozzles, and 100 to 115 in. lb (11.30-12.99 N.m) on earlier nozzles.

6 Check for nozzle chatter by first isolating the tester gauge and operating the handle to 60 strokes a minute. The pencil nozzle should chatter. If it does not, the valve is probably bent or is sticking, If readjustment still does not give chatter, replace the nozzle.

7 Seat leakage. Check seat leakage by raising pressure to 300 psi (21 kg/cm²) below opening pressure. No drops of fuel should appear on the tip.

8 Back leakage. Check back leakage by raising nozzle tip up higher than return (Figure 17.45). Raise

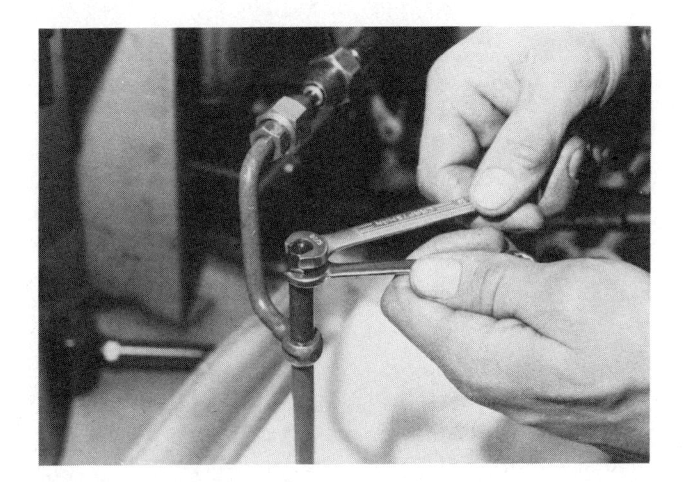

Figure 17.43 Setting opening pressure.

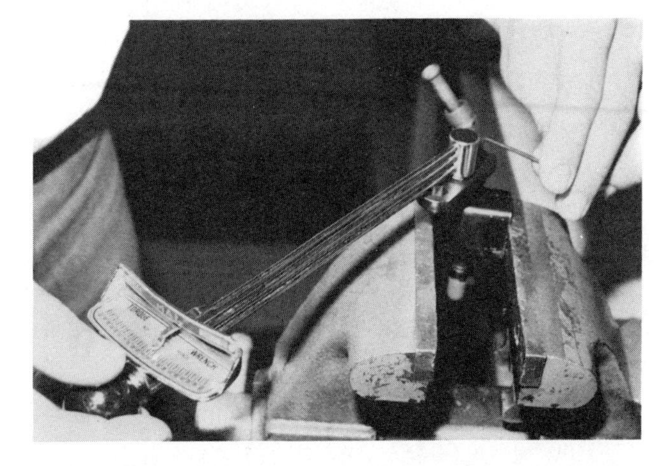

Figure 17.44 Torquing the lock screw after adjusting the lift.

Figure 17.45 Testing for back leakage.

pressure to 1500 psi (105 kg/cm²) and count the number of drops from return in 30 seconds. There should be 3 to 10 drops. If back leakage is excessive, the nozzle must be replaced.

D. Procedure for Installation of Pencil Nozzles

1 Install compression seal and carbon dam. A special tool is used to install the carbon dam (Figure 17.46).

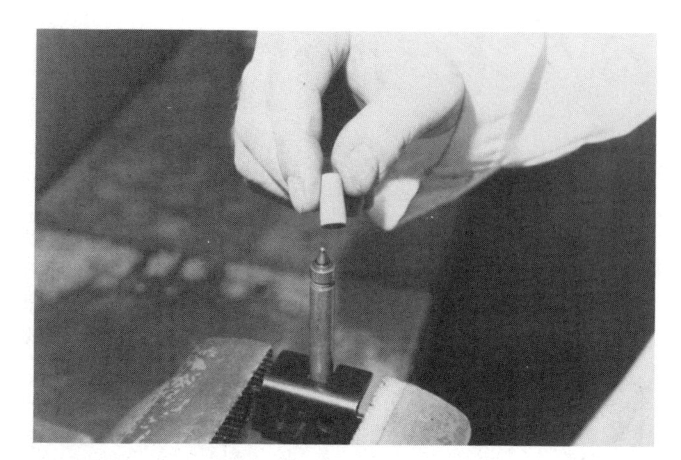

Figure 17.46 Tool used to install the carbon dam.

2 Install nozzles in cylinder head without lubricant and attach the holding clamp. The clamp should be torqued to 20 ft lb (27.12 N.m).

3 When installing fuel leak-off lines, always use new plastic boots on the nozzles to prevent leaking.

SUMMARY

If the procedures outlined in this chapter are followed, injection nozzle servicing is an easy task.

When working with any type of nozzle not listed in this chapter, always refer to the manufacturer's technical manual. It will give the correct torques, opening pressures, operation, and any other pertinent data. If a question still exists, consult your instructor or contact your nearest fuel injection service shop for information.

REVIEW QUESTIONS

1 Name the two basic types of nozzles.

2 What is the purpose of the Pintaux nozzle?

3 Explain the difference between a standard pintle nozzle and the throttling pintle.

4 What does each of the following numbers and letters stand for: DN 12 SD 12?

5 State the difference between a nozzle and an injector.

6 Why do most pintle nozzles require centering on the nozzle body?

7 Explain in detail the procedures for removing and installing nozzles in the engine.

8 Why are orifice nozzles used with direct injection engines?

9 List several reasons why nozzles should be cleaned regularly.

10 Why are retaining nut torque and nozzle hold-down torque so critical?

11 What is the purpose of dowel pins and timing lines in reference to nozzles?

12 Why shouldn't the Roosa Master pencil nozzles be soaked in a strong parts cleaner?

13 What five tests are made on the nozzle test stand?

14 What additional test is performed on Roosa Master pencil nozzles and how is it accomplished?

15 List all steps required in the cleaning of a regular throttling pintle nozzle.

16 Explain how a faulty nozzle can be located in the engine.

18

American Bosch Fuel Injection Systems

American Bosch is one of the oldest manufacturers of diesel fuel injection equipment in the world with more than 40 years' experience in the field. American Bosch is a division of AMBAC Industries, Inc., formed in 1968. The main office at Springfield, Massachusetts, produces fuel injection, hydraulic and electrical equipment. Additional fuel injection manufacturing plants are located in Breda, Holland, and Brescia, Italy.

Throughout the years American Bosch has produced many models of pumps used on various diesel engines in all types of applications. Two of the most common American Bosch pumps used on mobile diesel engines have been the PSB pump and the APE pump. Thousands of these pumps have been applied to farm tractors, heavy equipment, and motor trucks. In recent years the PSB pump has been replaced by the 100 series pump and a V8 model has been developed that is very similar to the APE in-line pump.

OBJECTIVES

Upon completion of this chapter the student will be able to:

1. List two different types of pumps available from American Bosch.
2. Explain and demonstrate the proper procedure for pump disassembly (pump model of student's choice).
3. Discuss and point out on a wall chart the fuel flow in a fuel pump of the student's choice.
4. Explain the reason for having an intravance mechanism in a 100 series pump.
5. Demonstrate the ability to calibrate an American Bosch pump (student's choice).
6. Demonstrate the ability to set port closure on an American Bosch pump (student's choice).
7. Demonstrate the ability to correctly install and time an American Bosch pump to an engine (student's choice).

Figure 18.1 Model APE injection pump (courtesy American Bosch—United Technologies Corporation, Springfield, Mass.).

GENERAL INFORMATION

The APE series Bosch pump (Figure 18.1) is a multiplunger in-line pump built for four, six, or eight cylinders. These pumps use the common port and helix fuel control, are operated by an internal cam, and are available in three sizes: A (small), B (medium), BB (large). A VEE pump also uses the APE design. This pump has eight cylinders, four on each bank, and is mounted in the VEE on some engines. The APE pump can use several different governors but the GV variable speed type is the most common. Optional equipment for the pump includes supply pump, timing advance coupling, and hand priming pump.

The APF pump (Figure 18.2) is a single cylinder pump used for medium and low speed engines up to 1000 hp per cylinder. It is simply one plunger and barrel in an individual cast housing. Many of these pumps can be controlled by a single governor to give accurate fuel control. A common use of this type of pump would be for locomotive engines and the John Deere two cylinder engine for farm and industrial use.

The PSB injection pump uses a reciprocating and rotating motion to deliver fuel. One pumping plunger and delivery valve serves all outlets, insuring even delivery. The PSB operates at crankshaft speed and can be run at 3200 rpm maximum. This pump is available with an external spring governor (Figure 18.3) or an internal spring governor (Figure 18.4).

The PSJ, PSM, and Model 100 pumps are updated and improved versions of the PSB. They are capable of higher speeds (3600 rpm), are governor

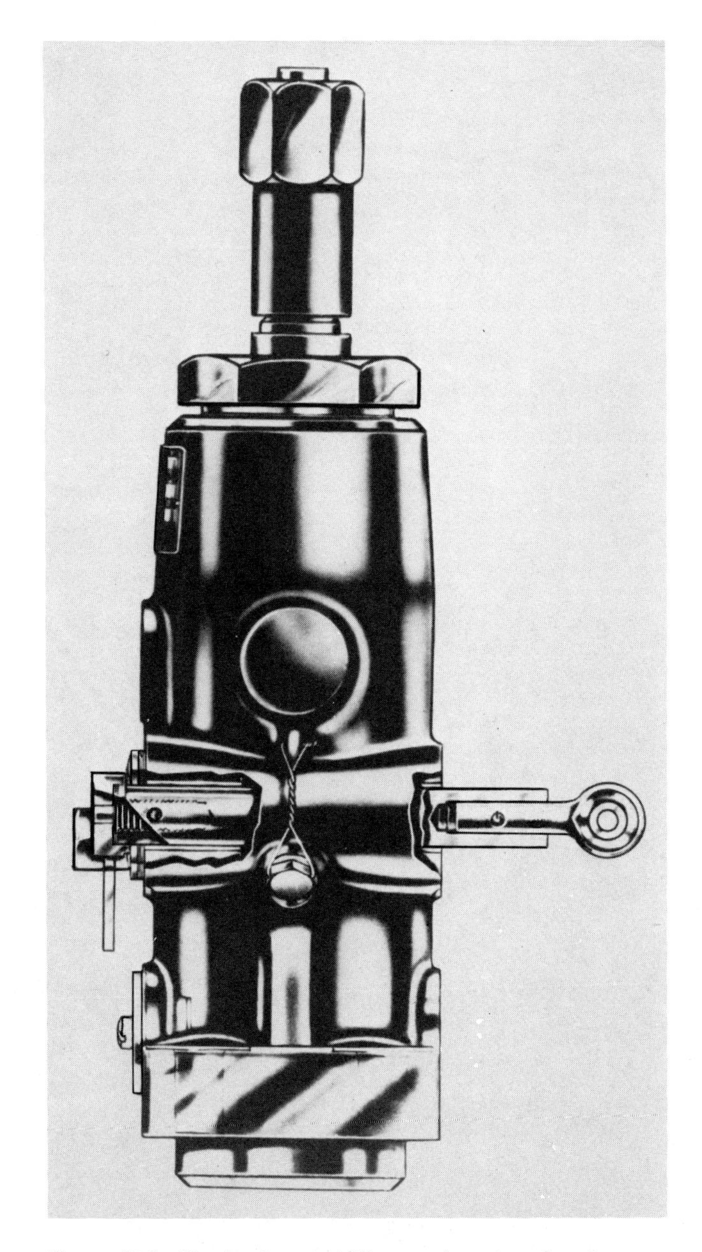

Figure 18.2 Single plunger APF pump (courtesy American Bosch—United Technologies Corporation, Springfield, Mass.).

controlled, and have a gear type supply pump mounted on the rear. An advance unit, either external (extravance) or internal (intravance) is available. These pumps are all lubricated by engine oil, are driven at crankshaft speed, and are single plunger units. The Model 100 (Figure 18.5) will be covered in this chapter.

I. Identification of APE Pump

The pump shown in Figure 18.6 is a model APE pump. Important information about this pump is listed on the pump nameplate. The nameplate provides infor-

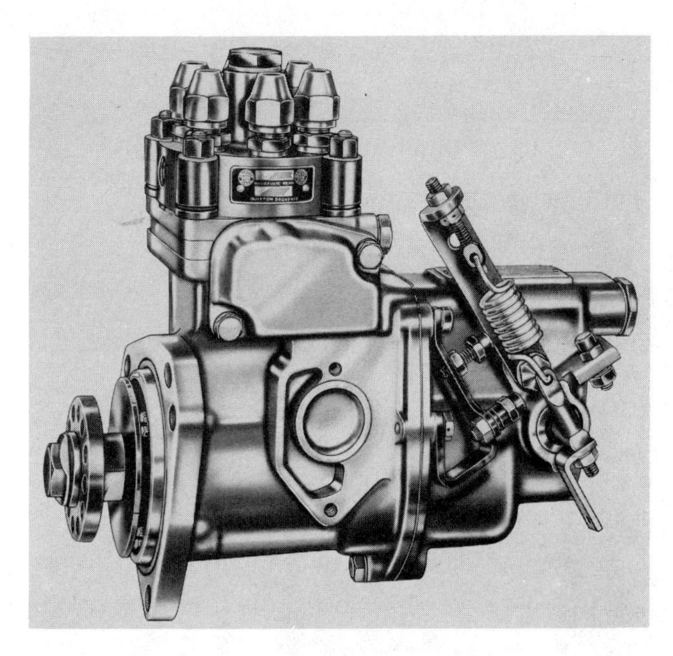

Figure 18.3 PSB external spring governor (courtesy American Bosch—United Technologies Corporation, Springfield, Mass.).

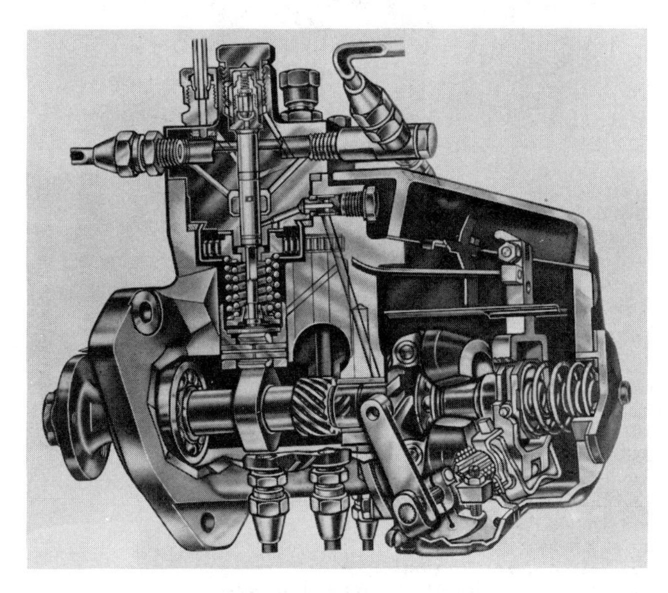

Figure 18.4 PSB internal spring governor (courtesy American Bosch—United Technologies Corporation, Springfield, Mass.).

Figure 18.5 Model 100 pump (courtesy American Bosch—United Technologies Corporation, Springfield, Mass.).

Figure 18.6 Model APE 8 VBB pump (courtesy American Bosch—United Technologies Corporation, Springfield, Mass.).

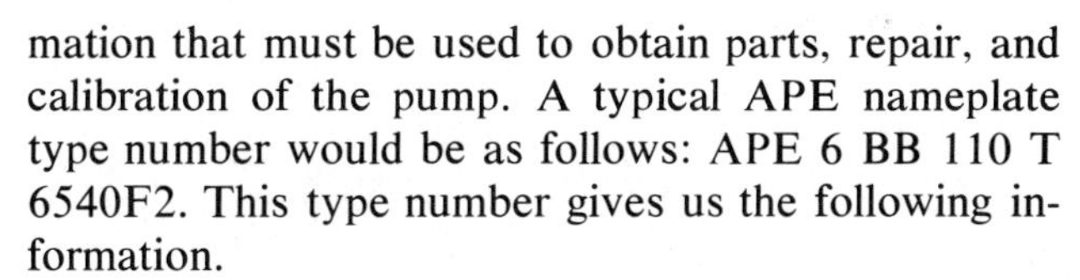

mation that must be used to obtain parts, repair, and calibration of the pump. A typical APE nameplate type number would be as follows: APE 6 BB 110 T 6540F2. This type number gives us the following information.

APE—type of injection pump
6—number of cylinders
BB—size of pump
110—plunger, diameter in tenths of millimeter (11.0 mm)
T—execution letter denotes change that affects interchangeability of parts.
6540—specification number denotes variations from base pump, also the identification number

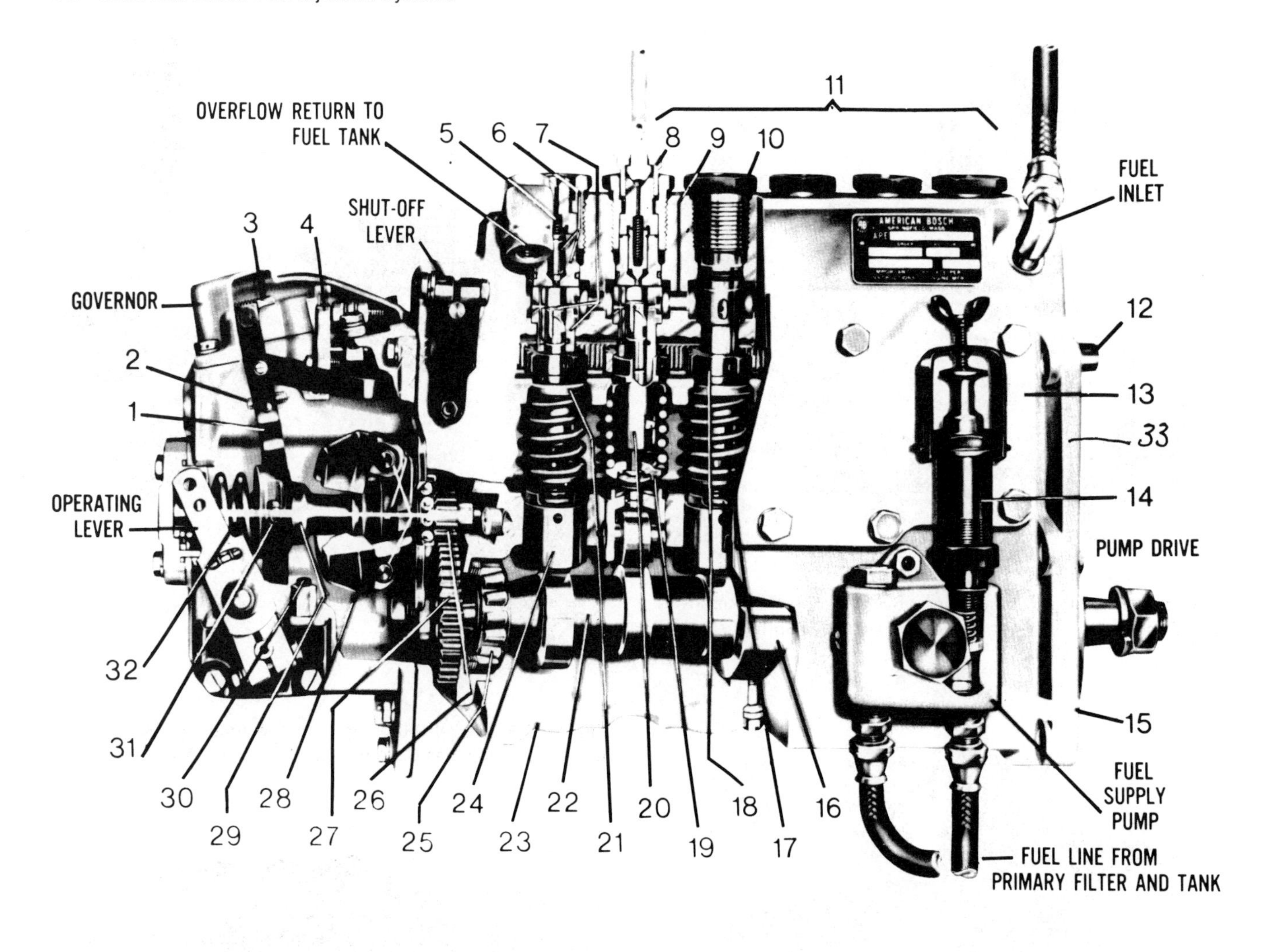

Nomenclature

1. **Fulcrum Lever Assembly**
2. **Droop Screw**
3. **Torque Control Cam**
4. **Adjustable Stop Plate**
5. **Delivery Valve Spring and Holder**
6. **Delivery Valve Assembly**
7. **Plunger and Barrel Assembly** **Turned 90° from standard for illustration purposes**
8. **Tubing Union Nut**
9. **Fuel Sump**
10. **Retaining Nut**
11. **Fuel Discharge Outlets (6)**
12. **Control Rack**
13. **Inspection Cover**
14. **Hand Primer (Optional)**
15. **Lubricating Oil Outlet**
16. **Camshaft Center Bearing**
17. **Plunger Spring**
18. **Control Sleeve Gear Segment**
19. **Lower Spring Seat**
20. **Control Sleeve**
21. **Upper Spring Seat**
22. **Camshaft**
23. **Closing Plug**
24. **Tappet Assembly**
25. **Camshaft Bearing**
26. **Driven Gear, Governor**
27. **Drive Gear and Friction Clutch Assembly**
28. **Flyweight Assembly**
29. **Sleeve Assembly**
30. **Speed Adjusting Screw**
31. **Fulcrum Lever Pivot Pin**
32. **Inner/Outer Governor Springs**
33. **Pump Housing**

Figure 18.7 Model APE component parts cross section (courtesy American Bosch—United Technologies Corporation, Springfield, Mass.).

for service manual use.
F – edition letter, denotes minor change.
2 – code number variation of specification number usually denotes speed or calibration change.

To fully understand the operation of the pump we must have a knowledge of the location of various parts and their correct names.

II. APE Component Parts (Figure 18.7)

A *Housing.* (33) The housing supports all other pump parts and has provision to mount the pump to the engine.

B *Camshaft.* (22) This imparts a reciprocating motion to the pumping plungers; drives the supply pump and mechanical governor.

C *Barrel and plunger assembly.* (7) High pressure is developed within the barrel by the plunger. The reciprocating motion pumps the fuel. Rotation of the plunger determines the quantity of fuel delivered.

D *Delivery valve.* (6) This valve serves to seal the line, retain fuel in the line between injections, and retract a certain amount of fuel from the line to insure immediate closing of the nozzle.

E *Control rack (or rod).* (12) This rotates all of the plungers simultaneously via linear force from the governor.

F *Roller tappets.* (24) They transmit the force of the camshaft to the pumping plungers, forcing them upward. The plunger springs then return the plunger to BDC (bottom dead center) following the contour of the cam.

III. Operation of APE Pump and GV Governor

A Operating cycle.

The APE pump is a multicylinder in-line type pump. It has an individual pumping element for each cylinder enclosed in one housing.

Fuel from the supply tank flows to the supply pump (either by gravity or suction). The supply pump then pressurizes the fuel in the main pump gallery to approximately 21 psi (1.47 kg/cm^2). With the cam lobe in the down position (BDC), fuel enters the intake port of the barrel (Figure 18.8) and fills the area above the plunger. As the cam starts to rise, the top of the plunger covers the spill port (Figure 18.9).

NOTE This is the start of injection and is known as port closure. The pump is timed at this point.

As the plunger continues upward, high fuel pressure opens the delivery valve and is forced through the line to the injection nozzle. Injection ceases when the lower edge of the helix uncovers the spill port. Fuel then rushes down the vertical slot at the edge of the plunger and back into the fuel gallery. This is the end of injection as the delivery valve then snaps shut and the nozzle stops injecting fuel. Even though the injection has stopped, the plunger will continue to rise until the top of the lobe is reached. The stroke of the cam is constant and the delivery stroke is variable, depending on plunger rotation.

Metering. As the governor rotates the pumping plunger via the control rack and plunger control gear, delivery will increase or decrease, de-

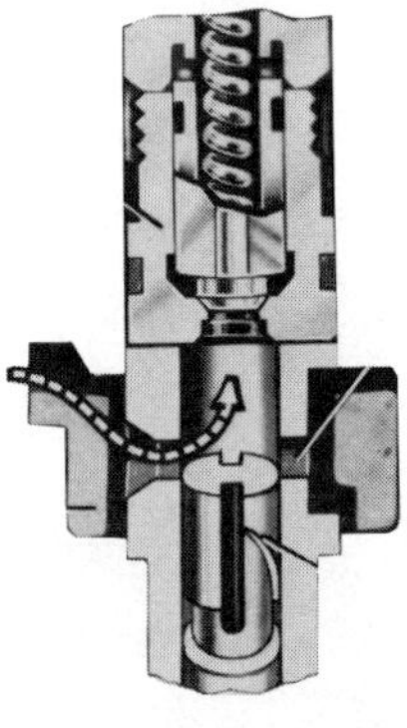

Figure 18.8 Cross section of pumping element on intake (courtesy American Bosch – United Technologies Corporation, Springfield, Mass.).

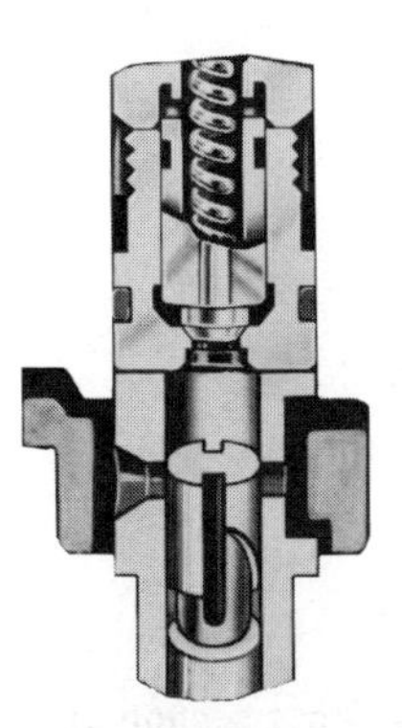

Figure 18.9 Cross section of pumping element at port closure (courtesy American Bosch – United Technologies Corporation, Springfield, Mass.).

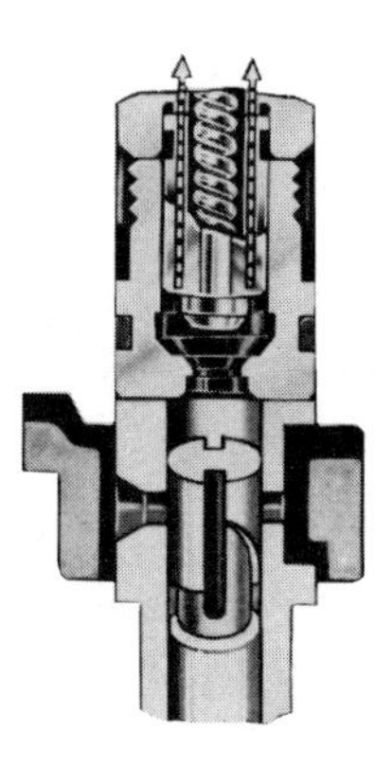

Figure 18.10 Cross section of pumping element during injection or fuel delivery (courtesy American Bosch—United Technologies Corporation, Springfield, Mass.).

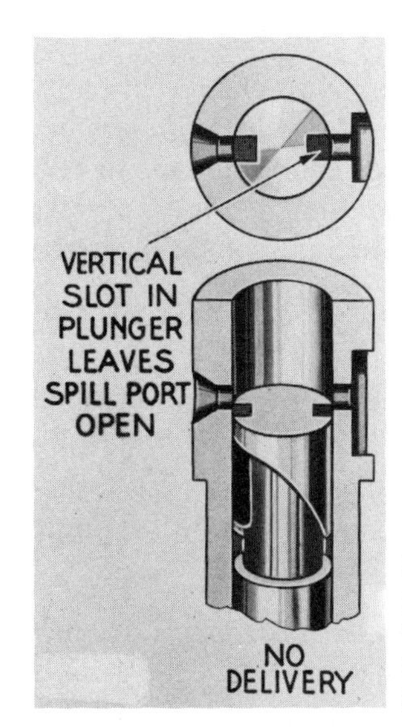

Figure 18.12 Cross section of pumping element in shutoff position (courtesy American Bosch—United Technologies Corporation, Springfield, Mass.).

pending on the plunger position. When set to deliver maximum fuel (Figure 18.10), the longest part of the helix is adjacent to the spill port. In this position the plunger delivers fuel for a long period before the fuel is spilled back to the gallery. Injection is then at the longest duration. When rotated for normal delivery (Figure 18.11), the effective stroke is much shorter than at full fuel and injection is therefore less.

Shutoff occurs when the vertical slot in the plunger lines up directly with the spill port in the barrel. The plunger can then develop no high pressure because all the fuel is pushed back into the fuel gallery (Figure 18.12).

This common fuel injection and metering control is known as port and helix. For more information, refer to Chapter 15.

B *Operation of delivery valve.*

The APE and APF pumps use one delivery valve per cylinder; each valve sits directly above the pumping plunger and barrel. The valve itself (Figure 18.13) is of conventional design, having a tapered seat and a retraction piston. As mentioned earlier, the valve serves three purposes:

1 When closed (Figure 18.14, right), it seals the line and retains fuel in the line for the next injection.

2 While closing (Figure 18.15, center), the retraction piston seals the line from the pump and at the same time removes its volume from the delivery valve holder.

3 As the valve continues moving downward, the area that it occupied is filled with fuel out of the injection line. This causes a pressure drop at the nozzle of 200 to 300 psi (14 to 21 kg/cm^2) that enables the nozzle to close quickly.

Figure 18.11 Cross section of pumping element at normal delivery (courtesy American Bosch—United Technologies Corporation, Springfield, Mass.).

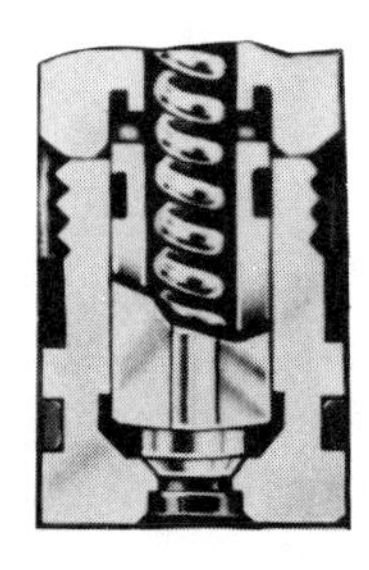

Figure 18.13 Delivery valve (courtesy American Bosch—United Technologies Corporation, Springfield, Mass.).

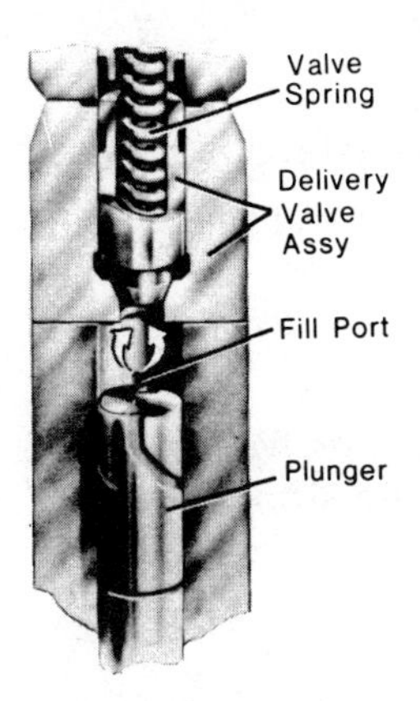

Figure 18.14 Delivery valve closed (courtesy American Bosch–United Technologies Corporation, Springfield, Mass.).

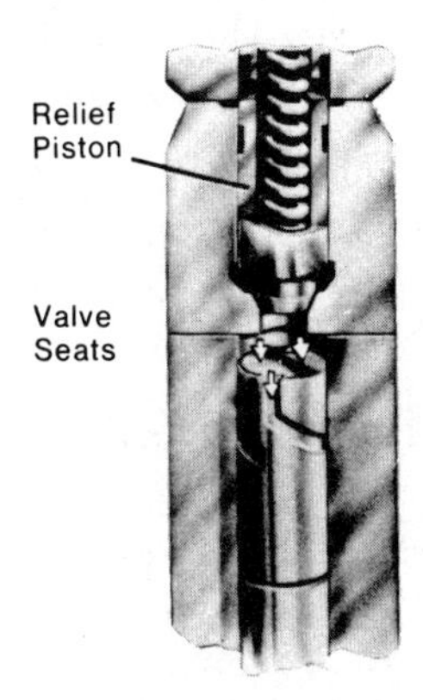

Figure 18.15 Delivery valve showing line retraction (courtesy American Bosch–United Technologies Corporation, Springfield, Mass.).

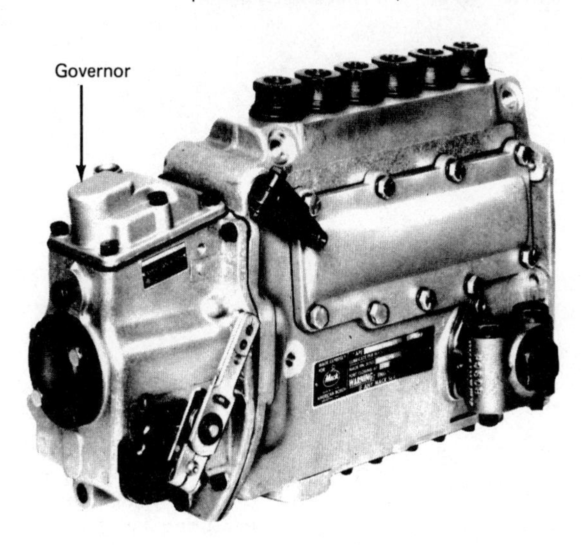

Figure 18.16 APE pump with mechanical governor (courtesy American Bosch–United Technologies Corporation, Springfield, Mass.).

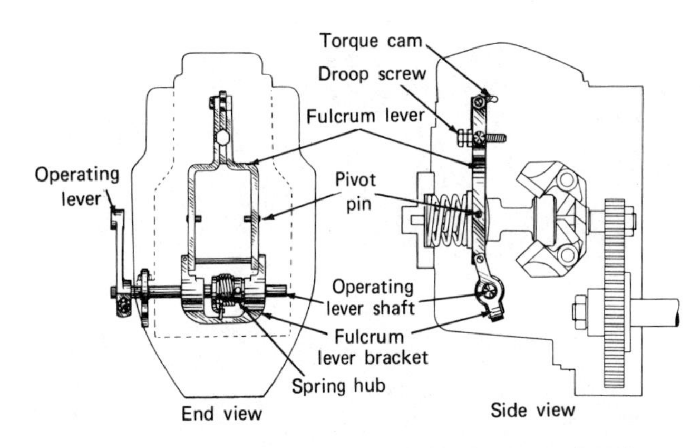

Figure 18.17 GV governor cross section (courtesy American Bosch–United Technologies Corporation, Springfield, Mass.).

C *Governor operation.*

The GV series governor is used on APE pumps in two types, the "GVA" for APE "A" pumps and the "GVP" for APE "B," "BB," and "VEE" pumps. Both mechanical governors are similar and are attached to the rear of the pump (Figure 18.16). A cross-sectional view of the GV governor is given in Figure 18.17, which lists the major components. Although the governor is similar to other mechanical governors, one item deserves special mention: The friction gear assembly is mounted on the end of the camshaft. It consists of spring discs that are pretensioned and a two-piece drive gear. This arrangement lets slippage occur between the governor weights and the camshaft when the speed of the camshaft changes suddenly. This allows the weights to run more smoothly, protecting the governor components.

1 *Engine stopped.*

NOTE When you study the operation of the GV governor, always remember that any movement, either rearward or forward, of the control sleeve (thrust sleeve) results in a direct movement of the fuel control rack.

With the engine stopped and the flyweights collapsed (Figure 18.18), the governor springs will force the thrust sleeve forward, and the fulcrum lever will pivot on its pins, allowing the control rack to move completely forward (with wide open throttle). This is the starting position, the position that allows the greatest injection of fuel.

2 *Medium speed.*

With the engine running at medium speed, the flyweights will push the thrust sleeve rearward, compressing the main governor springs (Figure 18.19). In this position the fulcrum lever is nearly vertical and the *droop screw* is contacting the contour of the stop plate.

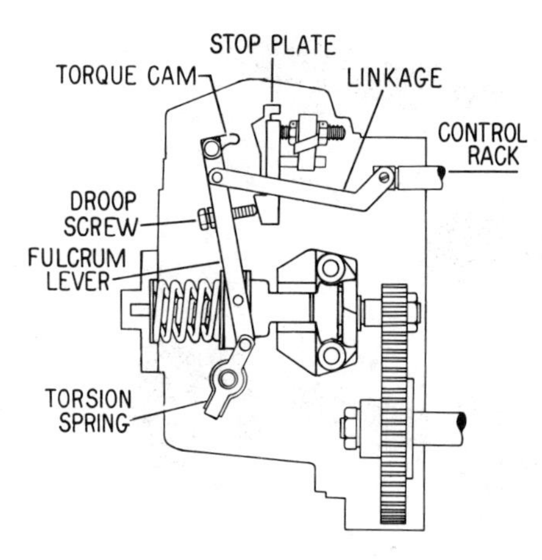

Figure 18.18 Flyweights all the way in (courtesy American Bosch—United Technologies Corporation, Springfield, Mass.).

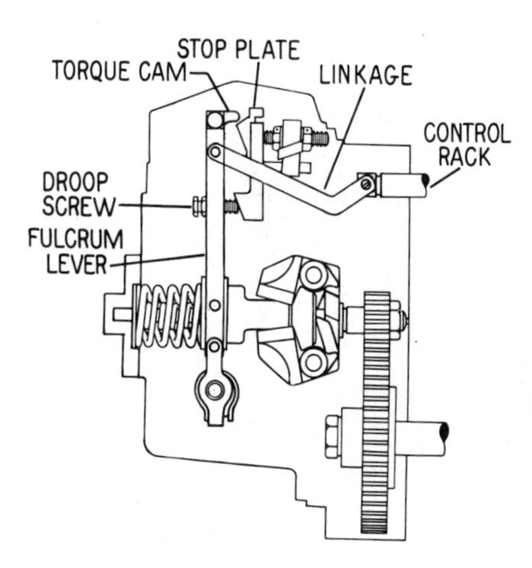

Figure 18.19 Main governor spring compressed (courtesy American Bosch—United Technologies Corporation, Springfield, Mass.).

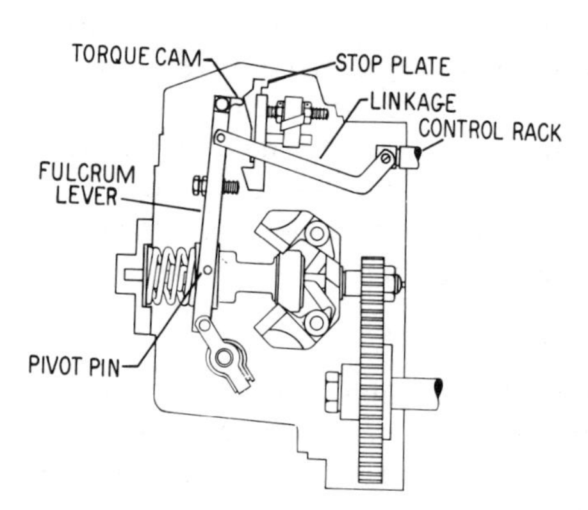

Figure 18.20 Torque cam contacting stop plate (courtesy American Bosch—United Technologies Corporation, Springfield, Mass.).

3 *Rated load.*

As the engine increases in speed up to rated load, the fulcrum lever is pushed back still further and tilts. This causes the torque cam nose to contact the stop plate (Figure 18.20).

NOTE Full load or rated load fuel delivery is adjusted at this speed by moving the stop plate either ahead or back.

4 *Breakaway.*

As speed increases over rated load, the weights develop enough force to push the fulcrum lever completely to the rear. The control rack also moves to the rear, cutting off fuel deliver.

For more information on governors, see Chapter 16.

IV. Disassembly of APE Pump with GV Governor

Before starting the disassembly, clean the exterior of the pump with solvent and blow dry. Also, record all model and serial numbers as they will be needed for ordering parts and testing the overhauled pump. Have at hand all special tools required for the task and start with a clean bench.

NOTE Various procedures that follow require that the front and rear of the pump be known. When working with any American Bosch pump, the drive end is considered the front of the pump.

A. Procedure for Pump Disassembly

1. Clamp the pump in a soft jawed vise or other suitable holding fixture.
2. Loosen and remove the delivery valve holders and valve springs.
3. Using the puller shown in Figure 18.21, remove the delivery valves.
4. Invert the pump and remove the closing plugs at the bottom of the pump (Figure 18.22).
5. Rotate the camshaft slowly and install the tappet holders as each plunger rises to the top (Figure 18.23).
6. Remove the drive coupling and key.

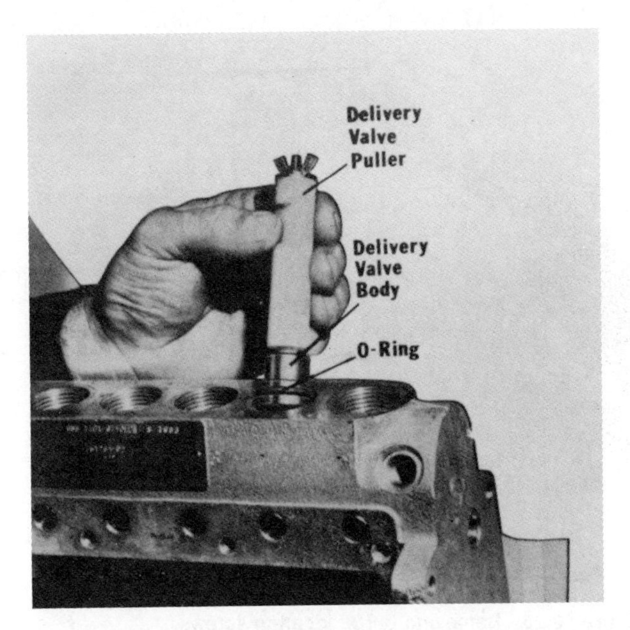

Figure 18.21 Delivery valve puller (courtesy American Bosch–United Technologies Corporation, Springfield, Mass.).

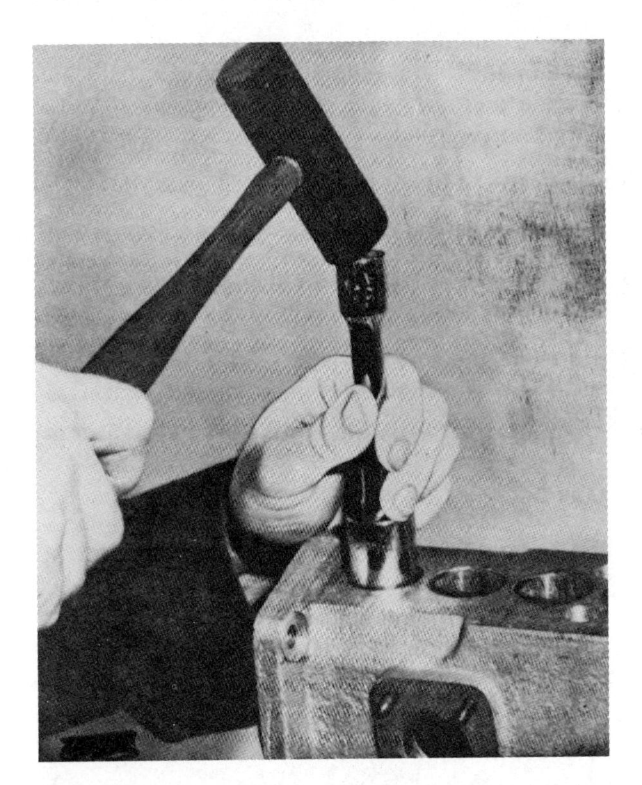

Figure 18.22 Removing closing plugs (courtesy American Bosch–United Technologies Corporation, Springfield, Mass.).

7 Remove the supply pump (if used) and the GV governor. Use care when unhooking the governor to pump linkage so it is not bent. (Refer to step B for governor disassembly.)

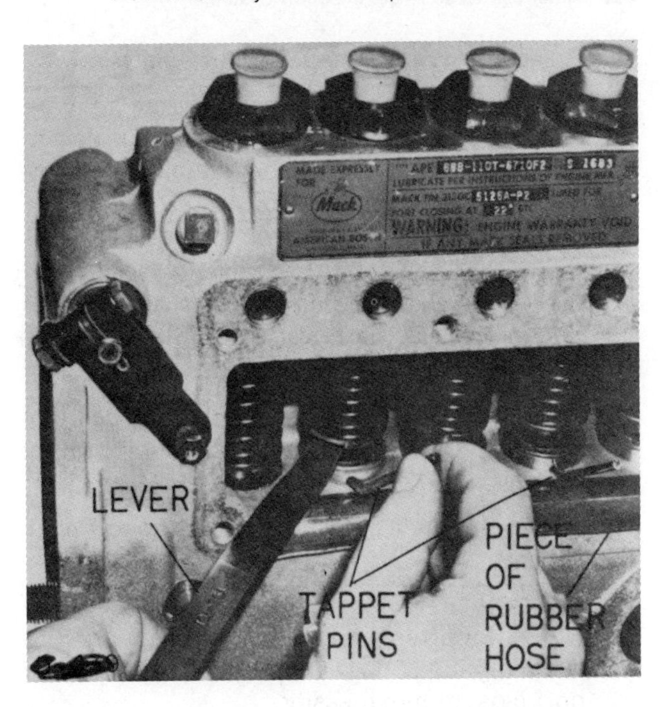

Figure 18.23 Installing tappet holders (courtesy American Bosch–United Technologies Corporation, Springfield, Mass.)

8 Remove the governor drive gear after removing the camshaft nut. The inner drive hub is held on the shaft by the taper fit only.

NOTE When removing the clutch springs, be certain to keep them in proper order in relation to the shims (Figure 18.24).

Figure 18.24 Shim and clutch spring location (courtesy American Bosch–United Technologies Corporation, Springfield, Mass.).

9 Remove the front and rear bearing plates.

CAUTION Note the position of the timing mark on the front plate so that it will be returned to the same position when the pump is reassembled.

10 Loosen and remove the two center bearing retaining screws and then pull the camshaft from the housing and place in a clean tray. Be careful of the bearings.

11 Push down on roller tappets with a wooden tool and remove the tappet holders. Release the spring pressure slowly.

12 Remove the roller tappets through the end plate opening.

CAUTION Keep the tappets in a numbered tray with the corresponding pumping plunger to avoid mix-up. This will make the adjustment of port closure much easier.

13 Using a magnet or mechanical fingers, withdraw the pumping plungers with lower spring seat.

CAUTION Keep the plungers in order in a numbered tray until they can be replaced in their respective barrels. The plunger and barrel form a matched set.

14 Remove the plunger spring, control sleeve, and upper spring seat.

15 Pull out the control rack.

16 Put the pump upright in the vise again and remove the barrel locating screws (Figure 18.25).

17 Carefully lift out each barrel and insert its respective plunger into it. Place the unit in a divided tray.

Figure 18.25 Removing barrel locating screws.

B. Procedure for Governor Disassembly

1 Remove top cover of governor and bumper spring.

2 Remove the external control lever after first scribing a location mark on the lever and the shaft.

3 *Do not* remove the stop plate or cam nose unless replacement is necessary.

4 Remove the operating lever shaft after first removing the cover plates and set screw (Figure 18.26).

5 Remove the screws holding all the internal rotating parts in place and pull the assembly from the governor housing.

6 Pull out the operating lever spring and hub.

7 Remove the end cap, bushing, and oil seal.

8 Remove the inner and outer governor springs with spring seat, shims, and thrust sleeve.

9 Lift out the fulcrum lever. Remove the driven gear and press the shaft from the ball bearing.

10 Remove keys and press the weight pins out of the shaft.

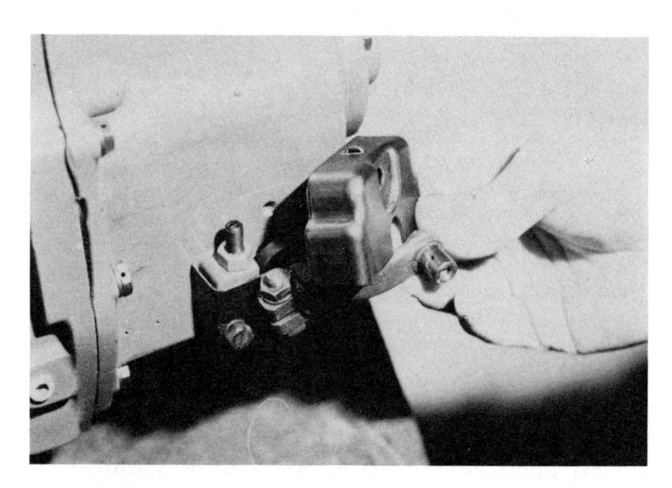

Figure 18.26 Removing cover plate.

V. Examination and Replacement of Parts

Thoroughly clean all parts in solvent or parts cleaner and blow dry. Spread them out on a clean bench for inspection.

A. Procedure for Examination of Pump Parts

1 Check all major housings for cracks, breakage, or thread damage.

2 The lobes of the camshaft should be smooth without evidence of flaking or pounding.

3 Replace cam bearings if the pump is being completely overhauled.

4 Check roller tappets, springs, and supply pump roller for flaking or chipping.

5 Carefully examine the pumping plungers for scoring and wear. If score marks can be felt with the fingernail, the plunger/barrel assembly should be replaced.

B. Procedure for Examination of Governor Parts

1 Check all governor parts for excessive wear, for example, flat spots, rough bearings, galled parts.

2 Replace pin bushings in the governor weights if wear is high.

CAUTION Flyweights are fitted with one needle bearing and one bushing *or* two bushings per pin. This determines the percentage of speed regulation. Be certain to replace bearings or bushings with identical parts.

3 Replace the flexible steel clutch discs to eliminate excess slippage of flyweights.

NOTE Be careful to arrange the discs and shims in their correct order to maintain the correct tension.

Always replace all gaskets, washers, and oil seals every time the pump/governor is disassembled. American Bosch has gasket kits available with all gaskets required for overhaul.

VI. Reassembly

Prepare the workbench for reassembly by cleaning it well. Discard all worn parts and any other unnecessary material.

A. Procedure for Pump Reassembly

1 Clamp the pump housing in the brass jawed vise in an upright position.

2 Install all of the barrels in their correct bore, making sure to align the slot that locates the barrel with the barrel locating screw hole. Install the screws.

3 Install the delivery valves, springs, and gaskets. Screw in the holders and torque.

CAUTION Torque values of different-size APE pumps will vary. Always refer to the torque chart that applies to your pump.

4 Push the control rack into the housing and secure with the screw.

5 Invert the housing in the vise.

6 Install the control sleeves so that the screw on the sleeve is parallel to the control rack (Figure 18.27).

7 Set the upper spring seat in place over the control sleeve.

8 Install plunger return springs.

9 Assemble the plungers with lower spring seat to the barrels, making sure the plunger is mated to the correct barrel. Also, the assembly marks on the control sleeve and plunger must match.

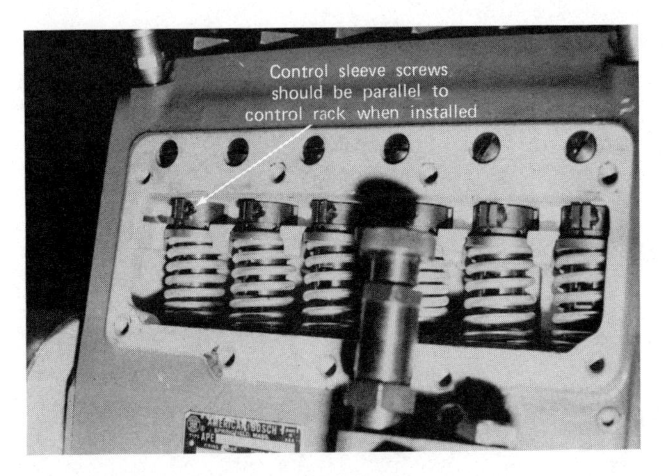

Figure 18.27 Installing control sleeves.

Figure 18.28 Checking camshaft end play (courtesy American Bosch—United Technologies Corporation, Springfield, Mass.).

10 Replace the roller tappets in their correct bore. Install the guide screws.

11 Compress the plunger springs with tool no. TSE76160 and install the tappet holders.

12 Apply a dab of Vaseline to the center bearing and place it on the camshaft. Slide the camshaft into the housing.

13 Screw in the center bearing screws and tighten to the correct torque.

14 Install both end plates and torque the screws.

15 The end play of the camshaft must now be checked. Place the tip of a dial indicator on the shaft as shown in Figure 18.28. Tap the camshaft back and forth to determine the end play. It should be 0.004 to 0.008 in. (0.1 to 0.2 mm). If incorrect, remove the end plate and add or delete shims until the end play is within limits.

Figure 18.29 Measuring governor clutch slipping torque.

16 Place the pump upright in the vise again.

17 While rotating the camshaft, remove the tappet holders.

18 Replace the base plugs. On cup plugs, use a sealer before installation.

B. Procedure for GV Governor Assembly

1 Install the stop lever in the governor housing.

2 Set the pump vertically in the vise. Assemble the drive hub and clutch components and secure with the retaining nut. Use a light coat of oil on all clutch parts.

3 The slipping torque of the clutch must be measured using tool no. TSE7947 (Figure 18.29). The gear should rotate at a pull of 3 to 3.5 lb (1.36 to 1.5 kg). If the tension is too great, shims must be added between the clutch discs. If the gear moves too easily, remove the shims.

4 Install the bearings on the governor shaft and secure with the retaining nut.

5 Slide the sleeves over the shaft, then the springs and retaining nut.

6 Fit the fulcrum lever to the weight shaft with the pins engaged with the thrust sleeve.
Upper holes = 10% regulation
Lower holes = 5% regulation

7 Install the hub and torsion spring so the spring ends engage the fulcrum lever bar (Figure 18.30).

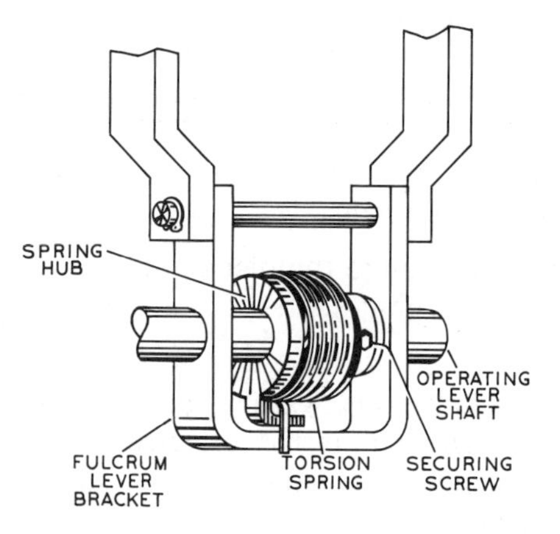

Figure 18.30 Correct assembly of governor torsion spring (courtesy American Bosch—United Technologies Corporation, Springfield, Mass.).

Figure 18.31 Locking governor shaft with small screw.

Figure 18.32 Governor spring gauge on housing.

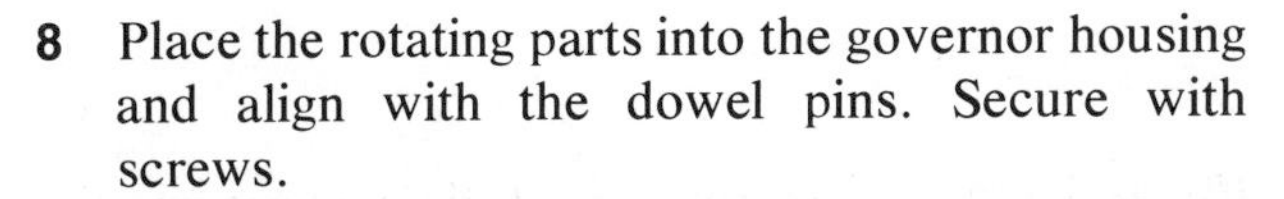

8 Place the rotating parts into the governor housing and align with the dowel pins. Secure with screws.

CAUTION The cutaway portion of the bridge must face down.

---

9 Install oil seals for lever shaft in the housing.

10 Slide the shaft into the housing and lock with the small screw (Figure 18.31).

11 Attach the governor to the injection pump, making certain to hook the pin in the control rack of the pump.

12 Before calibration, the tension of governor spring must be checked with tool no. TSE7939.

13 Remove the rear cover, locknut, ball bearing, spring seat, and *outer* spring.

NOTE Remove idle and full load adjusting screws.

14 Reinstall spring seat and place gauge on governor housing as shown in Figure 18.32.

15 Push the prongs of the gauge in until the spring seat contacts the inner spring. Tighten set screw.

16 Remove gauge and measure the distance the prongs protrude with a depth micrometer.

17 Add or remove shims under spring seat to obtain the specified gap.

NOTE If the spring is to be precompressed, add shims (or remove) until spring is flush and then add a shim of the thickness indicated.

18 Remove the inner spring and check the setting of the outer spring in the same manner.

19 Reinstall the ball bearing, locknut, and end cap. Screw in the idle and full load screws.

VII. Calibration of APE Pump with GV Governor

If an injection pump is received in the shop while still under warranty or after being overhauled, it must be tested and calibrated on a test bench. The following is the procedure to follow when calibrating an APE pump equipped with a GVA governor.

A. Procedure for Internal Timing (Port Closure or Port Opening)

This test is required to set all pumping elements in time so that the spill port for each element is covered at the correct time and in the correct relation to each other. Also, the external, or static, timing is set from this position.

1 Mount pump on test stand, using the correct adapter.

2 Connect the fuel supply to the gallery of the pump. Plug the outlet.

3 Place the control rack at the midposition.

4 Remove the delivery valve holder, delivery valve, and spring from no. 1 cylinder (cylinder nearest drive end). Replace holder.

Figure 18.33 Curved injector line installed on no. 1 pump outlet.

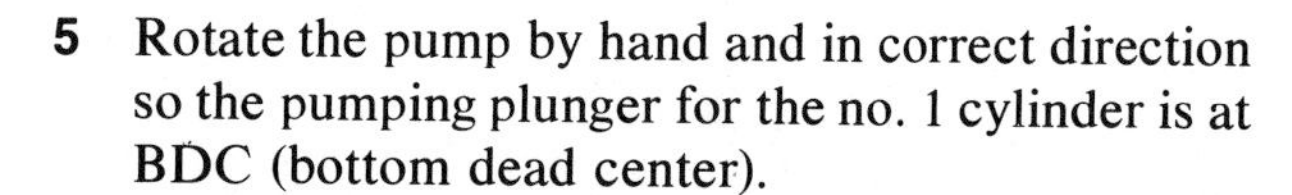

5 Rotate the pump by hand and in correct direction so the pumping plunger for the no. 1 cylinder is at BDC (bottom dead center).

6 Connect a curved injection line (Figure 18.33) to the no. 1 outlet.

7 Turn on the fuel supply and regulate to 2 psi (0.14 kg/cm^2). Fuel will flow from the curved line.

8 Rotate the pump by hand and in the direction of rotation until the flow of fuel slows to a drop every 10 seconds. *This is port closure.*

9 At this time the scribed mark on the camshaft *must* align with a similar mark on the end plate (Figure 18.34).

Figure 18.34 Camshaft and hub mark in alignment.

Figure 18.35 Tappet adjusting screw.

10 If the marks are not aligned, the plunger position must be changed, using the tappet adjusting screw (Figure 18.35).

NOTE On pump model APE-BB, adjusting shims are used on the roller tappets.

11 After the adjustment is made, lock the adjusting screw and recheck the marks.

12 With no. 1 cylinder correctly set, shut off the fuel supply, remove the line and holder, and replace the delivery valve, spring, and holder. Remove the holder from the no. 5 cylinder and take out the valve and spring. Replace the holder and curved line.

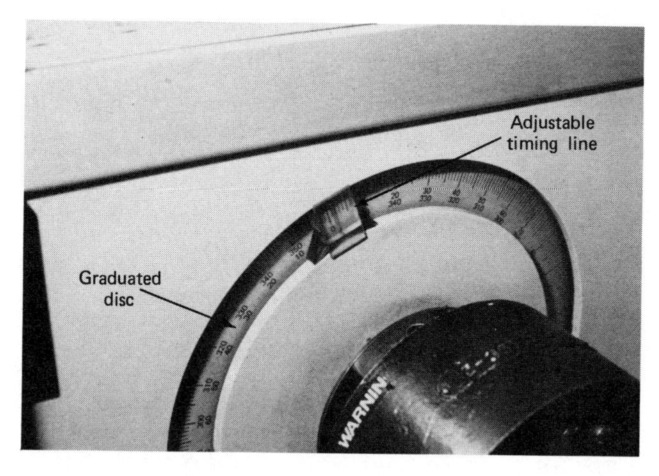

Figure 18.36 Setting degree wheel.

13 Set the graduated disc (Figure 18.36) to 0.

14 Start the fuel supply and rotate the pump by hand as follows:
90° – 4 cylinder pumps
60° – 6 cylinder pumps
45° – 8 cylinder pumps

15 When the disc shows the correct degree, the fuel flow must cease. If not, adjust the roller tappet.

16 Repeat this procedure in turn for the remaining cylinder. The usual firing order is as follows:
4 cylinders – 1,3,4,2
6 cylinders – 1,5,3,6,2,4

17 Pumps without external timing mark.

a Some APE pumps were made without timing marks. On these pumps it will be necessary to measure the distance from the top of the holder to the plunger (at BDC).

NOTE Delivery valve and spring must be removed as usual.

b Then flow test the pump and ascertain the point of port closure. At this point, again measure the distance as in item *a* above. The difference in the two readings is the plunger lift for port closing. It must be as stated on the pump specification. If not, adjustment of the roller tappet is required. Replace the springs and delivery valves when completed.

B *Equal fuel delivery.*
All pump outlets must deliver exactly the same quantity of fuel to give an even running engine.

1 Pour approximately 0.13 gallon (0.5 liter) of lubricating oil into the pump camshaft chamber.

2 Move the control rack to midposition.

3 Run the pump at 500 rpm for 5 to 10 minutes to remove air and stop any leaks.

4 Run the pump at the full load setting speed and take three draws of fuel.

5 Check the delivery of the third draw. All outlets must be even within 1 cc. If they are not even, loosen the clamping screw of the control sleeve (Figure 18.37) and rotate the sleeve (and plunger). To increase delivery, move the sleeve toward the control rack stop.

6 After every adjustment, be certain to lock the clamp screw.

C *Full fuel delivery.*

1 Run the pump at the speed listed on the specification sheet.

NOTE Throttle lever must be in wide open position (against stop).

2 Collect and record the delivery for the stroke count listed (usually 500 or 1000 strokes).

3 If delivery is incorrect, adjust the high idle stop screw (Figure 18.38) until the correct delivery is obtained.

Figure 18.37 Control clamp screw.

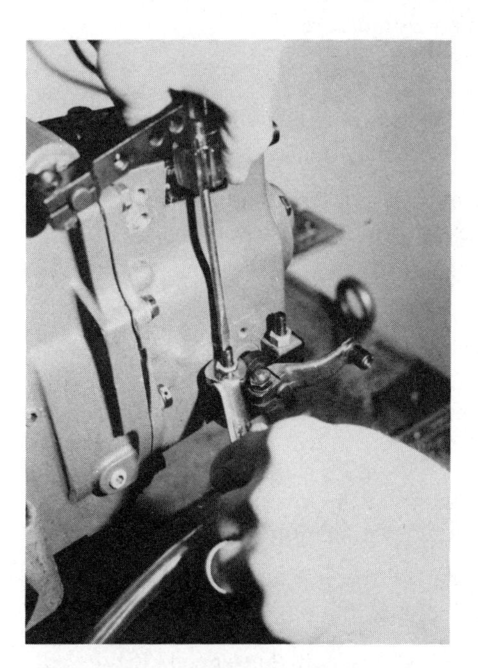

Figure 18.38 High idle stop screw.

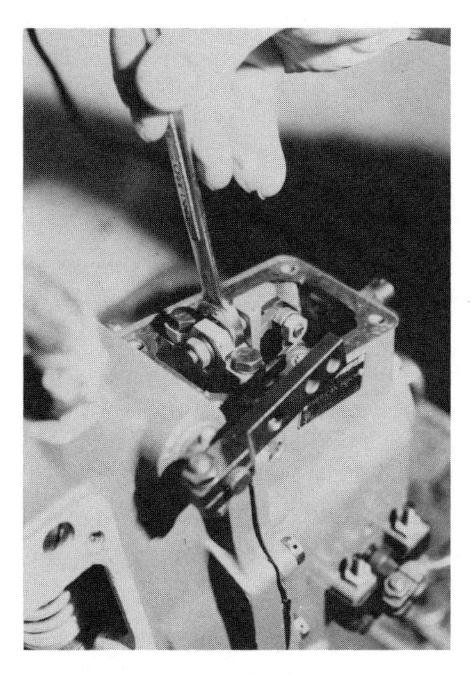

Figure 18.39 Adjusting stop plate.

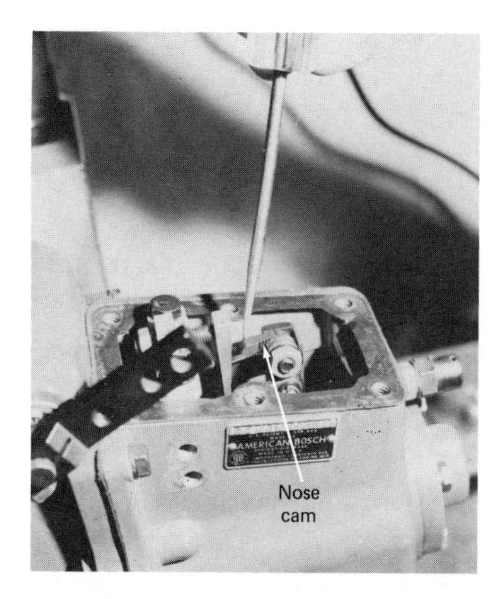

Figure 18.40 Cam leaving stop plate.

4 Next, adjust the stop plate (Figure 18.39) until it just contacts the nose cam; lock the stop plate, adjusting the screw.

D *Torque control.*

Reduce pump speed to $\frac{2}{3}$ of full load speed. Fuel delivery will increase.

E *Breakaway.*

1 At speed listed on the test sheet, the cam nose must leave the stop plate.

2 If the cam does not leave, adjust the high idle stop screw down until it does (Figure 18.40).

F *Bumper screw.*

With the pump running at high idle speed, turn in the bumper screw (Figure 18.41) until the governor oscillations are almost eliminated. Lock the screw in place.

G *Low idle.*

1 Run the pump at the lower speed for low idle given on the specification sheet.

2 Then place the throttle lever in low idle position and increase the pump speed 20 percent.

3 Delivery must cease at that speed. If not, adjust the low idle stop screw (Figure 18.42).

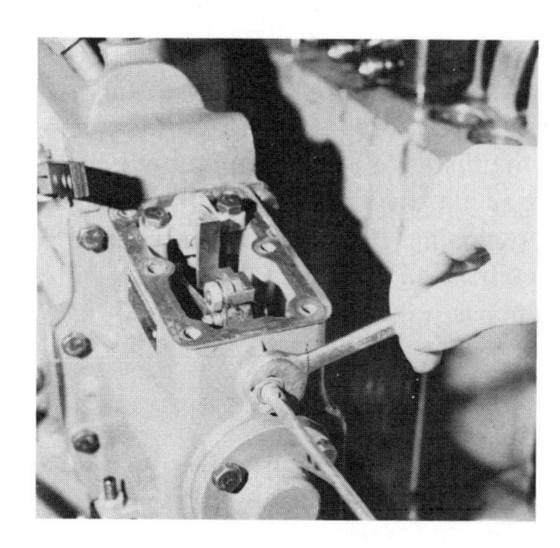

Figure 18.41 Adjusting bumper screw.

After completion of all adjustments, place the required seals on the high idle and full load stop screws. Remove the pump from the test bench, clean, and paint.

VIII. Identification of American Bosch PSJ, PSM, and 100 Pumps

The American Bosch pump shown in Figure 18.43 is a Model 100 injection pump. It is a flange mount, single plunger design of the constant stroke, distributing plunger, sleeve control type. Important information about this pump is listed on the pump nameplate. Found on the nameplate will be pump model number, serial number, and part number. A typical model num-

Figure 18.42 Adjusting low idle stop screw.

ber might be 100 6A 100A 9164A1, which stands for the following:

100—basic pump model
6—number of cylinders
A—size of pump
100—diameter of plunger in tenths of millimeters (100 = 10.0mm diameter)
A—execution of pump (denotes a major change)
9164—variation specification from a basic pump
A—specification edition (denotes a minor change)
1—specification code (may indicate a change in calibration)

IX. Component Parts—PSJ, PSM, and 100 Pumps

The following is a listing of the major sections of the pumps and their function.

A *Supply pump.* Mounted on the rear of the main pump, the supply pump pulls fuel through the primary fuel filter and then forces it through the secondary filter to the hydraulic head at low pressure.

B *Drive section.* The drive section includes the camshaft (with intravance on Model 100), roller tappet, and governor drive gears. These pumps run at the same speed as the engine. The drive section causes the plunger to reciprocate as well as rotate and drives the governor and supply pump.

Figure 18.43 100 series cross section (courtesy American Bosch—United Technologies Corporation, Springfield, Mass.).

C *Pumping section.* The pumping section includes the pumping plunger and control sleeve, hydraulic head, and delivery valves. The hydraulic head and plunger generate high pressure fuel and deliver it to the nozzles in the correct firing order.

Delivery valves serve to keep fuel in the line between injections and lower line pressure to insure immediate closing of the nozzle valve.

D *Control section.* The control section includes the governor and linkage components. The Model 100 uses a conventional mechanical flyweight governor to raise and lower the control sleeve in relation to the engine speed and load conditions.

X. Operation of PSJ, PSM, and 100 Pumps

A *Fuel flow and operation* (Figure 18.44). The American Bosch PSJ, PSM, and 100 pumps operate on the same principle with the exception of the automatic timing device. Only the 100 pump will be covered here. The Model 100 is a flange mount, single plunger design of the constant stroke, distributing plunger, sleeve control type. Fuel flows from the supply pump to the fuel inlet on the housing. This is the low pressure supply and is regulated by a relief valve located in the supply pump housing. An orifice located in the outlet fitting limits the amount of return fuel.

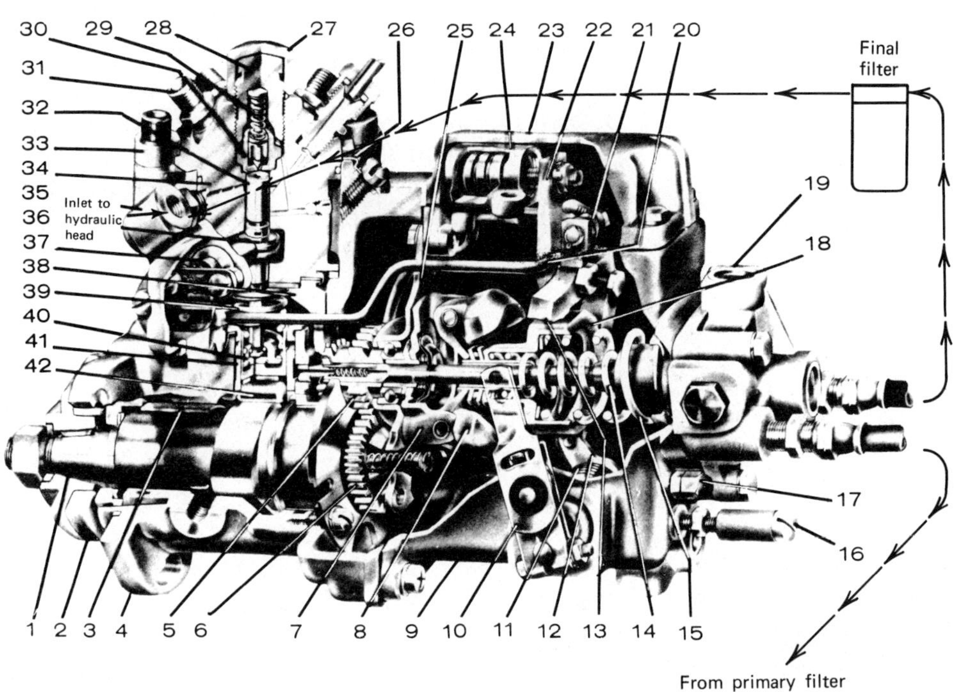

Nomenclature

1. Camshaft
2. Drive Plate
3. Camshaft Bearing
4. Pump Mounting Flange
5. Governor & Hydraulic Drive Gear
6. Camshaft Gear
7. Governor Weight Spider
8. Governor Weights
9. Governor Housing
10. Operating Lever
11. Operating Shaft Spring
12. Fulcrum Lever Bracket
13. Fulcrum Lever
14. Inner Governor Spring
15. Outer Governor Spring
16. Low Idle Screw (Spring Loaded)
17. High Idle Screw
18. Governor Sliding Sleeve
19. Fuel Supply Pump
20. Control Rod
21. Torque Cam
22. Stop Plate
23. Governor Top Cover
24. Excess Fuel Starting Device
25. Ball Bearing Support Plate
26. Head Indexing Plate
27. Delivery Valve Cap Nut & Gasket
28. Delivery Valve Holder
29. Delivery Valve Spring
30. Delivery Valve & Spring Guide
31. Fuel Discharge Outlet
32. Hydraulic Plunger
33. Hydraulic Head Clamping Screw & Holder
34. Hydraulic Head Assembly
35. Overflow Valve
36. Fuel Metering Sleeve
37. Control Unit Assembly
38. Face Gear
39. Plunger Return Spring
40. Plunger Button & Spring Seat
41. Tappet Guide
42. Tappet Roller

Figure 18.44 100 series fuel flow (courtesy American Bosch—United Technologies Corporation, Springfield, Mass.).

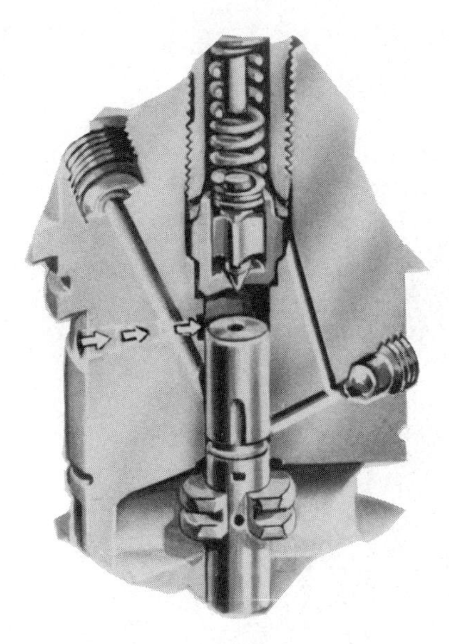

Figure 18.45 Pumping plunger area filling (courtesy American Bosch—United Technologies Corporation, Springfield, Mass.).

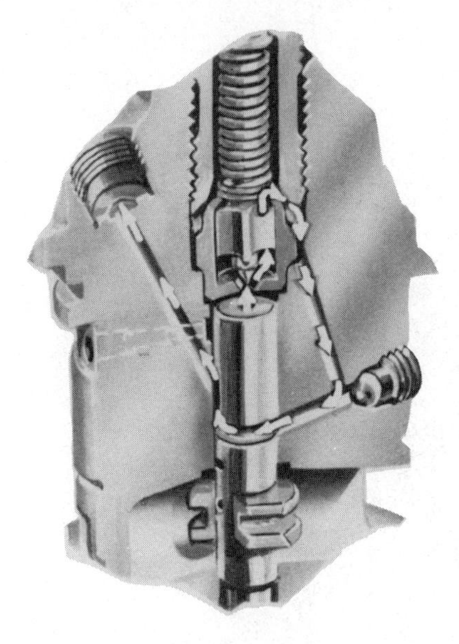

Figure 18.47 Hydraulic head fuel flow (courtesy American Bosch—United Technologies Corporation, Springfield, Mass.).

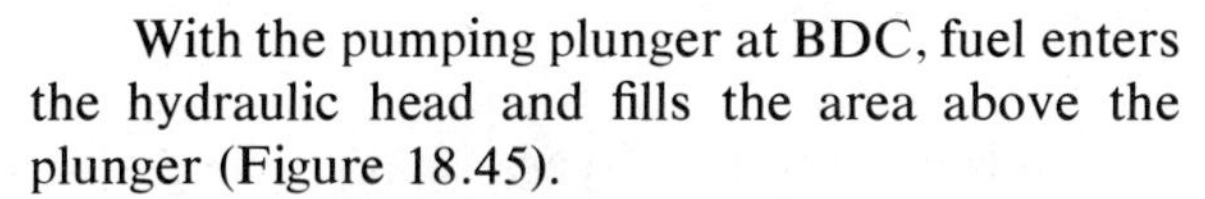

With the pumping plunger at BDC, fuel enters the hydraulic head and fills the area above the plunger (Figure 18.45).

The pumping plunger has a vertical hole extending down to the spill port. It also has a distribution slot that will distribute fuel to each cylinder in its correct firing position. As the rotating plunger moves upward, it covers the fuel intake. At this point the fuel above the plunger is trapped and ready for delivery (Figure 18.46). This is *port closure* and is very important to internal and external timing of the pump. As the plunger continues to move upward, the fuel will be forced by the delivery valve and back down into the hydraulic head (Figure 18.47). Fuel then flows to a horizontal annulus around the plunger, to the vertical slot, and up to the head outlet. Fuel then moves to the injection nozzle.

When sufficient fuel has been injected, as determined by the governor, the spill port in the plunger will be uncovered (Figure 18.48). This is the end of delivery, as the high pressure fuel above

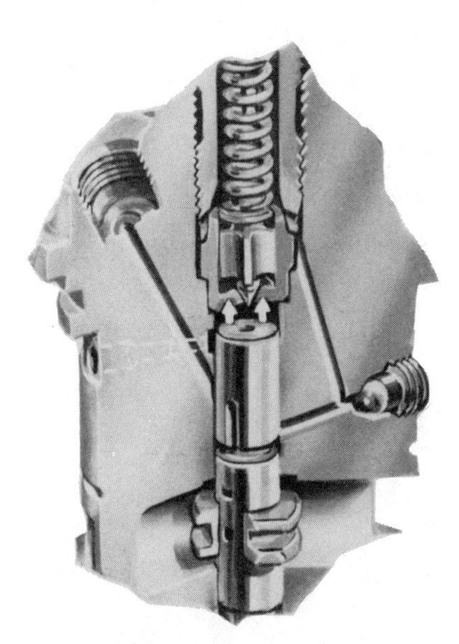

Figure 18.46 Port closure (courtesy American Bosch—United Technologies Corporation, Springfield, Mass.).

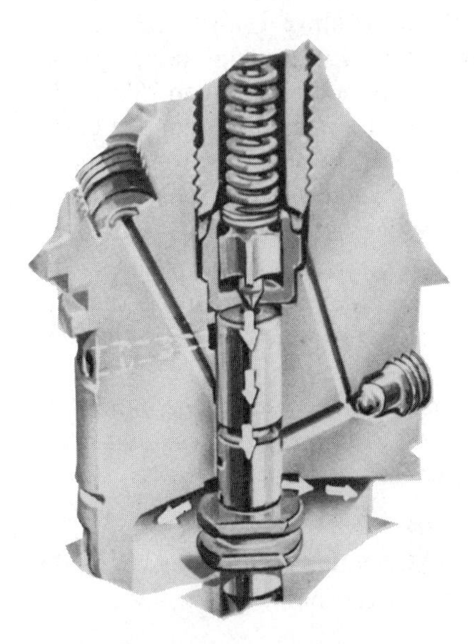

Figure 18.48 Spill port uncovered (courtesy American Bosch—United Technologies Corporation, Springfield, Mass.).

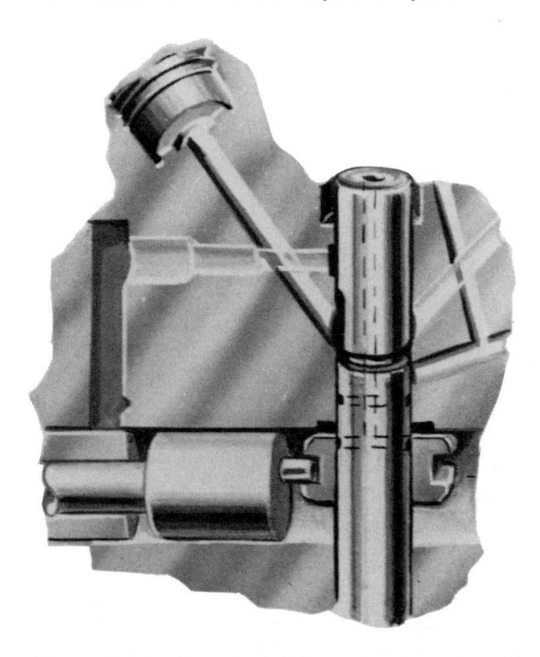

Figure 18.49 More fuel delivery (courtesy American Bosch—United Technologies Corporation, Springfield, Mass.).

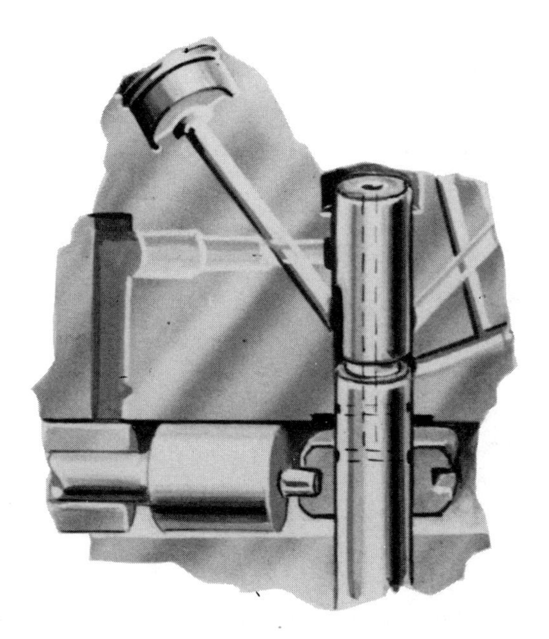

Figure 18.50 Less fuel delivery (courtesy American Bosch—United Technologies Corporation, Springfield, Mass.).

the plunger can now go down the center of the plunger and empty into the low pressure sump area. The delivery valve will close.

The amount of fuel delivered is controlled by the sliding control sleeve (thus the name, sleeve controlled metering). The vertical movement of this sleeve, which is controlled by the governor via an eccentric stud and control lever, will uncover the spill port in the plunger sooner or later in its cycle. L *later* = more delivery (Figure 18.49), *earlier* = less delivery (Figure 18.50).

This control allows for a completely variable control of fuel from no load to full load. *No delivery* (Figure 18.51) occurs when the sleeve is at the extreme lower position. In this position no pressure can be built above the plunger because the spill port is already open at BDC.

An operational diagram showing how the plunger is rotated and reciprocated in correct sequence is shown in Figure 18.52. Notice that the plunger is forced up and down at engine speed and rotated at one-half engine speed. One complete pump cycle (firing all cylinders) requires two revolutions of the camshaft.

B *Operation of governor.*

Refer to Part II, Section C, for operational details of mechanical governors.

C *Operation of supply pump.*

Late model distributor pumps use a positive displacement gear type supply pump (Figure 18.53). The function of the supply pump is to supply fuel to the hydraulic head at low pressure and at all speeds.

Fuel is drawn from the tank to the supply pump and forced out to the hydraulic head. When the correct operating pressure is reached, a relief

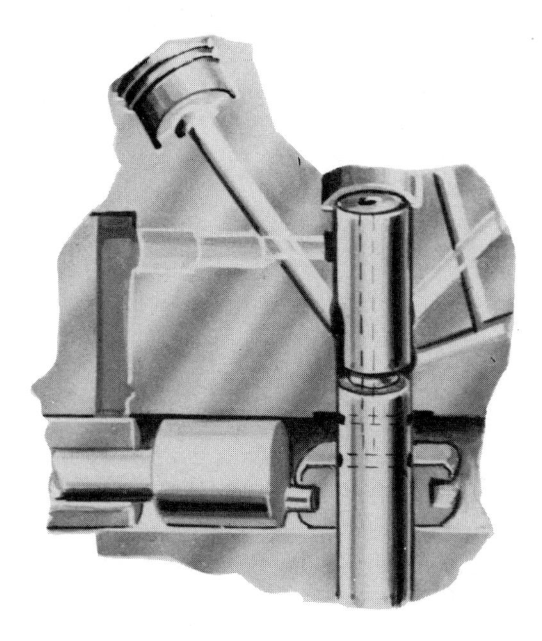

Figure 18.51 No delivery (courtesy American Bosch—United Technologies Corporation, Springfield, Mass.).

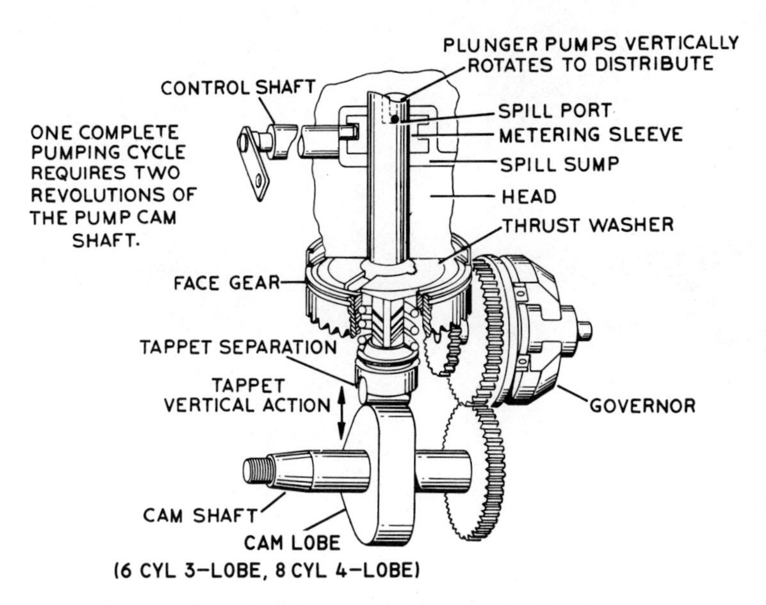

Figure 18.52 Operational diagram (courtesy American Bosch – United Technologies Corporation, Springfield, Mass.).

valve opens and bypasses fuel to the inlet.

When a hand primer is attached (optional), fuel is drawn in and forced out by the hand plunger (Figure 18.54).

The supply pump is driven by the governor shaft extending out the rear of the pump.

D *Operation of timing advance.*

The intravance timing unit takes the place of the standard camshaft and is actuated by oil pressure from the engine. The unit can provide up to a 20° advance.

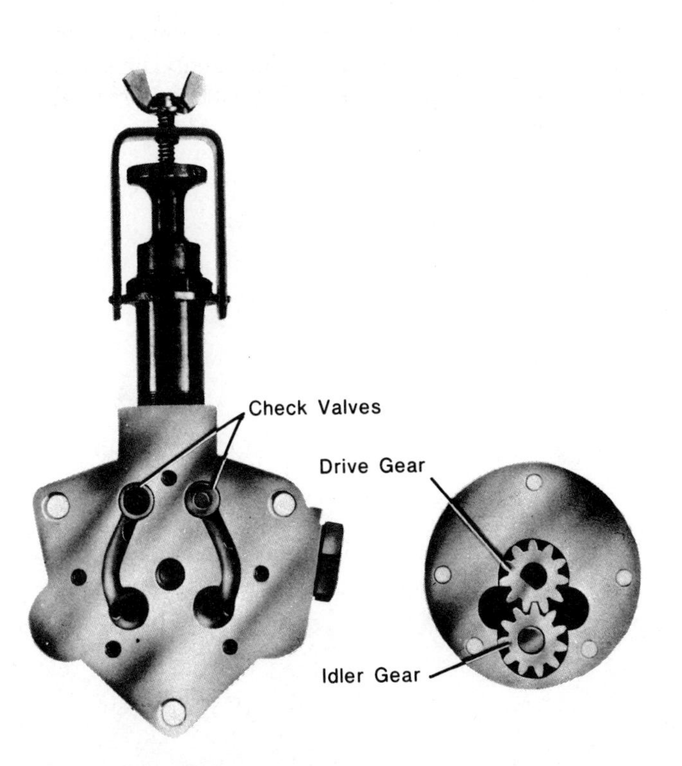

Figure 18.53 100 series gear pump (courtesy American Bosch – United Technologies Corporation, Springfield, Mass.).

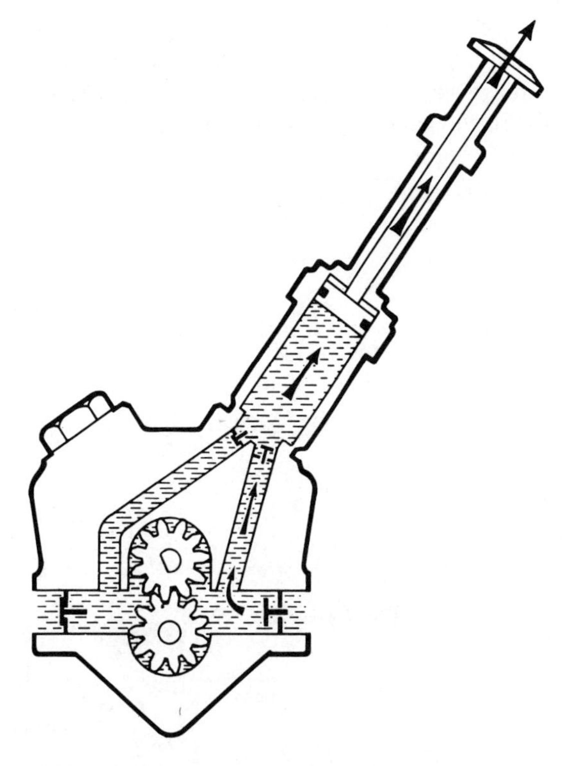

Figure 18.54 Hand primer (courtesy American Bosch – United Technologies Corporation, Springfield, Mass.).

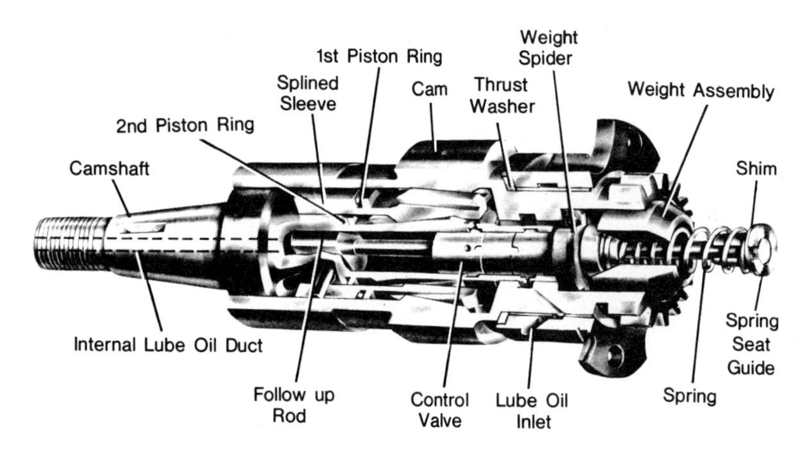

Figure 18.55 Cutaway view of intravance (courtesy American Bosch—United Technologies Corporation, Springfield, Mass.).

A cutaway view of all intravance parts is shown in Figure 18.55.

The intravance is classified as a speed advance unit because it responds to speed changes only. At low speed the control weights of the unit are collapsed and the control valve does not allow oil to enter the unit. There is no advance in this position. As engine speed rises, the control weights fly outward, pulling the control sleeve to the rear of the pump. Oil under pressure now passes into the unit and moves the splined sleeve inside the advance unit, causing it to rotate (Figure 18.56). It will rotate against the direction of rotation up to as much as 20°.

As engine speed falls, the weights will collapse and the control valve will be closed by the spring behind the weights. The cam will then move back as the oil drains from the piston area.

XI. Disassembly, Inspection, and Reassembly of Model 100

Lay out all special tools needed for pump repair on a clean bench. Clean the pump thoroughly in solvent and blow dry. Record model and serial numbers on the work order.

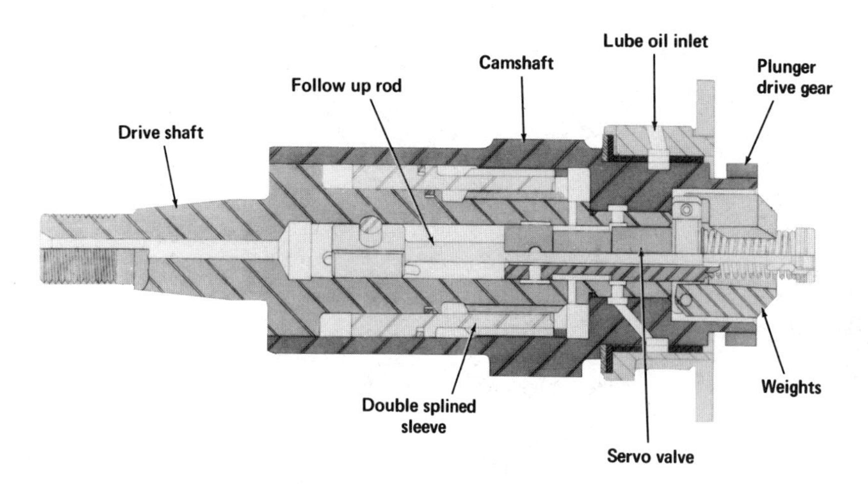

Figure 18.56 Intravance operation (courtesy International Harvester Company).

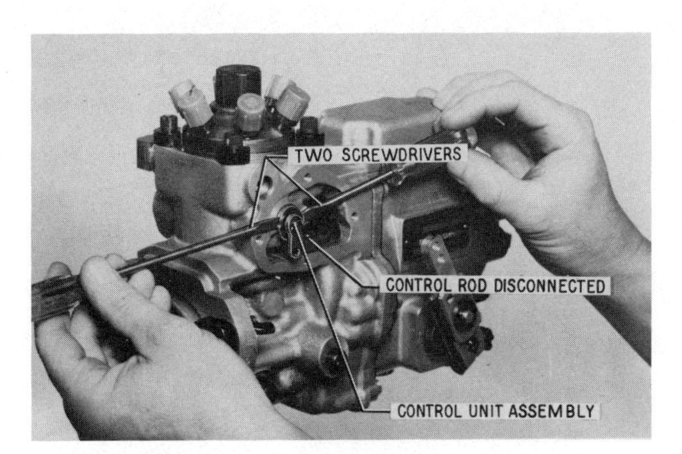

Figure 18.57 Removing control unit and spacers (courtesy American Bosch—United Technologies Corporation, Springfield, Mass.).

A. Procedure for Disassembly

1. Clamp pump in holder and place in a vise.
2. Using a three leg puller, remove the drive hub and back-up plate.
3. Remove the governor top cover and side inspection cover.
4. Remove the control unit plate and spacers (Figure 18.57). Disengage the control rod and pull out the control unit from the hydraulic head.

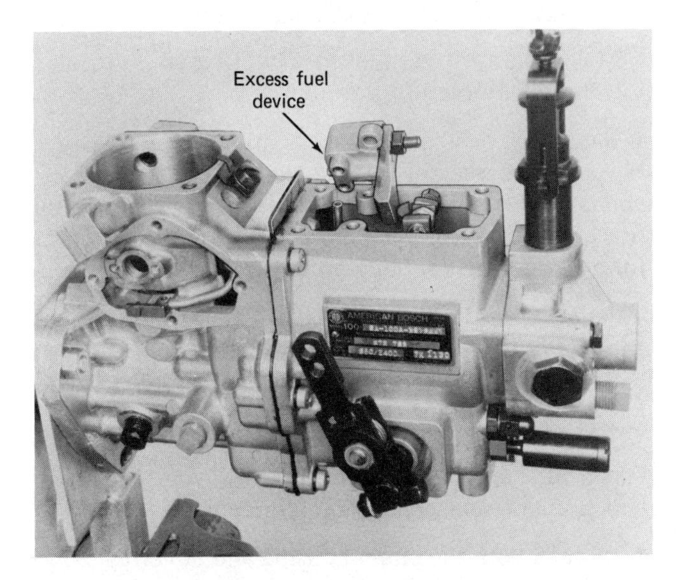

Figure 18.58 Removing excess fuel device (courtesy International Harvester Company).

5. Remove the delivery valve cap and loosen the delivery valve.
6. Remove the four head retaining screws and pull the head straight out of the housing.

CAUTION *Do not remove discharge fittings.*

7. Lift out the tappet and roller.
8. Remove excess fuel device (Figure 18.58).
9. Remove supply pump.
10. Carefully lift out the governor springs and spring seat with shims.
11. Tap governor housing to remove it from the main pump housing.
12. Remove camshaft rear bearing plate and pull the camshaft from the rear of the pump. Excess fuel device supply tube will come out with rear bearing plate and thrust washer (Figure 18.59).
13. Remove the front bearing plate with thrust washer and shims.
14. *Do not* disassemble the hydraulic head as general practice. The head is checked on the test bench when the pump is assembled. Remove the hydraulic head face gear and inspect the thrust washer.

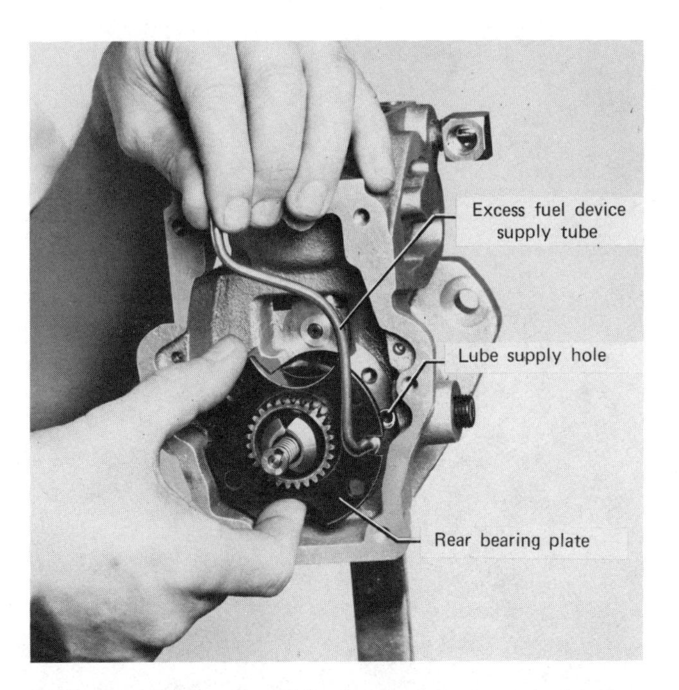

Figure 18.59 Removing oil line that leads to excess fuel device (courtesy American Bosch—United Technologies Corporation, Springfield, Mass.).

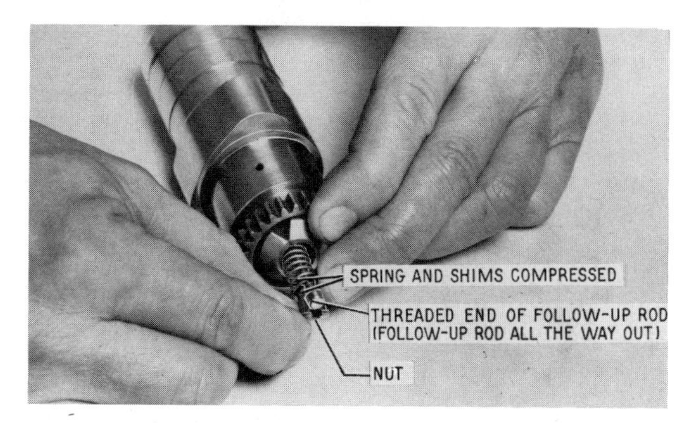

Figure 18.60 Disassembly of intravance (courtesy American Bosch—United Technologies Corporation, Springfield, Mass.).

15 Disassemble the intravance assembly (Figure 18.60) by first rotating the nut on the follow-up rod counterclockwise. When nut is past the threads, compress the spring and shims toward the drive end of cam and slide the nut sideways off the cam follow-up rod.

16 Remove weights, spiders, and servo piston.

17 Slide the cam apart.

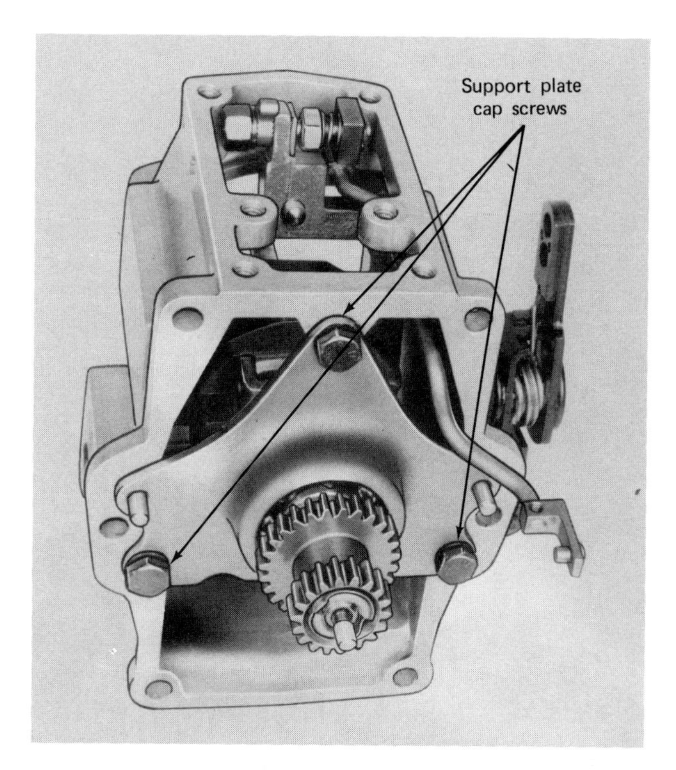

Figure 18.61 Removal of governor weight carrier (courtesy International Harvester Company).

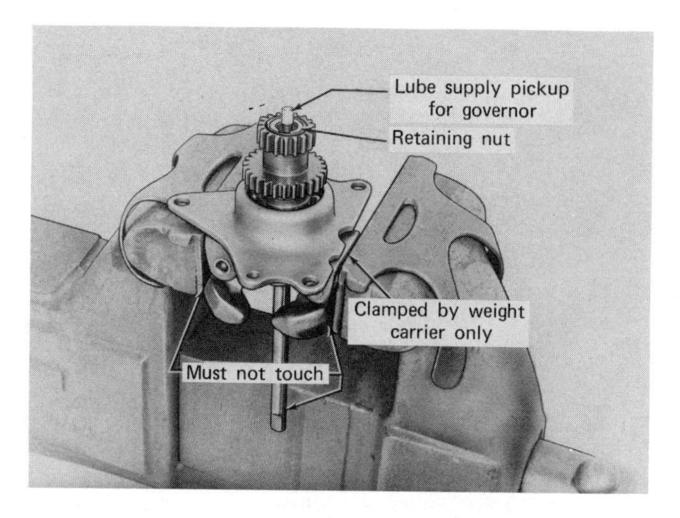

Figure 18.62 Weight and plate assembly installed in vise (courtesy International Harvester Company).

18 Remove the cross pin, rod, and inside helix.

19 Remove the screws from the support plate and pull out the governor weight and shaft assembly (Figure 18.61).

20 Remove the governor sleeve assembly.

21 Remove the screws from the stop lever and spring plate. On later pumps, pull the thrust return spring. Pull the shaft out of the governor housing.

22 Lift out the fulcrum lever assembly and the control rod.

23 Lock the weight and plate assembly in the vise as shown in Figure 18.62. Using a spanner socket with an inch pound torque wrench, rotate the assembly to check the slippage torque.

NOTE This torque must be checked while rotating the assembly (Figure 18.63).

If the torque is 40 in. lb (4.62 N.m), the weight assembly can be reassembled to the pump. If not, the spacer shims must be moved from one side to the other on the friction disc assembly. Before changing shims, check for a worn thrust surface on the spider bushing.

B. Procedure for Parts Inspection

Clean all parts in solvent and blow dry. Clean the bench area for inspection of pump parts.

1 Check both the pump housing and governor housing for damage, that is, thread damage, cracks, or wear.

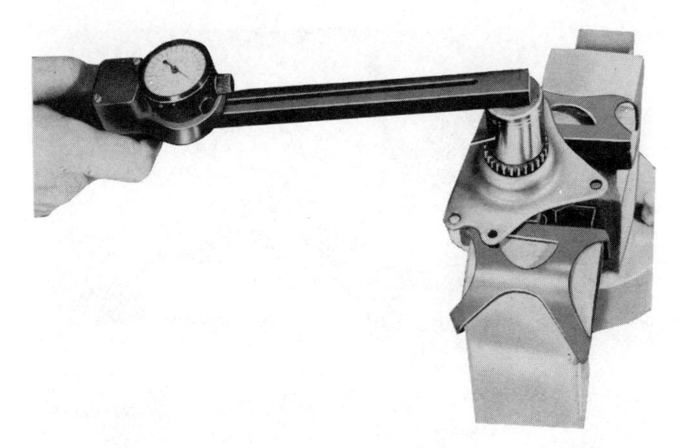

Figure 18.63 Checking governor drive slippage (courtesy American Bosch—United Technologies Corporation, Springfield, Mass.).

2 Check governor weights and pivot pins for excessive looseness. Replace if necessary.

NOTE Acceptable clearance between pins and weights is 0.007 in. (0.18 mm) before replacement is necessary.

3 Check governor slip clutch operation as outlined in step 23 above. Repair as required.

4 Check governor drive gear and head face gear.

5 Check all advance components for wear and breakage.

6 Inspect the cam lobes for flaking or pounding of material.

7 Inspect all springs for rust or breakage and replace as required.

C Parts replacement.
A normal parts replacement list for a 100 series pump would be:

1 Complete overhaul gasket set and seals.

2 Governor clutch friction springs.

3 Intravance follow-up rod spring.

4 Delivery valve cap nut seal ring.

5 Governor weights and pins.

6 Hydraulic head wear washer.

7 Governor spider bushing.

8 Governor slip clutch adjusting nut lock.

D. Procedure for Reassembly of American Bosch 100 Pump

Inspect end of plunger and plunger button. Inspect roller and tappet. When reassembling any injection pump, *always* replace all gaskets and seals as well as any needed parts.

NOTE Any assembly reference made to pump rotation will be as follows: front of pump—drive end, rear of pump—transfer pump end.

1 Mount pump on correct holding fixture and lock in vise.

2 Slide the camshaft and rear bearing plate with thrust washer into the housing so that the large portion of the plate is at twelve o'clock. Excess fuel tube must be behind the plate.

3 Place the correct O ring on the excess fuel supply tube.

4 Torque the screws in the rear bearing plate to the recommended torque.

5 Install the thrust washer over the front of the camshaft, bronze surface first, and then the shims that were under the bearing plate. Tighten the screws to the correct torque.

6 Using a dial indicator at the end of the camshaft, check end play by pushing and pulling on the shaft (Figure 18.64). End play must be 0.001 to 0.005 in. (0.025 to 0.13 mm). If the end play is not correct, add or remove shims under the front bearing plate until correct end play is achieved.

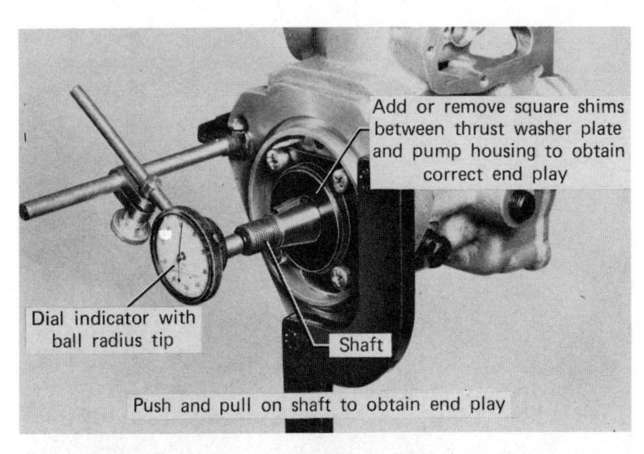

Figure 18.64 Checking camshaft end play (courtesy American Bosch—United Technologies Corporation, Springfield, Mass.).

7 Install the tappet roller.

8 Check for proper clearance between the roller and the tappet by first placing the camshaft at twelve o'clock. Place plastigage on the bronze section of the tappet. Insert tappet and press down lightly. Remove the tappet and check clearance. It must be 0.004 to 0.013 in. (0.10 to 0.33 mm).

NOTE If this clearance is not correct, replace the tappet and/or roller and recheck.

9 Install the shutoff control shaft if used.

10 Check the governor slip clutch as outlined in Part A, step 23. Adjust as required.

11 Install the fulcrum assembly, operating shaft, and sleeve to governor housing.

12 Position the governor lever sleeve over the fulcrum lever pivot pins.

13 Install the governor thrust sleeve on the governor shaft, then install the governor assembly into the governor housing and torque the support plate to specifications.

14 Place the camshaft so that the keyway is at the twelve o'clock position.

15 The chamfered tooth of the governor shaft gear (Figure 18.65) must be at twelve o'clock also.

16 Assemble the governor housing to the pump housing carefully.

CAUTION Do not damage the excess fuel supply tube.

Figure 18.65 Chamfered governor shaft gear.

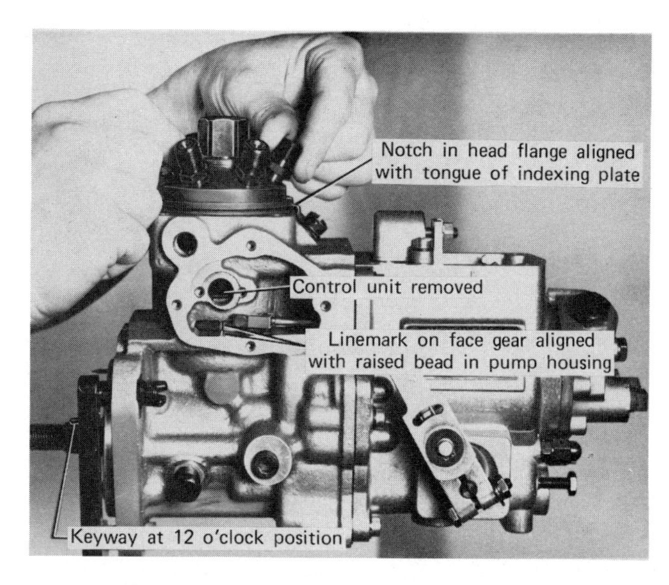

Figure 18.66 Installing hydraulic head (courtesy American Bosch—United Technologies Corporation), Springfield, Mass.).

17 Torque the attaching screws to the specified torque.

18 Install the excess fuel device over the oil supply tube.

19 Position the camshaft keyway at twelve o'clock.

20 Install the head, making certain the line mark on the face drive gear (Figure 18.66) is in alignment with the raised mark in the housing timing window.

21 Align the index notch of head with the plate or notch on later pumps.

22 Install screws and torque to the specified torque.

NOTE Delivery valve assembly should be checked at this time. Install delivery valve into the special tool and attach it to the nozzle tester. Opening pressure should be 1150 to 1450 psi (80 to 102 kg/cm^2). If the delivery valve opening pressure is not correct, a different spacer can be installed to change its opening pressure. After checking delivery valve, install it in the hydraulic head and torque to specifications.

23 Check to make sure fuel control sleeve is in its lowest position.

24 Rotate the sleeve pin on the control assembly (Figure 18.67) until it faces up. Install O ring or control unit, lubricate, and insert the control unit assembly.

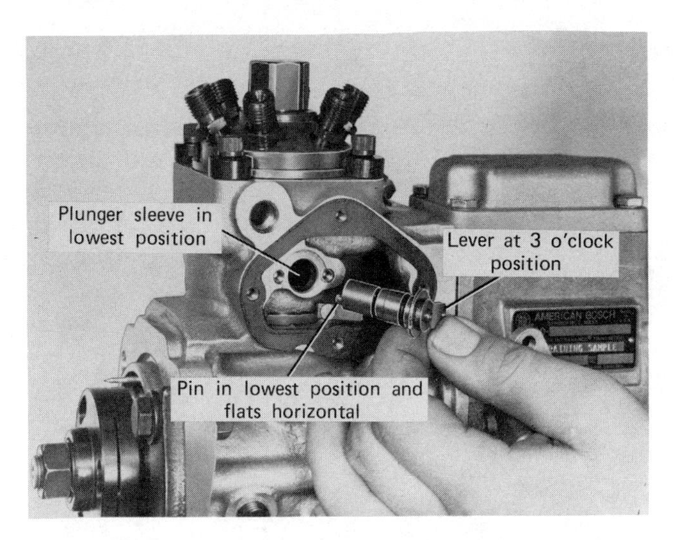

Figure 18.67 Installing control unit (courtesy American Bosch—United Technologies Corporation, Springfield, Mass.).

CAUTION After inserting the control unit, rotate the arm one full turn. If this cannot be done, the sleeve pin is not engaged.

25 Install the governor control rod. Assemble the control unit plate, spacers and screws.

26 Press the control assembly against the control unit plate and measure the clearance between the control rod and plate (Figure 18.68). This clearance must be 0.001 to 0.017 (0.025 to 0.43 mm).

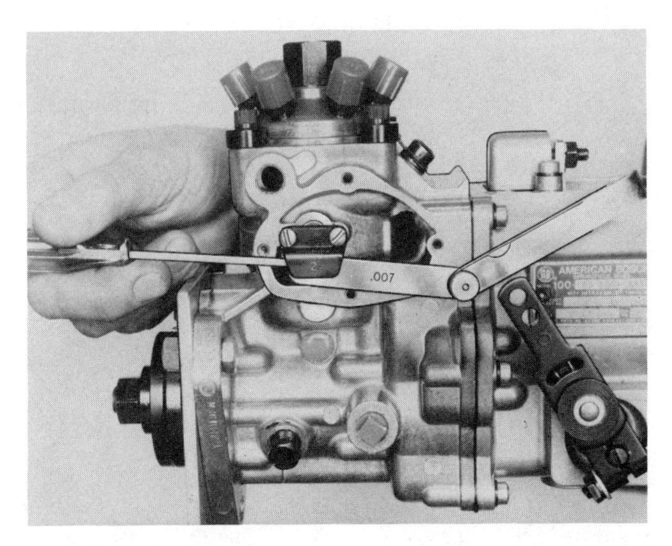

Figure 18.68 Measuring clearance between control rod and control retainer plate (courtesy International Harvester Company).

If clearance is not correct, change the plate. Three plates are available, nos. 1, 2, and 3.

27 Lock the control unit with wire so that the screws cannot loosen.

28 Assemble drive hub and torque.

29 Place the pump vertically in the vise with the governor end up. Install the inner governor spring with spacers.

30 Place the spring adjusting gauge into the housing (Figure 18.69a). The outer spring gauge should be firmly seated against the plate while the inner spring gauge should rest lightly on the inner spring guide.

31 Read graduations from zero to the last one visible. This figure is the inner spring gap. Add or

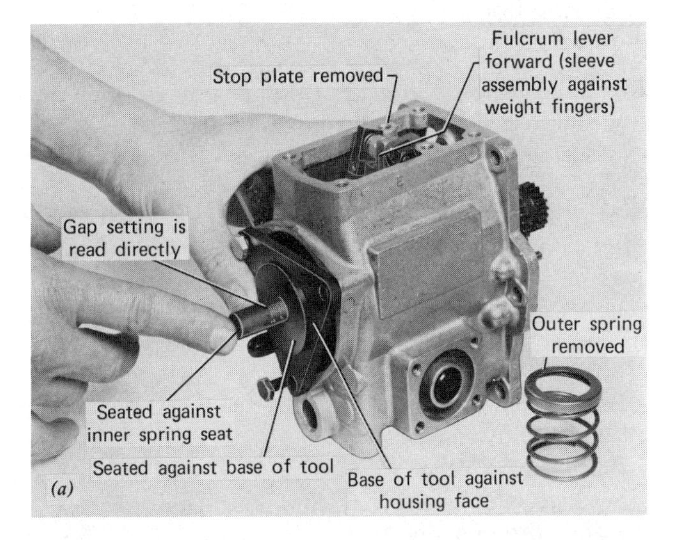

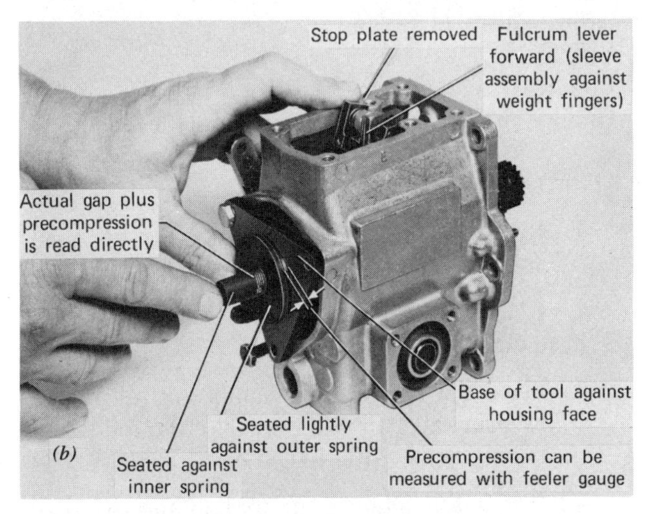

Figure 18.69 Governor spring adjusting gauge in housing (courtesy American Bosch—United Technologies Corporation, Springfield, Mass.).

subtract shims to bring the gap up to specifications.

32 Next, install the outer spring, spacers, and spring guide.

33 Reinstall the gauge and press down the outer spring gauge until it is seated. The amount of travel (as read on the center tool) is the outer spring precompression (Figure 18.69b). Shims can then be added or subtracted to obtain the correct precompression. Remove the adjusting gauge.

34 Install the outer spring guide, and install a new O ring and shaft seal on the supply pump.

NOTE Before installation of the T pump, install a new seal in the pump body. To remove the seal, place a $\frac{5}{8}$ in. coarse capscrew in a vise and turn the pump body and seal onto it. The seal can then be pulled from the housing.

35 Install a new seal in the T pump body and then install the T pump onto the governor housing. Insert and torque the screws to the correct torque.

XII. Bench Testing of Model 100 Pump

After pump overhaul or the replacement of hydraulic head or governor parts, the complete injection pump must be run on the test bench. The American Bosch procedure for testing is as follows:

Port closure setting
Set cam nose angle
Full load delivery
Set high idle, breakaway (cam nose leaves plate)
Adjust droop screw
Adjust overload setting
Cutoff
Low idle
Quality check (hydraulic head internal leakage)

A *Port closure.*

1 Remove the delivery valve, spring, and spacer.

2 Rotate the camshaft keyway to twelve o'clock.

3 Place the intravance assembly in full retard by using a small rod to push in the follow-up rod figure.

4 Install the plunger lift tool in the delivery valve holder bore. Set up a dial indicator so that it contacts the tool as shown in Figure 18.70.

Figure 18.70 Placing dial indicator so it contacts plunger lift tool.

5 Place a curved injection pipe on the no. 1 fuel delivery outlet.

6 Supply fuel to the hydraulic head at 0 psi, (0 kg/cm²). Fuel should run from the curved tube.

7 Place the operating lever in idle position and rotate the pump in the direction of rotation until the flow stops.

NOTE This position is port closure and is extremely important to pump timing.

8 If port closure is not as specified, the plunger button (Figure 18.71) must be exchanged with a different size. Recheck port closure.

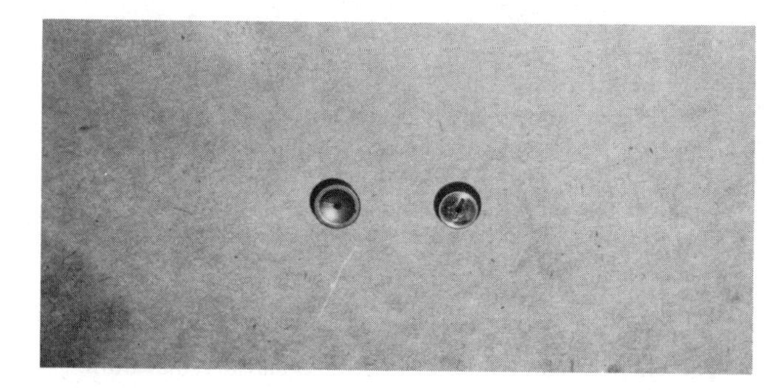

Figure 18.71 Plunger buttons.

Figure 18.72 100 series pump correctly installed on the test stand.

9 After setting port closure, remove the curved tube and install the delivery valve and spring. Torque the holder.

10 Complete the hookup of the pump to the test bench, attaching the lube oil supply and return lines, the fuel supply and return lines. Hook up the injection lines to the right nozzles or orifice plates (Figure 18.72).

B *Preparation of pump for testing.*
Before starting the tests, do the following:

1 Remove the governor cover and back out the high idle adjusting screw and droop screw.

2 Using the tool shown in Figure 18.73,1, set the angle of the cam nose (Figure 18.73,2) as stated on the specification sheet. Lock the adjusting screw.

Figure 18.73 Setting cam nose angle (courtesy International Harvester Company).

3 Install the correct overflow valve or orifice fitting.

4 Turn on the lube oil supply and adjust to 40 psi (2.8 kg/cm^2).

5 Run the pump for 10 minutes to stabilize the calibrating oil temperature.

NOTE FOR TESTING Since this pump runs at engine speed, two revolutions of the pump camshaft are required to fire all the cylinders. For this reason, the test bench counter must be set at twice the number of strokes given in the specification sheet to get the correct fuel delivery.

C *Full load delivery (stop plate position).*

1 With the operating lever in the high idle position, take a draw and record delivery.

2 If delivery is incorrect, adjust the stop plate screw (Figure 18.74) clockwise for decreased delivery and counterclockwise for increased delivery.

CAUTION Always use the stop plate support tool when doing adjustments to avoid bending the plate guide.

D *High idle.*

1 Raise the pump speed approximately 80 rpm over rated speed and adjust the high idle screw (Figure 18.75) until the cam nose just leaves the stop plate.

2 Recheck full load delivery. It must be the same as before. Readjust the stop plate if necessary.

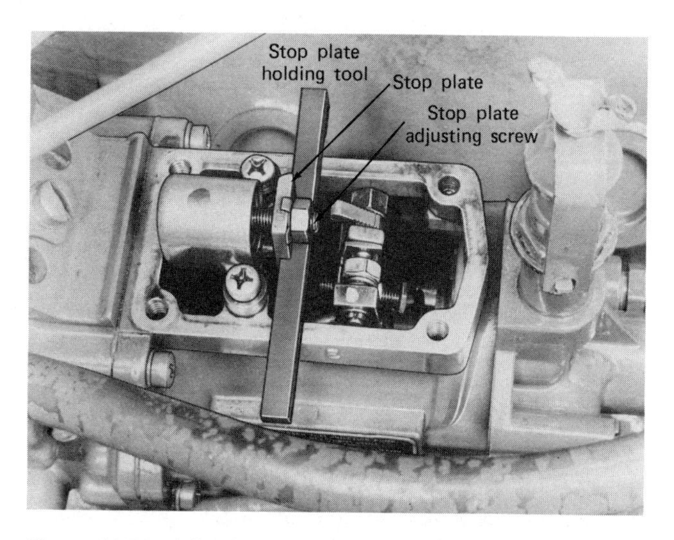

Figure 18.74 Adjusting stop plate screw (courtesy International Harvester Company).

Figure 18.76 Adjusting droop screw (courtesy International Harvester Company).

E *Governor cutoff.*

1. Increase the pump speed 17 percent over the rated rpm. Delivery must cease.
2. If delivery does not stop, recheck governor springs and shim thickness.

F *Droop setting.*

1. Run pump at required speed for droop setting and adjust droop screw (Figure 18.76) to obtain the correct delivery.

NOTE Droop should be set before overload to make sure that droop screw is not affecting it. If droop screw is against the stop plate at overload, there will not be enough fuel for overload.

2. Lock droop screw locknut before proceeding with further tests.

G *Overload.*

1. Run at specified speed for overload and check delivery.

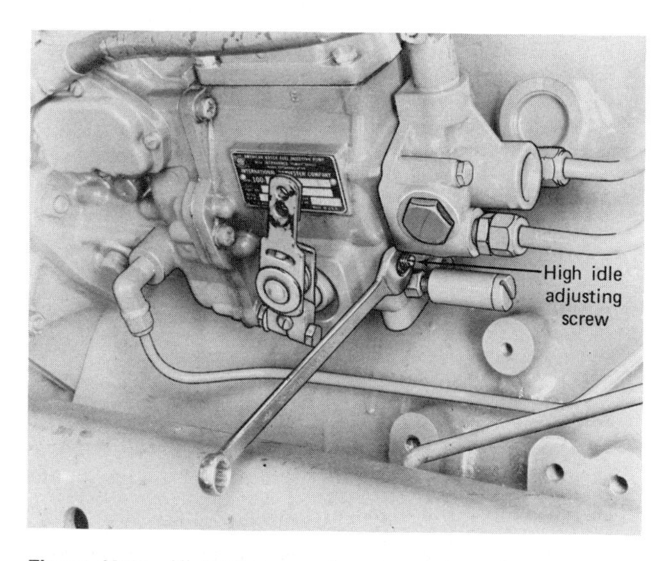

Figure 18.75 High idle screw (courtesy International Harvester Company).

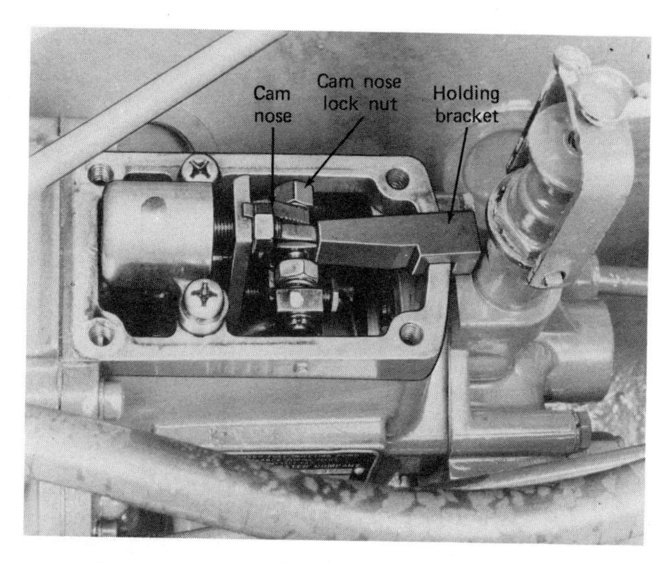

Figure 18.77 Cam nose angle (courtesy International Harvester Company).

2 Change cam nose angle if delivery is incorrect (Figure 18.77).

NOTE If the nose cam is adjusted, high idle must be rechecked and adjusted if necessary.

H *Recheck of rated load delivery.*

1 Recheck the delivery obtained for rated load, high idle, and cutoff. Correct if necessary.

2 Move the control lever to the stop position and check for a positive shutoff of fuel.

I *Low idle.*

1 Run pump at low idle speed and check delivery.

2 If necessary, adjust the low idle stop screw (Figure 18.78).

J *Quality check of hydraulic head.*

Run pump at 150 rpm. Delivery should be 40 cc for hydraulic heads with 10 mm plungers per 1000 strokes and 30 cc for hydraulic heads with 9 mm plungers.

K *Assemble and seal.*

1 Check all adjusting screws for tightness. Reinstall the governor cover and seal wires.

2 Remove pump from the test bench and prepare it for air pressure check by plugging the overflow orifice with a pipe plug. Then attach the air supply to the inlet side of the hydraulic head. Supply line pressure and immerse pump in cleaning fluid.

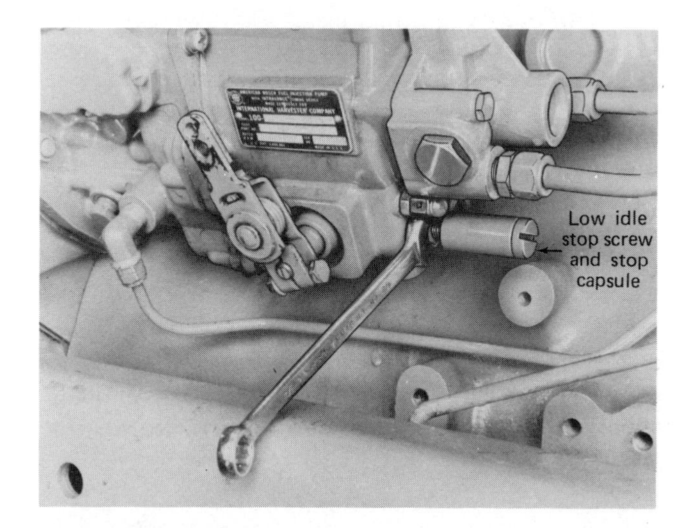

Figure 18.78 Adjusting low idle stop screw (courtesy International Harvester Company).

a Check for leaks around hydraulic head.

b Check for excess leaks at control shaft.

NOTE A small stream of bubbles escaping from the control shaft is considered normal.

3 Remove fittings from the inlet fitting and overflow valve.

4 Before installing control unit cover, make sure mark on face gear is in the control unit opening.

NOTE This makes sure that the pump is set to fire on the no. 1 cylinder.

5 Attach any special accessories that may have been used, such as a shutdown valve.

6 Align timing pointer with scribed mark on drive hub.

7 Rinse pump in cleaning solvent to remove all oil and grease.

8 Mask off with masking tape all nameplates and plug all openings with plastic caps.

9 Paint pump.

XIII. Troubleshooting the American Bosch 100 Pump

Be certain to make the following checks before blaming the pump for a specific problem:

A Are the correct lines and orifice plates being used?

B Is the pump being tested in the correct sequence?

C Does the test stand contain enough calibrating oil and is it at the correct temperature?

D Is the feed pressure correct?

E Is lube oil being supplied to the pump?

F Troubleshooting on the test stand.

1 Incorrect advance movement.

a Wrong transducer or timing light.

b Worn intravance parts.

c Low oil pressure.

d Intravance assembled incorrectly.

2 Uneven fuel delivery.

a Low supply pressure.

b Worn distributor head.

- c Worn delivery valve.
- d Wrong overflow valve on orifice fitting.

3 Low cranking fuel output.

- a Incorrect speed.
- b Excess fuel device binding.
- c Oil pressure still applied to the device at cranking speed.
- d Worn distributor head.
- e Worn roller tappet or roller.
- f Low T pump pressure.
- g Wrong overflow valve or orifice fitting.

G Troubleshooting on the engine.

1 Dilution of engine oil.

- a Hydraulic head O rings leaking.
- b Transfer pump seal leaking.
- c Excessive plunger wear.
- d Cold weather operation.

2 Erratic governor or loss of governor control.

- a Governor cover screws installed incorrectly, causing governor to bind.
- b Sticking control sleeve.
- c Slippage of governor clutch springs.
- d Incorrect governor springs and/or shims.
- e Plugged filters.
- f Collapsed suction line.

3 Low power.

- a Incorrect adjustment of full load screw.
- b Intravance not advancing enough or at the right time.
- c Incorrect adjustment of governor springs or wrong spring.
- d Droop screw not adjusted correctly.
- e Plugged filters.
- f Collapsed suction line.

4 Excessive smoke.

- a Loose delivery valve.
- b Governor slip clutch not set correctly.
- c Full load stop not properly adjusted.
- d Low oil pressure, causing excess fuel device to move forward.

5 Hard starting.

- a Fuel filters plugged.
- b Excess fuel device stuck in "run" position.
- c Hydraulic head or delivery valve worn or broken.
- d Throttle not in wide open position (WOT).

6 Pump cannot be timed.

- a Pump assembled incorrectly.
- b Timing gears worn.
- c Engine drive gear out of time.

7 Engine runs rough or irregular.

- a Pump timed to no. 6 cylinder.
- b Pump timed incorrectly on no. 1 cylinder.
- c Nozzles sticking or at uneven pressure.
- d Air in injection lines.
- e Loose or broken control unit.
- f Loose or binding governor linkage.

SUMMARY

The information contained in this chapter is intended to be used in conjunction with the service manual and specification sheet. From time to time, procedures and products are updated. Do not attempt to overhaul and calibrate a pump without referring to manufacturer's specifications for fuel delivery and other critical pump data. Remember, the pump overhaul job is only as good as the calibration it receives. If you are in doubt about a particular procedure or operation, see your instructor.

REVIEW QUESTIONS

1 The 100 series employs what type of fuel control?

a Inlet metering.

b Port bypass.

c Sleeve control.

d None of the above.

2 The pump turns at:

a Camshaft speed.

b Crankshaft speed.

c One-half camshaft speed.

d None of the above.

3 What type of supply pump does the pump employ?

a Vane.

b Rotor.

c Gear.

d Gerotor.

4 The plunger in the hydraulic head turns at:

a Camshaft speed.

b Crankshaft speed.

c One-half camshaft speed.

d None of the above.

5 Maximum fuel is adjusted by:

a Moving the screw in the hydraulic head.

b Adjusting the nose cam stop.

c Tension on the governor spring.

6 Low idle is adjusted by:

a Adjusting low idle stop screw.

b Length of throttle linkage.

c Screw in top cover.

d Two of the above.

7 The advance mechanism used on the Model 100 pump is:

a Load advance.

b Speed advance.

c Intravance.

d Two of the above.

8 Internal timing is accomplished by:

a Putting the camshaft keyway at twelve o'clock and lining up the mark on the face gear with the mark in the housing.

b Installing the head with the keyway in the camshaft at three o'clock.

c Both of the above.

d Neither of the above.

9 The slot or slack in face gear is to prevent:

a Engine overspeed.

b Excess fuel.

c Engine running backwards.

d Two of the above.

10 The delivery valve is used to:

a Prevent excess fuel.

b Prevent nozzle dribble.

c Maintain line pressure.

d Two of the above.

19

Robert Bosch Fuel Injection Pumps

In 1923 Robert Bosch produced the first practical diesel pump. This design enabled the newly developed diesel engine to become a viable engine for many applications. The method of fuel metering on this initial pump, designated as port and helix (high pressure metering), is still used on most modern injection pumps. Today Robert Bosch Corporation, located in Stuttgart, West Germany, and Charleston, South Carolina, produces injection nozzles, filters, pumps, and diesel testing equipment for a wide range of applications. This chapter will discuss in detail some of the most popular injection pumps and governors produced by Robert Bosch (Figure 19.1).

OBJECTIVES

Upon completion of this chapter the student will be able to:

1 Identify two models of Robert Bosch fuel injection pumps from a given list of various models.

2 Explain the fuel flow in one Robert Bosch pump (selected by student).

3 Correctly disassemble a Robert Bosch pump (selected by student), using the service manual.

4 Give two reasons for calibration of a Robert Bosch injection pump.

5 Calibrate a Robert Bosch pump (student's choice), using a service manual and specification sheet.

6 Explain the action of a delivery valve and why it is used.

7 List two advantages of an in-line injection pump.

8 List two advantages of a rotary injection pump.

9 Give three reasons for poor (slow) governor action in an in-line Robert Bosch injection pump.

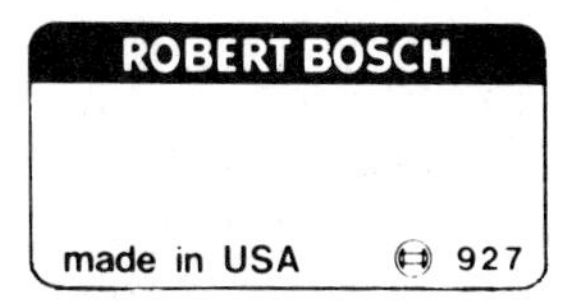

Figure 19.1 Bosch trademarks.

GENERAL INFORMATION

The A series pump (Figure 19.2) employs the design that originated in 1923. This design was the world's first injection pump for diesel engines. Up to 12 pumping elements may be used in one housing. Two sizes are used, the 2000 series and the 3000 series, which is somewhat heavier. The A pump may be equipped with a fuel lift pump, smoke limiter (aneroid), injection advance unit, and several different governors, including pneumatic, centrifugal, and electrical. There are countless applications. Some popular ones include John Deere, Case, Deutz, IHC, and Daimler-Benz.

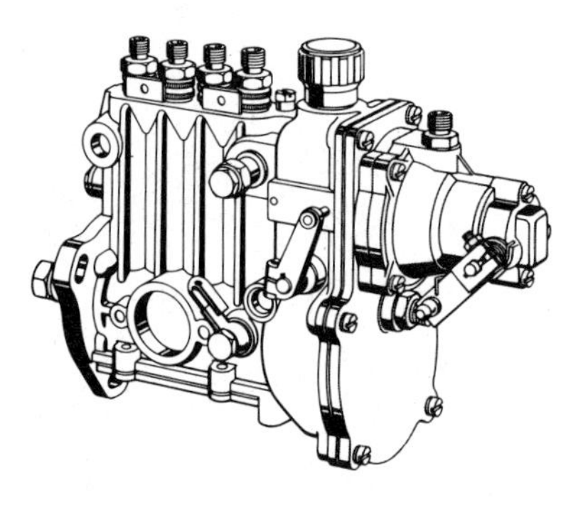

Figure 19.3 M series Bosch pump. (Courtesy Robert Bosch Corp.)

The M series pump (Figure 19.3) is very similar to the A series. It is smaller and lighter and uses a three-hole mounting flange. Shimless type roller tappets allow this pump to be shorter in overall height. Accessories for the M pump include fuel lift pump, automatic injection advance unit, and pneumatic governor. This pump was specially designed for Mercedes-Benz diesel automobiles, including models 170D to 240D. The pneumatic governor controls both fuel quantity and air flow to the engine.

One of the larger pumps in the Bosch line, the PE-P series (Figure 19.4), has many heavy-duty features, making it suitable for high output engines. Larger camshafts, plungers, and nonadjustable roller tappets enable this pump to be used with nozzle opening pressures of 10,000 psi (700 kg/cm²). Like the smaller pumps in the Bosch line, the P series can be fitted with a fuel lift pump(s), smoke limiter (aneroid), injection advance unit, and several different governors. Because of its increased size, the P pump is generally used on engines having more than 200 hp (149 kW).

Figure 19.2 A series Bosch pump.

Figure 19.4 PE-P series Bosch pump.

Figure 19.5 EP/VA series Bosch pump.

Figure 19.7 VE Bosch pump.

The Bosch EP/VA injection pump (Figure 19.5) is a single plunger, distributor type incorporating a spill-piston or hydraulic governor. This pump differs from the conventional in-line pumps in that the single plunger supplies fuel to all engine cylinders one after the other. Since the pump is lubricated by diesel fuel, it may be mounted either vertically or horizontally. A very compact unit, the EP/VA pump incorporates into the housing a fuel transfer pump, automatic timing advance, hydraulic governor, and hydraulic head assembly. Designed primarily for low horsepower output, high speed diesel engines, or where space is limited, the EP/VA pump is found on IHC, Deutz, Peugeot, and many other engines.

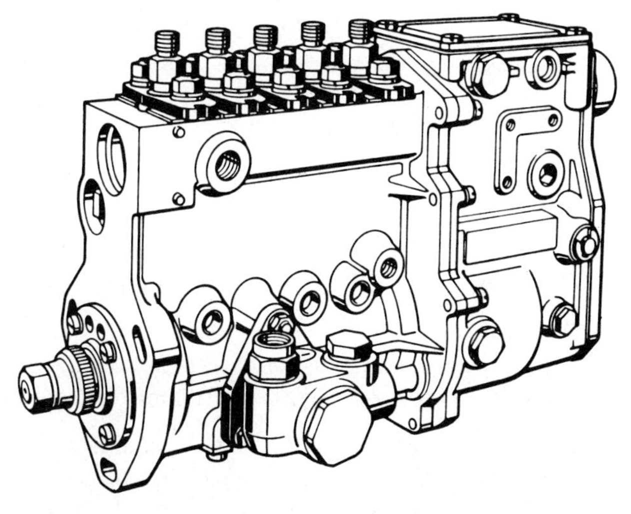

Figure 19.6 MW series Bosch pump. (Courtesy Robert Bosch Corp.)

To meet increasingly stringent emissions requirements, manufacturers of injection equipment are using much higher nozzle opening pressures than previously. Robert Bosch introduced the MW series pump (Figure 19.6) to handle this increase. The MW pump can handle higher pressures in part because of the incorporation of parts similar in design to the heavy-duty P series. These include a heavy-duty housing, larger camshaft, unitized delivery valve and barrel assemblies, and nonadjustable roller tappets. One of the first uses of the MW pump was on the Mercedes-Benz 300D five cylinder diesel engine and later on the 240D.

Like the EP/VA pump, the VE pump (Figure 19.7) is a single plunger, distributor type where the single plunger serves all outlets in the hydraulic head. The main difference between the EP/VA and the VE is in the governing system. Where the EP/VA uses an hydraulic governor, the VE employs a centrifugal mechanical governor. Features of the VE pump include smoke limiter (aneroid), load and speed sensitive timing, excess fuel starting device, and part load torque control. The VE is found generally on low output, high speed engines, such as the Volkswagen Rabbit Diesel.

I. Identification of Bosch In-Line Injection Pumps

The following information applies to all Bosch in-line pumps, including the A, P. M, and MW series pumps. Any minor differences between the various pumps will be noted.

The Bosch pump shown in Figure 19.8 is an in-line, multiplunger type. Important information about the pump is listed on the pump nameplate (Figure

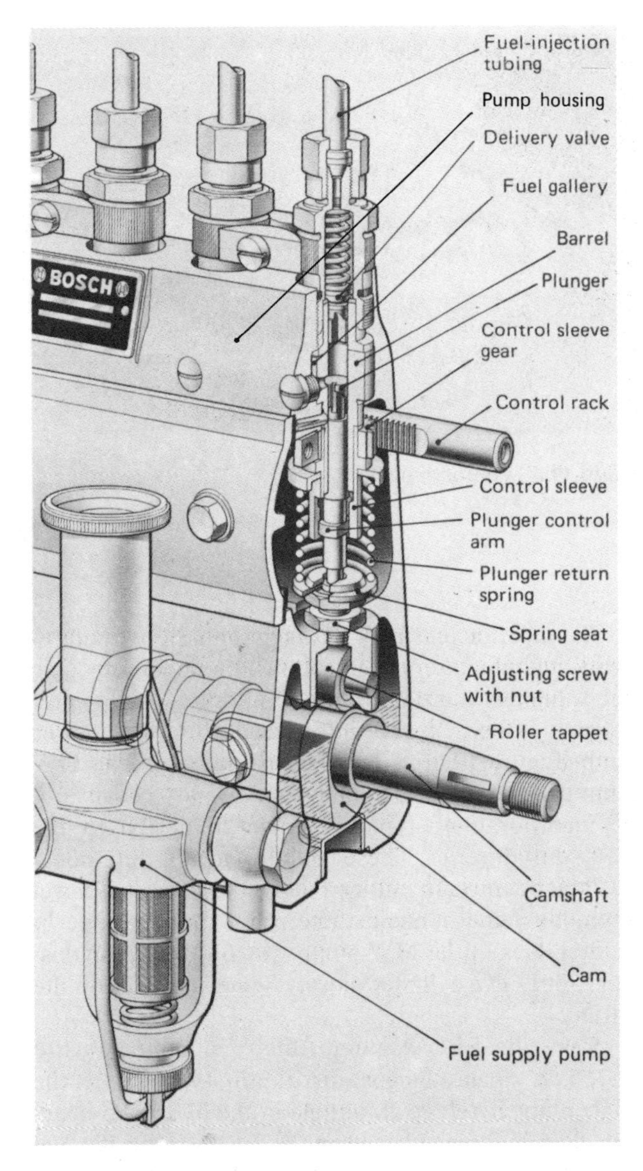

Figure 19.8 A series in-line pump (courtesy Robert Bosch Corporation).

19.9). This plate will list, among other items, pump serial number, pump model, and part number. Explanation of the pump code is listed below:

PE = pump model
S = flange cast on housing (no letter indicates a base mounted pump)
4 = number of pumping plungers
A = size of pump (A, P, M, etc)
80 = diameter of pumping plunger (80 mm)
B = letter denoting design change
420 = assembly number
L = denotes rotation (L = left hand/counterclockwise
R = right hand/clockwise)

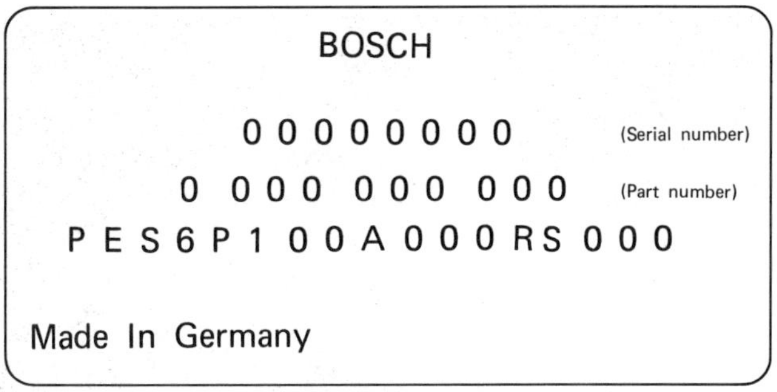

BOSCH

0 0 0 0 0 0 0 0 (Serial number)

0 000 000 000 (Part number)

P E S 6 P 1 0 0 A 0 0 0 R S 0 0 0

Made In Germany

Figure 19.9 Pump nameplate.

NOTE Pump rotation will be determined while looking at or facing the drive end of injection pump, the drive end being the front end of the pump.

S = production pump
445 = pump identification number (will vary depending on application)

II. Component Parts and Their Function

The pump shown in Figure 19.8 is typical of most Bosch in-line pumps.

A The *pump housing* contains all parts, acts as a lubricating oil reservoir, and has a low pressure fuel gallery surrounding the pumping plungers.

B The *camshaft* is coupled to the engine drive train. The camshaft causes reciprocating movement of the pumping plungers.

C The *pumping plunger and barrel assembly* performs two functions. It forces fuel past the delivery valve, into the injection line, and to the nozzle by way of its reciprocating action. It also controls the quantity of fuel and duration of injection by rotating action.

D The *roller tappets* ride directly on the camshaft and transmit its motion to the pumping plungers.

E The *plunger springs* keep the roller tappets pressing on the camshaft.

F The *control rack* transmits the action of the governor to the pumping plunger via the control sleeves.

G The *delivery valves* seal the delivery line from the barrel during the intake stroke and also relieve the pressure in the injection line to prevent nozzle dribble.

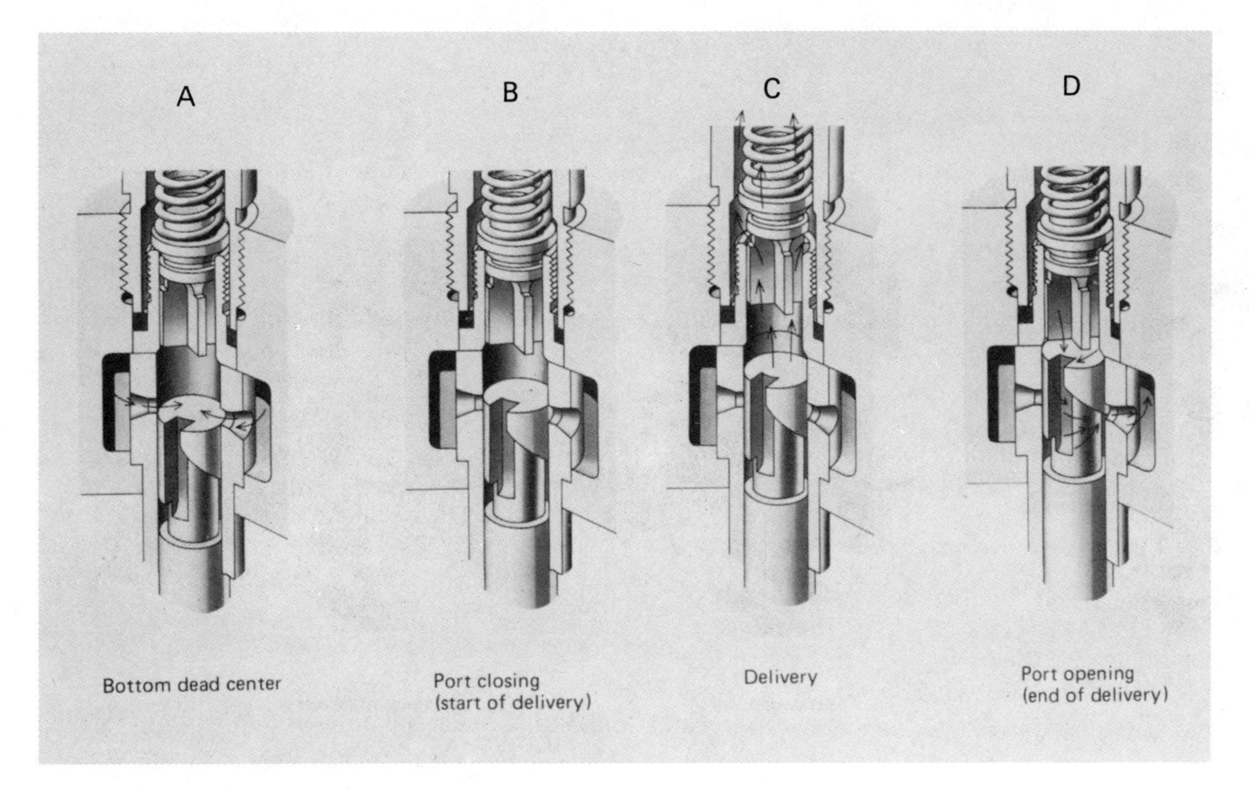

Figure 19.10 Plunger and barrel assembly showing port closure and port opening (courtesy Robert Bosch Corporation).

III. Pump Operation

The heart of the in-line injection pump is the plunger and barrel assembly (Figure 19.10). At this point all pumping and control of fuel output takes place. Because the plunger fits so precisely in the barrel (approximate clearance is only 0.002 mm or 0.00008 in.) there are no sealing rings to retain the injection pressure as the plunger pumps fuel.

NOTE Pumping plungers and barrels are lapped together to provide the necessary seal. Never interchange a plunger from one barrel to another. Plunger and barrel are sold as a matched set.

The pump camshaft provides a constant mechanical stroke of the pumping plunger, regardless of pump speed. The plunger is rotated indirectly by the governor to provide changes in fuel delivery. The upper edge of the pumping plunger has a vertical groove. Near the top is a helix (or control edge). The barrel may have either one or two control ports (Figure 19.10), depending on the pump size. Any fuel leaking by the plunger is collected in a groove cut into the barrel. From this point the fuel can return to the fuel inlet gallery.

When the pumping plunger is at BDC (bottom dead center), fuel is forced by the lift pump into the pump gallery. It enters the intake port (Figure 19.10, A) and floods the area above the plunger. The plunger is now forced upward by the camshaft. After approximately 2 mm ($\frac{3}{32}$ in.) upward travel, the plunger edge covers the inlet port (Figure 19.10, B). This is known as port closure and is very critical to the timing of the entire pump. (See Section XI, "Adjustment of Complete Pump," for complete details on adjustment.) Continued upward movement will force fuel past the delivery valve (Figure 19.10, C), open the injection nozzle, and inject fuel into the combustion chamber.

Injection will continue until the plunger has risen far enough to enable the lower edge of the helix to uncover the inlet port (Figure 19.10, D). At this time, pressurized fuel will rush down the vertical groove on the plunger and exit through the port. This is the end of delivery. After the end of fuel delivery, the plunger will continue to be forced upward by the camshaft, but this movement will not cause any further injection. The effective stroke of the pumping plunger is the time when fuel is being injected.

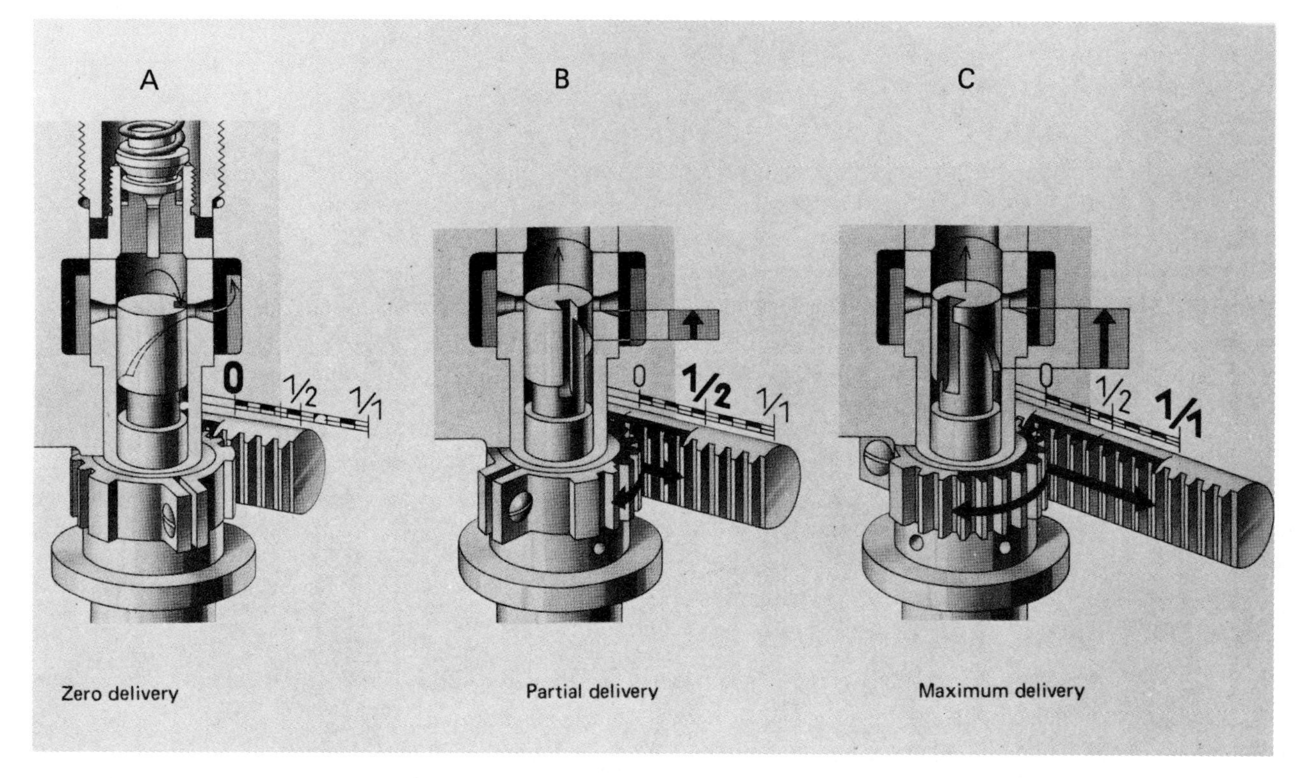

Figure 19.11 Plunger and barrel with control rack in various stages of delivery (courtesy Robert Bosch Corporation).

The amount and duration of fuel injection, as mentioned earlier, is controlled by a helix or control edge, the position of which is controlled by the governor. When the plunger is turned, as in Figure 19.11, A, the vertical groove is aligned with the inlet port and the upward movement of the plunger will not cause any pressure build-up above the plunger, since fuel can bypass back out the inlet ports. Fuel delivery at this point will be zero.

As the governor rotates the plunger via the control sleeve, the inlet port will be sealed off as the plunger moves upward (Figure 19.11, B). Since the short area of the helix is now being used, the injection amount will not be great. Injection will continue until the lower edge of the helix uncovers the inlet port. This is partial delivery.

Finally, the governor will rotate the plunger to the maximum fuel position (Figure 19.11, C). At this point the helix is at its longest point and injection will continue the full length or almost the full length of plunger travel. This position is usually reserved for providing excess fuel for starting. Some plungers have an additional cut on the edge of the plunger directly in line with the excess fuel position (Figure 19.12). This is known as a retard notch and will give delayed injection and maximum fuel for starting.

Another notch may be machined into the pumping plunger. This is a cylinder shutoff notch (Figure 19.13) for idle conditions. If this notch is aligned with the inlet port, fuel can be effectively cut off to run an eight cylinder engine on four cylinders. This design feature will allow the V8 engine to run more efficiently at idle although slightly rougher. For a more detailed expla-

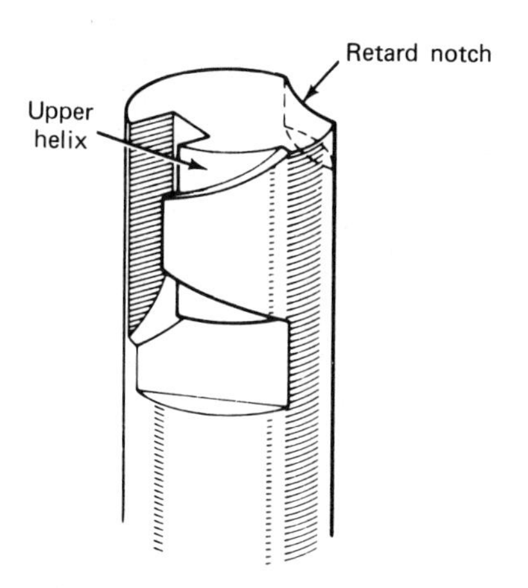

Figure 19.12 Plunger with retard notch (courtesy Robert Bosch Corporation).

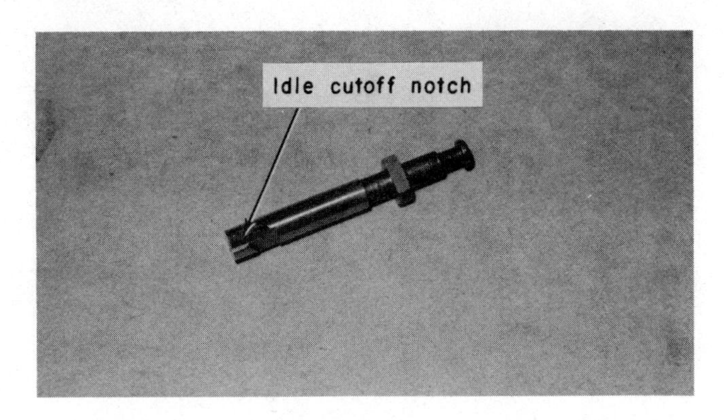

Figure 19.13 Plunger showing idle cutoff notch.

nation of plunger design and port helix operation see Chapter 15.

Located directly above the pumping plunger and barrel is the delivery valve (Figure 19.14). The delivery valve acts as a one-way check valve to help keep fuel in the injection line between injections. In addition, the delivery valve serves to relieve pressure in the injection line after injection to insure the instantaneous closing of the injection nozzle. This helps emission characteristics of the engine by doing away with nozzle dribble, which will cause smoke.

After injection is completed, pressure drops and the valve will close with the aid of the delivery valve spring. As the valve is closing and before it seats completely, the relief piston or retraction land will enter the bore of the delivery valve body (Figure 19.15), effectively sealing the injection line. As the valve continues the downward stroke and finally seats (Figure 19.16), it will increase the volume of the injection line an amount equivalent to the size of the retraction land. This volume increase in the injection line quickly lowers the line pressure below the nozzle opening pressure and insures nozzle closure.

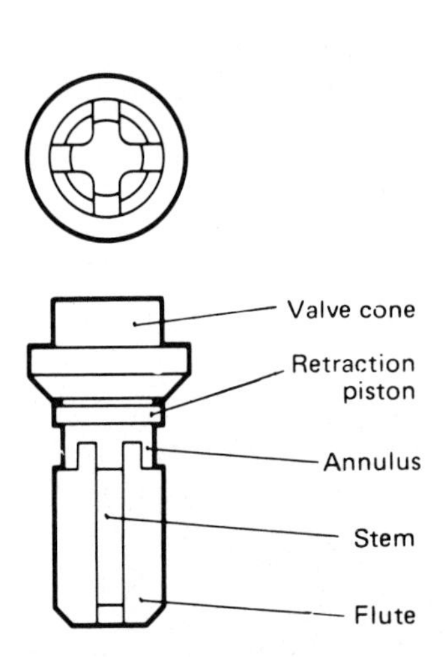

Figure 19.14 Delivery valve (courtesy Robert Bosch Corporation).

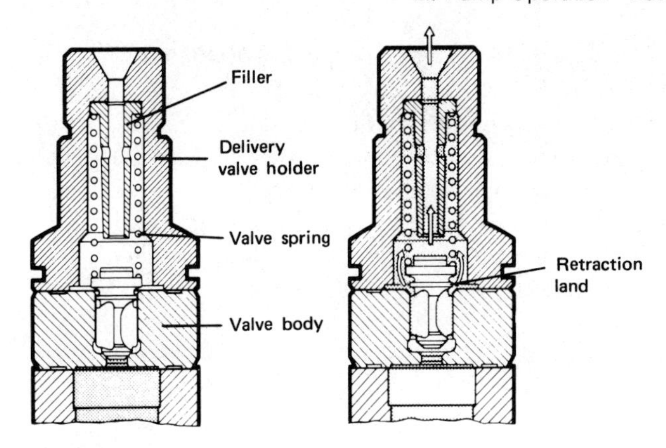

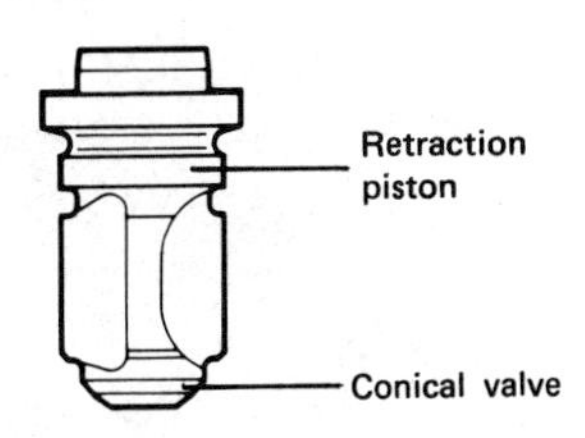

Figure 19.15 Delivery valve entering its bore (courtesy Robert Bosch Corporation).

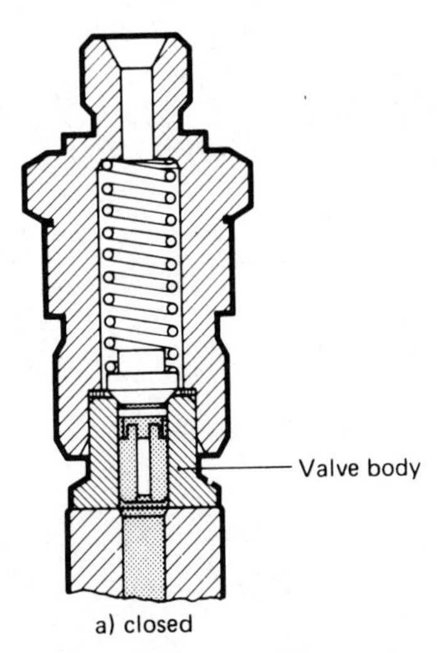

Figure 19.16 Delivery valve seated (courtesy Robert Bosch Corporation).

IV. Governors for In-Line Injection Pumps

In-line Bosch pumps can be fitted with various types of governors. The mechanical type will be discussed here. Bosch mechanical governors are of two basic types: (1) EP/RSV (variable speed) governors and EP/RSUV governors for slow speed engines, and (2) EP/RQV (limiting and variable speed) governors.

A *EP/RSV and EP/RSUV governors.*

1 The variable speed EP/RSV governor (Figure 19.17) is mounted on the rear of PE injection pumps. The governor flyweight assembly, mounted on the pump camshaft, acts together with the governor linkage to regulate the speed of the engine. The governor may be of single control lever design with the shutoff incorpo-

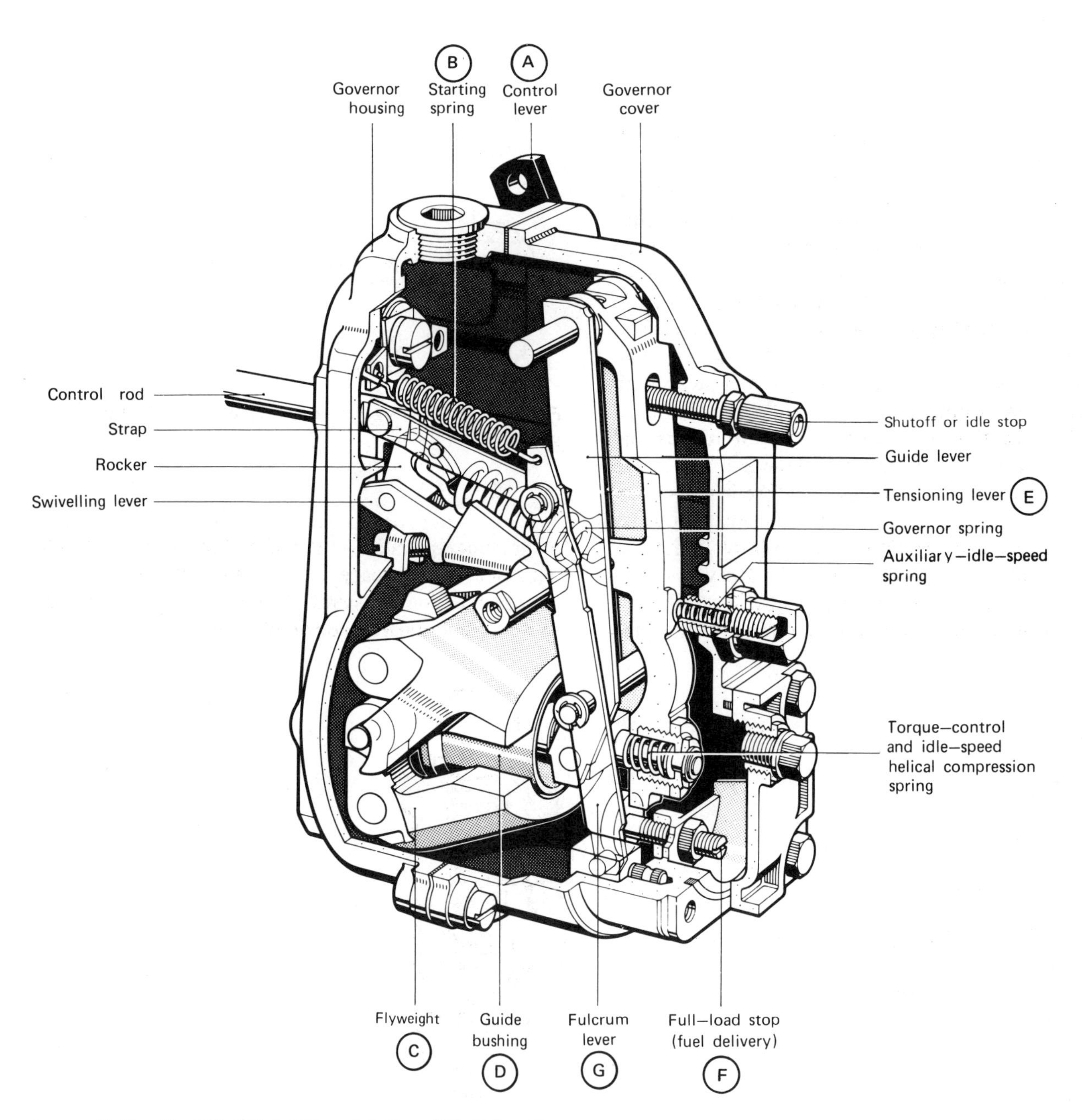

Figure 19.17 The EP/RSV governor (courtesy Robert Bosch Corporation).

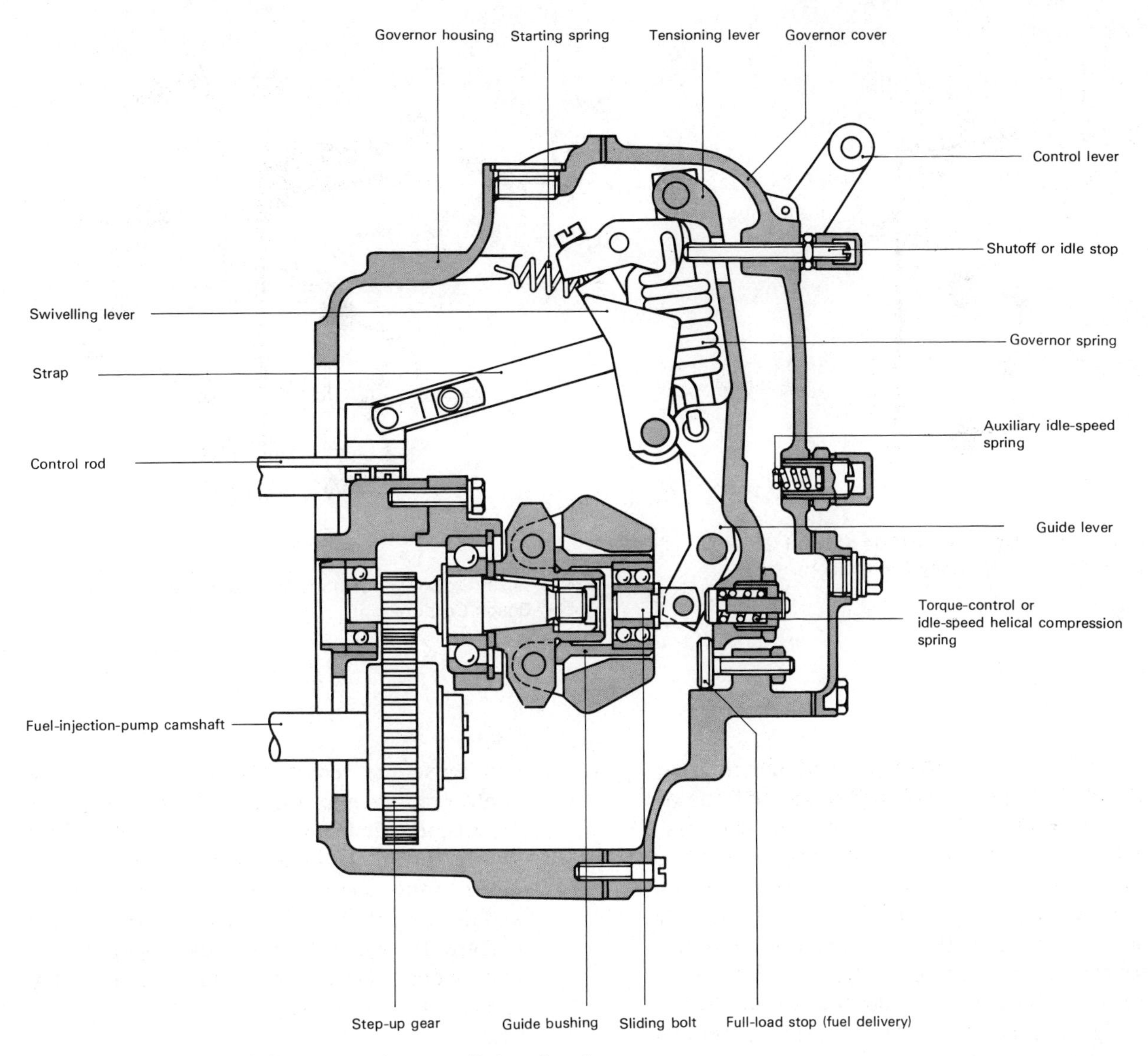

Figure 19.18 The EP/RSUV governor (courtesy Robert Bosch Corporation).

rated into the speed control lever, or it may have two control levers, one for shutoff and the other for speed regulation. This feature is dependent on engine manufacturer requirements.

2 For governing of very slow speed engines, the EP/RSUV governor was devised (Figure 19.18). The main operational difference between the EP/RSUV and the ER/RSV is the speed increase provided by the two gears that drive the governor, the common increase ratios being 1:1.86, 1:2.15, 1:2.75, 1:3.29, and 1:4.00. Other than this, operation of the two governors is identical as explained below.

3 Operation: Starting the engine. As the control lever (Figure 19.17,A) is moved to the maximum fuel position, the small starting spring (Figure 19.17,B) will pull the control rack to the starting fuel position (approximately 21 mm or $\frac{13}{16}$ in.) for excess fuel. This causes the pump to deliver more fuel to the combustion chamber for easy starting. At the same time, the governor weights (Figure 19.17,C) are collapsed (all the way inward) fully against the thrust sleeve (Figure 19.17,D) and the tensioning lever (Figure 19.17,E) is up to the full load stop screw (Figure 19.17,F).

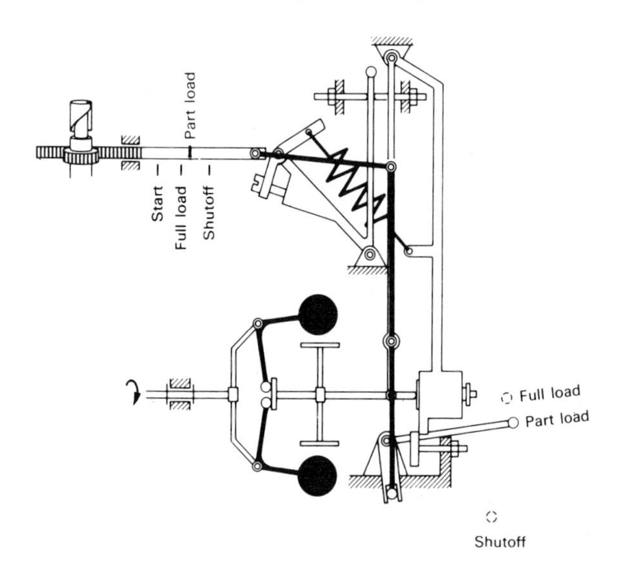

Figure 19.19 Control lever part throttle position (EP/RSV governor) (courtesy Robert Bosch Corporation).

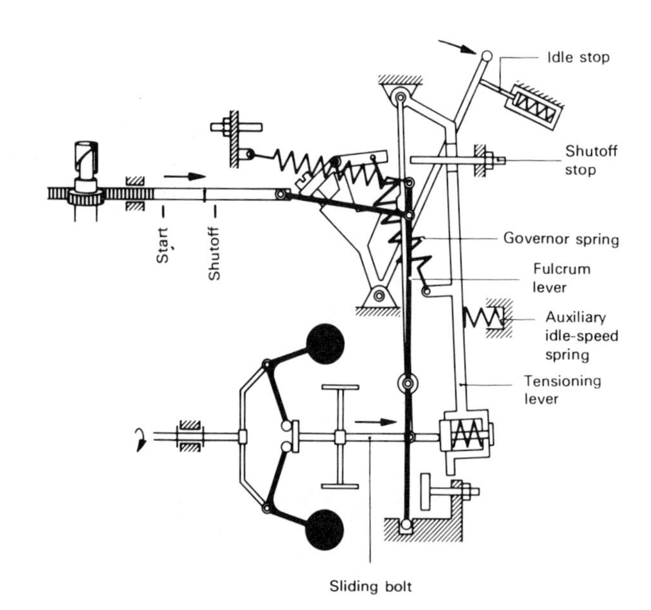

Figure 19.20 Low idle position (EP/RSV governor) (courtesy Robert Bosch Corporation).

When the engine starts and the rotating speed of the pump shaft increases, the force of the flyweights will quickly overcome the pressure exerted by the starting spring by forcing the thrust sleeve to the rear, causing the fulcrum lever (Figure 19.17,G) to pivot and pull the control rack in the same direction. With the control lever (throttle) still in the maximum position, the engine would quickly accelerate to high idle; therefore, the control lever should be moved to part throttle position (Figure 19.19) as soon as the engine is running.

At part throttle, the swivel lever has moved the main governor spring to a nearly vertical position where it exerts very little tension on the flyweights. As a result, the flyweights will swing out freely. As the flyweights swing out, they force the thrust sleeve rearward, moving the guide lever, which will pull the control rack to the low idle position (rear) (Figure 19.20).

NOTE Governor balance = idle flyweight force vs. slight extension of starting spring + slight tension of main spring + slight compression of torque capsule spring.

Movement of the speed control lever (throttle) from low idle to maximum speed will upset the delicate balance being held at idle and cause the tensioning lever to move the control rack to maximum delivery position. As the engine speeds up, the increased force of the flyweights will again cause the control rack to move to the rear, giving a lower fuel delivery, and the governor will balance at a higher speed (Figure 19.21). As long as the engine is not overloaded, the governor will maintain any speed within its range.

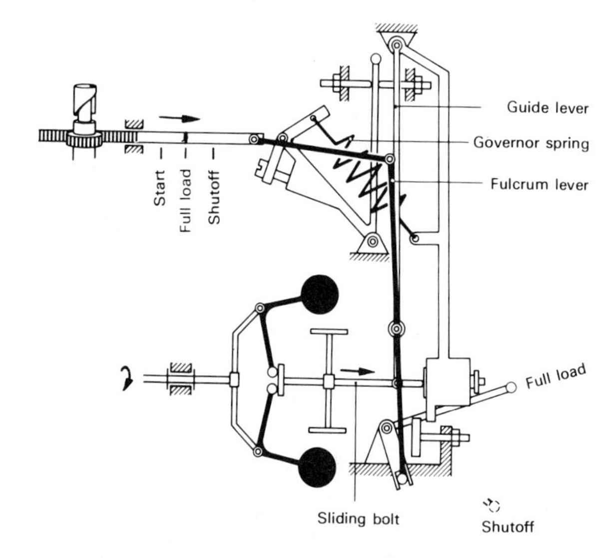

Figure 19.21 Rated speed position (EP/RSV governor) (courtesy Robert Bosch Corporation).

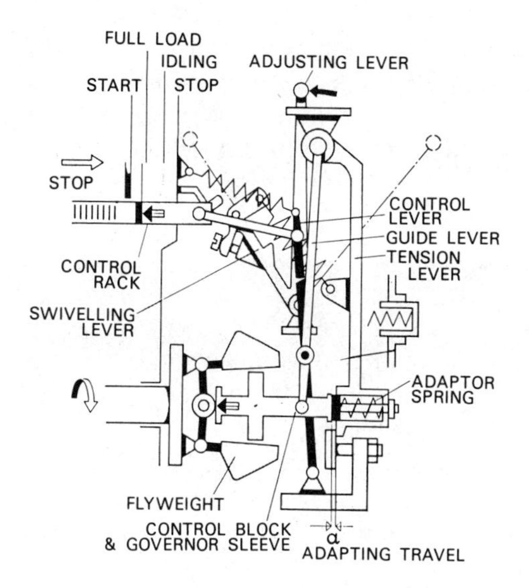

Figure 19.22 Overload position (EP/RSV governor); fuel controlled by torque capsule (courtesy Robert Bosch Corporation).

Figure 19.23 EP/RSV with separate stop lever.

At the maximum rated speed, the control lever is flush against the high speed stop screw, the swivel lever tensions the main governor spring fully, and the tensioning lever is held against the full load stop (torque capsule compressed).

NOTE The torque capsule is a small capsule with a spring in it that provides for excess fuel during engine overload.

In this position, the engine runs at rated speed and horsepower. Fuel setting is adjusted with the full load stop screw.

As engine speed is reduced from maximum rated speed, that is, in overload situations, the governor flyweights will move inward slightly. This allows the previously compressed torque capsule spring to expand and push the thrust sleeve to the front of the pump (Figure 19.22). When this happens, the control rack is also moved to the front of the pump, thereby increasing fuel slightly. This slight movement of the control rack by the torque capsule (usually 1 to 2 mm, 0.40 to 0.080 in.) provides sufficient fuel for overload conditions and yet gives good economy at rated load.

4 Torque control (EP/RSV governor) is used to give economical operation of a diesel engine at rated speed and load, yet still provide for an increase in fuel output as engine speed decreases, that is, when the engine is overloaded or "lugged" down.

The full load output of an injection pump is set at rated speed. This is the "normal" fuel setting and provides sufficient fuel for most conditions. When the engine slows down because of overload, it is capable of burning more fuel because of an increase in volumetric efficiency. This added fuel is obtained by the use of a torque control device (torque capsule) that automatically increases fuel pump output as engine speed drops and decreases it as engine speed rises, thereby preventing excess smoke and waste of fuel.

The EP/RSV governor may be equipped with a separate stop lever (Figure 19.23) that pulls the control rack to the stop position by pivoting the fulcrum lever on the guide lever. This action can take place at any speed or load setting, thereby causing the engine to stop.

If the governor has no special stop devices, shutoff is accomplished when the operator moves the speed control lever to the shutoff position (Figure 19.24). The small projections on the swiveling lever press against the guide lever, which moves the control rack to the stop position.

B *EP/RQV governors.*

The RQV (speed limiting) governor (Figure 19.25) differs from the RSV (variable speed) design in several ways. The governor weights, although still attached directly to the camshaft, operate radially on pins connected to the flyweight

Figure 19.24 EP/RSV governor with no stop lever.

hub and exert a pulling force with increasing speed rather than with the pushing force of the RSV governor. The sliding block, in combination with the S plate, forms an adjustable fulcrum for the governor levers that provide for the variable speeds in this governor. Torque control with the RQV-K governor is variable and more closely follows the fuel requirement curve of the engine by allowing control rack travel over a wider range, providing both increased and decreased delivery.

The RQV design can also be set up as a speed limiting governor, which would then be referred to as an RQ governor. The RQ governor controls only low idle speed and high idle speed of the engine.

1 *Operation of RQV governor (start position).*

As the control lever (Figure 19.25,A) is moved to the start position, the rocker (Figure 19.25,B) moves below the full load stop (Figure 19.25,C) and allows the control rack to move into the starting fuel position. When the engine starts, the flyweights quickly overcome the force of the governor springs (Figure 19.25,D) and pivot the fulcrum lever (Figure 19.25,E) to pull the control rack to low idle position.

2 *Operation of RQV governor (rated full load speed).*

At rated speed the flyweights swing out, compressing the governor spring and moving the fulcrum lever to a nearly vertical position. In this position the rocker is at its highest point on the V angled full load stop plate. This provides the correct amount of fuel to the engine to give rated horsepower with minimum smoke.

3 *Operation of RQV governor (torque control)* (Figure 19.26).

A decrease in engine speed due to an increase in load applied will allow the flyweights to swing inward, which, by means of the bellcranks (Figure 19.26,A) will push the fulcrum lever slightly rearward. This change of angle causes the rocker (Figure 19.26,B) to slide down the ramp of the full load stop plate and allows the control rack to move ahead, increasing fuel delivery for the overload condition.

V. Operation of Fuel Supply Pumps

To insure complete filling of the barrel assembly above the pumping plunger, the fuel gallery of the injection pump must be pressurized. A fuel supply pump is used to pump fuel from the fuel tank to the pump gallery (Figure 19.27).

A *The FP/K series fuel supply pump.*

This is a single acting plunger type pump usually mounted on the side of the main injection pump and driven off the pump camshaft. The pump can be equipped with a preliminary filter enclosed in a sediment bowl and also a hand primer as shown in Figure 19.27. The hand primer is used to purge (bleed) air from the system if it has run dry or if the fuel filters have been changed.

B *Suction/discharge stroke of fuel supply pump.*

On the suction stroke, the roller of the supply pump follows the camshaft inward, because of the force of the plunger spring (Figure 19.28*a*). As the plunger is moved inward, a low pressure area is created. Atmospheric pressure then pushes fuel through the preliminary filter, past the suction valve, and into the suction chamber. At the same time, the opposite side of the plunger pushes fuel from the pressure chamber into the outlet line. The pressure in this line, varying from 14 to 28 psi (1 to 2 kg/cm^2), depending on engine application, will close the pressure valve.

C *Intermediate stroke position.*

As the injection pump camshaft continues to revolve, it forces the roller tappet of the supply pump outward, away from the injection pump, also

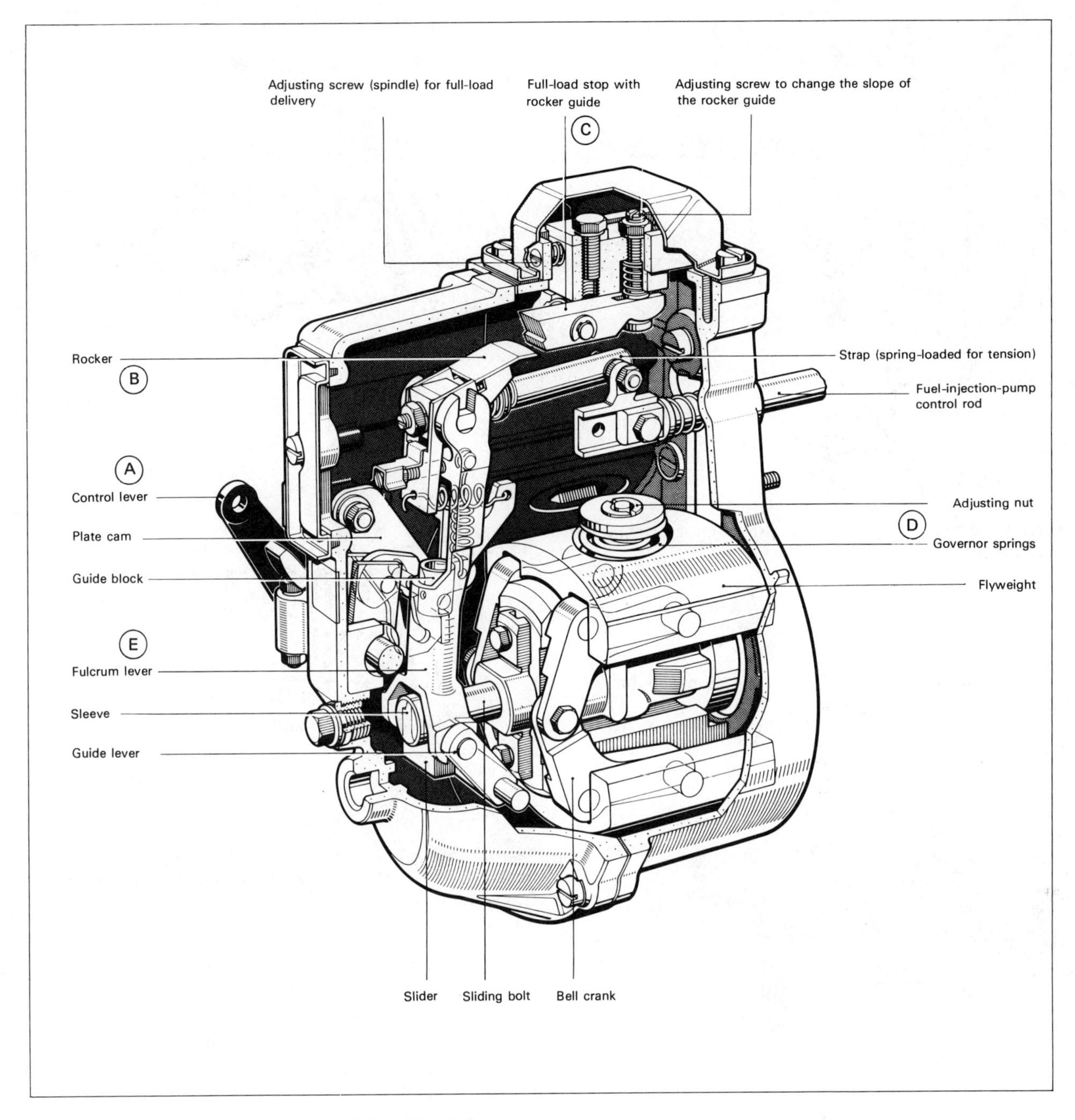

Figure 19.25 RQV governor (courtesy Robert Bosch Corporation).

pushing the plunger out (Figure 19.28*b*). Fuel trapped in the suction chamber will open the pressure valve and enter the pressure chamber. This fuel will also close the suction valve on the inlet line. This stroke completely fills the pressure chamber so that it can empty on the discharge stroke.

VI. Aneroid Operation (Smoke Limiter)

With increased emphasis and concern over emissions standards, methods must be used to reduce the quantity of visible smoke from diesel engines. One method of control used on Robert Bosch pumps is known as an aneroid or smoke limiter (Figure 19.29). The aneroid is a diaphragm fuel control usually mounted on the

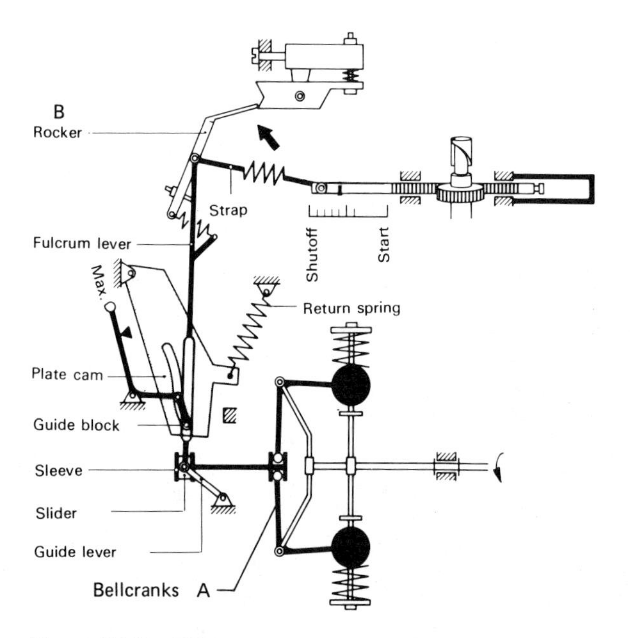

Figure 19.26 RQV governor torque control (courtesy Robert Bosch Corporation).

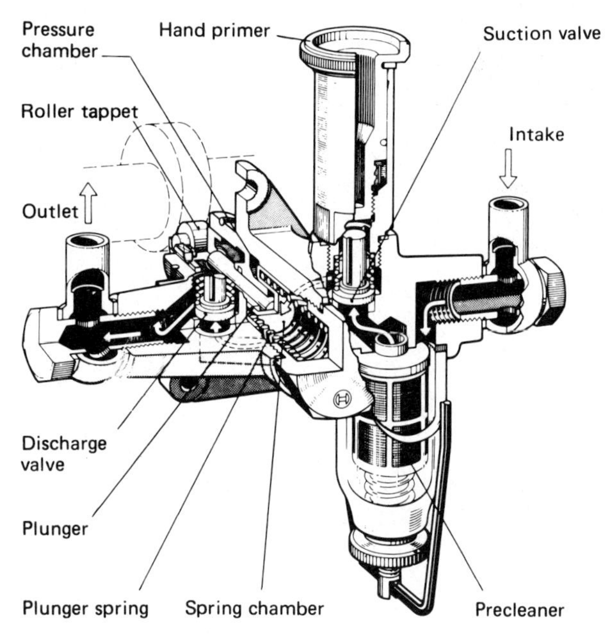

Figure 19.27 FP/K fuel supply pump (courtesy Robert Bosch Corporation).

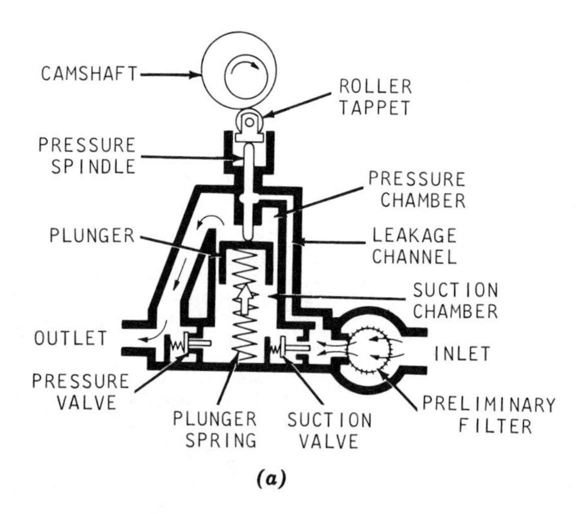

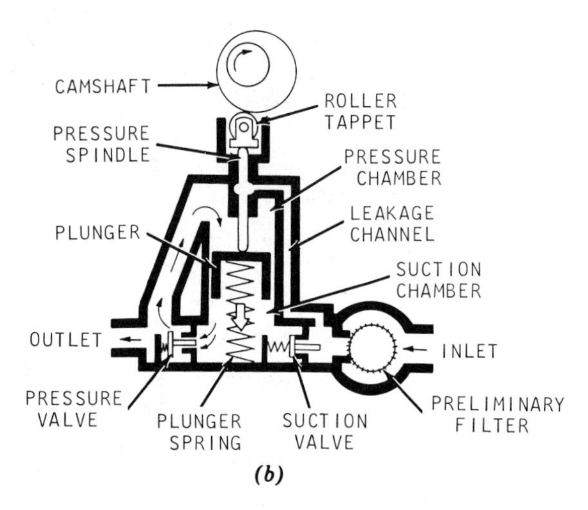

Figure 19.28 (*a*) FP/K transfer pump on inlet stroke. (*b*) FP/K transfer pump in intermediate position (courtesy Deere and Company).

governor housing, but it can be mounted on the front of the pump as well. It is operated by boost pressure (turbocharged engines). The aneroid eliminates unnecessary smoke during acceleration of turbocharged engines by limiting control rack movement until sufficient manifold pressure is built up by the turbocharger (full air charge) to effectively burn full load fuel.

With the engine stopped, the starting fuel control shaft must be moved, either manually through levers or hydraulically, to release the control rack so that the starting spring can pull the rack to the starting fuel position (Figure 19.30). After the engine starts, the governor weights will move the rack rearward, allowing the fuel control shaft (Figure 19.31) to spring back to normal "run" position. With the engine running without load at 800 RPM and higher, boost pressure is insufficient to overcome the spring pressure of the aneroid diaphragm. Thus the vertical adjusting shaft (Figure 19.32) is held upward, limiting control rack

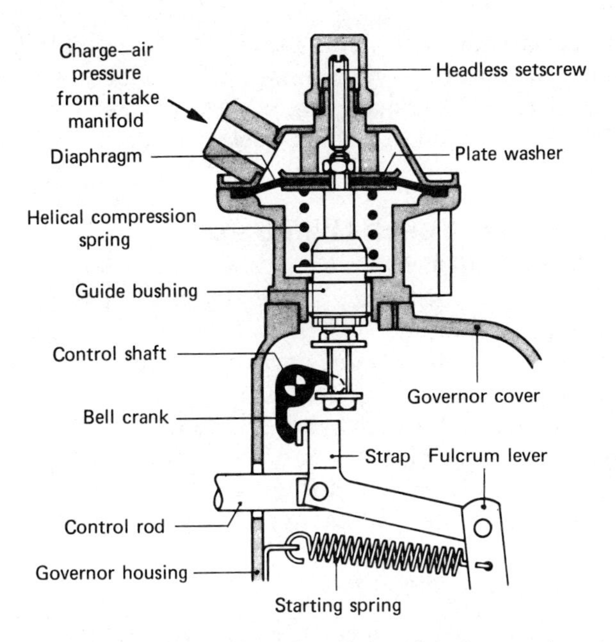

Figure 19.29 Aneroid used on Robert Bosch pump (courtesy Robert Bosch Corporation).

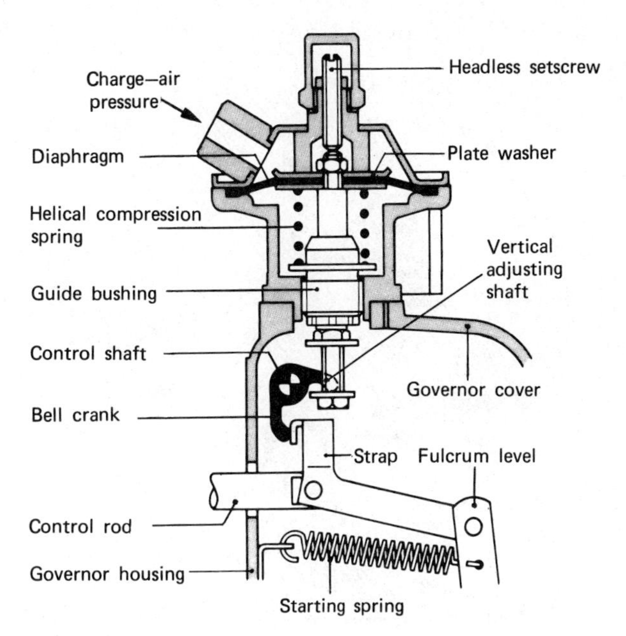

Figure 19.32 Aneroid limiting fuel delivery or rack travel (courtesy Robert Bosch Corporation).

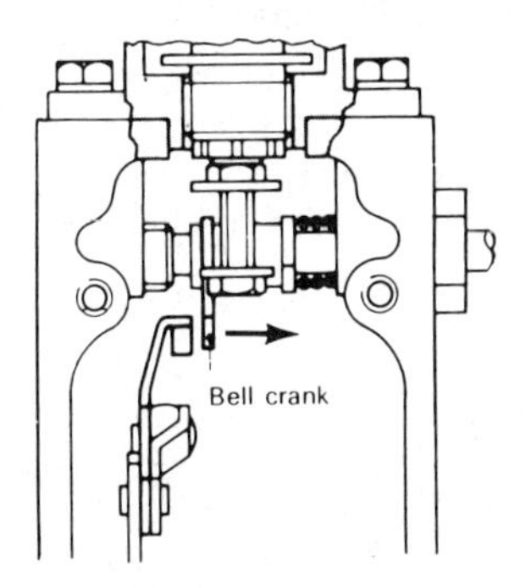

Figure 19.30 Rack in full position (RSV governor with aneroid) (courtesy Robert Bosch Corporation).

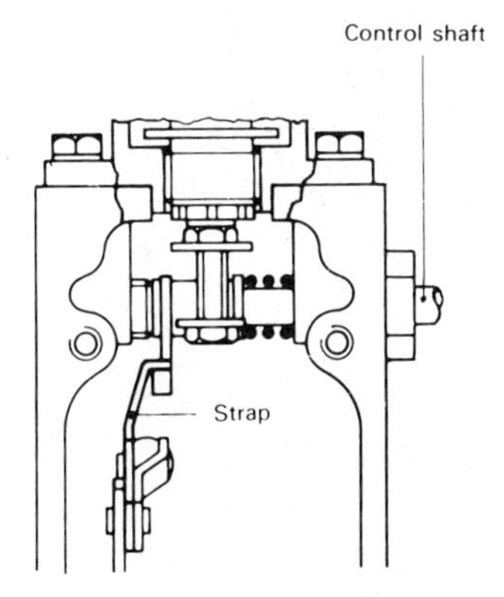

Figure 19.31 Normal run position (RSV governor) (courtesy Robert Bosch Corporation).

travel. As engine speed rises to 1200 rpm or more, the boost pressure increases to 11 psi (0.8 kg/cm^2), which moves the diaphragm down and allows the control rack to move forward, increasing fuel. This action delay allows the turbocharger time to come up to speed before full injection, thus reducing smoke.

VII. Timing Devices (In-Line Pumps)

In the combustion process, diesel fuel takes a certain amount of time to ignite and burn. As the engine runs faster, the burn time remains the same, and much of the burning takes place after TDC (top dead center). This is called ignition lag and almost always results in lowered performance. To offset this ignition lag, fuel must be injected sometime before TDC to give good performance at rated speed. However, with this fixed advance of injection, engine performance is optimum at rated speed only. Engines that vary speeds over a wide range, that is, automotive vehicles, need injection timed correctly at all speeds. This is the function of the timing device.

The Bosch EP/S..DR(L) timing device is used on in-line camshaft driven pumps (Figure 19.33). The EP/S..DR(L) is classified as an external, flyweight-operated device. Mounted at the front of the injection pump on the camshaft, the timing device is connected to the driving gear of the engine. Through the action of centrifugal force, the flyweights swing outward with increasing speed. Rollers mounted on the flyweights

Figure 19.33 EP/S . . DR(L) timing device used on in-line Bosch pumps (courtesy Robert Bosch Corporation).

push against the cam plate (Figure 19.34), which is connected to the pump camshaft. This causes the camshaft to rotate a maximum of eight degrees, providing proper timing in relation to engine speed.

VIII. Robert Bosch Distributor Injection Pumps

Robert Bosch distributor type injection pumps, models EP/VA..C and EP/VE, differ from the in-line PE series by having only one pumping plunger to serve all engine cylinders. PE pumps, as already shown, have one plunger for each cylinder. The EP/VA..C incorporates an hydraulic governing system while the VE uses a mechanical centrifugal governor. Because of the distributor design, these pumps can be mounted in any position, as the interior of the pump is completely filled with diesel fuel.

A *EP/VA..C Operation* (Figure 19.35).

NOTE EP/VA..A and EP/VA..B models are similar.

After passing through the fuel filters, diesel fuel enters the injection pump and then flows (Fig-

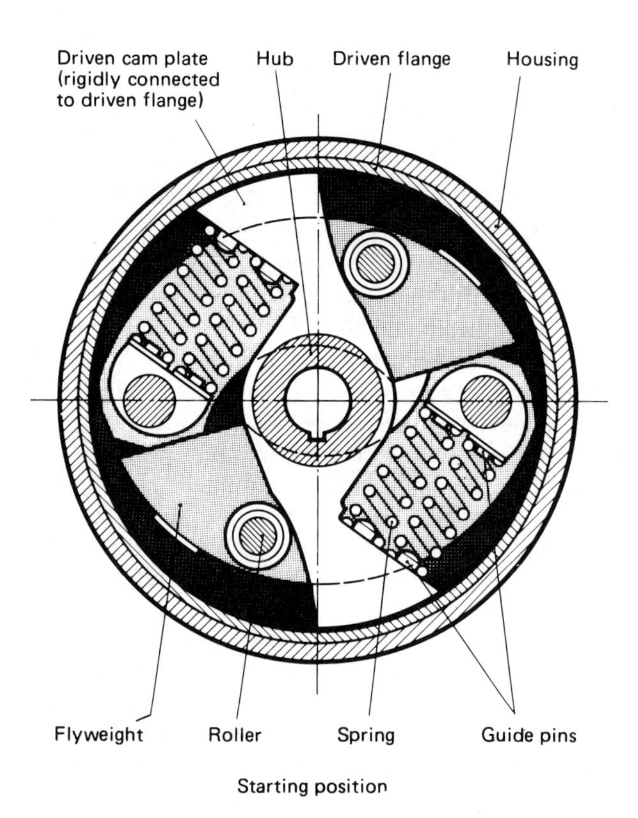

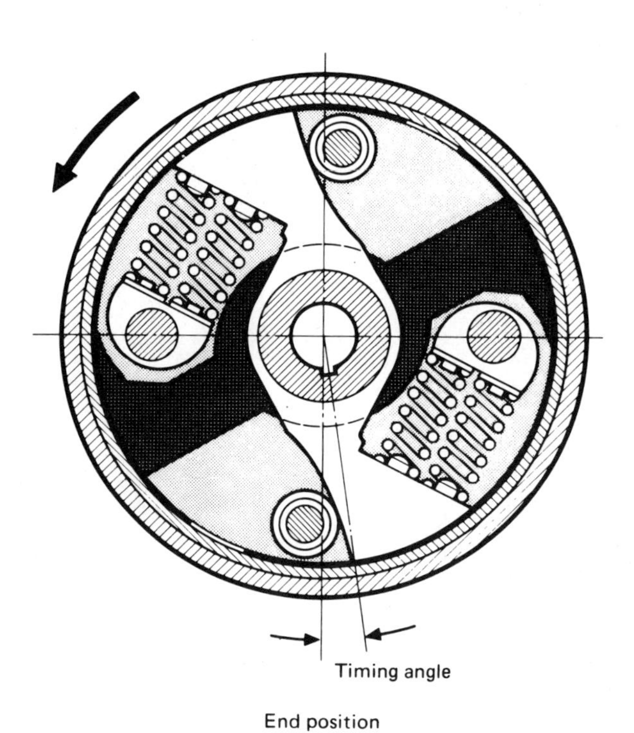

Figure 19.34 Operation of automatic timing device (courtesy Robert Bosch Corporation).

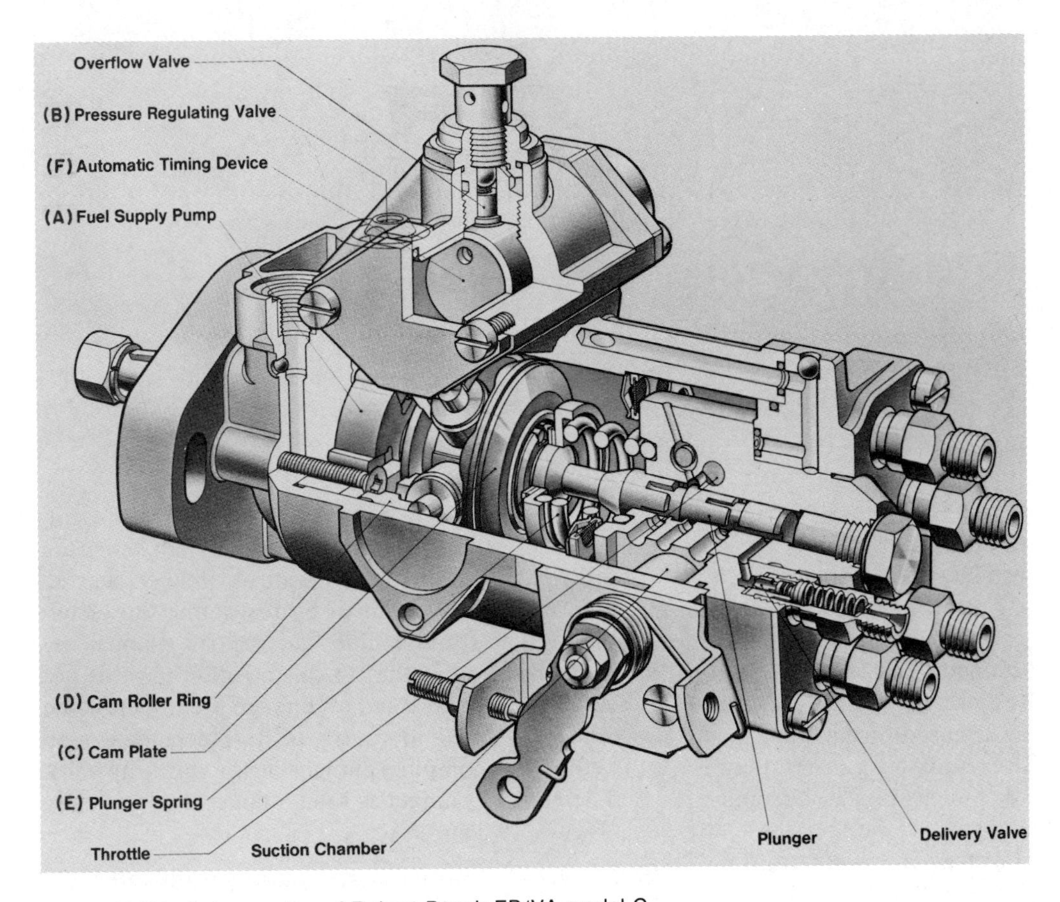

Figure 19.35 Cutaway view of Robert Bosch EP/VA model C rotary injection pump (courtesy Robert Bosch Corporation).

ure 19.35,A) directly into the fuel supply pump. Coming from the supply pump, the fuel passes to the pressure regulating valve (Figure 19.35,B) and is regulated to approximately 110 to 120 psi (7.5 to 8.5 kg/cm²), depending on speed. Bypass fuel from the pressure regulating valve is routed to the inlet side of the fuel supply pump. Pressure fuel from the supply pump is directed into the main pump housing where it completely fills the interior and lubricates the moving parts. Overflow from the pump housing and excess fuel from around the front seal rise to the top of the pump and flow back to the fuel tank.

The same shaft that rotates the fuel supply pump also rotates a face cam, which has a lobe for each engine cylinder (Figure 19.35,C). The rotation of the face cam riding on the cam ring assembly (Figure 19.35,D) causes reciprocating motion of the pumping plunger. Also, the face cam has a pin in its face coupled to the plunger that causes the plunger to rotate while reciprocating. The plunger return spring (Figure 19.35,E) keeps the plunger foot pressed tightly against the face cam, which in turn presses against the cam ring assembly.

A variation in pump timing can be achieved by moving the stationary cam ring assembly. This is done by supplying fuel at housing pressure to the advance piston (Figure 19.35,F), which causes the cam ring assembly to move via a small pin. The amount of advance varies with engine speed and fuel transfer pump pressure.

B *Fuel delivery and governing.*

With the plunger at B D C. (Figure 19.36), fuel at housing pressure enters the hydraulic head at ports A and B and fills the area above the plunger and enters the governing system around the lower annulus on the plunger at point C. As the plunger moves upward, it seals off the inlet fuel ports and no more fuel can enter the high pressure chamber (Figure 19.36,D). This is commencement of delivery and is the point where static timing of the engine is set.

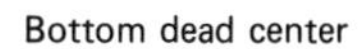

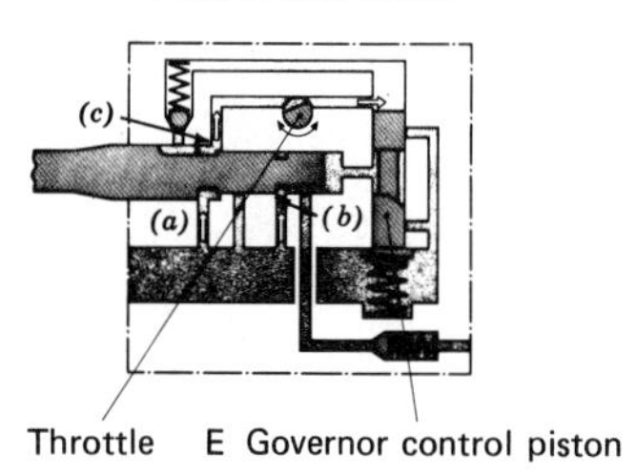

Figure 19.36 EP/VA fuel delivery and governing (courtesy Robert Bosch Corporation).

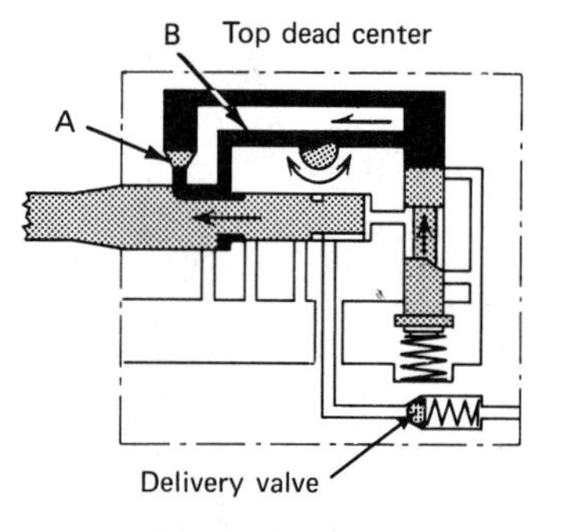

Figure 19.38 Fuel trapped behind plunger (courtesy Robert Bosch Corporation).

Fuel from the top of the plunger goes to the injection nozzle and also to the control plunger (Figure 19.36,E), which determines the amount of fuel injected per stroke. Fuel pressure moves the control plunger outward against the spring pressure until the helix on the plunger (Figure 19.37,A) uncovers the spill port (Figure 19.37,B). Pressure above the plunger is now released to the housing, and injection ceases. The pumping plunger may continue moving upward, but no fuel is injected because the spill port is open.

As the pumping plunger goes downward, the check valve in the governing circuit (Figure 19.38,A) closes and traps fuel behind the control plunger. Fuel must now be bled off from behind the control plunger by the throttle (Figure 19.38,B), the opening of which is controlled by the external throttle lever on the injection pump. If the fuel is bled off quickly from behind the control plunger, it will move back to its mechanical stop.

NOTE Mechanical stop is when the control plunger is stopped mechanically by contact with the hydraulic head.

Thus, when the pumping plunger again moves upward, it must move the control plunger a further distance outward before the spill port is uncovered. More fuel will be injected, this being the full load position of the control plunger. Partial load position is obtained by restricting the bleed-off of fuel from behind the control plunger by means of the throttle. In this position there is not sufficient time to bleed off enough fuel to allow the control plunger to return to its mechanical stop before the pumping plunger again starts upward. The control plunger is then stopped hydraulically (Figure 19.39).

NOTE When plunger is stopped hydraulically, it is stopped by fuel trapped behind it.

As the pumping plunger moves upward and forces fuel into the governing circuit, the control plunger doesn't move as far before opening the spill port. Less fuel is injected so partial load is obtained. The throttle, by controlling leak-off fuel, controls the distance the control plunger reciprocates. As a result, the throttle controls the engine speed. By rotating the control plunger (externally), the helix can be positioned for more or less fuel per stroke since rotation of control plunger controls full load quantity. As the pumping plunger rotates, a slot will line up with a different head outlet port

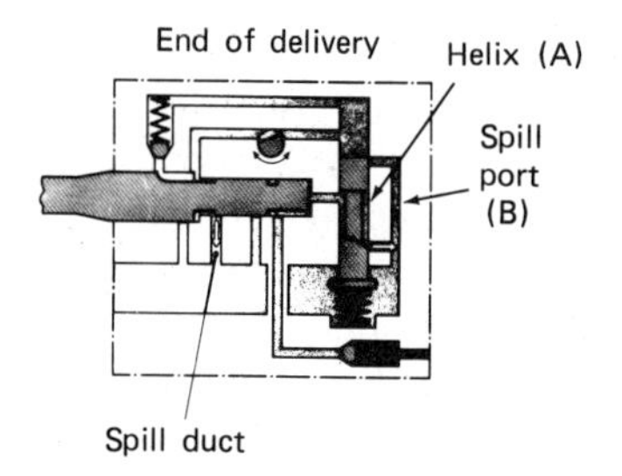

Figure 19.37 EP/VA fuel delivery (courtesy Robert Bosch Corporation).

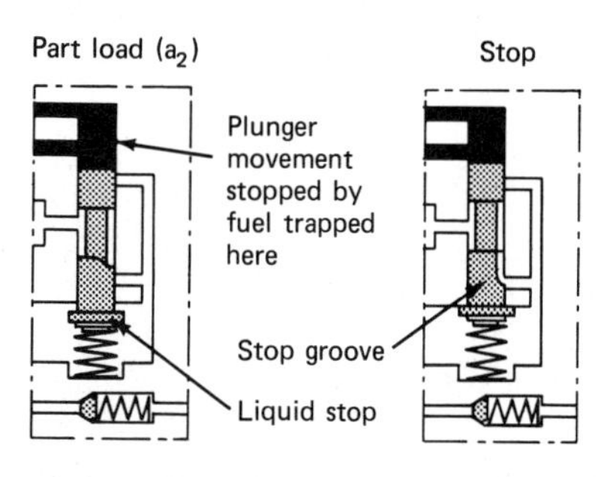

Figure 19.39 Hydraulic stop of control plunger (EP/VA model C) (courtesy Robert Bosch Corporation).

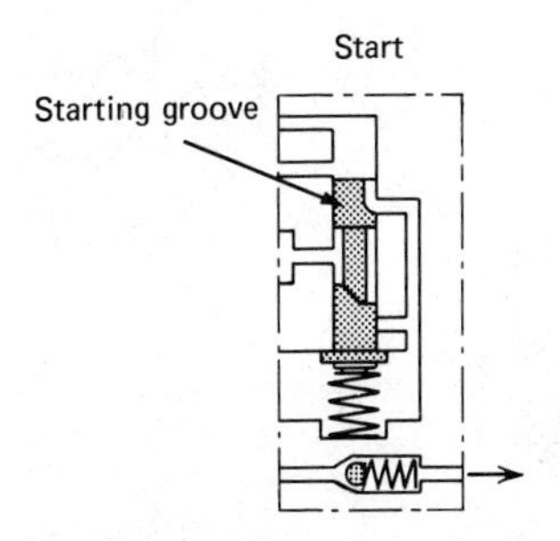

Figure 19.40 EP/VA starting fuel position (courtesy Robert Bosch Corporation).

and distribute fuel to each engine cylinder in turn. Governor regulation (response of the governor under varying load conditions) is determined by the pretension of the control plunger spring and can be adjusted by shims.

C *Starting fuel position.*

Excess fuel for starting is obtained by not allowing the control plunger to reciprocate at all. When this happens, all the fuel above the pumping plunger will be injected, as the control plunger never uncovers the high pressure spill port. This is accomplished by bleeding off the governor circuit fuel through the starting fuel groove (Figure 19.40) on the control plunger. The control plunger is rotated by the external lever until this groove is lined up. There are no rotating mechanical parts in the governor of the EP/VA..C and the adjustments of full load fuel, high idle, low idle, starting fuel, and shutoff are all done externally.

IX. Repair Instructions (In-Line Pumps)

Fuel injection pumps are precision units and will give long, dependable service if properly lubricated and clean fuel is used.

If trouble is apparent in engine operation, make certain that clean fuel filters are installed, all nozzles are working properly, and the injection pump is correctly timed. Also make certain the engine itself is in good mechanical order, for example, the compression is good and there is proper tappet clearance. See Chapter 26 on engine tune-up.

If the problem still exists, the injection pump must be removed and checked.

A *Removal from engine.*

1. Clean exterior engine, especially around the injection pump and nozzles.
2. Rotate engine to fire no. 1 cylinder and align correct timing marks (see engine manual).
3. Shut off fuel supply.
4. Remove injection lines, fuel feed lines, and aneroid linkage (if used).
5. Disconnect throttle linkage, shutoff linkage, and aneroid linkage (if used).
6. Remove drive coupling bolts if used.
7. Remove mounting bolts and carefully remove injection pump from engine.

CAUTION Do not attempt to service an injection pump without the proper tools and service information.

When received for service, an injection pump must be cleaned, visually examined, and the model and serial numbers recorded on the work order. With injection pumps still under factory or shop warranty, the pump should be mounted on the test bench and checked for correct adjustments and proper operation before disassembly. Pumps in the shop for general overhaul must first be disassembled and checked for worn parts, which may not be visible in a preliminary check on the test bench.

B. Procedure for Disassembly

Because of the many different styles of pumps in the PE series, it is impossible to list the exact disassembly for each pump. If a question exists, consult the technical manual for the particular pump being worked on or consult your instructor.

NOTE All repair work on injection pumps must be done in a room that is spotlessly clean to protect the precision parts.

To aid in disassembly, all pump parts must be placed in a divided plastic or metal tray similar to the one in Figure 19.41.

NOTE A plastic tray (called the organizer) is available directly from the Robert Bosch Corporation.

1. Mount pump on the proper holding fixture (Figure 19.42).
2. Remove fuel supply pump(s), side cover (A and M pumps), and aneroid linkage.
3. Remove screws that hold the governor to housing

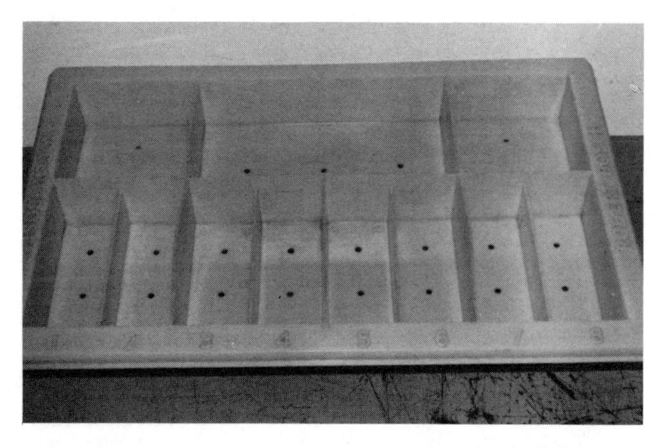

Figure 19.41 Tray for holding pump parts (courtesy Robert Bosch Corporation).

Figure 19.42 Pump holding fixture with pump mounted on it.

Figure 19.43 Unhooking governor control rack linkage.

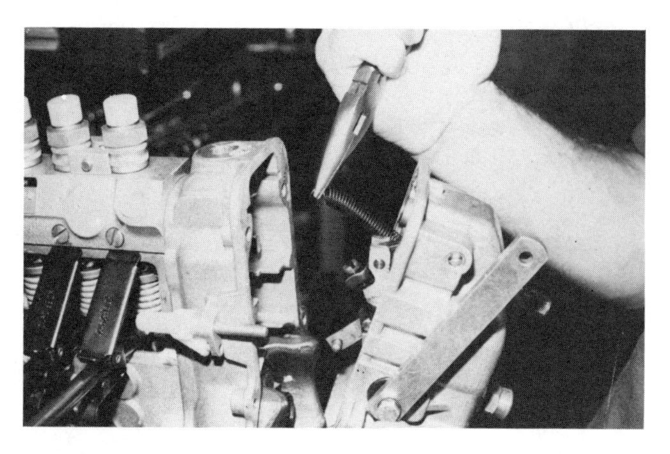

Figure 19.44 Unhooking starting spring.

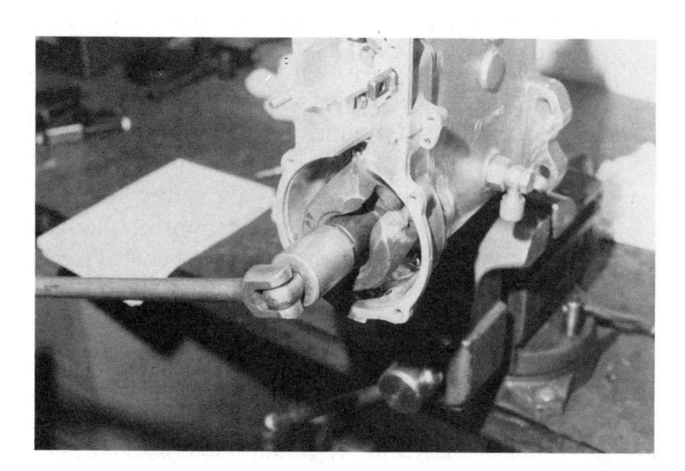

Figure 19.45 Removing flyweight nut.

Figure 19.46 Removing flyweight assembly using special puller.

and tap governor loose just far enough to unhook the linkage to the control rack (Figure 19.43).

4 On pumps using the RSV governor, unhook the small starting spring before removing the governor (Figure 19.44).

5 Using the flyweight nut wrench (Figure 19.45), loosen and remove the flyweight nut. The drive end of the camshaft must be held to prevent rotation.

6 The flyweight assembly can then be removed, using the special puller shown in Figure 19.46.

7 Rotate pump camshaft until no. 1 plunger is at TDC and install tappet holder to raise the tappet off the camshaft. Various types of holders are used, depending on the pump type (Figure 19.47). On a P pump the tappet holder must be rotated to raise the tappet to its highest point. By rotating the camshaft, all of the tappet holders can be installed in the same manner.

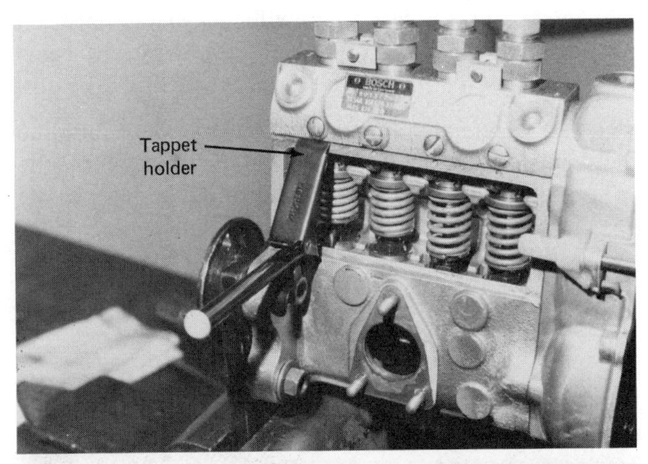

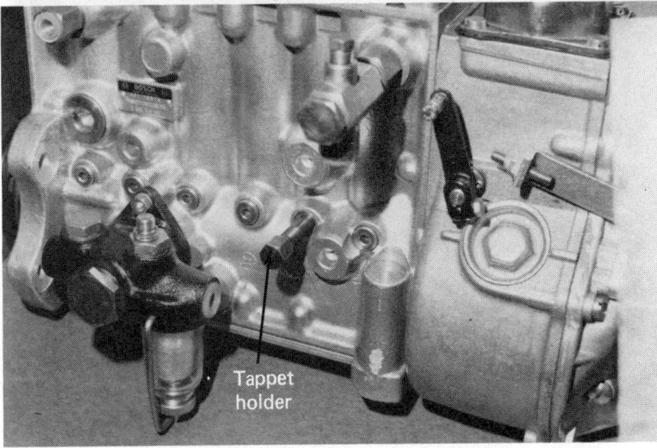

Figure 19.47 Various types of tappet holders.

Figure 19.48 Pulling drive hub or gear.

8 Using a suitable puller, remove the drive hub or gear, being careful not to damage it (Figure 19.48). Invert pump.

9 Remove the screws retaining the front bearing plate to the housing and remove housing. Note the position of the oil drain holes and timing pointer (or line) so that reassembly will be correct (Figure 19.49). On pumps of six cylinders or more, it will be necessary to remove the screws holding the half shell center bearing. On P pumps the bottom plate must first be removed (Figure 19.50).

10 Carefully remove the camshaft (Figure 19.51). Do not drop the center bearing shell as the camshaft is removed.

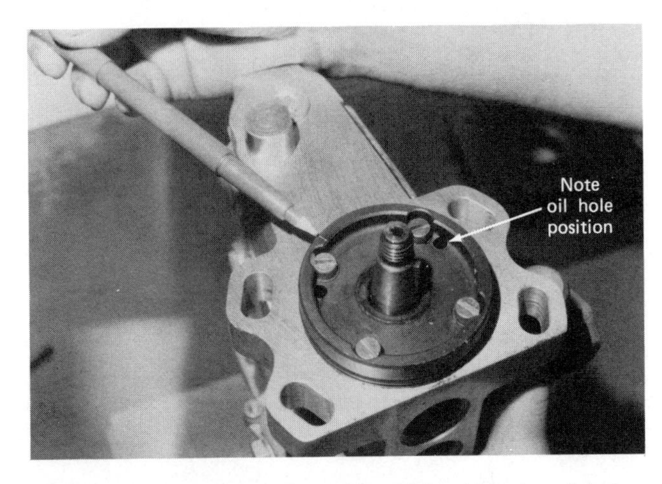

Figure 19.49 Oil drain hole position of front bearing plate to front housing.

Figure 19.50 Removal of center bearing screws.

Figure 19.51 Camshaft removal.

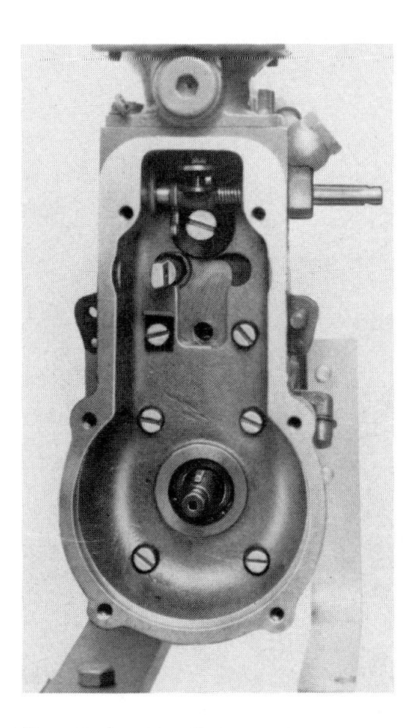

Figure 19.52 Removal of intermediate governor housing (courtesy Deere and Company).

Figure 19.53 Spacer between main housing and intermediate housing on P pumps (courtesy Deere and Company).

11 Remove the screws retaining the intermediate governor housing to the pump housing (Figure 19.52). Note the position of the bracket for the starting spring, if used. Gently tap the housing loose. On P pumps there will be a spacer between the intermediate housing and the camshaft bearing (Figure 19.53).

12 On A pumps knock out the closing plugs (cuplike plugs made from brass) on the bottom of the pump (Figure 19.54) by driving them into the housing. Discard the closing plugs.

13 Release the pressure on the tappet holders by pressing the tappets down with the correct tool

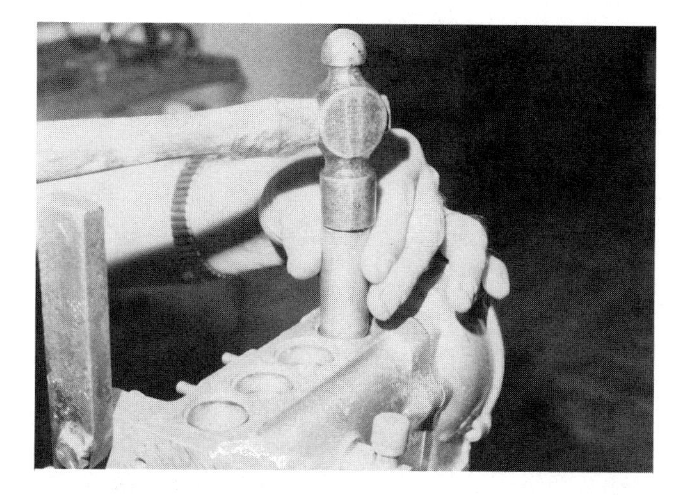

Figure 19.54 Removing the knock-out closing plugs from the pump housing.

Figure 19.55 Releasing pressure on tappet holders with releasing tool.

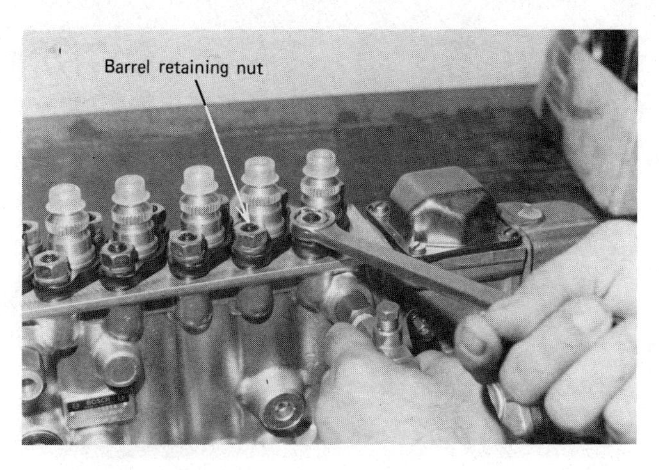

Figure 19.57 Removing barrel retaining nuts on P pumps.

recommended in the pump service manual (Figure 19.55).

14 Remove the roller tappets and place them in the divided tray.

NOTE Make certain that roller tappets, plunger, and barrel assemblies and associated parts are kept in the divided tray according to the cylinder number. This will aid reassembly and make calibration much easier.

15 Carefully remove plungers with lower spring seat (Figure 19.56).

NOTE Avoid touching the plungers with fingers and do not burr the surface in any way.

16 Remove the plunger springs, upper spring seats, and control sleeves.

17 On A and M pumps, remove the clamping pieces between the delivery valve holders. On P pumps remove the pressed steel cover.

18 On P pumps remove the two retaining nuts on each barrel assembly (Figure 19.57). Then use two screwdrivers to pry up the barrel assembly. Upon removal, insert the plunger back into its respective barrel to prevent mix-up.

19 On A, B, and M pumps remove the delivery valve holders, springs, and valves and discard the delivery valve gaskets. Then push up on the barrel as-

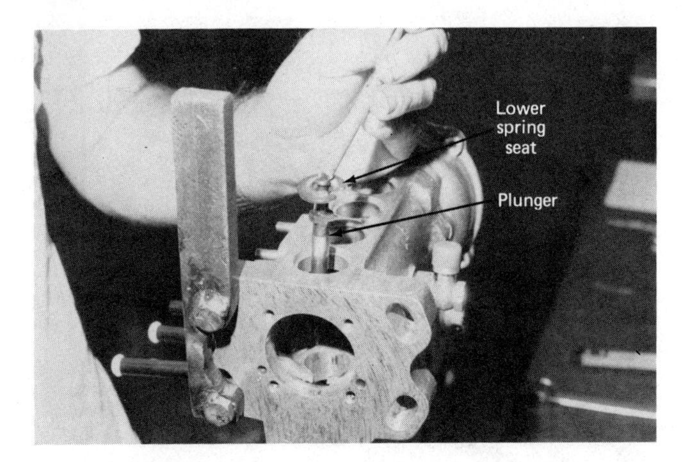

Figure 19.56 Removing plungers and lower spring seat.

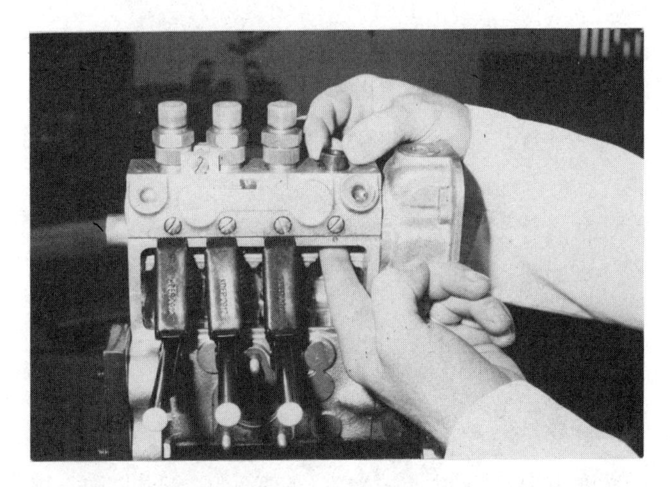

Figure 19.58 Removing barrel assembly on A series pump.

Figure 19.59 Removing rack retaining screw on A series pump.

sembly and remove it from the housing (Figure 19.58). Insert plungers into the barrel.

20 Remove the rack retaining screw and remove the control rack (Figure 19.59)

21 On P pumps the barrel assembly can be disassembled by placing it in the holding tool and using the special serrated socket to loosen the delivery valve holder (Figure 19.60). All parts of the assembly are shown in Figure 19.61.

CAUTION On A pumps the fuel baffle screws directly above the side cover must not be removed as they are epoxied in place. Damage to the housing will result if the screws are removed (Figure 19.62).

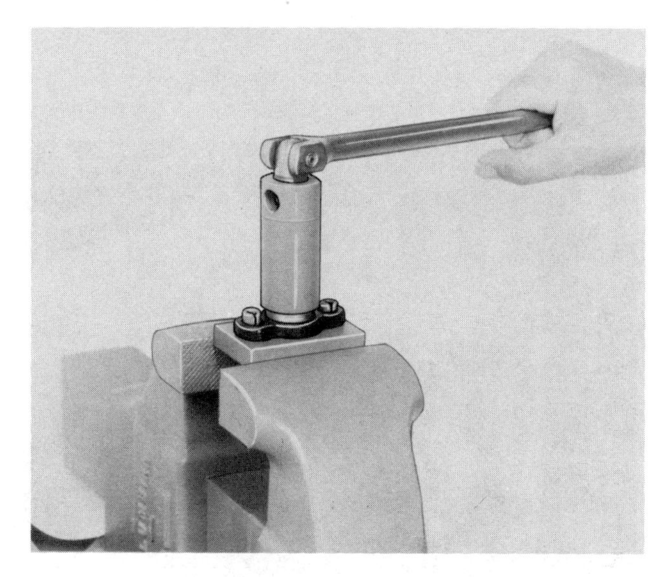

Figure 19.60 Removing delivery valve holder on P pumps (courtesy Deere and Company).

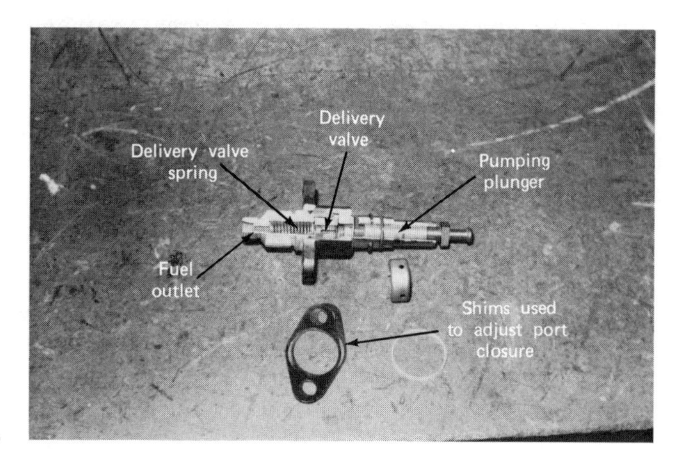

Figure 19.61 Components of P pump barrel assembly.

C. Procedure for Inspection and Repair of Component Parts

Thoroughly clean all pump parts in parts cleaner and rinse with clean solvent. Inspect all parts for wear or damage as indicated below.

1 *Barrel and plunger assembly.* Inspect plunger carefully for scoring. If this scoring can be felt with the fingernail, the assembly should be replaced. Avoid touching the lapped surface of the plunger with the fingers! (Figure 19.63)

2 *Pump housing.* Check the pump housing for thread

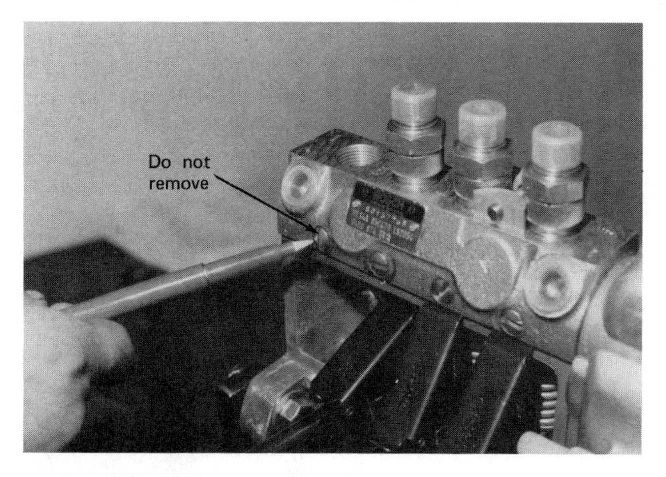

Figure 19.62 A pump baffle screws.

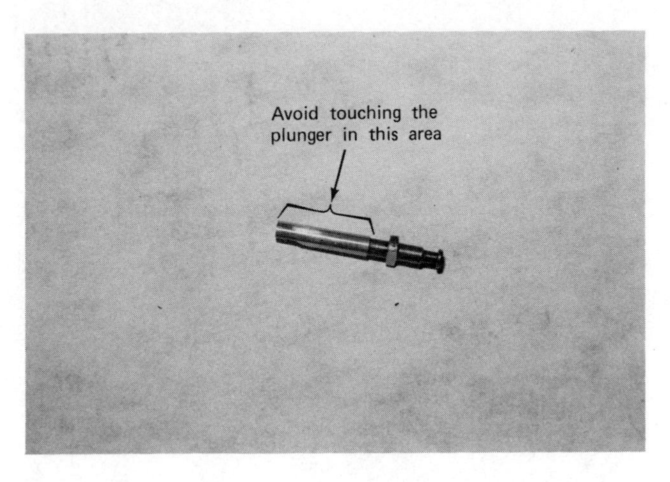

Figure 19.63 Do not touch this area of the lapped surface.

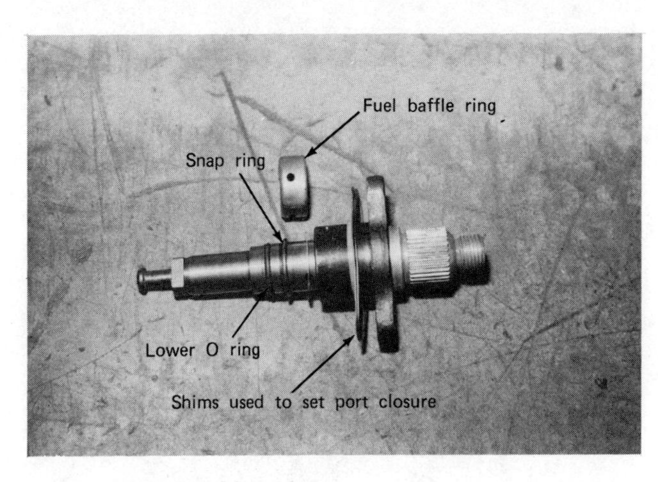

Figure 19.64 Position of shims, snap rings, and O rings on pump barrel.

damage, cracks, or aný other damage that would render it unserviceable.

3 *Camshaft.* Check the camshaft for flaking of metal on the lobe surface, thread damage, and out-of-roundness. Replace as necessary.

4 *Camshaft bearings.* Check bearings for galling and excessive wear. Check the half shell center bearings for scoring. As a general rule, all bearings are replaced during a major overhaul of the pump or if any contamination is found.

5 *Roller tappets.* Inspect all rollers for scoring and flaking of the surface metal. Any rough edges on the tappet should be removed.

6 *Delivery valves.* Make certain the valves are not sticking in the holder. Check the lapped surface of the delivery valve body. It must be absolutely clean and damage free. Lap the body before reusing.

7 *Delivery valve and plunger springs.* If the springs show any sign of rust or corrosion, they should be replaced. Also check for broken or bent springs and replace as necessary.

8 *Control rack and control sleeves.* Check the tooth segment of rack and sleeves for excessive wear. Replace if needed. The slots in the control sleeves that guide the plunger vane should be checked for wear and spreading apart.

NOTE Always replace all gaskets, sealing rings, oil seals, and packings when overhauling a pump. It is recommended that a gasket package supplied by Robert Bosch be used when overhauling a pump.

D. Procedure for Pump Reassembly

As the pump is assembled, lubricate all O rings and moving parts with test oil.

1 On P pumps assemble the barrel assembly and torque the delivery valve holder on the barrel to specified torque. Install the shims, snap rings, and O rings, as shown in Figure 19.64.

2 Install barrel assemblies in the pump housing and torque the retaining nuts. Be careful not to damage the O rings as the assemblies are being installed (Figure 19.65).

NOTE On P pumps install the plungers in the barrels and screw the tappet holders all the way in for leakage check. Plug outlet of fuel gallery and apply air to the inlet at 35 psi (2.5 kg/cm^2). Immerse the pump in test oil or solvent and check for air bubbles. Disregard small bubbles coming from the pumping plungers. Repair any leaks from barrel O rings. After the leak test, carefully remove the plungers and continue the assembly.

3 On A and M pumps lap the top surface of the barrels and install them in the housing, *making certain to align the groove in the barrel with the pin in the housing* (Figure 19.66). On B pumps a barrel locating screw is used. This must be installed after the barrel.

4 Using new delivery valve gaskets, install the delivery valve, spring, and holder. Torque the holders to specifications.

Figure 19.65 Installation of barrel assemblies on P pump (courtesy Deere and Company).

Figure 19.67 Installing threaded rack bushing (P pumps) (courtesy Deere and Company).

5 Install the control rack and rack locating screw. On P pumps press the guide block for the control rack into the housing, making certain that the vertical groove is parallel to the pump housing (Figure 19.67). Screw in the threading bushing and torque.

6 Place the rack in center position and install the control sleeves, making sure that they are properly timed to the control rack (Figure 19.68).

7 Invert pump and install upper spring retainers and plunger springs.

8 Rinse off the pumping plungers with test oil and install them in their respective barrel (putting on the lower spring seat as the plunger is inserted into barrel).

NOTE The vertical mark on the plunger vane or shoulder must face away from the control rack on A, M, and B pumps (Figure 19.69).

9 Install roller tappets, push completely down with the wooden tappet tool, and install tappet holders.

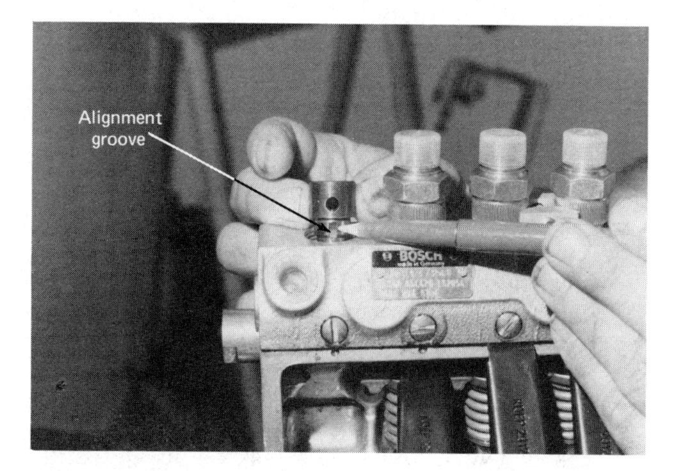

Figure 19.66 Align barrel groove with locating pin (A and M pumps).

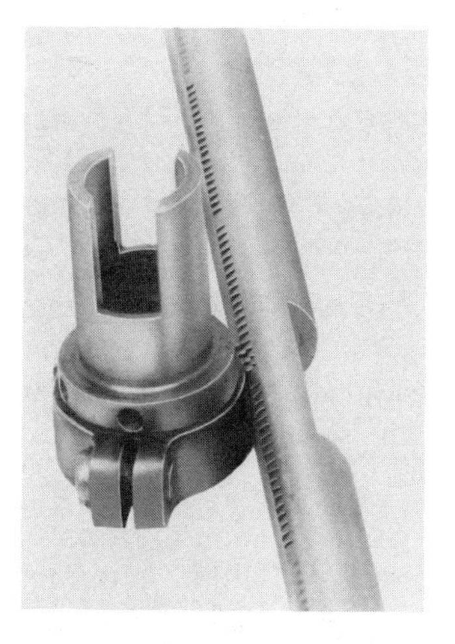

Figure 19.68 Timing control sleeves to the control rack (courtesy Deere and Company).

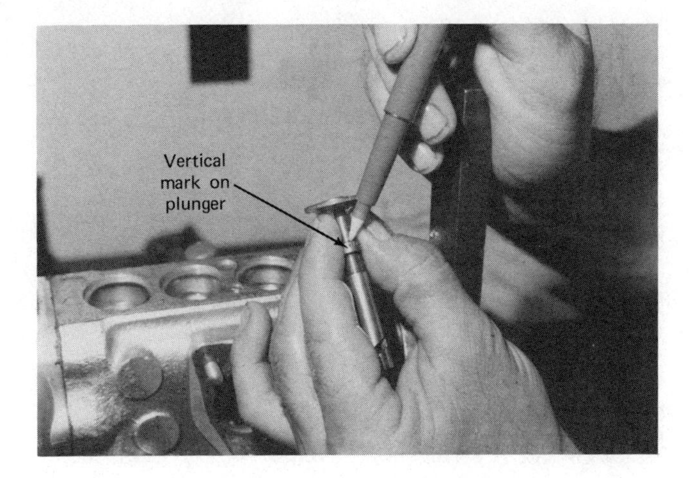

Figure 19.69 Vertical timing mark on plunger vane (A pump).

10 Install new camshaft bearings if needed. Oil these lightly and install the camshaft into the housing without the center bearing. Install the bearing end plate, noting the position of oil drain holes (if used) and torque to specified torque (Figure 19.70).

11 The camshaft protrusion should now be checked using the camshaft taper gauge. Measure the clearance between the pump housing flange and the taper gauge (Figure 19.71). This measurement should be as follows:

A pumps = 9.5 ± 0.5 mm (0.374 ± 0.020 in.)
B pumps = 9.5 ± 0.5 mm (0.374 ± 0.020 in.)
M pumps = 9.5 ± 0.5 mm (0.374 ± 0.020 in.)
P pumps = 13-14 mm (0.511 − 0.551 in.)

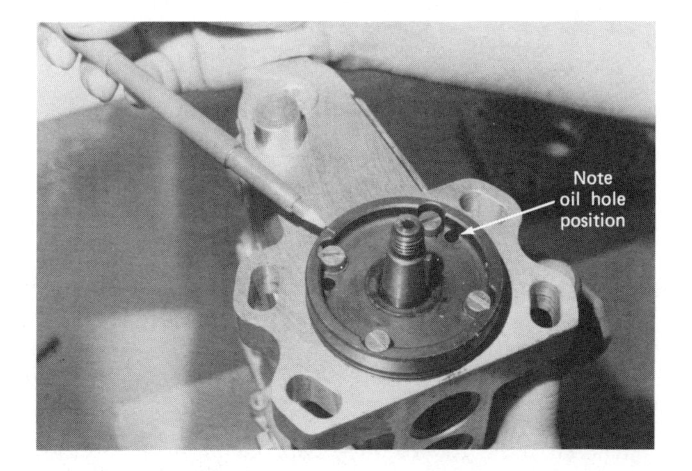

Figure 19.70 Bearing retainer plate correctly installed.

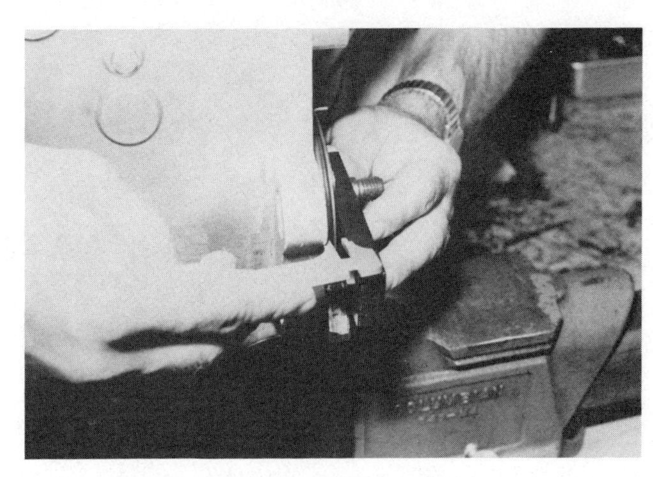

Figure 19.71 Checking camshaft protrusion.

12 If protrusion is incorrect, shims under the camshaft bearings must be varied from one end of the camshaft to the other until the correct protrusion is obtained.

13 After the correct protrusion is obtained, check the camshaft end play with a tool similar to the one shown in Figure 19.72. Pull and push steadily on the tool to get an accurate reading. If end play is not as listed in the specification sheet, remove the camshaft and add or delete shims on the front bearing to correct the end play.

14 When the end play is correct, remove the camshaft, mount the center thrust bearing, and reinstall the camshaft. Torque the bearing end plate screws and the center bearing screws to specification.

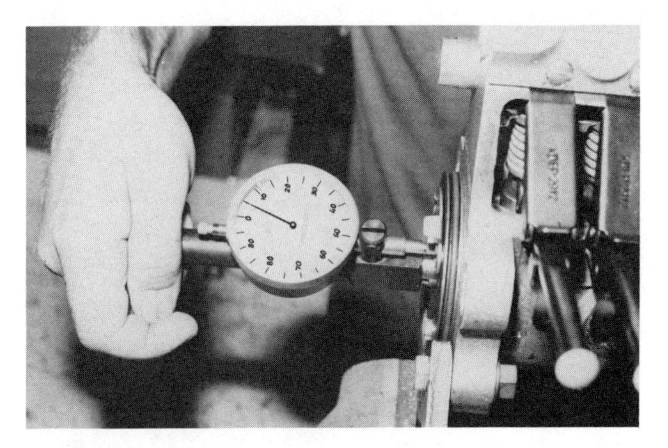

Figure 19.72 Checking camshaft end play.

Figure 19.73 Removing cross pin (EP/RSV governor).

15 Drive in new closing plugs on A pumps or install the base plate on P pumps.

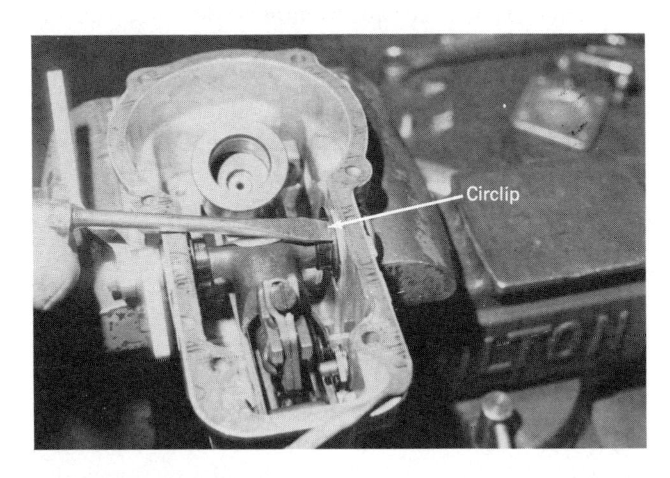

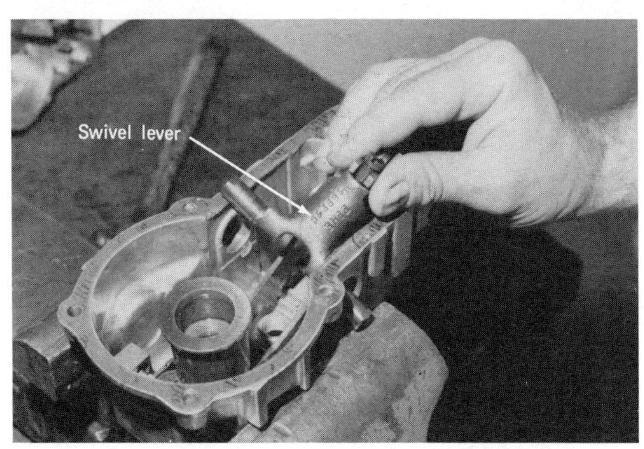

Figure 19.74 Removing swivel lever (EP/RSV governor).

Figure 19.75 Removing guide lever.

16 Remove tappet holders, allowing tappets to set back down on camshaft.

17 Install the fuel supply pump.

18 Install the governor.

19 Mount pump on test bench for adjustments.

E. Procedure for Governor Disassembly (EP/RSV)

1 Remove the two closing plugs at the top of the governor and drive out the cross pin (Figure 19.73). Remove the external control lever.

2 Remove the two retaining circlips on the swivel lever. Tap out the lever bushings and remove the swivel lever (Figure 19.74).

3 Remove the tensioning lever with main governor spring.

4 Remove the guide lever with the thrust sleeve and starting spring (Figure 19.75).

5 If a separate stop device is used, remove it.

6 Drive the guide lever from the thrust sleeve, as shown in Figure 19.76.

F. Procedure for Inspection and Repair of Governor

Wash all parts and remove all old gasket material. Inspect all parts for wear.

Figure 19.76 Removing thrust sleeve from guide lever.

1 The governor cover must not be worn where the swivel lever bushings fit.

2 Check swivel lever shaft for excessive wear. Replace as necessary.

3 Inspect the cross shaft and tensioning lever for wear.

4 Replace the main governor spring if it is rusted, bent, or shows signs of fatigue.

5 Replace all gaskets, seals, and O rings.

G. Procedure for Assembly (EP/RSV Governor)

Before assembling the governor, check the fitting dimension of the knuckle and thrust sleeve. This involves checking the distance from governor housing to rear of the thrust sleeve (removed from the guide lever and installed between the governor weights) with shims installed (Figure 19.77). If this dimension isn't as specified on the specification sheet, add or delete shims until the correct measurement is obtained. Reinstall thrust sleeve (with shims) to the knuckle on guide lever.

1 Check guide lever for distortion as shown in Figure 19.78. Both ends of lever must contact the flat surface together.

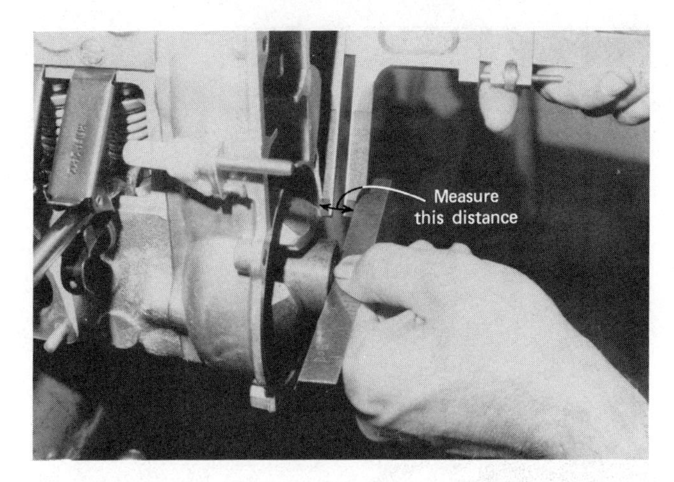

Figure 19.77 Checking governor thrust sleeve dimension.

2 Install the stop device (if used), using new O rings.

3 Install the guide lever with thrust sleeve into the housing and insert the cross pin into one leg.

4 Insert swivel lever and bushings. Lock the bushings in place with the retaining circlips (keepers).

5 Maneuver the tensioning lever with governor spring under the swivel lever and hook the spring to both levers.

6 Insert the cross pin through the tensioning lever and the second leg of the guide lever. Install the two closing plugs.

7 Attach the external control lever to the swivel lever, using shims to take up excess play.

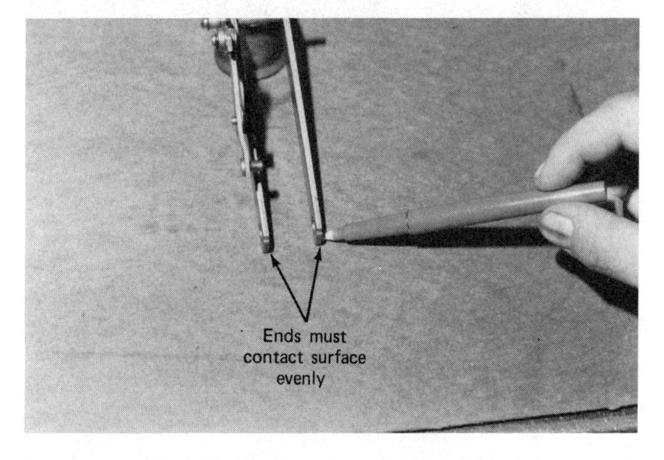

Figure 19.78 Checking guide lever for distortion.

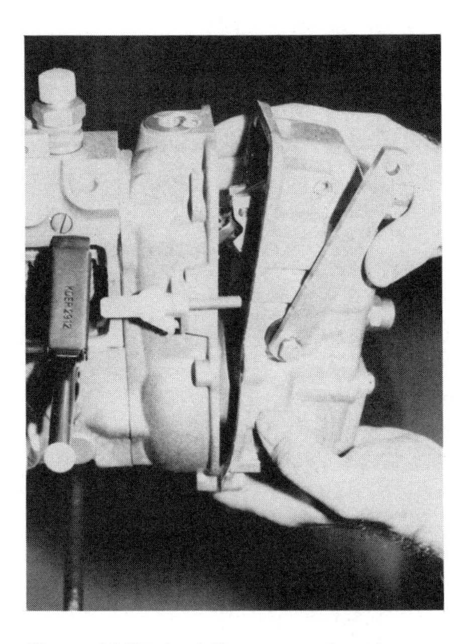

Figure 19.79 Install governor housing.

8 Using a new gasket, install the governor housing to the injection pump (Figure 19.79).

NOTE Pump with governor must be adjusted on the test bench after governor repair.

X. Repair Instructions (Distributor Pumps)

Before starting disassembly, clean the exterior of the pump and mount it in the proper holding fixture (Figure 19.80).

Figure 19.80 EP/VA pump mounted on holding fixture.

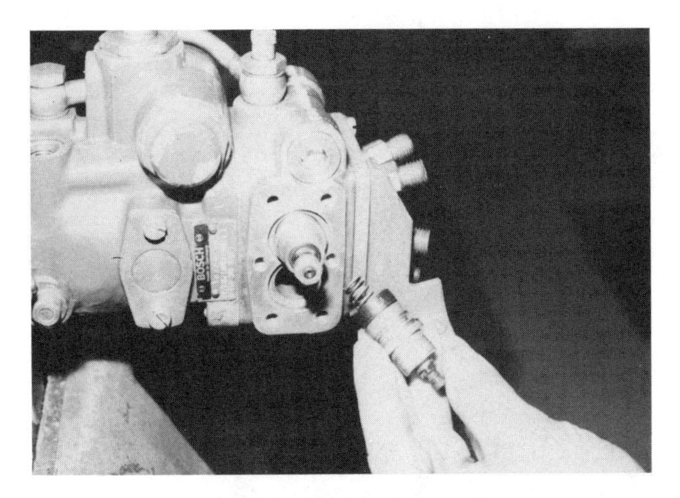

Figure 19.81 Removing bushing and shaft of delivery rate control.

A. Procedure for Disassembly

1 Remove the throttle lever and delivery rate lever. Remove torsion springs.

2 Remove stop plate. Remove bushing and shaft of delivery rate control (Figure 19.81). Remove spring and piston.

Figure 19.82 Removing delivery rate control shaft from bushing (EP/VA).

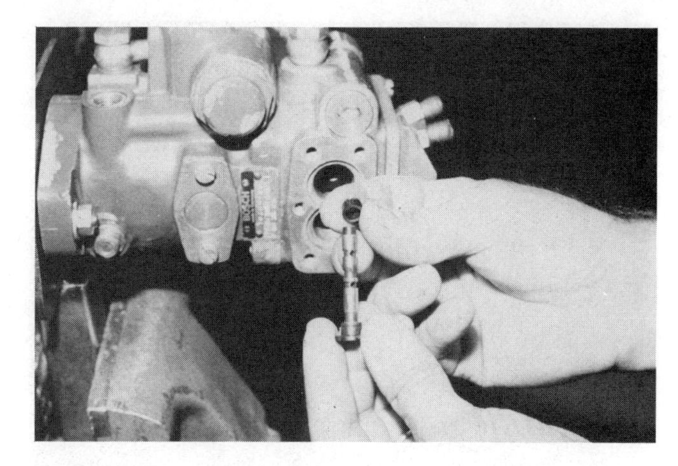

Figure 19.83 Removing throttle and throttle parts (EP/VA).

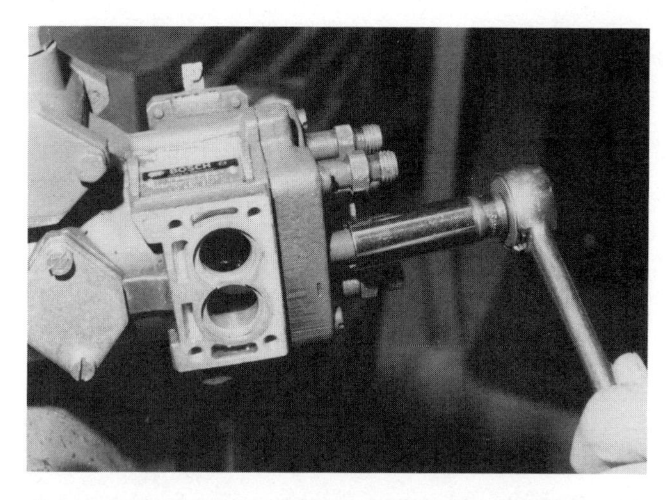

Figure 19.84 Removing delivery valve holder and delivery valve (EP/VA).

3 Remove shaft from bushing, as shown in Figure 19.82.

4 Take out the bushing and shaft for throttle, throttle, and throttle spacer ring (Figure 19.83).

CAUTION Do not lose the spacer ring because it is matched to the throttle and the hydraulic head during production.

5 Remove the delivery valve holders, springs, and valves (Figure 19.84), noting the letters stamped on the hydraulic head.

NOTE All delivery valve parts must be returned to the same bore. Discard the old delivery valve gaskets.

6 Set the pump vertically in the vise. Loosen the hydraulic head retaining screws and remove the head with caution so that pumping plunger does not fall out. Remove the plunger, plunger spring, and washers (Figure 19.85).

7 Remove the face cam.

8 Remove the side cover and the timing pointer (Figure 19.86).

9 Remove the covers on the automatic timing device and take out the spring (Figure 19.87).

10 Remove the overflow valve. Lift out the timing pin clip (Figure 19.88).

11 Remove the cross disc (Figure 19.89).

12 Take out the timing member lock pin and clip.

13 Push the connecting pin of the advance unit into the pump.

Figure 19.85 Removing plunger and plunger spring (EP/VA).

Figure 19.86 Removing side cover and timing pointer (EP/VA).

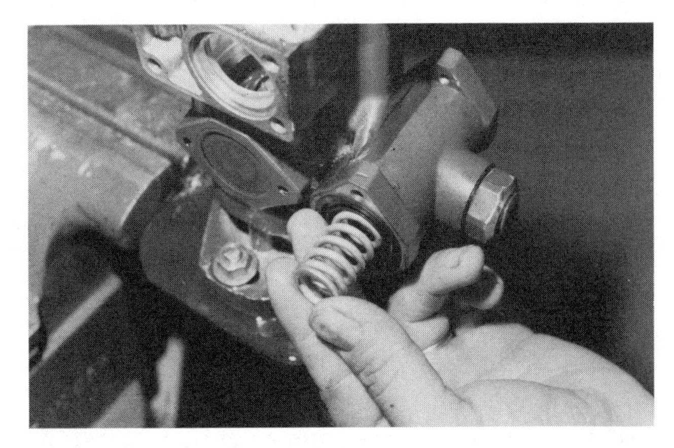

Figure 19.87 Removing automatic timing device, covers, and springs.

Figure 19.90 Removing roller ring assembly.

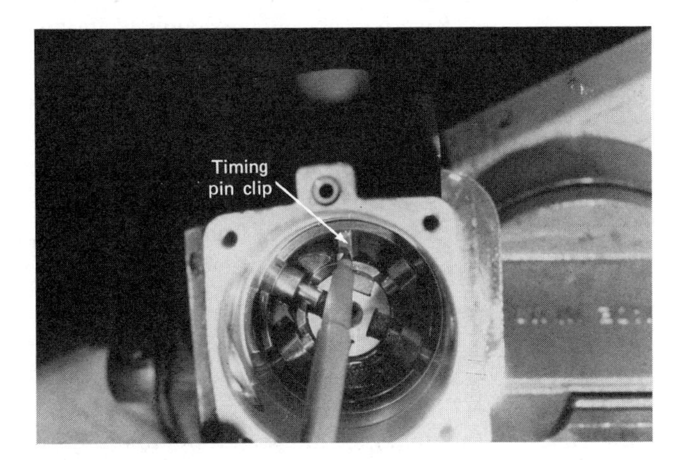

Figure 19.88 Removing timing pin clip.

14 Push out the timing piston.

15 Pull out the roller ring assembly (Figure 19.90).

CAUTION Do not drop or interchange the cam rollers.

16 Remove the fastening screws for the fuel supply pump. Invert the pump while holding the drive shaft and push out support ring, supply pump, and shaft as a unit. Remove the supply pump from the shaft. Do not interchange the pump blades (Figure 19.91).

17 Remove pressure regulating valve for the supply pump.

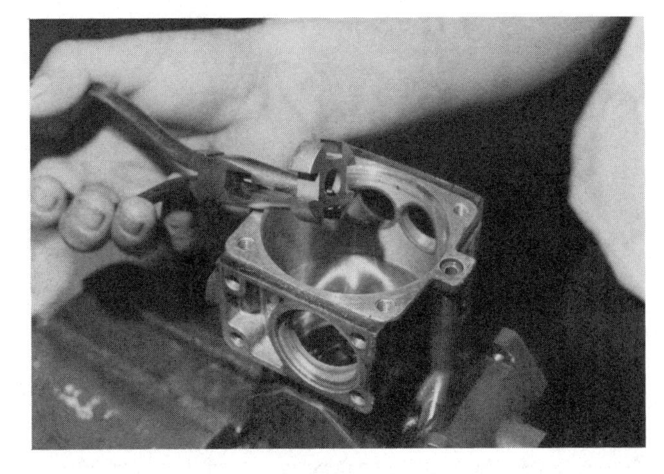

Figure 19.89 Removing cross disc.

Figure 19.91 T pump blades shown in T pump.

B. Procedure for Inspection and Repair of Parts

Replace all worn or broken parts. When replacing parts, the following must be treated as units:

1 Hydraulic head with pumping plunger, control plunger, and throttle with spacer.

2 Pump housing with timing piston.

3 Cam roller ring with rollers and washers.

4 Supply pump with vanes and liner.

Always replace all gaskets, O rings, and seals. Be sure to dip them in test oil before assembly.

Figure 19.93 Installation of supply pump and drive shaft.

C. Procedure for Assembly

1 Place the pump housing in the holding fixture and replace the front seal.

2 Assemble the drive shaft and supply pump. The bevel on the impeller must face the retainer.

3 When assembling the supply pump liner, note the hole farthest from the inside diameter (Figure 19.92). If the pump runs clockwise, the hole must be on the right, looking at the threaded end of the drive shaft. If the pump turns counterclockwise, the hole must be on the left. The third hole must always be located on top.

4 Install the supply pump, drive shaft and support ring as shown in Figure 19.93. Install the retaining screws and torque.

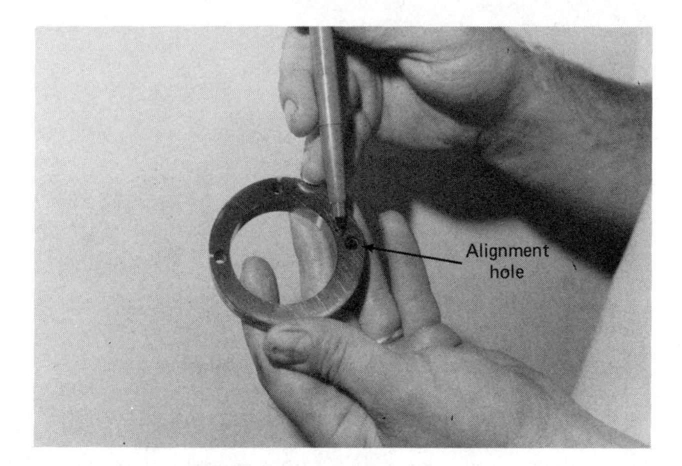

Figure 19.92 Hole in T pump liner.

5 Insert the connecting pin in the cam roller ring and install the ring assembly into the pump with the pin facing the advance device (Figure 19.94).

6 Install the timing piston, with the spring on the left for clockwise pumps and the spring on the right for counterclockwise pumps.

7 Push the connecting pin outward into the advance piston and install the locking pin and clip (Figure 19.95). Insert the cross disc.

8 Install overflow valve, advance spring, and shims.

9 Install timing pointer. Locate the face cam in the pump.

NOTE The drive pin in face cam must be aligned with the key in the drive shaft.

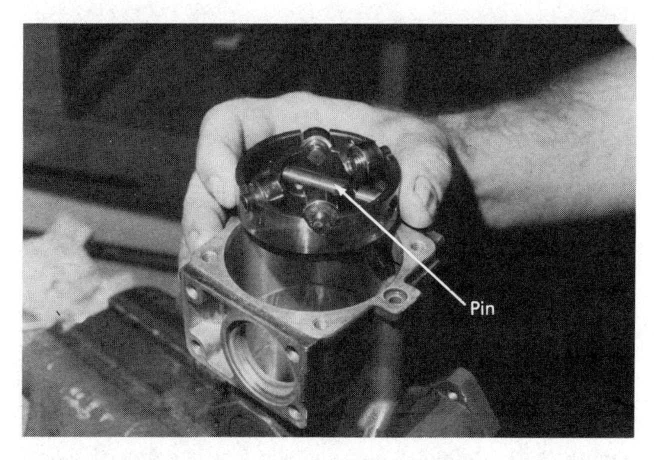

Figure 19.94 Installing connecting pin in cam roller ring.

Figure 19.95 Installing locking pin and clip.

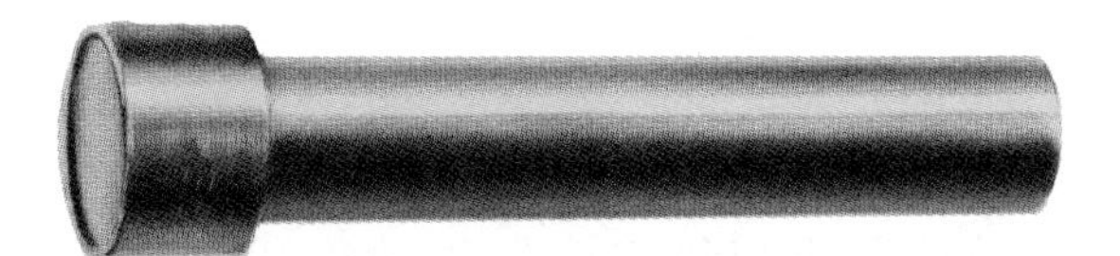

Figure 19.96 Tool used in aligning the hydraulic head with the pump housing.

10 Place the plunger with the shim on the cam.

11 Align the hydraulic head with the housing, using the tool shown in Figure 19.96.

12 Some EP/VA..C pumps employ O prestroke. On these pumps a code number is stamped on the hydraulic head near the top plate. The specification sheet will explain the code number. Set the pumping plunger to BDC. Zero the dial indicator on a lapping block (Figure 19.97). With the dial indicator and mounting device, measure the distance to the plunger. This value is subtracted from 30 mm to give the actual reading. If the reading is incorrect, the shim under the plunger foot must be changed.

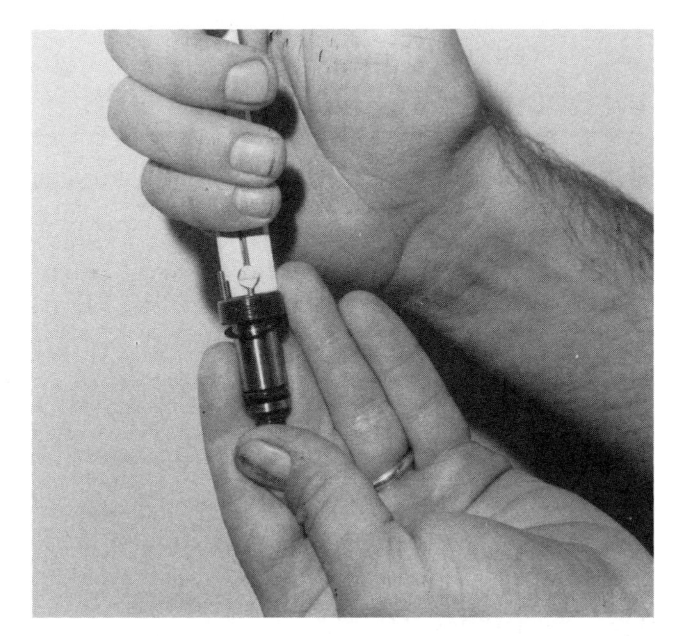

Figure 19.98 Measuring inside governor control shaft.

13 Measure inside the governor control shaft from end face to spring seat (dimension 4) and record (Figure 19.98).

Figure 19.97 Zero dial indicator before setting on hydraulic head.

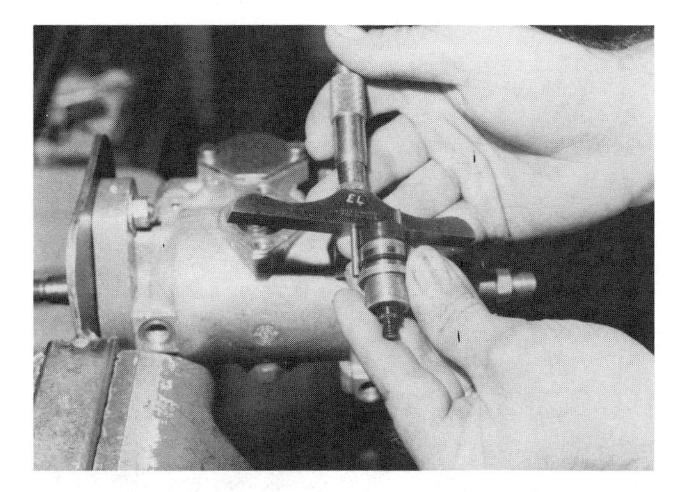

Figure 19.99 Measuring distance from end face of shaft to bushing shoulder.

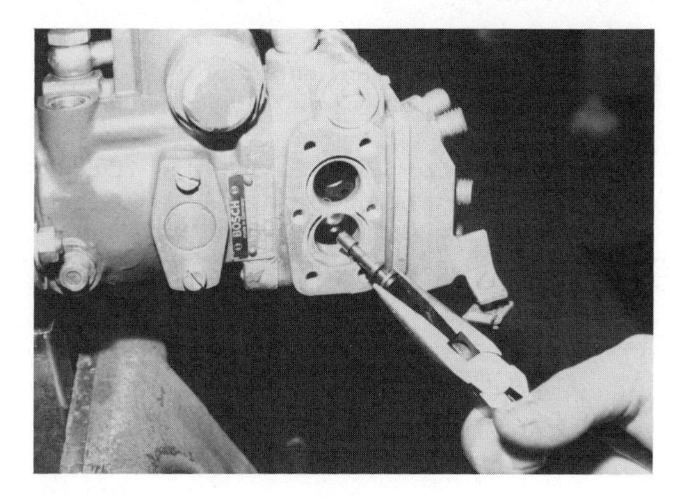

Figure 19.100 Installing control plunger in head.

14 Insert the shaft into the bushing and measure the distance from end face of the shaft to the bushing shoulder (Figure 19.99). This is dimension 5.

Dimension $1-2=3$
$4-5=6$
$3+6=7$
$7-$dimension V (on specification sheet) $=$ thickness of shim required

15 Assemble throttle and delivery rate control plungers to head (Figure 19.100).

16 Install bushings and cover plate onto control shafts (Figure 19.101).

17 Install delivery valves, springs, and holders. Torque to specifications.

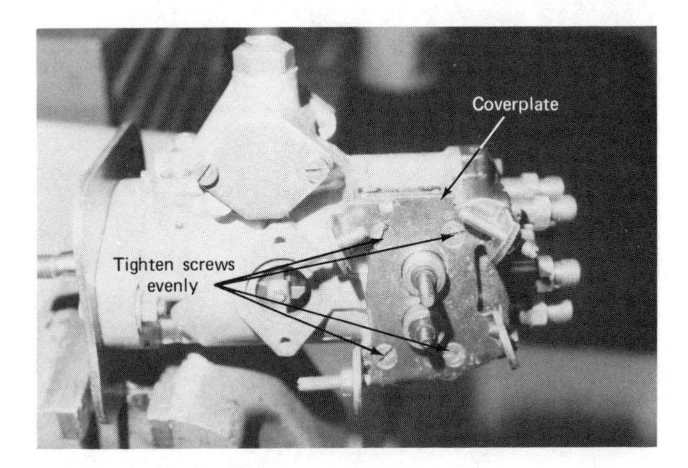

Figure 19.101 Installing cover bracket.

D. Procedure for Setting Port Closure (Except on O Prestroke pumps)

Mount the assembled pump to the test bench. Mount a dial indicator and prestroke measuring device to the pump and supply fuel at 3 psi (0.2 kg/cm^2) to the pump inlet. Port closure occurs when the fuel running from the plunger area is just stopped as the pump is rotated from BDC. The indicator reading should correspond to the one listed on the specification sheet. If it does not, remove the hydraulic head and apply a shim of different thickness under the plunger foot. Reassemble and check once again.

E. Procedure for Setting Cam Stroke

Remove the hydraulic head and take out the plunger spring.

1 Reinstall the plunger with the spacer sleeve (Figure 19.102).

2 Measure and record the distance from the plunger foot shim to the seating surface of the hydraulic head (Figure 19.103. This is dimension *a*).

3 Set cam plate to BDC.

4 Measure from housing surface to center of face cam (This is dimension *b*).

5 Compare *a* to *b*.

6 Compensate by varying shim size between spring seat and plunger.

Figure 19.102 Installing plunger in hydraulic head.

Figure 19.103 Measuring distance from plunger foot to seating surface of hydraulic head.

7 Reassemble pump and set up on test bench to complete pump adjustment.

XI. Procedure for Testing In-Line Pumps on Test Bench

The following is the test bench procedure for a PE..A (2000 series) pump with an EP/RSV governor. Other models will be quite similar; however, when testing a pump, always use the specification sheet that corresponds to the pump model.

When calibrating any injection pump, always check the governor and pump designation to get the correct specification sheet. Mount the pump with the governor on the test bench with injection lines, rack travel indicator, and plunger lift gauge. There are 10 basic adjustments to be made on most in-line pumps. These are:

1 Port closure or plunger lift.

2 Control rack movement—equal delivery.

3 Adjust main governor spring.

4 Full load.

5 Torque control adjustment.

6 Low idle.

7 Breakaway.

8 High idle.

9 Starting fuel quantity.

10 Aneroid adjustment.

Before testing, remove the rear cover plate on the governor, the closing screw on top of the governor, bumper spring assembly, torque capsule, and aneroid. Attach rack travel gauge and set to 0 when the control rack is in the stop position. Add lube oil to the pump and governor (approximately $\frac{3}{4}$ pint or 0.35 liter). Run in the full load adjustment screw until only three to four threads show beyond the locknut. Place a pan under the pump to catch waste oil.

A. Procedure for Port Closure or Plunger Lift

There are two methods used to check and adjust port closure. These are the low pressure and high presure methods explained below:

1 *Low pressure method.* To correctly set port closure, we must determine the exact point at which the pumping plunger covers the spill port because this is the start of injection and the pump will be timed externally at this point. Remove the delivery valve and spring from no. 1 cylinder and reinstall the holder. Position plunger lift gauge on no. 1 cylinder and rotate the pump so that the tappet is on BDC and zero indicator. Attach a curved tube to the no. 1 holder (Figure 19.104). Set control rack to the 12 mm position, plug fuel gallery outlet, and supply fuel to the pump at 3 psi (0.210 kg/cm²). Fuel will run from the curved tube. Rotate the pump (in nor-

Figure 19.104 Curved tube attached to no. 1 outlet.

Figure 19.105 Adjustment of roller tappet.

mal direction) until the fuel flow is reduced to one drop every 10 seconds. The reading on the plunger lift gauge should correspond to the one in the test specification, usually between 1.7 and 2.5 mm. If it doesn't, the roller tappet must be adjusted (Figure 19.105) until the plunger lift is as specified $\pm \frac{1}{2}$ degree.

When the correct plunger lift is obtained, set the degree wheel on the test bench to 0. Shut off the fuel supply to the pump and reinstall the delivery valve and spring for no. 1 cylinder. Remove the valve and spring for the next cylinder in the *firing order* and reinstall the holder. Mount the curved tube. Start fuel supply and make sure that fuel runs from the tube. Rotate the pump in the normal direction until the fuel slows to a drip. This will occur at 60° intervals for 6 cylinder pumps, 90° intervals for 4 cylinder pumps, etc. Readjust if necessary. Follow through the firing order for the remaining cylinders, making sure all plungers are in *PHASE*. meaing $\pm \frac{1}{2}°$ of each other. This is extremely important.

When all cylinders have been adjusted, make certain all valves and springs are in place and all delivery valve holders correctly torqued. Proceed to step B.

2 *High pressure method.* If the test bench is equipped with a high pressure pump for setting prestroke, considerable time can be saved when calibrating the pump because the removal and installation of delivery valves are eliminated. Close the fuel outlet from the pump gallery and connect the high pressure hose to the inlet. Also, install the high pressure bleed-off valves in the injection lines (Figure 19.106). Install plunger lift gauge and rotate the pump until gauge reads 0 at BDC. Turn on the high pressure pump and open bleed-off valve for no. 1 cylinder. Fuel should flow at this time.

Figure 19.106 High pressure bleed-off valves for checking port closure.

CAUTION Do not allow the high pressure fuel to strike your arm or hand, as it may penetrate the skin and cause infection.

Rotate the pump until the fuel flow is reduced to a *dribble*. The reading on the plunger lift gauge should be as stated in the test specification. If not, adjust the roller tappet to obtain the correct lift ± 0.1 mm (0.004 in.). Close the bleed-off valve for no. 1 cylinder and open valve for the next cylinder in the *firing order* and rotate pump to check the prestroke on that cylinder. Continue to do this with the remaining cylinders, setting all cylinders to $\pm \frac{1}{2}°$. When finished, shut off the high pressure pump and disconnect the hose.

B. Procedure for Checking Control Rack Movement—Equal Delivery

Move the control lever forward and backward several times to make sure it is free. The rack should travel at least 20 mm (as indicated on rack travel gauge). If it does not, the pump is incorrectly assembled or the rack travel gauge is not properly set to zero.

Figure 19.107 Adjusting control sleeve.

NOTE If the pump is equipped with an aneroid, it must be removed at this time. On governors not having a separate stop lever, move the control lever to the shutoff position and turn in the rack stop screw until a travel of 0.3 to 1.0 mm (0.012 to 0.040 in.) is indicated on rack travel dial indicator.

Governors with a separate stop lever use the rack stop screw to set low idle, which we shall explain later. Check cylinders for equal fuel delivery by running the pump at approximately 750 rpm. Take a draw of 1000 strokes and record the delivery from all cylinders. Delivery must be within ± 2 cc. If delivery is uneven, adjust the control sleeve (Figure 19.107) to get the correct delivery. Tighten locking screws.

NOTE Take at least three draws of 1000 strokes before adjusting the control sleeves. This insures that no air is present in the injection lines, nozzles, and pump gallery.

C. Procedure for Adjustment Main Governor Spring

Place the control lever at the angle listed in the specification. Back off the high idle adjusting screw. Run the pump at the first speed listed for setting the main spring. Adjust the main governor spring (Figure 19.108) until the correct rack travel, as listed in the specification, is obtained.

NOTE Fine adjustments can be made by repositioning the control lever slightly.

Run the high idle adjusting screw up against the lever. Drive the pump at the other speeds listed under "Adjusting Main Governor Spring" on calibration sheet and see that the rack travel valves correspond to those listed. For the remaining tests, install the aneroid.

Figure 19.108 Adjusting main governor spring.

D. Procedure for Full Load Adjustment

Run the pump at speed listed for full load adjustment with control lever up against high speed stop screw. Turn the full load adjusting screw (Figure 19.109) counterclockwise until the specified cubic centimeter (cc) delivery is obtained. Tighen the locknut. Note the exact position of the rack in millimeters and record for torque control adjustment.

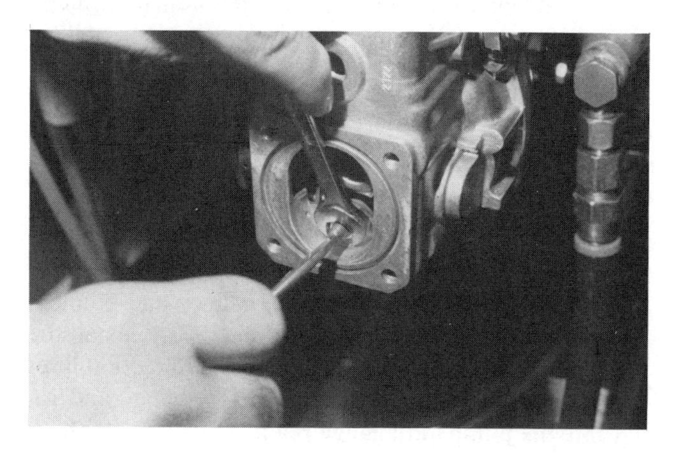

Figure 19.109 Full load adjusting screw.

NOTE If an aneroid is used, a full boost pressure must be used to test the full load quantity.

E. Procedure for Torque Control Adjustment

With pump still running at the speed listed for full load adjustment, turn the torque capsule into the tensioning lever until the rack travel increases by 1 mm (0.040 in.). Lock the capsule in place with the locknut. Back out the full load adjusting screw until rack position is the same as recorded in Step D. Tighten the locknut. Check the torque control by running the pump at all the speeds listed under torque control. Rack travel must correspond to the valves given.

F. Procedure for Low Idle Adjustment

1. *Separate stop lever type.* A rack travel will be in the specification sheet for low idle adjustment. Set the control lever to obtain a rack travel of *1 mm (0.040 in.) less* than given. Screw in the bumper spring assembly (Figure 19.110) to get the rack position specified. This preloads the bumper spring to prevent governor surge. Tighten the bumper spring assembly with its locknut. Adjust the low idle stop screw (Figure 19.111) until the specified fuel delivery is obtained and tighten the locknut.

2. *Single lever pump types.* Adjust the bumper spring assembly exactly as listed for governors having a separate stop lever. Since low idle is set with the linkage on the engine for pumps having a single lever, no low idle fuel delivery is given. Simply set the bumper spring assembly.

Figure 19.110 Adjusting bumper spring assembly.

Figure 19.111 Low idle stop screw.

G. Procedure for Breakaway Adjustment

Run the pump at the speed listed for breakaway and observe the control rack indicator. At the correct speed it should start to flutter (small, rapid vibrations of the indicator needle) and move to the 0 position. If it flutters too early or too late (according to specification sheet), adjust the high speed stop screw (Figure 19.112) until the speed is correct. Tighten the locknut.

H. Procedure for High Idle Adjustment

With the control lever against stop screw, run the pump at the speed listed for high idle. Check quantity of fuel delivered at this speed. If incorrect (does not meet specifications), adjust with the main governor spring screw.

Figure 19.112 High speed stop screw

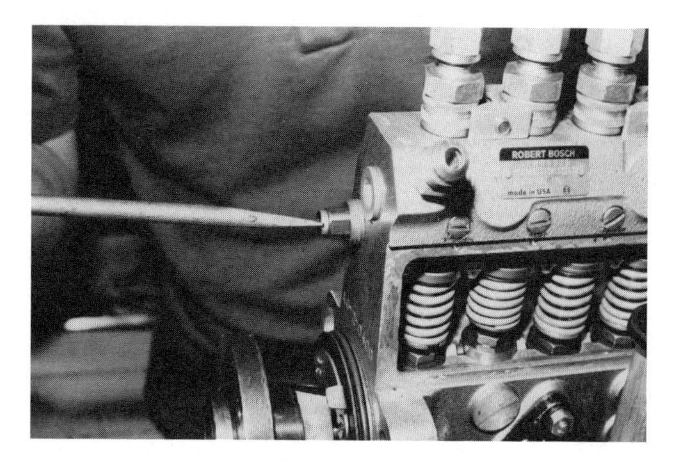

Figure 19.113 Rack stop screw.

NOTE When high idle is adjusted, breakaway must be rechecked.

I. Procedure for Starting Fuel Delivery Adjustment

Remove rack travel indicator from pump and reinstall rack screw (with shims) if present and rack cap. Run pump at the speed listed for starting fuel. Put control lever in full load position and check the quantity delivered. Adjust by removing shims under the rack stop screw to increase fuel delivery (Figure 19.113).

J. Procedure for Aneroid Adjustment

Run pump without aneroid pressure and see that rack travel and fuel delivery correspond to those listed for 0

Figure 19.114 Aneroid adjusting screw.

Figure 19.115 Aneroid diaphragm spring screw.

psi (0 kg/cm²). If incorrect, adjust with the top screw on the aneroid to obtain the correct travel (Figure 19.114). Run the pump at the speed listed for aneroid adjustment with air pressure and see when the aneroid just starts to move the control rack and when it finishes. Adjust with the aneroid diaphragm spring screw (Figure 19.115) to obtain the correct values.

After all adjustments have been completed, check the pump for leaks, correct torques, and seal the pump to prevent unauthorized adjustments.

XII. Procedure for Calibration (Distributor Pumps)

Mount pump on test bench with appropriate fittings. Test nozzles must be set to 2130 psi (149 kg/cm²) with one nozzle (on some models) set to 2840 psi (198 kg/cm²). The order of calibration for EP/VA. .C and B series pumps is as follows:

A *Set supply pump pressure.* Adjust the supply pump pressure by tapping the adjusting plug on the pressure regulator with a punch and small hammer (Figure 19.116). Drive the plug in to raise pressure to required limits.

NOTE If plug is driven too far and pressure exceeds the specifications, the regulating plug assembly must be removed by turning it counterclockwise. After removal, drive the adjusting plug out until it is flush with plug retainer. Then reinstall plug and check pressure.

B *Set timing advance.* Check advance with the gauge shown in Figure 19.117. If incorrect, add or remove shims for the spring side to obtain the correct advance.

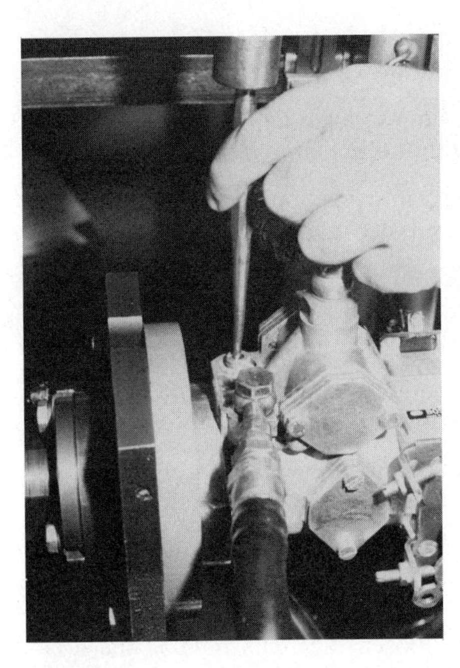

Figure 19.116 Adjusting T pump pressure regulator.

C *Full load quantity.* Adjust full load quantity with delivery screw as shown in Figure 19.118.

D *Breakaway.* Adjust the high idle stop screw to give fuel delivery required for breakaway (Figure 19.119).

E *Recheck full load quantity.* To correctly check adjustment of delivery rate control spring, recheck the full load delivery. If delivery is less than in step C, shims must be added to the delivery rate control spring housing or a different spring applied.

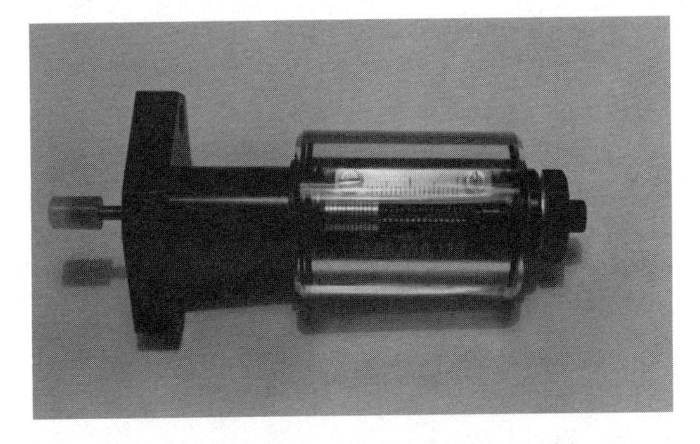

Figure 19.117 Timing gauge for checking advance movement.

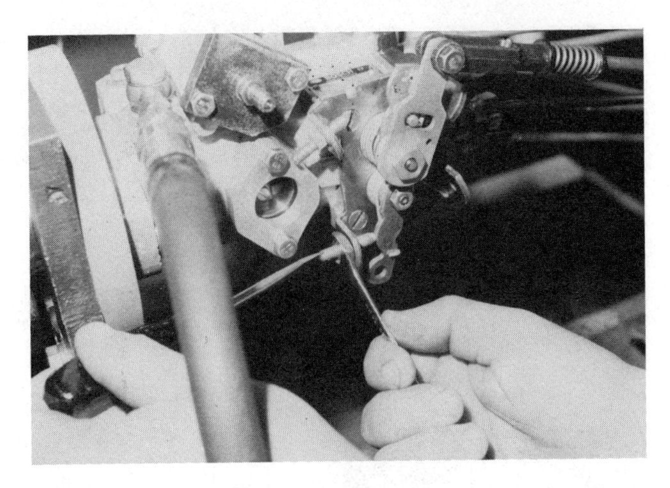

Figure 19.118 Full load delivery screw.

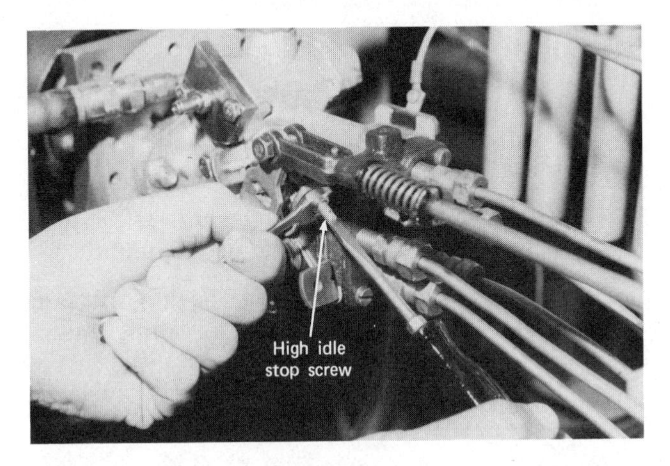

Figure 19.119 High idle stop screw.

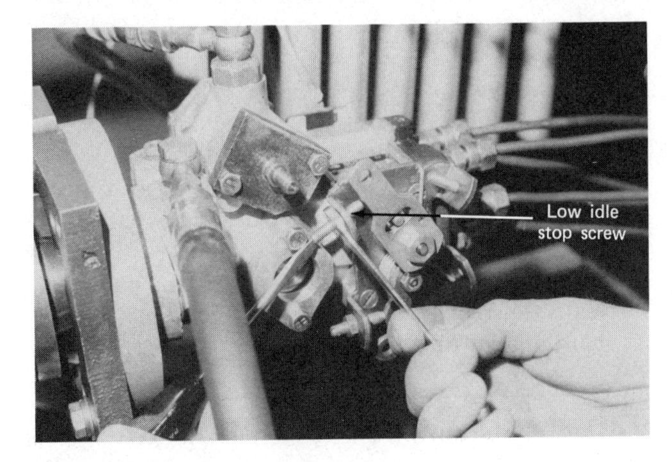

Figure 19.120 Low idle stop screw.

F *Low idle.* Run pump at the speed listed for low idle. Adjust low idle screw to obtain the correct fuel delivery (Figure 19.120).

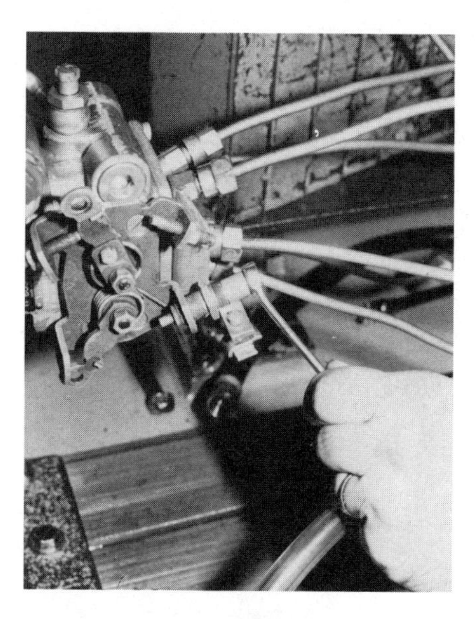

Figure 19.121 Adjusting starting fuel quantity.

G *Start quantity.*

NOTE On B pumps the throttle lever must be in low idle position when checking start quantity.

Run pump at speed listed for start quantity. Rotate delivery rate lever to just contact the spring loaded stop.

CAUTION *Do not compress spring!*

Record the fuel delivery obtained. If necessary, adjust the stop (Figure 19.121) to obtain the start quantity. When the stop is compressed, fuel delivery must cease.

H *Operate pump at the speed listed for overflow check.* Collect the test oil in a graduate and record quantity. If not as listed in the specification sheet, the overflow valve should be replaced. If a problem arises, consult the Bosch technical manual or ask your instructor.

After completion of tests, check the pump for leaks, correct torques, and seal the pump to prevent unauthorized adjustments.

XIII. Installation of Pump to Engine

1 Set no. 1 cylinder on compression stroke and bring it to the correct timing mark.

NOTE If the engine was placed in this position before pump removal it saves locating and repositioning engine to correct timing mark.

2 Rotate the pump to align the timing mark on the drive hub with the timing pointer.

3 Slide the pump into position on the engine and attach mounting bolts (nuts) finger tight.

4 Install the bolts that hold the pump drive gear to the drive hub (on most pumps).

5 If the pump flange has slotted mounting holes, position the pump midway in the slot and tighten the mounting bolts (nuts).

6 Make certain pump timing marks are still in alignment and tighten the bolts holding the drive gear to the drive hub.

7 Rotate the engine two revolutions in the direction of rotation and recheck the timing marks. If they do not align, loosen the bolts holding the drive gear and reposition the drive hub until the marks align.

NOTE If these marks are continually incorrect, check the pump drive gears and/or chains for excessive wear.

8 Attach the high pressure injection lines and tighten at the pump. Leave the lines loose at the nozzles at this time.

9 Attach fuel inlet and return lines and any electrical wires used (for example, on the shutoff solenoid).

10 Attach the aneroid pressure line if used.

11 Most in-line pumps are pressure oil lubricated. *Do not forget to hook up this line.*

12 Connect the throttle linkage and adjust.

13 On new or overhauled pumps it is wise to install new fuel filters before starting the engine.

14 Using a hand primer, force fuel up to the fuel transfer pump and into the main pump.

15 Provide a means of emergency shutdown and start the engine.

16 After engine has reached operating temperature check the following:

a Low idle }
b High idle } using an accurate tachometer

c Fuel leakage

d Horsepower—using a dynometer if available.

XIV. Troubleshooting the Robert Bosch PE Pump

The following procedures should be used as a guideline when troubleshooting the Robert Bosch PE pump.

A After overhaul, problems may occur on the test stand that are not covered in the testing section. Before blaming the pump, check the following procedures.

- **1** Make certain the pump is mounted correctly and that gauges are being read correctly.
- **2** Check for leaking lines.
- **3** Are test nozzles at the correct pressure?
- **4** Are all test conditions (oil temperature, feed pressure, and line size) being met?
- **5** Are tests being made in the correct order?

B If the preceding procedures are satisfactory, determine if all of the following conditions are being met.

- **1** If you are unable to determine exact port closure, check the following items.
  - **a** Is the control rack in correct position?
  - **b** Is the supply pressure too high?
  - **c** Are the pumping plungers worn?
  - **d** Is the pumping plunger correct?
- **2** If there is uneven delivery, check the following as the cause of the problem.
  - **a** Are the control sleeves out of adjustment?
  - **b** Are the plungers worn?
  - **c** Is the outlet restriction valve broken or worn?
  - **d** Is the lift pump defective?
  - **e** Is the supply line the correct size?
  - **f** Is the test stand low on oil?
  - **g** Is the delivery valve spring broken or the valve worn?
  - **h** Is there fuel leakage between delivery valve and barrel?
- **3** Check the following items if you are unable to obtain exact control rack travel when adjusting the main governor spring.
  - **a** Is the gauge being read correctly?
  - **b** Is the external control lever tight against the stop screw?
  - **c** Are the control sleeves out-of-time?
  - **d** Is the control lever set at the correct angle?
  - **e** Is the main spring weak or the correct spring for application?
  - **f** Is the camshaft centered in the housing?
  - **g** Is the governor thrust sleeve adjusted correctly?
  - **h** Is the rack binding?
- **4** Check the following if you are unable to obtain torque control rack travel.
  - **a** Is the torque capsule adjusted according to the procedure listed under testing?
  - **b** Is the torque capsule spring weak or incorrect for application?
  - **c** Is the main governor spring correct?
- **5** If you are unable to obtain specified starting fuel delivery, check the following items.
  - **a** Are there too many shims under the rack screw (if so equipped)?
  - **b** Are the pumping plungers worn?
  - **c** Is the calibrating oil too hot?
  - **d** Is the test nozzle pressure too high?
  - **e** Is the supply pressure too low?
  - **f** Are the delivery valves worn?
  - **g** Is there fuel leakage at injection lines?

C After installation to engine, other problems may surface that will require adjustment to the fuel system. This information is given on the assumption that the engine and subassemblies are in good mechanical condition.

- **1** If the engine will not start, the following items should be examined.
  - **a** Are the fuel filters dirty?
  - **b** Are there loose connections on the supply side that lead to air in the system?
  - **c** Is the fuel tank too low or empty?
  - **d** Is the pump timed correctly to the engine?
  - **e** Have the lines been properly bled?

f Is the stop lever in the shutoff position?

2 Check the following items if the engine is low on power.

a Are the fuel filters dirty?

b Are the shutoff cable and control properly adjusted?

c Is the full load delivery adjusted correctly?

d Is the torque capsule adjusted correctly?

e Is the pump timed correctly to the engine?

f Is the lift pump defective?

g Is the main governor spring weak or incorrectly adjusted?

h Is the throttle lever tight against the high idle stop screw?

i Is the high idle adjusted correctly?

j Is the port closure adjusted correctly?

3 Check the following items if there is low idle surge (hunting).

a Is the bumper spring adjusted properly?

b Is the idle speed too low?

c Is the fuel delivery uneven?

d Are the control rack or governor parts binding?

4 Check the following items if the acceleration is slow.

a Is the governor spring weak?

b Is the aneroid adjustment correct?

SUMMARY

This chapter has covered the most popular Robert Bosch injection pumps. Robert Bosch also makes other pump models that were not included because they are not widely distributed or are used for specific purposes.

The information given is intended to be used in conjunction with current test specifications and torque charts.

From time to time certain questions or problems may arise that are not fully covered in the chapter. If that is the case, the Robert Bosch service manual should be consulted, if available, or ask your instructor.

Diesel pumps are continually changing; even though they may appear to be about the same as earlier models, important changes have been made. This chapter provides all the latest information available at the time of printing.

Engine dealers using Robert Bosch equipment and authorized Robert Bosch service shops periodically receive service bulletins issued by the factory, which include all the latest changes and improvements that have been incorporated or will be incorporated into Robert Bosch pumps. The wise mechanic will keep abreast of these changes.

REVIEW QUESTIONS

1 What type of fuel control is employed on Robert Bosch in-line injection pumps?

2 What is the purpose of the delivery valves?

3 What do each of the following numbers and letters stand for?
PES 4 A 80 B 420 LS 54

4 Why is pressure maintained in the fuel gallery of in-line pumps?

5 Define port closure and its effect on engine operation.

6 How is port closure adjusted on the following pumps?
PE..A
PE..P
PE..M

7 How is the high pressure fuel sealed within the pumping barrel?

8 What type of fuel control is employed on Robert Bosch distributor injection pumps?

9 What type of governor is used in EP/VA pumps?

10 On what applications would the EP/RSUV governor be used?

11 For what purpose is torque control used?

12 How is torque control adjusted on the EP/RSV governors?

13 What is the purpose of the aneroid or smoke limiter?

14 List all steps required in adjusting a pump with an EP/RSV governor.

15 How is the start of injection varied on an in-line pump?

16 What is the major difference between an EP/VA pump and the VE pump?

17 How is lift pump pressure adjusted on the EP/VA pump?

18 What do each of the following numbers and letters stand for?
EP/RSV 350-950

19 What adjustments on the EP/RSV governor can be made on the engine?

20 List all steps required to properly adjust the EP/VA..C pump on the test bench.

20

CAV–Simms

CAV–Simms of England has been producing fuel injection equipment for many years. Up until 1956 their products included nozzles, filters, and in-line injection pumps. In that year an agreement was reached with Roosa Master, enabling CAV to produce the DPA distributor pump, which was designed after the A model Roosa Master pump. Some of CAV's in-line pumps (AA, NN, and BPE) are no longer in production. They were conventional in design, employing individual pump elements for each cylinder, and capable of using mechanical and pneumatic governors.

OBJECTIVES

Upon completion of this chapter the student will be able to:

1. Explain fuel flow and pump operation in one CAV pump (student's choice).
2. Demonstrate the ability to name all the parts in one CAV pump.
3. Demonstrate the ability to disassemble correctly and inspect a CAV pump.
4. Select from a list of pump components the parts that are normally replaced during a pump overhaul.
5. Demonstrate the ability to test correctly and calibrate a CAV pump.

GENERAL INFORMATION

In most cases the CAV–Simms pumps have been installed on engines built in foreign countries.

One of the major users of CAV pumps over the years has been Perkins Engine Company. Early Perkins engines used the BPE and later engines used the DPA. Some current models use the DP-15.

In recent years a pump similar to the DPA has been built for certain companies. This pump, called the ROTO-Diesel, looks like the DPA but employs some different features. Information provided in this chapter concerning operation, disassembly, assembly, and testing of the DPA model pump can be directly applied to the ROTO-Diesel pump.

I. The DPA Pump

One of the most popular pumps produced by CAV is the DPA, a single cylinder, twin plunger, distributor type pump (Figure 20.1). It uses inlet metering and is

Figure 20.1 DPA pump.

lubricated by the diesel fuel that it pumps. Either mechanical or hydraulic governors can be fitted to the DPA.

As an aid to understanding the operation of the DPA pump, a list of the component parts and their function follows.

A *Component parts* (Figure 20.2).

1 *Pump housing.* The housing holds all pump parts and has a flange for mounting the pump to the engine.

2 *Governor weight retainer.* It holds the governor weights in correct relation to each other during operation.

3 *Governor thrust sleeve.* This part transfers the motion of the governor weights to the governor arm.

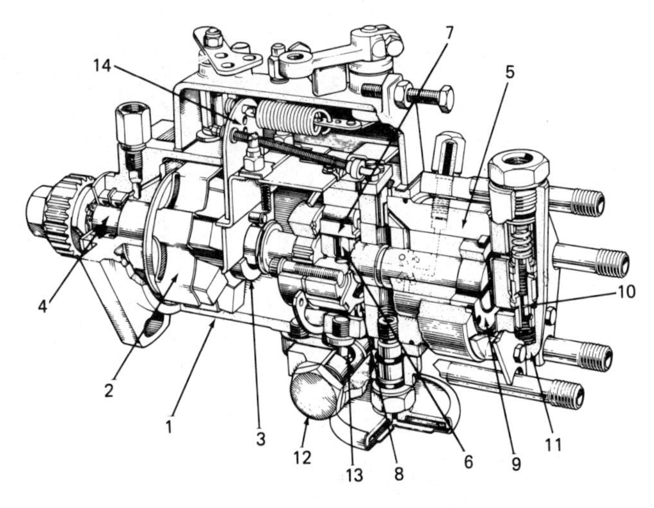

Figure 20.2 Diagram of a DPA pump (courtesy CAV–Lucas).

4 *Drive shaft.* The drive shaft supports the governor assembly and drives the rotor and transfer pump.

5 *Hydraulic head and rotor.* This is the main pumping element in the injection pump. The rotor carries the pumping plungers, rollers and shoes, and adjusting plates.

6 *Pumping plungers.* The plungers are fitted to the rotor and are pushed inward by the rollers and shoes to pump the fuel.

7 *Rollers and shoes.* The rollers fit into the shoes and contact the cam much like a cam follower in an engine.

8 *Adjusting plates.* Plates that are mounted on the rotor and limit the outward travel of the rollers and shoes. By limiting the outward travel of the rollers and shoes the maximum fuel delivery is controlled.

9 *Transfer pump.* This pump consists of liner and blades. It delivers fuel at low pressure to the metering valve.

10 *Pressure regulating valve.* Situated in the end plate, the regulating valve controls the transfer pump pressure throughout the speed range.

11 *End plate.* Provides a cover for the transfer pump.

12 *Advance mechanism.* Located on the bottom of the pump. A hydraulic piston rotates the cam ring via the cam advance stud *against* the direction of pump rotation.

13 *Cam advance stud.* A stud threaded into the cam that connects it to the cam advance mechanism.

14 *Governor.* Responds to engine load changes to maintain a preset speed.

B *Pump operation—charge cycle* (Figure 20.3).

Fuel enters the inlet connection, passes through a prefilter, and is routed to the top of the transfer pump. As the transfer pump rotates, it forces fuel through a drilled passageway at the bottom of the hydraulic head. From this point fuel is routed to the advance unit and around the rotor to the underside of the metering valve. The metering valve has a vertical groove machined into it. This groove aligns with the diagonal passageway. The extent of this alignment is determined by the governor, which rotates the metering valve to allow more or less fuel to pass the valve. The fuel then enters one of the charge ports in the hydraulic head (one port for each engine cylinder). When this charge port aligns with the single charge port in the rotor, the fuel flows into the rotor, goes to the front of the rotor (the rear is blocked), and enters the pumping cylinder. This forces the pumping plungers, and rollers and shoes outward against the cam ring. During the charge cycle the rollers fall into the low area on the cam ring. The pump cylinder is then completely charged with fuel.

C *Discharge cycle.*

As the rotor revolves, the charge port goes out of alignment and a single discharge port comes into line (Figure 20.3). At the same time, the rollers, shoes, and pumping plungers are forced inward as they contact the raised portions of the cam ring. This action forces fuel from the pumping cylinder down the center bore of the rotor to the discharge port (Figure 20.3), which is aligned at this time. (The charge port is now blocked.) The fuel flows through the head outlet and to the nozzle. (Remember that the fuel leaving the pumping cylinder during discharge is not injected from the nozzle on that same cycle, but displaces fuel at the head outlet, causing an exact amount to be forced out at the nozzle.) The rotor continues revolving and another charge cycle begins. The complete charge and discharge cycles are shown in Figure 20.3.

D *Pressurizing valve operation.*

To keep the injection lines full of fuel for the next injection, pressurizing valves (one for each cylinder) are installed in the head outlet bolts (Figure 20.2). In addition to these valves, the cam ring is ground so that it provides a relief of pressure in the injection lines at the end of the discharge cycle. This relief allows line pressure to drop below the nozzle opening pressure far enough so that the nozzle snaps shut quickly and completely. This prevents nozzle "dribble," which causes excessive smoke.

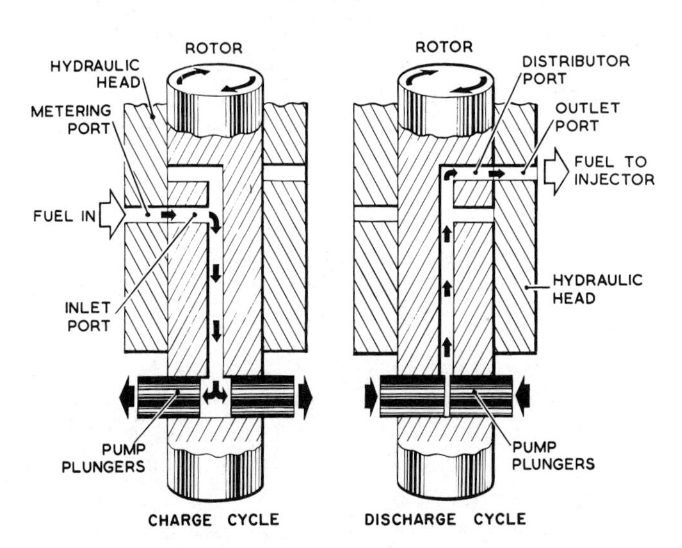

Figure 20.3 Pump charging cycle (courtesy CAV–Lucas).

E *End plate operation.*

The end plate (Figure 20.4) serves three functions:

1. Provides a fuel inlet to the pump.
2. Seals the transfer pump.
3. Contains the pressure regulating valve for the transfer pump.

   The pressure regulating valve has two different functions:

   a. When the pump is being primed (before rotation), fuel is *forced* into the inlet connection through the nylon mesh filter. It enters the regulating sleeve at the upper port and forces the regulating piston downward, compressing the priming spring. When the piston has moved down far enough to uncover the lower port in the sleeve, fuel flows directly into the hydraulic head. The pump is then primed.

   b. When the pump is rotating (engine running), fuel is *pulled* into the end plate by the transfer pump. Again it passes through the nylon prefilter and enters the transfer pump at port 9 (Figure 20.4). As the transfer pump revolves, it forces the fuel into the hydraulic head and also into the end plate. When the transfer pump builds up pressure, it forces the piston upward against the regulating spring (Figure 20.4). As the correct pressure is reached, the piston uncovers the regulating port and bypasses just enough fuel back to the inlet side of the transfer pump to maintain pressure at the preset point.

      Transfer pump pressure is adjusted by one of two methods:

      (1) On some pump models the spring guide (Figure 20.4) can be replaced with one of a different size that will compress the regulating spring more or less, thus changing the pressure.

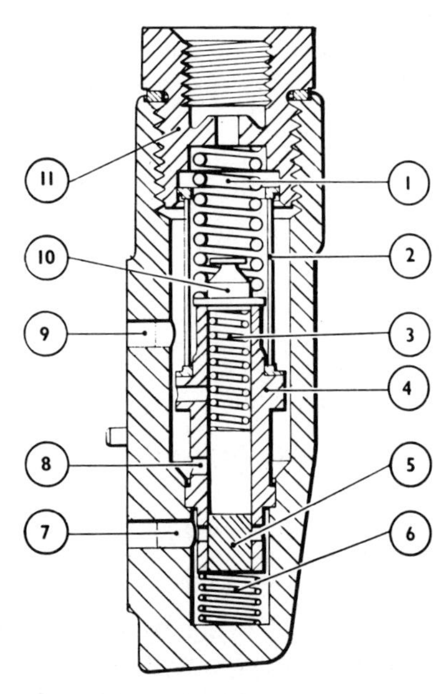

1. *Retaining Spring.*
2. *Nylon Filter.*
3. *Regulating Spring.*
4. *Valve Sleeve.*
5. *Piston.*
6. *Priming Spring.*
7. *Fuel passage to Transfer Pump Outlet.*
8. *Regulating Port.*
9. *Fuel passage to Transfer Pump Inlet.*
10. *Spring Guide.*
11. *Fuel Inlet Connection.*

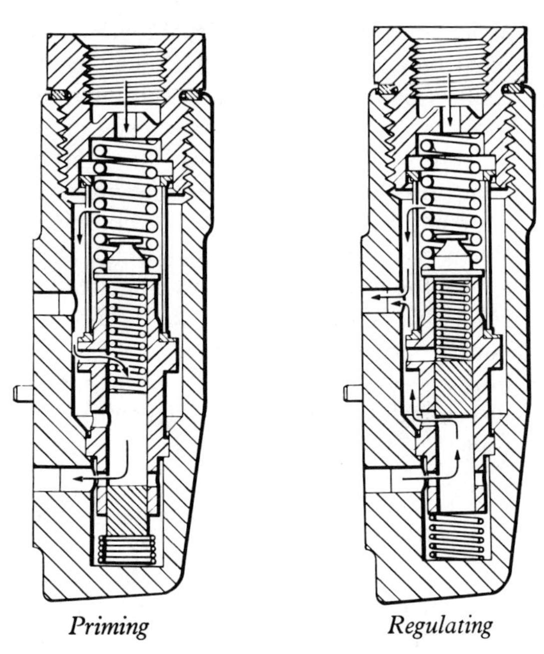

Figure 20.4 Cross section of end plate (courtesy CAV–Lucas).

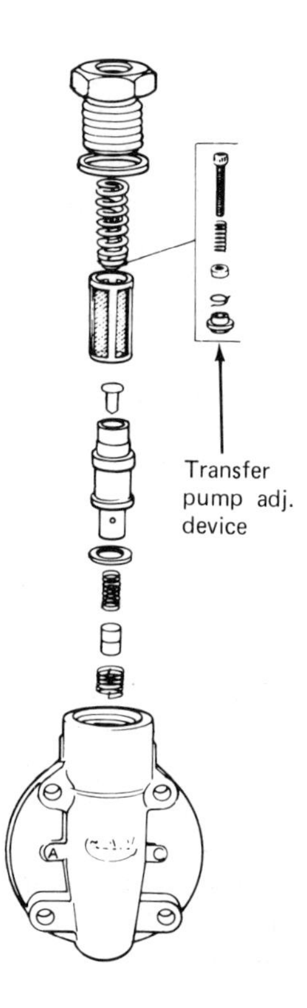

Figure 20.5 Transfer pump adjustment device (courtesy CAV–Lucas).

(2) On other models an adjuster (Figure 20.5) is used. With the correct tool, this adjuster allows the transfer pump pressure to be adjusted while the pump is running on the test bench.

F *Pump lubrication and return oil circuit.*

As stated earlier, the entire DPA pump is full of diesel fuel. The rotor is supplied with fuel directly from the transfer pump. In addition to meeting the requirements of the pumping plungers and the advance mechansim for fuel, the transfer pump supplies additional fuel to lubricate the entire pump. From the annular ring of the rotor (Figure 20.6), a groove leads to the front of the rotor. From this point fuel (and air) flows into the main pump housing. At the top front of this housing a return fitting allows the fuel (and air) to return to the main supply tank. This fuel lubricates all governor components and at the same time bleeds the system of any air that may have entered.

G *Governor operation.*

The DPA pump is available with two governors, a mechanical governor or a hydraulic governor. The mechanical governor will be covered here.

A diagram of the mechanical governor is given in Figure 20.7. The flyweights (B) transmit

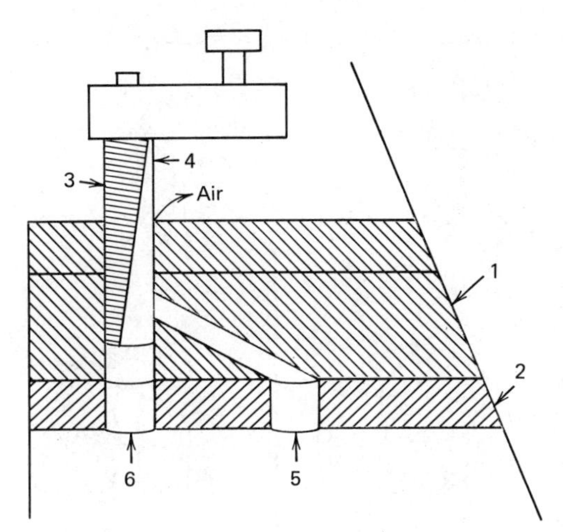

Figure 20.6 Return oil circuit. (1) Barrel (2) Sleeve (3) Metering valve (4) Air bleed passage (5) Delivery passage (6) Tranfer pressure.

force through the thrust sleeve (A), causing the governor lever (C) to pivot. This movement causes rotation of the metering valve (G), which then admits more or less fuel to the pumping cylinder as required by the engine. The flyweight force is opposed by the main governor spring (F) and the idling spring (E). These forces balance out to maintain a set engine speed. When required, a mechanical shutoff bar rotates the metering valve to the shutoff position regardless of engine speed.

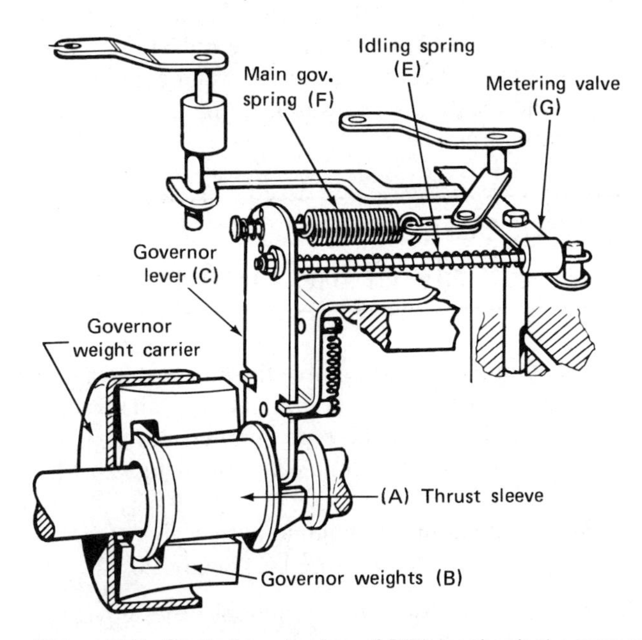

Figure 20.7 Control mechanism of DPA mechanical governor (courtesy CAV–Lucas).

1 *Idle operation.* At idle the flyweights provide little force against the governor lever, but since the main governor spring is now slackened the weights fly outward, causing the metering valve to rotate to the idle position. Within the idling speed range, the idling spring gives sensitive speed control.

2 *Full load speed.* At full load speed, the throttle applies maximum spring pressure on the governor lever. Because of the load applied to the engine, it can only run fast enough to balance the heavy spring load with flyweight force. At full load, the metering valve has rotated to allow the maximum amount of fuel to enter the rotor.

3 *High idle and cutoff.* With the throttle shaft still applying maximum spring force to the governor lever and the load on the engine suddenly removed, the engine will speed up. The increased force of the flyweights will then overcome the spring pressure and rotate the metering valve to the closed position. As the valve closes completely, governor cutoff is reached. This limits maximum engine speed. The engine will slow down until the flyweights and governor spring reach a balance that allows only a small amount of fuel into the rotor. This is known as high idle speed.

H *Advance operation.*

If the pump is fitted with an advance device (Figure 20.8), fuel is supplied from the transfer pump to the head locating screw, which is hollow in this case. Fuel is then routed to the power piston. The piston pushes the advance stud, which is screwed to the cam ring. The cam is forced against rotation, which advances injection. A return check valve keeps the cam stationary during impact of the rollers against the cam lobes. The cam is allowed to return by the normal leakage between piston and cylinder as transfer pressure falls.

Two advance units are used on DPA pumps: speed advance and load advance.

1 *Speed advance.* Speed advance is used to advance the injection timing over the complete operating range of the engine to give the best power and economy. Speed advance, as the name implies, is sensitive to speed changes. This speed change causes transfer pump pressure to rise or fall, which causes the advance piston to advance more or less as required. Changes in load have no effect on the advance if speed is kept constant. The speed ad-

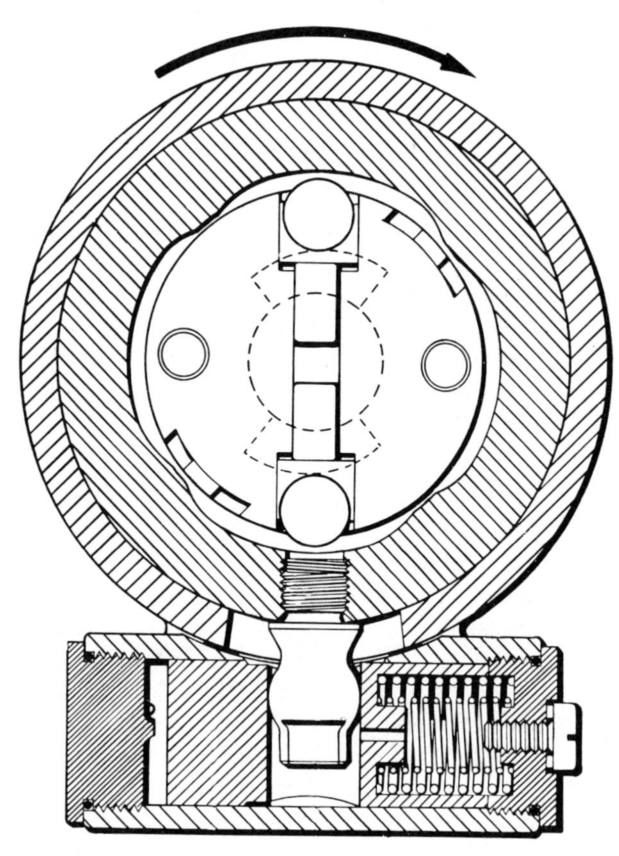

Figure 20.8 Automatic advance device (courtesy CAV–Lucas).

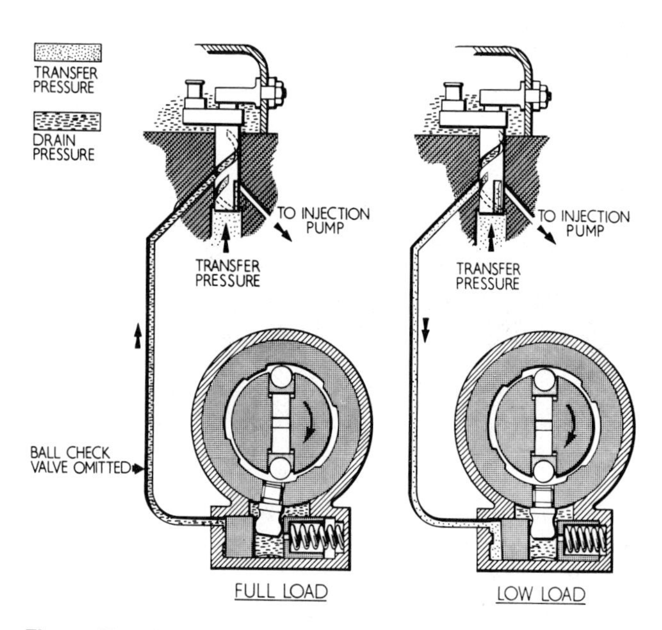

Figure 20.9 Load advance mechanism (courtesy CAV–Lucas).

vance unit is adjusted by changing the advance springs and also with shims.

2 *Load advance.* Without an advance unit and at light loads, the metering valve doesn't allow enough fuel into the pumping cylinder to push the pistons and rollers all the way out. Since they are not all the way out against the cam, they will contact the cam lobe later in the revolution, making fuel injection occur later. In some engines, notably precombustion chamber engines, this late injection causes misfiring. To counteract this condition, a load advance mechanism (Figure 20.9) is used to advance the cam at light loads. An additional groove is machined into the metering valve to allow fuel to flow by it and into the advance circuit.

This groove allows the proper amount of fuel to enter the advance circuit, depending on the adjustment. At full load the governor rotates the valve so that the groove no longer is aligned with the passage leading to the advance circuit; thus there is no advance. As the governor rotates the valve to light load, the groove again lines up and the cam will advance. In this position the metering valve is adjusted to obtain the correct advance. The valve is adjusted by raising or lowering it in the hydraulic head by means of an eccentric screw at the rear of the governor cover.

Certain pumps, load or speed advance, use a damper attachment (Figure 20.10) to smooth out transfer pressure pulsations. This is simply a diaphragm that is connected to the head locating screw and is in the transfer pressure circuit.

I. Procedure for Disassembly of the DPA pump.

Before beginning to disassemble the pump, thoroughly clean the exterior in solvent and dry with air. Record all serial numbers and the type number. An explanation of the nameplate code is given in Figure 20.11.

1 Mount the pump on holding fixture no. 7044/888 supplied in the CAV tool kit (Figure 20.12).

2 Loosen the top cover, unhook the governor spring from the throttle lever, and remove the cover. Discard the gasket.

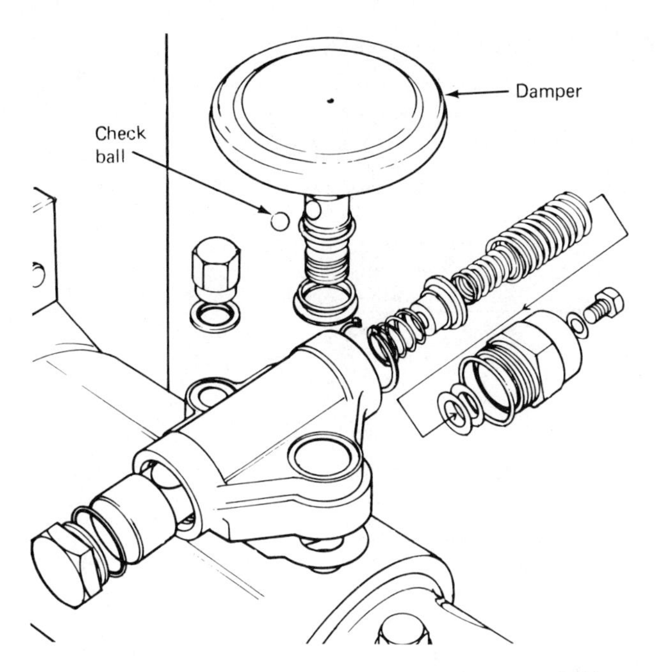

Figure 20.10 Damper used on DPA pump (courtesy CAV–Lucas).

Figure 20.12 DPA pump mounted in holding bracket (courtesy CAV–Lucas).

EXAMPLE
A/75/800/1/220

A = pump tested with BDN12SD12 nozzles
75 = maximum fuel setting, CC's/1000 strokes
800 = maximum fuel setting speed
1 = governor spring position (numbers under "Code" below list different spring positions)
2220 = maximum no load speed (engine)

Code	Control Arm	Throttle Link
1	1	1
2	1	2
3	1	3
4	2	1
5	2	2
6	2	3
7	3	1
8	3	2
9	3	3
0	Hydraulic governor	

Figure 20.11 Table with nameplate explanation and governor spring position.

3 Remove the governor springs by unhooking the main spring from the link. Remove the link.

4 Remove shutoff bar (Figure 20.13).

5 Loosen and remove the three screws holding the governor bracket to the pump housing.

6 Remove the bracket straight up with the linkage, governor arm, and metering valve.

7 Remove the metering valve from the linkage.

8 Loosen the end plate screws and inlet connection.

Figure 20.13 Shutoff bar.

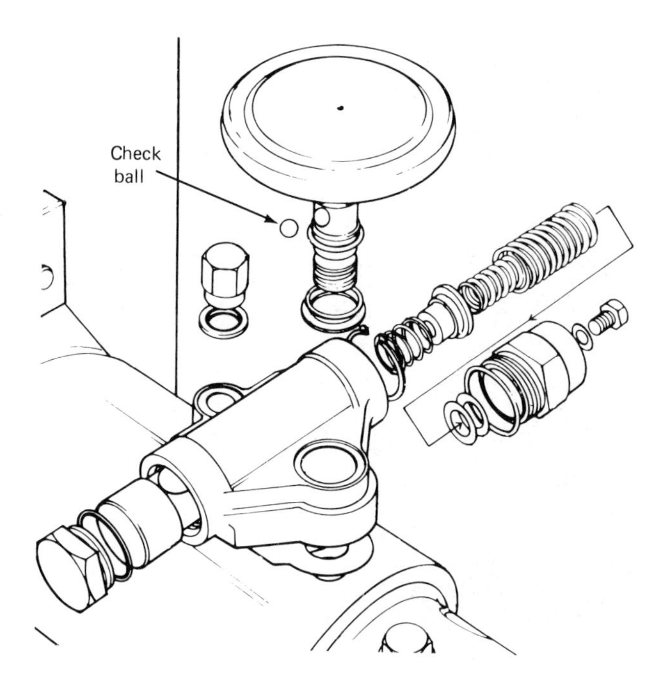

Figure 20.14 Check ball in damper bolt (courtesy CAV–Lucas).

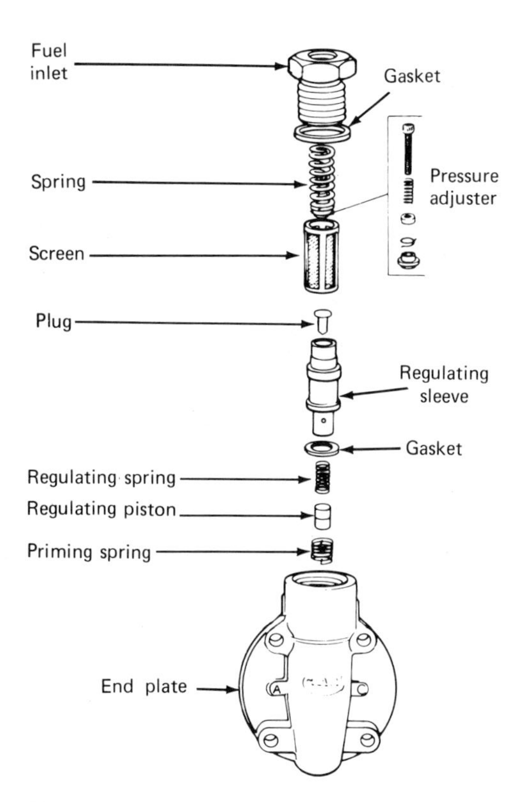

Figure 20.15 Pressure regulator and components (courtesy CAV–Lucas).

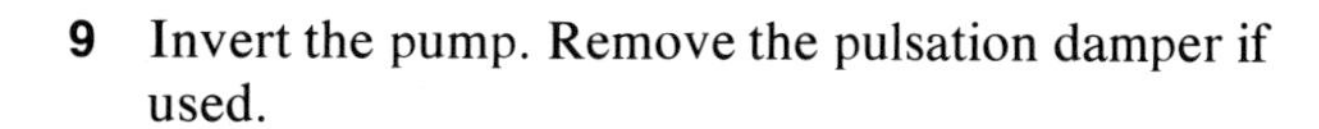

9 Invert the pump. Remove the pulsation damper if used.

NOTE When removing damper or head locating screw, do not lose the small check ball that fits into a hole in the screw (Figure 20.14).

10 Remove advance piston plugs (if so equipped). Remove nut that retains advance housing to the pump. Lift off advance unit.

11 Using tool no. 7144/14, loosen and remove the cam advance stud from the cam ring.

12 Remove fuel inlet connection from the end plate. Remove the end plate and remove all pressure regulating components (Figure 20.15).

13 Lift out the transfer pump blades and then the pump liner.

14 Hold the drive hub and loosen the rotor nut with tool no. 7044/889. *Do not remove.*

CAUTION The rotor nut is marked with an arrow. It must be loosened in this direction or it will break (Figure 20.16).

15 Remove head locking screws and carefully withdraw the hydraulic head from the housing.

16 Unscrew the rotor nut from the rotor and carefully remove the rotor from the head.

17 Lift off the cam rollers and shoes. The pumping plungers should be kept in the rotor to prevent any damage to them.

18 Remove the cam ring.

19 Compress the timing ring (circlip) with snap ring pliers and remove.

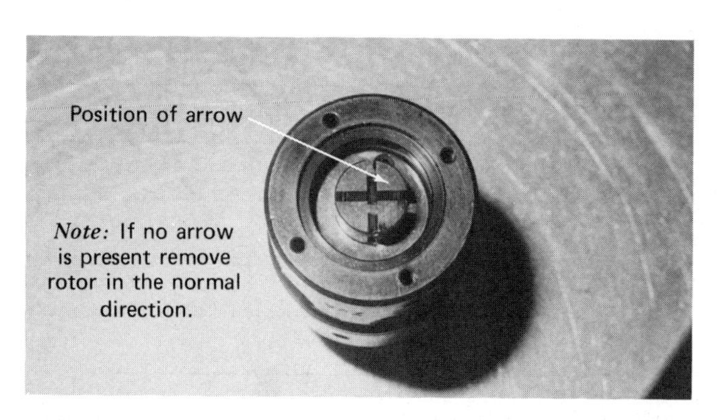

Figure 20.16 Arrow on rotor nut.

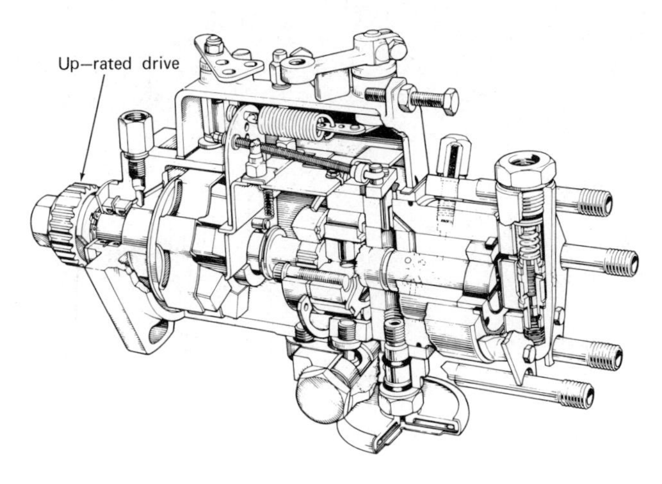

Figure 20.17 Up-rated drive pump (courtesy CAV–Lucas).

Figure 20.18 Removal of drive shaft before removal of drive gear and key (courtesy CAV–Lucas).

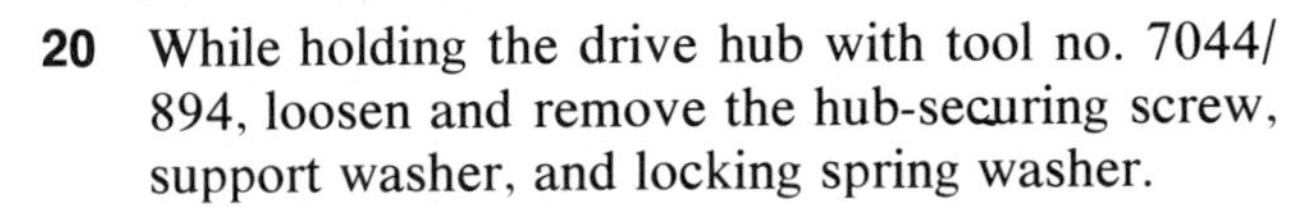

20 While holding the drive hub with tool no. 7044/894, loosen and remove the hub-securing screw, support washer, and locking spring washer.

21 Pull out the splined drive shaft and governor weight assembly.

22 Remove the O ring from the drive shaft and take off the weight retainer, weights, and thrust sleeve.

23 Pull the drive hub from the housing and remove the oil seal.

NOTE On CAV DPA pumps using the heavy duty up-rated drive (Figure 20.17), the following steps apply:

(1) Remove drive shaft nut and lock.
(2) Using tool no. 7244/105, remove pump drive gear and key (Figure 20.18).
(3) Remove the snap ring and thrust washer.
(4) Withdraw the drive shaft assembly.
(5) Remove the shaft seals.

J. Procedure for Reassembly of the DPA Pump

NOTE Before reassembly of the injection pump, check all parts for wear marks, scoring, and chipping. Replace any parts that are in doubt; also replace all gaskets and seals.

1 With the pump housing mounted on the assembly fixture, install the drive shaft seal with tool no. 7144/260.

2 Place drive shaft in governor weight assembly tool no. 7144/894 (Figure 20.19) with threaded inside end up. Place thrust sleeve and washer over the shaft. Place the governor weights on the

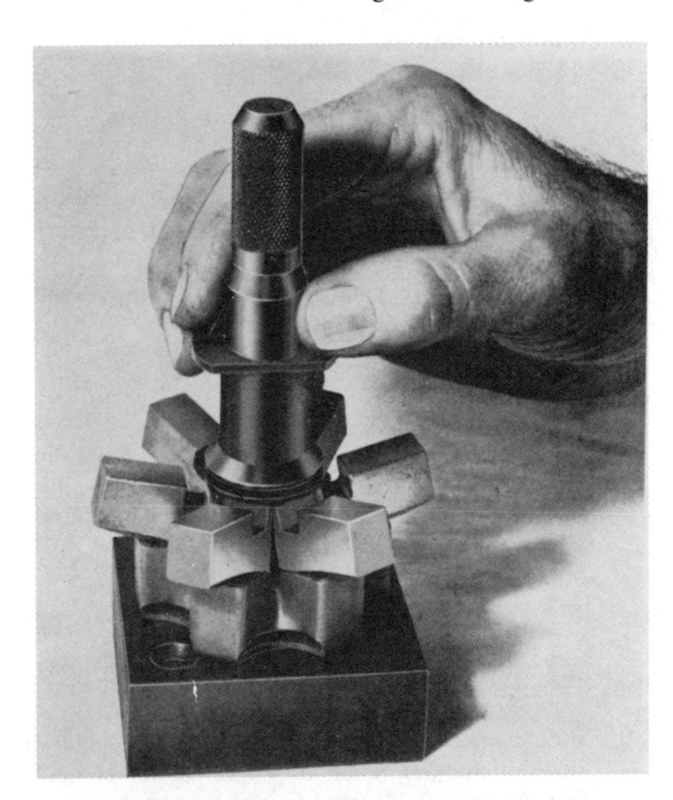

Figure 20.19 Drive shaft in governor weight assembly tool (courtesy CAV–Lucas).

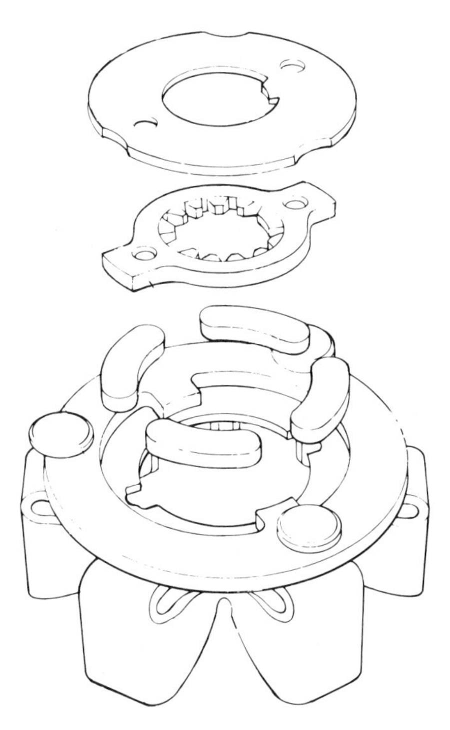

Figure 20.20 Assembling cush-type weight retainer (courtesy CAV–Lucas).

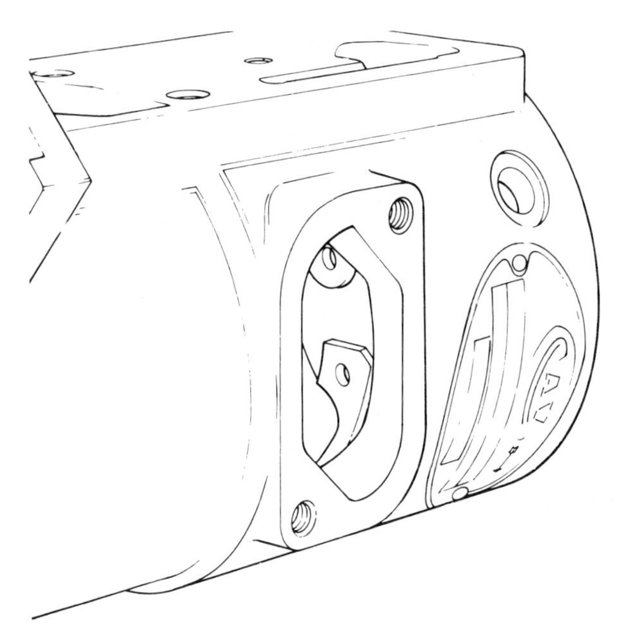

Figure 20.21 Timing ring position (courtesy CAV–Lucas).

fixture so that the lip of the weight sits on the thrust washer.

3 Install the weight retainer over the weights, making certain they slip into the pockets.

NOTE When a cush-type weight retainer is used, assemble it first as shown in Figure 20.20.

4 Install the drive shaft O ring and remove the complete assembly from the fixture.

5 Install the drive hub through the front seal carefully.

6 Insert the drive shaft weight assembly into the housing and engage the splines of the drive hub. Install, in correct order, the support washers, spring washer, and drive shaft screw. Torque to 285 in. lb (32.20 N·m).

7 Install the timing ring against the shoulder in the pump housing with the flat end of the ring positioned as shown in Figure 20.21.

8 Slide the cam ring into the housing and up against the timing ring. The directional arrow on the cam ring (Figure 20.22) must face the same direction as the arrow on the nameplate.

9 Invert the pump and install the cam advance screw in the cam ring and torque with adapter no. 7144/14. On pumps without an advance unit, install the cam locating screw and torque.

10 Make certain pumping plungers are free in their bore. Place the bottom roller adjusting plate over the rotor and insert the rotor in the hydraulic head. Rotate the lower plate until the tangs are aligned with the shoe slots in the rotor.

11 Install rollers and shoes, making certain that the ramp on the shoes coincides with the ramp on the lower plate.

Figure 20.22 Directional arrow on cam ring.

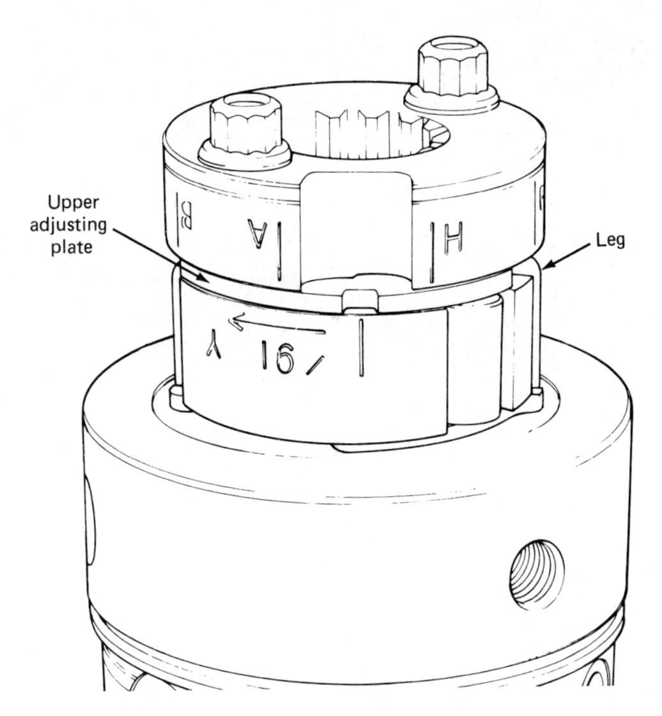

Figure 20.23 Installing upper adjusting plate (courtesy CAV–Lucas).

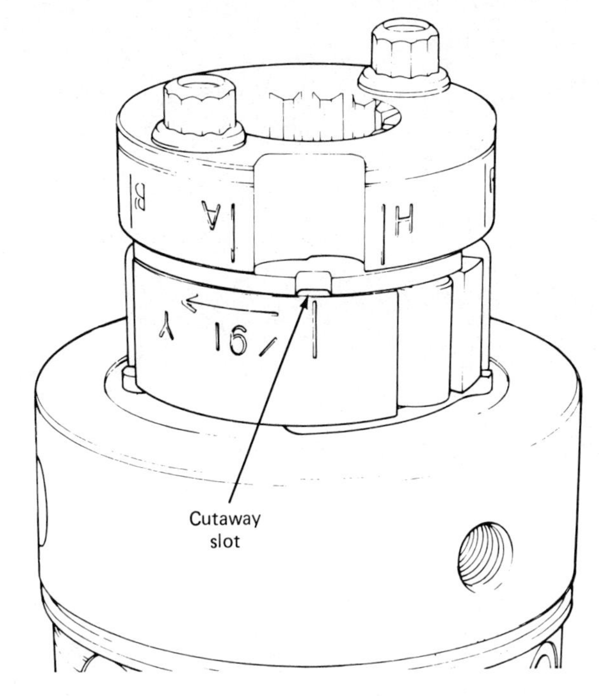

Figure 20.24 Top plate alignment with scribe mark on rotor (courtesy CAV–Lucas).

12 Install the upper adjusting plate, engaging the legs into the slots of the lower plate (Figure 20.23).

NOTE The cutaway slot on the top plate *must align* with the scribed mark on the rotor (Figure 20.24).

13 Place the drive plate with machined recess down so that the slot of the plate aligns with the vertical scribe mark on the rotor (Figure 20.25). Screw the drive plate retaining screws in finger tight.

14 Set roller-to-roller dimension (this determines maximum fuel output of the pump) by attaching tool no. 7144/262 to the hydraulic head (Figure 20.26). Supply fuel to the fitting at 450 psi (31.5 kg/cm^2) so that rollers will be forced outward. Measure the dimension with a micrometer. Set by rotating the upper adjusting plate (Figure 20.27). The dimension must match that listed in the CAV test plan.

15 Torque the drive plate screws.

16 Install and lightly tighten the rotor nut.

NOTE Check direction of rotation for proper tightening.

17 Lubricate the hydraulic head O ring and slide the

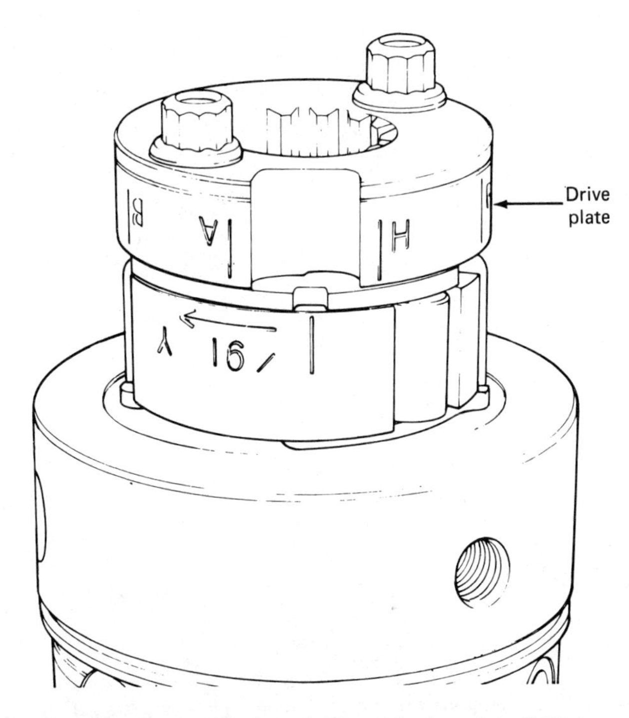

Figure 20.25 Installation of drive plate (courtesy CAV–Lucas).

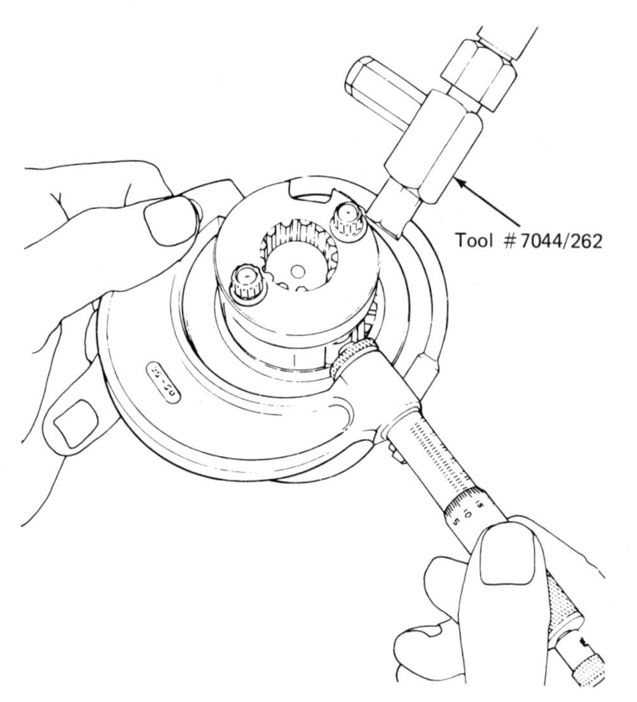

Figure 20.26 Setting roller-to-roller dimension with no. 7044/262 tool (courtesy CAV–Lucas).

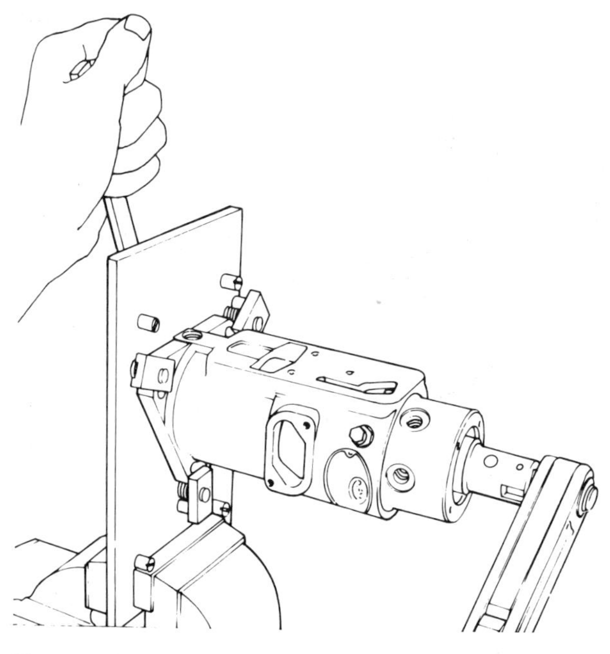

Figure 20.28 Tightening rotor nut (courtesy CAV–Lucas).

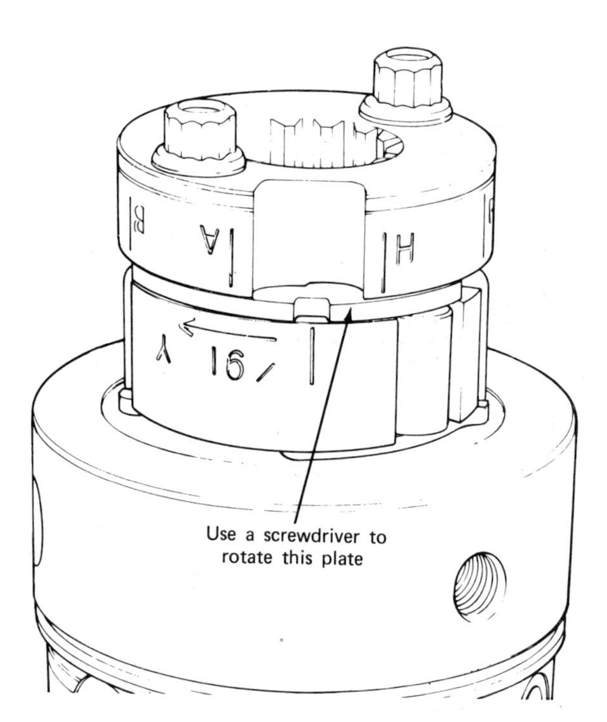

Figure 20.27 Roller-to-roller dimension adjusting plate (courtesy CAV–Lucas).

head into the pump housing, engaging the spline on the drive shaft.

CAUTION Rotate the head as it is inserted to avoid cutting the seal ring.

Install the two head locking screws finger tight.

18 Assemble the advance unit, making certain the solid end of the advance piston is on the same end as the port supplying the advance with fuel.

19 Using a new gasket, set the advance unit over the advance screw. Secure it with the lock cap and the head locating screw. Torque the head locking screws, locating screw and cap nut.

20 Using tools no. 7144/773 and no. 7044/889, tighten the rotor nut to the correct torque (Figure 20.28).

21 Install the transfer pump liner with the locating notch at 3 o'clock for a clockwise rotation pump and at 9 o'clock for a counterclockwise pump (Figure 20.29).

22 Install the pump blades and the end plate seal.

23 Assemble the end plate components as shown in Figure 20.30, and install the end plate to pump, making certain to correctly locate the dowel pin.

24 Install and torque the four end plate screws.

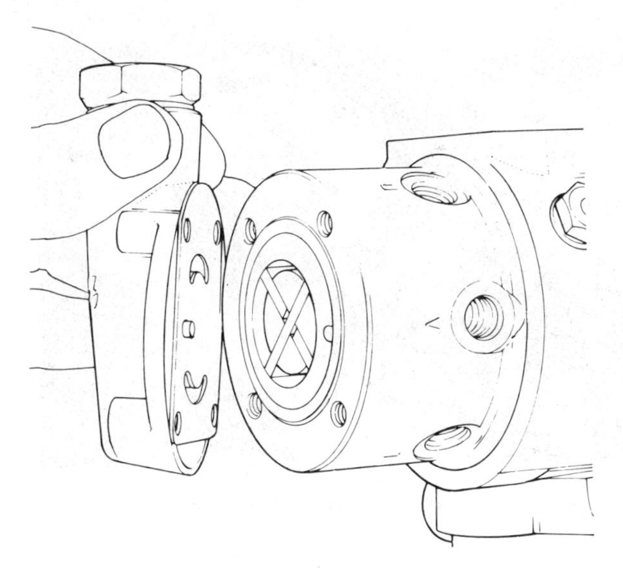

Figure 20.29 Installing T pump liner (courtesy CAV–Lucas).

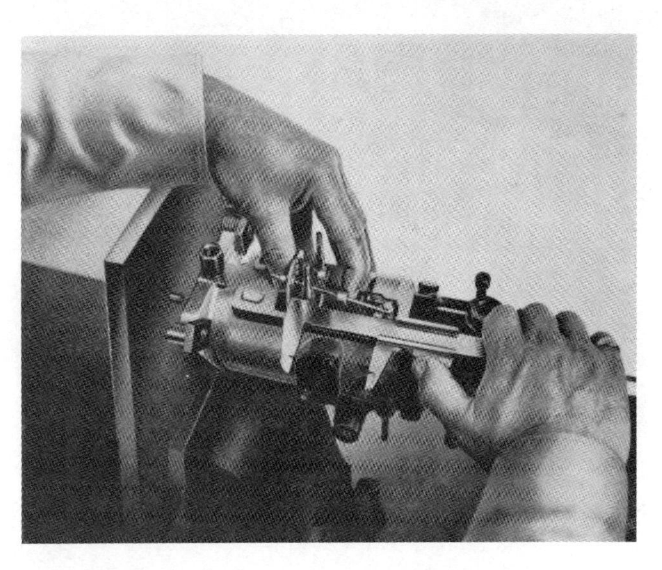

Figure 20.31 Setting the governor link length (courtesy CAV–Lucas).

25 Assemble the governor arm to the control bracket and attach arm spring. Put the spring retainer on the linkage hook and then the long spring, fiber washer, governor arm, pivot ball, ball washer, linkage nut, and lock washer. Snap the metering valve on the linkage hook.

Figure 20.30 Correct assembly of end plate parts (courtesy CAV-Lucas).

26 Place this assembly into the pump housing by setting the tips of the governor arm on the steps of the thrust sleeve and starting the metering valve in its bore.

27 Install keep plate, tab washers, and long control cover studs finger tight.

28 Install the bracket screw with a new tab washer. Torque the cover studs and bracket screw.

29 Set the governor link length as shown in Figure 20.31. Push rearward on the linkage, adjust hook, and measure the distance from the control cover stud (large diameter) to the large end of the spring retainer contacting the metering valve. Measure this distance with a vernier caliper held parallel to the axis of the pump. If it is incorrect, adjust the nut on the linkage hook and tighten the locknut.

30 Install a new housing gasket and insert the stop bar.

31 Install the shutoff shaft in the governor cover. Insert the idle spring and retainer into the correct hole in the governor arm as determined by the code number on nameplate (see Section I, Part I) or in test plan.

32 Attach the main governor spring and hook the other end into the throttle link (again as determined by the pump code).

33 Place the governor cover and the throttle shaft on the cover studs.

Figure 20.32 DPA pump on test bench.

Figure 20.33 Pressure gauge installed on pump.

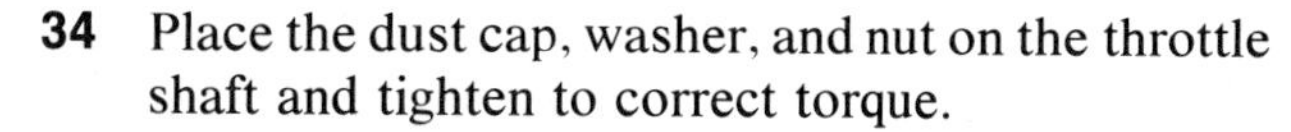

34 Place the dust cap, washer, and nut on the throttle shaft and tighten to correct torque.

35 Secure the governor cover with seal washers and acorn nuts. Torque to specified value.

36 Attach the return connection.

37 Attach the inspection hole plate.

K. Procedure for Testing

After overhaul, all injection pumps *must* be mounted to a test bench and checked for correct operation. These tests include: correct fuel delivery, governor operation, transfer pump pressure, and advance movement. Failure to make these adjustments could result in engine and/or pump damage.

The CAV–DPA pump can be tested on several different test benches. Mounting and operating instructions vary from brand to brand, so consult the test bench manual before mounting the DPA pump. Following is the correct testing procedure after the pump has been mounted on the test bench (Figure 20.32).

1 *Initial run.* Supply calibrating oil to the pump inlet and start test test.

CAUTION Make sure the pump is rotated in the correct direction as indicated on the nameplate.

Drive the pump at 100 rpm until the fuel bleeds from all lines at the nozzle end. Tighten the lines. Make sure the pump has transfer pump pressure. Run the pump up to 500 rpm and check for external leaks. Correct all leaks before starting the tests. Run the pump at this speed for about 10 minutes to warm it up and purge air from the system.

2 *Transfer pump pressure.*

a Attach a 0 to 150 psi (0 to 10.5 kg/cm^2) gauge to the transfer pressure adapter installed in place of the vented head locking screw (Figure 20.33).

b Run pump at the speed listed in the test plan (all speeds are pump speeds) for transfer pressure adjustment. Check pressure.

c If pressure is incorrect, adjust on certain pumps by changing the end plate sleeve plug (Figure 20.34). On other pumps, simply screw the transfer pressure adjuster in or out as required. (*In = higher pressure, out = lower pressure.*) It is best to set the pressure to the high side of the specification.

3 *Automatic advance.* To check the advance, first attach gauge no. 7244-59 with pin no. 7244-70 to pump, as shown in Figure 20.35. Zero the pointer. Run the pump at the speed(s) given for the advance position on the test plan. If the advance is incorrect, adjust as follows:

a *Speed advance.* Add or subtract shims between the piston and spring cap.

b *Load advance.* With the shutoff lever adjusted as stated on the test plan, set the advance with the eccentric screw at the rear of the governor cover.

4 *Maximum fuel delivery.* Run pump at the speed in-

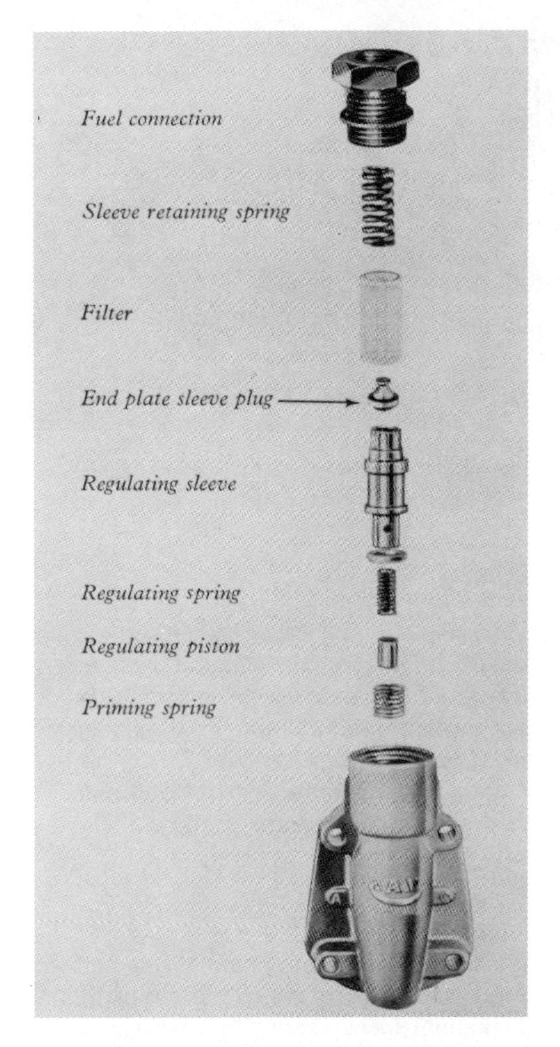

Figure 20.34 End plate adjusting plug (courtesy CAV–Lucas).

dicated on test plan or on the nameplate. Check fuel delivery at full throttle. Measure the volume of fuel collected for 200 strokes on the test bench counter.

NOTE All fuel delivery checks on DPA pumps are for 200 strokes.

If fuel delivery is incorrect, adjust as follows:

a On internal adjust pumps, remove the side inspection cover. Rotate the pump until the slot in the adjusting plate is visible (Figure 20.36). Loosen the two drive plate screws. Move the adjusting plate to get the appropriate amount of fuel. The direction the drive plate is turned to raise or lower the output depends on the type of adjusting plate used.

NOTE Only a *slight* movement of the plate is required to change fuel output.

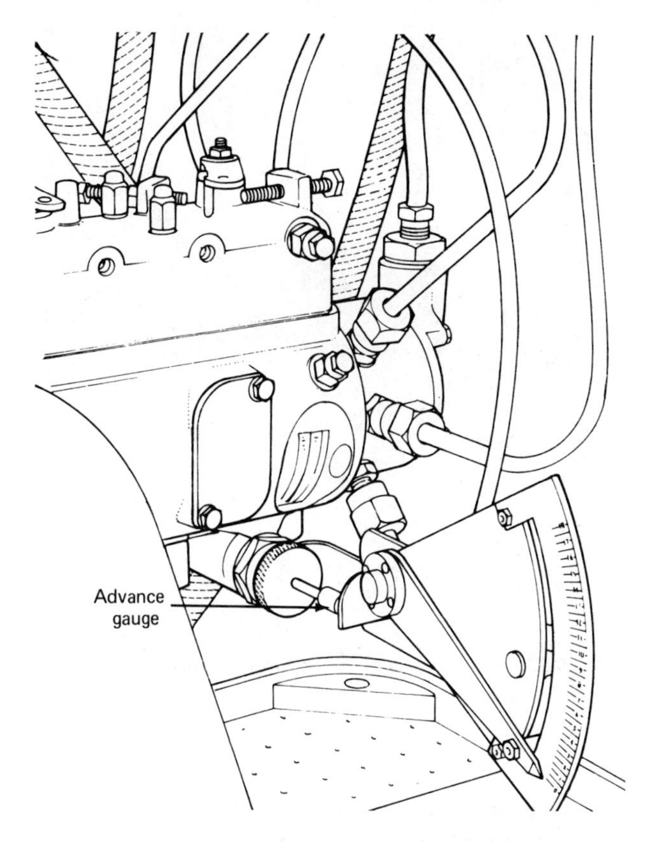

Figure 20.35 Advance gauge installed on pump (courtesy CAV–Lucas).

Torque the drive plate screws evenly using tools no. 7144/482 and no. 7144/511A and a torque wrench. Correct torque when using a spanner wrench is:

Pumping plungers up to and including 7.5 mm = 115 in. lb (12.99 N·m), 8.0 mm and above = 180 in. lb (20.34 N·m).

Replace side inspection cover.

Figure 20.36 DPA pump with side cover removed.

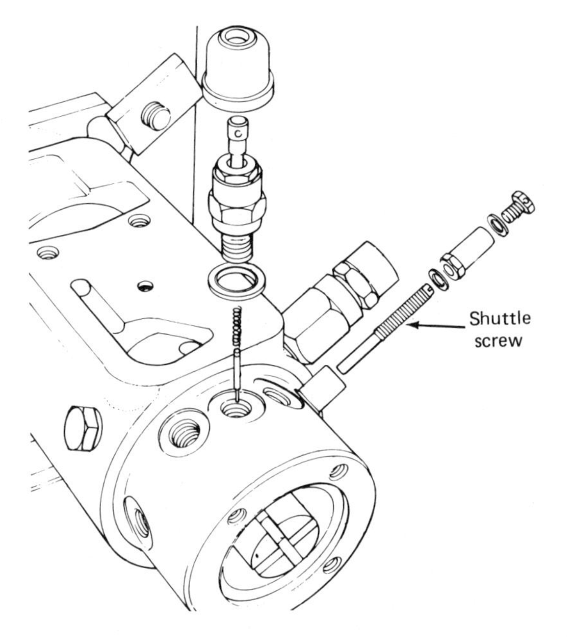

Figure 20.37 Shuttle stop screw (courtesy CAV–Lucas).

Figure 20.39 Head outlet lettering (courtesy CAV–Lucas).

b With external adjusting pumps, screw the shuttle stop screw (Figure 20.37) inward to reduce delivery or outward to increase. Retighten shuttle nut and replace screw.

5 *Cranking fuel delivery check.* This test determines the condition of the hydraulic head and rotor and is critical to pump performance. Run the pump with the throttle in wide open position at 100 rpm and check delivery. If insufficient quantity is delivered, the hydraulic head and rotor must be replaced. Any delivery *equal to or over* the required amount is acceptable.

6 *Shutoff operation.* Run the pump at the required speed and move the shutoff lever to the off position. Fuel delivery must cease.

Spring Position Code	First RPM	Second RPM	Third RPM
1, 2, 5, 8, 9	RPM listed	RPM listed	RPM listed
3, 6	RPM listed	RPM listed	RPM listed
4, 7	RPM listed	RPM listed	RPM listed

Figure 20.38 Spring code and rpm table (courtesy CAV–Lucas).

7 *Governor spring check.* Three rpm test settings are given (Figure 20.38). Run the pump at the first rpm and simply record the delivery. Then run the pump at the second rpm and set the maximum speed screw to get an average delivery of 2.0 cc (no line over 3.0 cc). Go back to the first rpm and recheck delivery. It should be the same or 0.4 cc below.

NOTE If delivery is too low, the governor spring must be replaced.

8 *Maximum speed check.* Run pump at one half the speed listed on nameplate (or check test plan) and set the maximum speed screw to give 2.0 cc delivery. Lock stop screw and seal.

9 *Timing.* After testing, all DPA pumps must be timed both internally and externally. To time internally, remove the pump from the test bench and assemble to the holding fixture. Remove the side inspection plate. Attach tool no. 7144-262 (which was used to set roller-to-roller dimension) to correct outlet as indicated on the test plan (Figure 20.39) and supply fuel to the hydraulic head at 450 psi (31.5 kg/cm^2). The outlets are lettered on the hydraulic head. Rotate pump (in running direction) until resistance is felt. With pump in this position, work through the side cover and position the flat of the timing ring (or mark) (Figure 20.40) so that it lines up with the correct letter on the drive plate as indicated on the test plan under timing. After internal timing is completed, install the flange marking gauge no. 7244/27 (Figure 20.41) as shown, using the correct drive shaft insert. With the pump still on the internal timing position, rotate the scale on the flange marking tool until the correct degree mark (as listed on test plan under tim-

Figure 20.40 Positioning timing ring.

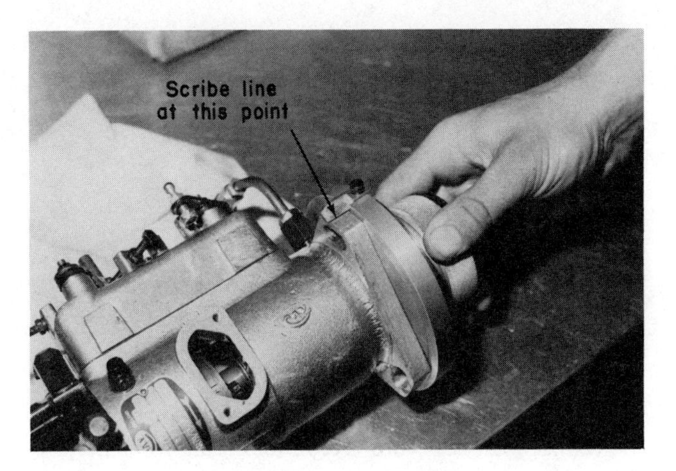

Figure 20.42 Checking housing timing mark.

ing) lines up with pointer (Figure 20.42). At this time check the scribed timing on the housing flange. If it lines up, the pump timing is correct. If it doesn't line up, file off the old mark and scribe a new one in the correct location. Remove timing tools. Replace the side cover.

L *Sealing and final checks.*

After all tests are complete, the drive shaft screw must be retorqued (except on up-rated drive pumps). To do this, loosen and retorque the drive shaft screw *three* times to insure seating of the spring washer. Correct torque is:

1.13 in. (28.5 mm) = 285 in. lb (32.20 N·m).
1.25 in. (31.7 mm) = 320 in. lb (36.15 N·m).

CAUTION This torque is extremely critical as a loose screw will cause a total loss of governor control, which could damage the pump and/or engine.

Figure 20.41 Flange marking gauge on pump.

Cap all outlets with plastic caps. Blow off the exterior of pump with compressed air. Place masking tape over the nameplate in preparation for painting. Paint pump with appropriate color.

II. Procedure for Installation to Engine (CAV-DPA)

A Be certain engine is still on no. 1 cylinder compression stroke and at the correct timing mark.

B Rotate pump until discharge port is lined up with no. 1 outlet.

C Mount pump and tighten the bolts finger tight.

D Align the mark on pump flange with mark on engine. Tighten mounting bolts.

E Rotate engine two revolutions and recheck timing. Adjust as necessary.

F Install injection lines, supply lines and fuel return lines.

CAUTION Be careful not to bend the injection lines.

G Supply fuel to pump inlet and bleed out all air. Tighten the connection.

H Attach throttle and shutoff linkage and adjust.

I Provide a means of emergency shutdown.

J Start engine and run at 800 to 900 rpm.

K Locate and correct any fuel leaks.

L After engine has reached operating temperature check high and low idle speeds with an external tachometer.

III. The DP-15 Distributor Pump

The latest addition to the CAV line is the DP-15. The DP-15 is a distributor pump capable of serving up to eight cylinders. It uses a small mechanical governor and a rotating metering valve.

The DP-15 is completely filled with diesel fuel and can be mounted in any required position. The hydraulic head and rotor serve as the main structural component of the pump with the governor housing and transfer pump being attached to it.

A *Component parts* (Figure 20.43).

1. *Hydraulic head/rotor.* The hydraulic head is the main component of the pump with the governor and transfer pump fitted to it. It serves to pump and distribute fuel to the injection nozzles in correct order.
2. *Drive housing.* It serves to mount the pump to the engine. It also contains the governor and the drive shaft, which carries the rollers and shoes.
3. *End cover.* The end cover contains the transfer pump and throttle components. The fuel inlet and outlet are mounted to this cap.

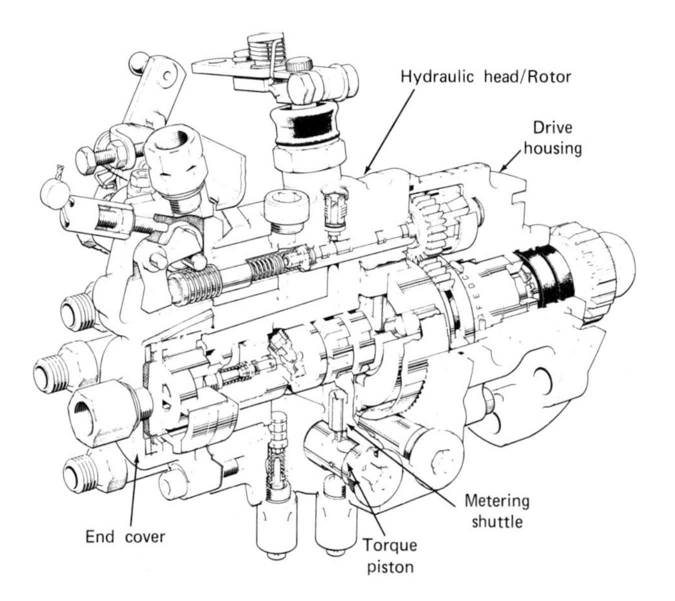

Figure 20.43 Component parts of DP-15 (courtesy CAV–Lucas).

4. *Metering shuttle(s).* The metering shuttles are charged by metering pressure from the metering valve. The shuttle(s) then discharges to the pumping plungers.
5. *Torque piston.* This piston determines the movement of the shuttle, and thereby the amount of fuel pumped for a given load or speed condition.

B *Pump operation.*

1. *Charge cycle.* Diesel fuel enters the inlet connection at the end cap and goes to the transfer pump (Figure 20.44,1). From the transfer pump, fuel is routed to the metering valve (Figure 20.44,2), the cam advance device (Figure 20.44,3), and to the underside of the left-hand shuttle (Figure 20.44,4).

NOTE Only the twin shuttle design is shown here.

The fuel going to the metering valve is lowered in pressure slightly as it goes by the rotating valve. The opening of this valve is controlled by the movement of the mechanical governor that operates directly on its end. Fuel from the metering valve flows to the fuel shutoff valve (Figure 20.44,5) and then to a slot in the rotor. It then forces the right-hand shuttle downward (Figure 20.44,6) until it contacts the torque piston (Figure 20.44,7). Fuel that is beneath the R.H. shuttle is discharged to the housing and returns to the supply tank. While the R.H. shuttle is being charged by metering fuel, transfer pressure is forcing the L.H. shuttle upward. The fuel above the L.H. shuttle then goes to a slot in the rotor, opens the inlet check valve (Figure 20.44,8), travels up the center of the rotor, and forces the pumping plungers apart (Figure 20.44,9).

NOTE The displacement of the shuttle is equal to the displacement of the pumping plungers.

This completes the charge cycle, in which one shuttle is charged and the cylinders of the pumping plungers are also charged with fuel.

2. *Discharge cycle.* As the rotor continues to revolve, the rollers are forced inward by the cam lobes (Figure 20.44,9). This forces fuel out of the cylinders and down the center of the rotor. The fuel opens the delivery valve (Figure 20.44,10) and goes out the discharge port (Figure 20.44,11) which is in register with a discharge port in the hydraulic head. The fuel

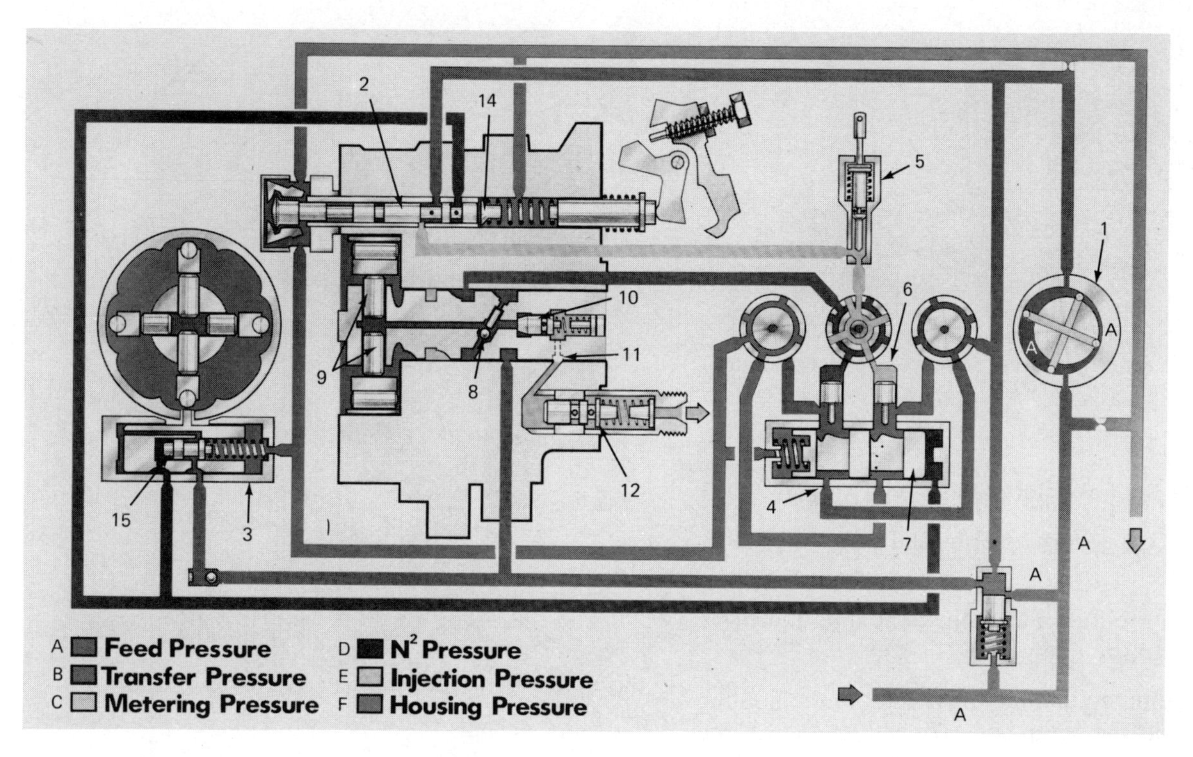

Figure 20.44 DP-15 fuel injection pump fuel flow (courtesy CAV–Lucas).

then opens the pressurizing valve (Figure 20.44,12) and goes to the injection nozzle.

NOTE The inlet check valve (Figure 20.44, 8) is closed at the end of the shuttle stroke by the small piston above it.

As the shuttle ends its stroke, no more fuel can be supplied to the check ball. Transfer pressure behind the small piston then can force the ball on its seat. It is held on its seat until injection is completed.

3 *N^2 pressure.* N^2 pressure is regulated at the metering valve by a plate valve (Figure 20.44,14). N^2 pressure operates the torque piston (Figure 20.44,7) against a return spring. The cam profile on the torque piston can be adjusted to give the correct fuel delivery at varying speeds. A slot in the piston also allows for excess fuel at starting. N^2 pressure also operates a small servo valve in the advance piston (Figure 20.44,15). The advance of the cam ring depends on pump speed, and the servo valve tends to eliminate any advance timing changes caused by changes in fuel temperature, that is, viscosity.

4 *Metering shuttles.* Twin shuttles are used when greater fuel outputs are required. Using two shuttles gives a greater time period in which to fully charge the pumping plungers. A single shuttle is used on pumps for lower horsepower engines.

C. Procedure for Disassembly (DP-15)

Before beginning disassembly, always record all type and serial numbers from the pump name tag to assist in ordering parts. When working on any injection pump, always keep the work area spotlessly clean. Always have all required special tools available.

1 Mount the pump on fixture no. 7244-331.

2 Remove drive gear, using tool no. 7244-24.

3 Remove the end cover screws and lift off the end cover.

4 Remove transfer pump components (Figure 20.45).

CAUTION Do not lose the rollers on the ends of the blades.

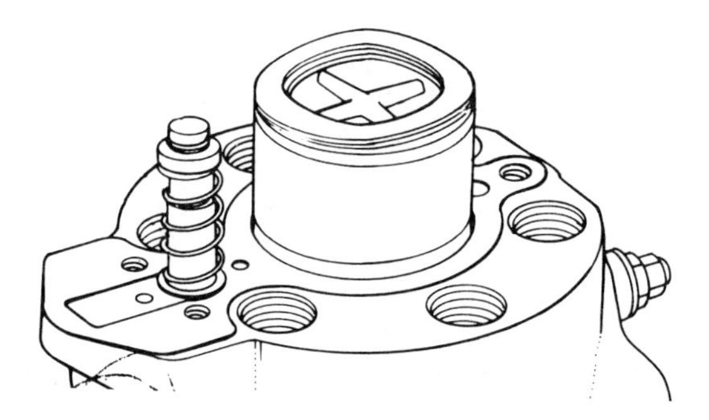

Figure 20.45 Removing transfer pump components (courtesy CAV–Lucas).

5 Lift out the governor plunger with spring.

6 Remove the governor spring and plate valve (Figure 20.46).

7 Remove the four allen screws holding the drive housing to the pump body.

8 With the pump mounted vertically in the holding fixture, lift the pump body straight up and off the rotor. Remove the metering valve (Figure 20.47).

9 Remove the stop control unit from the pump body.

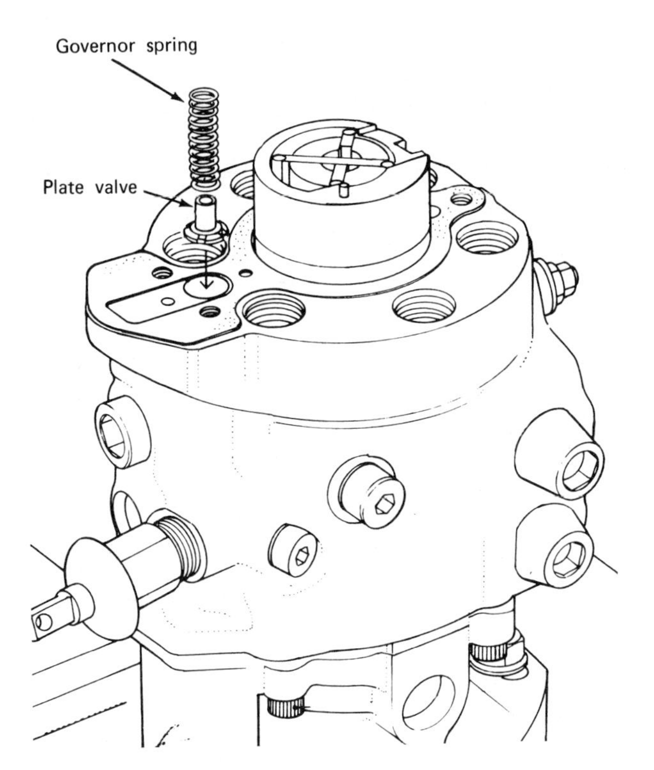

Figure 20.46 Removing governor spring and plate valve (courtesy CAV–Lucas).

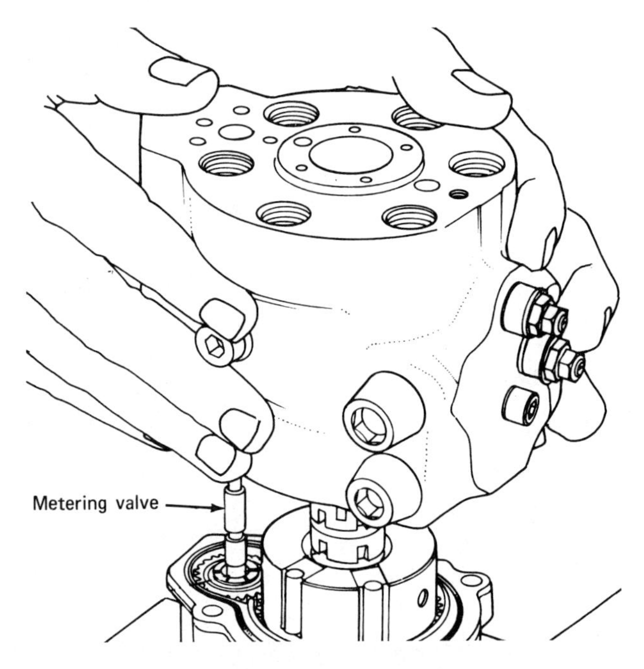

Figure 20.47 Removing metering valve (courtesy CAV–Lucas).

10 Unscrew the lock screw (1) and remove the metering valve sleeve adjuster (2) and sleeve from pump body (Figure 20.48).

11 Remove the coned advance plug (1) from the pump body (Figure 20.49). Lift out the stem as-

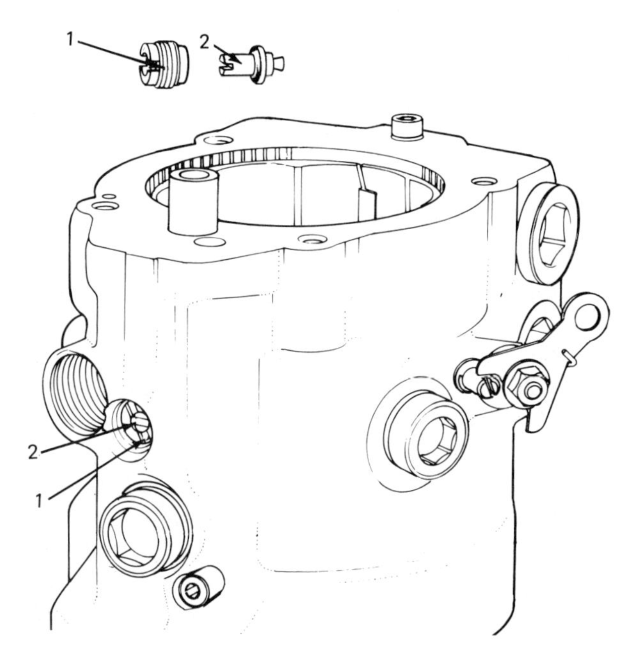

Figure 20.48 Removing metering valve sleeve adjuster (courtesy CAV–Lucas).

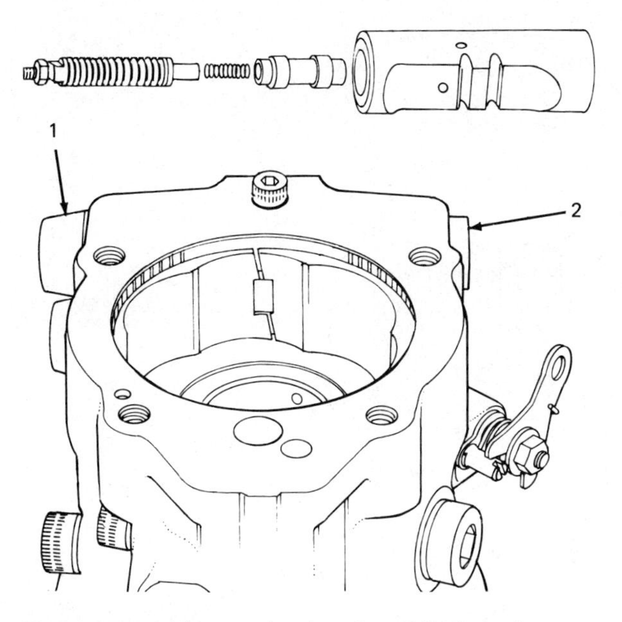

Figure 20.49 Advance plug (courtesy CAV–Lucas).

sembly. Remove the plug opposite the coned plug (2).

12 Push the advance piston to the blank end and remove the cam ring.

13 Remove the advance piston with the servo valve and spring.

14 Remove the excess fuel device by removing the control lever (1) and the stop screw (2) (Figure 20.50). Pull the shaft from the pump body.

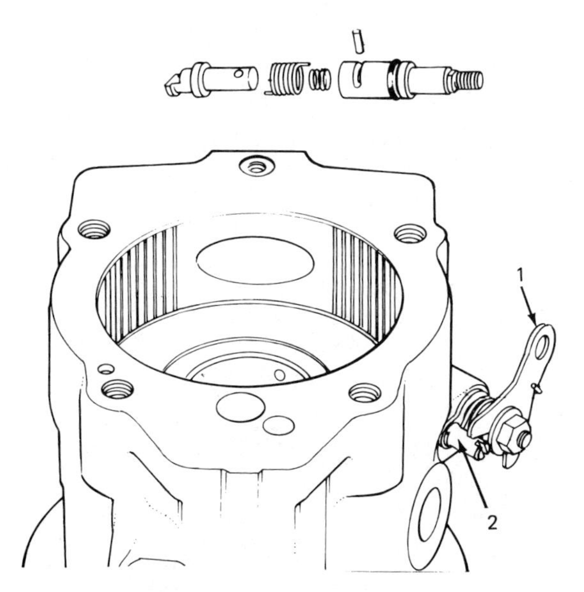

Figure 20.50 Removal of excess fuel device (courtesy CAV–Lucas).

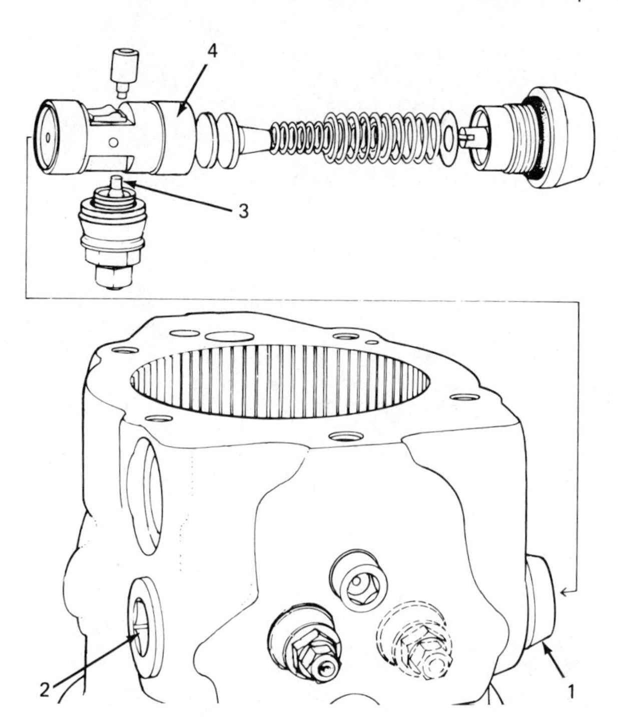

Figure 20.51 Plug of torque piston assembly (courtesy CAV–Lucas).

15 Remove the coned plug of the torque piston assembly (1) and also the plug on the opposite side (2) (Figure 20.51). Remove the maximum fuel adjustment screw and then the torque piston from the body (3).

16 Remove the shuttle(s) (1) and transfer the pressure regulator (2) with the spring (3) (Figure 20.52).

17 Remove the throttle shaft and lever pivot from the end cap.

18 Carefully remove rollers and shoes.

CAUTION Rollers and shoes are a matched set and must not be interchanged. Remove the rotor from the drive shaft, being careful not to drop the pumping plungers.

19 Remove the drive shaft (1), the gear ring with rubber buffers, and the governor unit (Figure 20.53).

20 Disassemble the governor unit by lifting the weight carrier off the governor gear. Remove the weights (Figure 20.54).

21 Remove the drive shaft seals from the drive housing and discard.

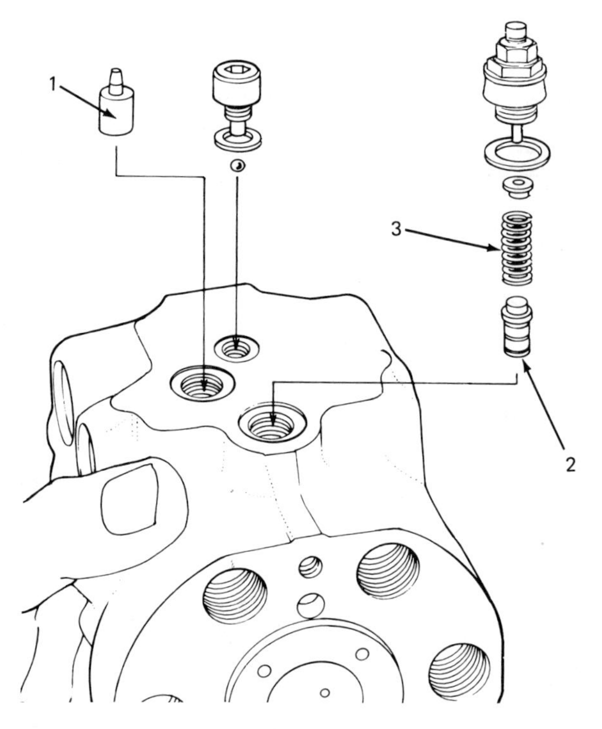

Figure 20.52 Transfer pump pressure regulator (courtesy CAV–Lucas).

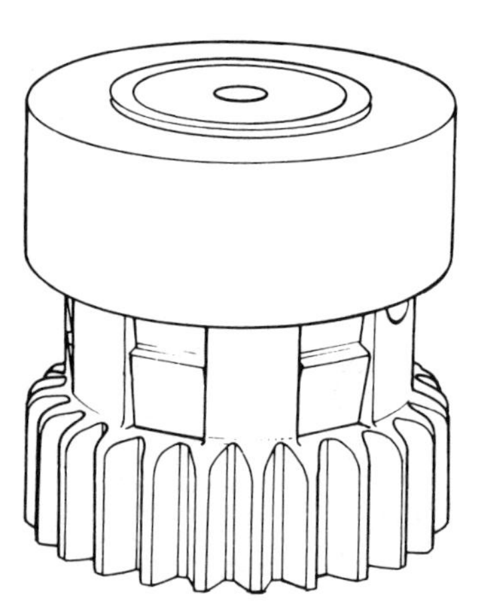

Figure 20.54 Removing the governor weights (courtesy CAV–Lucas).

D. Procedure for Inspection

Carefully examine all pump components for scoring, rust, corrosion, and thread damage. Checks to be made on specific items are given below:

1 *Rotor*. Inspect rotor for deep scoring. Make certain the pumping plungers move freely.

a *Lift*. Using tools no. 7244-279 and no. 7244-278, attach a dial indicator to the rotor as

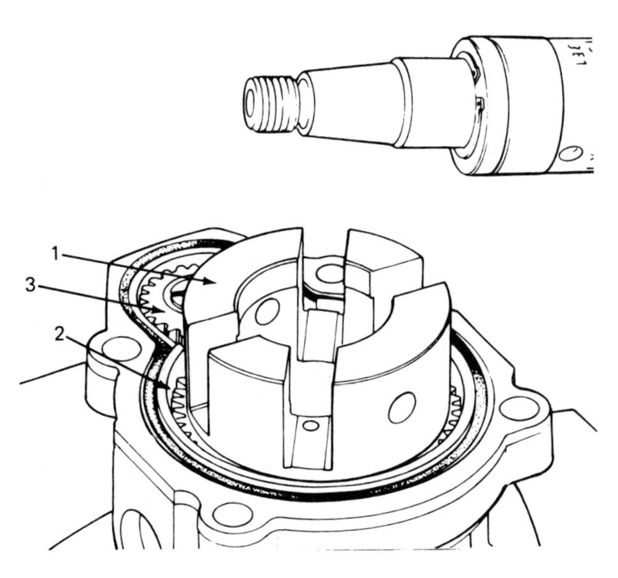

Figure 20.53 Drive shaft with governor (courtesy CAV–Lucas).

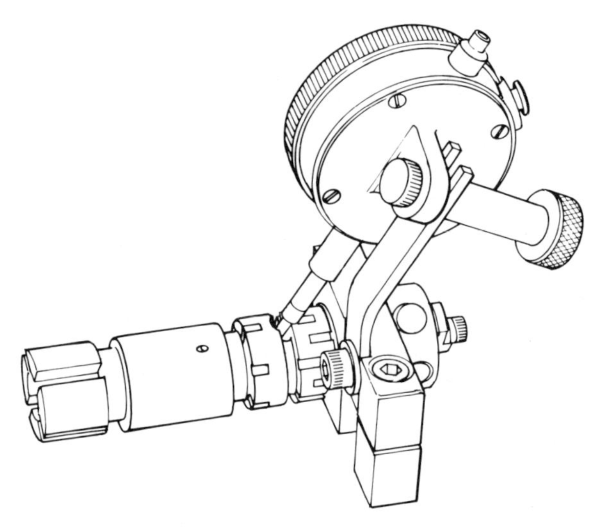

Figure 20.55 Checking the lift of the inlet check valve (courtesy CAV–Lucas).

shown in Figure 20.55. This will check the lift of the inlet check valve. The lift must be between 0.5 and 0.85 mm (0.020 and 0.34 in.). If the lift is incorrect, the rotor/pump body must be replaced as a unit.

CAUTION *Do not remove the check valve assembly. Distortion of the rotor will result.*

b *Leakage.* Using tool no. 7244-218 and a nozzle hand stand, check the inlet check valve and delivery valve stop for leakage (Figure 20.56). Apply pressure to the rotor at 3000 psi (210 kg/cm²) and check the inlet check valve and delivery valve stop for leakage. If any leakage is found, replace the rotor/pump body as a unit. *Do not remove inlet check valve or delivery valve.*

2 *Drive shaft end play.* Assemble the drive shaft and drive plate into drive housing without the shaft seals. Attach tool no. 7044-634 onto the drive shaft. Attach the drive housing to pump body (Figure 20.57). Push in on the tool and tighten the thimble until it touches the drive housing. Record the reading (reading A). Next pull out on the tool and again turn the thimble in until it touches the drive housing. Record the reading (reading B). Subtract reading A from reading B to obtain drive shaft end play. It should be 0.1 to 0.45 mm (0.004 to 0.018 in.) If incorrect, the drive shaft or drive housing must be replaced. Remove the tool from the drive housing and remove the drive housing from the pump body.

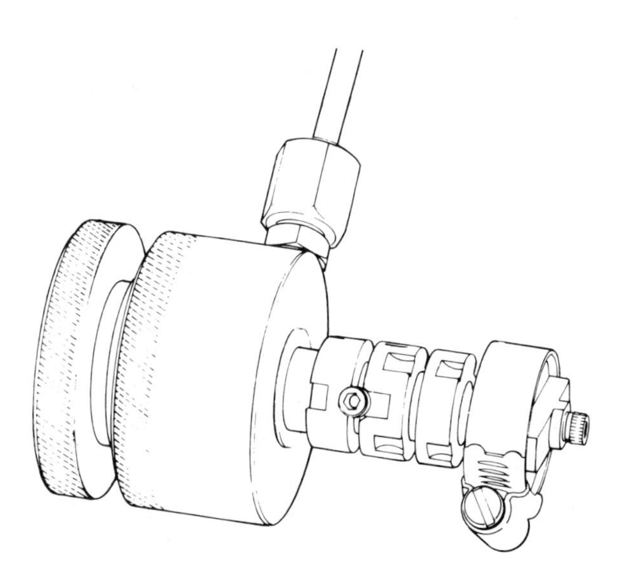

Figure 20.56 Checking inlet check valve for leakage (courtesy CAV–Lucas).

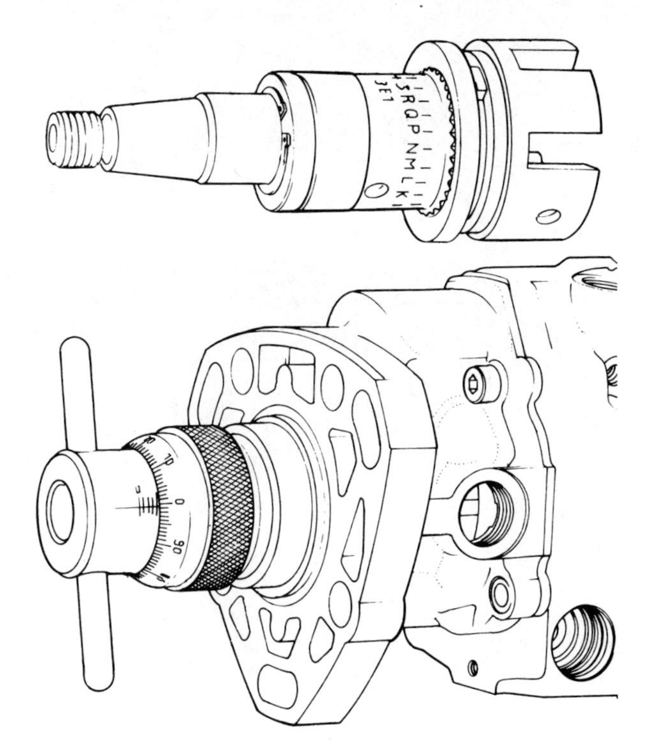

Figure 20.57 Attaching drive housing to pump body (courtesy CAV–Lucas).

E. Procedure for Assembly

NOTE Always replace *all* O rings, gaskets, and seals when overhauling any injection pump.

1 Assemble excess fuel shaft to pump body as shown in Figure 20.50.

2 Install the transfer pump regulator valve into the pump body and install the regulator adjusting plug. Tighten to 27.12 N·m (240 in. lb)

3 Insert the metering shuttle(s) into the pump body (stem portion last).

4 Install the torque piston into its bore in the pump body (Figure 20.58). The groove in the block must be in the center of the maximum fuel adjusting hole. Screw the plug of the maximum fuel screw into the body. Tighten to 27.12 N·m (240) in. lb). Then screw in the maximum fuel adjusting screw until it engages the groove of the torque piston. Tighten both plugs of the torque piston to 27.12 N·m (240 in. lb).

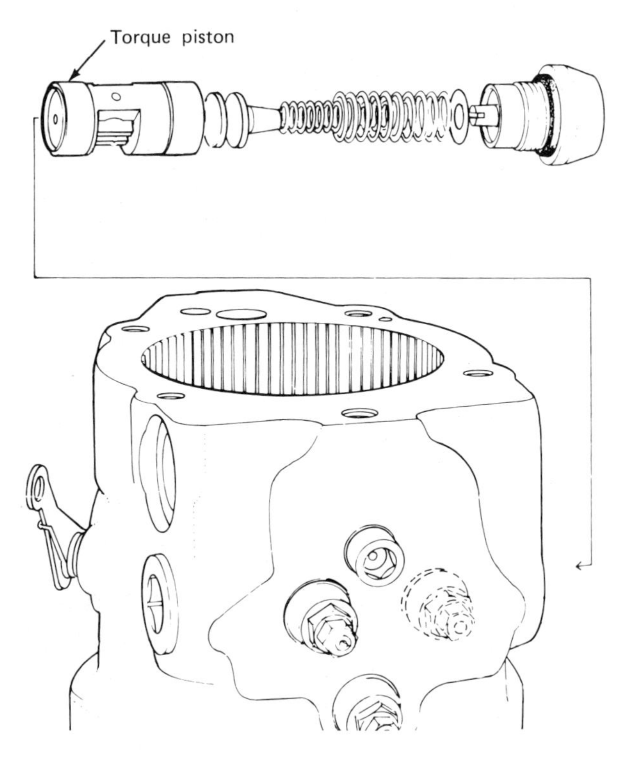

Figure 20.58 Installing torque piston (courtesy CAV–Lucas).

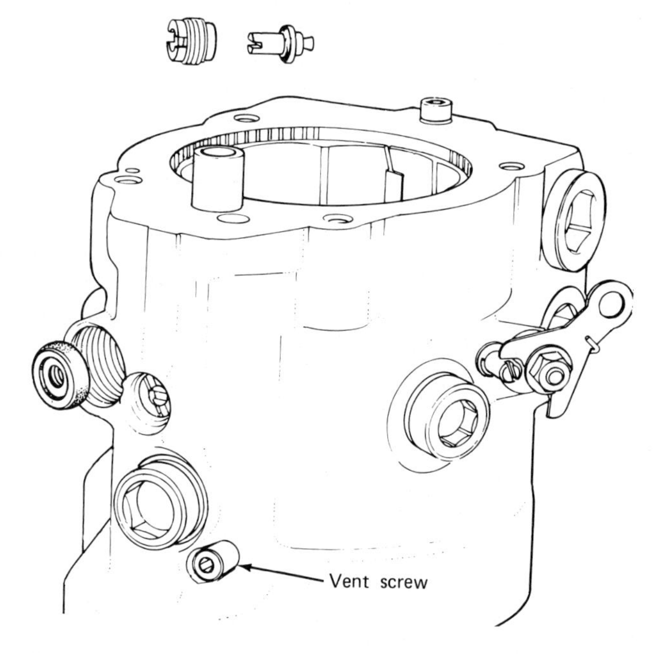

Figure 20.60 Installing vent screw (courtesy CAV–Lucas).

5 Assemble the servo valve to advance piston and install piston to pump body. Install cam ring so that the slot in ring engages the tooth of the advance piston (Figure 20.59).

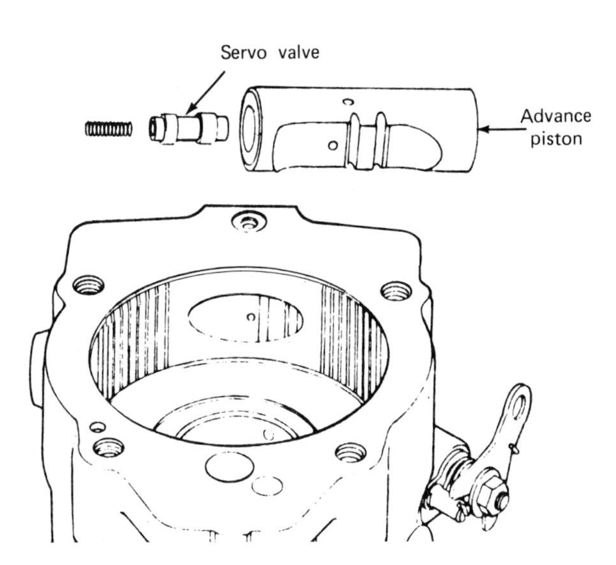

Figure 20.59 Assemble cam ring to advance piston (courtesy CAV–Lucas).

NOTE The part numbers and directional arrow on the cam ring must face downward.

6 Install and tighten vent screw to correct torque (Figure 20.60).

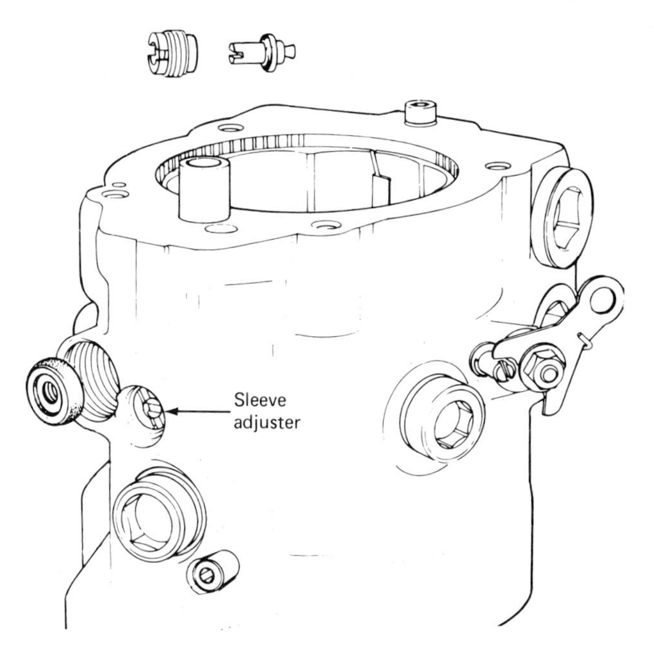

Figure 20.61 Installing metering valve sleeve adjuster (courtesy CAV–Lucas).

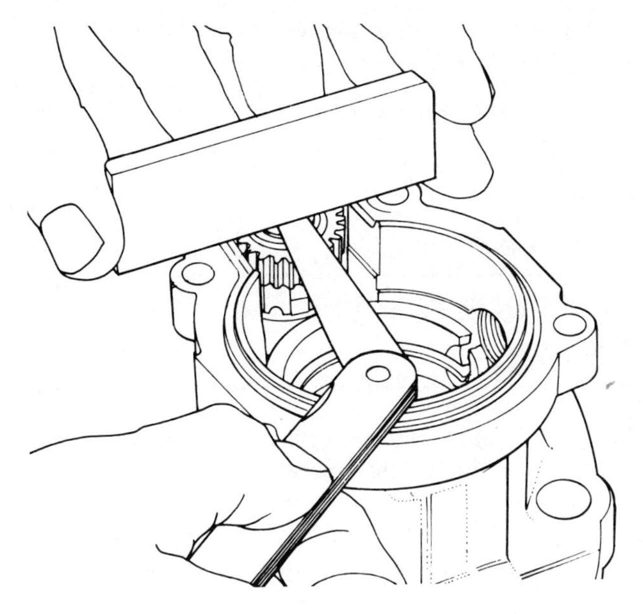

Figure 20.62 Setting end play of governor assembly (courtesy CAV–Lucas).

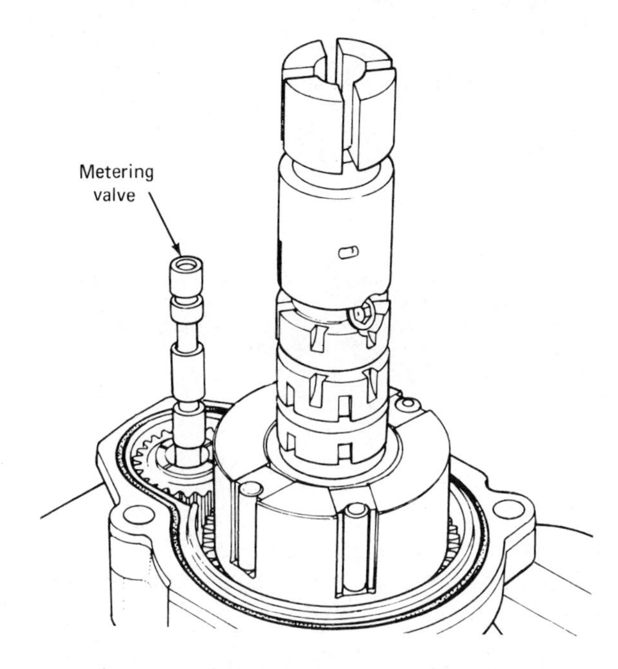

Figure 20.63 Installing the metering valve in the governor assembly (courtesy CAV–Lucas).

7 Install metering valve sleeve until it protrudes 10 mm (0.400 in.) above the body. Insert the metering valve sleeve adjuster (Figure 20.61).

8 Install the drive housing seals in the following manner: the outer seal (lip side first) from the outside of the drive housing and the inner seal (lip side first) from the inside.

9 Place shims and thrust pad for the governor assembly into the drive housing. Set the governor in position and measure end play as shown in Figure 20.62. It should be 0.05 to 0.10 mm (0.002 to 0.004 in.). Change shims under nylon pad to obtain correct end play.

10 Install the governor drive gear ring and rubber inserts.

11 Install protection cap no. 7144-820 over drive shaft and install shaft into housing. Remove protection cap.

12 Install seal ring in drive housing. Place the rotor in the drive shaft with rollers and shoes in the proper position. Insert the metering valve in the governor assembly, engaging the thrust sleeve (Figure 20.63).

13 Place the pump body over the rotor and metering valve. Install and torque the retaining screws to 7.34 N·m (65 in. lb).

14 Adjust metering valve, as shown in Figure 20.64, until tool plunger is level with the tool no. 7244-227.

15 Install the stop control unit.

16 Place the rotor retaining rings in the rotor slot. The notch in the edge of the retaining ring must face away from the metering valve.

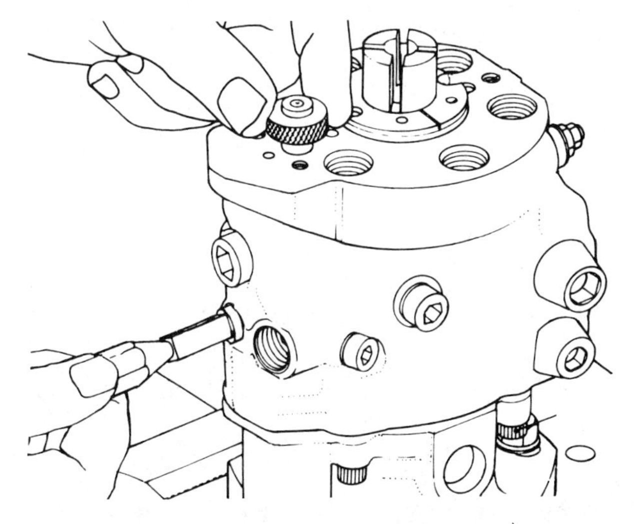

Figure 20.64 Adjusting metering valve (courtesy CAV–Lucas).

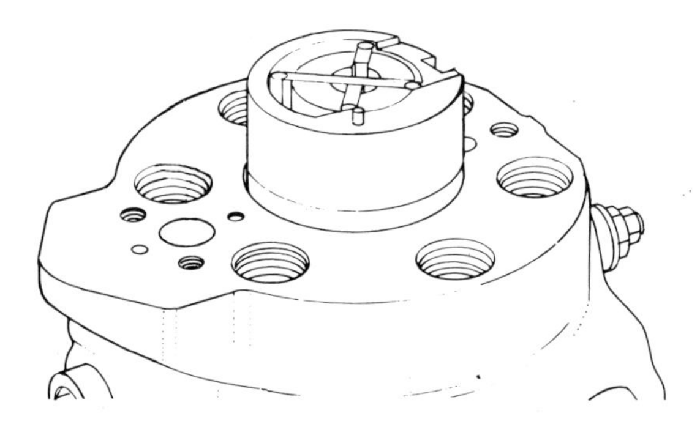

Figure 20.65 Transfer pump liner in position (courtesy CAV–Lucas).

17 Place the transfer pump liner and lock pins in position (Figure 20.65) and then the pump blades and blade rollers.

18 Place the governor plunger in the pump body. Place tool no. 7244-336 over the plunger and adjust until the inner end just touches the top of the peg. Remove the tool and measure from the screw to the end of tool and compare it with the dimension given on the test plan. Add or subtract shims as required.

19 Install the governor plunger spring, the plunger with shims and peg. Place wave washers over the transfer pump liner.

20 Assemble throttle lever and shaft to the end cap and install the end cap on the pump body.

21 Install the pressurizing valves and springs in order, as shown in Figure 20.66. Apply sealer (such as Loctite) to the high pressure outlets and tighten to 45.19 N·m (400 in. lb).

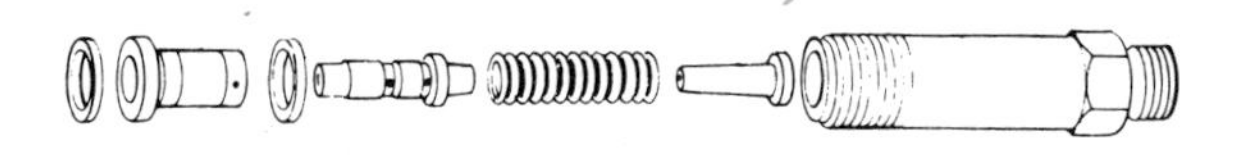

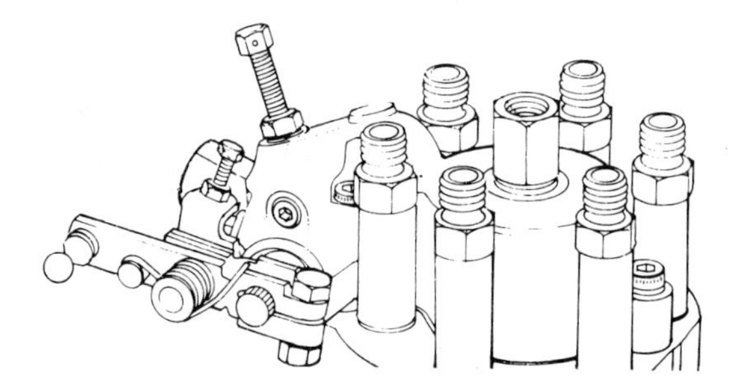

Figure 20.66 Installing pressurizing valves (courtesy CAV–Lucas).

22 Install the drive hub and torque to 81.35 N·m (720 in. lb).

23 Remove pump from fixture and install it on the test bench.

F. Procedure for Testing

The DP-15 pump can be tested on a number of test benches. Mounting and operation of these benches vary so always consult the test bench manual before testing the DP-15. The following procedure is to be followed after the pump is correctly mounted to the bench:

1 *Mount tool no.* 7244-217 to check the transfer pump pressure.

2 *Run pump for 10 minutes* to remove air and stabilize the temperature at 40°C (105°F). Repair leaks.

NOTE Some fuel will leak from the many adjusting screws during testing. At this time stop only those leaks at other points.

3 *Set transfer pump pressure.* Run pump at the speed specified on test plan and observe the pump pressure. If it is incorrect, adjust the pressure adjusting screw. Turn the screw in to raise the pressure.

4 *N^2 pressure check.* Calculate this pressure by subtracting the housing pressure from the pressure observed at the N^2 pressure plug (Figure 20.67).

5 *Maximum fuel set.* Run pump at specified speed and check fuel delivery. If incorrect, adjust the screw (or screws on an eight cylinder pump) to get the correct amount (Figure 20.68). Screwing inward will decrease delivery.

6 *Fuel delivery set.* Adjust with the screw in the center of the torque plug (Figure 20.69).

7 *Excess fuel device.* By changing the shims under the outer torque piston spring, the speed at which the torque piston moves out of the excess fuel notch can be changed. Add more shims to increase the speed at which the torque piston moves.

8 *Advance set.* Attach the advance gauge no. 7244-

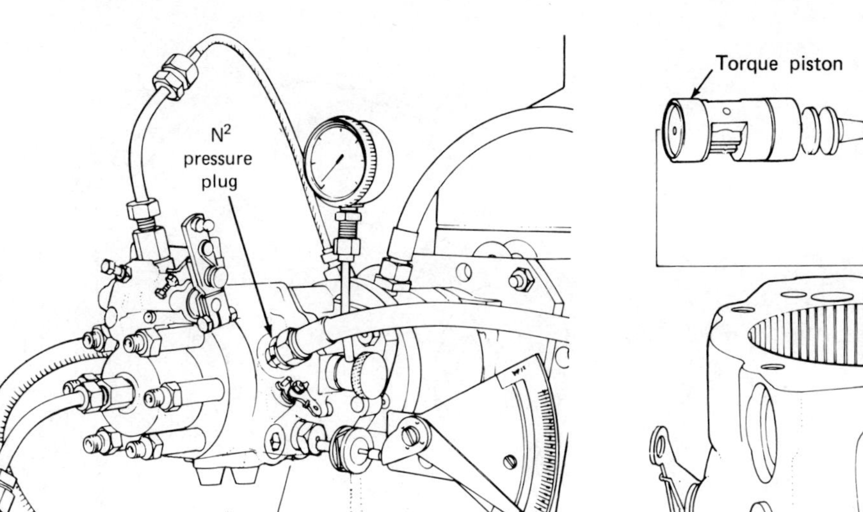

Figure 20.67 N^2 pressure plug (courtesy CAV–Lucas).

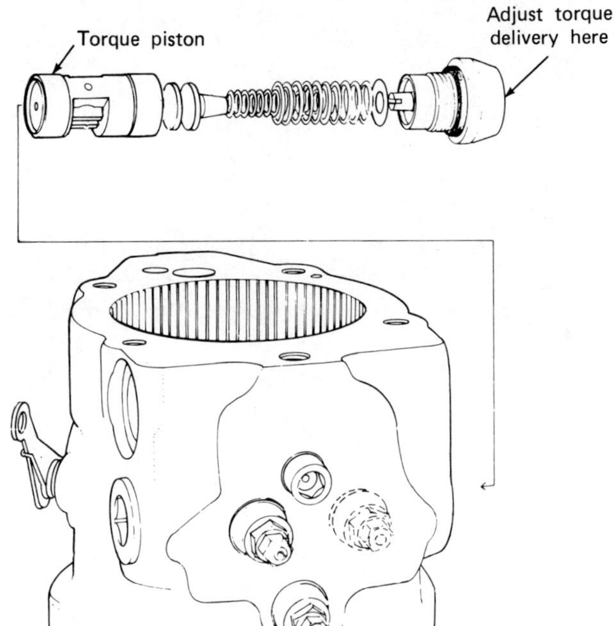

Figure 20.69 Setting torque fuel delivery (courtesy CAV–Lucas).

190 to the advance unit. Run the pump at the indicated speed for advance operation. If advance is incorrect, adjust the set screw in the advance piston test plug (Figure 20.70).

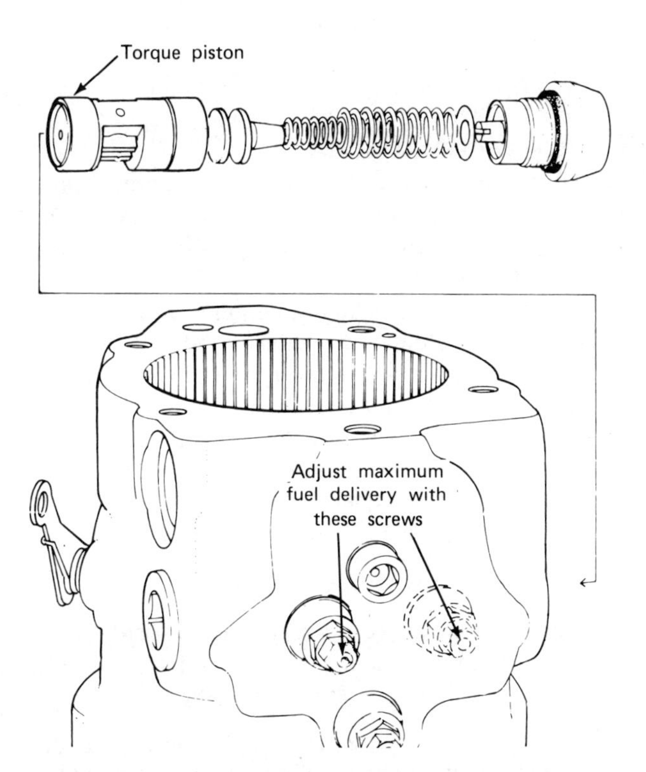

Figure 20.68 Setting maximum fuel delivery (courtesy CAV–Lucas).

9 *Maximum speed set.* Run pump at the speed listed in the test plan and adjust the high idle stop screw (Figure 20.71) to obtain the correct fuel delivery. Screw inward to reduce delivery.

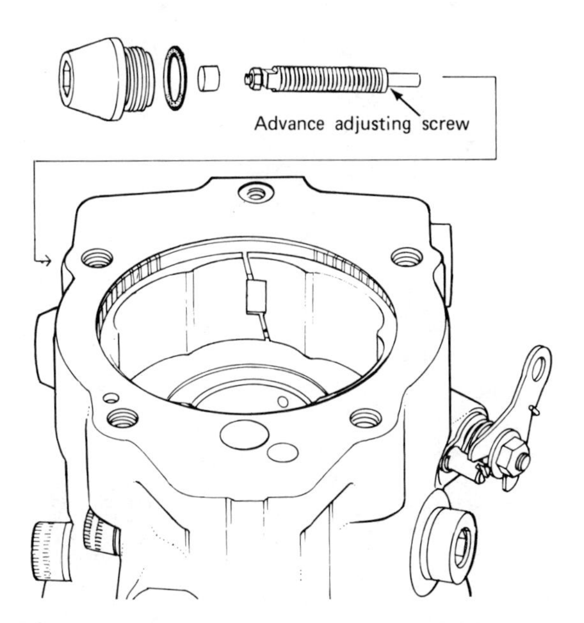

Figure 20.70 Adjusting advance movement (courtesy CAV–Lucas).

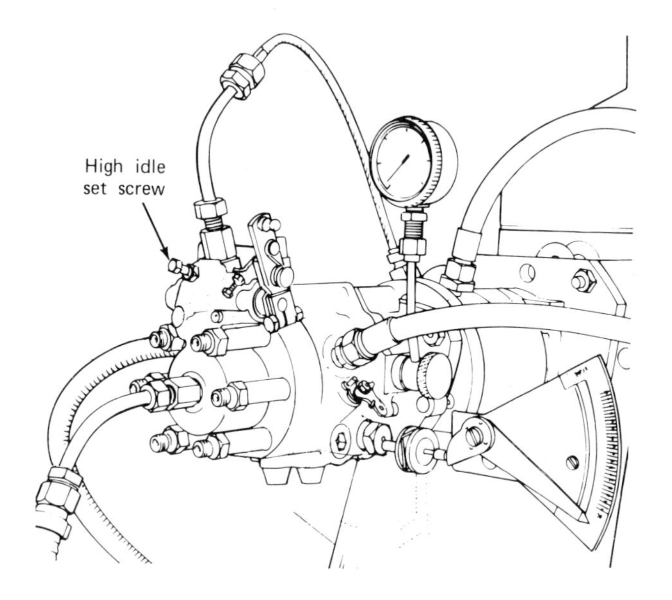

Figure 20.71 Setting high idle (courtesy CAV–Lucas).

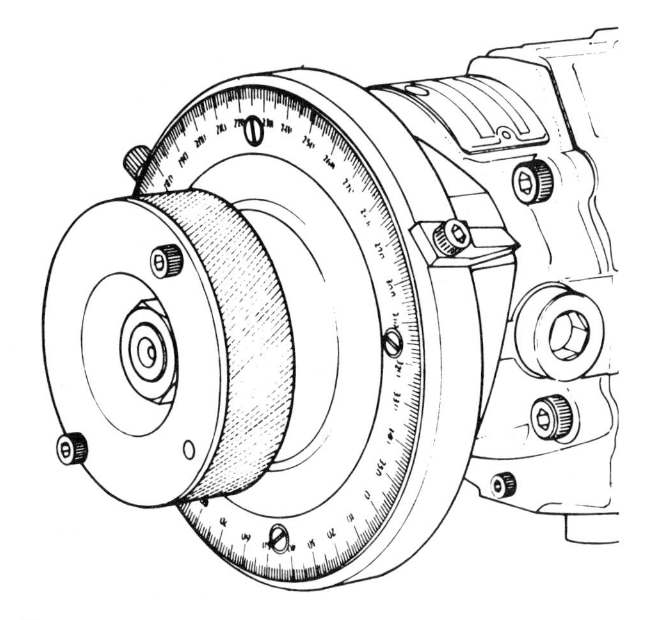

Figure 20.73 Timing the pump (courtesy CAV–Lucas).

10 *Low idle set.* Run pump at idling speed and set fuel delivery with the low idle screw (Figure 20.72). Screw in to increase delivery.

11 *Sealing.* Attach all plastic and aluminum seals with O rings. Run the pump for 10 minutes at 700 rpm and check for leaks.

12 *Timing the pump.* Place tool no. 7244-27 on the drive shaft (Figure 20.73) and set the degree wheel to the degree listed on the test plan. Supply fuel to the pump at 5 psi (0.35 kg/cm^2). Rotate the pump until a resistance is felt. At this point pencil mark a line above the pointer of the degree wheel. Recheck the timing mark and scribe a permanent line across the flange.

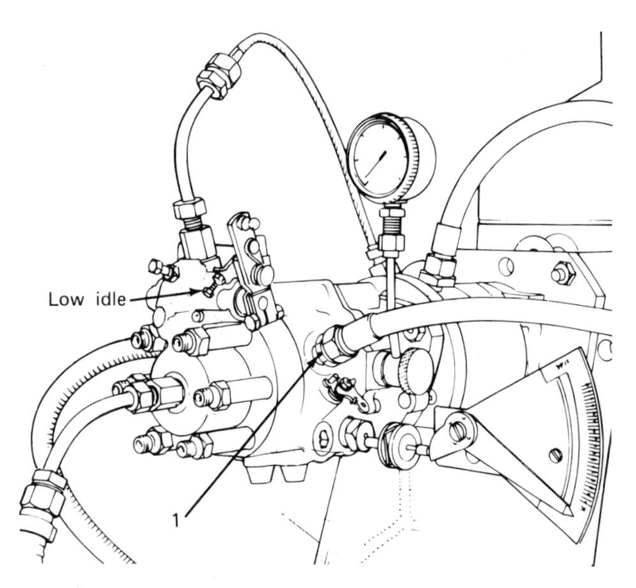

Figure 20.72 Setting low idle (courtesy CAV–Lucas).

With compressed air, dry pump thoroughly. Then plug all holes with plastic caps and paint the pump with the correct color.

IV. CAV-Simms Pumps

The CAV Minimec and Simms fuel injection pumps are very similar. The Simms will be covered here. The same information can be directly applied to the Minimec pump (Figure 20.74).

The Simms pump is an in-line pump having an individual pumping element for each cylinder. These elements are cam-operated and spring-returned. A mechanical governor is mounted at the front (drive end) of the pump and directly controls the fuel delivery of each pump element through a control rod. The pump can be fitted with a boost control, automatic timing advance, and an excess fuel device to aid starting.

A *Component parts* (Figure 20.75).

1 *Pump housing* (Figure 20.75,26). Contains all working parts, acts as a reservoir for lube oil, and has an engine mounting flange.

2 *Camshaft* (Figure 20.75,28). Actuates the pumping plungers up and down and carries the governor weights and the drive mechanism.

Figure 20.74 Simms in-line pump.

3 *Pumping barrel and plunger* (Figure 20.75,60). The matched barrel and plunger assembly does the high pressure pumping, controls fuel quantity delivered, and keeps high pressure fuel sealed from the pump reservoir.

4 *Tappet assembly* (Figure 20.75,59). The tappet assembly transfers the motion of the camshaft to the pumping plunger. It also carries a pad (shim) that can be exchanged to correct port closure (beginning of injection).

5 *Delivery valves*. The delivery valves (one for each element) serve two purposes. These are to seal the injection line between injections and to lower line pressure sufficiently to eliminate nozzle dribble.

6 *Feed pump*. The feed or lift pump pulls fuel from the main supply tank and sends it to the fuel gallery of the injection pump. This serves to keep the gallery free of air and at a slight pressure (about 22.5 psi) (1.57 kg/cm^2).

7 *Governor* (Figure 20.75,56). Various governors (mechanical and hydraulic) keep the engine running at the proper speed as dictated by load conditions and throttle position.

8 *Boost control (aneroid)*. A boost control is used to limit or prevent excessive smoke during engine acceleration (on turbocharged engines). The boost unit is controlled by intake manifold pressure.

B *Principles of operation*.

With in-line injection pumps, all pumping and fuel control occurs at the plunger and barrel assembly. Since these two parts are precisely ground, they do not require any sealing rings. They must be replaced as a set.

The camshaft provides a constant stroke of the plunger regardless of speed. The plunger is rotated indirectly by the governor to provide changes in fuel delivery. Through the center of the plunger (and approximately one-third of its length) is a drilled passage. Near the top is a helix. The barrel has a control port where fuel enters from the pump gallery (Figure 20.76).

When the pumping plunger is at bottom dead center (BDC), fuel is forced by the lift pump into the pump gallery. It enters the intake port (Figure 20.77,A) and fills the area above the plunger. The plunger is now forced upward by the camshaft. After a slight upward movement, the top edge of the plunger covers the intake port (Figure 20.77,B). This is known as port closure. Continued upward movement will force fuel past the delivery valve (Figure 20.77,C) and to the nozzle.

Injection will continue until the plunger has risen far enough to enable the lower edge of the helix to uncover the inlet port (Figure 20.77,D). At this time pressurized fuel will rush down the center of the plunger and exit through the inlet (now called bypass) port. This is the end of delivery. No more fuel will be injected even though the plunger is still being forced upward.

The amount and duration of injection, as mentioned earlier, is controlled by a helix, the position of which is controlled by the governor. The governor causes the control rack to move back and forth (linear movement). This rotates the plungers via the fork on the end of the plunger. When the plunger is turned, as in Figure 20.77,C, fuel delivery is at maximum since the plunger must move the greatest distance from port closure to uncover the lower helix. In low idle position (Figure 20.77,D), movement upward of the plunger is much less before the helix uncovers the bypass port. As a result, less fuel is delivered.

NOTE For a more detailed explanation of port and helix fuel control and different types of helixes used, see Chapter 15.

When no fuel is required (shutoff), the plunger is rotated to position (Figure 20.77,F). In this position the top of the plunger and the helix both are open to the inlet (bypass) port. No pressure can be built up and no fuel injected.

The delivery valve (Figure 20.78) acts as a check valve to help keep fuel in the line between injections. In addition, the valve relieves pressure

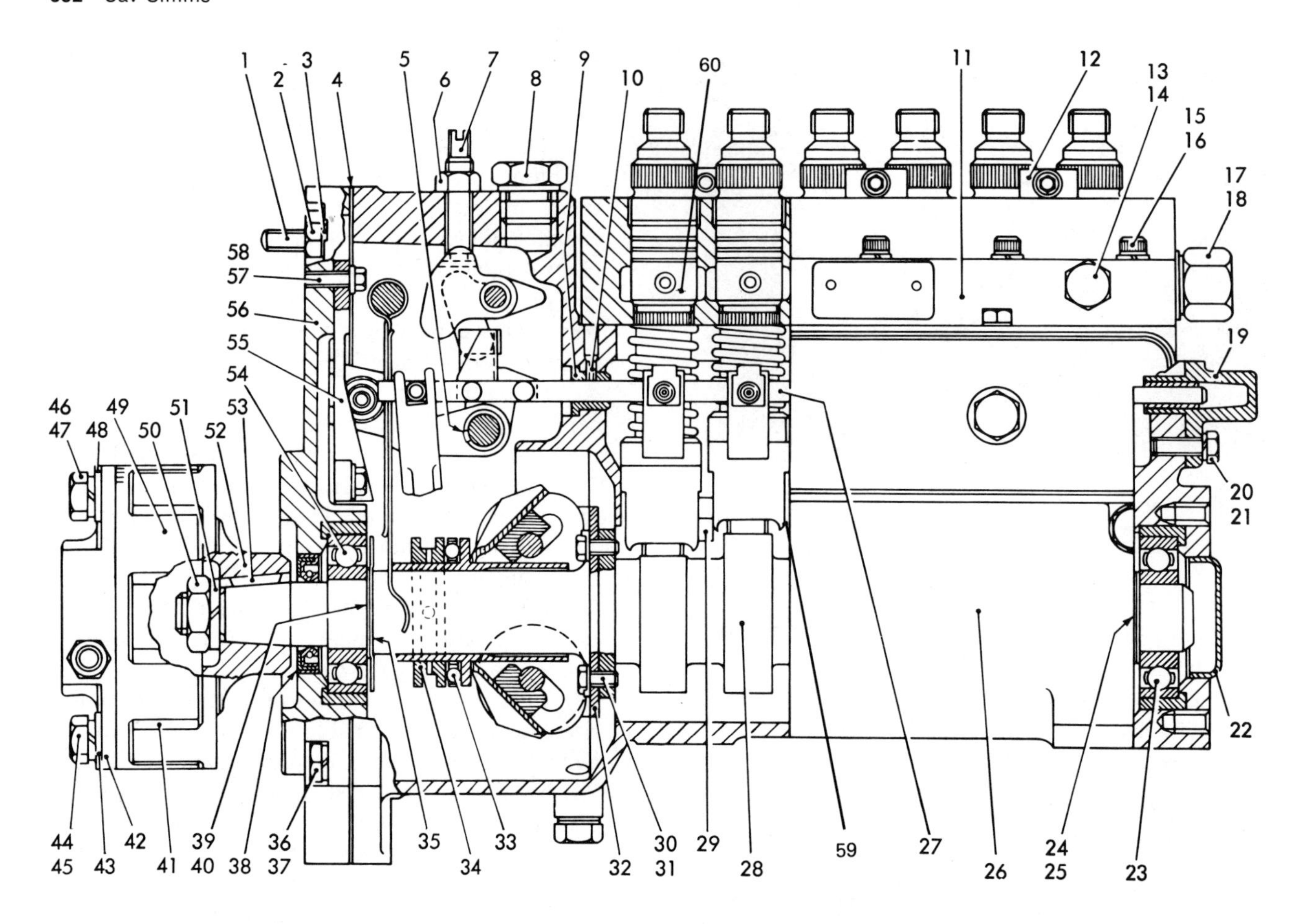

1 Stud
2 Nut
3 Spring Washer
4 Gasket
5 Woodruff Key
6 Nut
7 Max. Fuel Stop Screw
8 Oil Filler Plug
9 Control Rod Bush
10 Groverlok Pin
11 Pump Body
12 Clamp
13 Air Vent Screw
14 Joint Washer
15 Screw
16 Spring Washer
17 Fuel Inlet Adaptor
18 Joint Washer
19 Control Rod Cover
20 Screw
21 Spring Washer
22 Expansion Plug
23 Ball Bearing
24 Shim (0.1 mm.)
25 Shim (0.2 mm.)
26 Pump Unit Housing
27 Control Rod
28 Camshaft
29 Tappet Locating T-Piece
30 Backplate Screw
31 Tab Washer
32 Backplate
33 Thrust Bearing
34 Thrust Pad
35 Baffle Washer
36 Screw
37 Spring Washer
38 Oil Seal
39 Shim (0.1 mm.)
40 Shim (0.2 mm.)
41 Insert
42 Driving Flange
43 Clamp Plate
44 Dowel Screw
45 Spring Washer
46 Screw
47 Spring Washer
48 Clamp Plate
49 Dog Flange
50 Camshaft Nut
51 Spring Washer
52 Pump Flange
53 Woodruff Key
54 Ball Bearing
55 Ramp
56 Governor Cover
57 Screw
58 Lock Washer
59 Tappet Assembly
60 Pumping Plunger/Barrel

Figure 20.75 Component parts of Simms pump (courtesy CAV–Lucas).

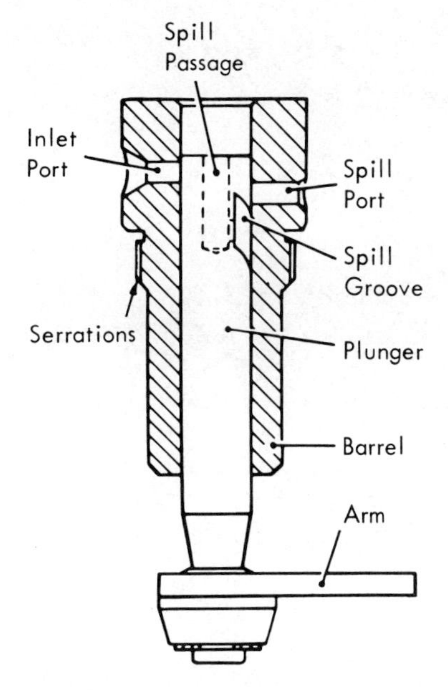

Figure 20.76 Pumping plungers with barrel (courtesy CAV–Lucas).

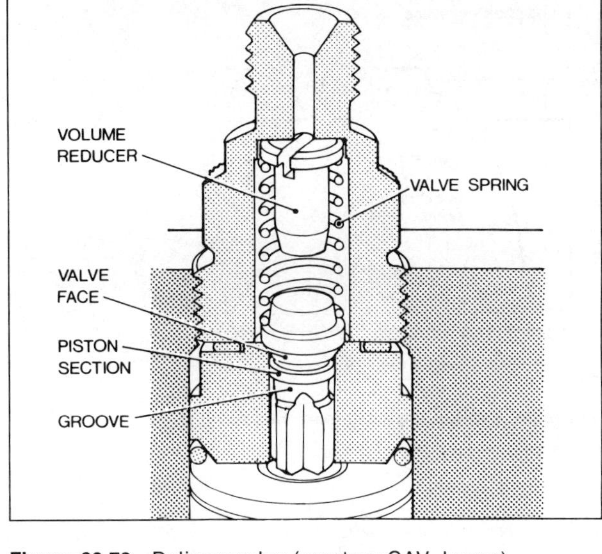

Figure 20.78 Delivery valve (courtesy CAV–Lucas).

in the line after each injection to insure immediate closing of the nozzle. After injection is completed, pressure drops and the valve will close with the aid of the delivery valve spring.

A graph of pressure changes in the injection line with and without a delivery valve is given in Figure 20.79.

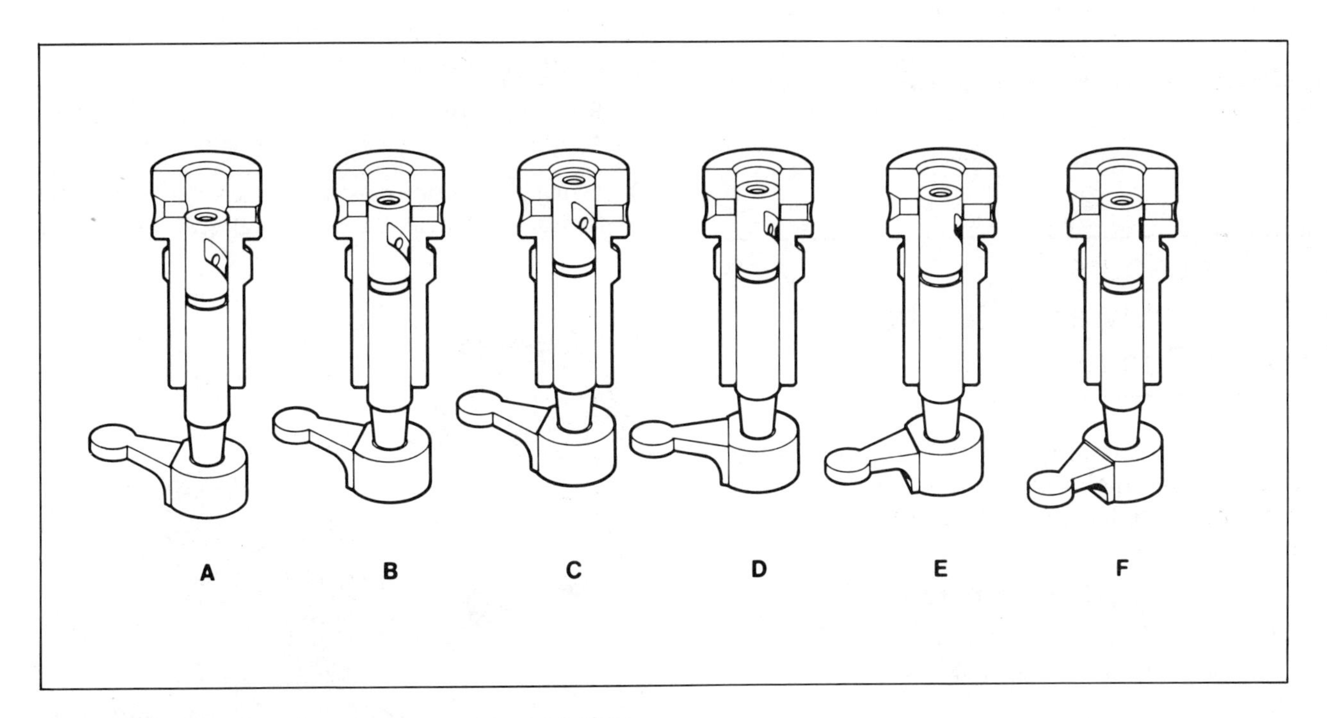

Figure 20.77 Operation of pumping plunger (courtesy CAV–Lucas).

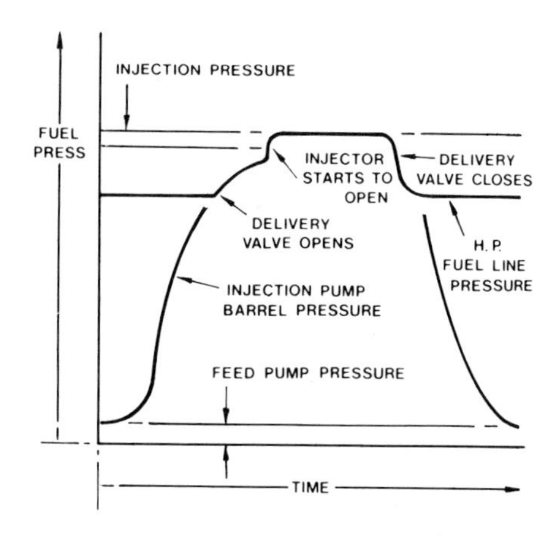

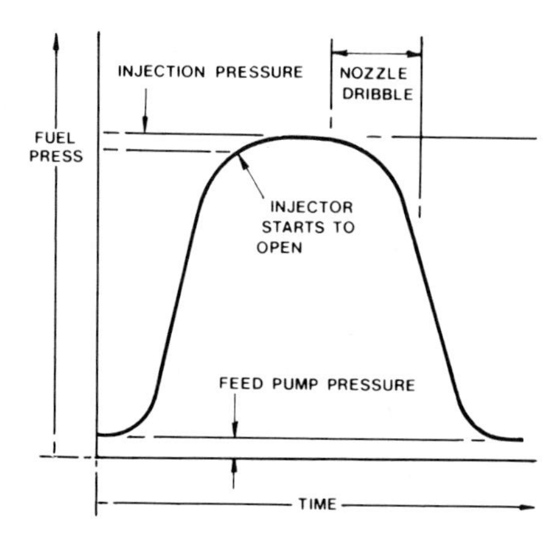

Figure 20.79 Pressure changes during operation (courtesy CAV–Lucas).

C. Procedure for Disassembly and Overhaul of In-Line Pumps

Before starting disassembly of the injection pump, thoroughly clean the exterior and drain the lube oil. Also record all model and serial numbers on the work order.

1 Lock the pump in a vise that has brass jaws to avoid marring the pump.

2 Remove side inspection cover.

3 Remove the screws securing the pump body to the pump housing (Figure 20.80).

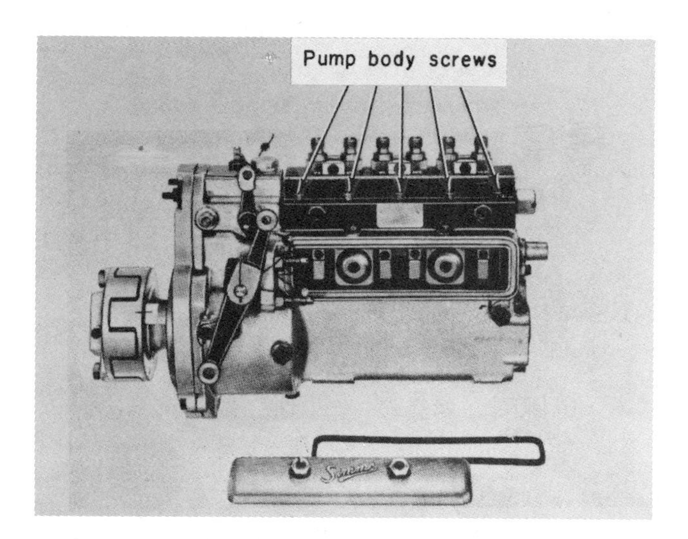

Figure 20.80 Removing pump body screws (courtesy CAV–Lucas).

4 Lay the pump on its side and carefully remove the pump body (Figure 20.81).

5 Place all the plungers, springs, and spring discs in a divided tray to keep all matched parts together. This will aid in reassembly and testing.

6 Clamp the pump body in the brass jawed vise and remove the delivery valve holder clamps and holders.

NOTE A special serrated socket must be used to remove the holder or it will be damaged.

7 Remove springs and volume reducers (Figure 20.82).

8 Using a rubber or plastic hammer, tap on the bottom of the barrel. Catch the delivery valves and

Figure 20.81 Removing pump body (courtesy CAV–Lucas).

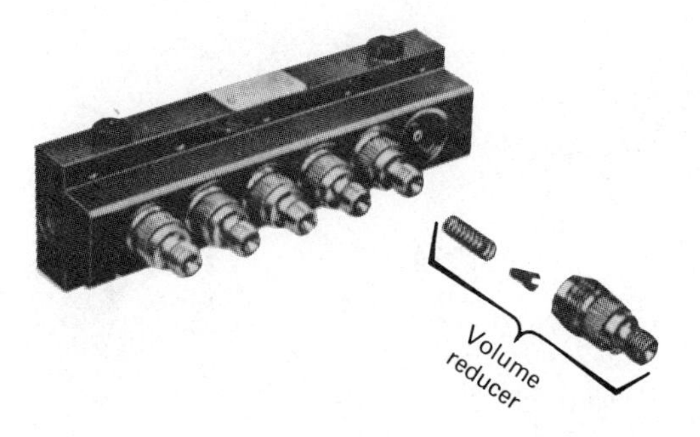

Figure 20.82 Removing volume reducers (courtesy CAV–Lucas).

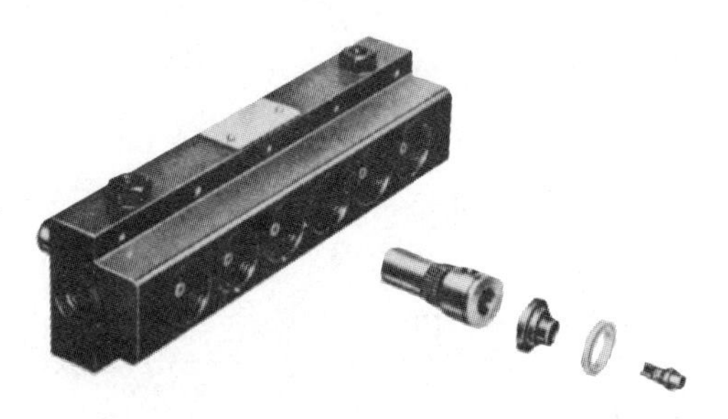

Figure 20.83 Delivery valves and barrels out of housing (courtesy CAV–Lucas).

barrels as shown in Figure 20.83. Keep them in the divided tray.

9 Lift out the tappets and place in the tray.

10 Remove the drive coupling by attaching the special holding wrench and removing the camshaft nut. Use a puller as shown to remove the coupling (Figure 20.84).

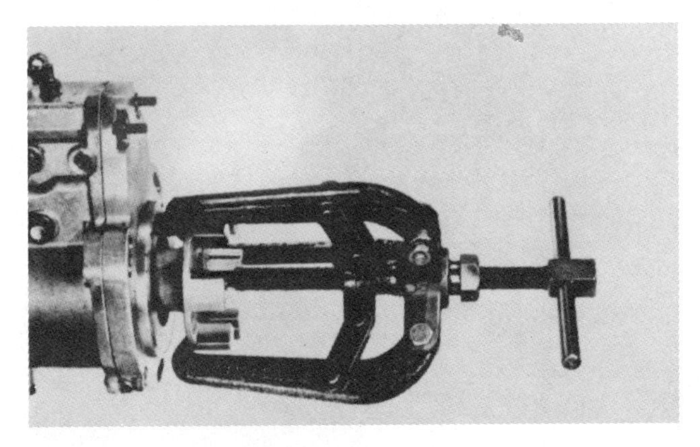

Figure 20.84 Removing drive coupling (courtesy CAV–Lucas).

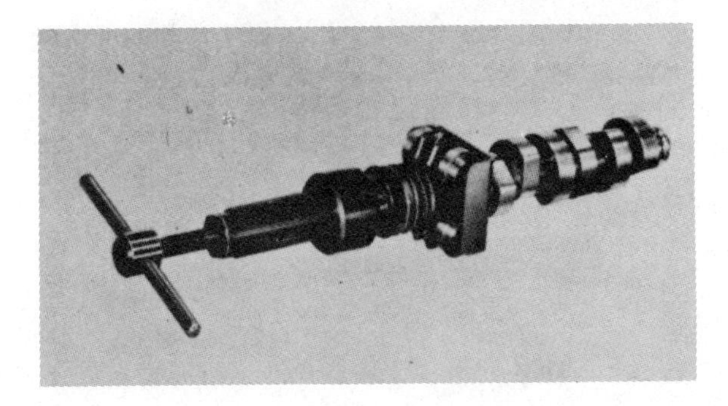

Figure 20.85 Removing bearing races (courtesy CAV–Lucas).

11 Loosen and remove the governor cover and roller assembly from the control lever.

12 Remove the fulcrum pin and governor springs.

13 Remove rocking lever and thrust washer.

14 Pull out the camshaft and governor weight assembly.

15 Pull off the bearing ball cages.

16 Using an appropriate tool, pull the inner ball races off the camshaft as shown in Figure 20.85.

17 Slide off the baffle and thrust washers and governor weight assembly.

18 Remove the speed control shaft.

19 Remove maximum stop lever and excess fuel shaft.

20 Drive out the sealing plug (from pump housing) if it shows signs of leaking.

21 Unless obviously damaged, do not remove the control rack or forks.

22 Remove oil seal from the front governor cover.

23 Remove outer ball bearing races from the governor cover and pump housing.

D. Procedure for Inspection of Pump Components

After disassembly is complete, clean all pump parts in a parts cleaning solution or in solvent to remove all paint, varnish, and dirty oil. Rinse the parts in clean solvent and blow dry. Inspect all parts for excessive wear as outlined below:

1 *Pump housing*. Check the housing for cracks or thread damage.

2 *Plunger/barrel assemblies.* Check the plunger for scoring, which can be felt with the fingernail. If it can be felt this way, the element should be replaced.

3 *Camshaft.* Check the lobes carefully for flaking of material, which indicates that the hardened surface is breaking up.

4 *Camshaft bearings.* The camshaft bearings are normally replaced during a major overhaul of the pump.

5 *Roller tappets.* Inspect the tappets for excessive wear on the rollers and pin. Remove any burrs with emery cloth.

6 *Delivery valves.* Check the tapered seat of the valve for pitting and excessive wear. Also check the relief piston of the valve for deep scoring.

7 *Springs.* Check the plunger and delivery valve springs for rust and warping. Replace as needed.

E. Procedure for Reassembling the In-Line Pump

1 *Always* replace all gaskets, seals, and O rings when overhauling an injection pump.

CAUTION Keep all parts spotlessly clean during assembly. Rinse parts and O rings in clean calibrating oil before using.

2 Install new outer ball bearing races in the pump housing and governor cover.

3 Assemble the ramp to the governor cover.

Figure 20.87 Installing shims and camshaft bearings (courtesy CAV–Lucas).

NOTE Use a sealing compound on the threads.

4 Install stop control shaft and excess fuel shaft.

5 Install the roller control lever in the pump housing (Figure 20.86).

6 Place the governor weight assembly and thrust sleeve on the camshaft. Then install the shims and bearings (Figure 20.87).

7 Place the camshaft in the pump housing. Place the governor cover on and tighten the screws.

8 Check the end play of the camshaft as shown (Figure 20.88). The end play should be 0.05 to 0.127 mm (0.020 to 0.050 in.). Add or deduct shims to obtain the correct play.

NOTE Always divide the shims equally from end to end.

9 Remove the governor cover and install the pivot pin, governor springs, and fulcrum pin.

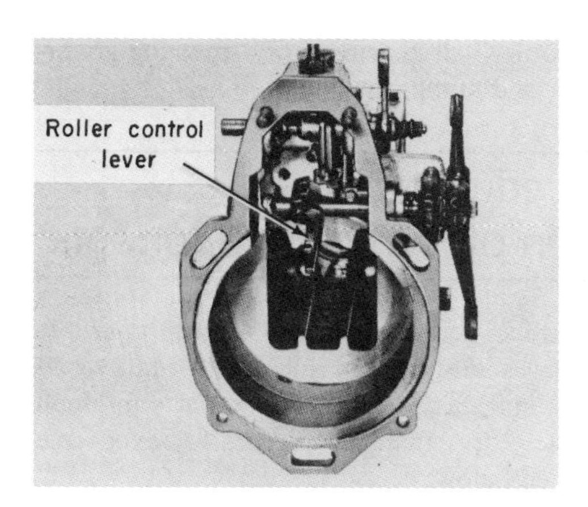

Figure 20.86 Installed roller control lever (courtesy CAV–Lucas).

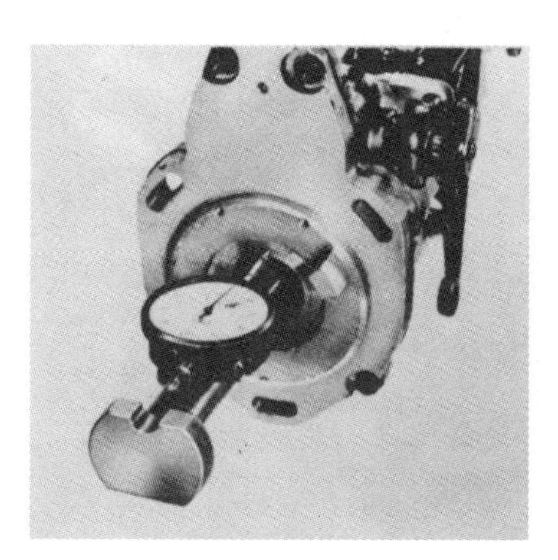

Figure 20.88 Checking camshaft end play (courtesy CAV–Lucas).

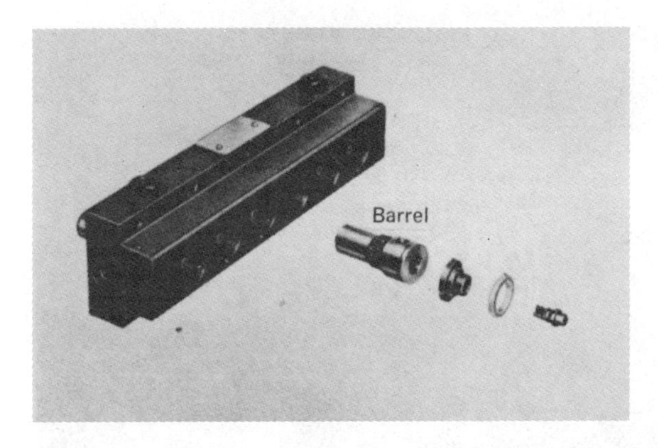

Figure 20.89 Locating the barrel in the pump body (courtesy CAV–Lucas).

Figure 20.90 Phasing the pump (courtesy CAV–Lucas).

10 Install the roller assembly on the roller control lever. Install the governor cover without the oil seal. After the cover is on, tap the oil seal into the governor cover.

11 Install the drive coupling and torque to 59.66 N·m (44 ft lb).

12 Install the roller tappets into the pump housing.

CAUTION Make certain the tappets are returned to their original bore.

13 Fit the barrels back into their correct position in the pump body. The missing tooth must line up with the missing tooth in the pump body (Figure 20.89).

14 Install the delivery valves, joint rings, springs, volume reducers, and delivery valve holders. Torque the holders to 54.23 N·m to 61.01 N·m (40–45 ft lb).

15 Install the pumping plungers in the pump body and install the body in the housing, making sure to engage the plunger arms with the control forks.

NOTE Further assembly will include phasing and port closure.

16 Mount the pump to the test bench. Remove the delivery valve, spring, and volume reducer from the no. 1 element. Install the special plunger lift gauge in place of the delivery valve holder (Figure 20.90).

17 Rotate the pump until the no. 1 plunger is at BDC. Zero the dial indicator. Supply calibrating oil to the pump at 0.14 kg/cm^2 (2 psi).

18 Continue to rotate the pump in the direction of rotation until fuel flow ceases.

NOTE This is port closure or commencement of delivery. The travel shown on the lift gauge must correspond to the value listed on the fuel setting data sheet.

19 If the lift is incorrect, the shim of the roller tappet must be exchanged. Record the variation so the shim may be exchanged when the pump is disassembled. If correct, set the degree wheel of the test bench to zero.

20 Shut off the fuel supply. Remove the lift gauge and install the delivery valve components and torque holder.

21 Remove the delivery valve and spring from the next element in the firing sequence. Install a curved tube to the delivery valve holder.

22 Turn on the fuel supply and rotate the pump. Watch the degree wheel. The fuel must stop running when the wheel indicates 45° (8 cylinder), 60° (6 cylinder), 90° (4 cylinder), 120° (3 cylinder). Record any variation.

23 Check the remaining elements in a similar manner.

24 Remove the pump body and install shims of the correct thickness in the roller tappets. Changing shim thickness 0.1 mm is equal to $\frac{1}{2}$° of pump rotation. Use the recorded degree values of obtain the correct shim.

25 Lay the pump body on its side and install the plungers with springs and spring seats.

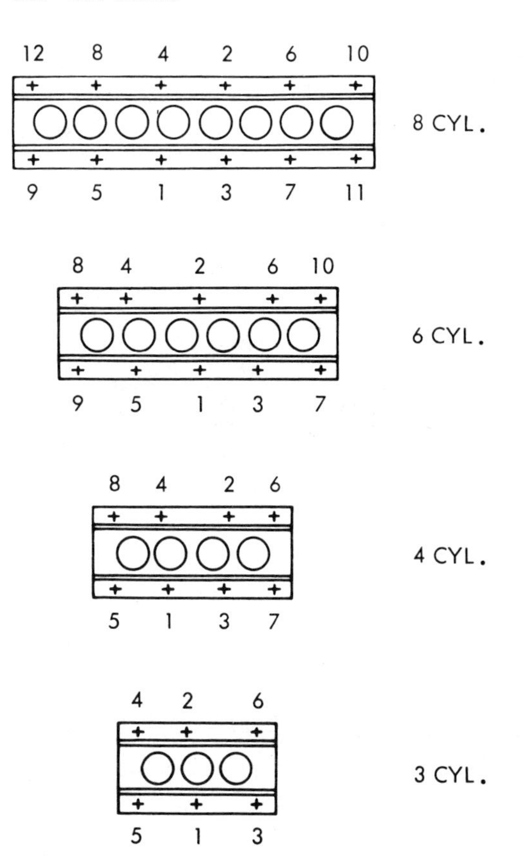

Figure 20.91 Torque sequence for pump body screws (courtesy CAV–Lucas).

26 Apply the sealer to the mating surface of the pump housing and install the body.

CAUTION Make certain to engage the plunger arms in the control forks.

27 While pushing down on the pump body, install and snug up the retaining screws.

28 Torque the screws in correct sequence to 6.78 N·m (5 ft lb) (Figure 20.91).

29 Install the delivery valve holder clamps.

F. Procedure for Calibration of In-Line Pumps

After overhaul, all pumps must be mounted on a test bench for calibration.

Since these pumps are lubricated by engine oil, make sure to add oil to the pump before operation.

Figure 20.92 Setting equal fuel delivery.

Run the pump for 10 minutes to remove air from the lines. Repair all leaks.

The following tests are to be performed on Simms CAV in-line pumps:

1 *Equal fuel delivery.* Set the rearmost control fork to the dimension given on the data sheet. This is usually 0.5 mm (0.020 in.) and is measured from the square section of the control rod (Figure 20.92). Tighten the fork screw. Run the pump and record delivery from this element. Adjust all remaining control forks to give exactly the same delivery.

NOTE All fuel checks are for 200 shots (strokes).

2 *Maximum fuel delivery.* Move the speed control lever to high speed position. Run the pump at the speed specified on the data sheet for maximum fuel. Adjust the maximum fuel screw (Figure 20.93) until the correct amount of fuel is obtained.

3 *Low idle.* Run the pump at specified speed with speed control lever in the low idle position. Adjust the low idle stop screw to get the correct delivery (Figure 20.94).

4 *High idle.* Place speed control lever in maximum speed position. Run pump at high idle and adjust delivery with the high speed stop screw (Figure 20.95).

5 *Excess fuel (starting delivery).* Move the excess fuel shaft (Figure 20.96) to give increased fuel. The quantity delivered must equal or exceed that given on the data sheet.

CAUTION This test is for 100 shots (strokes).

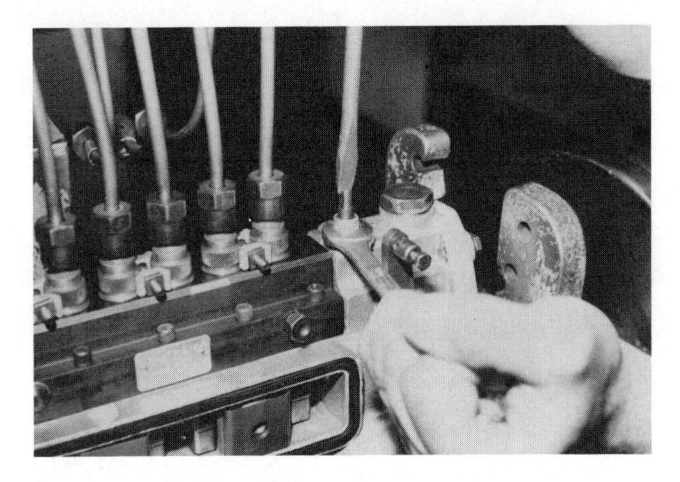

Figure 20.93 Adjusting maximum fuel.

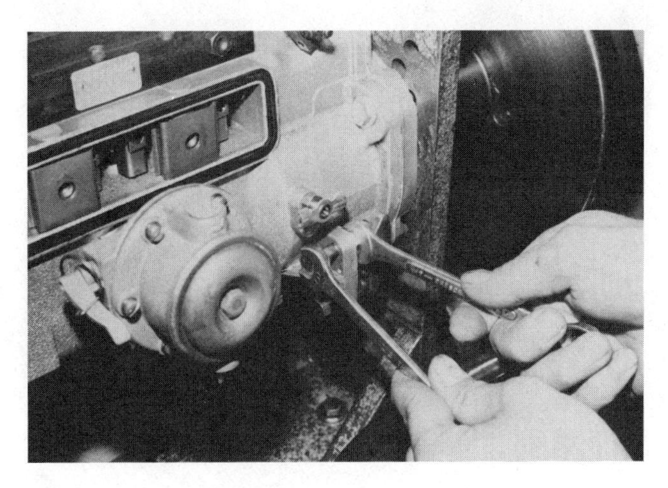

Figure 20.95 Setting high idle.

6 *Governor cutoff/breakaway.* The control rod should begin moving to the stop position at the speed listed. Delivery should then cease at a somewhat higher speed.

> **CAUTION** To prevent engine damage, delivery must cease at speed listed in test data.

7 *Leak test.* After all calibration tests are completed, install the side cover and torque the screws to 9.49 N·m (7 ft lb). Run the pump at 500 rpm and check all connections for leakage of fuel or oil.

Install sealing wires, remove from test bench, clean, plug all openings with plastic caps, and paint with the correct color.

V. Troubleshooting the CAV and Simms Pumps

A While the pump is on the test bench, certain problems may arise because of incorrect testing conditions or equipment. So before troubleshooting the injection pump, check the following:

1 Are the test nozzles at the correct opening pressure?

2 Are the correct nozzles being used?

3 Is the calibrating oil at the correct temperature?

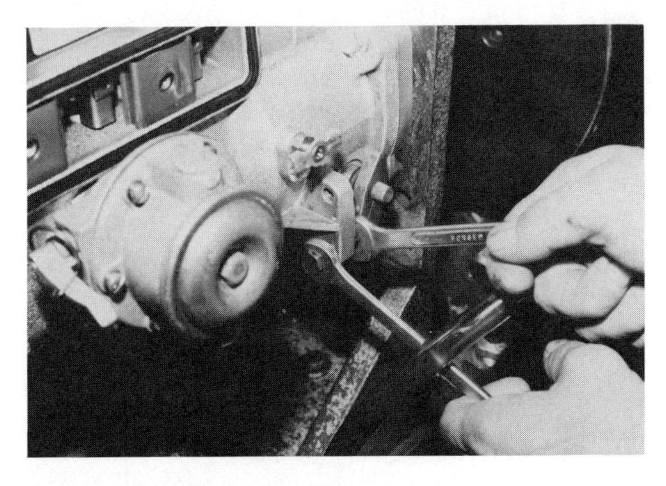

Figure 20.94 Adjusting low idle.

Figure 20.96 Excess fuel control shaft.

4 Are the correct injection lines being used?

5 Are the correct gauges and adapters for the pump being tested in use?

B If a problem still exists with the pump on the test bench, check the following when:

1 Full load delivery is low or uneven after setting high idle (DPA):

- **a** Governor link length incorrect.
- **b** Wrong governor spring being used.
- **c** Governor spring weak.
- **d** Metering valve worn.

2 Governor cutoff cannot be achieved:

- **a** Governor link length incorrect.
- **b** Worn metering valve.
- **c** rive hub screw loose (except up-rated drive).
- **d** Hydraulic head worn.

3 Maximum fuel delivery at cranking speed cannot be obtained:

- **a** Transfer pump pressure low.
- **b** Maximum fuel delivery low.
- **c** Control lever not in wide open position.
- **d** Hydraulic head worn.

4 Correct advance values cannot be obtained:

- **a** Low transfer pump pressure.
- **b** Checking device not correctly zeroed.
- **c** Defective or incorrect advance spring.
- **d** Worn advance piston.
- **e** Incorrect number of adjustment shims.

C If a problem exists in engine operation after the pump has been installed, check the following when:

1 Engine will not start:

- **a** Tank valve off.
- **b** Fuel filters plugged.
- **c** Hand throttle in idle position or shutoff lever in off position.
- **d** Pump timed incorrectly.

2 Engine starts hard:

- **a** Fuel filters plugged.
- **b** Throttle in idle position.
- **c** Pump timed incorrectly.
- **d** Hydraulic head of pump worn excessively.

3 Low power:

- **a** Dirty air or fuel filters.
- **b** Shutoff lever in partial off position.
- **c** Advance unit not working.
- **d** Pump timed incorrectly.
- **e** Roller-to-roller dimension incorrect.
- **f** Shuttle valve sticking (DP-15).
- **g** Worn governor components.

4 Engine overspeeds:

- **a** Metering valve sticking.
- **b** Governor adjusted incorrectly.
- **c** Drive hub screw loose or broken (DPA).
- **d** Incorrect metering valve.

5 Engine smokes black:

- **a** Roller-to-roller dimension incorrect.
- **b** Dirty injection nozzles.
- **c** Dirty air cleaner.
- **d** Pump timing incorrect.

6 Engine smokes white or blue:

- **a** Advance unit not working.
- **b** Pump timed to engine incorrectly.
- **c** Low compression.
- **d** Engine burning oil.
- **e** Port closure incorrect (Simms).

7 Engine runs rough or irregular:

- **a** Air in supply or injection lines.
- **b** Sticking shuttle valve or metering valve.
- **c** Unequal delivery due to adjustment (Simms pumps) or worn hydraulic head (distributor pump).
- **d** Dirty or sticking nozzles.

SUMMARY

This chapter has presented a complete study of CAV and Simms fuel pumps. With this information and careful work, the mechanic should realize much success in repairing injection pumps. It is wise to proceed, however, only when the required special tools and spare parts are available. Always work in a clean area. Use common sense and good work habits. After you have worked on several pumps, you will know and understand the basic overhaul procedures without referring to the chapter. However, always keep up to date by reading service bulletins and attending factory authorized service schools.

REVIEW QUESTIONS

1 The CAV–DPA pump uses:

- **a** Inlet metering.
- **b** Sleeve control.
- **c** Port and helix.
- **d** None of the above.

2 Testing transfer pump pressure is done by: (DPA)

- **a** Connecting the pressure gauge to the bleed port on the side of the pump.
- **b** Connecting the gauge to the fuel inlet.
- **c** Connecting the gauge to the bottom of the transfer pump.
- **d** None of the above.

3 High idle is adjusted by: (DPA)

- **a** Changing the control linkage.
- **b** Adjusting the transfer pump pressure.
- **c** Adjusting the high idle stop screw.

4 *True or false:* The CAV DP-15 has one delivery valve for line retraction.

5 *True or false:* The set code stamped on the pump name tag gives needed information about calibration of the DPA pump.

6 The main purpose of the shuttle valve on the DP-15 is:

- **a** Even delivery.
- **b** More fuel delivery at low idle.
- **c** More time to fully charge the plungers.
- **d** All of the above.

7 The roller-to-roller dimension is adjusted on a DPA pump by:

- **a** The leaf spring.
- **b** The shoe adjusting plates.
- **c** The torque screw.
- **d** The twin shuttles.

8 What type of fuel control does the in-line Simms pump use?

- **a** Port and helix.
- **b** Inlet metering.
- **c** High pressure inlet.
- **d** Sleeve metering.

9 Equal delivery on Simms pumps is obtained by:

- **a** Changing control segments.
- **b** Adjusting the full load screw.
- **c** Changing the plunger lift.
- **d** Adjusting the control forks.

10 *True or false:* All DP-15 pumps use a twin shuttle arrangement.

11 Describe N^2 pressure as it is used on the DP-15 pump.

12 If the inlet check valve in the rotor of the DP-15 pump leaks:

- **a** It should be removed and lapped with a fine paste to restore the seating surface.
- **b** It should be replaced.
- **c** Replace the pump body/rotor as an assembly.
- **d** None of the above.

13 What component stops the downward travel of the shuttle valve in the DP-15 pump?

- **a** The advance piston.

- **b** The torque piston.
- **c** The cam ring.
- **d** The end cap.

14 True or false: The Simms pump uses no delivery valves.

15 The drive adapter screw torque on the DPA is critical because:

- **a** A loose screw could lower the transfer pressure.
- **b** A loose screw can cause the loss of the governor control.
- **c** A loose screw reduces fuel delivery.
- **d** None of the above.

21

Roosa Master Injection Pumps

Stanadyne/Hartford (Roosa Master) has been producing distributor type fuel injection pumps since 1952. The original design idea came from Vernon Roosa, who was a diesel mechanic working primarily with generator sets.

Instead of using the in-line injection pump that had been in common use, Roosa devised a single pumping element to supply fuel to all cylinders. This was the first single cylinder, opposed plunger, distributor type injection pump. Today Stanadyne/Hartford produces many different distributor pumps, pencil nozzles, single and dual element fuel filters, water separators, and electric fuel pumps.

OBJECTIVES

Upon completion of this chapter the student will be able to:

1. Explain fuel flow and pump operation in two Roosa Master pumps (student's choice).
2. Demonstrate the ability to correctly name all parts in two Roosa Master pumps (student's choice).
3. Demonstrate the ability to correctly disassemble and inspect a Roosa Master pump (student's choice).
4. Select from a pump component parts list the parts that are normally replaced during a pump overhaul (student's choice of pump).
5. Demonstrate the ability to correctly test and calibrate two Roosa Master pumps (student's choice).

GENERAL INFORMATION

The A model pump (Figure 21.1) was the first from Roosa Master. It used a single rotor turning within a hydraulic head, inlet metering (measuring the fuel on the inlet of the pump), and a single cylinder with two opposed pistons. The A pump went into full scale production in 1952 on Hercules equipment. Although later replaced by the B and D models, the basic design of the A pump is still used by CAV, Ltd., of England.

Figure 21.1 A model Roosa Master pump (courtesy Stanadyne).

CAV signed an agreement with Roosa Master in 1953 that gave CAV the right to produce and market pumps of Roosa Master design. In 1956 CAV introduced the DPA pump. The basic design of the A pump and its updated versions have been very popular around the world.

Introduced in 1956, the D model Roosa Master (Figure 21.2) was the first pump that showed the design of the current production DB. The D pump used a .750 in. (19 mm) rotor, horizontal throttle shaft (rather than the vertical shaft used on earlier models), and a simpler governor mechanism. In addition to the mechanical governor, a load advance mechanism was available.

Figure 21.2 D model Roosa Master pump (courtesy Stanadyne).

Figure 21.3 DB model Roosa Master pump.

The DB pump (Figure 21.3), first produced in 1958, used the design of earlier Roosa Master pumps and added many improvements. The rotor diameter was increased to 0.920 in. (23.35 mm). The transfer pump rotor was made as a part of the distributor rotor with a central delivery valve located in the distributor rotor to serve all cylinders. Accessories such as automatic advance and electric shutoff could be an integral part of the DB housing. The DB could be mounted either horizontally or vertically, since the housing was filled with diesel fuel. Still in use today, the DB can be used on engines of up to eight cylinders and can be fitted with automatic speed or load advance; torque control; electric shutoff; mechanical, electric, and variable droop (percent of regulation); governors; and aneroid control.

NOTE Roosa Master manufactures a slightly modified DB pump for John Deere called the JDB.

To meet the fuel requirements of high horsepower engines, Roosa Master introduced the DC pump in 1962. Comparable to the DB in many ways, the DC (Figure 21.4) used two cylinders and four opposed plungers to supply more fuel and an increased rate of injection. Because of increased driving torque of this pump, the DC used a splined drive shaft and distributor rotor.

In 1969 Roosa Master started supplying a new, compact pump model to industry. Still employing the rotary, single cylinder design used on earlier models, the C (and later the CB) model (Figure 21.5) was made extremely compact in part because of a smaller gover-

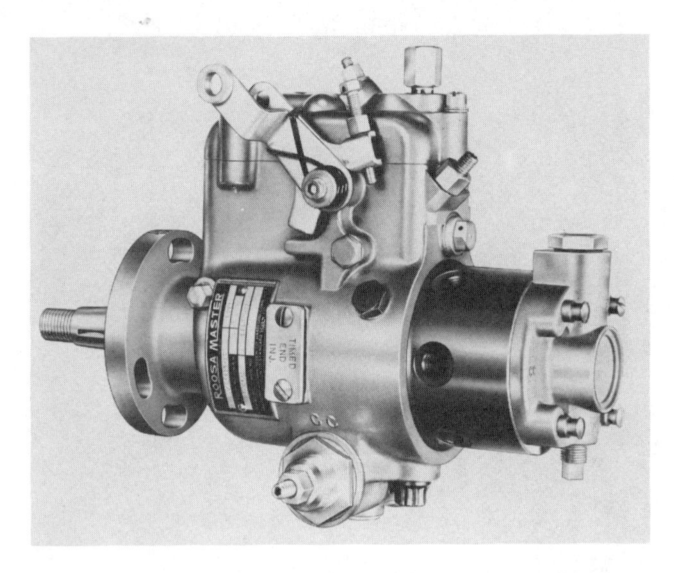

Figure 21.4 DC model Roosa Master pump (courtesy Stanadyne).

nor assembly and a rotating metering valve. The C and CB models were used mainly on John Deere equipment and are no longer produced.

The need for a heavier duty pump was met in 1972 when the DM pump was introduced. The DM (Figure 21.6) uses a highly modified hydraulic head with flared fuel line fittings on the rear of the head. This arrangement eliminates the extra parts of a banjo (a round connection at the end of the fuel injection line) connection. Also, a larger drive shaft and metering valve are used. A bearing rather than a pilot tube is used in conjunction with a three or four hole mounting flange. At present the DM is the largest pump produced by Roosa Master and is capable of over 8000 psi (560 kg/cm^2) line pressures. Like the DC model, the

Figure 21.5 CB model Roosa Master Pump.

Figure 21.6 DM model Roosa Master pump (courtesy Stanadyne).

DM is available in a four plunger version.

The latest pump to be introduced by Roosa Master (1977), the DB2, is already in widespread use on engines from GM. The DB2 (Figure 21.7) combines features of the older DB and the DM. Pumps for automotive use will feature a limiting speed governor (controls only low and high idle speeds). The DB2 is also available for other applications to engines having up to eight cylinders using many different governors.

I. Component Parts (DB, DC)

The Roosa Master DB is a single cylinder, the DC a twin cylinder opposed plunger, inlet metering, distributor type pump. The main components of the pump are:

Figure 21.7 DB2 model Roosa Master pump.

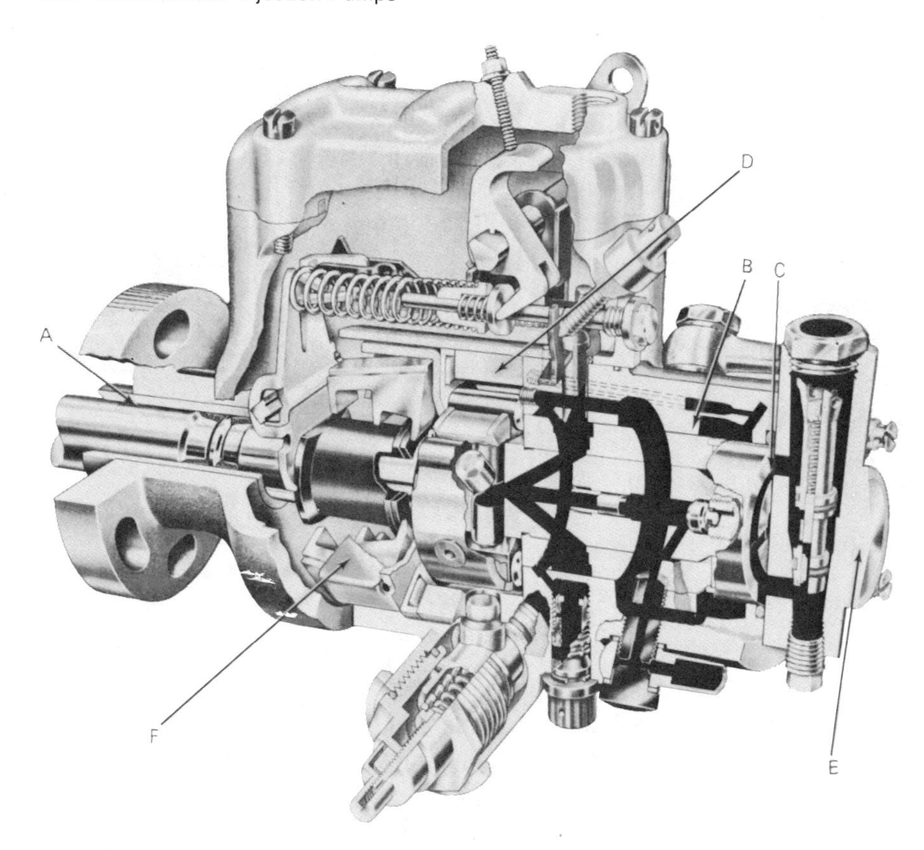

Figure 21.8 Cutaway view of DB pump (courtesy Stanadyne).

A *Drive shaft* (Figure 21.8,A). The drive shaft connects to the engine drive gear train and rotates the rotor of the hydraulic head, governor weight retainer, and transfer pump.

B *Hydraulic head and rotor assembly* (Figure 21.8,B). The hydraulic head is the main unit of the pump, containing the lapped and fitted rotor that has the pumping cylinder and drilled passages to distribute the fuel. Fuel is metered, distributed, and discharged by the hydraulic head.

C *Transfer pump* (Figure 21.8,C). Located at the rear of the hydraulic head, the transfer pump supplies fuel to the metering valve at approximately 80 psi (5.60 kg/cm^2). Additional fuel is fed to the automatic advance mechanism. Remaining fuel fills the pump housing and is returned to the fuel tank via the fuel return line.

D *Cam ring* (Figure 21.8,D). The cam ring has internal lobes, one for each engine cylinder, which force the pumping plungers inward at the correct moment, thus pressurizing and injecting fuel. If an automatic advance device is used, it is connected to the cam ring.

E *End plate* (Figure 21.8,E). The end plate not only seals the back of the pump but houses the transfer pump pressure regulating valve and the fuel inlet to the pump.

F *Governor* (Figure 21.8,F). The governor assembly consists of the weights, weight retainer, thrust sleeve, pivot shaft, governor arm, governor springs, and linkage. The sole function of the governor is to rotate the metering valve to the correct opening, depending on engine speed and load.

II. Principles of Operation

NOTE The front of the pump is the drive shaft end and the rear is the end plate end.

A *Fuel flow (charge cycle).* Fuel enters the inlet connection and passes through the strainer (Figure 21.9,1) to the transfer pump (Figure 21.9,2). Fuel from the transfer pump is forced into a drilled passageway (Figure 21.9,3) at the bottom of the hydraulic head. It then flows into an annulus (a ring formed by two half circles) within the hydraulic head (Figure 21.9,4). This annulus allows fuel to

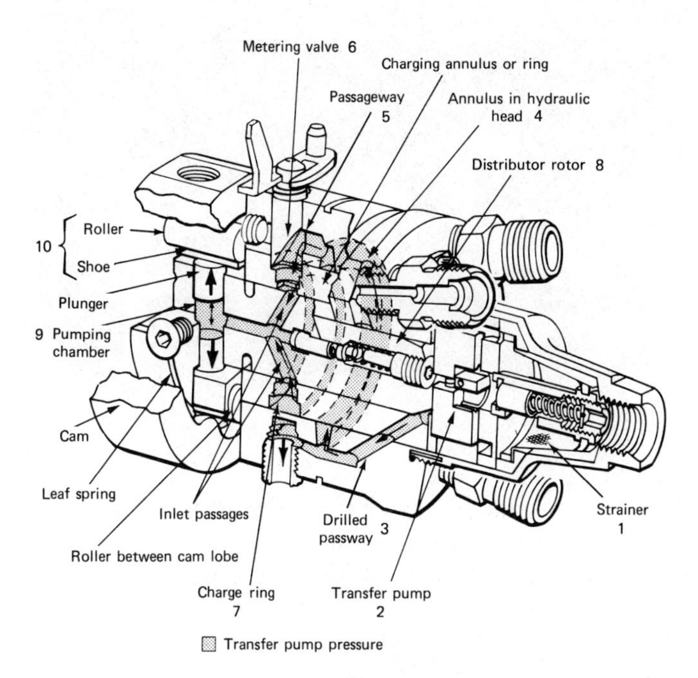

Figure 21.9 Fuel flow in hydraulic head during charging (courtesy Stanadyne).

flow from the bottom of the hydraulic head to the top where it passes through the short connecting passage (Figure 21.9,5) leading to the metering valve. The metering valve (Figure 21.9,6) measure fuel at transfer pump pressure, which is much lower than injection pressure, hence, the name inlet or low pressure metering. The helical cut on the valve uncovers more or less of the inlet port as determined by governor action. Fuel metered by the valve enters the charge ring directly below it (Figure 21.9,7). The charge ring has a port for each engine cylinder. The rotor (Figure 21.9,8) has a centrally drilled bore leading to the pumping cylinder. Leading from this central bore to the charge ring is a diagonal bore (two bores on some pumps). As the drive shaft rotates it also rotates the rotor. This rotation causes the diagonal bore(s) in the rotor to line up (register) with one of the ports in the charge ring. Fuel then moves into the diagonal bore(s) and then to the pumping cylinder (Figure 21.9,9). Since this fuel is under transfer pump pressure, it forces the two opposed plungers (four plungers on DC and DM4 models) outward until their movement is stopped by the leaf spring or cam ring. As the rotor continues to revolve, the charge port(s) goes out of register, thus trapping the fuel in the pumping cylinder. Complete outward movement of the pumping plunger is possible because the cam rollers and shoes (Figure 21.9,10) are betweeen the cam lobes of the internal cam.

B *Fuel flow (discharge cycle)* (Figure 21.10). As the rotor continues to revolve, the rollers and shoes (and pumping plunger) are forced inward by the lobes of the internal cam ring. This action pushes the fuel from the pumping cylinder back into the central bore of the rotor. However, because the rotor has turned slightly, the charge port(s) is now out of register. Fuel is blocked from returning through the charge port(s), and therefore continues down the central bore and enters the bore of the delivery valve. After passing by the delivery valve, the fuel enters the spring area of the valve. In the spring area is the discharge port. There is one discharge port in the rotor that lines up (registers) with the head outlet bores (one at a time) that lead to the outside of the hydraulic head. At this point fuel is transferred to the injection lines and finally to the nozzles.

NOTE When fuel is being discharged, the charge port(s) is out of register, and vice versa.

C *Delivery valve operation.* The delivery valve (Figure 21.11) is hollow and uses no seat. As fuel enters the valve, it presses against the top of the valve and forces it backward against the spring. This movement, A, is known as the *retraction value* and is given in millimeters.

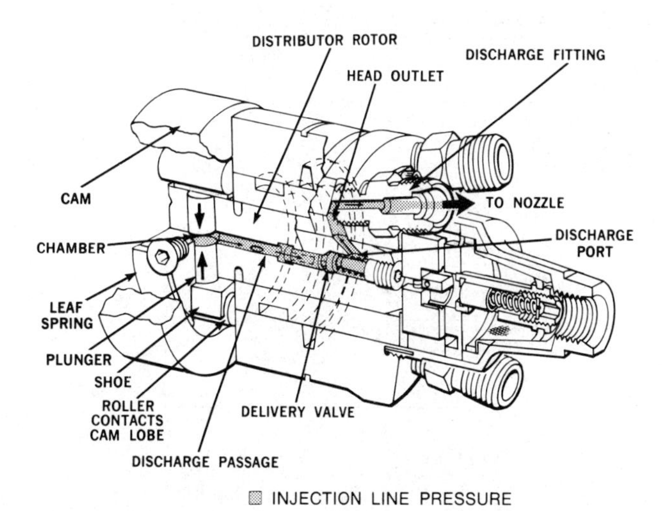

Figure 21.10 Complete discharge cycle (courtesy Stanadyne).

The delivery valve has two functions:

1 To retain fuel in the injection line so that injection can start quickly on the next stroke.

2 To lower pressure in the injection line so that the nozzle valve will close instantaneously after injection to prevent smoke and waste of fuel. *Line retraction,* as it is called, is accomplished as the delivery valve moves out of its bore. This adds the volume of the valve to the spring cavity (Figure 21.11). After injection ceases, the delivery valve spring forces the valve to the closed position. Sealing of the spring cavity is accomplished before the valve stops moving into its bore.

Since the cavity is sealed, no more fuel can leak out, but as the valve continues moving inward it removes its volume from the cavity. As the volume is decreased, line pressure is lowered as fuel returns from the injection line. This single valve handles flow for all cylinders and retracts the same amount for each cylinder.

Additional line retraction is accomplished on some models by using cam lobe retraction. A retraction step is added to the cam lobe (Figure 21.12). When injection stops, pressure on the plunger side of the delivery valve is quickly dropped as the rollers fall into the retraction step of the cam. Retraction value of the cam is usually equal to or slightly greater than delivery valve retraction.

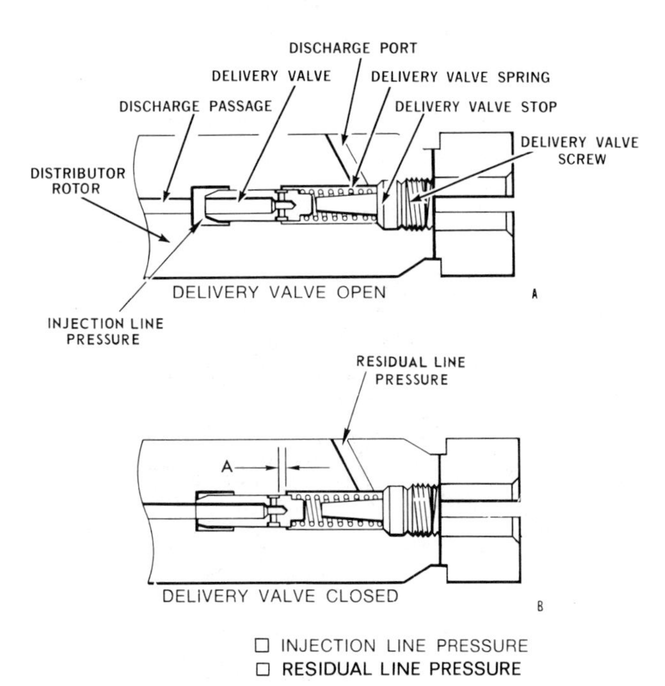

Figure 21.11 Delivery valve operation (courtesy Stanadyne).

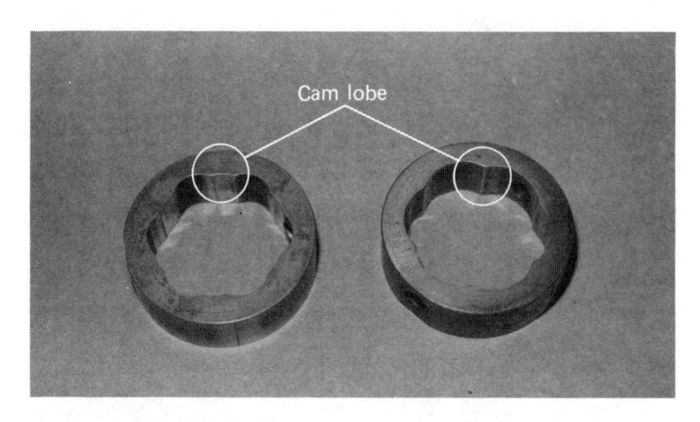

Figure 21.12 Standard and retraction cams.

D *End plate operation.* The end plate serves to cover the transfer pump and houses the pressure-regulating valve. End plates used on DB and DC models are of two designs: (1) the early design and (2) the late design.

1 The early design (Figure 21.13) has the inlet at the top and the pressure regulating valve running diagonally through the end plate. During *hand priming* of the pump (Figure 21.13*a*), fuel is forced into the inlet by the priming pump. It enters the top of the regulating piston downward until port A is uncovered. This depresses the priming spring. Fuel then flows around the lower end of the sleeve and enters the transfer pump. When hand priming stops, the priming spring will force the regulating piston up to cover again port A.

With the pump in operation (Figure 21.13*b*), fuel is drawn into the inlet by the transfer pump and is routed to the hydraulic head. Fuel also flows to the underside of the pressure regulating sleeve through the lower half-moon slot of the end plate. The fuel forces the regulating piston upward, compressing the regulating spring against the stop plug. With increasing pressure port B is eventually uncovered, spilling fuel back to the inlet side of the pump. Pressure is adjusted externally by exchanging the step plug with one having a larger or smaller step (Figure 21.14).

2 The late model end plate (Figure 21.15) has the inlet and pressure regulating valve built as one unit. Fuel is drawn through the inlet A and

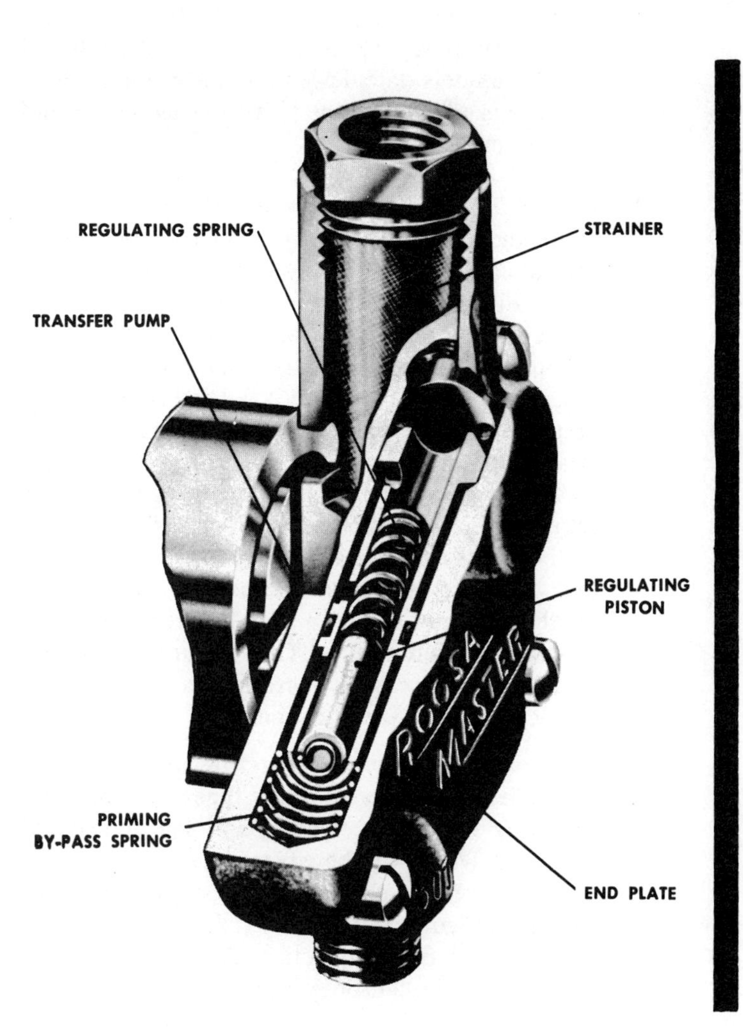

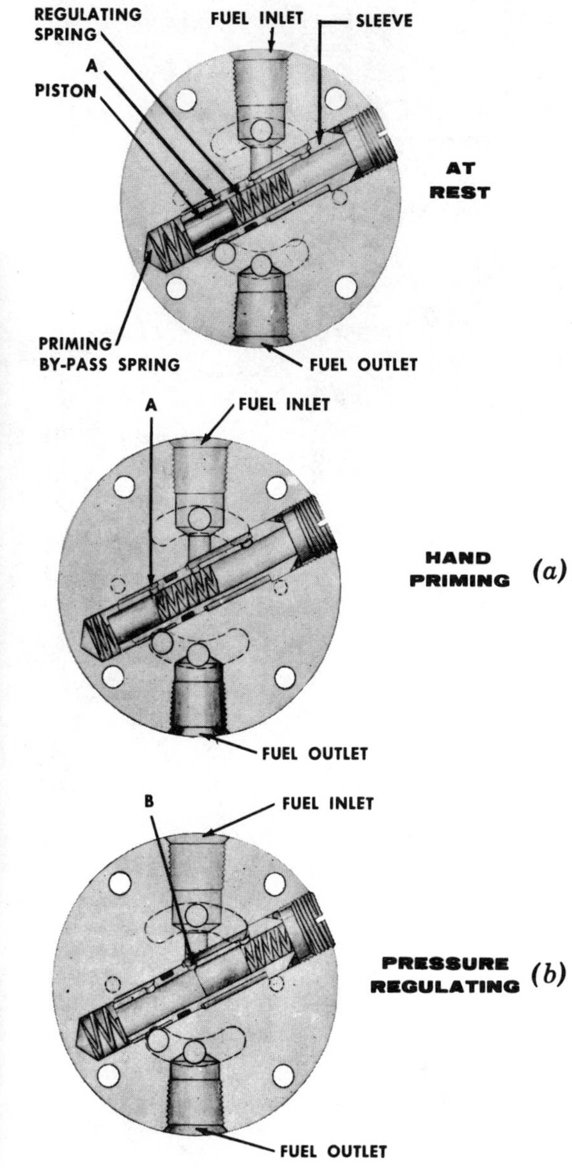

Figure 21.13 Operation of early style end plate (courtesy Stanadyne).

enters the transfer pump. Fuel from the outlet of the transfer pump enters the underside of the pressure regulating piston and forces it upward until port B is uncovered. Fuel pressure is then regulated by bleeding fuel back to the inlet. Port C prevents excessively high fuel pressures if engine speed rises too high.

Running pressure is adjusted with the adjusting plug (Figure 21.15*b*) through the fuel inlet. A small orifice in this plug provides viscosity compensation to eliminate any pressure

Figure 21.14 Step plugs used to change transfer pump pressure.

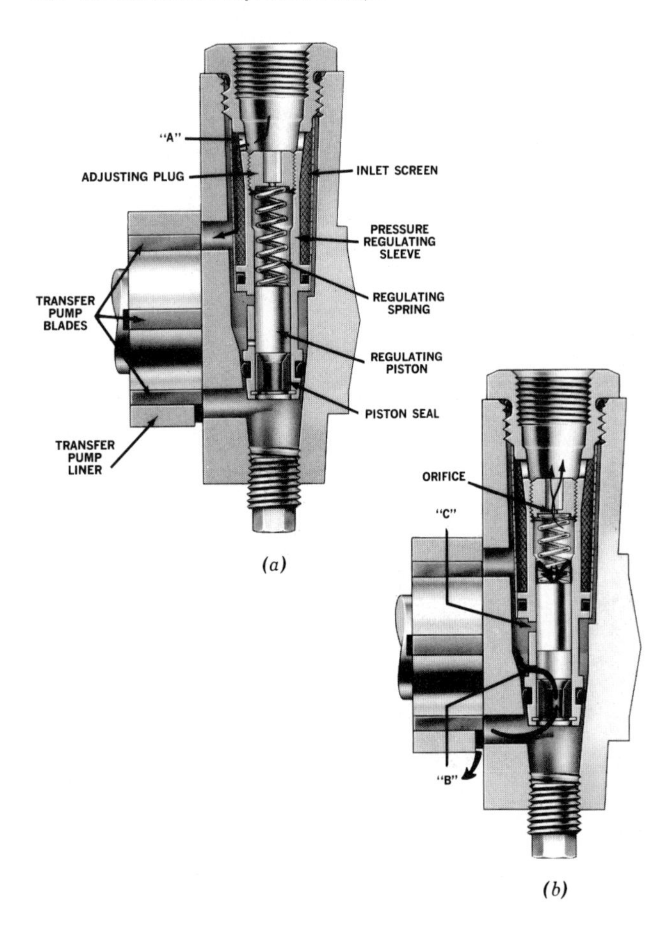

Figure 21.15 Late model end plate in operation (courtesy Stanadyne).

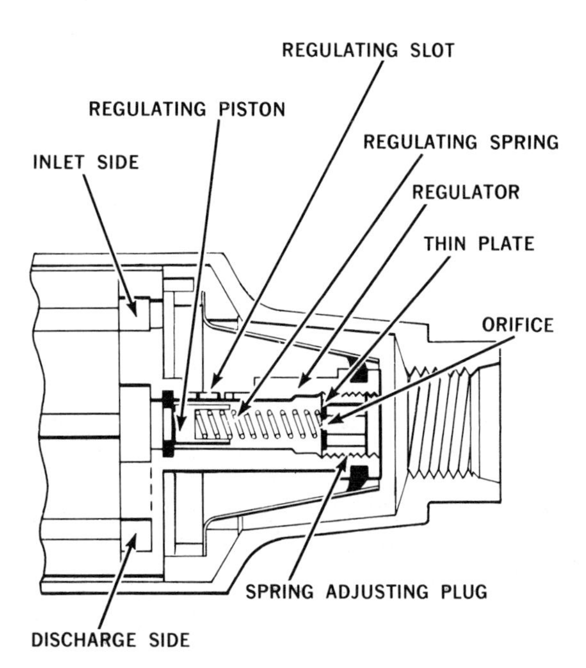

Figure 21.16 End plate used on model DM and DB2 pump (courtesy Stanadyne).

differences caused by varying fuel viscosity. Both style end plates have a pipe plug on the bottom for attaching a pressure gauge. With this gauge the transfer pump pressure can be checked at all pump speeds.

End plates used on DM and DB2 pumps are of the design shown in Figure 21.16. Operation is very similar to that of the DB and DC models (late design). Transfer pump pressure is adjusted in the same way as on DB and DC models.

E *Return fuel circuit.* To prevent the possibility of air being forced into the pumping cylinder(s), a special cavity is built into the pump to collect and return this air back to the fuel tank. This cavity is located at the top of the transfer pump on DB and DC models and leads to a small opening in the housing interior (Figure 21.17). Any air that may be present in the transfer pump is routed through this circuit and returned to the tank. A vent wire, located in the air vent passage, restricts the flow of return fuel, and also breaks up large bubbles of air so they pass easily out of the return. On the DB2 this vent wire assembly is accessible after removing the governor cover and can be exchanged with a wire of different size. This feature allows the mechanic to increase or decrease return flow quantity according to specifications.

F *Governor operation.* Roosa Master pumps are available with several different types of governors, including hydraulic, electric, and the centrifugal governor, which will be covered here.

Components of the centrifugal governor are shown in Figure 21.18.

The function and action of this governor is similar to other centrifugal governors:

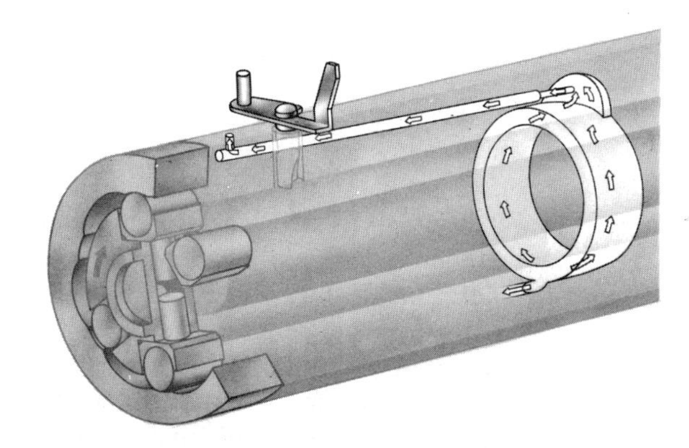

Figure 21.17 Return fuel circuit (courtesy Stanadyne).

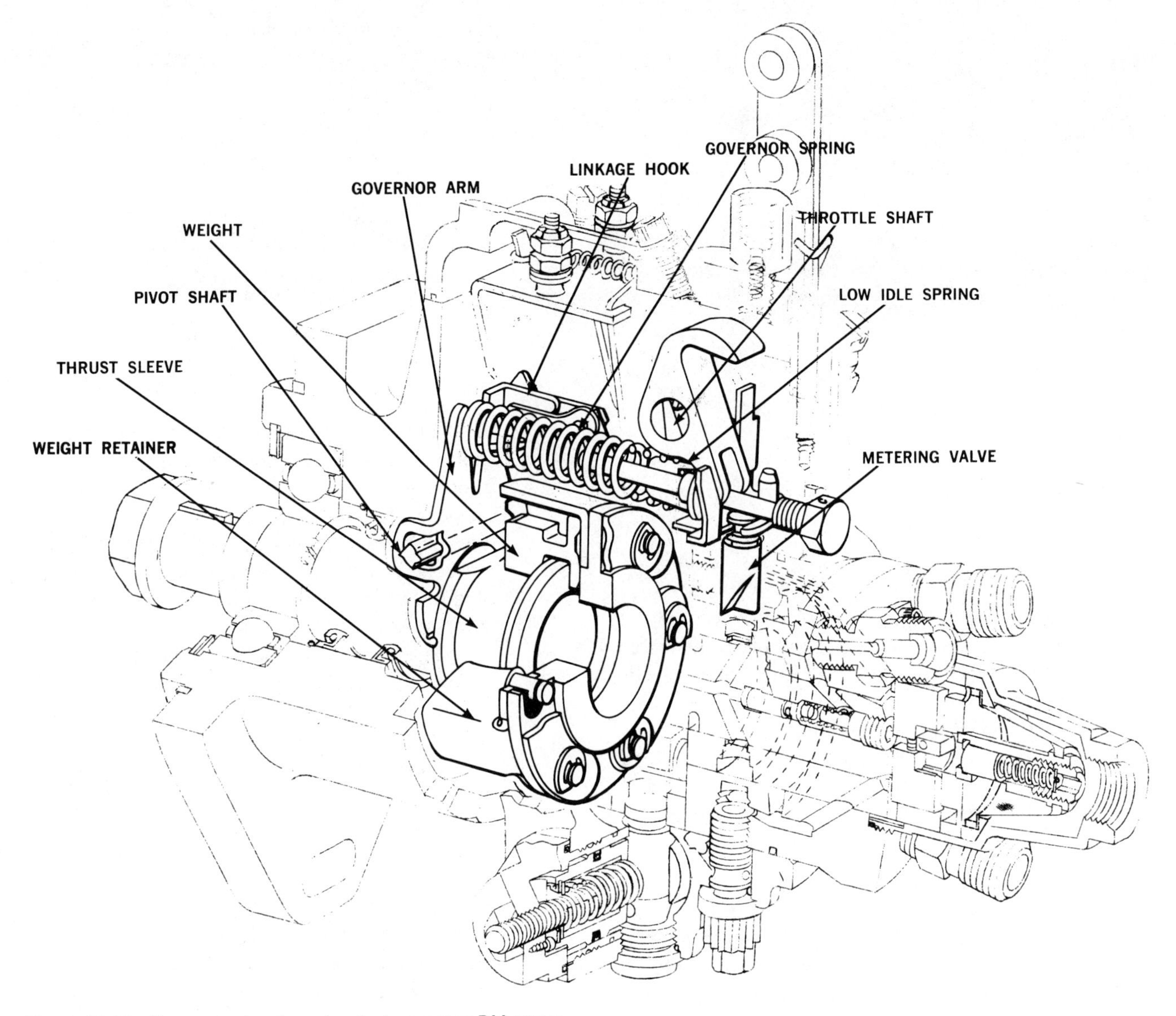

Figure 21.18 Components of mechanical governor DM pump (courtesy Stanadyne).

1. *Idle operation.* In idle position the flyweights move out part way, carrying the thrust sleeve to the front of the pump. This causes the governor arm to pivot on the pivot shaft and force the governor linkage to the rear. This rotates the metering valve to deliver less fuel to the engine. The manually controlled throttle shaft applies little spring pressure to the governor arm in low idle position. Because of this, the small force of the weights at idle can rotate the metering valve to the nearly closed position. During idle operation the light idling spring provides more sensitive speed regulation.

2. *Full load speed.* At full load speed the throttle shaft applies maximum spring pressure on the governor arm. Because of the load applied to the engine, it can only run fast enough to balance the heavy spring load applied to the governor arm. At this point the metering valve has rotated to allow the maximum amount of fuel to enter the charge ring.

3. *High idle and cutoff.* With the throttle shaft still applying maximum force to the governor arm and the load on the engine suddenly removed, the engine will speed up. The increased

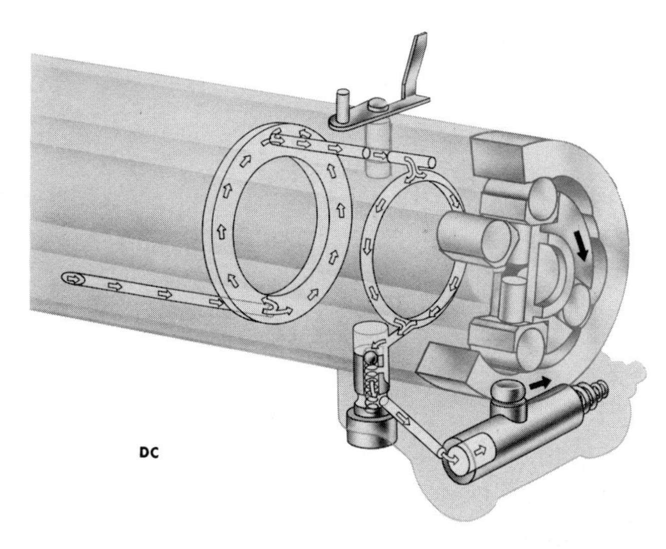

Figure 21.19 Advance circuit of DB and DC models (courtesy Stanadyne).

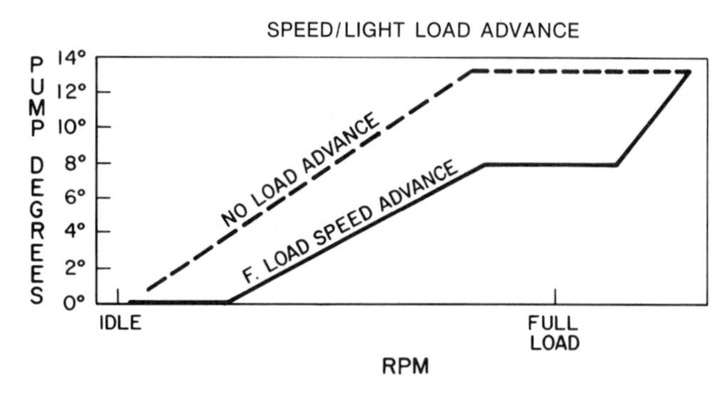

Figure 21.20 Load advance graph (courtesy Stanadyne).

force of the flyweights will then overcome the spring pressure and rotate the metering valve to the closed position. As the valve closes completely, governor cutoff is reached. This limits maximum engine speed. The flyweights and governor spring reach a balance at no load speed, allowing only a small amount of fuel into the charge ring. This is known as high idle/ no load speed.

G *Automatic advance operation.* If the pump is fitted with an automatic advance unit, fuel is forced by the metering valve through a groove in the valve. The fuel then enters a second annular ring (Figure 21.19) that connects to the piston (known as the power piston) of the advance unit. Fuel pressure forces the piston against the cam advance screw. This screw, in turn, advances the cam by rotating it against the direction of rotation. Fuel reaches the servo piston through the hollow center of the hydraulic head locating screw. This screw has a return check valve that bleeds fuel pressure off slowly, allowing the servo piston to return normally to its original position. A second piston on the opposite side of the cam advance screw has springs that force the cam in the direction of rotation back to the start position. This is called the spring piston.

Roosa Master pumps employ two types of automatic advance units: (1) load advance and (2) speed advance.

1 Without an advance unit and at light loads, the metering valve doesn't allow enough fuel into the pumping cylinder to push the pistons and rollers all the way out. Since they are not all the way out against the cam, they will contact the cam lobe later in the revolution, making fuel injection occur later. In some engines, especially precombustion chamber engines, this late injection at light loads causes missing and sputter.

The purpose of the *load advance* is to advance the cam slightly at light loads. The graph (Figure 21.20) shows when the load advance is in operation. A groove is cut into the metering valve to allow fuel to flow by it and into the advance circuit. The metering valve used for load advance is shown in Figure 21.21. This V-shaped groove allows more or less fuel to enter the advance circuit, depending on how it is adjusted. At full load the governor rotates the valve so that the groove no longer is aligned with the passage leading to the advance circuit (no advance). As the governor rotates the valve to light load, the groove again lines up and the cam advances. In this position the metering valve is adjusted to obtain the correct advance. The valve is adjusted by raising or lowering it in the hydraulic head by the tapered section of the guide stud (Figure 21.22).

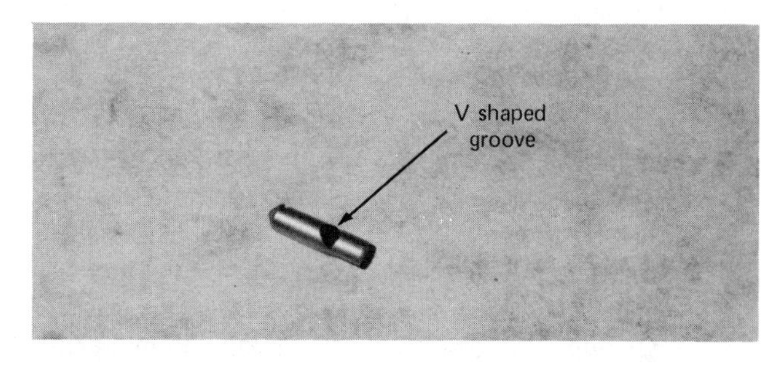

Figure 21.21 Load advance metering valve.

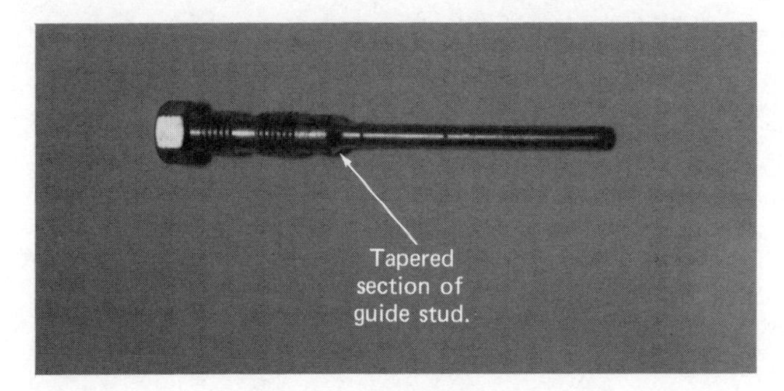

Figure 21.22 Load advance guide stud.

2 *Speed advance* is used to advance the injection timing over the complete operating range of the engine to give the best power and economy. Speed advance also uses a metering valve with a groove cut in it (Figure 21.23). This valve, however, allows the same flow of fuel to enter the advance circuit at all times. Speed advance, as the name implies, is sensitive to changes in speed. This speed change causes transfer pump pressure to rise or fall. The pressure causes the power piston to advance more or less as required. Changes in load have no effect on the advance if speed is kept constant. The speed advance unit is adjusted by shims placed on the spring piston side or by a trimmer screw (Figure 21.24) that can be located in either piston.

With either load or speed advance, the cam is in the retard position during engine cranking. This provides for easier starting, since injection takes place near top dead center (TDC) when the compressed air is at its higher temperature. Total advance available from either load or speed advance is dependent on the length of the advance pistons. On DB2 and DM pumps, only one annulus is used, and the fuel required to move the advance servo piston is routed directly to the top of the hydraulic head locating screw (Figure 21.25).

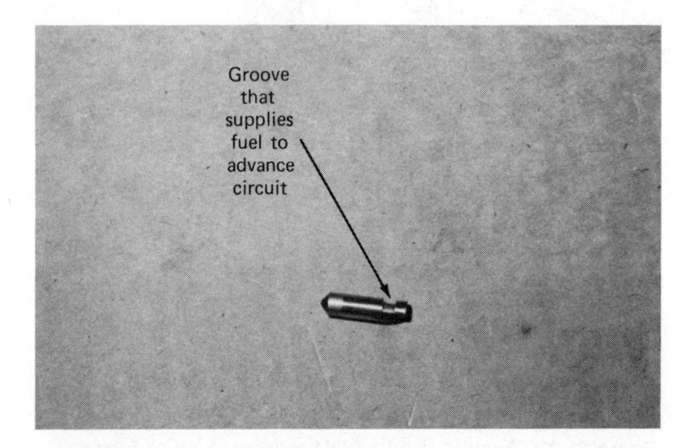

Figure 21.23 Speed advance metering valve.

Figure 21.24 Trimmer screw adjustment of advance.

H *Torque control.* Torque control or torque "backup" on Roosa Master pumps is controlled by metering valve opening, charge time, and transfer pump pressure. A torque control screw is installed in the pump housing to limit metering valve open-

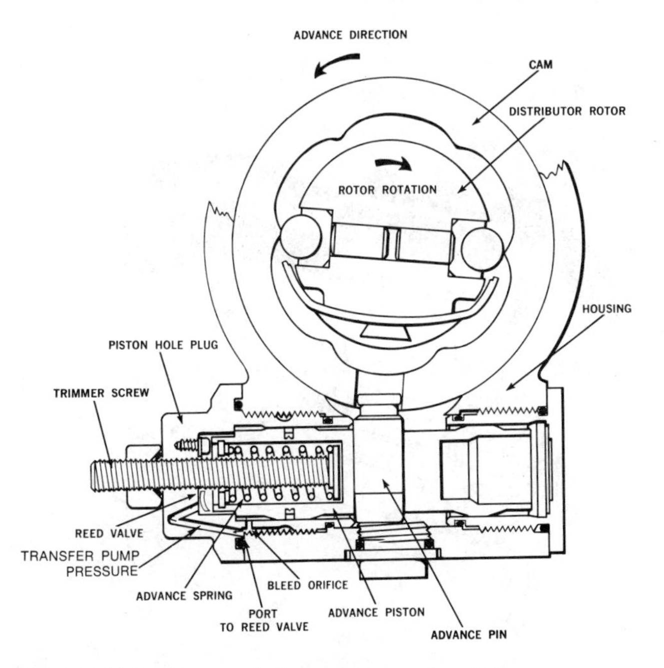

Figure 21.25 Advance fuel passages used on DM.

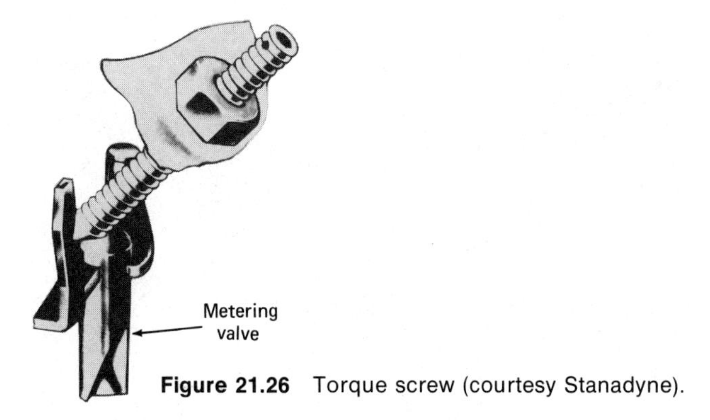

Figure 21.26 Torque screw (courtesy Stanadyne).

ing (Figure 21.26). At rated load and speed, the valve arm will be contacting the torque screw. As the engine speed drops, because of overload conditions, the engine requires more fuel. This fuel is obtained because of the slower pump speed. As pump speed drops, more time is available for fuel to charge the pumping cylinders. Fuel delivery then increases (even though the metering valve does not rotate). This torque control provides for less fuel consumption while at the same time allowing extra fuel for lugging.

I *Electrical shutoff.* On some engine installations it is an advantage to be able to shut fuel off with the starter key rather than use a pull cable. If this is the case, an electrical shutoff solenoid will be mounted in the governor cover (Figure 21.27). With the energized-to-run (ETR) solenoid, current is supplied to the solenoid whenever the engine is in operation. This allows the governor linkage and metering valve to operate normally. When the key is turned to shutoff, the return spring of the solenoid will push the governor linkage to the rear of pump, rotating the metering valve to the closed position. The engine will then stop.

J *Aneroid control.* On some turbocharged engines, an aneroid is fitted to reduce smoke when the engine is accelerated. The aneroid (Figure 21.28) consists of a flexible diaphragm that is spring loaded. With the engine at idle, the spring keeps the shutoff pushed upward, which partially cuts off the fuel supply. As the engine is accelerated, the pressure developed by the turbocharger gradually overcomes the spring force against the diaphragm and pushes the shutoff lever to the run position. This delay in the opening of the shutoff lever controls fuel supply to the engine while the turbocharger picks up speed. When the turbocharger is at full speed, maximum fuel can then be delivered to the engine.

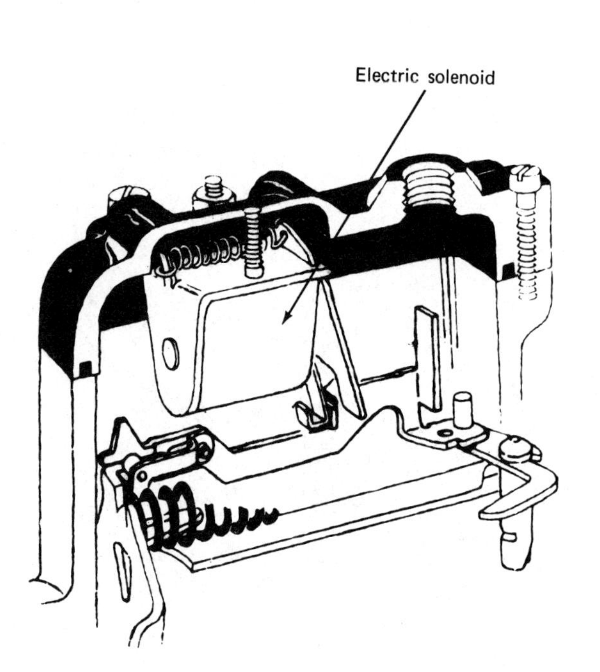

Figure 21.27 Electric shutoff solenoid (courtesy Stanadyne).

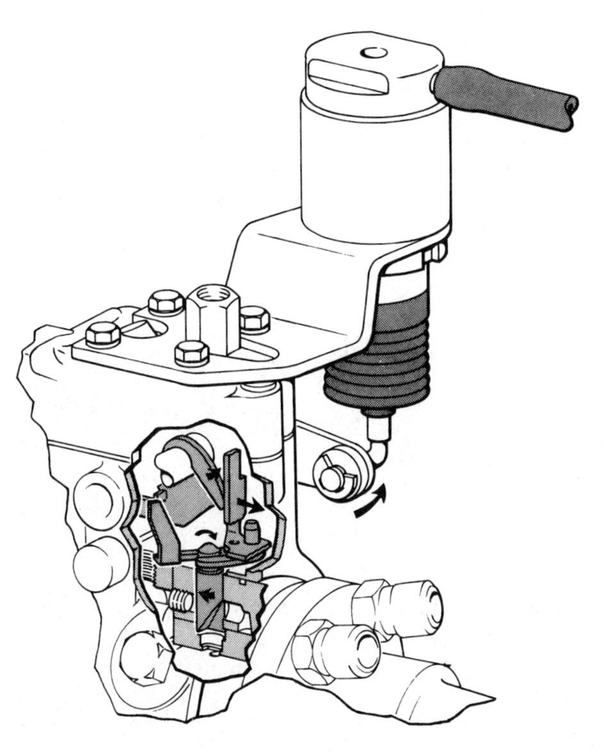

Figure 21.28 Aneroid used on DM model (courtesy Stanadyne).

K. Procedure for Removal from Engine

1. Clean the entire side of the engine that the pump is located on.
2. Turn off fuel supply.
3. Remove the small timing window cover on the side of the pump and rotate the engine in the normal running direction until the two small marks line up in the window.

Note At this time the engine will be on no. 1 cylinder, compression stroke.

4. Remove inlet and fuel return lines.
5. Remove fuel injection lines completely.

CAUTION Do not bend injection lines in order to remove the pump.

6. Remove throttle linkage and shutoff cable (if used).
7. Tie throttle lever in the wide open position so flyweights do not move out of place within the pump.
8. Remove mount nuts and washers.
9. Slide pump straight back off the drive shaft (flange mount) or straight up with the shaft (on vertical mount pumps).
10. Place cap plugs in all pump openings and in all fuel lines.
11. Drain all diesel fuel from the pump.
12. Wash the pump thoroughly in solvent and blow dry.
13. Set the pump in a clean, protected place if it is not going to be worked on right away.

III. Disassembly, Inspection, and Reassembly of Models DB, DC, and DB2

Before disassembly, secure the proper tools and specifications. To aid in locating the correct specifications, carefully read the code on the pump name tag. This tag contains valuable information about the operation and application of the pump. A complete breakdown of the various name tags is given below.

DBGVC631–5AJ (example name tag on DB pump)

- DB pump model
- G type of governor (mechanical)
- V method of mounting (vertical) (F = flange mounted)
- C direction of rotation from drive end (clockwise) (CC = counterclockwise)
- 6 number of cylinders (2,3,4,6, and 8 cylinders)
- 31 plunger diameter (0.310 in.)
  - 25 = 0.250 in. 33 = 0.330 in.
  - 27 = 0.270 in. 35 = 0.350 in.
  - 29 = 0.290 in. 37 = 0.370 in.
  - 31 = 0.310 in. 39 = 0.390 in.
- 5AJ engine application or engine use code

DB2825PC3145 (example name tag on DB2 pump)

- DB2 pump model
- 8 number of cylinders (available in 2,3,4,6, and 8)
- 25 plunger diameter (0.250 in.) (6.35 mm)
- PC accessory code. Include this code in any reference to pump.
- 3145 specification number. Determines fuel adjustment for a given application.

Prior to disassembly, clean the exterior of the pump and mount it on the correct holding fixture (Figure 21.29), using flat washers against the pump housing.

DB and DC–fixture no. 13363
DM–fixture no. 19965
DB2–fixture no. 20029

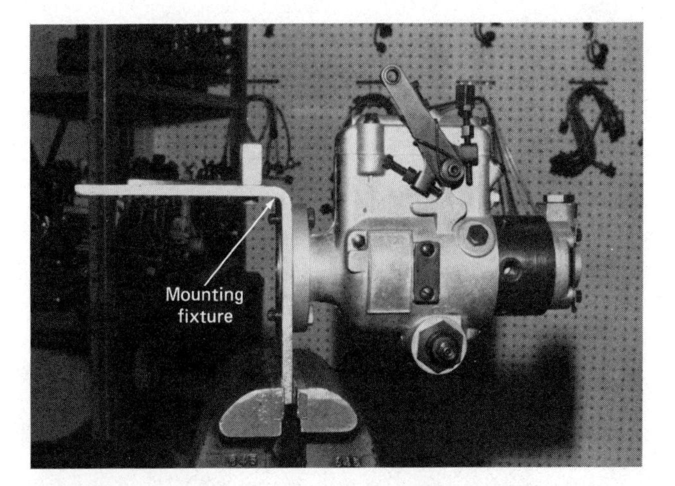

Figure 21.29 Fixture used to disassemble and reassemble the pump.

A. Procedure for Disassembly

NOTE As parts are removed, place them in a pan of clean calibrating oil or diesel fuel.

1 Remove and discard all lead seals and mounting flange O rings.

2 Remove the three screws holding on the governor cover. Discard cover seal.

3 Using tool no. 20992, remove the cam shutoff and cam stop (if used) from the throttle lever shaft. Discard the cam shutoff.

4 Remove the throttle shaft assembly and throttle lever. Noting the position of these parts during disassembly will aid in the correct positioning during reassembly.

5 Loosen guide stud. Hold the governor spring, idle spring, and spring retainer while removing the guide stud. Remove the springs.

6 Push down on the metering valve and lift the linkage assembly off the metering valve arm. Pull the linkage back to unhook it from the governor arm and hang it over the side of the housing.

7 Remove the metering valve assembly with spring.

8 On DB2 pumps, remove the vent wire screw assembly with wrench no. 16336 (Figure 21.30).

9 Remove the head locking screws (Figure 21.31).

10 On pumps without automatic advance, invert the pump in a vise and remove the head locating screw and cam locking screw (Figure 21.32).

Figure 21.31 Removing head locking screws.

Figure 21.32 Head locating screw and cam locking screw.

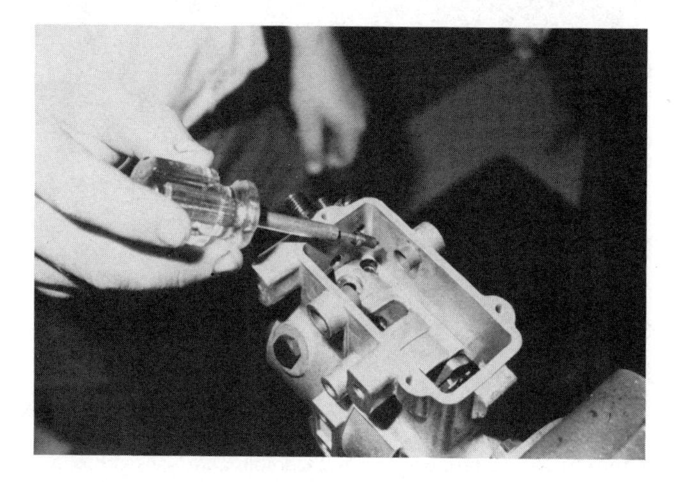

Figure 21.30 Removing vent wire assembly (DB2).

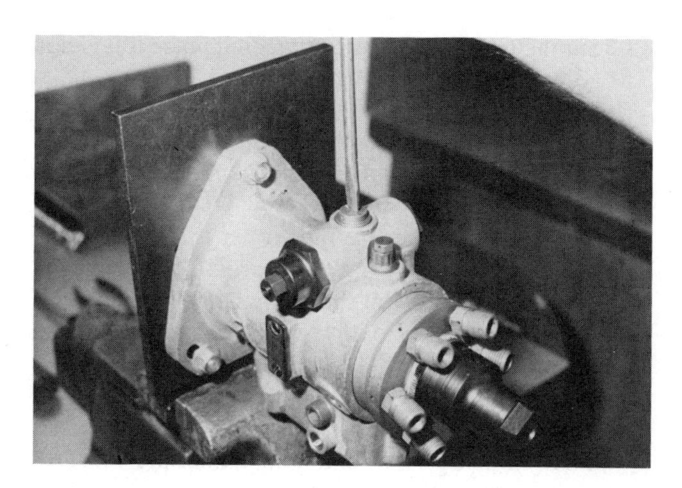

Figure 21.33 Removing the advance plug.

Figure 21.34 Removing the advance pin.

Figure 21.36 Installing advance stud bushing for removal of stud.

Proceed to step 16. On pumps with automatic advance, invert the pump in a vise and remove the head locating screw.

11 Tap the advance plug to loosen it. Then remove the plug using tool no. 14490 (Figure 21.33).

12 If the pump has an advance pin, use tool no. 13301 and remove the pin (Figure 21.34). If the pump has an advance stud, proceed to step 13.

13 Remove the power piston plug and spring piston plug using tool no. 14490 (Figure 21.35).

NOTE On pumps with an advance pin, the power piston will come out with the power piston plug. On pumps using an advance stud, slide washers must be removed after the plugs are removed.

Figure 21.35 Removing advance piston plugs.

CAUTION Keep all advance springs, seats, and seals in correct position or advance may not operate correctly when reassembled.

14 Remove the power piston on DB and DC pumps by pulling it out with advance pin or by unscrewing the trimmer screw through the lower piston plug on DM pumps. If the pump has an advance stud, remove the power piston by rapping the power piston plug against the hand. Remove and discard the piston ring and seal from the power piston.

15 If an advance stud is used, remove it using the correct tool (no. 15499) and bushing (no. 15500). Screw the bushing into the housing over the stud (Figure 21.36). Insert the tool into the stud and remove stud using a $\frac{5}{8}$ in. wrench (Figure 21.37).

16 Return pump to original position in vise. *Carefully* remove the hydraulic head by grasping and pulling it rearward, rotating it slightly as it is pulled out (Figure 21.38).

CAUTION When the head O ring clears the pump housing, the head will come out quite easily. Be certain not to drop any parts as the hydraulic head is removed.

17 Disassemble the governor by letting the weights, thrust sleeve, and thrust washer fall into the hand.

18 Set the hydraulic head assembly in the holding fixture, supported by the governor weight retainer as shown in Figure 21.39.

Figure 21.37 Removing advance stud.

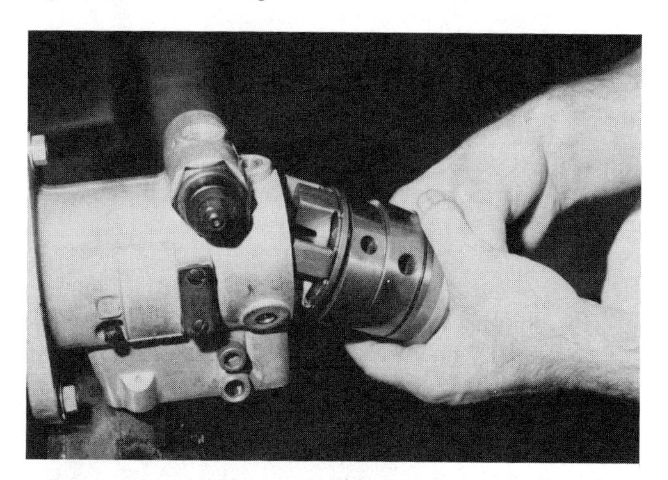

Figure 21.38 Removing the hydraulic head.

Figure 21.39 Hydraulic head in holding fixture.

Figure 21.40 End cap locking screw on DM and DB2.

19 Remove end plate and transfer pump components.

On DB and DC pumps, remove the four end plate screws and remove end plate. Remove steel insert on new style end plates. Remove pressure regulating valve components by removing the step plug and using tool no. 13301 to remove the sleeve on old style end plates. On new plates simply remove the inlet fitting with regulating sleeve. Lift out the end plate seal, transfer pump blades, and liner.

On DB2 and DM pumps, remove the end cap locking screw and plate (Figure 21.40). Unscrew the end cap. Disassemble the regulator assembly as shown in (Figure 21.41). Remove the transfer pump blades and liner. Remove the end cap seal.

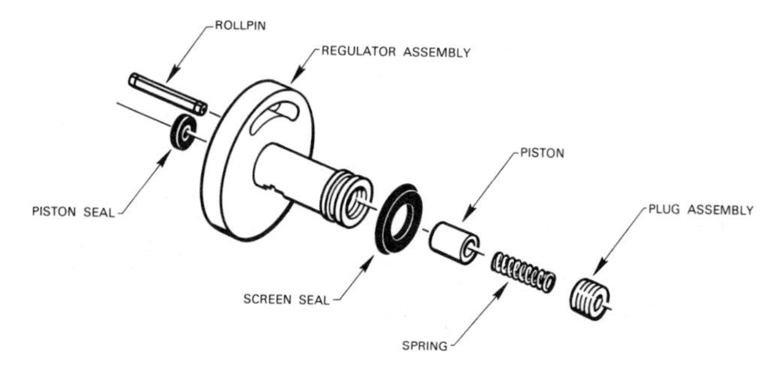

Figure 21.41 Regulating components (courtesy Stanadyne).

Figure 21.42 Removing rotor retainer snap ring.

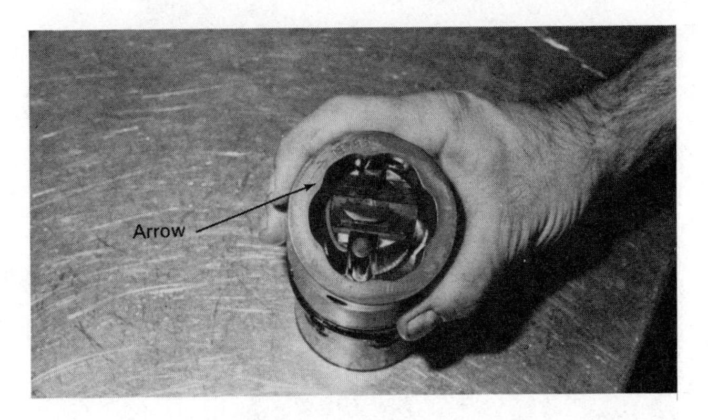

Figure 21.44 Directional arrow on cam ring.

20 Remove the rotor retainer snap ring (Figure 21.42).

21 Remove rotor retainers.

CAUTION After rotor retainers are removed, rotor is not held in the hydraulic head. Do not let the rotor fall out when handling the hydraulic head.

If working on a DM pump, proceed to step 28.

22 Invert hydraulic head and set it down on a clean bench. With a snap ring pliers, remove the weight retainer snap ring (Figure 21.43). Then remove weight retainer.

NOTE On DC pumps the weight retainer must be pressed off.

23 Remove cam ring, noting the direction of rotation arrow (Figure 21.44).

24 Remove the leaf spring, rollers and shoes, and pumping plungers.

25 Withdraw the rotor from hydraulic head bore.

CAUTION To avoid contamination to the precision surface of the rotor, do not handle the rotor with bare hands. Always rinse off the rotor with clean fuel before inserting it in the head bore. Leave the rotor in the head bore whenever possible.

---

26 Set the rotor in tool no. 16313 and lock in a vise. Remove the delivery valve retaining screw and stop. Remove the delivery valve with tool no. 13383.

27 Remove the metering valve from the arm by inserting it in the bore of fixture no. 19965. Tap the arm off (Figure 21.45).

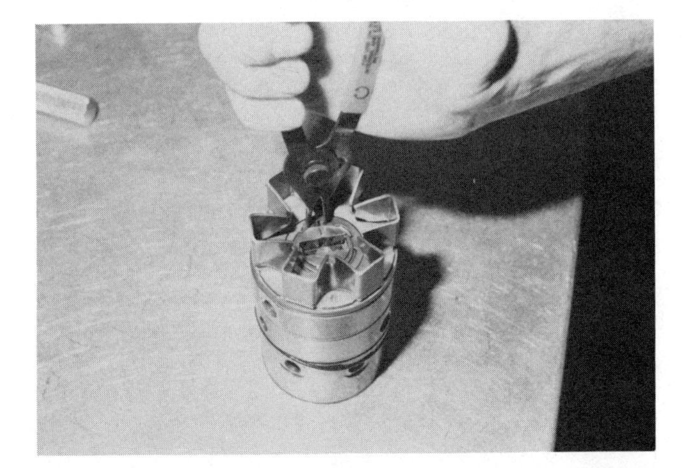

Figure 21.43 Removing weight retainer snap ring.

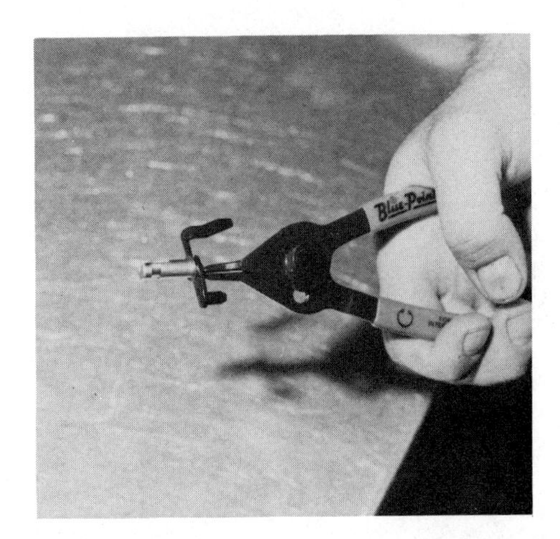

Figure 21.45 Removing metering valve from arm.

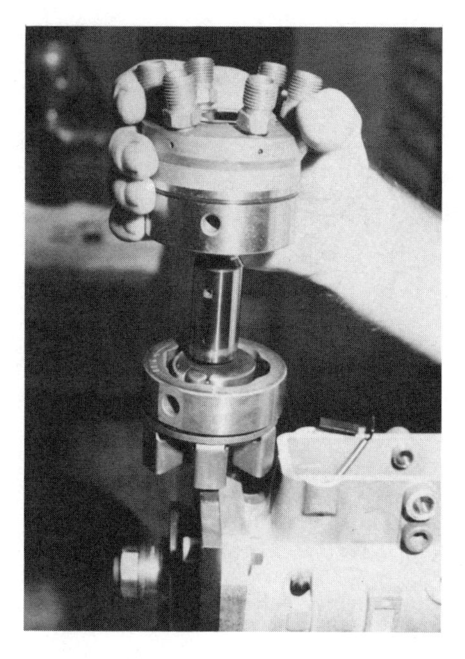

Figure 21.46 Removing hydraulic head from rotor.

Steps 28 to 32 apply to DM pumps only.

28 Carefully lift the hydraulic head off the rotor (Figure 21.46).

29 Remove cam ring, leaf spring, rollers and shoes and pumping plungers.

CAUTION Pay particular attention to the location of parts on DM4 pumps.

30 Remove delivery valve retaining screw, delivery valve stop, and extract valve using tool no. 13383.

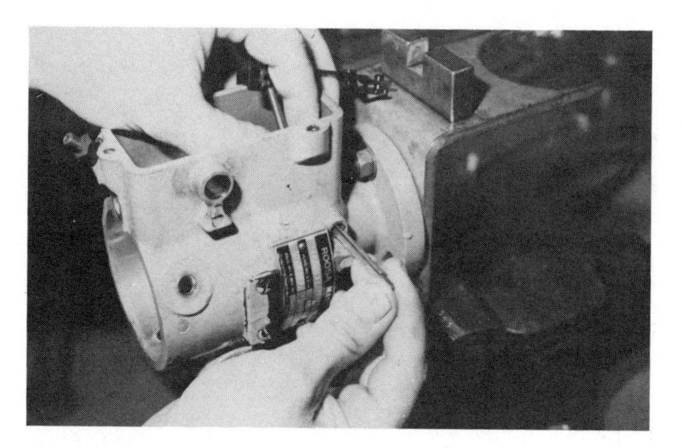

Figure 21.48 Pushing the pivot shaft from housing.

31 Remove retaining rings, cushion retainers, and rubber cushions from the weight retainer (Figure 21.47).

32 Remove metering valve from arm.

33 Remove pivot shaft nuts and push pivot shaft out of housing (Figure 21.48).

34 Remove governor arm and linkage assembly.

The following steps apply to DM pumps only.

35 Remove drive shaft hub.

36 Using tool no. 20043, remove shaft retaining ring (Figure 21.49). *Use caution.*

37 Remove spring washer and drive shaft with bearing.

38 Press seals from housing after removing the seal retaining ring. Discard the seals.

Figure 21.47 Removing the weight retainer rubber cushions.

Figure 21.49 Removing drive shaft retaining ring.

Painted pump parts can be soaked in a parts cleaner to remove loose paint, grease, and dirt. All internal parts should be rinsed in clean solvent or fuel and inspected for wear or breakage. Clean the work area in preparation for assembly.

B. Procedure for Parts Inspection

After all parts are rinsed with solvent or fuel, they should be inspected for wear according to the following list:

1 Check the pump housing for thread damage or breakage. On DB and DC models the pilot tube may require replacement if it shows signs of wear from the drive shaft seals. Refer to page 384 for instructions on replacing the pilot tube.

2 Check for excessive wear on the knife edge of the pivot shaft.

3 Check components such as flyweights and weight retainer for wear on their pivot points and replace as required.

4 Make certain the control linkage is free of wear at all pivot points and does not bind.

5 Check metering valve for scoring, sticking, or wear. Replace if needed.

6 Closely examine the transfer pump blades and blade springs.

NOTE Usually two-piece blades will require replacement because the spring loading increases the chance of scoring. Measure one-piece blades end to end with a micrometer. They must be at least 1.0930 in. (27.75 mm) long to be put back into service.

7 Examine the cam ring at the points shown in Figure 21.50 for flaking or chipping. Replace the cam ring if required.

8 Check wear on rollers and shoes by trying to pull the roller straight out the shoe. If this can be done, replace both pieces. Also check rollers for scoring and shoes for wear in the contact area of the leaf spring.

9 Make a thorough examination of the hydraulic head and rotor assembly. Make certain the vent wires are free by shaking the head and listening for the sliding of the wires. Carefully check rotor for excessive scoring and abrasive wear. The pumping plungers must slide freely in their bore and show no signs of scoring.

NOTE The only positive check on hydraulic head and rotor wear is the cranking fuel delivery test performed on the test bench.

10 Check delivery valve for scoring around the main body of the valve and scuffing of the retraction collar. Replace if required.

CAUTION Delivery valves are sold in standard size and oversize. If the rotor nas been machined to accommodate the oversize valve, it will be marked OV on the front 180° opposite the leaf spring (Figure 21.51). Oversize valves will be blackened on both ends. Make certain the correct valve is used.

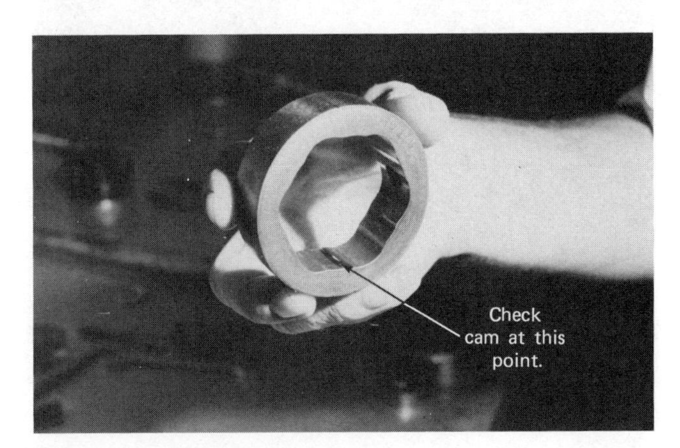

Figure 21.50 Examination of cam ring.

Figure 21.51 Oversize delivery valve marking.

Figure 21.52 Checking drive shaft tang for wear.

11 Check the drive shaft tang (if shaft is with pump) as shown in Figure 21.52.
Minimum values:
DB and DB2 – 0.305 in. (7.75 mm)
DM – 0.430 in. (10.9 mm)

12 If an electrical shutoff solenoid is used, check it for operation and cracking of the sealing material.

13 On pumps with load advance, check the adjustable guide stud for wear on the taper directly above the metering valve. Replace if small dents are found on the taper (Figure 21.53).

14 On pumps with automatic advance of any type, check the power piston and piston ring for excessive wear. Any wear in this area can cause the advance to function incorrectly.

NOTE *Always* replace all O rings, gaskets, and seals when an injection pump is overhauled. Overhaul gasket packages are supplied, which include all needed gaskets.

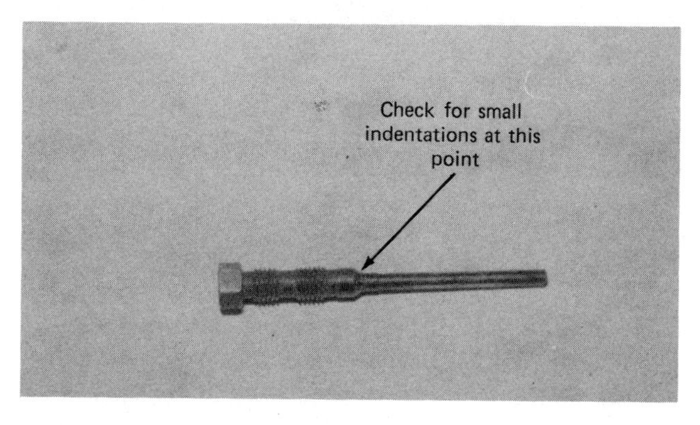

Figure 21.53 Worn guide stud.

C. Procedure for Reassembly

When reassembling the injection pump, always observe the torque values listed for the pump being worked on.

1 If it is determined that the pilot tube should be replaced on DB, DC, or DB2 pumps, install it as follows:

a If pilot tube has been removed previously, proceed with step **b;** if not, remove pilot tube using the special driving tool shown in Figure 21.54.

b Coat pilot tube with the epoxy that can be obtained from Roosa Master.

c Install pilot tube using the positioning tool shown in Figure 21.55.

d Inspect inside the housing, making sure no epoxy has been left in the housing near the pilot tube.

e Place housing under a heat lamp and allow to dry.

NOTE The epoxy will dry under the heat lamp in about 2½ hours; at room temperature it will take about 12 hours.

2 On DM pumps, press new drive shaft seals into the housing with tool no. 20268. The seal lips must oppose each other when installed. Apply a light coat of Vaseline to the seals and insert the drive shaft into the front of housing. Install spring

Figure 21.54 Tool used to remove pilot tube.

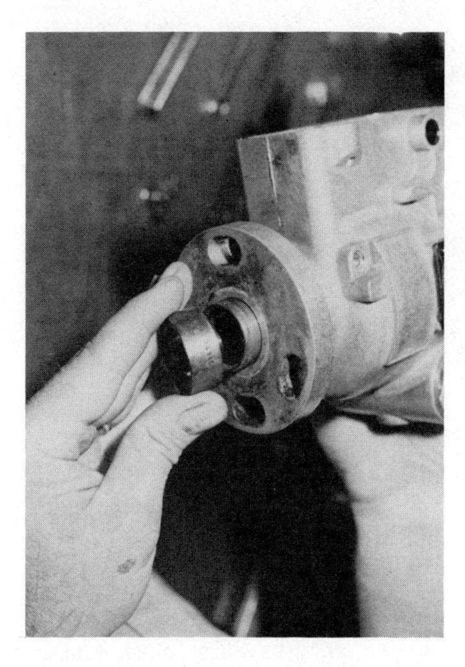

Figure 21.55 Positioning pilot tube to correct protrusion.

washer and retaining ring. Install gear hub to the shaft and torque to the specified torque.

3 Install delivery valve, spring and stop into rotor bore using sequence shown in Figure 21.56. Torque stop plug to 85 to 90 in. lb (9.60 N·m to 10.17).

CAUTION This is a very critical torque since overtightening may cause head seizure and under torque will cause fuel leakage (hard starting).

4 Rinse rotor in clean fuel and install pumping plungers, rollers, and shoes. Mount the leaf spring(s) to the rotor and screw in the adjusting screw far enough so that the shoe cannot be pulled out from the rotor.

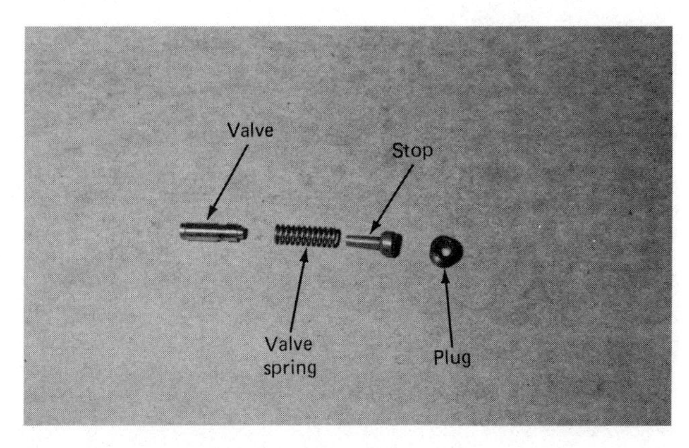

Figure 21.56 Correct assembly sequence for delivery valve.

Figure 21.57 Method used to set roller-to-roller dimension.

5 Clamp setting fixture no. 19969 carefully into vise (on the flat side) and install the rotor carefully. The rollers must be held inward when transferring the rotor so that they do not fall out. Attach an air line to the fixture inlet and regulate pressure between 40 to 100 psi (3 to 7 kg/cm^2). This will force the rollers and shoes to their maximum outward travel as controlled by the leaf spring(s). *Check roller-to-roller dimension* as shown in Figure 21.57. If it is incorrect, adjust it by turning the leaf spring adjusting screw clockwise to increase the dimension, counterclockwise to decrease the dimension.

CAUTION On all four plunger models, each set of rollers must be adjusted to within 0.003 in. (0.076 mm) of each other and both must meet specifications.

NOTE The roller-to-roller dimension gives a completely accurate setting of maximum fuel output on a new injection pump.

6 On all four plunger models and certain DB2 models (refer to appropriate specification sheet) the centrality of the rollers *must* be checked. (This assures that all rollers will contact the cam ring at the same instant.)

a Rotate the rotor until one roller is aligned with the dial indicator on setting bracket no.

19969. Slide the indicator plunger inward until it contacts the roller and depresses the plunger at least 0.010 in. (0.25 mm). Lock the indicator and set the pointer on zero.

b Rotate the rotor in either direction until the next roller contacts the indicator plunger. It should read ±0.002 in. (0.050 mm) of zero. Check all rollers before you correct any one roller.

c If any roller is beyond the specification, rollers and/or shoes may be interchanged to obtain the correct centrality.

CAUTION This check must be made accurately. Failure to do so could cause seizure of the hydraulic head.

d Always recheck roller-to-roller dimension after correcting centrality.

7 While holding inward on the rollers, carefully remove the rotor from the bracket and rinse it again in clean calibrating oil. Then, on DB, DC, and DB2 models, install the rotor in the hydraulic head, and on DM pumps install the weight retainer as shown in Figure 21.58. Install the cushion retainer and the small retaining circlips (Figure 21.59) with tool no. 20045. When installed, these clips should not rotate when slight pressure is applied to them. If they do rotate, replace all of the circlips.

Figure 21.58 Installing weight retainer on DM model.

Figure 21.59 Installing retaining circlips.

Figure 21.60 Installing cam ring on DM model.

8 On DM pumps, set the rotor on top of fixture no. 19965 (weight retainer down). Place the cam ring over the rotor with the directional arrow pointing in the *opposite* direction of rotation (Figure 21.60). Rinse hydraulic head in calibrating oil and slide it over the rotor as shown in Figure 21.61.

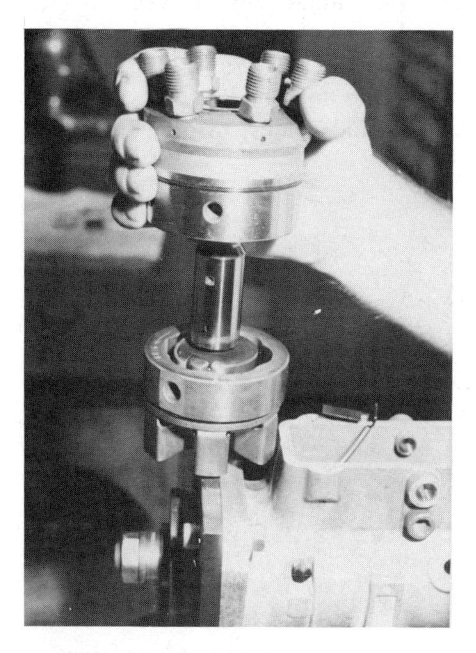

Figure 21.61 Installing hydraulic head to rotor.

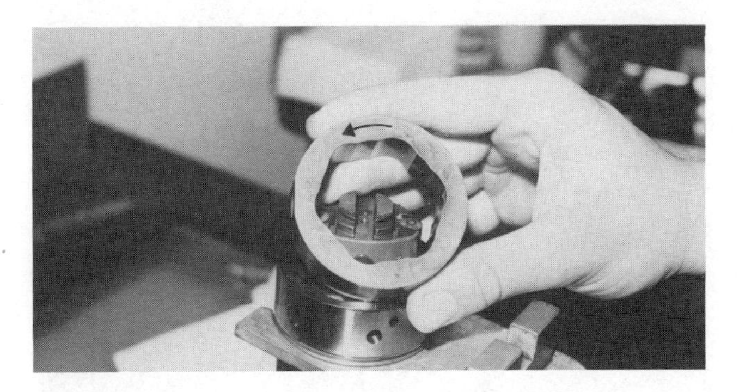

Figure 21.62 Cam ring rotation (DB, DB2, and DC).

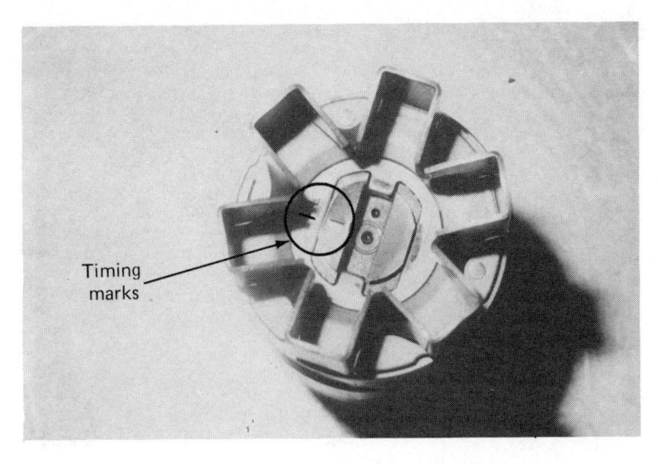

Figure 21.64 Correctly timing the weight retainer to rotor.

9 On DB, DB2, and DC pumps, install the cam ring with directional arrow pointing *in the direction of* rotation (Figure 21.62). Attach a new, flexible retaining ring between the weight retainer and the retainer hub (if so equipped) by stretching the ring over the rivets, as shown in Figure 21.63. Install this assembly on the rotor, being certain to align the timing marks (Figure 21.64). Install the retaining snap ring.

NOTE On DC pumps the weight retainer assembly must be pressed onto the rotor and then the ring attached. Align the marks as shown in Figure 21.65.

10 Carefully invert head assembly (except DM pumps) to prevent rotor from falling out.

11 Install the rotor retainers and, then, on DB and DC pumps install the retaining snap ring with tool no. 13375 as shown in Figure 21.66. On DM and DB2 pumps install the transfer pump liner-locating ring (Figure 21.67) approximately three-fourths down.

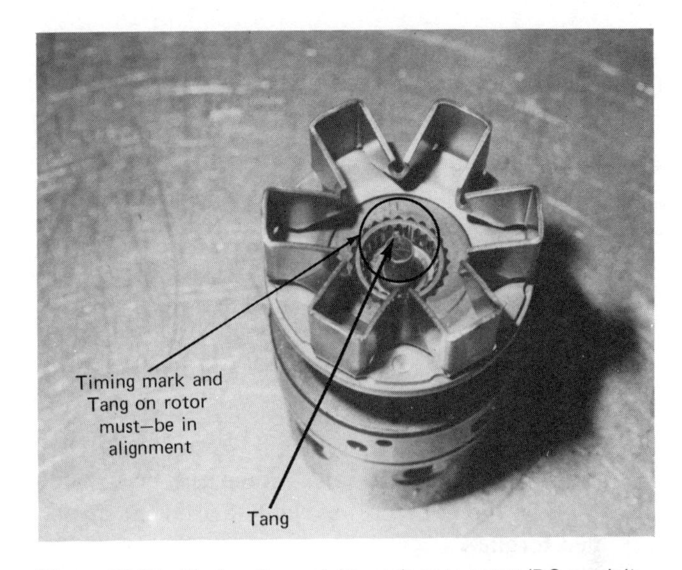

Figure 21.65 Timing the weight retainer to rotor (DC model).

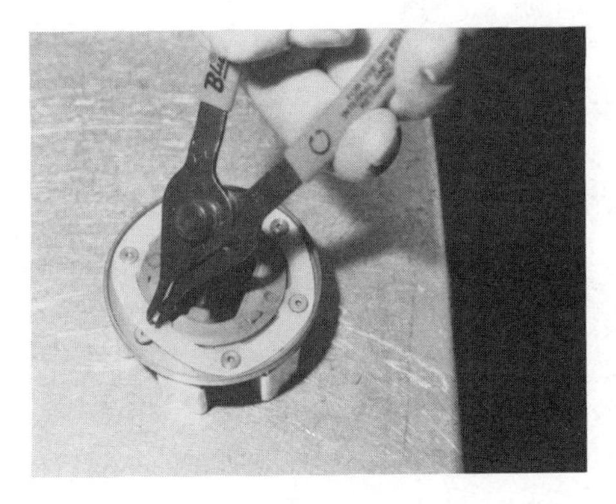

Figure 21.63 Installing flexible retaining ring.

Figure 21.66 Installing the rotor retainer snap ring.

Figure 21.67 Positioning transfer pump liner locating ring.

12 Install the transfer pump end cap seal. Make certain the ring is down all the way.

13 Install the transfer pump liner and align the slot with the rollpin in the regulator. On DB and DC pumps, the rotational marking c (clockwise) or cc (counterclockwise) must face outward when installed (Figure 21.68).

14 Install the blade springs in the transfer pump blades and compress the blades together with light pressure. Install the blades into the rotor slots.

15 On DB and DC pumps, install the steel thrust insert in the end plate and mount the end plate. Tighten the four screws evenly and to the specified torque.

16 On DM and DB2 pumps, attach the regulator assembly and screw on the end cap.

Figure 21.68 Rotational marking of liner.

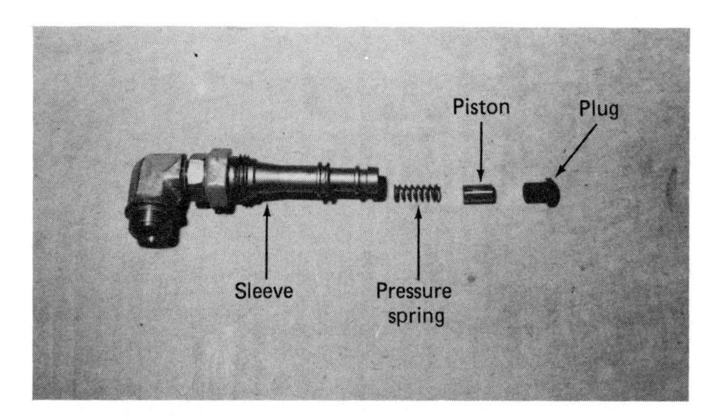

Figure 21.69 Pressure regulator components.

CAUTION Be certain the cap is started straight by turning the end cap counterclockwise until a slight "click" is heard. This will aid in starting the end cap without crossing the threads.

17 On DB and DC pumps, assemble the pressure regulator as shown in Figure 21.69 and screw it into the end plate.

18 Insert the pivot shaft (knife edge to the rear) (Figure 21.70) into the housing while holding the governor arm in the housing. Install O rings and pivot shaft cap nuts and tighten to the specified torque.

19 Invert hydraulic head and install the governor weights, thrust washer, and thrust sleeve.

CAUTION On DB and DC pumps the thrust washer has a chamfered edge that must face the thrust sleeve when installed. All weights must be of even height and collapsed against the sleeve when installed.

Figure 21.70 Knife edge of pivot shaft.

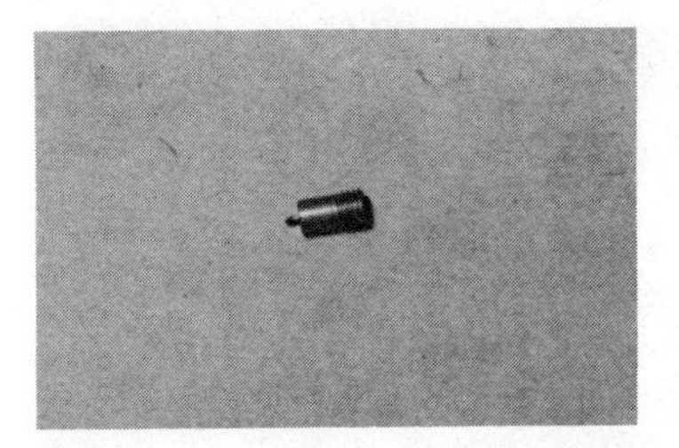

Figure 21.71 Vent wire assembly (DB2).

20 The hydraulic head assembly can now be installed in the pump housing. Install a new head seal ring and apply a light film of grease to the inside of the housing. Insert the head with the metering valve bore UP. Use a slight twisting motion when installing the head to avoid cutting the head seal.

21 Install the two head-locking screws finger tight.

22 On DB2 pumps install the vent wire assembly (Figure 21.71).

23 Install the metering valve into its bore and attach the governor linkage over the valve arm pin.

24 Install the guide stud with a new washer by depressing the metering valve far enough to allow the guide stud to enter the housing.

NOTE On DB2 models, make sure guide stud goes under the spring clip to hold the metering valve up in its bore.

Attach the main governor spring, spring seat, and low idle spring over the stud. Hook the spring over the projection on the governor arm and tighten the guide stud to correct torque.

25 Invert the pump in the vise. Install the head-locating screw and torque.

26 Tighten the two head-locking screws to the correct torque.

27 If a cam advance screw is used, assemble the tool no. 15499/15500. Screw the bushing (Figure 21.72) into the housing over the cam advance screw. Then insert screw wrench and tighten to specified torque. Proceed to step 29.

NOTE Check cam advance screw after installation to make sure the head of the screw did not crack upon torquing it.

28 If an advance pin is used, first assemble the advance components, that is, insert the advance piston into the power piston plug and screw the assembly into the housing. Rotate the cam ring until the unthreaded hole is aligned with the advance piston hole. Insert the advance pin (Figure 21.73). Install the spring piston plug with advance springs. Torque both plugs. Install the advance screw hole plug and tighten.

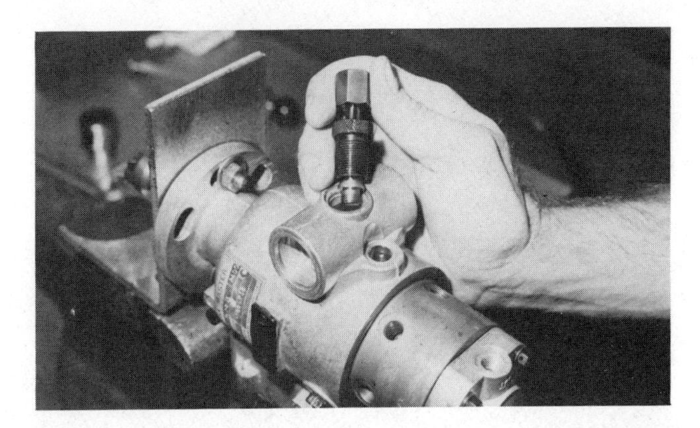

Figure 21.72 Tool used to install advance screw.

29 When a cam advance screw is used, assemble both plugs and stick slide washers to the pistons with vaseline. Install power piston plug according to rotation of the pump. Marks are present on the housing designating clockwise (c) or counterclockwise (cc) rotation. The power piston must go on the correct side depending on rotation.

30 Install spring piston plug, advance screw hole plug, and torque.

31 Install throttle shaft, forked throttle lever (facing guide stud), and shutoff shaft (if used) into housing. Install a new shutoff cam and cam stop.

Figure 21.73 Inserting the advance pin.

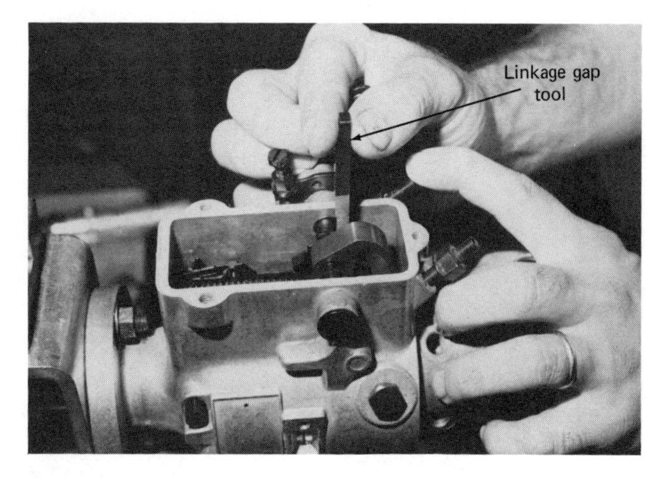

Figure 21.74 Checking governor linkage gap.

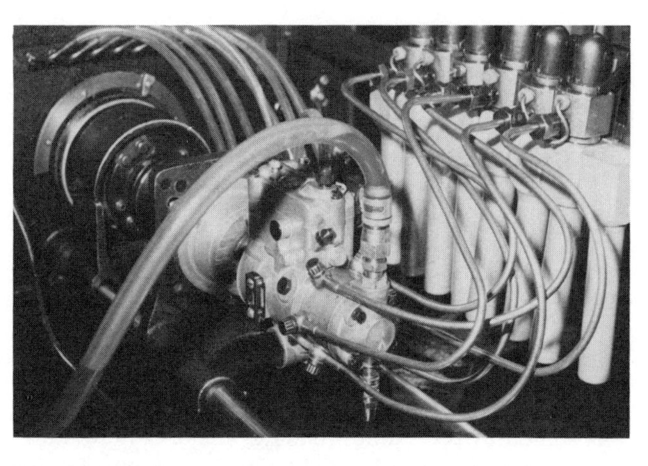
Figure 21.75 Pump correctly mounted to test bench.

NOTE On some pump models, only the shutoff cam or the cam stop will be used.

32 Insert the gauge as shown between the throttle shaft and governor linkage (Figure 21.74). The small portion of the gauge should fit, but the second, larger step should not.

33 Adjust linkage screw until adjustment is correct.

34 Check all governor parts for freedom of movement and install cover.

35 On DM and DB2 pumps, install the end cap locking plate and screw.

36 The pump can now be removed from the holding fixture and mounted on the test bench.

IV. Procedure for Calibration and Testing

After overhaul or any parts replacement, all injection pumps must be mounted to a test bench and checked for correct operation. These tests include: correct fuel delivery, governor operation, transfer pump pressure, and advance movement. Failure to do these adjustments could result in engine and/or pump damage.

Roosa Master pumps can be tested on a variety of test benches. Mounting and operating instructions vary from brand to brand. Consult the instructions manual supplied with the test bench before attaching the Roosa Master pump.

Following is the correct testing procedure after the pump is mounted on the test bench (Figure 21.75):

A *Initial run.* Supply calibrating oil to pump inlet and start test bench.

CAUTION Make certain pump is rotated in the correct direction as indicated on the specification sheet.

Drive pump at 100 rpm until fuel bleeds from all injection lines (at nozzles). Tighten the lines. Make sure the pump has transfer pump pressure. Run pump up to 500 rpm and check for any leaks. Correct all leaks before starting tests. Run at this speed for about 10 minutes to warm up pump and purge air from the system. Repair any leaks.

B *Transfer pump vacuum and pressure test.*

1 Attach a 0 to 150 psi (0 to 11 kg/cm^2) gauge to the plug on the bottom of the end plate on DB and DC pumps, (Figure 21.76) to the head locating screw on DM pumps, or to the end cap locking screw on DB2 pumps.

2 Run pump at one half the speed indicated on test specification (engine speeds) for checking vacuum. Restrict the pump fuel inlet by shutting off the boost pump and closing the inlet fuel line valve. Vacuum should be 18 in. at the speed indicated on test plan.

3 Open fuel inlet valve, turn on the boost pump, and adjust boost pressure.

NOTE Boost pressure on all models 1 to 3 psi (0.07 to 0.20 kg/cm^2), except on the DB2 which requires 5 psi (0.35 kg/cm^2) boost, and check transfer pump pressure.

Figure 21.76 Transfer pump pressure gauge attached to pump.

If the pressure is incorrect, adjust pressure on DB and DC as shown in Figure 21.77. On DM and DB2 pumps adjust pressure as shown in Figure 21.78. Clockwise adjustment will increase the pressure.

CAUTION At no time should 130 psi (9 kg/cm^2) be exceeded. This could result in head seizure.

4 Check return fuel as indicated in the specifications at this point; if not listed there, a standard specification is 150 to 250 cc per minute. A

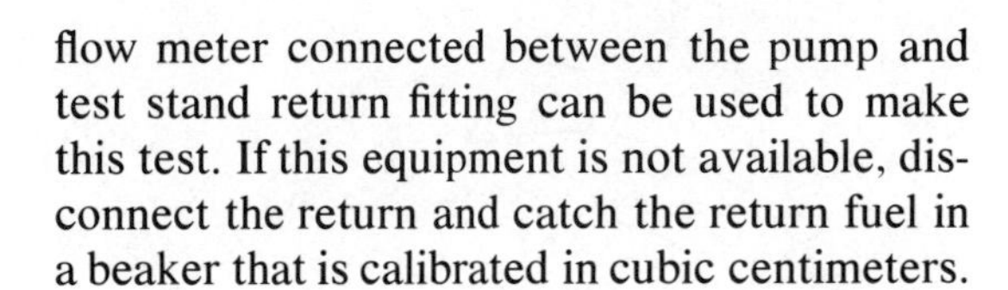
flow meter connected between the pump and test stand return fitting can be used to make this test. If this equipment is not available, disconnect the return and catch the return fuel in a beaker that is calibrated in cubic centimeters.

C *Maximum delivery check.* Before making this check, back out torque screw and high idle screw approximately five turns. Run pump at speed specified for full (rated) fuel and divert fuel flow to graduates. Collect fuel for 1000 strokes. Run at least three tests before recording delivery. (This wets the glass of the graduates and removes air from the lines.) If the delivery is incorrect, it must be adjusted by removing the following parts:

1 Governor cover.

2 Main governor spring and guide stud.

Rotate the pump until the leaf spring adjusting screw can be seen through the cam ring (Figure 21.79). With an allen wrench, adjust the leaf spring. Replace the parts removed and recheck the fuel delivery.

CAUTION Do not adjust DC or DM4 pumps on the test stand. If deliveryх is incorrect, the pump must be disassembled, the rollers readjusted, and roller centrality checked.

D *Torque control* (if used). Run pump at speed specified for torque control delivery and check delivery.

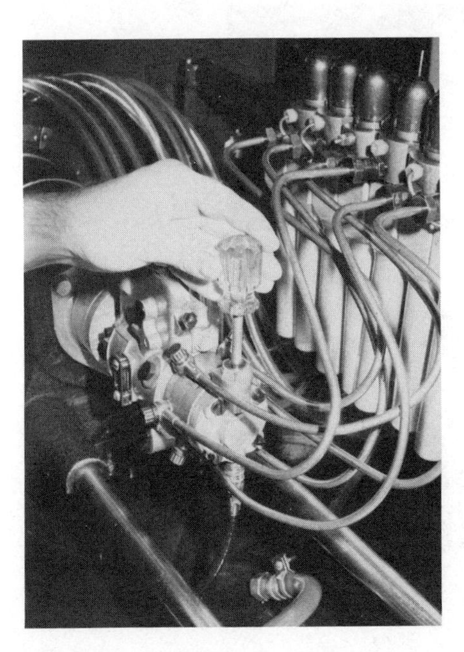

Figure 21.77 Adjusting transfer pump pressure on DB.

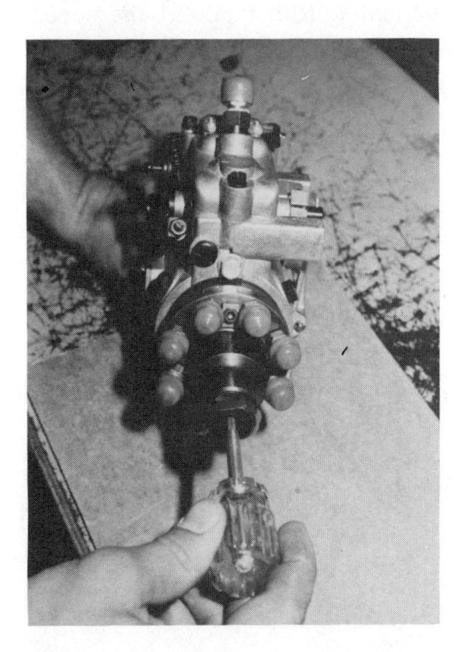

Figure 21.78 Adjusting transfer pump pressure on DB2.

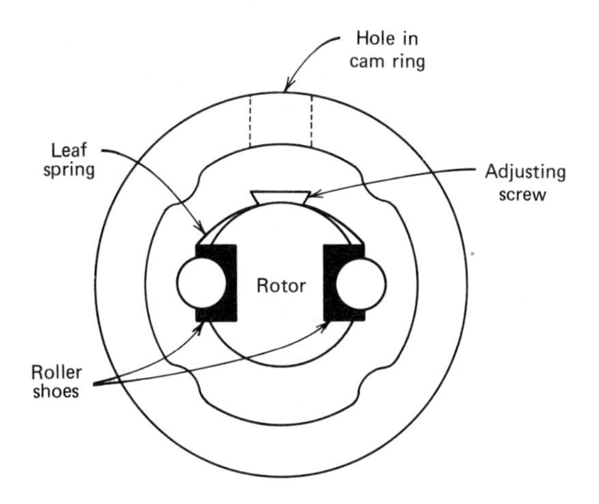

Figure 21.79 Leaf spring adjusting screw.

Figure 21.81 Adjusting high idle.

If incorrect, adjust the torque screw (Figure 21.80) to obtain the correct delivery. Clockwise adjustment decreases fuel delivery.

E *High idle adjustment.* Check the fuel delivery at the speed listed for high idle (WOT—wide open throttle). If incorrect, adjust with the high idle screw (Figure 21.81). Turning the screw inward decreases delivery.

F *Recheck torque control.* To make sure that high idle adjustment does not interfere with torque backup, torque fuel delivery must be rechecked. If this delivery is less than before, the main governor spring must be replaced.

G *Check governor cutoff.* Run pump at the speed listed for governor cutoff. At this speed, fuel delivery must not exceed that listed. If delivery is too high, recheck the governor linkage gap and wear of the metering valve. Adjust or replace as needed.

CAUTION Fuel must not exceed the quantity listed, as engine runaway will occur if not corrected. Do not run pump at cutoff speed any longer than necessary, as the rotor is insufficiently lubricated at this speed.

H *Advance adjustment.*

1 *Speed advance.* Attach timing window to pump, as shown in Figure 21.82 (if used).

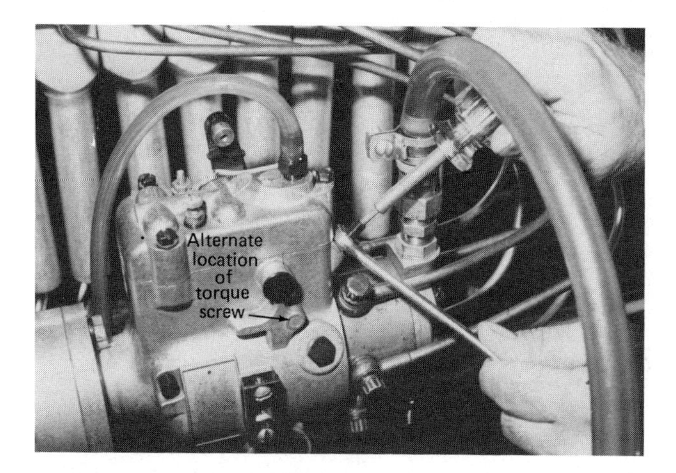

Figure 21.80 Adjusting torque screw.

Figure 21.82 Timing window used to check advance.

Figure 21.83 Adjusting advance movement.

Figure 21.85 Adjusting low idle on throttle lever.

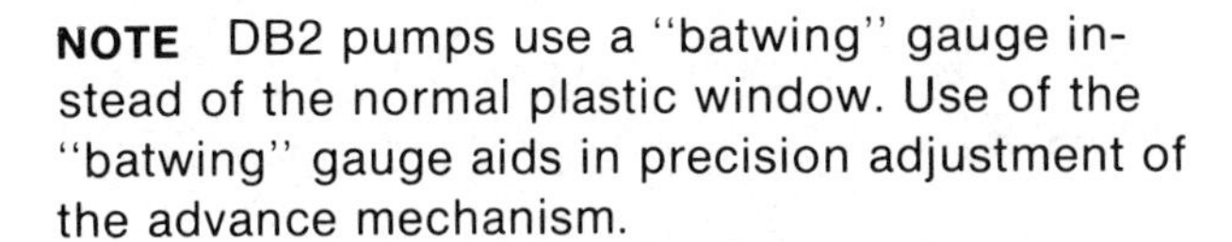

NOTE DB2 pumps use a "batwing" gauge instead of the normal plastic window. Use of the "batwing" gauge aids in precision adjustment of the advance mechanism.

Run pump at speed given for initial advance. Every line on the timing window is two degrees. If advance is incorrect, adjust the trimmer screw or shims as required (Figure 21.83).

NOTE Transfer pump pressure is critical to advance operation. Be certain pressure is correct before starting advance check. Check remaining check points.

2 *Load advance.* Attach timing window as shown in Figure 21.82. Adjust pump speed within the range given until the first delivery is obtained. Adjust advance to degree given by turning the guide stud (Figure 21.84). Turning the guide stud in will increase advance. Next, increase pump RPM, to obtain second delivery. When delivery is correct, check advance. Check the third setting in a similar manner. When adjusting delivery be sure to keep the RPM in the range given.

EXAMPLE:

All speeds are between 900 to 980 rpm.

900 rpm = 38 cc = 0° advance
940 rpm = 24 cc = 2° advance
980 rpm = 10 cc = 4° advance

I *Low idle.* Run pump at low idle speed given. Move the throttle control lever from WOT to low idle. Check delivery for 1000 strokes. If incorrect, adjust with the low idle screw on the throttle lever (Figure 21.85) or on top of the governor cover (Figure 21.86).

NOTE On model DB2, the throttle low idle position must be set before placing the pump on the

Figure 21.84 Adjusting load advance.

Figure 21.86 Adjusting low idle on top cover.

test stand. To set low idle delivery, adjust the min/max governor assembly to increase or decrease the governor spring tension on the control arm. This adustment (low idle delivery) must be rechecked after all the other settings are completed.

J *Cranking fuel delivery.* Cranking fuel delivery is critical because it is an excellent indicator of hydraulic head condition. Make certain pump is at 110 to 115° F (43 to 46° C) before running this check. Run pump at the speed listed and at WOT. Fuel delivery must meet or exceed quantity listed. If not, the hydraulic head must be replaced.

K *Leak test.* Run pump at 500 rpm and examine closely for leaks. Correct as necessary.

Remove pump from the test stand and cap all outlets with plastic caps. Connect a regulated air supply line to the outlet fitting of the pump and supply 10 psi (0.70 kg/cm^2) to the pump while immersing it in fuel oil or solvent. Any bubbles indicate a leak that must be repaired. Blow off the exterior of pump with compressed air. Place masking tape over the name tag in preparation for painting. Paint pump the same color as the engine it will be used on.

V. Troubleshooting the Roosa Master Pump on Test Bench

A *General.*

After the injection pump has been overhauled and mounted on the test bench, test the pump in the correct sequence as stated in this chapter. If a problem develops when the pump is on the test bench, check the following items first:

1. Correct injection lines being used.
2. Inlet line not leaking.
3. Test oil at correct temperature.
4. Inlet line full of calibrating oil.
5. Gauges mounted correctly and being read correctly.

B *Specific.*

Then check the following items if a specific problem continues:

1. Unable to set correct transfer pump pressure.
   - a Worn or missing transfer pump blades.
   - b Incorrect pressure regulating spring.
   - c Regulating valve sticking.
2. Full load delivery cannot be obtained.
   - a Transfer pressure too low.
   - b Test oil temperature too high.
   - c High idle screw not backed out.
   - d Cam ring in backwards.
   - e Insufficient boost pressure (DB2).
   - f Return line plugged.
   - g Hydraulic head and rotor worn out.
3. Unable to set high idle.
   - a Worn governor components.
   - b Governor linkage incorrect length.
   - c Throttle lever not against stop.
   - d Test stand not maintaining high idle speed.
   - e Guide stud bent (DB2).
   - f Incorrect assembly of governor components.
4. Housing pressure too high.
   - a Return fuel flow gauge connected.
   - b Defective return fuel connection.
   - c Plugged return line.
   - d Vent wire assembly bypassing too much fuel (DB2).
5. Pump does not deliver sufficient cranking fuel.
   - a Transfer pressure low.
   - b Throttle not in wide open position.
   - c Full load delivery not set correctly.
   - d Hydraulic head worn.
   - e Delivery valve stop loose or worn.

VI. Procedure for Installation of Pump to Engine

1. Make certain engine is on no. 1 cylinder and at the correct timing position. For example: TDC, 10° BTDC.

2 Align the two small timing marks on the pump.

3 Install new drive shaft seals with an installation tool.

4 Lubricate the seals and pilot tube of pump with vaseline or light grease.

5 Before installing the pump double check the timing by aligning the punch marks on the drive shaft and the hydraulic head.

6 Using the seal compressor, compress the seals and slide the pump onto the drive shaft.

CAUTION Be very careful not to turn the seals over when installing the pump. This will allow fuel to enter the engine crankcase.

7 Be certain the pump slides all the way up to the mounting. Do not pull the pump up with the mounting nuts.

8 Rotate the pump slightly to line up the timing marks *exactly*. Tighten the mounting bolts.

9 Bar the engine in direction of rotation two revolutions and recheck timing. Reposition the pump if necessary.

NOTE If timing changes time after time, inspect the gear train for excessive wear.

10 Install fuel injection lines being careful not to bend them. On DB and DC pumps make sure to use one washer on each side of the fuel line.

CAUTION If one washer is missing or falls out of position, the line bolt will go in too far and seize the hydraulic head.

11 Install fuel supply and return lines, throttle, and shutoff linkage.

12 Turn fuel supply on and bleed supply line up to pump inlet.

13 Provide a means of emergency shutdown.

14 Start engine and complete the job.

VII. Troubleshooting on the Engine

1 Pump cannot be timed.

- a Engine timing position not correct.
- b Timing mark was not scribed on governor weight retainer.
- c Gear train worn.
- d Timing mark on governor weight retainer scribed at incorrect location.

2 Hard starting or will not start.

- a Pump timed incorrectly to engine.
- b Fuel filters dirty.
- c Transfer pressure low.
- d Hydraulic head worn.
- e Electric shutoff solenoid defective.
- f Delivery valve stop loose or worn.
- g Delivery valve spring broken.

3 Engine runs rough or irregular.

- a Defective delivery valve.
- b Advance piston sticking.
- c Load advance not adjusted correctly (if so equipped).
- d Defective nozzle.

4 Erratic governor operation.

- a Flexible governor retainer ring broken.
- b Metering valve sticking.
- c Governor linkage set incorrectly.
- d Worn governor parts.

5 Engine starts but will not continue to run.

- a Pieces of flexible governor retainer ring plugging fuel return.
- b Plugged fuel filter.
- c Defective supply pump.
- d Air in supply line.

6 Low power.

- a Advance not working (could be caused by a partial restriction of the fuel return line).
- b Pump not timed correctly.
- c Dirty fuel filters.
- d Shutoff partially closed.
- e Engine not running at correct speed.
- f Engine camshaft gear worn.

7 Engine will not shut off.

- a Electric solenoid defective.
- b Shutoff cable broken.

c Governor linkage screw loose.

8 Dilution of engine oil.

a Drive shaft seals worn.

b Drive shaft seal turned over during shaft installation.

c Defective nozzle.

d Pilot tube worn or grooved.

9 Excessive smoke.

a Full load adjustment too high.

b Advance not working correctly.

c Pump timed incorrectly.

d Torque screw incorrectly adjusted.

e Delivery valve sticking.

f Worn governor parts.

SUMMARY

Although all diesel injection pumps are designed to do the same basic job, there are great differences between brands. This chapter has presented a complete listing of Roosa Master fuel injection pumps. The service procedures are quite different from an in-line pump, so always follow these instructions carefully. This rule applies regardless of any previous experience with other brands or types of pumps. The mechanic is urged to keep abreast of all new service information that is available. When working, problems will occasionally arise that are not covered in this chapter. In that case, consult the Roosa Master service manual or ask your instructor.

REVIEW QUESTIONS

1 What type of fuel control does Roosa Master use?

a Inlet metering.

b Pressure inlet.

c Port and helix.

2 A delivery valve is used as:

a A device to keep pressure in the pump.

b An antidribble device.

c Neither of the above.

d Both a and b.

3 Explain what each unit of the following serial no. means:
DCGFC831-9AJ. (Write on back of answer sheet.)

4 What is the difference between speed advance and load advance on a Roosa Master pump (DB)?

a Load advance is sensitive to speed and speed advance is sensitive to load.

b Load advance changes when load changes and speed advance changes when speed changes.

c Opposite of b.

d Load and speed advance work the same.

e None of the above.

5 How many pumping plungers does the DB model have?

a 2.

b 1.

c 4.

d None of the above.

6 How is the governor in the DB pump lubricated?

a Fuel oil.

b Engine oil.

c Lubriplate.

d None of the above.

7 What is the purpose of a transfer pump (T pump)?

a To deliver fuel to the filters.

b To deliver pressurized fuel to the metering valve.

c To prevent overfueling.

d None of the above.

8 Why must the metering valve be changed when changing a pump from no advance to load or speed advance?

a To eliminate overfueling.

b To increase fuel delivery.

c To eliminate governor surge.

d To develop more power.

e To get fuel to the advance mechanism.

9 How is the load advance mechanism adjusted?

a By increasing spring tension.

b By adjusting governor spring guide stud.

c By adjusting T pump pressure.

d By changing advance pistons.

10 In what location can transfer pump pressure be checked on a Roosa Master pump?

a On top of pump.

b Port of fuel filter.

c Port on side plate.

d Port on end plate.

e Two of the above.

11 Why do the plungers move completely outward during operation? (DB and DC)

a Centrifugal force.

b Fuel pressure.

c Air pressure.

d b and c.

e None of the above.

12 By what is maximum fuel delivery set on a pump that does not have a torque screw?

a T pump trimmer.

b Advance screw.

c Nonadjustable.

d None of the above.

13 The transfer pump pressure on a Roosa Master pump can be adjusted:

a With an adjusting screw.

b By adjusting screw in end plate.

c By changing filter.

d By turning cam ring around.

14 Erratic engine operation surge, misfiring and poor governing may be caused by: (on a DB)

a Cam backwards in housing.

b End plate regulating piston sticking in prime position.

c Passage from T pump to metering valve clogged with foreign matter.

d Fuel lines clogged or restricted.

15 If the T pump has pressure but no fuel is being delivered to the nozzle, the first check would be:

a Governor: not operating, parts or linkage worn, sticking or binding, or incorrectly assembled.

b Torque screw adjustment.

c Shutoff device at stop.

d Metering valve sticking or closed.

22

Cummins Diesel Fuel System (PT)

The PT diesel fuel system (Figure 22.1) is manufactured by Cummins Engine Company for use on its engines. The system is not used on any other engine. The PT fuel system was introduced on Cummins engines in 1951. It replaced another Cummins fuel system, which was called the double disc fuel system. The newer PT system was much simpler in design and application and also contained fewer moving parts. Because of this, it has become the standard for all Cummins engines manufactured.

OBJECTIVES

Upon completion of this chapter the student will be able to:

1. Explain the operating principle of a Cummins PT fuel pump.
2. List and point out on a chart the fuel flow in a typical PT fuel pump.
3. Explain with a diagram the fuel flow in a Cummins PTD injector.
4. List two differences between a PT flange type injector and a PTB injector.
5. List two differences between a PTB injector and PTD injector.
6. State one purpose of the balance orifice.
7. List two functions of the Cummins fuel injector.
8. List two problems that may result from loose injector adjustment.
9. Explain what information is contained in the injector cup number.
10. Explain what information is contained in the injector body code.
11. List two functions of the injector plunger in the PTB, PTC, and PTD injector.
12. List two problems that may result from tight injector adjustment.
13. List the two most common governor types found on Cummins engines.

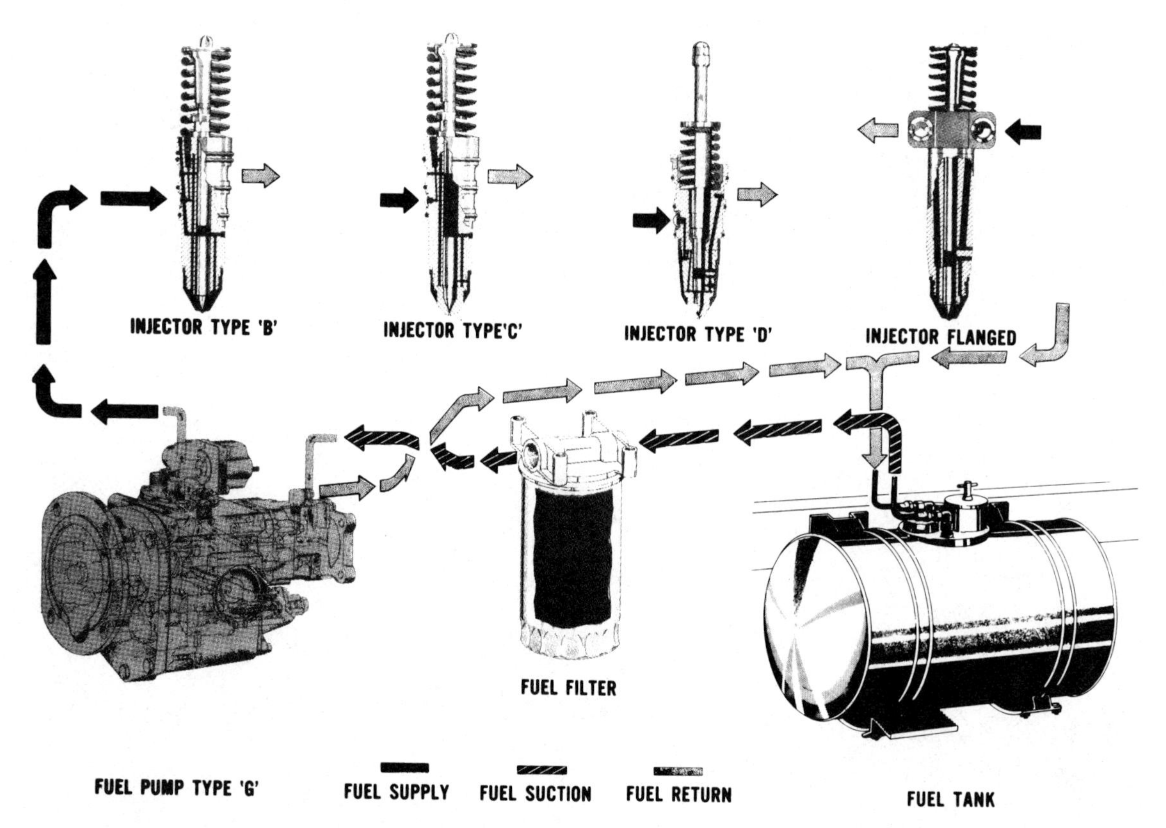

Figure 22.1 Cummins PT fuel system (courtesy Cummins Engine Company).

14 Name the component part of a PTG fuel pump that controls the amount of bypassed fuel.

15 Demonstrate the correct procedure for idle speed adjustment.

16 List one purpose of throttle leakage and how it should be adjusted.

17 Demonstrate and explain how the fuel manifold pressure may be changed in small amounts.

18 Explain how the idle spring plunger can be identified.

19 Explain how engine speed is controlled between low and high idle on an automotive type governor.

20 Explain how engine performance will be affected if the gear pump is worn.

21 List two differences between a PTD injector and an American Bosch injector.

22 Demonstrate and explain the procedures that should be used to install an injector in the engine.

23 Demonstrate and explain the following injector adjustment methods:
a Torque wrench method.
b Dial indicator method.

24 List the way in which a top stop PTD injector differs from a regular PTD injector.

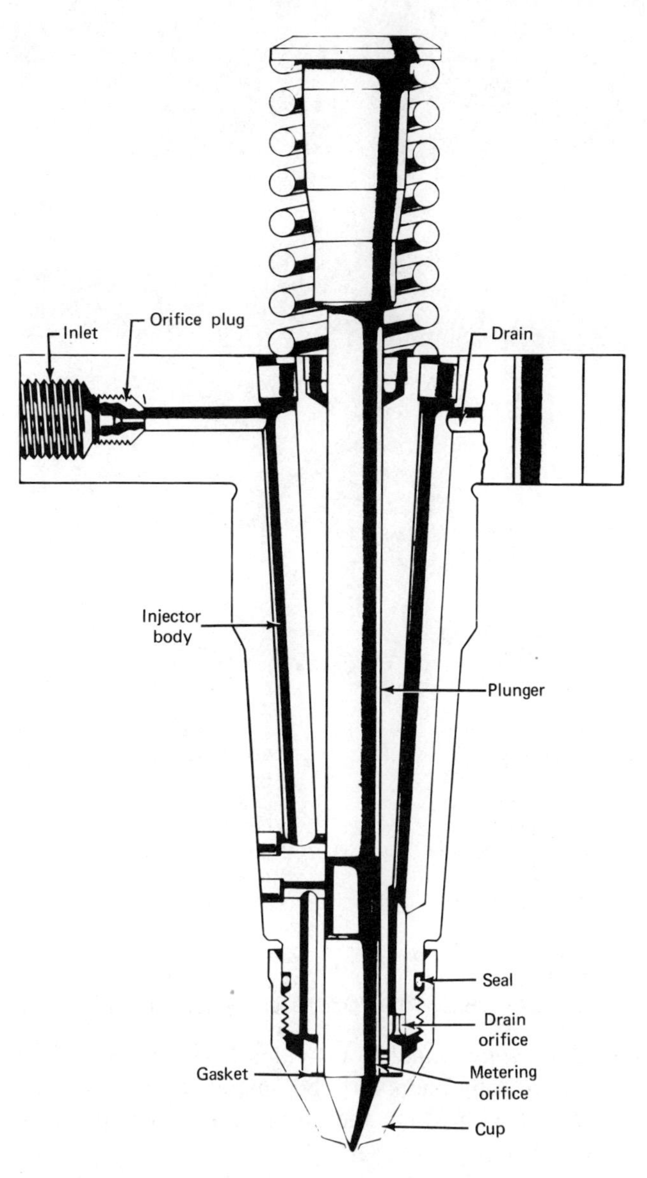

Figure 22.2 PT flange type injector (courtesy Cummins Engine Company).

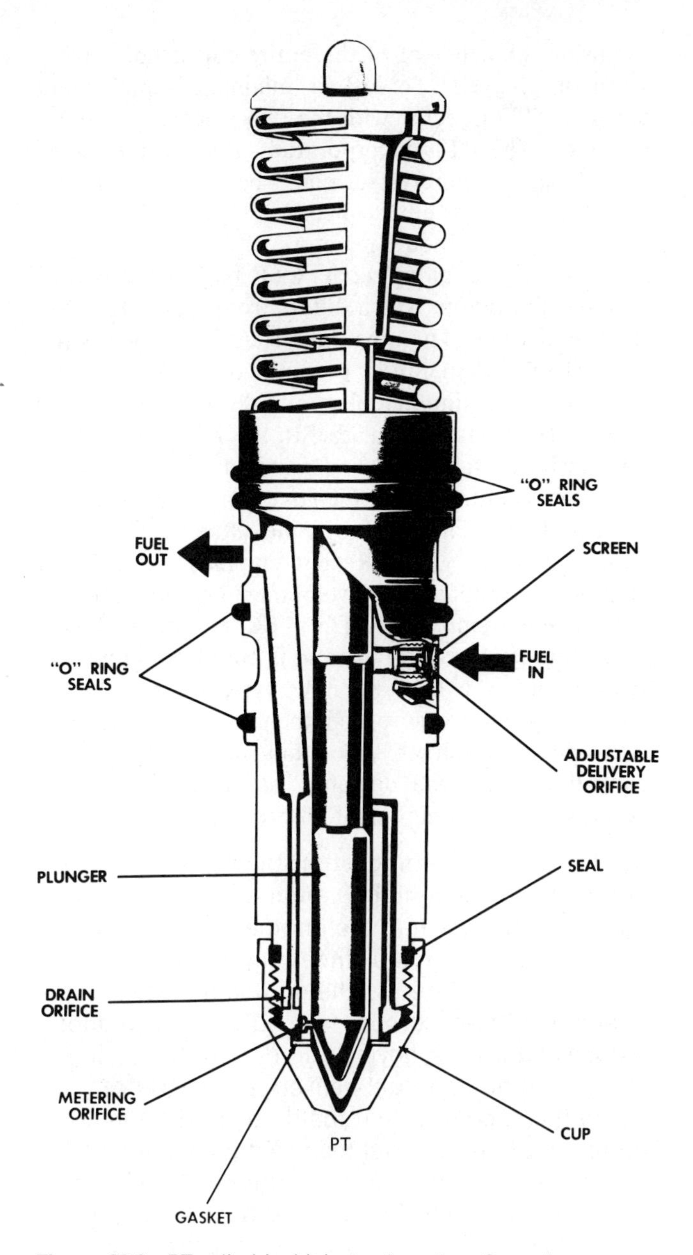

Figure 22.3 PT cylindrical injector (courtesy Cummins Engine Company).

GENERAL INFORMATION

The Cummins PT (pressure-time) fuel system's first model was composed of the PT flange type injector (Figure 22.2) and a PTR (pressure-time regulator) type pump. This PT fuel system uses a principle that is based on pressure and time. The pressure supplied to the injector is supplied by a low pressure gear pump. Time allowed for fuel measuring is controlled by the injector plunger, which opens and closes the metering orifice. This time is regulated by the speed of the engine since the injector plunger is camshaft driven. By varying the two elements of fuel pressure and time, the engine speed and horsepower are controlled.

Injectors and pumps have changed over the years as engine horsepower and exhaust emission requirements changed. From the initial flange type injector other injector models were developed. The PT cylindrical injector (Figure 22.3), first introduced with the internal fuel line engines, was a cylindrical (round) injector and utilized the same basic principle as the flange type PT. Following the PT cylindrical were the PTB and PTC injectors. The PTC was an improvement over the PTB since it used a two-piece cup or injector tip that decreased the cost of the cup. Early cups used on the PTand PTB were larger and of one piece, requiring replacement of the entire cup if holes were worn or plugged. To further minimize replacement costs, a PTD injector with the two-piece cup was introduced. The PTD incorporated a replaceable barrel and plunger assembly that could be changed without replacement of the entire injector body as had been the case in previous model injectors.

The original pump model PTR has been replaced by the PTG pump, which differs from the PTR pump in several ways. The most important difference is the method of fuel manifold pressure regulation. In the PTR pump, maximum fuel manifold was controlled by a separate pressure regulator. In the PTG the regulator has been eliminated and maximum fuel manifold pressure is controlled by the governor, hence the designation PTG. The PTG was the standard Cummins pump for many years. Recent emphasis on exhaust emissions prompted the development and utilization of the current pump, the PTG AFC pump. A device somewhat similar to an aneroid is built into the pump. An aneroid is a flow–no flow bypass valve that is operated by air pressure from the intake manifold. The AFC device differs by providing flow and pressure control to meet the demands of the engine during periods of low intake manifold pressure.

NOTE One of the biggest problems involving aneroid and turbo equipped engines over the years has been the fact that the engine acceleration was adversely affected since only a small amount of fuel was delivered to the engine until turbo or intake manifold pressure was built up. (The aneroid was an on–off device and did not have the ability to modulate fuel delivery.) The PTG AFC pump has a control unit that is built into the pump and controls fuel manifold pressure under various engine operating conditions. Since this device controls the fuel with respect to intake manifold or turbo pressure, the model designation AFC, which stands for air-fuel control, was adapted.

The PT fuel system has proven to be a reliable and rugged fuel system for Cummins engines.

I. Component Parts (PTR, PTG, and AFC)

The basic PT pump (Figure 22.4) is composed of the following component parts:

A The *gear pump* is an external type that is mounted on the rear of the pump housing. Fuel is drawn from the tank through a fuel filter and then supplied to the throttle and governor within the pump housing.

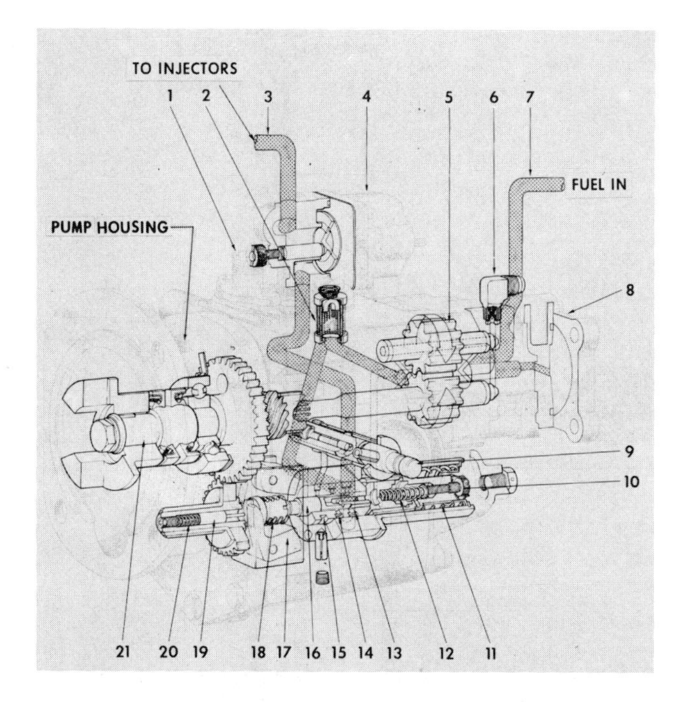

1 TACHOMETER SHAFT
2 FILTER SCREEN
3 FUEL TO INJECTORS
4 SHUT-DOWN VALVE
5 GEAR PUMP
6 CHECK VALVE ELBOW
7 FUEL FROM TANK
8 PULSATION DAMPER
9 THROTTLE SHAFT
10 IDLE ADJUSTING SCREW
11 HIGH SPEED SPRING
12 IDLE SPRING
13 GEAR PUMP PRESSURE
14 FUEL MANIFOLD PRESSURE
15 IDLE PRESSURE
16 GOVERNOR PLUNGER
17 GOVERNOR WEIGHTS
18 TORQUE SPRING
19 GOVERNOR ASSIST PLUNGER
20 GOVERNOR ASSIST SPRING
21 MAIN SHAFT
22 PUMP HOUSING

Figure 22.4 Basic PTG Pump (Courtesy Cummins Engine Co.)

B The *pump housing* provides a place for all the pump components to be installed and a flange mount with which the pump can be bolted to the engine.

C The *pulsation damper* is connected to the rear of the gear pump and contains a steel diaphragm that absorbs pulsations in the fuel that are created by the gear pump during operation.

D The *throttle* controls engine speed between low and high idle when a limiting speed governor is used. On pumps with variable speed governors, it allows the operator to adjust the engine speed to the operating conditions.

NOTE The throttle controls fuel flow to the injectors. It is composed of a shaft with a hole in it. The size of this hole can be adjusted during calibration for correct fuel flow.

E Two basic *governor* types, the limiting speed governor and the variable speed governor, are used on Cummins fuel pumps. Cummins sometimes calls them the automotive type governor and the mechanical variable speed (MVS) governor. The au-

tomotive type is used on Cummins engines in trucks, while the MVS governor is used on any application where governor action over the entire engine speed range is required. Both of the governors used are mechanical type governors. In either case the governor is mounted within or is directly connected to the main pump housing.
The following governor parts make up the governor:

1. Governor weights.
2. Governor control plunger.
3. Governor weight assist plunger and spring.
4. Torque spring.
5. Idle spring.
6. Idle spring plunger (button).

F The *pressure regulator* (type R pump only) regulates the maximum pressure that is supplied to the injectors from the pump. Its job is very important because if excessive pressure is supplied to the injectors, engine overfueling damage can result. This pressure regulator is a bypass valve that bypasses any excessive fuel flow. The PTG type pump does not use a separate pressure regulator. Pressure regulation in the PTG type pump has been incorporated into the governor (Figure 22.5).

G A special *tach drive* has been incorporated into the pump housing for the engine tachometer. The tachometer cable or electric tachometer sending unit can be connected directly to it.

H The *drive coupling* is mounted on the pump main drive shaft. The drive coupling has three tangs on it that are engaged into a rubber or plastic spider and installed on the engine, which has a similar drive coupling.

I The engine *shutdown solenoid* is an electrically operated solenoid valve, mounted on top of the pump, that allows the operator to start and stop the engine with the vehicle ignition switch. When electricity is supplied to it, the solenoid valve is open and allows fuel to flow to the injectors. When the electricity is shut off, the fuel flow to the injectors is stopped, causing the engine to shut down. The solenoid is equipped with a knob that allows it to be operated manually if the power supply should be lost. To operate the valve manually, the knob (mounted on the forward end of the solenoid) must be turned all the way clockwise. Electric operation from the key switch requires that it must be turned all the way out (counterclockwise).

J The *AFC (air-fuel control) device* controls the air-fuel ratio of the engine under varying load and speed conditions. It is operated by intake manifold pressure and spring pressure.

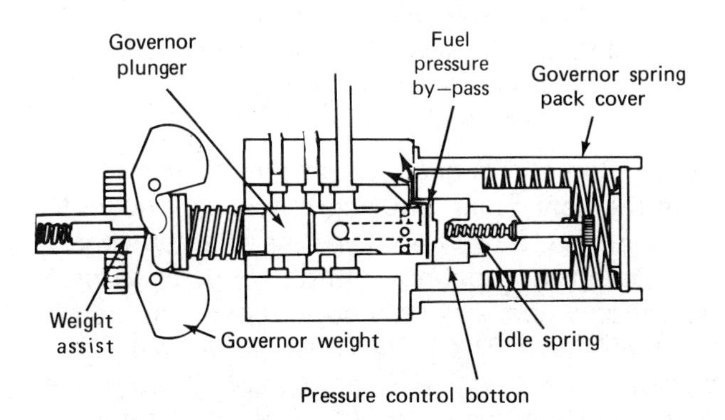

Figure 22.5 PTG governor (courtesy Cummins Engine Company).

II. Pump Identification

When pump servicing is required, identification of the pump must be made to insure correct repair and calibration procedures.

A *Pump nameplate explanation.*
All Cummins pumps have a nameplate attached to them when they are manufactured or rebuilt. This nameplate contains a code that is used for looking up the proper fuel pump setting in the fuel system data book. Several types of nameplates have been used over the years as pump models and identification procedures changed.

B *An example of the former marking procedure is illustrated in Figure 22.6.*

1. The number on the top line, 0156, represents the control parts list.
2. The next five spaces is the service shop order number.

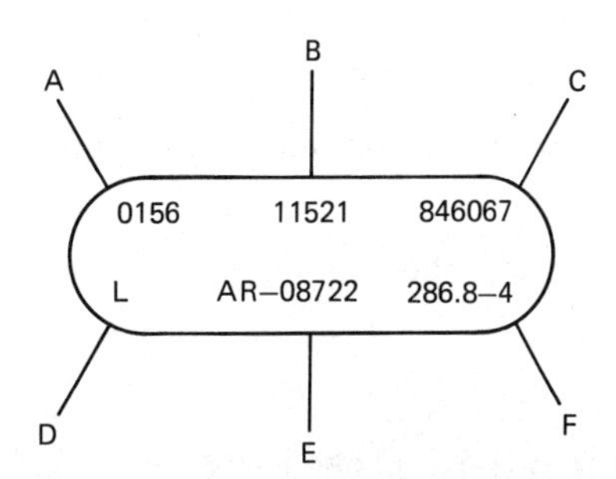

Figure 22.6 Old style identification plate. A. Control parts list. B. Shops order number. C. Pump serial number. D. "L" left-hand rotation. E. Fuel pump part number. F. Calibration card number and suffix letter (courtesy Cummins Engine Company).

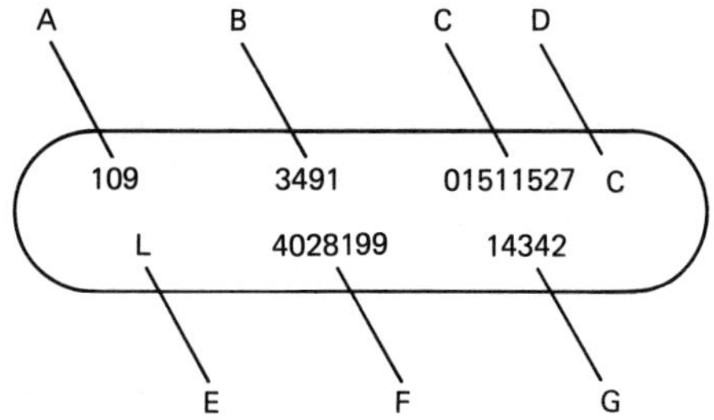

A B C D

109 3491 01511527 C

L 4028199 14342

E F G

Figure 22.7 New style identification plate. A. Control parts list. B. Fuel pump code. C. Pump serial number. D. Latest code revision. E. Left-hand rotation. F. Pump assembly number. G. Engine shop order number (courtesy Cummins Engine Company).

3 The next number represents the fuel pump serial number.

4 The L denotes pump rotation.

5 The next number is the pump part number.

6 The last number and letter indicates the calibration card number.

C *The present method of pump identification is shown in Figure 22.7.*

1 Starting with the top line from left to right, the numbers and letters represent the following information:

a CPL 109 is the control parts list number 109.

b 3491 is the fuel pump code number and must be used when looking up the pump specification sheet.

c 01511527 C represents the pump serial number and the revision letter.

2 The second line on the nameplate indicates the following information:

a The L indicates the pump rotation as viewed from the drive end.

b 4028199 is the pump assembly part number.

c 14342 represents the engine shop order number.

III. Pump Operation and Fuel Flow

As mentioned earlier in this chapter, PT stands for pressure-time, which is the basic principle of the Cummins system. Fuel control (delivery) in a Cummins PTG type fuel pump is accomplished by the utilization of a very simple principle: fuel delivery is directly related to the fuel pressure, the time allowed for delivery, and the size of the orifice through which it must flow. Fuel flow in the PTG fuel pump is shown in Figure 22.8.

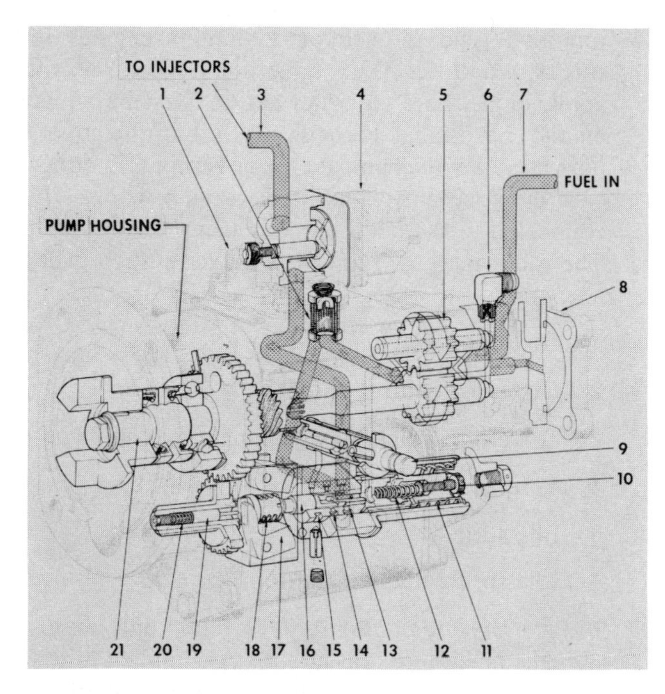

Figure 22.8 Fuel flow (PTG pump) (courtesy Cummins Engine Company).

A *Fuel flow (PTG fuel pump with automotive governor).*

The following fuel flow is given for a PTG fuel pump with an automotive governor. Its major application is on an engine that is installed in a truck.

1 Fuel is drawn by the gear pump from the fuel tank through the system fuel filter, which is located between the tank and pump.

2 Fuel is then supplied under pressure (this pressure is dependent on the speed of the gear pump and system restriction) to the filter screen within the pump housing.

NOTE Some PTG gear pumps have a bleed line from the gear pump housing that is connected to the injection return line. The bleed line prevents gear pump overheating during periods of pump operation when very little or no fuel is required to run the engine. The fuel that would normally be pumped to the engine is recirculated through the pump housing, which could cause overheating.

Throttle Leakage
Throttle Open
Throttle Closed
Idle Fuel
Fuel to Injectors
By—Pass
Weight Assist
Governor Weight
Idle Spring

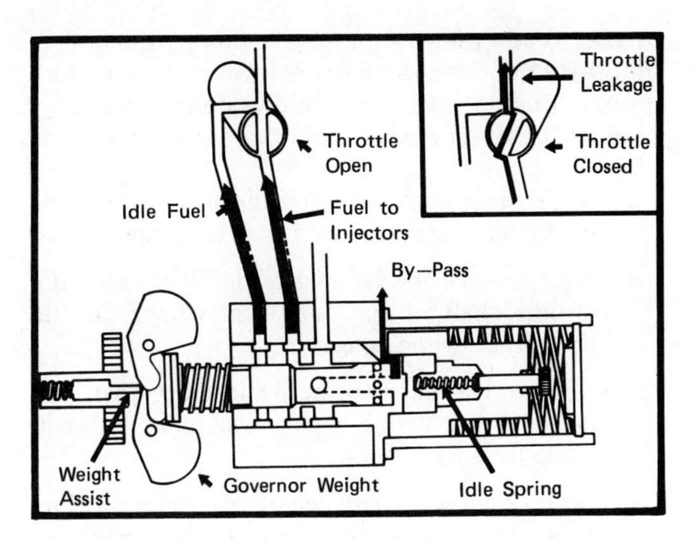

Figure 22.9 Governor plunger and barrel (courtesy Cummins Engine Company).

This operation occurs, for example, during downhill operation of a truck.

3 From the filter screen, fuel flows to the governor.

4 If the engine operating speed is somewhere between low idle and high idle, the governor will have no effect on the fuel flow. Fuel flow under those conditions is regulated by the throttle, which is controlled by the operator.

NOTE PTG type pump governors will control the maximum fuel pressure that is supplied to the fuel manifold through the shutdown valve when the throttle is in the full fuel position. This action is similar to a pressure regulator valve and effectively regulates fuel delivery to the engine.

5 If the engine reaches full speed, the governor will move the governor plunger to cover fuel passages that supply fuel to the injectors. This governor action will result regardless of the throttle position (Figure 22.9).

6 When the governor shuts off the fuel to the injectors, the fuel is bypassed back into the main pump housing.

7 Fuel in the main pump housing is used for lubrication of the governor and governor plunger. As the housing becomes full of fuel, fuel is returned to the gear pump inlet suction side to be used over again.

8 This completes the fuel flow within the pump housing. Fuel leaving the pump flows through the shutdown valve and then by a steel line to the passageways within the cylinder head.

B *Fuel flow (PTG AFC with automotive governor) (Figure 22.10).*

Fuel flow through the AFC pump is similar to the standard PTG pump with the following exceptions:

1 As the fuel leaves the throttle, it passes through the AFC unit.

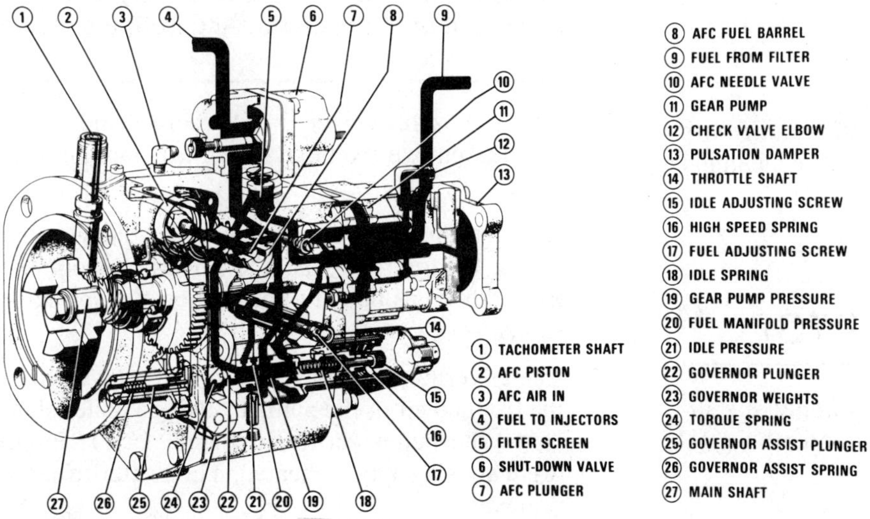

1 TACHOMETER SHAFT
2 AFC PISTON
3 AFC AIR IN
4 FUEL TO INJECTORS
5 FILTER SCREEN
6 SHUT-DOWN VALVE
7 AFC PLUNGER
8 AFC FUEL BARREL
9 FUEL FROM FILTER
10 AFC NEEDLE VALVE
11 GEAR PUMP
12 CHECK VALVE ELBOW
13 PULSATION DAMPER
14 THROTTLE SHAFT
15 IDLE ADJUSTING SCREW
16 HIGH SPEED SPRING
17 FUEL ADJUSTING SCREW
18 IDLE SPRING
19 GEAR PUMP PRESSURE
20 FUEL MANIFOLD PRESSURE
21 IDLE PRESSURE
22 GOVERNOR PLUNGER
23 GOVERNOR WEIGHTS
24 TORQUE SPRING
25 GOVERNOR ASSIST PLUNGER
26 GOVERNOR ASSIST SPRING
27 MAIN SHAFT

Figure 22.10 Fuel flow of PTG AFC with automatic governor (courtesy Cummins Engine Company).

2 When no intake manifold pressure is being supplied to the AFC unit, fuel cannot flow around the AFC plunger.

3 However, since fuel must be supplied to run the engine, during periods of low manifold pressure a bypass passage has been provided with a flow adjustment valve called a no-air screw.

4 Until intake manifold pressure rises and moves the AFC valve, fuel is supplied to the engine through the no-air valve passage.

5 This condition exists until turbocharger speed increases, resulting in increased intake manifold pressure, which is supplied to the AFC unit air chamber.

6 The AFC control plunger moves to uncover a passageway that allows fuel to bypass the no-air screw and passageway.

7 An additional drilling has been provided in the pump housing through which fuel can pass from the AFC control plunger to the shutdown valve.

8 From the shutdown valve, fuel is supplied to the engine fuel manifold.

It can be seen that the AFC unit will then regulate the fuel pressure and, consequently, the fuel delivery to the engine so that it corresponds to the amount of air pressure in the intake manifold. This delivery relationship allows the engine to develop full power at all engine speeds and load conditions. In addition, the exhaust emissions of the engine are greatly reduced.

IV. Fuel Pump Disassembly and Inspection

Before pump disassembly, clean the entire pump with cleaning fluid and compressed air. After the cleaning, proceed with the following steps:

A. Procedure for Disassembly (PTR and PTG)

The following disassembly procedures apply specifically to PTR and PTG pumps with an automotive governor.

1 Remove the shutdown valve.

2 Remove the pulsation damper and gear pump retaining bolts.

NOTE When removing the gear pump, it is not necessary to remove all the screws from the gear pump housing, since four screws hold it to the pump. The other screws hold the pump together.

3 Remove the gear pump from the main pump housing by tapping lightly with a plastic hammer.

4 Next remove the large snap ring that holds the throttle shaft in place and remove the throttle shaft from the housing.

5 Remove the four allen head capscrews that hold the governor spring pack cover onto the main pump housing.

6 With a snap ring pliers, remove the governor shim pack snap ring and the governor spring pack assembly.

7 Remove the capscrews that hold the front cover to the pump housing and remove the front cover.

CAUTION When removing the governor front cover, tip it back slightly as it separates from the main pump housing. This will prevent the governor plunger from dropping out.

8 Remove the governor plunger by pulling it from the main pump housing.

9 Remove the tachometer drive assembly retainer cap.

10 Using a brass drift and a hammer, carefully drive the tachometer assembly from the pump housing.

CAUTION The tach drive assembly must be driven out of the main housing from the inside out.

11 Using a screwdriver head socket and drive handle, remove the filter screen cap and the filter screen.

12 Remove the governor weight assist plunger and spring before removing the governor from the front cover, using care not to drop the spring or shims.

13 The governor assembly should be removed from the front cover of the pump by using a puller similar to the puller shown in Figure 22.11. Pump drive housing may be heated in hot water to aid in removal.

14 Before removal of the pump drive shaft and gear, a snap ring behind the drive gear on the inside of the drive housing must be removed.

Figure 22.11 Governor assembly puller (courtesy Cummins Engine Company).

15 The next step is to remove the drive shaft from the housing.

- **a** First install a long bolt in the drive hub retainer capscrew hole and support and drive housing in a press.
- **b** Force out the drive shaft by pressing on the bolt.

16 Remove the drive shaft seals from the front housing with a hammer and seal driver and discard.

17 Clean all pump parts thoroughly by soaking them in a cleaning solvent. Then rinse and blow off with compressed air.

B. Procedure for Disassembly (PTG AFC Type with Automotive Governor)

Disassembly of the AFC pump differs in several ways from the PTR and PTG pumps. Those procedures that differ from the PTR and PTG pumps are listed.

1 *Throttle assembly.* The throttle assembly cannot be removed on an AFC pump until the snap ring located inside the pump housing is removed. This requires that the front drive housing be removed. After removal of the internal snap ring, the cover plate on the outside of the housing must be removed. This plate is held in place with drive screws (screws that are driven in with a hammer). These screws and plate must also be removed before the throttle can be removed.

2 *"No-air" bleed screw.* This adjusting screw should be removed so that a new O ring can be installed on it. To remove, loosen the locknut and turn the screw counterclockwise.

3 *Air-fuel control.* Mount the pump housing on a suitable bracket so that it can be mounted in a vise for removal of the AFC assembly.

- **a** Remove the three cover plate screws.
- **b** Remove the cover plate.
- **c** Remove bellows, AFC, and plunger from the pump housing.
- **d** Lift out the bellows spring and shim.
- **e** If pump is to be completely disassembled, a snap ring must be removed before the AFC barrel can be removed.
- **f** Pull the barrel from the housing and remove the barrel O ring.

C. Procedure for Inspection (PTR and PTG)

All pump parts must be inspected carefully before reuse. The following inspection procedures must be closely followed:

1 The *pump housing* contains the throttle shaft bushing and the governor plunger (bushing). These two bushings must be closely inspected to determine if the housing is reusable.

2 The *throttle shaft bushing* should be inspected for scoring and wear. If the bushing is worn and some doubt exists about the fit of the throttle shaft into the bushing, an oversize throttle shaft must be fitted.

3 The *governor bushing or barrel* should be inspected closely for wear. It is replaceable in the housing, but in most cases the housing is replaced with a rebuilt housing, if the barrel is badly scored.

4 The *governor plunger* should be inspected closely for wear. If it is worn excessively or scored, the plunger should be replaced.

NOTE New governor plungers can be fitted to an old governor barrel if the plunger is worn. This is a selective fit process that is done by inserting progressively larger plungers into the barrel until the plunger will no longer enter the barrel. Then select a plunger two sizes smaller than the one that would not enter the governor barrel bore. This is the plunger to use. Plungers are identified by their color (Figure 22.12).

If the governor thrust washer is worn, it can be replaced by removing the roll pin in the plunger and then installing a new washer and roll pin.

NOTE A rebuilt pump housing is supplied with a new governor plunger and throttle shaft.

5 The *governor weight assembly* should be checked for wear at the weights and pins. If the weights and pins are worn, the weight carrier should be replaced.

6 The *governor carrier bushing* should be checked for wear and, if worn, should be replaced.

7 Check the *drive shaft* closely for wear at the point where the seals ride on it. Check the drive shaft ball bearing. If it is rough, replace the bearing.

8 Disassemble the *gear pump* and check for visual signs of wear. A worn gear pump should be replaced with a rebuilt one.

9 The *tachometer drive* should be checked for wear in its bushing. The normal clearance between the shaft and bushing is 0.002 to 0.003 in. (0.05 to 0.08 mm). If worn in excess of this, replace the bushing and/or shaft.

10 In most cases the *shutdown valve* is not disassembled unless a problem exists with it. If it works when power is applied to it, it can be used again. If it does not work, the solenoid part of the valve can be replaced easily by removing the four screws that hold it to the shutdown valve and installing a new one.

D. Procedure for Inspection (PTG AFC)

The inspection of the PTG AFC pump is similar to the PTR and PTG pumps with exception of the AFC plunger and bellows assemblies.

TABLE 22-1: CURRENT PT (TYPE G) GOVERNOR PLUNGERS

Code	Red	Blue	Green	Yellow	Orange	Black	Gray	Purple	Usage	Code Letter
Size	0	1	2	3	4	5	6	7		—
Part No.	169660	169661	169662	169663	169664	169665	169666	169667	Standard	J
Part No.	182530	182531	182532	182533	182534	182535	182536	182537	SVS. Gov.	—
Part No.	159320	159321	159322	159323	159324	159325	161586	161587	JT	G
Part No.	168630	168631	168632	168633	168634	168635	168636	168637	V12, H, J, V6, V8	H
Part No.	203350	203351	203352	203353	203354	203355	203356	203357	Lower VS	
Part No.	213240	213241	213242	213243	213244	213245	213246	213247	V-378	
									V-504, V-555	
Part No.	212350	212351	212352	212353	212354	212355			Upper VS	
Part No.	213610	213611	213612	213613	213614	213615			Upper VS	
Part No.	3009380	3009381	3009382	3009383	3009384	3009385	3009386	3009387	AFC-5° End	

TABLE 22-2: CURRENT PT (TYPE R) GOVERNOR PLUNGERS

Code	Red	Blue	Green	Yellow	Orange	Black	Usage	Code Letter
Size	0	1	2	3	4	5		
Part No.	102510	102511	102512	102513	115584	115585	Standard	A
Part No.	70810	151941	151942	151943	151944	151945	MVS. Gov.	D
Part No.	114690	114691	114692	114693	115604	115605	MVS. Gov.	C
Part No.	151970	151971	151972	151973	151974	151975	MVS. Gov.	E
Part No.	105030	105031	105032	105033	115554	115555	Torque Converter Governor	K
Part No.	159410	159411	159412	159413	159414	159415	Torque Converter Governor	M
Part No.	109960	109961	109962	109963	109964	115915	Road Speed Gov.	B
Part No.	161110	161111	161112	161113	161114	161115	MVS. Gov. with PT (type B) Injectors	F

Figure 22.12 Plunger identification (courtesy Cummins Engine Company).

1 The AFC bellows should be inspected for cracks or breaks and replaced if any doubt exists about their condition.

2 Check the piston for scoring and scratches.

3 Check the AFC barrel for scratches and scores.

4 Remove all O rings from the AFC barrel and discard.

V. Pump Assembly and Calibration

A. Procedure for Assembly (PTR and PTG)

Before attempting to reassemble the pump, make sure that all parts have been thoroughly cleaned and inspected. New gaskets and seals should always be used when reassembling the pump.

1 Mount the pump housing on a suitable bracket that can be clamped in a vise or in a pump work stand, as shown in Figure 22.13.

2 Install the governor idle spring washer over the idle adjusting screw in the governor spring plunger (Figure 22.14).

3 Install the idle spring over the idle adjusting screw in the governor spring plunger.

Figure 22.13 Pump housing mounted on holding bracket for reassembly.

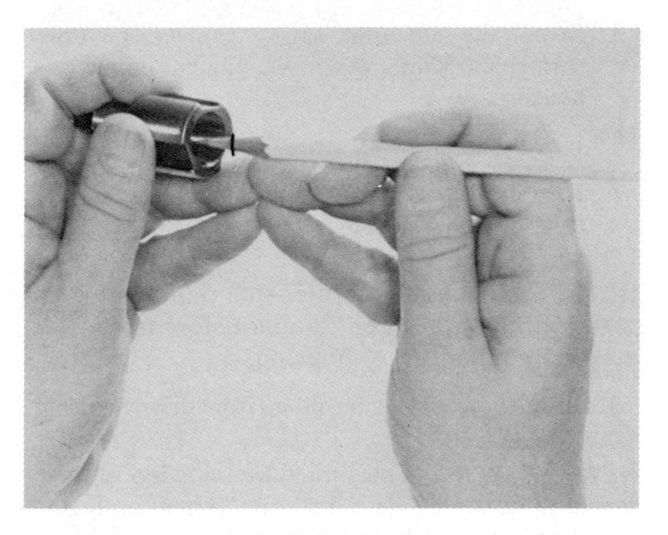

Figure 22.14 Installing idle spring washer on adjusting screw (courtesy Cummins Engine Company).

4 Install the idle plunger button over the idle spring and into the spring plunger (Figure 22.15).

NOTE The idle plunger button regulates maximum fuel pressure and, consequently, engine horsepower. It should be selected for the pump being worked on by referring to the pump code and calibration data.

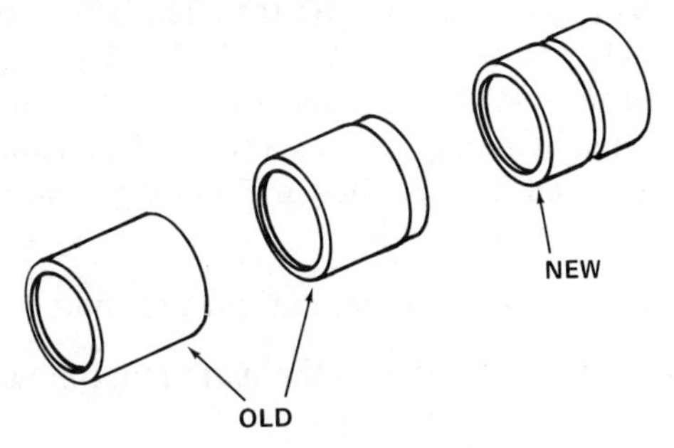

Figure 22.15 Installing the idle spring plunger (courtesy Cummins Engine Company).

5 Install the governor spring plunger, governor spring, and shim pack in the plunger bore in the housing.

6 Install the governor spring snap ring.

7 Lubricate the governor control plunger and install it in the pump housing.

NOTE Make sure the torque spring (if used) is installed on the governor plunger before it is installed in the governor housing.

8 Install new seals in the pump front drive housing.

NOTE Drive shaft seals must be installed in different directions. The outer drive seal should be installed with the lip to the outside of the pump housing. The inner seal should be installed with the seal lip toward the rear of the pump. The seals should not be pressed tightly together or the weep hole between them will be plugged. Install the seals flush to the housing shoulder; press the shaft and bearing into the housing.

9 If a seal protection tool is available, install it on the drive shaft. Place the shaft retaining snap ring on the shaft between the governor drive gear and bearing.

10 Install the shaft in the pump front housing and remove the seal protection tool (if used).

11 If a seal protection tool is not available, lubricate the shaft with engine oil and install it in the seals carefully, turning the shaft and working it from side to side. This will help prevent rolling of the seal lip.

12 Using a snap ring pliers, compress the retainer ring and insert it in the groove in the front housing.

13 Install the governor weight assembly and bushing into the front housing. On early pumps the governor weight assembly is held into the bushing with a snap ring that must be installed before the assembly can be pressed into the housing. Current governor weight assemblies are not held into the bushing by a snap ring; as a result the bushing can be pressed into the housing and then the weight assembly installed into it.

14 Place a new O ring on the throttle shaft.

15 Insert the throttle into the main pump housing.

NOTE Current PTG AFC pumps use two different throttle shafts. One has a threaded screw for adjusting fuel flow and the other has a shimmed plunger in the throttle shaft. The throttle shaft with the threaded fuel adjusting screw is held in place with a snap ring inside the pump housing. The shimmed type restriction plunger should be shimmed so that the hole in the throttle is wide open. The adjusting plunger in either throttle should not restrict the throttle opening during pump calibration. As a result, the screw in the throttle should be turned out four or five turns in order not to restrict the throttle opening.

16 Depending on which type throttle assembly you have, insert a small snap ring on the inside of the pump housing or insert a large snap ring on the outside of the pump housing.

17 Note the position of the oil groove in the tach drive bushing. This oil groove should line up with the fuel pump drive during installation.

18 Using a special bushing driver, ST-1032 or an equivalent, drive the tach bushing and tach drive into the pump housing.

19 Install the tach drive seal spacer and drive seal.

CAUTION Use a seal driver and do not drive the seal down too far or it will crush the drive seal spacer.

---

20 Install the dust seal with white side up and then install the retainer and screws.

21 Place the front housing gasket on the main pump housing.

22 Place the governor plunger drive tang in the horizontal position.

23 Install the weight assist plunger with spring and shims in the governor shaft bore.

24 Place the governor weight assembly in the horizontal position.

25 Install the front housing, making sure that the governor control plunger tang engages the slot between the governor weights.

26 Install bolts in front housing and torque capscrews to 10 ft lb (13.56 N m).

27 Install a new gasket on the gear pump.

NOTE Gear pumps must be correctly installed, dependent on the pump rotation. Right-hand rotation pumps require the notch in the housing to be in the upper right-hand corner and left-hand pumps require the pump housing notch to be in the left-hand corner. In-line pumps have pulsation dampers on the left side viewed from the drive end. V-8 engine pumps have the pulsation dampers on the right side viewed from the drive end.

28 Install gear pump.

29 Install the pulsation damper on the back of the gear pump and insert bolts

30 Install the gear pump, damper, and capscrew and tighten to 10 ft lb (13.56 N·m).

31 Determine what type fuel inlet fitting is required for the pump being repaired.

NOTE Some pumps require a fitting with straight thread and an O ring, while other pumps use a fitting that has a tapered pipe thread.

32 Install correct inlet fitting in gear pump.

33 Install fuel filter and cap in pump housing.

34 Install fuel shutdown valve and torque capscrews.

B. Procedure for Assembly (PTG AFC with Automotive Governor)

The assembly of the AFC pump is similar to the standard PTG pump with the following exceptions:

1 The AFC plunger must be installed in the pump housing using new O rings.

NOTE Early AFC barrels were made from cast iron. They have been replaced by a steel barrel that has better wear characteristics. Use the new steel barrel when rebuilding a pump. Steel barrels can be identified in two ways: either a groove is cut around the top of the barrel or there is an S before the part number (Figure 22.16). Before you install the AFC barrel, place torque spring no. 139585 behind the barrel and install a flat snap ring. The AFC pump was not originally equipped with a torque spring behind the AFC plunger, but field experience has proven that the barrel moves

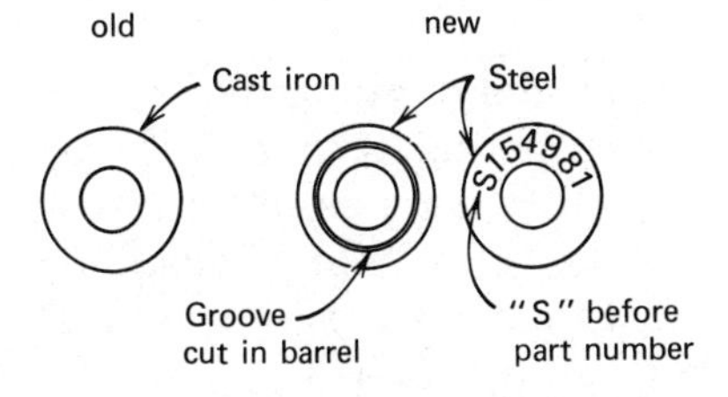

Figure 22.16 AFC barrel identification (courtesy Cummins Engine Company).

in the housing, causing incorrect engine operation.

2 Next install the bellow spring steel washer in the pump housing.

3 Now install the bellows spring on the steel washer in the AFC housing.

NOTE Bellows springs will have to be held in the housing unless the housing is in a horizontal position.

4 Make sure the O ring groove in the AFC plunger is clean, and then lubricate and slide the O ring on the plunger.

5 Use tool no. 3375146 to install the "glyd" ring on the AFC plunger and in the O ring groove.

NOTE The "glyd" ring sits on top of the O ring in the plunger groove and is made from teflon.

6 Install the bellows piston on the AFC plunger and tighten the nut to 30 to 40 in. lb (3.39 to 4.52 N·m).

7 Install the AFC plunger in the glyd ring-forming tool to form the "glyd" ring before the AFC plunger is installed in the barrel.

NOTE If the "glyd" ring is not formed before installation, it may stick in the barrel.

8 Carefully insert the AFC plunger in the AFC barrel and slide it out of the forming tool.

NOTE The forming tool will act much like a ring compressor when you install the AFC plunger in its barrel.

9 Push the AFC plunger into the pump and position the bellows in the bellow housing.

10 Make sure the holes in the bellows line up with the holes in the pump housing and then install the AFC cover plate.

11 After lubrication, install the O ring on the no-air needle valve and insert the valve into the pump housing, turning the valve in until it seats. Leave the locknut loose until the pump has been run on the test stand.

C. Procedure for Pump Calibration and Testing

Mount the pump to be tested on the test stand and prepare the pump for test stand operation. Make sure all connections are of the correct type and are in good condition. Any suction leaks on the inlet line would allow enough air to enter the injection pump so that proper calibration would be impossible. Before attempting to calibrate the pump, look up the calibration data by using the pump code number found on the pump nameplate.

Although the following calibration instructions are for a PTG AFC pump, the information can be used to calibrate a PTG pump without an AFC.

1 Remove the three capscrews that hold the AFC cover plate on the main pump housing.

2 Install AFC adjusting tool no. 3375137.

NOTE When installing the service tool, check to be sure that the bellows retainer nut fits in the large socket. The small socket should fit on the adjusting screw locknut. The allen wrench tool should fit in the AFC adjusting screw.

3 Install the capscrews and torque to 30 to 35 in. lb (3.39 to 4.52 N·m).

CAUTION *Do not overtighten!*

4 Supply 25 psi (1.75 kg/cm^2) air pressure to the AFC control unit. (The air line should have been connected during pump mounting and hookup.)

5 Screw the knurled knob on the pump shutdown valve all the way in. This opens the shutdown valve, allowing fuel to flow through it.

6 Open the manifold valve on the flow panel or test stand.

7 Start the stand and run the pump at 500 rpm with the throttle in the wide open position.

NOTE As you start the test stand, watch the plastic inlet line for fuel flow; if the pump does not pick up fuel soon after it is started, check to determine the reason. Extended operation of the pump without fuel may damage it.

8 Increase the pump speed to 100 rpm below rated speed.

9 Adjust the inlet restriction valve so that 8 in. of vacuum shows on the vacuum gauge.

10 Run the pump for 5 minutes to seat new bearings or bushings and bleed all the air from the pump.

11 Check the gear pump suction:

a Run the pump at 500 rpm.

b Close the suction fitting valve or restriction valve.

c The vacuum should increase to 25 in.

d If the vacuum is low, replace or rebuild the gear pump.

12 Adjust the flow control valve so that the flow meter indicates the flow specified in calibration data (pump throttle in wide open position).

13 Readjust the suction fitting valve with the pump running at 100 rpm below rated speed to show 5-8 in. of vacuum on the vacuum gauge.

NOTE Make sure no air is visible in the fuel passing the flow meter. If air is visible, check the pump carefully for vacuum leaks.

14 Check governor cutoff rpm by increasing the pump speed until the fuel pressure begins to drop. This drop should be at the rpm indicated in the pump calibration data. (Pump throttle must be in wide open position.)

NOTE If the cutoff speed does not match calibration data, the governor spring must be shimmed. Adding shims will increase the cutoff speed and removing shims will decrease the cutoff speed. The governor cutoff speed will be changed 2 rpm by every 0.001 in. (0.025 mm) of shim. To install or remove shims from the governor spring pack, the governor spring cover must be removed. Before installing the cover, make sure the gasket is in good condition or a suction leak may result.

15 Run the pump at 1500 rpm and move the throttle back and forth to remove all air from the system.

16 Increase the pump speed until the fuel pressure begins to drop. This should match the pump high idle rpm listed in the calibration data.

17 Open the throttle leakage valve and close the manifold or flow valve. Check the throttle leakage by placing the throttle in idle position and running the pump at rated speed.

NOTE Unless your test stand is equipped with a throttle leakage flow meter or bypass valve, the line that is connected to the fuel cutoff solenoid must be removed and directed into a beaker that will measure the amount of throttle leakage. Fuel is generally directed into the beaker for 1 minute.

CAUTION Do not run the pump in this condition for any extended period of time or it may overheat, causing pump damage. Pump temperature during the test should be checked by feeling the pump. Normally the pump will feel warm, but not hot, to the touch.

Figure 22.17 Adjusting rear throttle stop screw (courtesy Cummins Engine Company).

Figure 22.18 Checking throttle lever position with protractor (courtesy Cummins Engine Company).

18 Compare the amount of fuel in the beaker to calibration data sheet.

19 If the throttle leakage does not correspond to the amount specified, adjust the rear throttle stop screw on AFC pumps in or out until the correct flow is obtained (Figure 22.17). Standard PTG pumps without AFC have a different stop screw setup; throttle travel to the rear is adjusted by the front stop screw, and throttle travel to the front is adjusted by the rear stop screw.

NOTE Throttle leakage prevents under running of the engine during rapid deceleration, and engine stalling during rapid acceleration and the time required for the engine to decelerate. This is a critical setting, so make sure it is correct.

20 Check the throttle lever position using a throttle adjusting protractor (Figure 22.18). When the throttle is in the idle position (all the way back), the throttle lever center line should line up between the two drilled idle holes in the protractor.

21 If the throttle does not index properly with the idle holes in the protractor, loosen the throttle clamp screw and move the throttle accordingly.

NOTE Make sure the throttle shaft is all the way back against the rear throttle stop screw before adjusting the lever position. *Do not* adjust the rear throttle stop screw to obtain the correct lever position. This adjustment was made during the throttle leakage test and must not be changed.

22 Check the throttle lever forward movement. Total movement from idle to full fuel should be as listed in the calibration data (usually 28°). If total travel is not correct, adjust the front stop screw.

23 Check the pump idle speed setting with the pump running at the specified idle speed.

24 Close the main flow valve and open the idle valve.

25 Make sure the throttle is in the idle position and hold it there.

26 The pressure on the fuel pressure gauge should now match specifications.

27 If pressure is low or high, stop the pump. Remove the pipe plug from the governor spring pack cover and adjust the idle adjusting screw in for more fuel and out for less.

NOTE The idle adjusting screw is held in place by a spring lock that clamps on the outside of the screw head. When adjusting the screw, a slight click will be felt as you turn the screw. If the idle pressure cannot be adjusted, a second spring washer may need to be added to the idle adjusting screw. If this is necessary, the idle spring pack cover must be removed and the governor spring plunger and spring pack removed from the pump housing by taking out the retaining snap ring (Figure 22.19).

28 After the idle pressure setting is satisfactory, open the manifold valve and close the idle flow valve.

29 Run the pump at rated rpm and compare pressure and flow to specifications.

NOTE Recheck the vacuum gauge during the test to insure that it stays at 8 in.

30 If pressure on the fuel pressure gauge is not correct, the throttle restriction must be changed.

Figure 22.19 Governor spring retaining snap ring (courtesy Cummins Engine Company).

NOTE Throttle restriction is changed on some pumps by removing the throttle shaft and adding or subtracting shims under the throttle restriction plunger (Figure 22.20). Other pumps have a fuel flow adjusting screw that can be adjusted with an allen wrench. The allen wrench screw is located inside the outer end of the throttle shaft (Figure 22.21). At the end of the throttle shaft a steel ball has been inserted. This ball must be removed to gain access to the adjusting screw. If the ball cannot be removed, a new throttle shaft must be used and a steel ball inserted after calibration. Since throttle shafts are available in five oversizes, make sure you select a throttle shaft of the correct size.

CAUTION The throttle shaft is matched to the pump housing bore and cannot be replaced with just any throttle shaft. Follow the throttle shaft color code and size closely when replacing the throttle shaft.

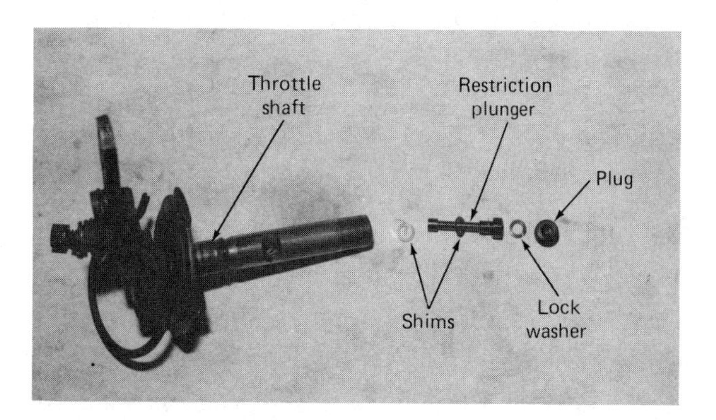

Figure 22.20 Throttle restriction plunger.

Figure 22.21 Throttle shaft with adjusting screw (courtesy Cummins Engine Company).

31 After adjustment has been made to the throttle shaft restriction, adjust the flow to specification and check pressure.

32 Reduce the pump speed to check the torque backup. This speed is given in calibration data and is listed as check point no. 1.

33 Adjust the manifold valve for correct flow listed at this rpm.

34 At this point the pressure on the pressure gauge should correspond to specifications.

NOTE If the torque backup fuel delivery is not correct, the pump must be disassembled and the torque spring checked.

35 To check the torque spring, disassemble the pump at the front drive housing.

36 Make sure the torque spring has the correct color code for the pump being calibrated.

NOTE After changing the spring it is necessary to recheck all the previously made checks, such as governor cutoff, fuel pressure, and fuel flow at rated speed.

37 If the rated speed check point is all right, set the pump speed to the speed indicated in check point no. 2.

38 If the pressure is not as specified, the pump must

be disassembled again and the weight assist plunger protrusion checked. The weight assist plunger protrusion can be changed by adding or subtracting shims between it and the spring.

NOTE The weight assist plunger protrusion must be decreased to lower pressure and increased to raise the pressure. After making any adjustment to the weight assist plunger, the pump high idle governor cutoff and idle fuel pressure must be checked.

39 If the pump is equipped with an AFC device, it must be checked at this time for correct operation.

40 Locate the no-air adjusting valve behind the throttle lever. Loosen the locknut and screw the valve all the way in until it is seated.

41 Run the pump at the AFC rpm (taken from the pump spec sheet) with the manifold valve open and the idle valve closed.

42 Shut off the air supply to the AFC diaphragm.

43 A pressure reading of 5 psi (0.350 kg/cm^2) or less should result. Make sure the flow meter and pressure gauge have stabilized before making the pressure reading.

44 Adjust the air pressure supplied to the AFC bellows to the pressure indicated in the specification sheet.

NOTE Make this pressure adjustment slowly so that you do not exceed the recommended setting.

45 After the pressure has been adjusted, adjust the manifold valve or flow meter to the flow called for in the specification sheet.

46 If the pressure does not correspond with the specification, the AFC plunger must be adjusted. This adjustment is made with the adjusting tool that was installed in place of the AFC housing during the pump mounting.

47 To adjust the plunger, turn off the air supply to the AFC diaphragm.

48 Push in on the larger diameter of the tool, turning it back and forth slightly to engage the AFC diaphragm locknut.

49 Next push in on the smaller diameter of the tool to engage the small socket head with the AFC plunger and then push the socket head wrench into the end of the AFC plunger.

50 Adjust the AFC plunger.

51 After adjusting the AFC plunger, run the pump with no-air pressure until flow and pressure stabilize.

52 Turn on the air supply and adjust to the specified pressure.

53 Check the flow meter and make adjustments as required.

NOTE The adjustment of the AFC unit may have to be repeated until the correct flow and pressure are obtained.

54 After adjustment of the AFC plunger, it should be checked for resistance to movement within its barrel.

55 Set the air pressure at 25 psi (1.75 kg/cm^2) and then decrease to the AFC psi listed in the specification sheet.

56 As the air pressure is changed from high to low, the fuel pressure should not change over 15 psi (1.05 kg/cm^2).

57 If the pressure exceeds 15 psi (1.05 kg/cm^2), check the AFC plunger for freedom of movement.

58 Recheck the pressure difference.

59 Shut off the air supply to the AFC diaphragm.

60 Start the pump and run it at the rpm specified for the no-air setting.

61 With the pump running at this rpm, adjust the no-air screw with tool no. 3375140 if available.

NOTE In the event that a tool is not available, the no-air screw can be adjusted with a deep, thin wall socket and a screwdriver. As the pressure setting is changed, the manifold or flow valve will have to be adjusted to maintain the proper flow.

After the correct pressure is obtained, lock the no-air screw locknut.

62 When the pump calibration is complete, the throttle shaft restriction adjustment should be sealed so that it cannot be tampered with. This is done by installing a steel ball in the end of the shaft.

63 The steel ball must be installed with special tool no. 3375204 to prevent changing the throttle shaft adjustments. *Do not use a hammer and a punch.*

64 Remove the AFC adjusting tool and install the AFC cover. Make sure the plunger spring is seated correctly in the bellows before installing the cover.

65 Install the seal wires to seal the AFC cover, governor spring pack, and front housing.

66 Plug all inlet and outlet ports with plastic caps.

67 Tape off the nameplate and paint pump.

VI. Troubleshooting the Cummins Fuel Pump

The following procedures should be used as a guideline when troubleshooting the Cummins fuel pump on the test stand (A to G) and on the engine (H to M).

A If the pump will not pump fuel after initial installation on the test stand (no flow shown in flow meter), the following procedures should be followed:

1. Loosen the line, recheck all fittings, and tighten.
2. Determine if the shutoff solenoid is in run position.
3. Check the pump nameplate to see if the pump rotation is correct.
4. Check the fit between idle spring plunger and governor plunger. It may be necessary to change one or both to obtain a good fit between them.
5. Check the gear pump suction to determine if the gear pump is worn out.
6. Make sure the gear pump is installed correctly on the main pump body.

B Check to see if there is aeration of fuel in the flow meter. This indicates an air leak somewhere on the suction side of the gear pump.

NOTE Any leakage at various places on the PTG housing can cause suction leaks, since the PTG housing is on the suction side of the gear pump during operation.

The following parts, if defective, will permit air leaks into the pump:

1. Determine if the tachometer drive seal is leaking. Check by putting a small amount of diesel fuel into the tachometer drive coupling with the pump running. It should not be sucked into the pump.
2. Check all pump housing gaskets and retighten all capscrews.
3. Remove the throttle shaft and check the O ring on the throttle shaft. Replace the O ring if necessary.
4. Check the drive shaft seals. Using an oil can, squirt a small amount of oil into the weep hole between the seals. The oil should not be sucked away. If it is, the inside or back seal is leaking.

C If the governor cutoff is not correct, each of the following should be examined:

1. Check or possibly replace the governor.
2. Check governor weights and see if the pins are worn.
3. Determine if the governor plunger is sticking in its barrel.
4. Determine if the proper number of shims are on the governor spring.

D If the throttle leakage cannot be adjusted so that it comes back to the same setting after being moved, examine the following items:

1. Determine if the throttle shaft is worn or scored.
2. Check the governor plunger to see if it is worn.

E If the fuel manifold pressure canot be adjusted correctly, determine if the following items are in proper working order:

1. See if the idle spring plunger is correct.
2. Check the suction and determine if the gear pump is worn.
3. Check the governor plunger to see if it is worn.
4. Determine if the throttle shaft is worn.
5. See if the adjustment of the throttle restriction is correct.

F If check points no. 1 and/or no. 2 (on calibration chart) do not meet specifications, check the following items:

1. Is the torque spring correct?
2. Is the weight assist plunger adjusted correctly?
3. Are the governor weights correct for application?
4. Is the gear pump worn?

G If the pump operation is noisy, determine which of the following parts is worn:

1. Is the governor worn?

2 Is the governor drive gear worn?

3 Is the gear pump worn or scored?

H If the engine will not run, check the following procedures:

1 Check for operation of the electric shutdown valve.

2 Change the fuel filter if any doubt exists about whether it may be plugged.

3 Make sure all lines leading to the pump are tight.

4 Check the fuel inlet line for restriction by blowing back through it with an air hose.

5 Remove the tach drive cable and crank the engine; the tach drive should turn at this time. This is a good indication that the pump is or is not turning.

6 If the pump is not turning, check the drive spider or splined sleeve.

I If the engine runs but is low on horsepower, examine the following items to detect the problem:

1 Check the fuel filter and change if necessary.

2 Check the snap pressure.

NOTE Snap pressure is checked by attaching a pressure gauge (Figure 22.22) to an extra outlet fitting on the shutdown solenoid and then accelerating the engine quickly from idle to full throttle. Read the maximum pressure on the gauge during acceleration. It should be very near the pressure listed on the pump calibration sheet under *engine fuel pressure.* Snap pressure checks will not be accurate on AFC pumps.

Figure 22.22 Checking snap pressure (courtesy Cummins Engine Company).

3 If the snap pressure is low, it may be necessary to change the throttle restriction or the idle spring plunger button.

4 If there is an incorrect high idle setting, change the governor spring shims.

5 If the throttle travel is not correct, check to make sure the throttle is in the wide open position with the accelerator pedal all the way down.

J The engine often does not decelerate properly.

This is a common complaint after a pump has been overhauled and calibrated on the test stand, since each engine may require a different amount of fuel during deceleration.

1 If engine deceleration is too slow, turn the throttle stop screw counterclockwise $\frac{1}{8}$ to $\frac{1}{4}$ in. turn.

NOTE Check the position of this screw before moving it and then move it a small amount. If no change occurs, put the screw back in its original position and lock the locknut.

2 Make sure the throttle return spring is returning the throttle to the idle position.

3 There should be no binding or sticking of the accelerator linkage.

K If the engine stalls or under runs as it slows down to idle, check the following items to determine the problems:

1 Is the idle adjusting screw correctly adjusted?

NOTE The idle screw is adjusted by removing the pipe plug in the governor spring pack cover and inserting a screwdriver through the opening to engage the screw. Turning the screw in increases idle speed and turning it out decreases idle speed.

2 Is there a suction leak at the fuel inlet line?

3 Is the fuel filter restricted?

4 Is the throttle leakage screw adjusted correctly?

NOTE When adjusting the throttle leakage on the engine, move the screw in small amounts and return it to its original position if no improvement is noticed.

L If the engine high idle is incorrect, check each of the following items:

1. Add or subtract shims from the governor spring as needed.
2. Check throttle travel.
3. Check the governor to determine if it is worn or incorrect.

M If there is excessive black smoke from engine under load, examine the following items:

1. Check the snap pressure and adjust the throttle restriction if required.
2. Determine if the correct governor idle spring plunger button has been used.

If additional information on Cummins PT pump overhaul and calibration is required, consult your instructor or Cummins fuel system service manual.

VII. Injectors

Cummins injectors are found in several different models (Figure 22.23). Early PT injectors were flange type. Later injectors were cylindrical (round) and were built in four models, PT, PTB, PTC, and PTD.

A *Injector identification.*

Each injector has information stamped on it that will be required during repair and calibration. This information can be found anywhere on the body. The marking system that you can expect to find on Cummins injectors is shown in Figure 22.24. (The numbers associated with each arrow correspond to items listed.)

1. This flow code refers to the amount of fuel in cubic centimeters (cc) that an injector should deliver in 1000 strokes on the injector test stand when it is properly adjusted.
2. This number indicates the number of holes that are in the injector cup. For example, the number 8 would indicate that there are eight holes in the injector cup.
3. This number represents the size of the hole in the injector cup in thousandths. For example, the number 7 would indicate 0.007 in. (0.178 mm) holes.
4. This number indicates the angle of the injector cup holes using the surface of the cylinder head or an imaginary line at a right angle to the injector as a base line. For example, the number 17 would indicate that the cup holes are at 17°.

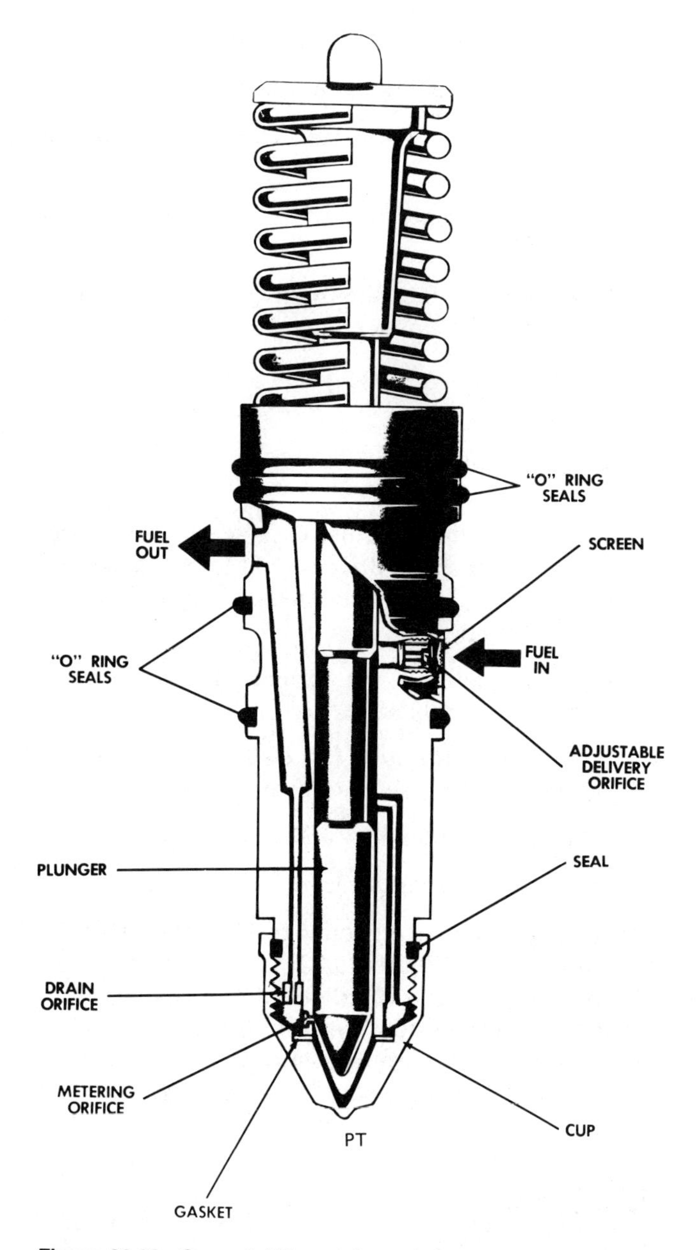

Figure 22.23 Several different Cummins injectors (courtesy Cummins Engine Company).

NOTE By using the information in items 2, 3, and 4 above, the cup on this injector (Figure 22.25) should have eight holes 0.007 in. (0.178 mm) in diameter and placed at a 17° angle. It is commonly called an 8-7-17 degree cup.

5. Plungers and bodies on PT, PTB, and PTC injectors are routinely rebuilt and matched together. It is necessary that they be kept together since they are a matched set. This

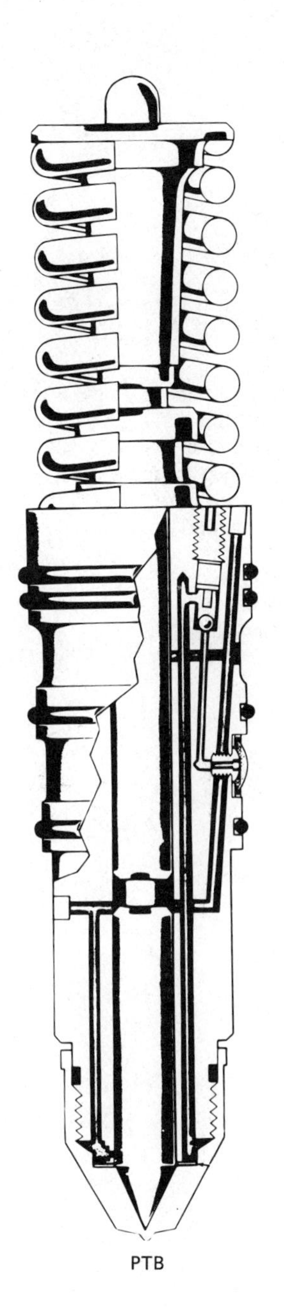

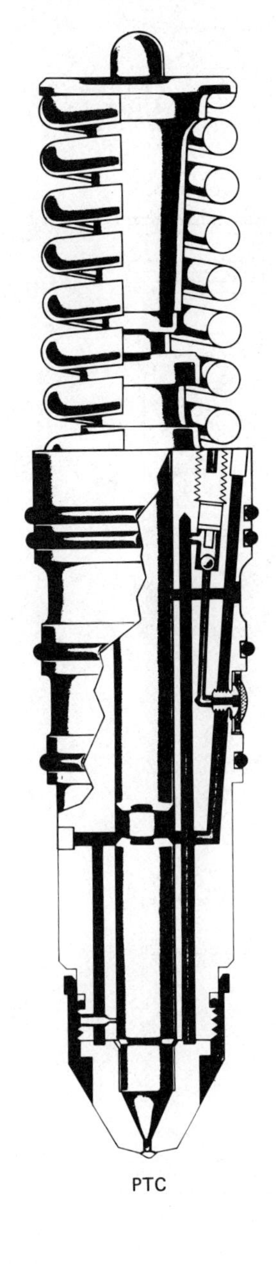

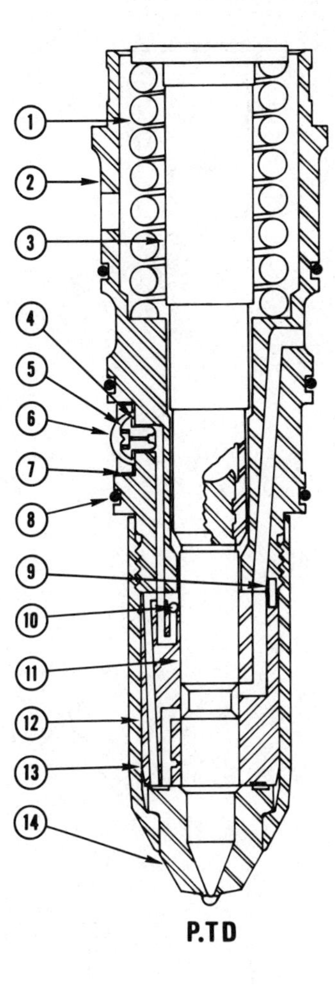

number represents the number oversize of the plunger and should correspond with the body. For example, if a plunger has a no. 1 stamped on it (Figure 22.26), the body it is installed in should also be a no. 1. Intermixing of bodies and plungers without respect to size cannot be done.

NOTE PTD injectors have a replaceable plunger and barrel assembly; therefore the plunger (and body) size is not stamped on the outside of the injector body, as it is for PTB and PTC injectors (Figure 22.24).

6 The body part number is a number used for parts replacement.

7, 8, and **9** These three numbers represent the month, day, and year that the injector was manufactured.

10 The injector style is stamped on the injector body.

B *Component parts.*

The Cummins PTB, PTC, and PTD injectors are composed of the following component parts (Figure 22.23):

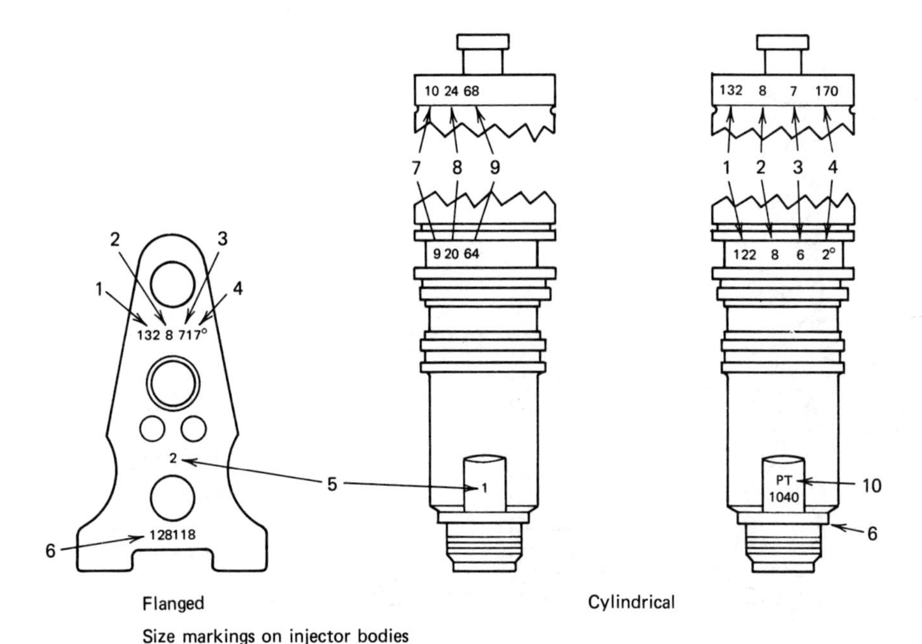

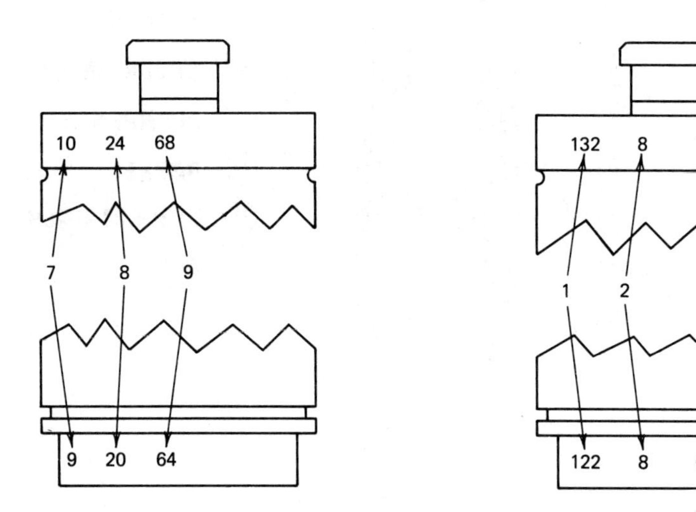

Figure 22.24 Injector identification (courtesy Cummins Engine Company).

1 Body.

2 Cup (single or two piece).

3 Plunger.

4 Plunger return spring.

5 Balance orifice.

6 Barrel and plunger (PTD only).

7 Injector link.

C *Injector operation and fuel flow (PT flange type, PTB, PTC, and PTD).*

The injector in the Cummins PT fuel system is operated by the engine camshaft via the cam followers, push tubes, and the injector rocker arm (Figure 22.27).

The injector's function is to time, meter, inject (pressurize), and atomize the fuel. Fuel is supplied to the injector from the passageways in the cylinder head. Fuel then flows through the injector in this order (fuel flow given is for PTD injector):

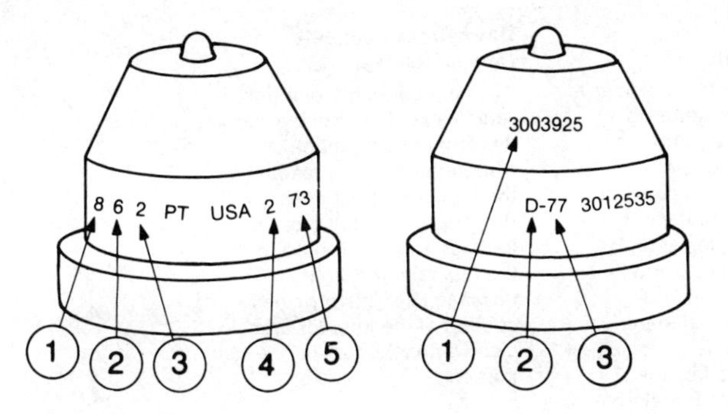

OLD
1. Number of Holes
2. Size of Holes
3. Degree of Holes
4. Month
5. Year

NEW
1. Cup Part Number
2. Year Quarter Made
   A.—First Quarter
   B.—Second Quarter
   C.—Third Quarter
   D.—Fourth Quarter
3. Year

Figure 22.25 Identification of injector cups (courtesy Cummins Engine Company).

1 Fuel is supplied to the injector balance orifice from the fuel passageways in the cylinder head.

2 Fuel then flows through the injector as shown and described in Figure 22.28.

VIII. Injector Disassembly, Cleaning, and Inspection

Assume that the injectors are removed from the engine and are ready to be disassembled, cleaned, and repaired.

NOTE Before disassembly of the injector, rinse it in solvent and blow it off with compressed air. Remove body O rings and discard them. Then proceed with the following steps:

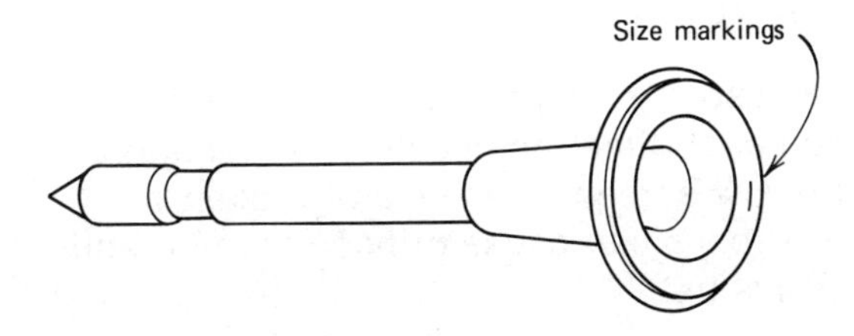

Figure 22.26 Plunger identification (courtesy Cummins Engine Company).

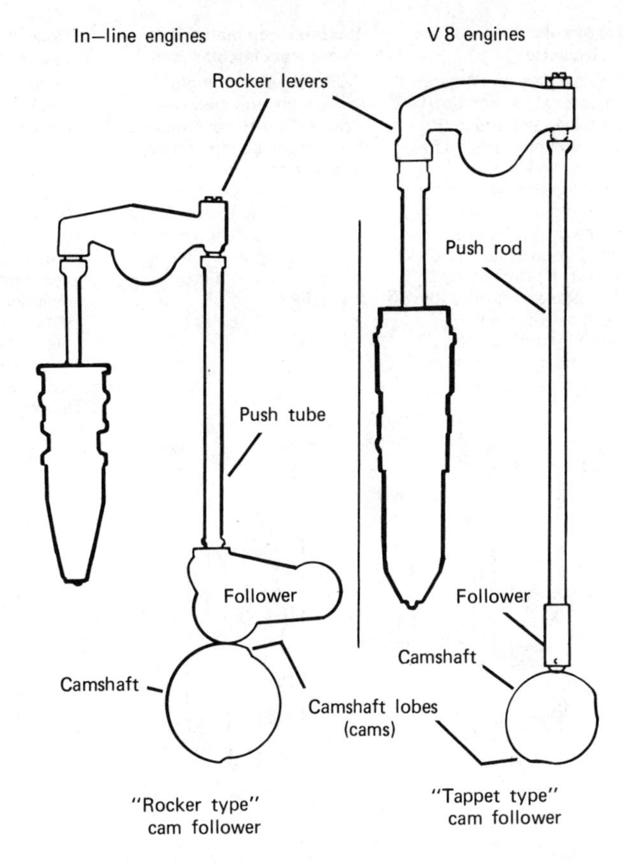

Figure 22.27 Injector operating train (courtesy Cummins Engine Company).

A. Procedure for Disassembly of the PT Flange Type Injector

1 Remove the plunger and plunger spring.

2 Place the injector body in the holding fixture if available.

NOTE If a factory-approved holding fixture is not available, a fixture can easily be made by inserting two capscrews upside down in a vise. The capscrews should be spaced so that the injector can be inverted and placed on them. Do not clamp the injector directly into the vise.

3 Remove the cup with a special socket and socket handle.

NOTE Cup wrenches or sockets look much like ordinary sockets but have splines in them that fit the corresponding spline on the injector cup. The cup socket fits a standard $\frac{1}{2}$ in. square drive

Start upstroke (fuel circulates)
Fuel at low pressure enters the injector at (A) and flows through the inlet orifice (B), internal drillings, around the annular-groove in the injector cup and up passage (D) to return to the fuel tank. The amount of fuel flowing through the injector is determined by the fuel pressure before the inlet orifice (B). Fuel pressure in turn is determined by engine speed, governor and throttle.

Upstroke complete (fuel enters injector cup)
As the injector plunger moves upward, metering orifice (C) is uncovered and fuel enters the injector cup. The amount is determined by the fuel pressure. Passage (D) is blocked, momentarily stopping circulation of fuel and isolating the metering orifice from pressure pulsations.

Downstroke (fuel injection)
As the plunger moves down and closes the metering orifice, fuel entry into the cup is cut off. As the plunger continues down, it forces fuel out of the cup through tiny holes at high pressure as a fine spray. This assures complete combustion of fuel in the cylinder. When fuel passage (D) is uncovered by the plunger undercut, fuel again begins to flow through return passage (E) to the fuel tank.

Downstroke complete (fuel circulates)
After injection, the plunger remains seated until the next metering and injection cycle. Although no fuel is reaching the injector cup, it does flow freely through the injector and is returned to the fuel tank through passage (E). This provides cooling of the injector and also warms the fuel in the tank.

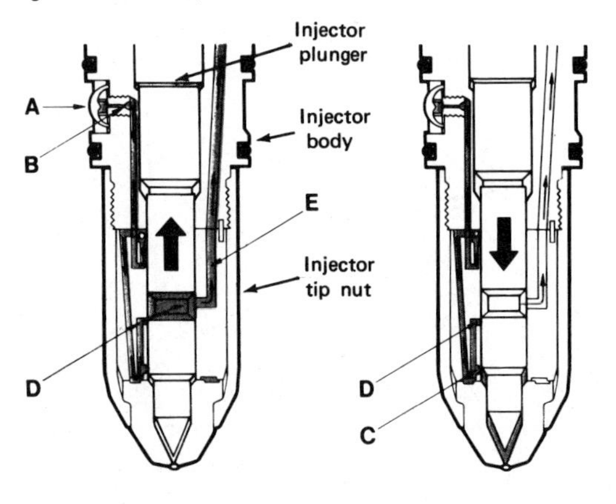

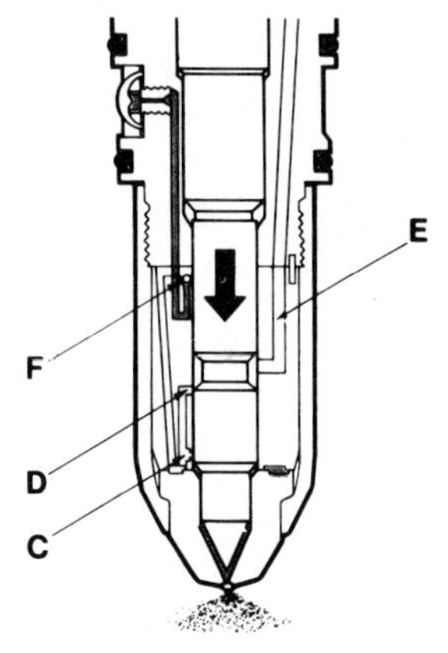

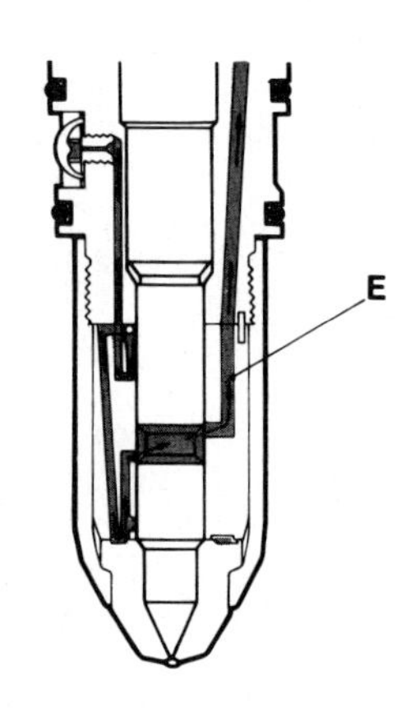

Figure 22.28 PTD Injector fuel flow (courtesy Cummins Engine Company).

ratchet or flex handle. Cup wrenches are available from any Cummins authorized dealer.

CAUTION Cup wrenches must be used! The use of any other type wrench to remove the cup will damage it and it will have to be replaced.

4 Remove and discard cup-to-body O ring.

5 Remove the cup gasket from the cup, noting the number of notches on the cup gasket. This number should correspond to the number stamped on the injector body plunger coupling. If an incorrect gasket has been used in the injector, it must be replaced with a new one.

NOTE It is a good practice to replace all cup washers when overhauling flange type injectors.

B. Procedure for Disassembly of the PTB and PTC Flange Type Injector

1 Remove the plunger and plunger spring from the injector body.

2 Remove the spring from the injector plunger.

NOTE If injector hold down plate was not removed previously, remove it at this time.

3 Place the injector plunger back in the injector body.

CAUTION Do not intermix injector plungers and bodies. Plungers and bodies are classified according to size and must remain as a matched pair.

4 Remove orifice screen retaining ring and remove and discard orifice screen.

NOTE It is also a good practice to replace the orifice screen each time the injector is overhauled. The cost of the screen is small, and broken or damaged screens can cause orifice plugging and a cylinder misfire.

5 Clamp the injector into a holding fixture if available (Figure 22.29).

6 If a holding fixture is not available, a $\frac{15}{16}$ in. wrench

Figure 22.29 Cummins injector holding fixture (courtesy Cummins Engine Company).

clamped in a vise can be used to hold the injector while the cup is loosened.

7 Use an injector cup wrench to loosen the cup.

8 After loosening the cup, remove it from the injector.

9 Remove the ball check plug from the injector body. The ball check retaining plug is located on the top end of the injector body.

10 Remove the ball check.

C *Dissasembly of PTD standard and "top stop" injectors.*

PTD injectors are disassembled in much the same way as PTB and PTC injectors except that they must be installed in a loading fixture before disassembly.

1 Remove the plunger link if not previously done.

NOTE The plunger link on the PTD injector is simply lifted out of the plunger coupling. No snap ring is used to hold it in place. Also on top stop injectors the plunger stop nut must be removed.

2 Remove the plunger.

3 Remove the spring from the plunger.

4 Install the injector in the loading fixture, and install the cup retaining nut wrench and body holding fixture (Figure 22.30).

5 Torque the hold down screw to the required torque.

6 Loosen the cup retainer nut.

7 Remove the injector from the holding fixture.

8 Complete the disassembly by removing the cup retainer and *catching the check ball* as barrel and plunger assembly come apart.

D. Procedure for Cleaning PT Flange (PTB, PTC, PTD, and PTD Top Stop) Injectors

If injectors and injector parts have not been rinsed with cleaning solvent, do so before placing them in the cleaning solution.

1 Place all injector parts in a basket and soak them in a cleaning solvent like carburetor cleaner. An alternate method of cleaning would be to use an ultrasonic cleaner.

NOTE Ultrasonic cleaners (Figure 22.31) use high frequency sound waves to clean the carbon from the injector parts. This type of cleaning usually

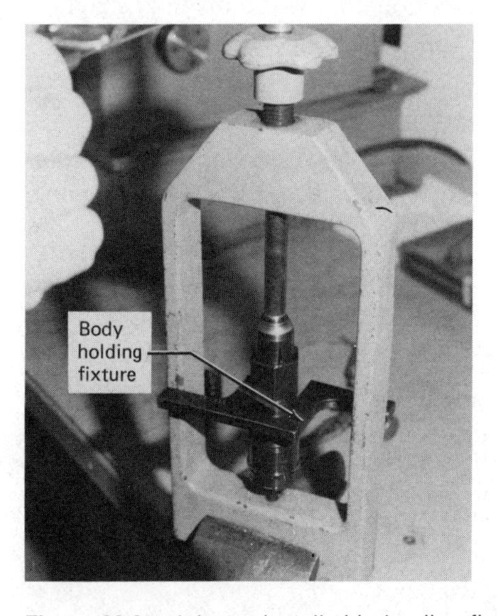

Figure 22.30 Injector installed in loading fixture.

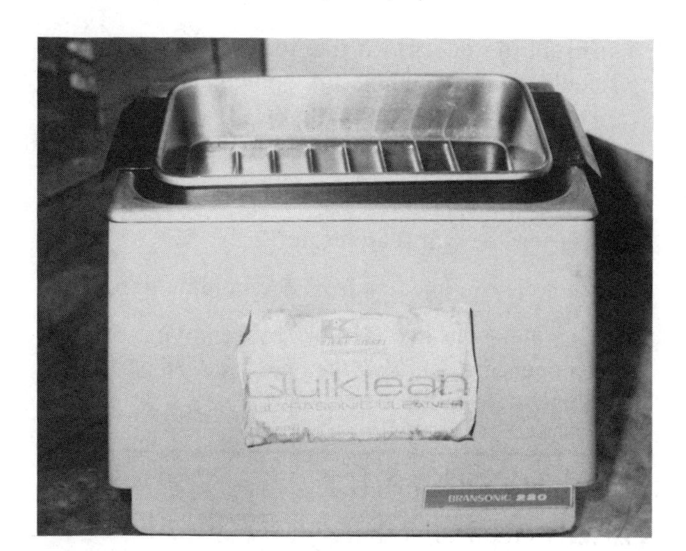

Figure 22.31 Ultra sonic cleaner.

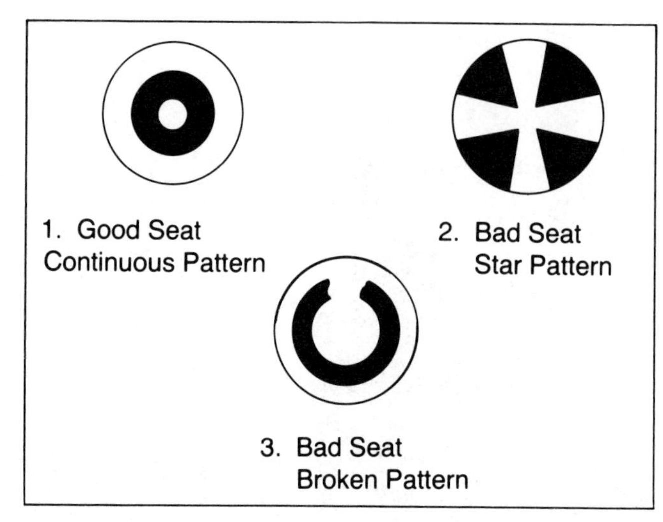

Figure 22.32 Check cup-to-plunger contact (courtesy Cummins Engine Company).

does a better job and is much faster than a cleaning solvent.

2 After the carbon has been removed or loosened by one of the methods mentioned above, flush the parts thoroughly in a clean parts washing solution.

3 Blow off all parts with clean, dry, compressed air.

E. Procedure for Parts Inspection (PTB and PTC)

Parts inspection is one of the most important procedures in rebuilding and repairing injectors.

1 Inspect the injection cup with a magnifying glass for burred, oblong, or plugged holes.

2 If a hole is plugged with carbon, the carbon can sometimes be removed by using a cleaning wire about 0.001 in. (0.025 mm) smaller than the hole. For example, if the hole is 0.007 in. (0.178 mm), use a 0.006 in. (0.152 mm) cleaning wire.

NOTE If the cup hole cannot be unplugged either by ultrasonic cleaner or with a cleaning wire, the cup may have to be replaced.

3 Inspect the cup plunger seating surface for pitting and blistering. Replace if pitting or blistering is excessive.

4 If the plunger seating surface is considered to be in good condition, the plunger-to-cup contact should be checked.

5 Generally the plunger-to-cup contact can be checked by looking at the cup. A darkened area where the plunger contacts the cup will be visible.

6 If the plunger-to-cup contact cannot be determined by looking at the cup, place a small amount of bluing compound on the plunger and place it firmly inside the injector cup.

NOTE Turn the plunger approximately 90° after placing the plunger in the cup.

7 Seat contact is considered good (Figure 22.32) if the plunger contacts 40 percent or more of the cup.

8 If seat contact is not acceptable, do not attempt to lap the plunger and cup together. Replace the cup.

CAUTION When replacing cups, do not attempt to intermix PTC and PTD cups, since the first has a tapered shoulder and the other one a straight shoulder. The shoulder is used for alignment; therefore, the cups cannot be switched from injector to injector.

NOTE If the cup is to be used over, make sure all bluing has been removed from the cup.

9 Check plunger for scoring. Generally, if the plunger is free in the injector body, the body and plunger can be used over again.

10 Check the plunger link for wear (Figure 22.33).

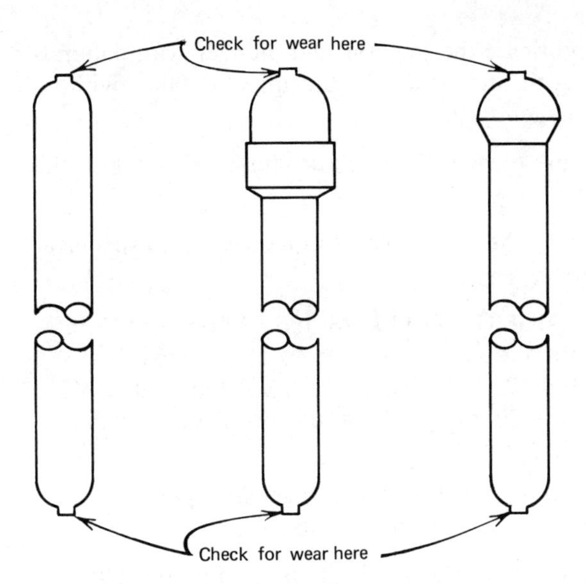

Figure 22.33 Worn plunger link (courtesy Cummins Engine Company).

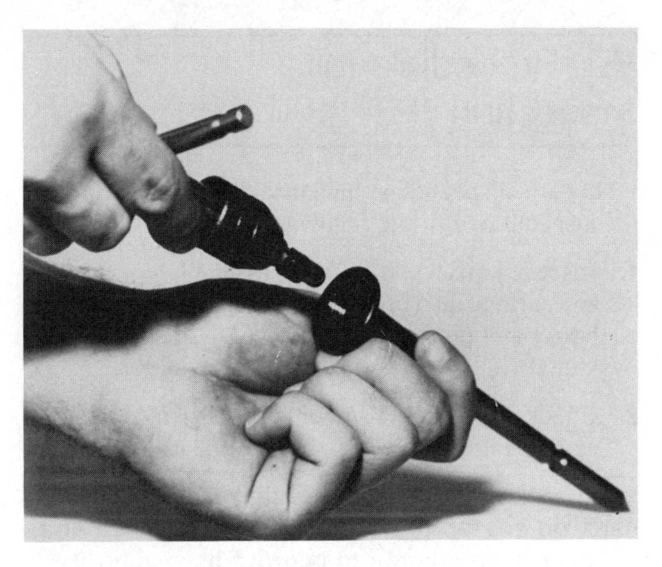

Figure 22.35 Injector plunger link removal (courtesy Cummins Engine Company).

11 Replace the plunger link if excessive wear is evident.

NOTE Replacement procedures depend on what type plunger link is used (Figure 22.34).

12 The E-ring type plunger and retainer must be placed in a block of wood and the E-ring ears broken off with a hammer. A new plunger and E-ring must then be installed.

13 The plunger link, which is held in place with a sleeve, can be removed by using a tap holder (Figure 22.35).

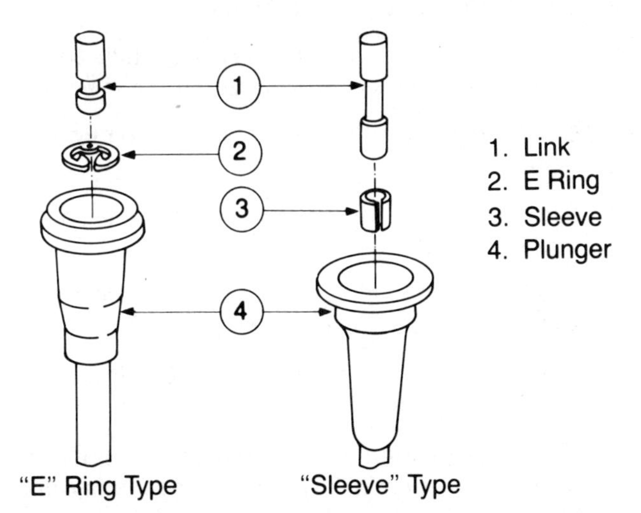

Figure 22.34 Injector plunger link (courtesy Cummins Engine Company).

14 Install the retainer sleeve on the plunger link and install the link by tapping lightly with a plastic hammer.

15 Check the injector body for scoring in the plunger bore. If the body is badly scored, the body will have to be replaced.

NOTE The injector body and plunger must be replaced as a matched set.

16 Check the body cup contact area closely for nicks and scratches.

17 If the body is nicked or scratched, it can be cleaned by lapping the body on a lapping block.

NOTE Make sure the body is held straight and not tipped sideways when lapping. Tipping the body sideways will cause the body surface to become rounded, which will prevent a good cup-to-body seal after the cup has been installed.

CAUTION Make sure all lapping compound is thoroughly cleaned from the injector body after the lapping process is finished.

18 If doubt exists about the squareness of the injector cup surface, it can be checked by using bluing and a flat metal surface.

NOTE The back side of a lapping block works well for this check.

F. Procedure for Inspection (PTD)

PTD parts inspection is similar to PTB and PTC injector inspection with the following exceptions:

1. Inspect barrel and plunger assembly for scoring and surface interruption. If unfit for further use, the barrel and plunger assembly can be replaced individually.
2. Check the check ball for any flattened areas or scratches on it. Replace the ball if it is damaged.

NOTE Since the ball and its seat are equally hard, do not tap the check ball with a punch and hammer in an attempt to improve its seating.

IX. Injector Assembly and Calibration

Assembly and calibration of injectors must be carried out in the following manner if the injectors are to function correctly:

A. Procedure for Flange Type Inspection and Assembly

1. Place the injector on a suitable holding fixture.
2. Install a new O ring on the plunger body.
3. Lubricate the O ring with calibrating fluid or Vaseline.
4. Install a new injector cup washer.

NOTE As outlined under disassembly, the cup washer must be selected according to the number on the injector plunger and body. The washer is identified by the number of notches on its rim (Figure 22.36). See the chart in Figure 22.36 for gasket selection.

5. Install the cup and cup washer finger tight on body.

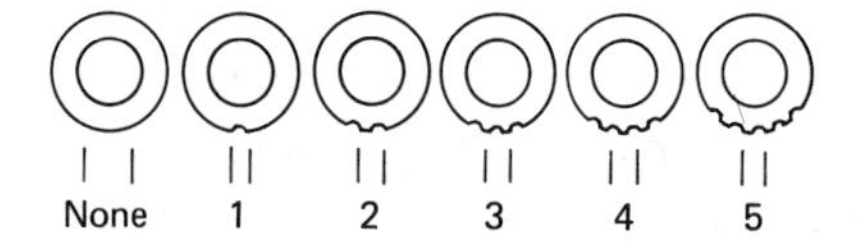

Figure 22.36 Cup gasket notches (courtesy Cummins Engine Company).

6. Lubricate the injector and plunger with calibrating oil and install in the injector body without plunger spring.
7. Press or hold the injector plunger firmly into the cup.
8. Using the cup socket, torque the cup to specifications.

NOTE All J, C, V6-140, V8-185, V-504, and V-555 injectors should be torqued to 45 ft lb (61.01 N·m). Injectors H, NH, V6-200, V8-265, V-903, V8-350, VT8-430, and V12 should be torqued to 55 ft lb (74.57 N·m).

9. Remove the injector plunger and pour the calibrating oil into injector housing.
10. Install the plunger in housing, forcing it in the cup.
11. Make sure fuel comes out of the holes in the injector cup.

B. Procedure for Assembly of PTB and PTC Injectors

The assembly of PTB and PTC injectors differs from flange injectors. It is recommended that the following procedure be followed:

1. Make sure the injector body is thoroughly cleaned with compressed air before attempting to reassemble the injector.
2. Install the cup O ring on the injector.
3. Select a new cup gasket (PTB only) that corresponds to the class size marked on the injector.

NOTE PTC injectors that use a two-piece cup do not use cup gaskets.

4. Screw the cup or cup and nut tightly on the injector body finger.
5. Dip the injector plunger in clean calibrating oil and insert it in the injector body.
6. If a special holding fixture is available, insert the injector in holding fixture.
7. If a holding fixture is not available, clamp a $\frac{7}{8}$ in. open end wrench in a vise and place the injector body and plunger in the wrench.
8. While holding the plunger in the cup to align it with the injector body, torque the cup to the rec-

ommended torque with a special cup wrench and torque wrench.

9 After torquing the cup, lift the plunger a few inches in its bore and force it into the cup. It should move up and down in the body and cup freely.

10 Remove the injector plunger from the body and pour in some calibrating oil. Push the plunger into its seat. Calibrating oil should flow out of all the cup holes.

11 Install the orifice plug in the injector body and a new gasket on the cup type.

12 Install new O rings on the injector body.

13 Install the check ball and retaining plug. Torque the plug to 50 in. lb (5.65 N·m).

14 Remove the plunger and install the spring.

C. Procedure for Assembly of PTD and PTD Top Stop Injectors

NOTE Top stop injectors differ from standard PTD injectors in that they have a stop nut that limits the upward travel of the injector plunger. When installed in an engine, this feature allows the injection train, push rods, and rocker arms to get better lubrication, since the upward spring pressure of the injector plunger spring is carried by the injector stop.

PTD injectors require the use of a special loading device during reassembly. It is recommended that disassembly and reassembly of PTD injectors not be attempted without a loading fixture.

NOTE Loading fixtures are available from Cummins Engine Company and Bacharach Instrument Company.

1 Make sure the body and barrel mating surfaces are free from nicks and scratches. If they are not, lap them as outlined in the inspection section.

2 Install the orifice plug with new gasket in the injector body.

3 Hold the barrel with plunger removed and install the check ball.

4 Install a gasket if you are working on a $\frac{5}{16}$ type injector. Other injectors do not use a gasket.

5 Place the injector body on the barrel, lining up the roll pins or spiral pins.

NOTE Many of the first injectors used roll pins, while injectors that were developed later used spiral pins for alignment.

6 Holding adapter and barrel tightly together, turn them over and place on workbench.

NOTE Adapter and barrel must be held together to prevent the check ball from falling out.

7 Place cup and cup holder on the injector barrel.

8 Coat the cup nut with lubricating oil and install over cup on the adapter.

9 Install the nut finger tight and then back off slightly.

10 Dip the injector plunger in calibrating oil and install in the injector body without a spring.

11 Install the cup retainer wrench and injector body.

12 Insert injector in the loading fixture.

NOTE Depending on the type of loading fixture you are using, the following procedures may vary somewhat. Follow the instructions provided by the fixture manufacturer.

13 After installation of the injector in the load fixture, torque the cup nut to the correct torque by using a crowfoot wrench and torque wrench.

14 Remove the injector from the loading fixture.

15 Check the plunger-to-cup alignment, remove the plunger from the injector body, and pour a small amount of fuel in the injector body.

16 Carefully install the plunger (without spring) in the injector body and force the plunger quickly into the cup.

17 Step 16 will tell you if all the cup holes are open, since fuel will squirt out of all the holes if they are open. The plunger should slide out easily when the injector is turned upside down. If it doesn't, then place the injector in the holding fixture and loosen and retorque the cup.

NOTE A special tool is available from Cummins Engine Company to check plunger sticking. If a large number of injectors are done by your shop, this would be a good investment.

18 Pull the plunger from the injector body, install the spring and reinstall into the injector body.

19 Injector assembly is now complete with the exception of the inlet screen. Do not install the screen at this time because it will need to be removed when the injector is tested and calibrated.

D. Procedure for Testing and Calibration

Injector testing and calibration occur after injector disassembly and repair. Cummins injectors are tested and calibrated for plunger and cup leakage and injector flow. If "top stop" injectors are being serviced, the top stop setting must not be made until after the injector is calibrated.

NOTE Testing of injector plunger and cup leakage requires the use of a special injector leakage tester (Figure 22.37) that is supplied by Cummins Engine Company. This check is recommended by Cummins and also highly recommended by the author. If you have a tester of this type, follow the instructions provided with the machine or in the Cummins shop manual to test the injectors. If the tester is not available, an estimation of plunger-to-body leakage can be made on the injection calibration stand. Consult your instructor for further instructions.

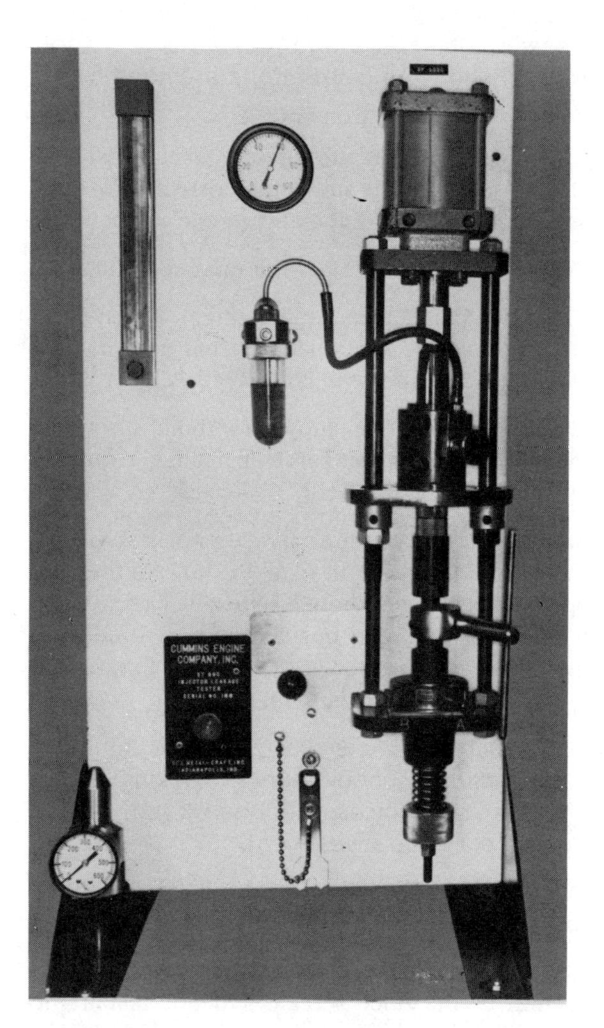

Figure 22.37 Injector leakage tester (courtesy Cummins Engine Company).

1. Consult injector calibration tables to obtain the correct restrictor orifice size.
2. Check the test stand for the correct restrictor orifice size.

NOTE Depending on what type of comparator or calibrator (Figure 22.38) you are using, different procedures must be used in mounting the injector in the machine. Follow the manufacturer's instructions for the machine being used. The following instructions are for a Bacharach comparator.

3. Change the restrictor orifice if necessary.

NOTE The injector seat orifice is installed in the discharge head of the test stand to simulate compression pressure on the injector.

4. Select the correct master injector for the injector to be calibrated. A master injector is one that has been calibrated at the factory for a given flow. This injector is used to calibrate or test the injector calibration machine before it is used.

NOTE A 0.020 in. restrictor orifice should be used when calibrating the injector test stand.

5. Select the correct adapter pot, lubricate the injector O rings, and install the master injector into the injector adapter pot.

Figure 22.38 Injector comparator.

Figure 22.39 Injector comparator in TDC position.

NOTE The injector may be difficult to push into the pot. Place the pot on a solid surface, such as the jaws of a vise, and push firmly, using hand pressure.

6 Select the correct plunger link. It should be marked H or NH when using the master injector.

7 Make sure the cam pulley drive and cam are in top dead center position (Figure 22.39).

8 Install the injector and adapter pot in the calibrator and clamp into place.

9 Connect the inlet and return lines to the injector adapter pot.

10 Turn the cam pulley drive over several times by hand before starting the drive motor to make sure it rotates all the way around.

11 Start the drive motor and adjust pressure to 120 psi (8.4 kg/cm^2).

12 Close the graduate drain valve.

13 Start the count by pressing the count button.

NOTE When calibrating the test stand, the counter should be set at 1000 strokes until the first draws are taken. After taking three draws it may be necessary to adjust the count so the correct amount of fuel is being delivered.

14 After the count has stopped, read the cubic centimeters of fuel in the graduate. It should read 132 cc for the flange type master injector.

15 If delivery is low, adjust the counter for more strokes. If delivery is high, adjust the counter for fewer strokes.

16 Continue running draws and adjusting the counter until the delivery is exactly 132 cc.

17 Remove the master injector from the test stand in reverse order of installation.

18 Remove the master injector from the adapter pot and install the injector to be tested.

19 Install the injector and adapter pot on the test stand.

20 Test the injector, following the same procedure utilized when the master injector was tested.

21 If the injector does not meet the flow specifications, it may be necessary to change the inlet orifice or enlarge it by burnishing with a burnishing tool.

22 To change the inlet orifice, remove the inlet fuel fitting. Using an allen wrench, turn the orifice counterclockwise to remove it.

23 If flow was low, replace it with the next larger size. If flow was high, replace it with the next smaller size.

24 To use the burnishing tool to enlarge the orifice, push in on the small knurled knob of the tool until the point makes contact with the orifice.

NOTE Leave the test stand running while using the burnishing tool.

25 After contact with the orifice is felt, turn the knob slightly counterclockwise to engage the slot in the large knurled knob.

26 To enlarge the orifice, turn the large knob clockwise with reference to the pointer and marks on the burnishing tool body.

27 After enlarging the orifice two or three marks, withdraw the burnishing tool by turning the large knob counterclockwise while holding the small knob. This will disengage the burnishing needle, which can be pulled back from the inlet orifice.

28 Run three more checks on the injector for flow. If flow meets specifications, remove the injector from the comparator.

29 If flow does not meet specifications, burnish the orifice until the specifications are met.

NOTE Do not attempt to burnish the orifice to increase delivery more than 10 cc; install a larger orifice.

30 After removing the injector and adapter pot from the comparator, remove the injector from the adapter pot. Install the hold down plate on the injector.

31 Install button or wrap-around screen on the injector body.

NOTE If top stop injectors are being serviced, the "top stop" adjustment must be made at this time.

X. Procedure for Injector Installation and Adjustment (PTD)

A *Correct installation of a rebuilt or reconditioned injector is mandatory if the engine is to operate correctly.*

1. Check the O rings on the injector to make sure they are not torn or pinched.
2. Lubricate the O rings with 30 weight engine oil.
3. Clean the injector seat in the cylinder head with a tapered wooden stick and a rag.
4. Set the injector into the copper sleeve.
5. Remove the injector plunger and spring.
6. Using a plastic hammer handle, tap the injector into the sleeve. Hold the hammer handle on the injector plunger and tap the other end of the hammer with your hand.

NOTE Use only a plastic or hard rubber hammer handle to prevent getting splinters or dirt into the injector plunger link seat.

7. Place the hold down plates on the injector.
8. Install the hold down bolts and torque to 11 to 12 ft lb (14.91 N·m to 16.277 N·m) on NH engines and 30 to 35 ft lb (40.67 N·m to 47.45 N·m) on V8 engines.
9. Turn rocker arm up into position and position the push rod.
10. Install the rocker arm adjusting screw.

B *Injector adjustment*

Two methods are commonly used to set the injectors in Cummins engines: (1) the torque wrench method and (2) the dial indicator method.

Torque wrench method

1. Turn or bar engine over by hand to line up "VS" (valve set) mark on the vibration damper with the mark on the engine.

NOTE In-line engines have the VS marks on the accessory drive pulley and the timing mark on the engine timing gear cover. V8 engines have the VS marks on the vibration damper with the timing mark on the front of the engine block. (Some engines use A, B, and C instead of the VS mark. In this case A is the same as 1-6 VS, B equals 2-5 VS, and C equals 3-4 VS.) A typical VS mark on a six cylinder engine would be 1-6 VS, which stands for a one or six cylinder valve set. It should be noted that valves and injectors on only one cylinder can be set when the VS and timing marks are lined up because the engine is firing either number one or number six cylinder. To determine which cylinder can be adjusted when the VS 1-6 mark is lined up with the mark on the timing cover, check the valve rocker arms. The cylinder that can be set is the one that has both valves closed.

CAUTION Make sure you are on the correct cylinder before setting the injector or you may bend a push rod as you turn the engine over to set the next cylinder.

---

2. Using a large screwdriver, turn the injector rocker arm adjusting screw until it bottoms the injector plunger in the injector.
3. Turn the injector adjusting screw an additional $\frac{1}{4}$ turn after it bottoms, to insure that all the diesel fuel oil has been forced from the injector cup.
4. Loosen the injector rocker arm adjusting screw about one turn.
5. Using a screwdriver adapter on a torque wrench, tighten the injector adjusting screw to the torque listed on the engine nameplate.
6. Torque the injector adjusting screw locknut to specifications.

Dial indicator method.

Cummins Engine Company highly recommends the dial indicator method of setting injectors because it eliminates the variation in adjustment that can occur when the torque wrench method is used. The engine that is set by the dial indicator method runs with less exhaust smoke and more power.

1. To adjust the injector using the dial indicator method, use Figure 22.40 to determine which cylinder you should adjust.
2. Turn the engine until the A or 1-6 VS mark on damper and flywheel are in line with the mark on the timing cover or front cover of the engine.

INJECTOR AND VALVE SET POSITION

Bar in Direction	Pulley Position	Set Cylinder Injector	Set Cylinder Valves (Intake and Exhaust)
Start	A or 1-6 VS	3	5
Adv. to	B or 2-5 VS	6	3
Adv. to	C or 3-4 VS	2	6
Adv. to	A or 1-6 VS	4	2
Adv. to	B or 2-5 VS	1	4
Adv. to	C or 3-4 VS	5	1

Figure 22.40 Table for adjusting injectors and valves with a dial indicator.

NOTE It is recommended that you start making adjustments with the engine in the A position so that you can proceed through all the adjustments in logical order.

3 Make sure which cylinder you should set by checking the valves on cylinder no. 5. If both valves are closed on cylinder no. 5, you must adjust injector no. 3. If both valves on no. 5 are not closed, you must adjust injector no. 4.

4 After determining which cylinder to set, mount the dial indicator on the engine rocker arm box by screwing the T bolt into a rocker arm cover bolt hole.

5 Make sure the dial indicator plunger is resting on the injector plunger collar.

6 Adjust the dial indicator so that it is in the center of its travel.

7 Rotate the rocker arm down two or three times to insure that all the diesel fuel has been forced from the injector cup.

8 Rotate the rocker arm down and hold it. While holding it in this position, zero the dial indicator.

9 Release the pressure on the rocker and allow the injector rocker arm to rotate back.

10 Read the total travel on the indicator. It should be as listed in the specifications. Most six cylinder in-line engines with cast iron rocker housing will have a cold setting travel of 0.175 in. (4.44 mm).

11 If the reading does not meet the specifications, adjust the rocker arm adjusting screw until it does.

12 Torque the adjusting screw locknut to 30 to 40 ft lb (40.67 N·m to 54.23 N·m) and recheck the injector travel.

13 After setting the injector on number three cylinder, set the valve tappets on number five cylinder.

14 Turn engine to the bar 2-5 VS and set the injector and valve listed on the chart.

15 Continue setting valves and injectors in this manner until all are set.

NOTE Top stop injectors must not be adjusted using the dial indicator method. The top stop adjustment method must be used. Consult your instructor or Cummins engine service manual.

After all adjustments are complete, install new rocker arm cover gaskets and covers.

XI. Troubleshooting the Cummins Injector

A *On the test stand.* The following troubleshooting covers only the injector. Make sure your test stand is in good condition before blaming the injector.

1 If the injector does not deliver fuel, determine which of the following items is not properly cleaned:

a Is the balance orifice plugged?

b Is the restrictor orifice plugged?

c Are the injector body passageways plugged?

2 If the injector delivery is low, check the following items:

a If the balance orifice is too small, enlarge or replace it.

b Clean the holes in the injector cup if they are plugged.

c Determine if the injector plunger and body are worn.

d Correct the clamping pressure.

3 If the injector delivery is high, determine which of the following is causing the problem:

a If the balance orifice is too large, replace with a smaller one.

b If the inlet pressure is set too high, reset to 120 psi (8.40 kg/cm²).

B *In the engine.*

NOTE A misfiring injector can be located by holding down the injector rocker arm while the engine is operating. As you bar or hold down an injector, notice if the operation of the engine changes. If it does, the injector is probably working properly. If it does not change engine operation, the injector may be faulty.

1 If the cylinder that the injector is in misfires, check the following for the possible cause:

a Check injector adjustment and be sure it is correct.

b Remove the injector and check orifice to determine if it is plugged.

2 If the injector sticks in the down position, check the following items:

a Determine if the injector hold down capscrew was overtorqued.

b Determine if the injector cup is aligned correctly on the body.

3 If there is excessive engine smoke, one of the following items may be the reason:

a If the cup holes are plugged, remove injector and clean or replace cups.

b Check the adjustment of injector rocker arm to be sure it is correct.

The troubleshooting procedures listed here by no means cover all the problems that you may encounter. Hopefully this section will help you to organize and effectively troubleshoot the fuel system on which you are working.

SUMMARY

This chapter has explained in detail the procedures to be used in the overhaul, calibration, and installation of Cummins pumps and injectors. Although the terms and procedures may seem confusing at first, they will all be commonplace after you have worked on a number of pumps and injectors. If you have further questions concerning Cummins fuel injection equipment, consult the Cummins service manual or your instructor.

REVIEW QUESTIONS

1 Discuss and point out on a chart the fuel flow through a Cummins PTG pump.

2 Explain the basic operating principles of a Cummins PT fuel system.

3 List two differences between a PTB injector and a PTD injector.

4 List and explain two possible problems that may result from loose injector adjustments.

5 List and explain what information can be found on the injector cup.

6 List and explain two possible problems that may result from tight injector adjustment.

7 Demonstrate the correct procedure for making an idle speed adjustment on the Cummins PTG pump.

8 Explain how the idle spring plunger is identified.

9 Demonstrate the ability to install and adjust an injector in the engine, using the dial indicator method.

10 Explain why the dial indicator method is a better method of injector adjustment than the torque wrench method.

11 Explain the term *throttle leakage*.

12 Explain the operation of a PTG AFC pump.

13 Explain how an AFC unit differs from an aneroid.

14 How can a PTG pump be checked for suction leaks?

15 Where is calibration information for a PTG pump found?

23

Detroit Diesel Fuel System

The Detroit Diesel unit injector type fuel system (Figure 23.1) was first introduced on the Detroit two stroke cycle engine. Since its introduction this system has proven to be simple, easy to maintain, and very dependable. The system is not used by any other diesel engine manufacturer.

OBJECTIVES

Upon completion of this chapter the student will be able to:

1 Describe the fuel flow in a Detroit Diesel fuel system.

2 Explain and demonstrate the proper procedures for injector removal and installation.

3 List and explain the injector identification found on the name tag and tip.

4 Explain and demonstrate the proper disassembly and reassembly of a unit injector.

5 List and explain the steps that should be incorporated in a complete Detroit Diesel tune-up.

GENERAL INFORMATION

All Detroit Diesel engines use the same basic fuel system. The only difference from engine to engine is the governor and injector fuel delivery.

Injectors used on early Detroit Diesel engines were very similar to the ones being used at present. The major change in injectors has been the tip design. Old style or early injectors used a spray tip that had an opening pressure of approximately 800 psi (56 kg/cm^2). Current injectors use a needle valve spray tip with an opening pressure of 2300 to 3300 psi (161 to 231 kg/cm^2) to help eliminate excessive exhaust smoke. Each engine cylinder has its own injector that is operated by the engine camshaft through a push rod and rocker arm assembly. Engine speed is controlled by a mechanical governor that regulates the amount of fuel by controlling the injector control rack. Injectors are easily removed and installed for service.

I. Fuel System Components and Function

A *Gear pump* (Figure 23.2). The gear pump is driven by the engine and supplies fuel at low pressure (30

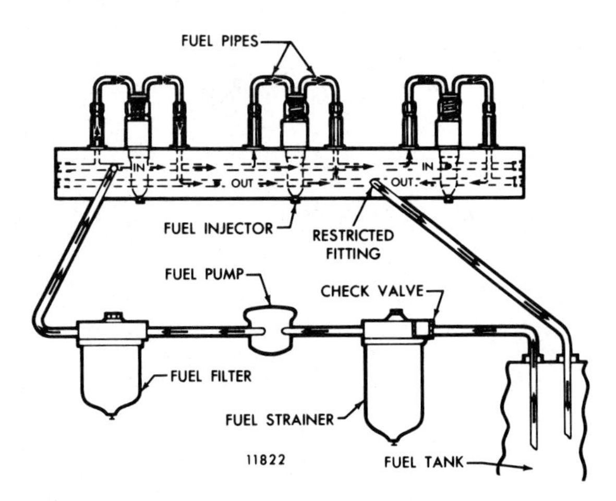

Figure 23.1 Detroit Diesel fuel system (courtesy Detroit Diesel Allison, Division of General Motors Corporation).

to 60 psi) (2.1 to 4.2 kg/cm²) to the injector.

B *Fuel filter* (Figure 23.3). It filters the fuel between the fuel tank and injector.

C *Injector* (Figure 23.4). The injector meters, pressurizes, times, and atomizes the fuel that is injected into the engine.

D *Control tube* (Figure 23.5). The control tube is connected to the injector control racks and the governor. It keeps the injectors balanced together and provides the right amount of fuel to run the engine under varying loads and speeds.

E *Fuel lines.* Steel or woven lines that serve as inlet and outlet lines.

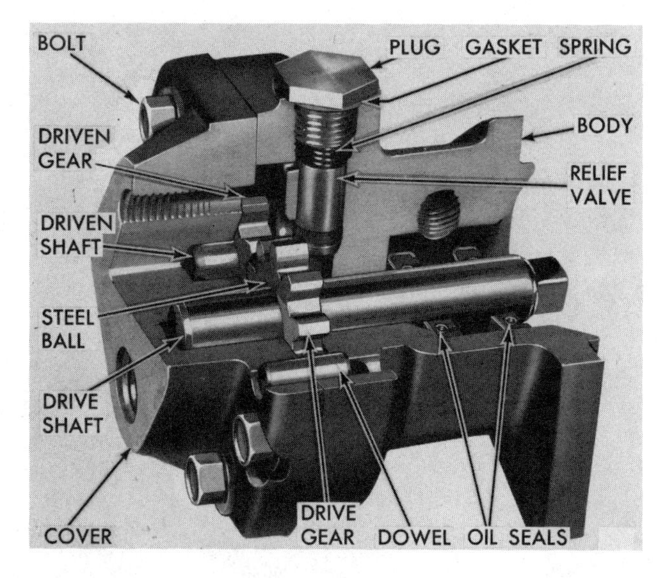

Figure 23.2 Gear pump (courtesy Detroit Diesel Allison, Division of General Motors Corporation).

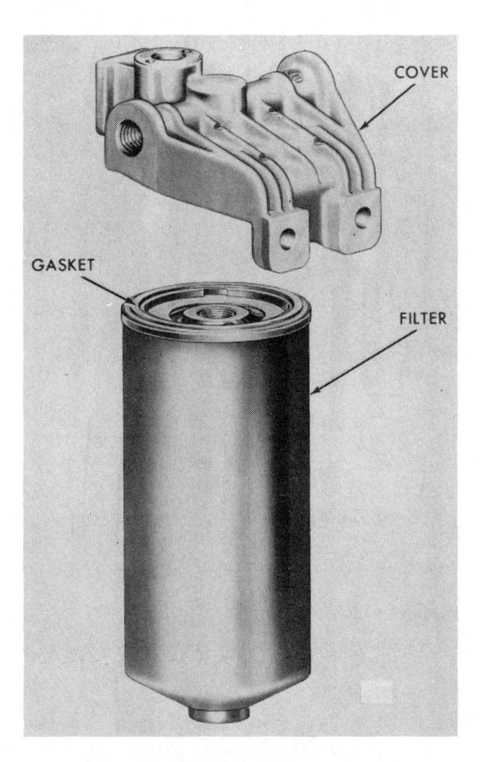

Figure 23.3 Fuel filter (courtesy Detroit Diesel Allison, Division of General Motors Corporation).

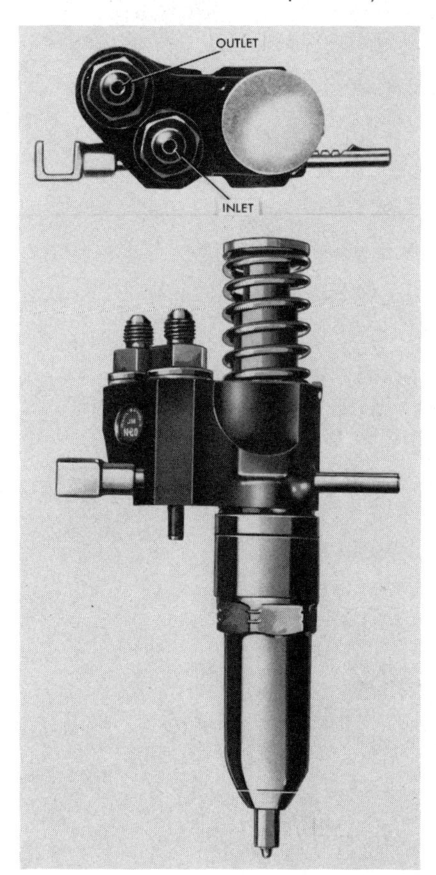

Figure 23.4 Injector (courtesy Detroit Diesel Allison, Division of General Motors Corporation).

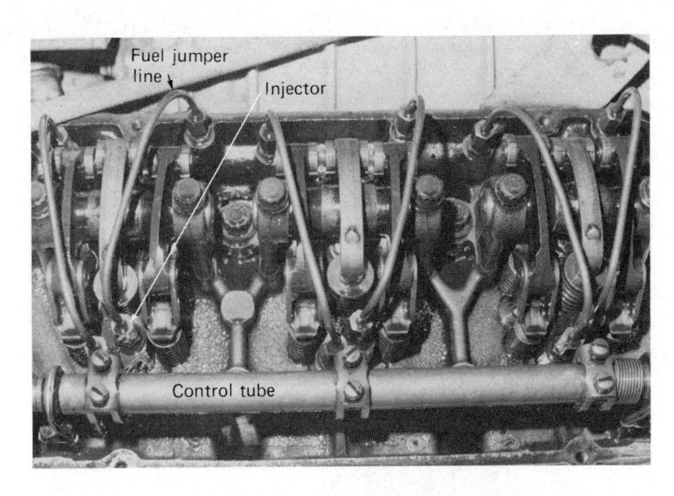

Figure 23.5 Control tube (courtesy Detroit Diesel Allison, Division of General Motors Corporation).

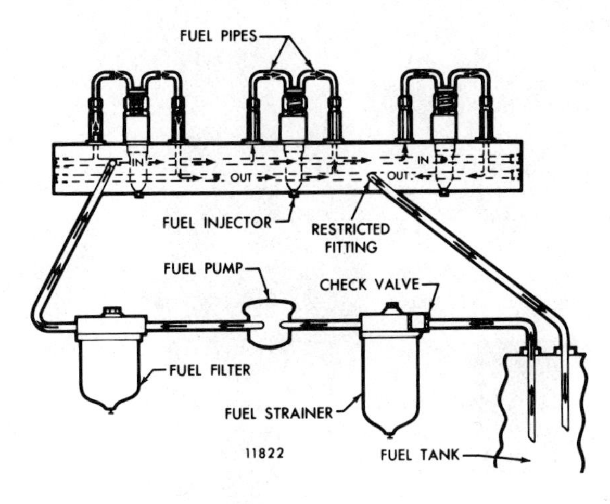

Figure 23.7 System fuel flow (courtesy Detroit Diesel Allison, Division of General Motors Corporation).

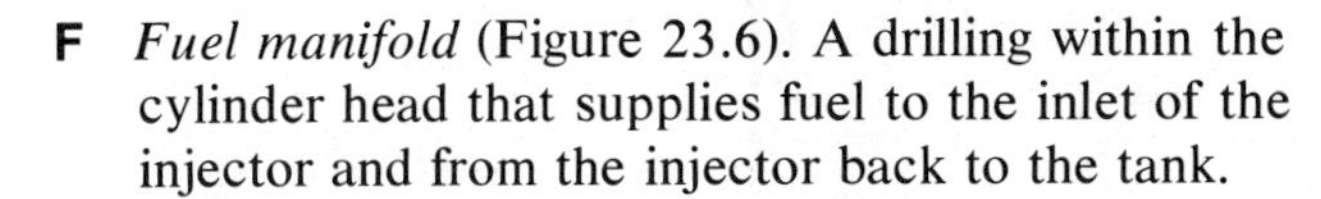

F *Fuel manifold* (Figure 23.6). A drilling within the cylinder head that supplies fuel to the inlet of the injector and from the injector back to the tank.

G *Fuel strainer.* A primary filter or metal type strainer through which the fuel passes before it reaches the gear pump.

H *Inlet check valve.* Located at the inlet side of the fuel strainer, this valve prevents the fuel in the system from running back to the tank.

II. Fuel System Fuel Flow

Fuel flow (Figure 23.7) in the Detroit Diesel system is somewhat different from a pump and nozzle type system, since most of the fuel flow in the system is at low pressure. Injection pressure is not developed until the injector plunger starts downward on its injection stroke. With reference to Figure 23.7, the fuel flow is as follows:

1. Fuel flows from the tank, drawn by the gear pump through a one way check valve located at the inlet side of the fuel strainer.
2. Fuel then flows through the fuel strainer or primary filter.
3. From the fuel strainer it flows through the gear pump and pressure is raised to approximately 60 psi (4.2 kg/cm^2).
4. After leaving the gear pump, fuel enters the final fuel filter.
5. Fuel then is directed to the inlet manifold in the cylinder head. The fuel inlet manifold is usually the top passageway in the cylinder head.

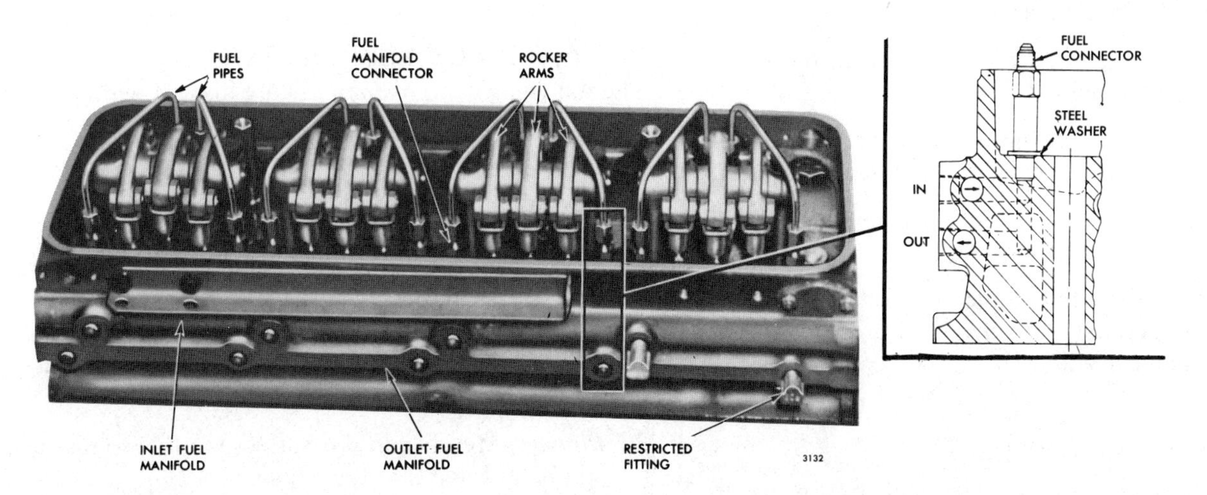

Figure 23.6 Fuel manifold (courtesy Detroit Diesel Allison, Division of General Motors Corporation).

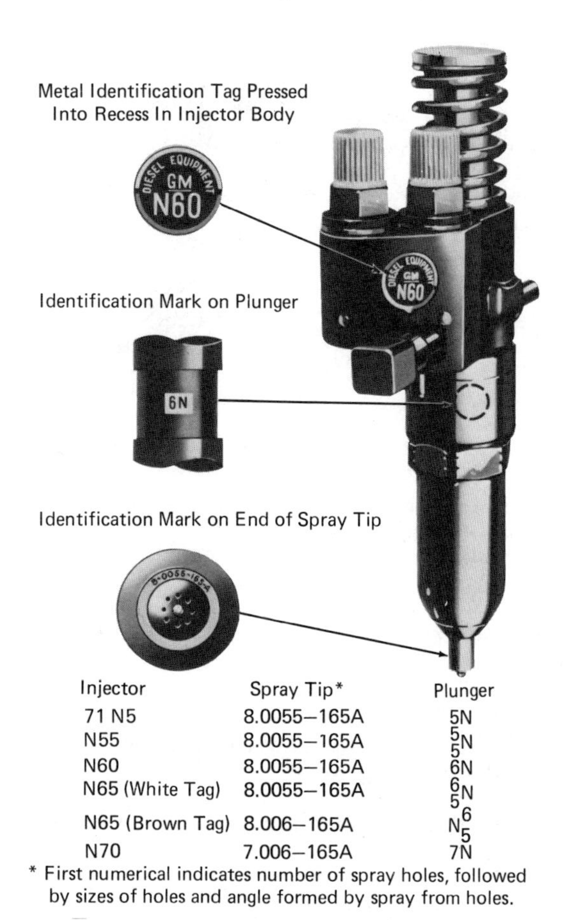

Injector	Spray Tip*	Plunger
71 N5	8.0055–165A	5N
N55	8.0055–165A	$\frac{5}{5}$N
N60	8.0055–165A	6N
N65 (White Tag)	8.0055–165A	$\frac{6}{5}$N
N65 (Brown Tag)	8.006–165A	N$\frac{6}{5}$
N70	7.006–165A	7N

* First numerical indicates number of spray holes, followed by sizes of holes and angle formed by spray from holes.

Figure 23.8 Injector and tip identification (courtesy Detroit Diesel Allison, Division of General Motors Corporation).

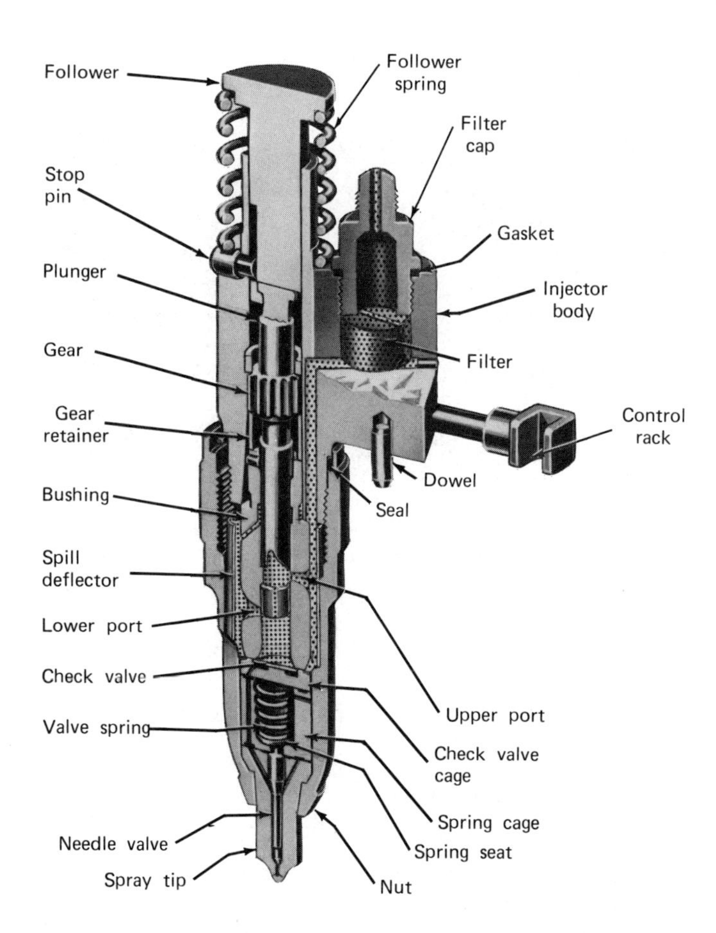

Figure 23.9 Injector components (courtesy Detroit Diesel Allison, Division of General Motors Corporation).

6 Fuel in the inlet manifold can flow through the manifold and out to each injector inlet via a jumper pipe or tube.

7 Fuel then flows through the injector where a certain amount, depending on engine requirements, is used to run the engine.

8 The remaining fuel circulates through the injector, helping to cool it.

9 Fuel then leaves the injector via the return jumper tube, which is connected to the return manifold.

10 Fuel flows through the return manifold and out through the restrictor fitting.

11 The restrictor fitting at the end of the return fuel manifold maintains the manifold pressure within the cylinder heads.

III. Injector Identification

Detroit Diesel injectors are identified by the round metal tag pressed into the injector body. This number represents the size of the injector plunger, which is installed inside the injector. In addition, the injector tip has stamped on it the number of holes, hole size, and hole angle (Figure 23.8). When installed in an engine, all injectors should be the same size. This insures that each cylinder will get the same amount of fuel, which makes for a smooth-running engine.

IV. Injector Component Parts

The following component parts are located within the fuel injector (see Figure 23.9).

A *Follower.* Provides a place for the rocker arm to activate the injector plunger.

B *Stop pin.* Prevents the injector follower from being pushed out of the body by the follower spring.

C *Follower spring.* Returns the follower and the plunger that is connected to the follower to the raised position.

D *Plunger.* Meters and pressurizes the fuel so that it can be injected into the engine.

E *Gear.* Attached to the plunger and meshed into a control rack, it helps control the fuel output of the injector.

F *Gear retainer.* Holds the gear in place when installed in the injector.

G *Bushing.* Closely fitted (lapped) to the plunger, the bushing (along with the plunger) controls the amount of fuel injected.

H *Spill deflector.* A stainless steel sleeve that prevents the high pressure, high velocity fuel from striking the injector body and eroding a hole in it.

I *Check valve.* A flat valve that prevents exhaust gases from flowing back into the injector if a small piece of carbon or dirt holds the needle valve open.

J *Check valve cage.* It provides a place for the check valve to be installed.

K *Valve spring.* The spring holds the needle valve on its seat until pressure builds up to overcome the spring tension.

L *Needle valve.* It opens and closes the tip. The valve is raised by fuel pressure and held shut by the valve spring.

M *Filter cap fitting.* Holds the inlet filter in place and provides a place to connect the inlet and return jumpers.

N *Filter cap gasket.* Seals the filter cap to the injector body.

O *Injector body.* Provides a place to mount all the other component parts of the injector.

P *Control rack.* Meshed with the gear on the pumping plunger, the control rack connects the injector to the control tube and the governor.

Q *Seal.* Provides the seal between the injector body and the injector tip nut.

R *Spring cage.* It provides a place for the needle valve spring and spring seat.

S *Nut.* Retains the injector valve and tip on the body.

V. Injector Fuel Flow (N Series Operation)

A *Fuel flow.*

All Detroit Diesel injectors have a similar fuel flow. One of these injectors, the N series injector, will be covered here.

1. As stated in the discussion on fuel system fuel flow, fuel reaches the injector from the inlet fuel manifold via a steel jumper line.

2 Fuel flow within the injector (Figure 23.10) starts at the injector inlet fitting.

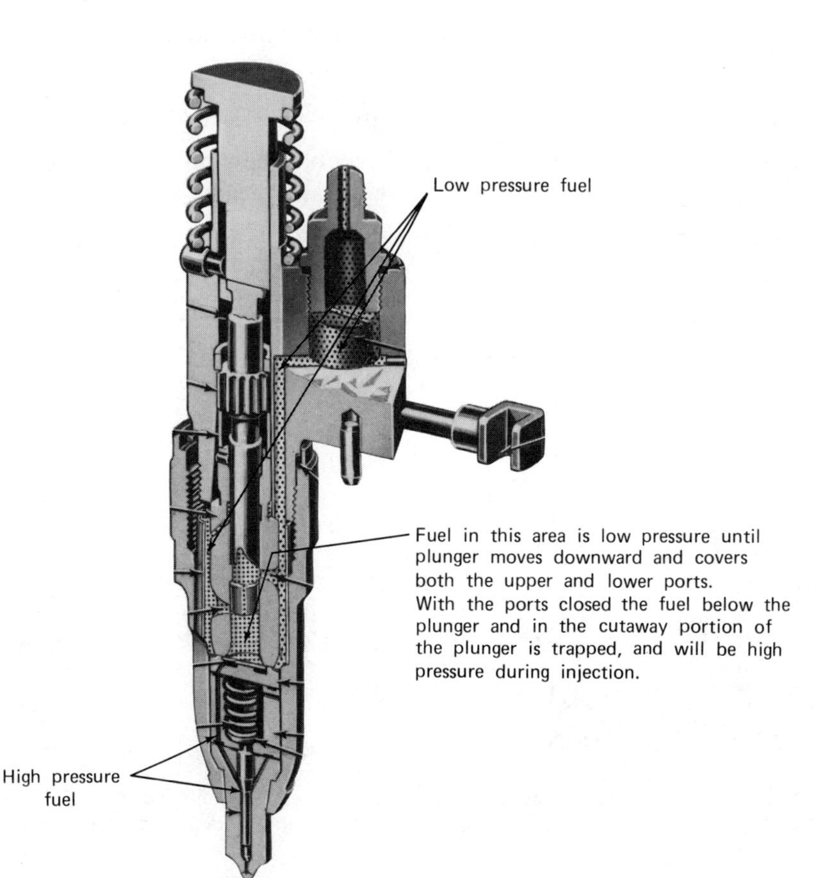

Figure 23.10 Injector fuel flow (courtesy Detroit Diesel Allison, Division of General Motors Corporation).

3. Fuel flows through the inlet fitting into the inlet filter.
4. From the inlet filter it enters a horizontal passageway that connects with a vertical drilling in the injector body.
5. Fuel flows into the injector body and surrounds the barrel assembly.
6. Fuel then flows through the inlet port of the barrel, filling the area below the pumping plunger.
7. As the plunger moves downward, it closes off the inlet and outlet ports in the plunger barrel.
8. The fuel then is trapped. As the plunger continues its downward movement, fuel is forced by the needle valve check valve into the spring cage.
9. Fuel then travels into a drilling in the tip and lifts the needle valve from its seat.
10. Fuel then is allowed to flow through the orifices in the tip and into the engine cylinder.
11. Not all of the fuel that is pumped into the injector body is used to run the engine.
12. The excess fuel circulates through the injector body to help lubricate and cool it.
13. This excess fuel is then returned to the return manifold and back to the tank.

B *Fuel metering.*

Fuel metering is controlled in the Detroit injector by the position of the helix in relation to the ports in the injector barrel (Figure 23.11). As can be seen from Figure 23.12, the amount of fuel delivered to the engine is controlled by the position of the helix in relation to the inlet and outlet ports in the plunger barrel. Turning the plunger causes the ports to close sooner or later, pumping more fuel into the engine cylinder.

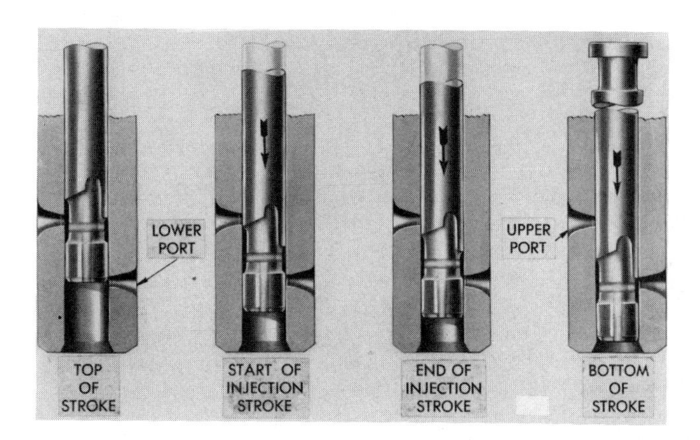

Figure 23.11 Barrel and plunger port and helix (courtesy Detroit Diesel Allison, Division of General Motors Corporation).

VI. Injector Disassembly, Inspection, and Reassembly

Servicing of the Detroit Diesel fuel system and injector is a relatively simple procedure. The injector disassembly and reassembly do not require a lot of special tools. In addition, the injector does not contain very many parts. Probably one of the most important tools required to successfully overhaul and test Detroit injectors is a comparator or calibrator. Many shops have machines of this type and routinely rebuild Detroit injectors. Shops that do not have this piece of equipment generally replace worn or misfiring injectors with a rebuilt unit. As stated previously, this is easily done and the price per unit is relatively cheap.

A. Procedure for Disassembly

1. Place the injector with tip pointed downward in a suitable holding device. A special holding fixture is available from Detroit Diesel, tool no. J22396.
2. Remove the inlet and outlet fittings, filters, and gaskets.
3. Push down on the follower and spring.
4. Insert a screwdriver between the spring and body near the follower retaining pin.
5. Lift the spring from the follower retaining pin and push the pin out of injector body.
6. Lift the follower, follower pin, and plunger from the injector body.

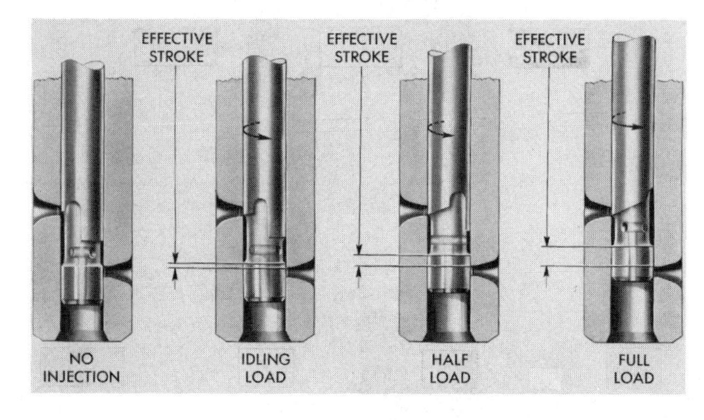

Figure 23.12 Fuel control by port and helix (courtesy Detroit Diesel Allison, Division of General Motors Corporation).

7 Turn the injector over and clamp it back into the fixture.

8 Using a deep socket or special GMC tool no. J4983-01, remove the injector nut.

CAUTION When lifting the injector nut from the body, lift it straight up to prevent losing the spray tip and valve parts.

9 Place the injector nut on a block of wood and drive the spray tip out with a driver, J1291-02, or other suitable tool and a hammer (Figure 23.13).

NOTE The tip driver must be something that would drive on the tip shoulder in order to prevent damage to the tip.

10 Lift the spill deflector from the injector body.

11 Remove the plunger barrel from the injector body.

12 Lift the injector body out of the holding fixture.

13 Tip the body upside down; this will cause the plunger control gear and retainer to fall from the body.

14 After removing the plunger control gear and retainer, the control rack can be removed from the body.

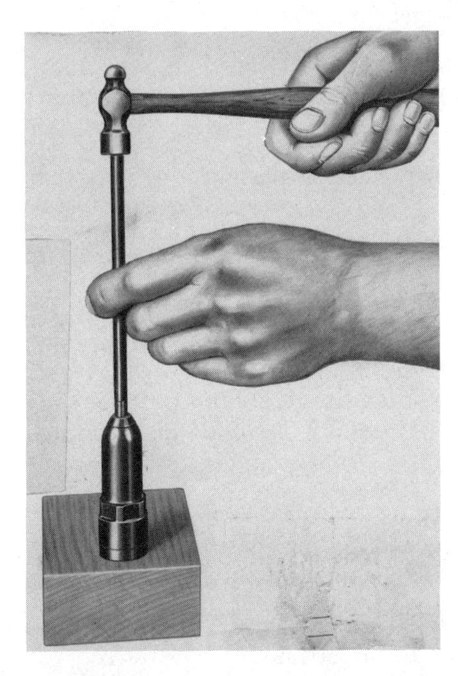

Figure 23.13 Removing injector tip from injector nut (courtesy Detroit Diesel Allison, Division of General Motors Corporation).

15 Place all the parts in a suitable cleaner that will remove all carbon and varnish buildup.

16 Allow parts to soak for approximately 1 hour. Then remove them from cleaner and blow until dry with compressed air.

B. Procedure for Inspection

Inspect all parts for wear or other damage as follows:

1 Check the injection spray tip with a magnifying glass for oblong or burred holes.

2 Check the holes in the injector tip with the correct size cleaning wire.

3 Clean the outside of the spray tip using a brass brush.

4 Inspect the needle valve body closely for scoring along the valve seat area.

5 Check the injector nut for carbon and clean thoroughly with a reamer if available.

6 Inspect the plunger and bushing assembly, looking closely at the plunger wear. Any plunger showing wear or the conditions indicated in Figure 23.14 should be discarded.

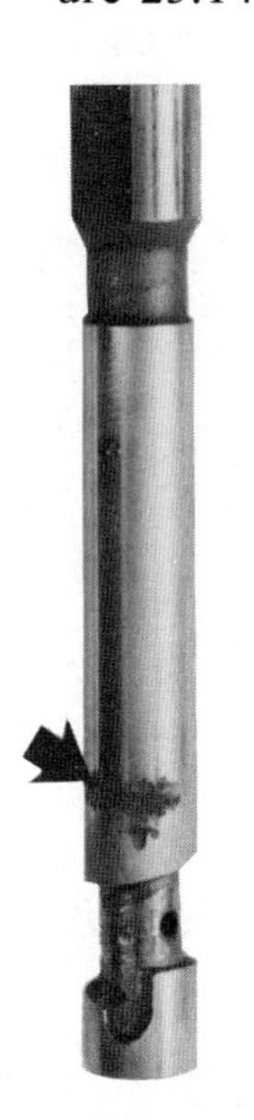

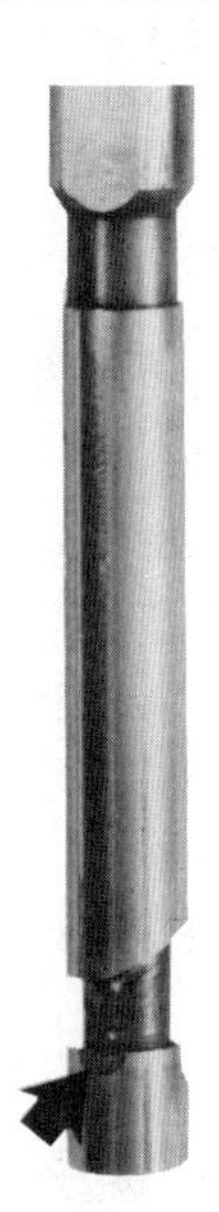

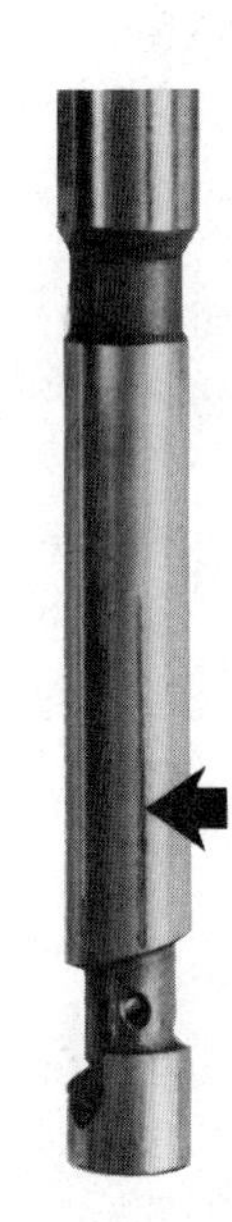

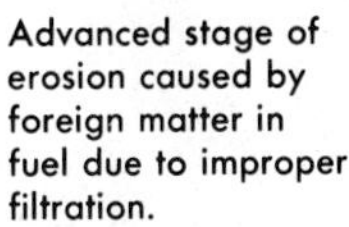

Advanced stage of erosion caused by foreign matter in fuel due to improper filtration.

Chipped at lower helix.

The above condition can be caused by either lack of fuel at high speeds or water in fuel.

Figure 23.14 Wear on pumping plunger (courtesy Detroit Diesel Allison, Division of General Motors Corporation).

NOTE Remember to keep all plunger and barrel assemblies matched together. *Do not* interchange them.

7 Check the plunger control gear and rack for worn or chipped teeth.

8 Check the follower spring with a spring tester.

NOTE The follower springs supplied currently have a wire diameter of 0.142 in. (3.65 mm). All injectors should be updated with this spring during an overhaul.

9 Check the injector body closely for nicks in the seal ring flat that encircles the injector body. Also check the body where the injector barrel contacts it. Ream the body with a special reamer if available.

10 Check all injector parts for burrs and roughness at the areas indicated in Figure 23.15.

11 If lapping is required, make sure the lapping block is clean and lap parts with 600 grit lapping powder.

12 Lap parts until nicks or burrs have been removed. Then clean with solvent and compressed air.

CAUTION Lap only enough to clean parts since excessive lapping can ruin them.

13 After lapping, polish the part on a lapping block without lapping compound.

C *Assembly.*

Before beginning assembly of the injector, make sure the work area is clean. Also, the injector assembly should only be done in an area that is free from airborne dirt and dust.

1 Insert the injector body in the holding tool with the top side up.

2 With the slotted side up, place a new filter into the inlet of the injector body.

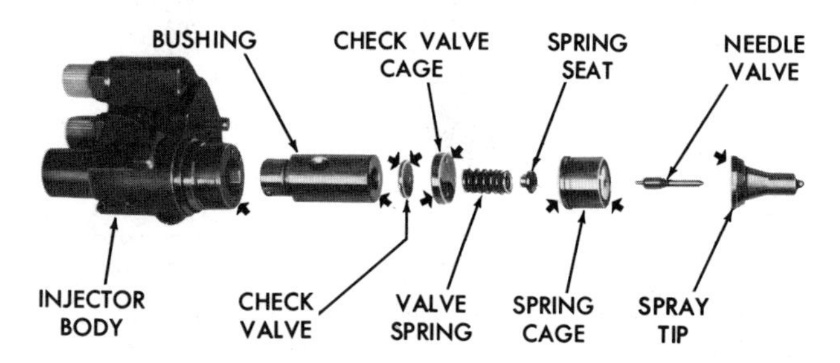

Figure 23.15 Areas to check closely for roughness or burrs (courtesy Detroit Diesel Allison, Division of General Motors Corporation).

NOTE The inlet side of the injector is immediately above the control rack. Offset body injectors use only one filter on the inlet side while standard or straight body injectors have an inlet as well as an outlet filter.

3 Install the inlet and outlet fittings with new gaskets and tighten to 65 to 75 in. lb (7.34 to 8.47 N·m).

4 After tightening the inlet and return fittings, blow compressed air through both of them to remove any foreign particles that may have been in them.

5 Place plastic caps on both inlet and return fittings to prevent the entry of dirt.

6 Remove the injector from the holding fixture.

7 Turn the injector body bottom end up and install the control rack.

8 Push the control rack into the body until the two timing marks can be seen when you look into the injector body bore.

9 Holding the rack in this position, insert the control gear into the body so that the timing marks on the rack and on the gear line up (Figure 23.16).

10 Install the injector body in the holding fixture and vise with the top end down.

11 Install the gear retainer.

12 Next install the plunger barrel onto the injector body by aligning the locating pin and inserting.

13 Install a new seal on the injector body and lubricate it and body threads with clean engine oil.

14 Install the spill deflector over the plunger barrel.

NOTE Before further assembly of the injector and spray tip, the spray tip lift should be checked using Detroit Diesel tool no. J9462-01 or its equivalent (Figure 23.17).

15 After the spray tip lift has been measured, the spray tip and spring assembly must be checked for opening pressure by installing it on the special adapters.

NOTE This test is not necessary if a tester is available to check the spray tip valve opening pressure after the injector has been completely assembled.

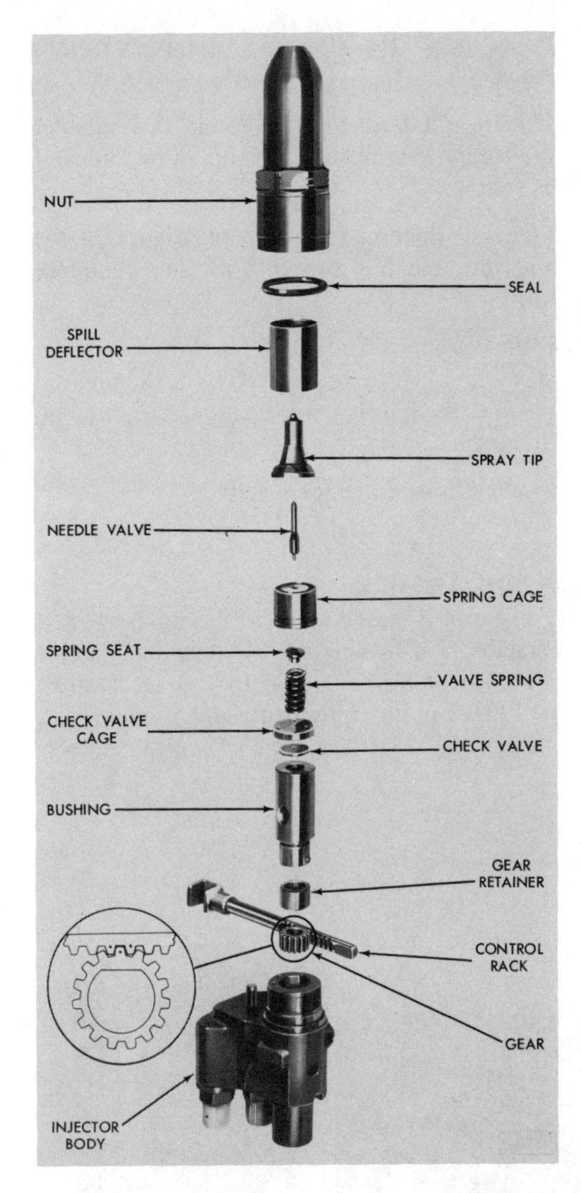

Figure 23.16 Timing marks on rack and control gear (courtesy Detroit Diesel Allison, Division of General Motors Corporation).

16 Assemble the spray tip and spring assembly on the adapter and attach the adapter to the nozzle test stand.

17 Operate the nozzle tester. Valve opening pressure should be 2300 to 3300 psi (161 to 231 kg/cm²).

18 If opening pressure is less than 2300, the valve spring may be weak or the needle valve may not be seated properly.

19 The spray tip should also be checked for leakage by pumping the pressure on the nozzle tester up to 1500 psi (105 kg/cm²).

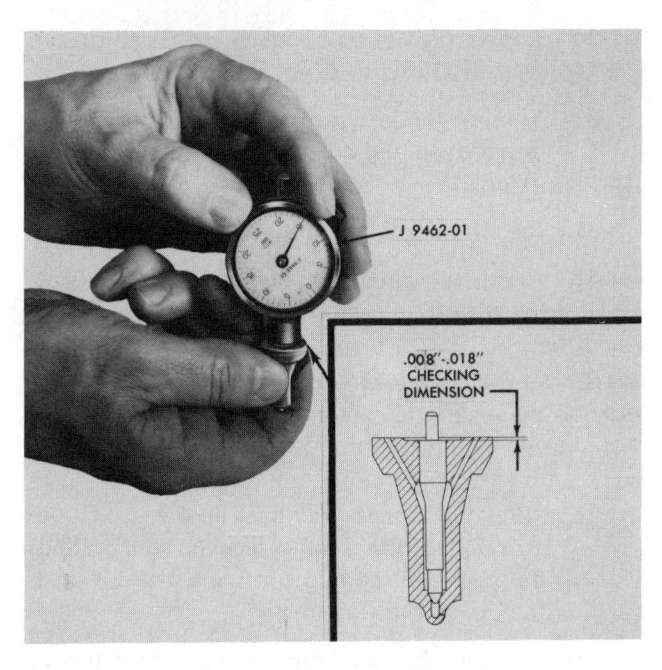

Figure 23.17 Checking spray tip lift (courtesy Detroit Diesel Allison, Division of General Motors Corporation).

20 Hold this pressure for at least 15 seconds. There should be no fuel droplets during this time.

NOTE A small amount of wetness on the tip is considered acceptable.

21 Remove the spray tip, spring, and spring cage from the adapter and install them in the injector body as follows.

22 Install the check valve.

23 Install the check valve cage.

24 Install the spring seat at the end of the spring.

25 Then insert the spring and spring seat into the spring cage, with the spring seat into the cage first.

26 Install the spring cage with spring and spring seat on top of the check valve cage, making sure that all the parts are in alignment after installation.

27 Install the needle valve into the spray tip, and then install the valve assembly on the spring case assembly.

28 Carefully place the injector nut over the parts that have been stacked on the injector.

29 Turn the nut by hand while holding down on the spray tip to prevent any dislodging of parts in the injector assembly.

30 Torque the nut with a deep socket or special Detroit Diesel tool, no. J49803-1.

31 Remove the injector from the holding fixture, turn it over, and insert it back into the holding fixture.

32 Install the follower onto the plunger.

33 Install the plunger, follower spring, and follower as a unit into the injector body.

NOTE As the plunger is inserted into the body, it may be necessary to move the control rack in and out slightly to get the plunger and gear to line up.

34 After the plunger has been inserted into the gear and injector body, place the stop pin into its groove in the injector body. Do not push the pin all the way in.

35 Position the stop pin so that the follower spring rests on the stop pin shoulder.

36 Align the slot in the follower with the stop pin and press down on it, depressing the spring.

37 Press in on the stop pin, snapping it into place.

38 If Detroit Diesel tool no. J5119 is available, the spray tip concentricity should be checked as follows.

39 Place the injector into the tool (Figure 23.18) and turn the injector 360° or one complete turn.

40 If the runout exceeds 0.008 in. (0.20 mm), the spray tip must be recentered by loosening the tip nut, recentering the tip, and retightening the nut.

41 Recheck the tip concentricity.

VII. Injector Testing

After the injector is assembled, it should be tested before installation in an engine. To test the injector properly, you must have an injector tester, Detroit Diesel no. J2310 or its equivalent, and a calibrator, Detroit Diesel no. J22410 (Figure 23.19) or its equivalent.

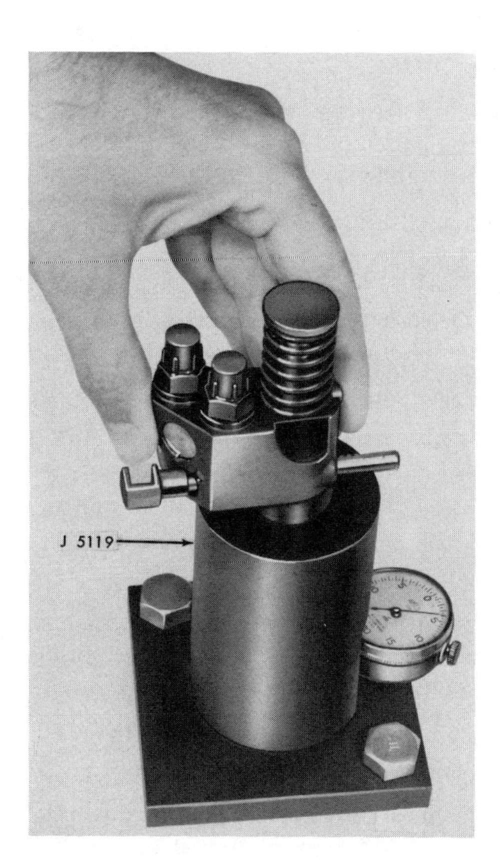

Figure 23.18 Injector concentricity tester (courtesy Detroit Diesel Allison, Division of General Motors Corporation).

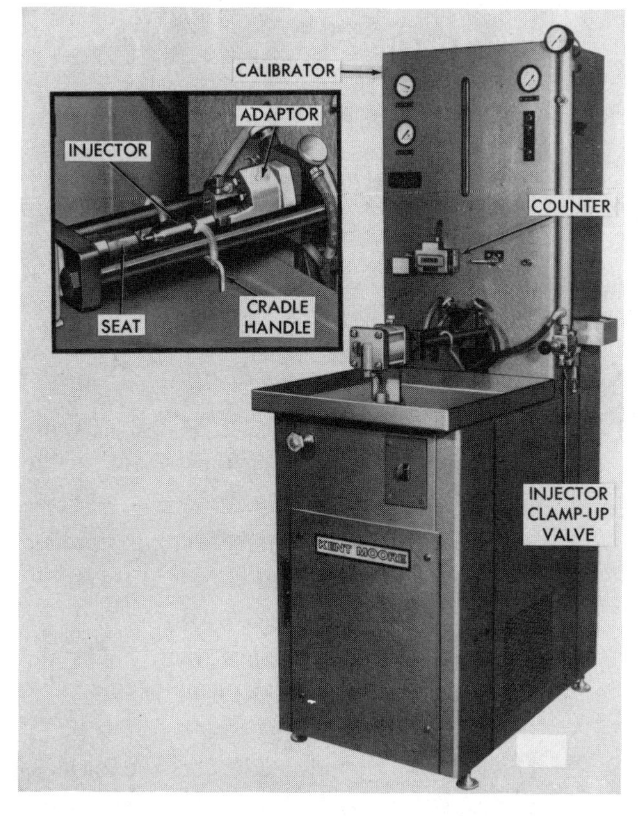

Figure 23.19 Injector calibrator (courtesy Detroit Diesel Allison, Division of General Motors Corporation).

A. Procedure for Testing the Control Rack and Plunger for Freedom of Movement

1 Place the injector in the special fixture available from Detroit Diesel or in another suitable device.

2 Push the plunger all the way down to the bottom of its stroke.

3 Move the control rack in and out of the injector body while allowing the plunger to move upward slowly.

4 If the rack is not free in the body, loosen and retighten the spray tip nut.

NOTE If this does not free the rack, the spray tip may have to be replaced.

B. Procedure for Spray Tip Opening Pressure

NOTE If the spray tip has been checked previously, it is not necessary to perform this test.

1 Install the injector in the tester and bleed all the air from the system.

2 Place the injector rack in the full fuel position.

3 Follow instructions for the type of tester you are using and build up pressure in the injector.

4 Observe the opening pressure and compare to specifications.

C. Procedure for High Pressure Test

The injector high pressure test is utilized to test the injector for leakage at the following points: spray tip nut seal ring, body plugs, and filter cap gaskets.

1 With the injector installed as in Section B, step 1 above, build up pressure in the injector 1600 to 2000 psi (112 to 140 kg/cm^2).

2 Check the injector closely for leaks.

NOTE Slight leakage at the control rack is considered normal.

D. Procedure for Pressure Holding Test

The pressure holding test is used to determine if the lapped surfaces are fitting together properly and the condition of the plunger and barrel assembly.

Since the procedure for making this test varies greatly from one tester to another, follow the test procedure for your tester in making this test.

E. Procedure for Fuel Delivery Test

Fuel delivery tests are made to determine if the injector is delivering the right amount of fuel. This test must be made on a machine called a calibrator or comparator, which simulates engine operation. Although several different makes of machine are available, the procedure for testing on all of them is very similar and is as follows:

1 Install the injector to be tested into the machine.

2 Connect the inlet and return lines.

3 Set the counter on the machine for 1000 strokes.

4 Place the injector control rack in the no-fuel position.

5 Start the machine, adjust the fuel pressure to 50 psi (3.5 kg/cm^2), and operate until all air bubbles disappear from the fuel return line.

6 With the machine running, start the counter. It is recommended that three draws be taken to insure accuracy.

NOTE The counter will stop automatically when 1000 strokes have been counted.

7 Read the fuel level in the graduate. The delivery should be as shown in the fuel output chart (Figure 23.20).

8 If injector delivery does not meet specifications, refer to the troubleshooting section of this chapter for possible causes.

NOTE Output of the Detroit injector can be changed only by changing the barrel and plunger assembly or tip assembly. If the injector does not need specifications for fuel delivery, try a different barrel and plunger or tip assembly. Output from injector to injector should be within 1 to 2 cc.

9 If the injector delivery meets specifications, remove the injector from the machine and place plastic caps on the openings.

Injector	Calibrator J 22410	
	Minimum	Maximum
71N5	50	54
N55	53	57
N60	57	61
N65(white tag)	64	68
N65(brown tag)	64	68
HN65	70	72
N70	71	75
N75	75	79
N80	81	85
N90	87	92
71C5	50	54
C55	53	57
C60	57	61
C65	64	68
C70	71	75
B55	53	57
B60	57	61
B65	64	68
71B5	50	54
B55E	53	57
7B5E	50	54

Figure 23.20 Injector fuel output chart (courtesy Detroit Diesel Allison, Division of General Motors Corporation).

VIII. Procedure for Injector Installation

To install the injector in the engine, the following steps must be followed:

1 Make sure that the injector you are installing has the same identification number as all other injectors in the engine if less than a complete set is being replaced.

2 Clean the copper injector seat in the cylinder head. A special reamer is available from Detroit Diesel to clean the carbon from the injector tube.

NOTE If a reamer is not available, a pointed wooden stick with a wiping towel on it may be used to clean the seat.

3 Place the injector in the cylinder head.

4 Make sure the injector control rack engages the control lever on the control tube.

5 Install the injector hold down clamp and bolt.

NOTE Make sure the injector hold down clamp is correctly placed on the injector to prevent interference with the valves and valve springs.

6 Torque the hold down bolt to 20 to 25 ft lb (27.12 to 33.89 N·m).

7 Lift the rocker arm assembly into position over the injector and valves.

NOTE If you are working on a four valve head engine, make sure the valve bridges are all in place before tightening the rocker assembly bolts.

8 Torque the rocker arm bolts to the recommended torque.

9 Remove the plastic caps from the injector inlet and return line connections.

10 Install the inlet and return fuel jumper lines.

11 Torque fuel jumper line nuts to 12 to 15 ft lb (16.27 to 20.34 N·m).

12 Adjust the injector timing and set tappets according to information given in the tune-up section of this chapter.

NOTE If a single injector is installed, it may not be necessary to complete all steps under engine tune-up. Injector timing, tappet setting, and injector control lever position are normally set when less than a complete set of injectors is installed.

IX. Engine Tune-up and Adjustment

Tune-up of the Detroit Diesel engine includes more operations than setting tappets and injectors. The following procedures must be followed in correct order. Since many different types of governors are used on Detroit Diesel engines depending on the engine application, the following tune-up will not cover all engine and governor types.

A. Procedure for Checking and/or Adjusting the Exhaust Valve Tappet Clearance

1 Remove rocker arm cover if not done previously.

NOTE Before checking and adjusting the valve tappet clearance, the rocker shaft bushings and shaft should be checked for wear. This is easily done by removing the hold down bolts and rotating the rocker arm assembly upward and tipping it back slightly. Remove the shaft, check for wear, and check bushings in the rocker arms. Replace if necessary.

2 Make sure the engine shutdown control is in the stop position.

3 Watch the injector rocker arm on the cylinder you wish to set and turn the engine with a bar or wrench in the direction of rotation.

NOTE If it is not convenient to use a bar or wrench to turn thé engine, you can use the starter, although this is not recommended by the author.

4 Turn the engine until the injector rocker arm is all the way down (injector depressed).

5 Check the clearance between end of the valve and valve rocker arm on an engine with two valve cylinder heads (Figure 23.21). On engines with four valve cylinder heads, check the clearance between the valve bridge and the rocker arm (Figure 23.22). If clearance is being adjusted on a four valve cylinder head with a spring-loaded bridge, the clearance should be checked between the bridge adjusting screw and the valve (Figure 23.23).

NOTE Valve clearance on two valve cylinder head engines is 0.009 in. (0.025 mm) hot (engine temperature 160 to 185° F), (75 to 85° C) and 0.011 in. (0.03 mm) cold. Valve clearance on four valve cylinder head engines is 0.014 in. (0.35 mm) hot (engine temperature 160 to 185° F), (72 to 85° C) and 0.016 in. (0.40 mm) cold.

6 If the clearance is incorrect (tight or loose), change the push rod length until the correct clearance is obtained.

CAUTION On engines using a bridge, the valve clearance is always corrected by adjusting the push rod length; do not adjust the bridge adjusting screw.

7 Adjust all the valves in the firing order sequence.

Figure 23.21 Checking clearance between rocker arm and valve stem (courtesy Detroit Diesel Allison, Division of General Motors Corporation).

Figure 23.22 Checking clearance between valve bridge and rocker arm (courtesy Detroit Diesel Allison, Division of General Motors Corporation).

B. Procedure for Checking and/or Adjusting the Injector Timing Dimension

1 Rotate the engine as outlined under tappet adjustment until the exhaust valve rocker arms are all the way down (exhaust valve open) on the cylinder to be adjusted.

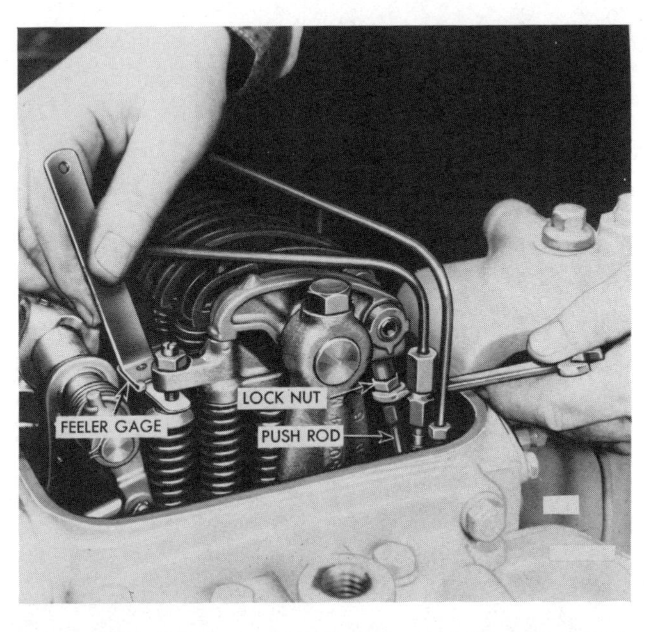

Figure 23.23 Checking clearance between bridge adjustment screw and valve (courtesy Detroit Diesel Allison, Division of General Motors Corporation).

Injector	Timing dimension	Timing gage	Camshaft timing
	Generator set applications		
All	1.460″	J 1853	Standard
	[a]All other applications		
7IN5	[b]1.460″	J 1853	[b]Standard
N55	[b]1.460″	J 1853	[b]Standard
N60	[b]1.460″	J 1853	[b]Standard
N60 (turbo)	1.484″	J 1242	Standard
N65 (white tag)	1.460″	J 1853	Standard
N65 (turbo brown tag)	1.484″	J 1242	Standard
N65 (non-turbo brown tag)	[c]1.484″	J 1242	[c]Advanced
HN65	1.460″	J 1853	Standard
N70 (turbo)	1.460″	J 1853	Standard
N70 (non-turbo)	1.460″	J 1853	Advanced
N75 (turbo)	1.460″	J 1853	Standard
N80 (turbo)	1.484″	J 1242	Standard
N80 (non-turbo)	[c]1.484″	J 1242	[c]Advanced
N90	1.460″	J 1853	Standard

[a] For automotive applications.
[b] Use 1.484 in. timing gage (J 1242) when engine has advanced camshaft timing. Correct to standard camshaft timing and 1.460 in. injector timing at first opportunity to be consistent with current production build.
[c] Use 1.460 in. timing gage (J 1853) when engine has standard camshaft timing. Correct to advanced camshaft timing and 1.484 in. injector timing at first opportunity.

Note: Advanced camshaft timing is indicated by "ADV–CAM–TIMING stamped on lower right-hand side of engine option plate.

Figure 23.24 Timing tool chart (courtesy Detroit Diesel Allison, Division of General Motors Corporation).

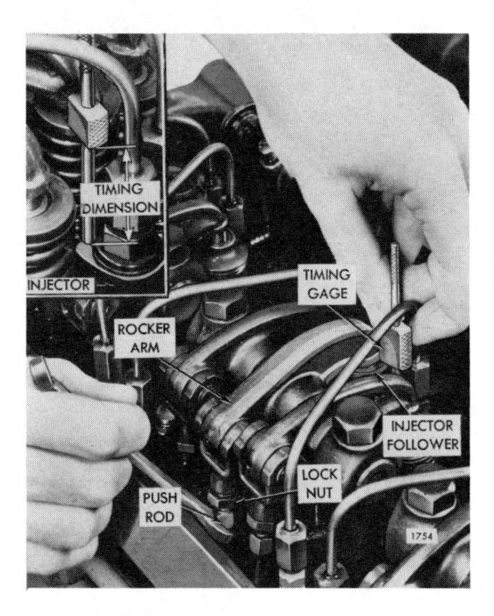

Figure 23.25 Injector timing tool in position to check injector (courtesy Detroit Diesel Allison, Division of General Motors Corporation).

2 Select the timing tool to be used from the chart (Figure 23.24).

3 Place the injector timing tool into the hole in the injector body (Figure 23.25).

4 Push down gently on the tool and attempt to twist it over the injector follower.

5 If the timing tool will just pass over the shoulder of the follower, the dimension is correct.

6 If the timing tool does not pass over the shoulder of the follower, the push rod should be adjusted.

7 Adjust the push rod by loosening the push rod locknut and turning the push rod with a wrench until the dimension is correct.

8 Lock the push rod locknut and recheck the dimension.

9 Continue turning the engine and adjusting the injectors in the firing order.

C. Procedure for Governor Gap Adjustment

1 Remove the governor high speed spring cover.

2 Loosen the buffer screw locknut and back out the buffer screw.

NOTE The buffer screw should be turned out until it protrudes approximately $\frac{5}{8}$ in. (16 mm) beyond the locknut.

3 With the engine running, adjust the engine low idle speed (Figure 23.26).

NOTE The idle speed on truck engines is usually set at 500 to 600 rpm.

4 With engine stopped, remove the governor cover by removing the hold down screws.

5 Select a 0.0015 in. (0.04 mm) feeler gauge to check the governor gap.

NOTE The governor gap check must be made while the engine is running. Since the governor cover and throttle have been removed, the engine speed must be controlled in some way other than with the throttle. One method of control is to clamp a vise grip wrench onto the control rack. Engine speed can then be regulated by the vise grip.

CAUTION Since the governor is not controlling engine speed at this time, make sure the engine is not overspeeded.

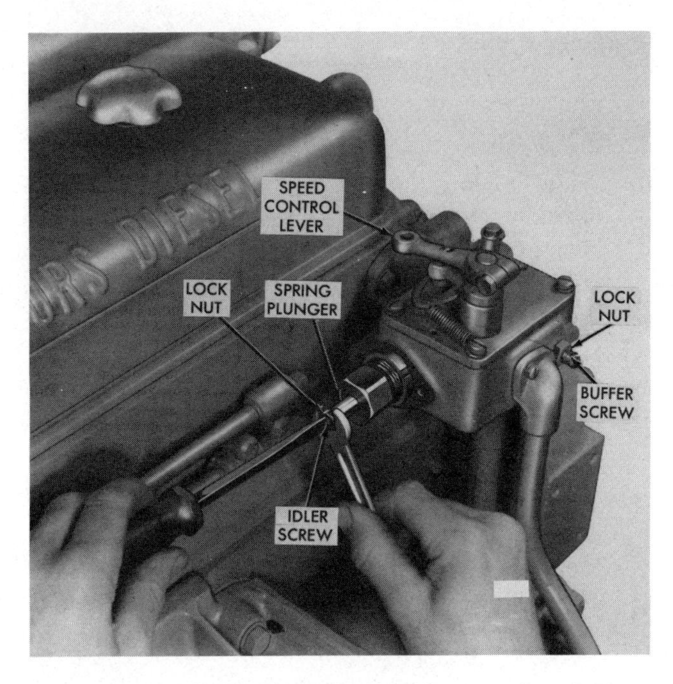

Figure 23.26 Adjusting low idle speed (courtesy Detroit Diesel Allison, Division of General Motors Corporation).

6 Run the engine at 800 to 1000 rpm and check the governor gap between the low speed spring cap and the high speed plunger with a 0.0015 in. (0.04 mm) feeler gauge (Figure 23.27).

7 If the gap is incorrect, adjust the gap adjusting screw.

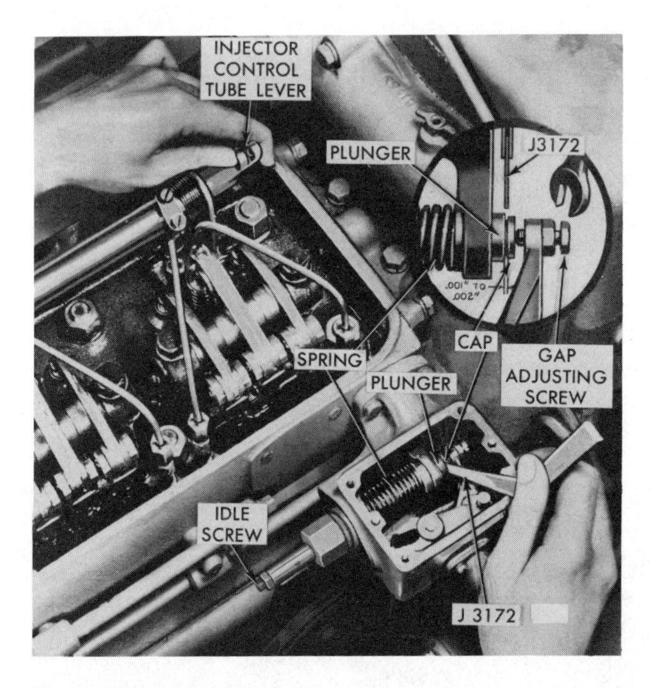

Figure 23.27 Checking governor gap (courtesy Detroit Diesel Allison, Division of General Motors Corporation).

8 Install the governor cover with a new gasket, making sure that the throttle lever is engaged in the differential lever.

D. Procedure for Injector Rack Control Lever Position V Engines

This adjustment is *very important* to the correct operation of the engine and should be done carefully.

1 Unhook the throttle linkage from the throttle at the top of the governor housing.

2 Back out the idle adjusting screw until 12 to 14 threads are exposed beyond the locknut.

3 Back out the buffer screw if not done previously.

4 Take out the right-hand clevis pin that connects the fuel control rod to the control tube lever.

5 Back out the inner and outer injector rack adjusting screws on both cylinder heads.

6 Place the speed control lever in the full speed position and hold it there.

7 With a screwdriver, turn in the inner adjusting screw of the first injector on the left bank until movement of the control tube is noticed. At this point the injector is in the full fuel position.

NOTE Right and left banks of the engine are always determined by looking at the engine from the rear.

8 Next turn the outer injector rack adjusting screw until it bottoms on the control tube.

9 Then adjust the inner and outer adjusting screws alternately until they are tight. The recommended torque is 24 to 36 in. lb (2.71 N m to 4.07 N m).

CAUTION To prevent stripping the threads in the injector control rack lever, do not overtighten screws.

10 With the speed control lever held in the full speed position, check the adjustment you have just made by pressing down on the injector rack clevis with a screwdriver (Figure 23.28). When you press on it, it will rock down. After you remove the screwdriver, the rack clevis should spring back to a horizontal position (Figure 23.29).

11 If the rack does not spring back up when you remove the screwdriver, the control rack adjustment is too loose. Loosen the outer screw on the

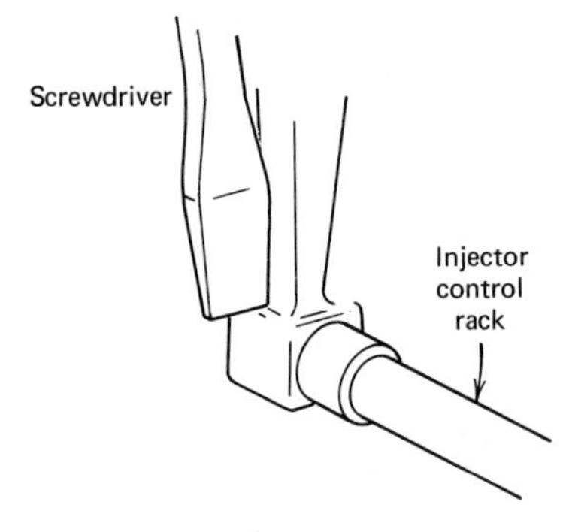

Figure 23.28 Checking rack clevis for movement (courtesy Detroit Diesel Allison, Division of General Motors Corporation).

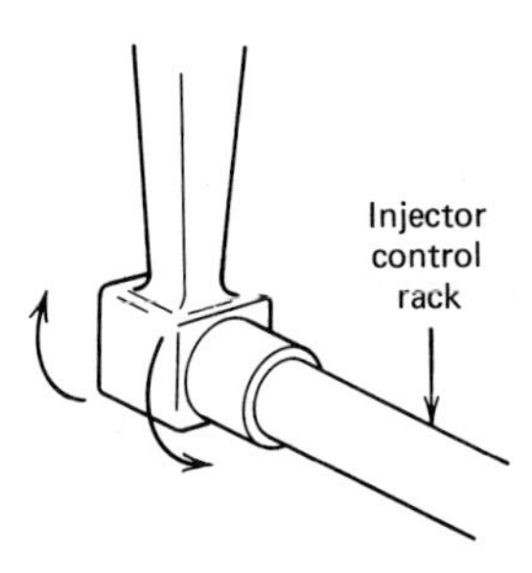

Figure 23.29 Typical rack movement (courtesy Detroit Diesel Allison, Division of General Motors Corporation).

injection control rack lever slightly and tighten the inner one. Recheck the rack by rocking it down with a screwdriver and releasing it. Continue adjusting and checking until the rack will rock up on its own.

12 Make sure the setting of the injector control rack lever is not too tight by watching the control tube as you adjust the lever. If the control tube moves before you have completed the adjustment, it is probably too tight.

13 Disconnect the left bank control tube from the fuel control tube lever by removing the connecting pin.

14 Connect the right bank injector control tube by inserting the pin into the control tube lever and governor fuel rod.

15 Now adjust the no. 1 right injector as you adjusted the no. 1 left injector.

16 Connect the left bank injector control tube lever and recheck the no. 1 left and no. 1 right injector racks for rock-back. Both injector racks should be set so they rock back up at this point.

NOTE Make sure the throttle is held in the full fuel position while making the checks.

17 Once you have the no. 1 left and no. 1 right injector rack adjusted correctly, disconnect both racks by removing the governor fuel rod and control tube pins.

18 While holding the control tube in the full fuel position, adjust the remaining injector control levers on both banks as outlined in step 10.

19 After all the injectors are adjusted on one bank, go back to the no. 1 injector and make a "rock check." If it does not rock back up, *do not readjust* the no. 1 injector control lever. Loosen and readjust the injectors on the bank until all check out as outlined in step 10.

20 After adjusting the injectors on both banks, reconnect the injector control tubes to the governor fuel rods by inserting the pins.

21 Make sure the pins have cotter pins in them to secure them in place.

22 Before starting the engine, turn the idle screw in so that about $\frac{3}{16}$ in. (4.7 mm) of it is visible beyond the locknut.

E. Procedure for Adjusting the High Idle No Load

1 Check the high idle no load by starting the engine.

NOTE Before making the high idle check, check to make sure that the engine emergency shutdown is working.

2 Move the throttle to full fuel position and check the rpm level.

3 If number of rpm's is incorrect, adjust the speed by loosening the spring retainer locknut and turning the high speed spring retainer in or out (Figure 23.30). If the engine speed is low, turn the retainer in; if the engine speed is high, turn the retainer out.

NOTE If the engine you are working on has a different type governor from the one in Figure 23.30, consult your Detroit Diesel service manual or your instructor.

F. Procedure for Idle Speed Adjustment

1 With the throttle in low idle position, check the engine low idle speed.

2 If it is incorrect, adjust as shown in Figure 23.26.

3 Turn the screw in to increase the idle speed and out to decrease it.

4 Lock the idle screw locknut.

G *Buffer screw adjustment.*
The buffer screw is used to eliminate engine speed change (roll) at low idle.

1 Loosen the buffer screw locknut and turn the screw in with engine running at idle until the engine roll has been eliminated.

2 Do not increase engine low idle speed more than 15 rpm with a buffer screw.

3 Recheck the engine at high idle. A change of 25 rpm or less in high idle is acceptable. If the speed change is greater than 25 rpm, the buffer screw must be reset and the rpm level checked again.

X. Troubleshooting and Testing

The following information should be followed closely when troubleshooting a Detroit Diesel fuel system.

A *Troubleshooting a Detroit Diesel injector that is installed in the comparator or calibrator.*

1 Low fuel delivery may be caused by the following conditions:

a Injection rack and control gear are not timed correctly.

b Worn barrel and plunger.

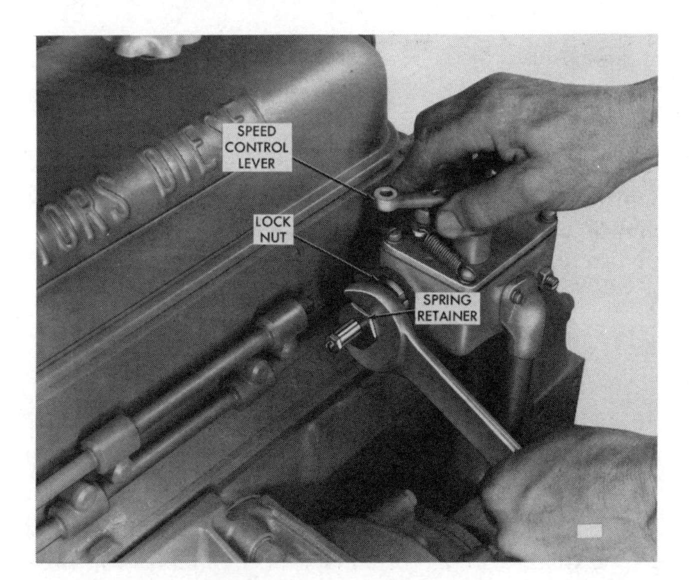

Figure 23.30 Adjusting high idle no load (courtesy Detroit Diesel Allison, Division of General Motors Corporation).

c Incorrectly fitted internal parts, valve seat, valve spring cage.

d Plugged holes in the spray tip.

2 High fuel delivery may occur because:

a The injector rack and control gear are not timed correctly.

b The machine counter is not set correctly.

3 Injector leaking fuel around tip retainer nut while running on the test stand may be caused by:

a Damaged tip nut O ring.

b Tip nut O ring missing.

4 Injector rack sticking may be caused by:

a An incorrectly assembled injector.

b A scored barrel and plunger assembly.

B *Troubleshooting the fuel injector when installed in the engine.*

1 If the cylinder misfires, the following conditions may be present:

a The injector timing dimension is set incorrectly.

b The injector control rack lever is incorrectly set.

c The injector delivery is low.

NOTE Isolating or finding a faulty injector should be done as follows: With the engine running and rocker arm cover removed, use a screwdriver to hold down the injector follower while noting the engine operation. If the engine sound changes, the injector probably is all right. If the sound does not change, the injector is probably faulty.

CAUTION Do not push sideways on the injector follower when pushing it down. Always push straight down on the injector or it may be damaged.

C *Troubleshooting the entire fuel system (engine).*

1 Low engine power may be caused by:

a An incorrectly set injector fuel control lever.

b Incorrect rack positioning.

c High idle engine speed set too low.

d Dirty fuel filters.

e Low gear pump pressure.

f A missing return line orifice.

g A worn governor.

2 A rough running engine may be caused by:

a A faulty fuel injector.

b Low fuel pressure.

c A restricted fuel filter.

d An improper injector timing adjustment.

3 A hard starting engine may be caused by:

a Low fuel pump pressure.

b Obstruction in spray tip holes.

c An incorrect timing adjustment.

d The shutoff control not adjusted properly.

4 Check for the following conditions if the engine does not idle without stalling:

a Low fuel pressure.

b Restricted fuel filter.

c Incorrect control rack setting.

d Idle adjusting screw not set correctly.

e Binding or sticking of injector linkage.

5 Check for the following conditions if the engine misses on one or more cylinders:

a Restricted fuel filter.

b Low fuel pressure.

c Return line orifice missing.

d Injector control rack levers incorrectly adjusted.

e Injector timing dimension incorrect.

6 Excessive engine smoke may be caused by:

a A plugged injector spray tip.

b Incorrect injector timing.

c A plugged return line orifice.

7 Crankcase dilution (fuel oil) may be caused by:

a A leaking injector tip nut O ring.

b A leaking jumper line.

c A leaking manifold fitting.

d The spray tip stuck open.

8 If the engine overspeeds, check for the following problems:

a A sticking injector control rack.

b Incorrect adjustment of control rack lever.

c The governor not adjusted correctly.

d A sticking accelerator linkage.

9 The following conditions may be present if the engine does not start:

a The shutoff control cable is incorrectly adjusted.

b The injector control rack is sticking.

c The injector timing dimension is incorrectly set.

SUMMARY

The chapter has dealt with a normal engine tune-up on a Detroit Diesel engine. Tune-ups are performed at regularly scheduled intervals or when an engine problem exists or when the injectors have been removed and replaced. This chapter provides you with the basic knowledge required to service, check, remove, and install injectors. For further information, consult your instructor or the Detroit Diesel service manual.

REVIEW QUESTIONS

1 What type of fuel system is the Detroit fuel system?

2 List five components of a Detroit Diesel fuel system and their function.

3 List four functions of the injector in the Detroit Diesel fuel system.

4 Explain with aid of a diagram the fuel flow in the injector.

5 In what way is the tag pressed into the body of a Detroit injector used by the mechanic?

6 Explain the fuel metering principle used in a Detroit Diesel injector.

7 Why should all injector barrel and plunger assemblies be kept as a matched unit?

8 Injector testing after assembly involves several tests. List and explain two of them.

9 Explain the general procedures involved in installing an injector.

10 A special timing tool is required to set injector timing. How is the correct tool selected?

11 The buffer screw is used to eliminate engine low speed surge or roll. What precautions must be used when adjusting it?

12 Explain why incorrect adjustment of the injector control rack levers can cause poor engine operation.

13 List and explain all steps that must be included in a Detroit Diesel tune-up.

24

Caterpillar Fuel Systems

Caterpillar is one of the few major engine manufacturers that also build its own fuel injection pumps. Caterpillar pumps are used only on Caterpillar engines and are not supplied for any other use. The first Caterpillar injection pumps, designated as forged body pumps (Figure 24.1), were the only pumps used by Caterpillar for many years. They used the common port and helix fuel control (scroll) and in many respects were similar to other in-line pumps of that period. Because these pumps are in very limited production, a thorough presentation will not be given here. As there are many of these pumps still in use, however, operational information and the basics on engine adjustments for forged body pumps will be given. If the situation should arise where more involved work is to be performed on a forged body pump, the mechanic should contact the appropriate service manual for the pump being worked on.

Later Caterpillar engines use an improved and streamlined version of the older forged body pump. These pumps are known as compact housing pumps (Figure 24.2). They use the same port and helix (scroll) fuel control as forged body pumps but are available with Woodward hydraulic governor, a hydra-mechanical governor, an air-fuel ratio control (aneroid), and variable timing advance unit. The compact housing is also offered in an in-line or V configuration.

All Caterpillar engines produced today use either the compact housing pump or the sleeve metering pump.

OBJECTIVES

Upon completion of this chapter the student will be able to:

1 Explain the operation of the scroll pumping element used in a Caterpillar pump.

2 Explain the operation of a sleeve metering pumping element used in a Caterpillar pump.

3 List and explain the function of two

Figure 24.1 An early forged body pump.

component parts of the scroll system pump.

4 List and explain the function of two component parts of the sleeve metering pump.

5 Describe and point out on a chart the operation of a hydra-mechanical governor.

6 Demonstrate the ability to remove and install a pump element in a scroll system pump.

7 Demonstrate the ability to set port closure on a compact housing pump.

8 Demonstrate the ability to set the maximum rack travel on a sleeve metering pump.

Figure 24.2 Compact housing pump.

9 Demonstrate the ability to remove and install a Caterpillar injection pump (student's choice).

GENERAL INFORMATION

All Caterpillar pumps, regardless of type (scroll or sleeve metering), utilize a single pumping element for each cylinder. All pumps used on four or six cylinder engines are in-line pumps. Pumps for V-8 engines, such as the 3208, use a pump designed in a V configuration. The principles utilized in the Caterpillar fuel system are very much like other systems that you have studied, such as American or Robert Bosch. Caterpillar pumps have been designed to be worked on with a minimum of equipment and special tools. Calibration of the injection pump is done right on the engine in contrast to other systems, which require that the pump be calibrated on a test stand.

I. Compact Housing Pumps

A *Operation of main pump.* (This information applies to forged body pumps as well.)

NOTE Throughout this section the illustrations show either the in-line or V compact housing pump. The operation of both is identical unless otherwise noted.

The component parts of the pump are shown in Figure 24.3. A camshaft (Figure 24.3,1) forces upward on a roller tappet (Figure 24.3,2), which then pushes the pumping plunger upward (Figure 24.3,3). Previous to this, while the plunger was at BDC, the transfer pump, which is mounted on the fuel filter adaptor on in-line pumps and on top of the V pump, supplies fuel at low pressure to the pump gallery (Figure 24.3,4).

Figure 24.4 illustrates the beginning and end of injection. From the gallery fuel flows into the area above the plunger (Figure 24.4,A), completely filling it. As the camshaft forces the plunger up more, it covers the inlet port (Figure 24.4,B). This is port closure and is the start of injection. Continued upward movement of the plunger will inject fuel until the lower edge of the helix (scroll) uncovers the spill port (Figure 24.4,C). This is the end of injection even though the plunger will continue to rise the full lift of the cam lobe before returning to BDC.

The reciprocating motion of the plungers then will create high pressure in the pump and open the nozzles, spraying fuel into the cylinder.

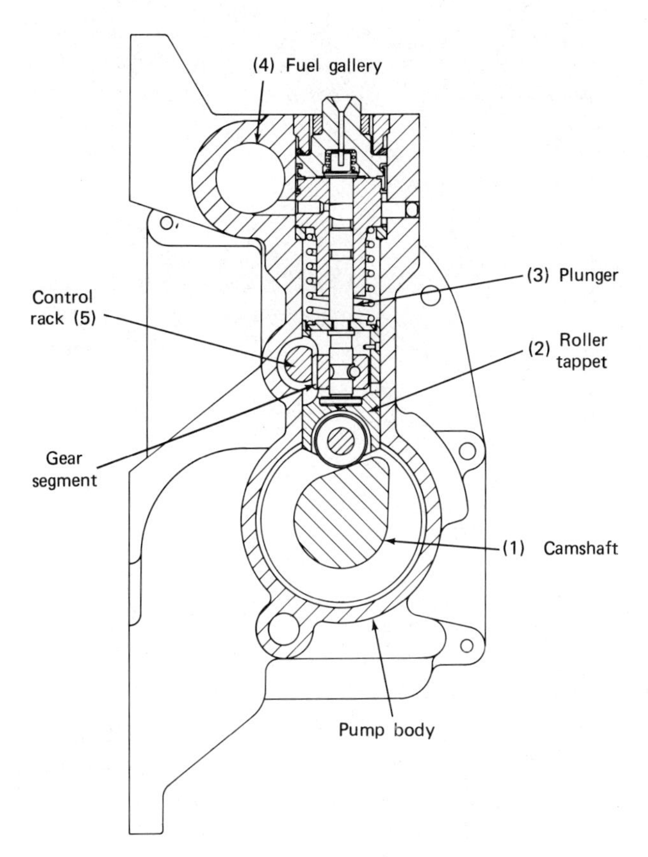

Figure 24.3 Cross-sectional view of a compact housing pump.

To control the quantity of injection, the plungers must be rotated to align different areas of the helix with the inlet and spill port. This is accomplished by placing a gear segment at the bottom of each pumping plunger (Figure 24.3,3). The gear segment is timed to a control rack (Figure 24.3,5). This rack is connected to the governor, which will rotate the pumping plungers to various

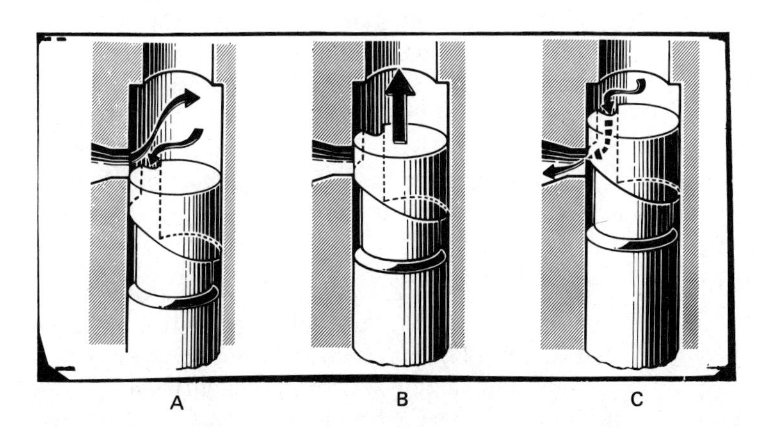

Figure 24.4 Beginning and ending of injection.

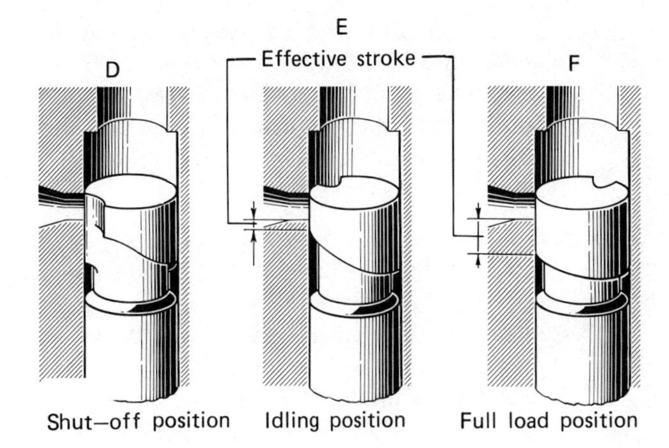

Figure 24.5 Various fuel delivery positions.

positions (Figure 24.5), depending on engine speed and load. These conditions are:

1 *Shutoff* (Figure 24.5,D). In this position the vertical groove machined in the plunger is in direct alignment with the fuel inlet port. When the camshaft pushes the plunger upward, the area above the plunger is full of fuel at the transfer pump pressure but, since the port is open, no injection pressure will be developed. This means that no injection will occur even though the plunger is stroking up and down the complete lift of the cam lobe.

2 *Idle position* (Figure 24.5,E). By moving the external control lever on the governor, the plungers will be rotated to the run position. Now when port closure occurs, fuel will be injected for a short while until the lower edge of the helix uncovers the spill port. This will provide sufficient fuel for low idle and high idle.

3 *Full load* (Figure 24.5,F). As the engine demands more fuel, the governor will rotate the plunger to the full fuel position. In this position injection will continue for a much longer time, resulting in increased fuel delivery. This position is for rated fuel delivery.

4 *Variable speed and load.* At speeds and loads other than those listed, the governor will position the plunger at a point somewhere between idle and full fuel delivery. If this would be for a motor truck application, the driver would control this position directly with the accelerator pedal. Also, for starting, the plunger will be so rotated that the inlet port will align with the longest section of the helix. This allows injection to continue for the longest possible time to aid in starting.

B *Operation of fuel transfer pump.*

The fuel transfer pump, mounted on top of the main pump V models, or on the fuel filter adapter on in-line pumps, or on the fuel pump auxiliary drive, is a positive displacement external gear type. It draws fuel from the tank and forces it to the injection pump fuel gallery at low pressure. Since this pump operates exactly as other external gear type pumps, no further explanation will be presented here. Refer to inspection and repair of this pump given later in this chapter.

C *Operation of hydra-mechanical governor* (Figure 24.6).

The Caterpillar hydra-mechanical governor is a hydraulically assisted mechanical governor. It will move the control rack (to which it is directly connected) which, in turn, rotates the pumping plungers. As mentioned earlier, plunger rotation causes more or less fuel injection. By the use of engine oil pressure assistance the total weight and size of the governor can be much less than a comparable mechanical governor. On in-line pumps the governor is mounted on the rear of the pump and on V pumps it is mounted on top of the pump.

1 *Starting position.* In the starting position the external control lever puts maximum pressure on the governor spring (Figure 24.6,1) which will push the control rack to the front or maximum fuel position. The governor weights are completely collapsed at this time.

2 *Run position.* As the engine starts, the flyweights (Figure 24.6,2) fly outward, compressing the governor spring. This flyweight movement will pull the oil control valve (Figure 24.6,3) to the rear. At the same time, engine oil under pressure will enter the governor at the front (Figure 24.6,4), flow around the sleeve (Figure 24.6,5), and force the piston (Figure 24.6,6) to the rear. This pulls the control rack to the rear, which provides less fuel to the engine, slowing it down. As the engine slows down, the flyweights will move inward because of the pressure of the main governor spring. This moves the control valve (Figure 24.6,3) to the front of the pump. Engine oil pressure is now routed to the large end of the piston (Figure 24.6,6) and moves it to the front of the pump. This movement results in more fuel delivery, causing the engine to speed up. When spring tension and oil pressure balance out, engine speed will stabilize.

A mechanical shutoff is provided by rotating the shutoff shaft (Figure 24.6,7) that contacts the collar (Figure 24.6,8) and pulls the control rack to the zero delivery position.

The same oil that moves the control valve provides for governor lubrication and eventually drains back into the engine crankcase through a line at the bottom of the governor housing.

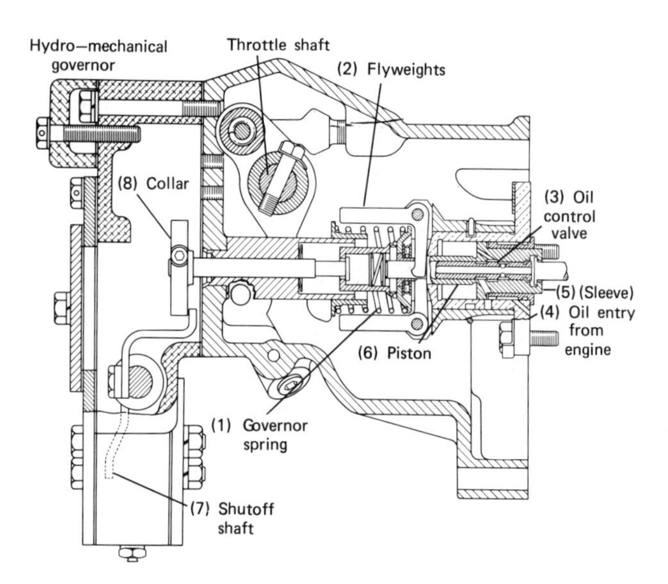

Figure 24.6 Cross-sectional view of governor for compact housing pumps.

D *Operation of air-fuel ratio control (aneroid).*

The Caterpillar air-fuel ratio control or aneroid (Figure 24.7) is used on turbocharged engines to limit or eliminate visible smoke during acceleration. The unit is a simple diaphragm and spring-type aneroid operated by intake manifold pressure.

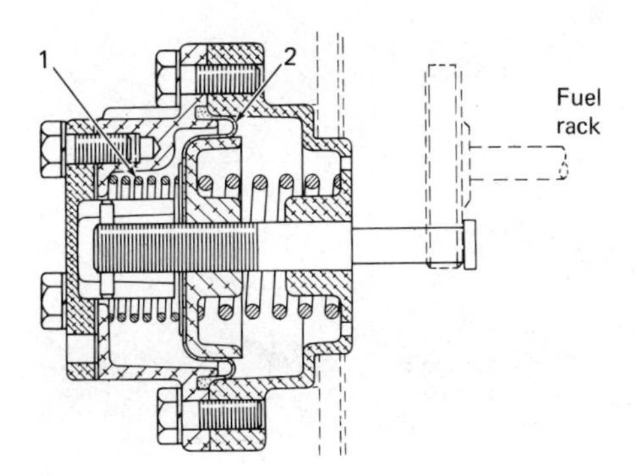

Figure 24.7 Air-fuel ratio control cross section as used on compact housing injection pumps.

The bolt of the aneroid connects with the collar that is used to shut off the engine (Figure 24.6,8). As the engine is accelerated, the collar will pull against the aneroid bolt that is under spring pressure (Figure 24.7,1). This will restrict control rack movement (which limits fuel) until the turbocharger develops sufficient intake manifold pressure to push the diaphragm (Figure 24.7,2) to the right. Then the control rack is free to travel to the full limit as set by the full load screw. This gradual increase in fuel delivery will limit the output of visible smoke during acceleration. However, as with all aneroids, acceleration will be slightly slower than engines that are not so equipped.

A hydraulic aneroid is used on some engines to achieve the same effect of reduced smoke and good performance. Both units are similar in appearance.

E *Operation of automatic injection advance.*

The advance unit (Figure 24.8) for Caterpillar engines (if so equipped) is a centrifugal type (some types being assisted by engine oil pressure) that is connected to the pump camshaft at the front and driven by the timing gears. As the engine speeds up and centrifugal force increases, the advance weights move outward and move the flange (Figure 24.8,1) a small amount as opposed to the drive gear. This flange is connected to the pump camshaft, which is then advanced. This causes injection to occur earlier in the cycle. Proper advance of injection is critical to engine operation.

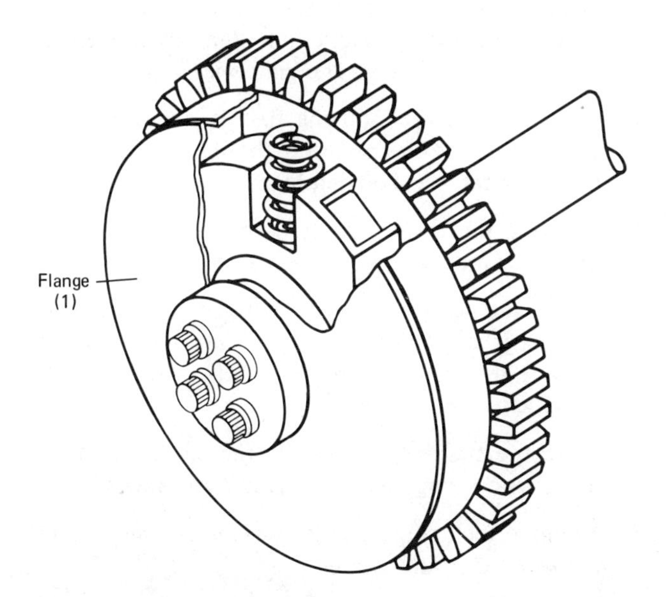

Figure 24.8 Centrifugal advance unit.

II. Removal and Disassembly of Compact Housing Pumps

A. Procedure for Removal

Before beginning, thoroughly clean the entire engine with a steam cleaner or high pressure washer. Cap all lines as they are removed.

1 Place engine's no. 1 cylinder on top dead center (TDC) compression stroke by installing the timing bolt in the threaded hole in the flywheel (Figure 24.9).

Figure 24.9 Installing timing bolt in the flywheel housing.

NOTE This timing bolt is stored in the engine near the installation hole.

2 When the bolt can be threaded into the flywheel, make certain no. 1 is on the compression stroke by checking the valves for that cylinder. Both rocker arms must be loose. If not, remove timing pin and rotate engine until the bolt again threads into the flywheel. This will now be no. 1.

NOTE On older engines this method may not be used to locate TDC no. 1. Always check the Caterpillar service manual for the specific engine being worked on.

3 With the engine correctly located at TDC no. 1 compression stroke, disconnect all fuel supply and return lines.

4 Remove all air pressure and lube oil lines leading to the pump.

5 Remove all fuel injection lines and cap.

6 Remove the bolts holding the drive gear to the pump camshaft (Figure 24.10).

7 Remove all mounting bolts and the slide pump out of drive gear.

CAUTION Since the pump is very heavy, always have an assistant help lift it from the engine or use a suitable hoist.

B. Procedure for Disassembly

Disassembly procedures are given for an in-line compact housing pump. The instructions for a V type are very similar. If a specific question is not answered

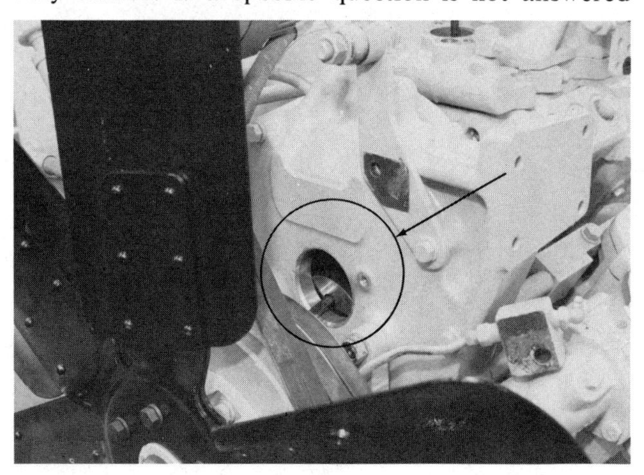

Figure 24.10 Removing pump camshaft drive bolts.

Figure 24.11 Compact housing pump governor housing removed.

here, consult your instructor or the Caterpillar service manual.

Before beginning disassembly, clean the pump and workbench. Have available clean lintfree rags and clean calibrating oil or diesel fuel. *Be clean!*

1 Begin by mounting the pump in a suitable holding fixture in a brass-jawed vise.

2 Remove the aneroid and electric shutoff solenoid if so equipped.

3 Remove the governor end cover (Figure 24.11) by removing the bolts. Dispose of the waste oil.

4 Remove the main governor housing by first removing the stop collar; this is done by loosening the set screw through the side opening.

5 Lift the governor housing from the pump body after removing the mounting bolts. Use a rubber mallet to tap the housing loose.

6 Remove the three mount bolts and pull off the governor mechanism.

7 Check the parts in the governor control cover and replace if any signs of wear are present. Always replace the oil seals on the throttle shaft (Figure 24.12).

8 Slide the main governor spring off the control shaft with spring seats, bearings, and races.

9 Remove lock ring cylinder weight assembly, and control valve and sleeve.

10 Pull out the felt washers around the pumping

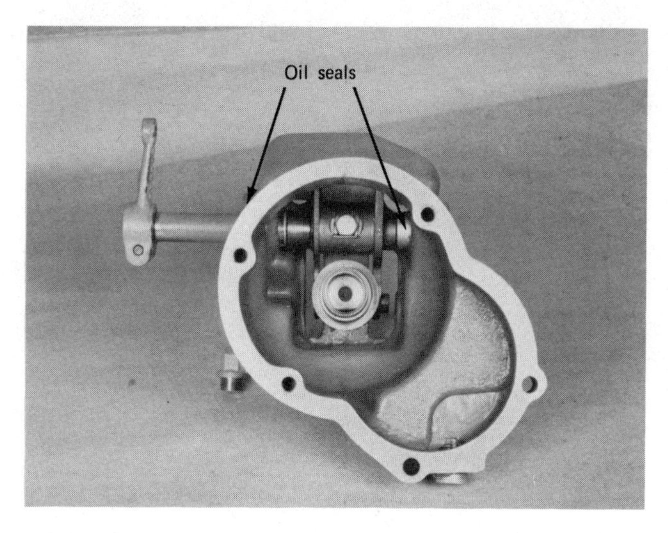

Figure 24.12 Governor housing with throttle shaft and seals.

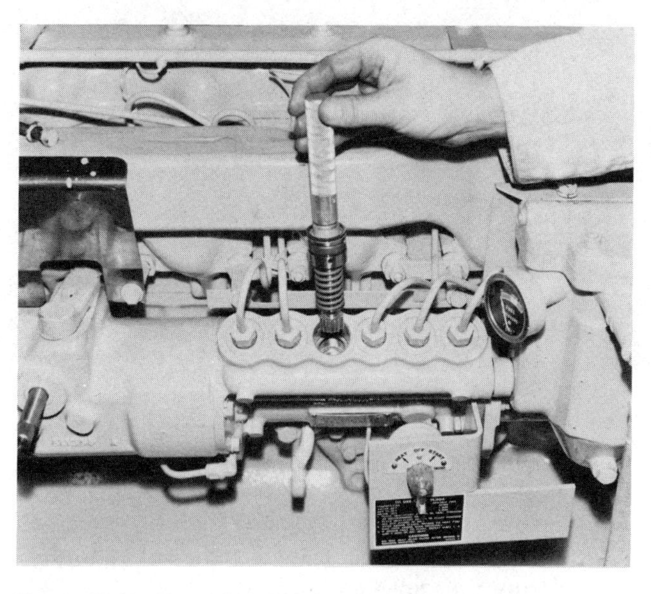

Figure 24.14 Removing the pumping element.

units. Remove the retaining bushing from the no. 1 cylinder with correct wrench (Figure 24.13).

11 Pull the pumping unit from the housing with the correct extractor (Figure 24.14).

NOTE This assembly will include the pumping plunger and barrel, check valve, return spring and seat, and bonnet. Be very careful not to drop any of these parts.

12 Remove the remaining pumping elements and place them in order in a clean tray.

CAUTION As these units are taken apart to inspect the plungers, do not mix the plungers from barrel to barrel. These are matched parts and must be kept together.

Figure 24.13 Removing retaining bushing from pump.

13 Pull the fuel control rack from the housing.

14 Carefully remove the spacers and lifters from each bore. Again, keep these parts together with their respective pumping element to avoid mix-up. A divided tray is very helpful during this operation.

15 Remove the bolt that retains the governor drive gear to the camshaft and lift off the plate and crescent-shaped clip. Take off the governor drive gear.

16 Carefully slide the camshaft from the pump housing from the drive end.

17 Place all painted parts in parts cleaner to remove all loose paint and grease. Rinse all internal pump parts in solvent and blow dry.

18 Lay the parts out for inspection and possible repair.

III. Procedure for Inspection of Compact Housing Pump Component Parts

After all parts have been cleaned, check them carefully for wear or damage.

A *Main housing.*

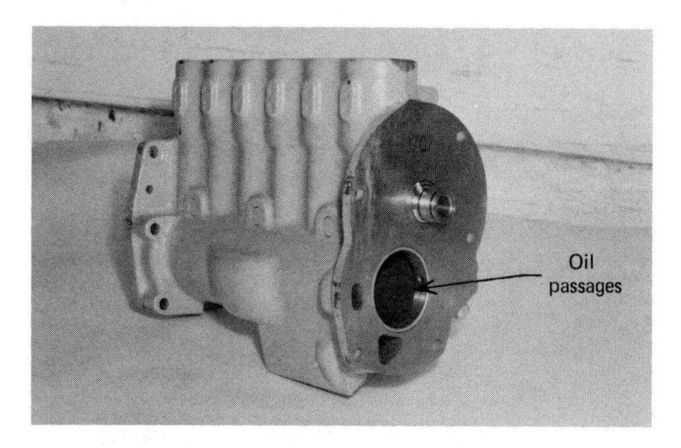

Figure 24.15 Oil passages correctly positioned.

1 Check for thread damage and cracks and measure the camshaft bushings.

2 If new bushings must be installed in the housing, make certain to locate correctly the oil passages indicated in Figure 24.15.

B *Camshaft.*

1 Measure the camshaft journals and compare them with Caterpillar specifications.

2 Inspect camshaft lobes for flaking or pitting. Replace as required.

C *Pumping elements.*

1 Check the helix area of the plunger closely for wear and scoring. If score marks can be felt with the fingernail, replace both the plunger and barrel.

2 Also, Caterpillar recommends measuring the plunger length (Figure 24.16). This length, plus the lifter washer (the yoke on forged body pumps) thickness, is critical to correct port closure or internal pump timing. Replace if out of specified limits.

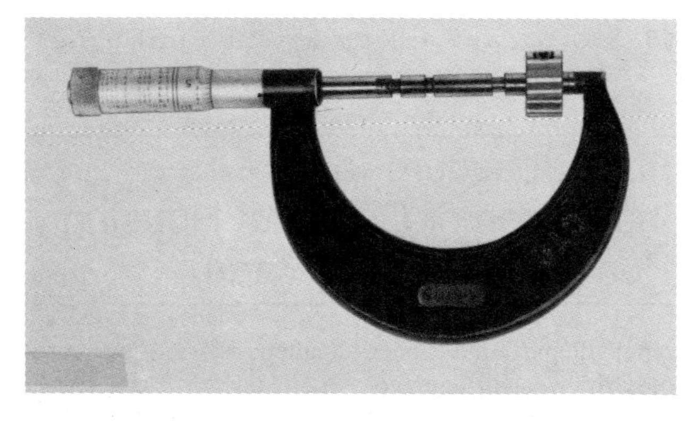

Figure 24.16 Measure overall length of the plunger.

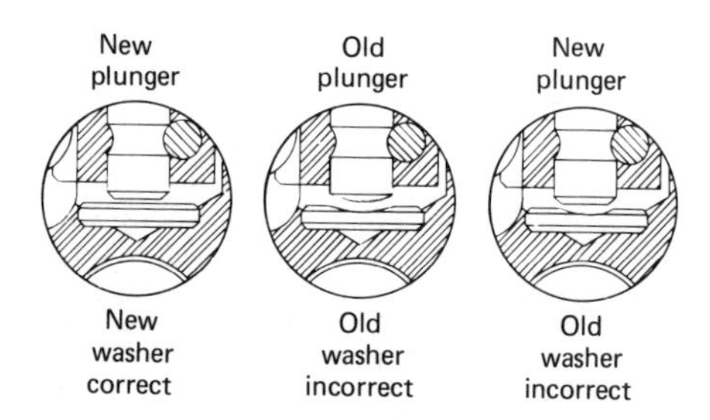

Figure 24.17 Lifter washer.

D *Lifter washers and roller tappets.*

1 When doing a major overhaul on a compact housing pump, carefully inspect the lifter washer (Figure 24.17) and replace if worn.

2 Also check the roller tappets for heavy score marks and wear of the roller.

E *Plunger springs.*

1 Test all springs for free length and pressure with a spring tester and compare to specifications.

F *Transfer pump.*

1 When checking the gear pump, check for excessive gear and housing wear.

2 Check for wear of the top plate and drive shaft.

3 Use Plastigage between the cover and gears to check clearance. Clearance should be 0.001 to 0.002 in. (0.02 to 0.05 mm).

4 Always carefully check the transfer pump drive shaft for grooves due to seal wear. Replace both the shaft and seal if necessary to prevent fuel leakage at this point.

G *Governor.*

1 Inspect all governor parts for excessive wear and breakage.

2 Test the springs for correct force.

3 On hydra-mechanical governors, make certain the oil pressure control valve works freely but without excessive clearance.

4 Always replace governor bushings and seals.

5 Be critical of any pivot point within the governor because any wear here will result in a poor responding engine.

Caterpillar service manuals list all clearances, spring pressures, and dimensions for the different sizes of pumps. Always be certain to refer to the correct manual for the pump being worked on.

IV. Reassembly and Installation to Engine

A. Procedure for Reassembly

When reassembling the injection pump, always replace all gaskets, seals, and O rings. As each part is added, dip it in clean fuel or light oil. If special tools are called for, make certain to use them. If these tools are not available at your shop, sometimes they can be fabricated. If not, the best course of action is to take the pump to a Caterpillar service station that has all the required tools.

1 Press in new camshaft bushings if required.

NOTE Make sure all oil supply and drain holes are in alignment.

2 Slide in the camshaft and install the governor drive gear with crescent-shaped clip and torque. Bend the lock tab (Figure 24.18).

3 Install the roller tappets (lifters) (Figure 24.19) with the cutout area facing the control rack (Figure 24.19,2).

4 Insert the timing spacers (Figure 24.19,1) on top of the lifters.

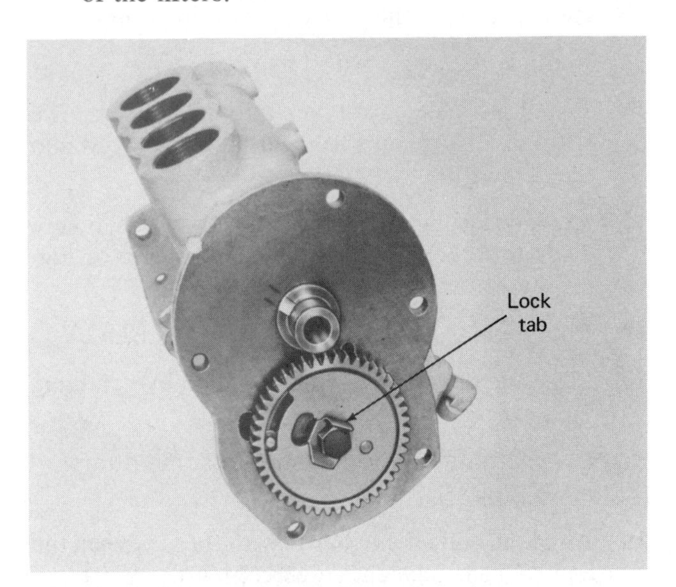

Figure 24.18 Correctly bent lock tabs on drive gear bolts.

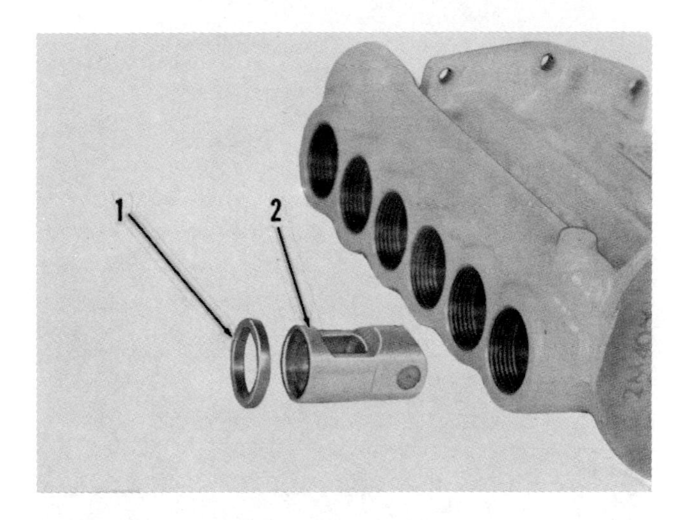

Figure 24.19 Roller tappet assembly.

NOTE At this time the timing dimension of each individual cylinder must be set (port closure). This will be set by changing the spacer inserted in step 4 with one of different thickness (if required). This procedure is very critical to pump timing on the engine and will greatly affect engine operation. Special tools required will be:
1. 5P1768 pointer
2. Micrometer depth gauge
3. Correct pump camshaft timing pin
4. 8S7167 gauge
5. 1P7210 gauge plate

5 Install the pointer to the pump housing (Figure 24.20).

6 Fasten the gauge plate to the pump camshaft (Figure 24.21) finger tight.

7 Install the pump timing pin and rotate the camshaft until it engages the notch (Figure 24.22).

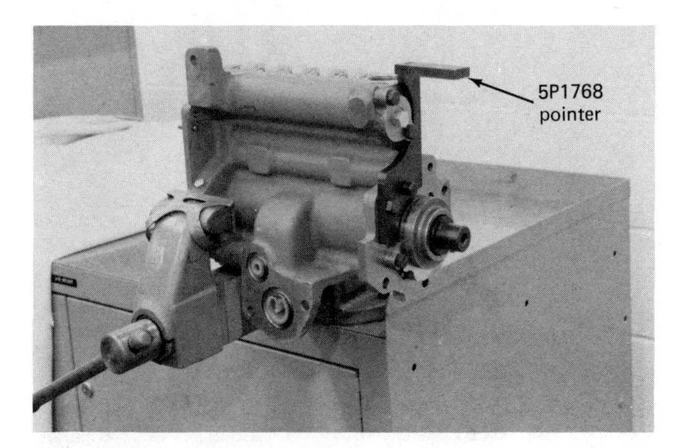

Figure 24.20 Phasing pointer mounted on pump.

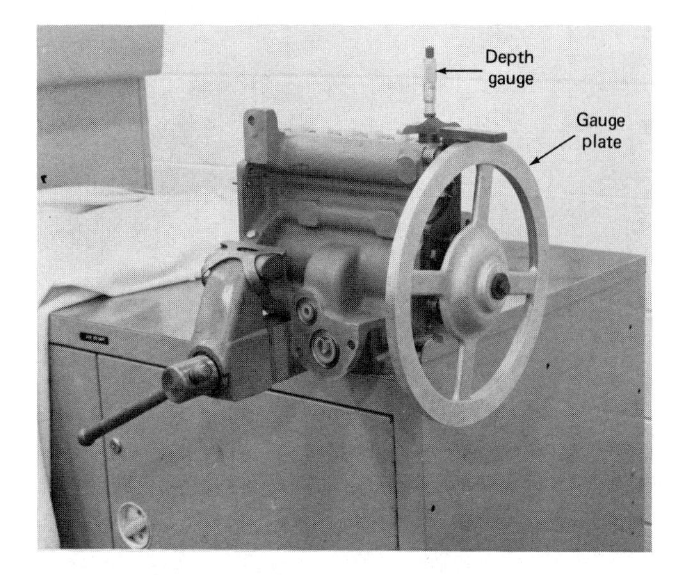

Figure 24.21 Gauge plate bolted to camshaft.

8 Rotate the gauge plate until the O mark aligns with the pointer. Tighten the bolt that holds the gauge plate to the camshaft.

9 Remove the camshaft timing pin.

10 Install the 8S7167 gauge in the no. 1 bore of the housing (Figure 24.23) and measure from the top of the gauge to the lifter wear washer.

11 The dimension recorded must correspond to the dimension given in the specification sheet. If it does not, the spacer must be exchanged until the correct reading is obtained.

NOTE This dimension will be different for direct injection engines and precombustion chamber engines.

Figure 24.22 Installing timing pin on pump.

12 The remaining lifters and spacers must be checked in the same manner. The gauge wheel must be rotated in normal rotation until the next cylinder in the firing order is in proper position. The charts below give the correct degree mark for six and eight cylinder engines. When the correct degree mark is lined up with the pointer, check the dimension. Change the spacer if necessary.

Six Cylinder

Lifter number	Degrees
1	0
5	60
3	120
6	180
2	240
4	300

Eight Cylinder

Lifter number	Degrees for engine type	
	(Precombustion)	(Direct injection)
1	354.5	346
8	39.5	31
4	84.5	76
3	129.5	121
6	174.5	166
5	219.5	211
7	264.5	256
2	309.5	301

13 To double-check the accuracy of the measurements, recheck the cylinders in firing order.

14 Remove the gauge wheel and pointer.

15 Install the fuel control rack into the housing. The left rack of V pumps has a groove machined into the face of the rack.

16 Assemble pumping elements as shown in Figure 24.24 for precombustion chamber engines or Figure 24.25 for direct injection engines.

17 Rotate pump until no. 1 cylinder is at BDC.

18 Place the rack in the center position by centering the gear teeth in the bore.

NOTE A centering stop is used on some pumps (Figure 24.26).

19 Align the bonnet and barrel with the groove in the control gear segment (Figure 24.27).

20 Carefully place the unit into the bore without ro-

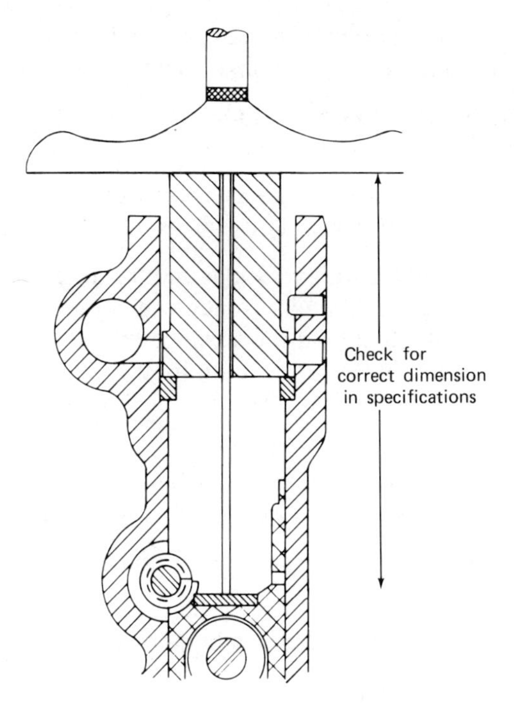

Figure 24.23 Measuring timing dimension.

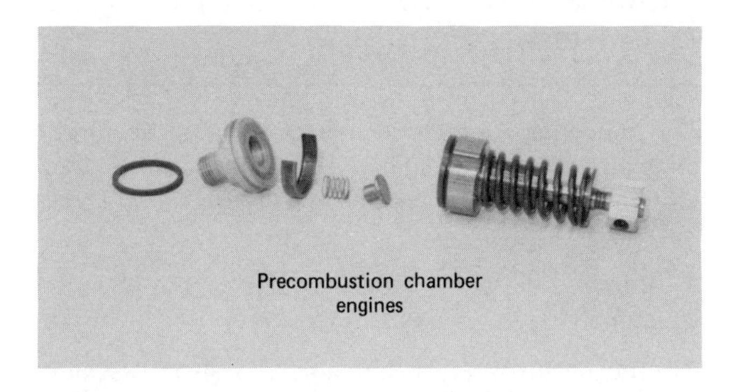

Figure 24.24 Assembly of pump element (precombustion chamber engines).

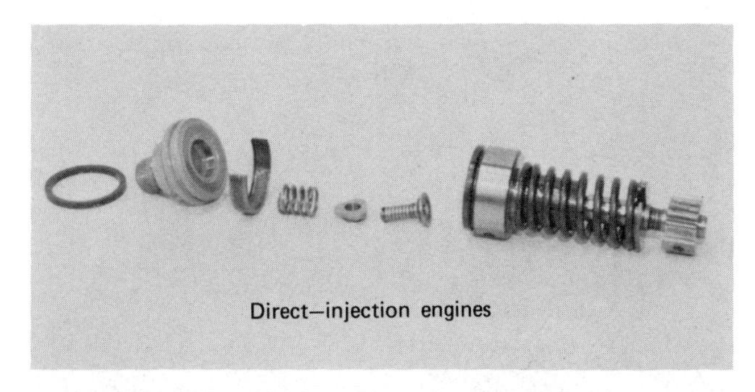

Figure 24.25 Assembly of pump element (direct injection engines).

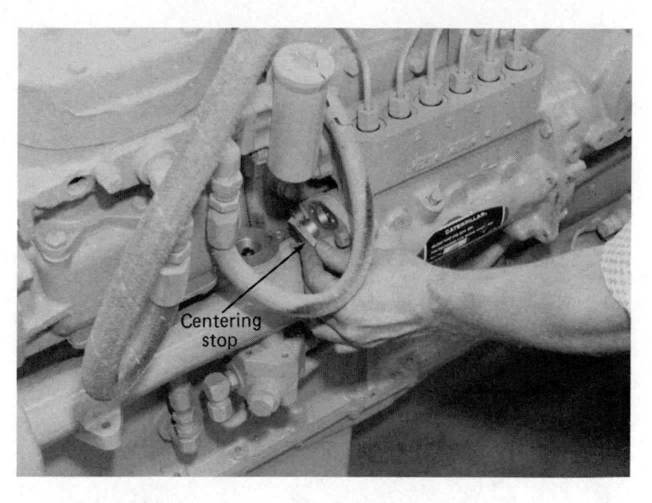

Figure 24.26 Stop used to center the control rack.

tating the plunger and make certain the notches line up (Figure 24.27, A and B).

21 Screw in the hold down bushing finger tight. Rack must still move freely.

22 After all elements are in place, check the total rack travel with a dial indicator or depth micrometer. It must be 0.800 in. (20.32 mm). If it is not, one or more of the gear segments are not properly aligned and must be removed and correctly installed.

23 If the correct travel is obtained, torque all the bushings to 150 ft lb (203.37 N·m). Install new felt covers over the bushings.

CAUTION If the correct travel is not obtained and the pump is installed to the engine, the engine could overspeed and cause great damage. Be very careful when installing the pumping elements and measuring rack travel.

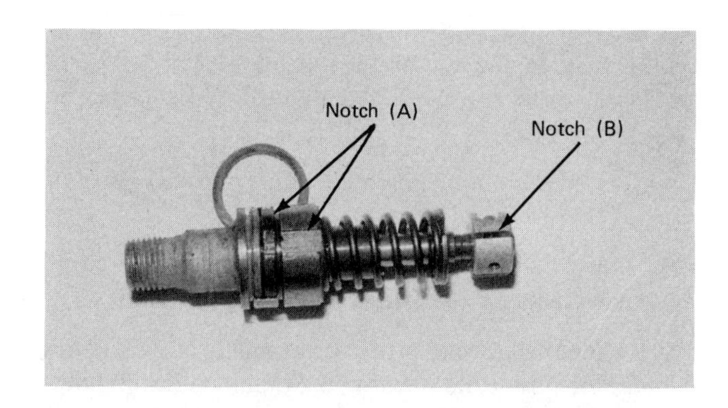

Figure 24.27 Alignment of grooves before assembly.

Figure 24.28 Installing the fuel transfer pump.

24 Install the control valve and sleeve, cylinder weight assembly and lock ring to governor mechanism.

25 Mount the control unit to the pump housing with the cap screws, engaging the governor drive gear.

26 Mount the governor housing and install the fuel stop collar.

27 Renew shaft seals, install levers, and mount end cover to the governor housing.

28 Attach the rear cover plate, aneroid, and shutoff solenoid.

29 On V pumps install the fuel transfer pump, engaging the gear as it is installed (Figure 24.28).

30 Mount pump to engine (if available) to finish adjustments.

B. Procedure for Installation

Make sure engine is still on the no. 1 cylinder compression stroke and the timing bolt (if used) is screwed into the flywheel.

1 Install the timing pin in the pump to lock camshaft at the no. 1 cylinder.

2 Hoist the pump into position on the engine, using new seals on the flange and fuel supply lines.

3 Align the bolt holes of the automatic advance unit to the pump and position the pump on the engine.

4 Install the bolts that hold the pump to the engine and tighten.

5 Place the bolts that hold the advance unit to the pump camshaft and tighten finger tight.

NOTE Tighten these bolts to the correct torque at this time.

6 Remove the timing pin from the pump camshaft and timing bolt from engine flywheel.

7 Rotate the engine two revolutions and try to reinstall the timing bolt and pin. If they will not go in, loosen the bolts that hold the advance unit to the pump camshaft and tap it gently until the pin will enter the pump camshaft. Torque the bolts and remove pins.

8 Install the front cover with a new gasket. Store the timing pin and bolt in their correct locations.

9 Install injection lines, supply lines, and throttle linkage.

10 Provide a means of emergency shutdown.

11 Supply fuel to pump and bleed lines.

C. Procedure for Installation and Timing of Forged Body Pump

The following timing procedures are for the forged body pumps used on older Caterpillar engines. The instructions that follow are specifically for the D9 tractor.

1 Remove the cover from the side of the injection pump (Figure 24.29).

2 Loosen the screws that hold the shaft assembly (Figure 24.29,3).

Figure 24.29 Injection pump with side covers removed.

3 Bar engine to TDC no. 1 cylinder on compression stroke.

4 Remove the dowel pin from storage hole (Figure 24.29,2) and insert pin in timing hole; press the pin in place.

5 Rotate pump camshaft until the dowel enters the slot (Figure 24.29,1).

6 Tighten the clamp screws on sleeve (Figure 24.29,3).

7 Replace dowel pin in the original hole and install the side cover of injection pump.

CAUTION Be certain to remove the dowel to prevent breakage.

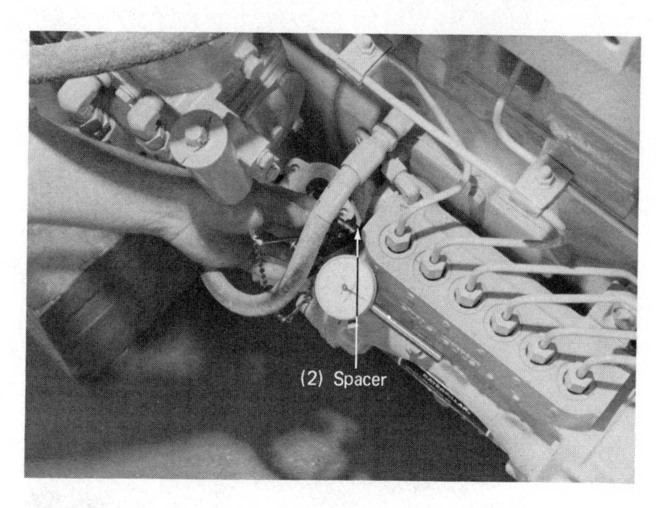

Figure 24.31 Rack travel bracket and dial indicator.

V. Procedure for Tests and Adjustments on Engine

The engine should be run up to operating temperature and all oil and fuel leaks stopped before doing any testing. The following tests are to be made:

A *Rack setting.*

The rack setting will determine the maximum amount of fuel injected during engine operation, thereby affecting horsepower. Be careful when making these adjustments.

1 Remove the stop and spacer from the front of the injection pump (Figure 24.30).

2 Install the 9S7350 bracket (in-line pumps) and dial indicator (Figure 24.31).

3 Move the control lever to the fuel on position. Next move the lever to off position and install spacer (Figure 24.31,2).

4 Zero the indicator.

5 Using an electric circuit tester, connect one end to the brass terminal on top of the governor housing (Figure 24.32). Ground the other end.

6 Rotate the governor control lever to the on position until the light comes on.

7 Read the dial indicator and record the measurement. Compare this reading to the rack setting listed in the Caterpillar service manual.

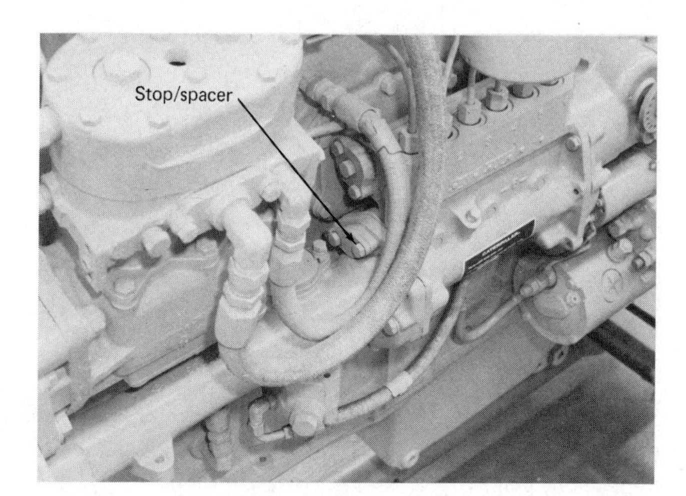

Figure 24.30 Removing rack stop prior to rack setting.

Figure 24.32 Connecting electrical tester to brass terminal.

Figure 24.33 Setting rack travel with stop collar screw.

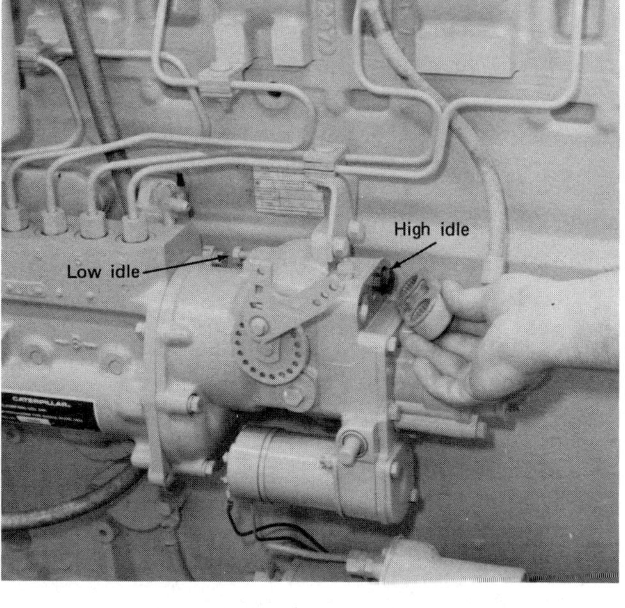

Figure 24.34 Removing high idle adjusting screw cover.

8 If there is a difference, adjust the control rack with collar adjusting screw (Figure 24.33). Recheck the adjustment several times to make sure it is correct.

B *High idle speed.*

1 Measure high idle rpm with the engine warmed up thoroughly.

2 If the high idle needs adjustment, remove the cover on compact housing pumps (Figure 24.34) or on forged body pumps (Figure 24.35).

3 Adjust the high idle by turning the correct adjustment screw.

4 Make certain the adjustment is correct and reinstall the cover.

C *Low idle.*

1 Proceed as when adjusting the high idle by removing the screw cover on forged body pumps. Turn adjusting screw until the correct low idle is obtained. Install cover.

2 On compact housing pumps simply turn the adjusting screw until the low idle speed is correct.

D *Air-fuel ratio control adjustment* (Figure 24.36).

NOTE Rack travel indicator used to set the fuel rack must be in place at this time.

1 Make sure the fuel rack setting is correct before proceeding with the air-fuel adjustment.

2 Remove the cover (Figure 24.36,2) from the aneroid and start the engine.

3 By hand push the valve (Figure 24.36,1) in until it stays by itself.

4 Move the throttle lever quickly to the wide open position (WOT) and read the dial. Record

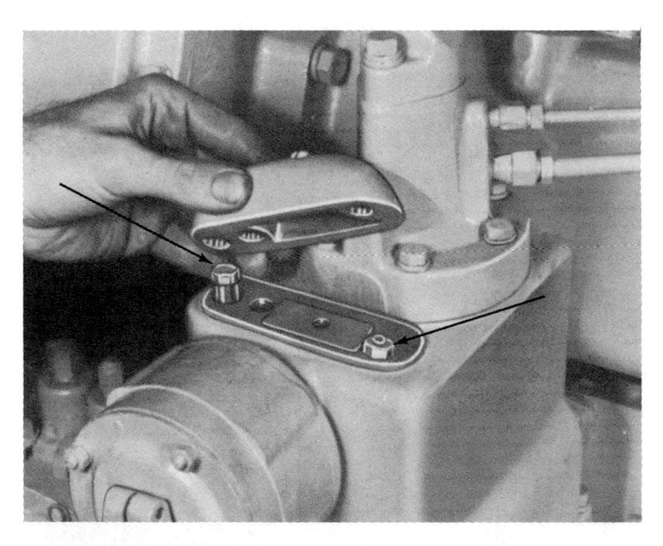

Figure 24.35 High and low idle adjusting screw cover on forged body pumps.

Figure 24.36 Air-fuel control.

Figure 24.37 Caterpillar sleeve metering injection pump.

this reading and compare with Caterpillar specifications.

5 If rack travel must be increased, turn the valve clockwise. If travel is to be decreased, turn the valve counterclockwise.

6 When the adjustment is correct, reinstall the aneroid cover.

The tests listed must be done on the engine and set according to Caterpillar specifications. Authorized Caterpillar dealers or shops selling equipment with Caterpillar engines will be the only sources of rack setting information.

It must also be remembered that Caterpillar truck engines sold in the United States have met Federal Emissions Standards. Any attempt to adjust injection pumps to specifications other than those listed by Caterpillar could result in severe fines for the service shop involved. Always proceed with caution in this area.

VI. Sleeve Metering Injection Pumps

The latest addition to the Caterpillar line of injection pumps is the sleeve metering pump (so named because of its method of fuel control). The sleeve metering pump is available in the in-line configuration as a four or six cylinder version and as an eight cylinder pump in the V formation.

The sleeve metering pump (Figure 24.37) uses a fuel metering principle similar to the American Bosch PSJ and 100 series pumps (Chapter 18); however, in the Caterpillar version, each cylinder has its own plunger and sleeve which, in turn, are controlled by one governor.

A *Main pump operation.*

Fuel flows from the main tank to the fuel filters and then to the transfer pump that is located on the front of the main pump (Figure 24.38). From this pump fuel flows into the main pump and is pressurized to 30 psi (2.1 kg/cm^2). The sleeve metering pump also differs from other Caterpillar pumps in that no engine oil is used for lubrication of the camshaft, roller tappets, and related parts. All these parts, as well as the plungers and sleeves, are lubricated by diesel fuel. Thus, at this time all internal pump parts are submerged in diesel fuel.

As the fuel moves to the plungers, it is free to enter the plunger and fill the area above it (Figure 24.39).

Figure 24.38 Fuel transfer pump (sleeve metering pump).

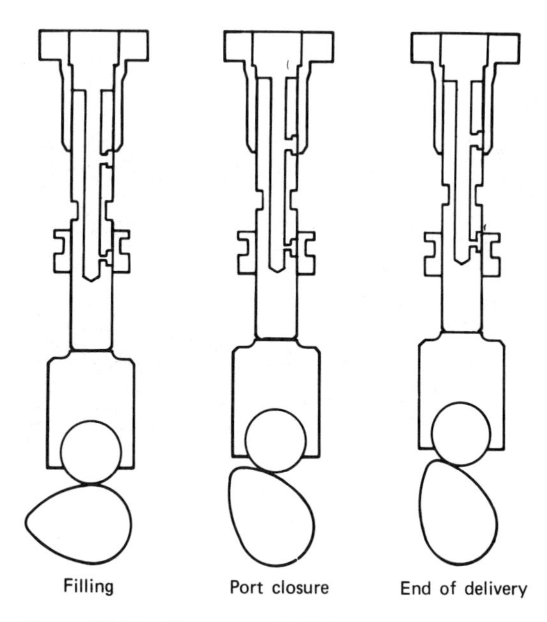

Figure 24.39 Plunger at BDC; barrel filling with fuel.

Injection will start when the camshaft lifts the pumping plungers far enough to close the fill port. At this time no more fuel enters the area above the plunger and the fuel is forced upward, opening the delivery valve, and is injected into the engine cylinder. Notice at this time that not only the fill port will be closed but also the spill port. A precision sleeve is fitted to the plunger and connected to the governor. As long as this sleeve covers the spill port, fuel will be injected. When the plunger has moved up far enough, the spill port will rise above the control sleeve and all high pressure fuel will spill back into the housing. This is the port opening and the end of injection. It is by the movement of the control sleeve that the amount of fuel injected per stroke is changed:

1 *Starting*. In the start position (Figure 24.40) the control sleeve will be moved to a position far away from BDC. In this position fuel will be injected for the longest possible time period before the spill port is uncovered. This is maximum delivery.

2 *Low idle*. When running at low idle the control sleeve will be at a position close to BDC. With the sleeve in this position the injection stroke of the plunger is quite short before the spill port is uncovered and injection ceases.

3 *Full load*. At full load delivery (and other deliveries between low idle and high idle) the control sleeve moves to a position farther away from BDC, but not as far as in the starting position. Here injection continues for a longer period than at low idle. This means increased delivery.

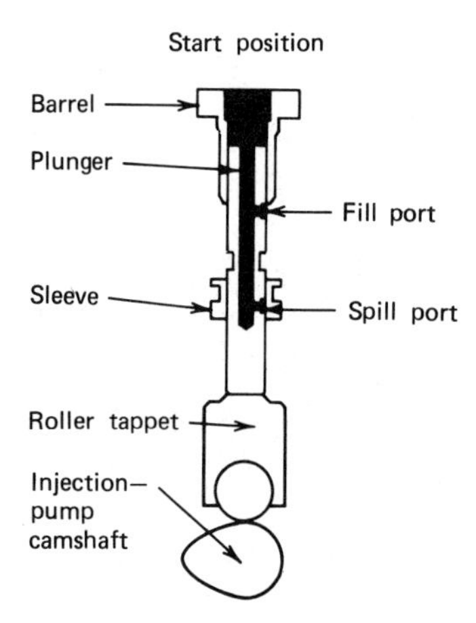

Figure 24.40 Start position. Sleeve is at the greatest distance from BDC.

4 *Shutoff*. When the control lever is moved to the shutoff position (Figure 24.41), the control sleeve is at its lowest position. As the plunger moves upward, the spill port is already uncovered and no injection occurs.

B *Transfer or lift pump*.

The Caterpillar injection pump uses an external gear type pump to supply fuel to the pump housing under pressure. Since the operation of this pump is identical to all external gear pumps, no special explanation is needed here.

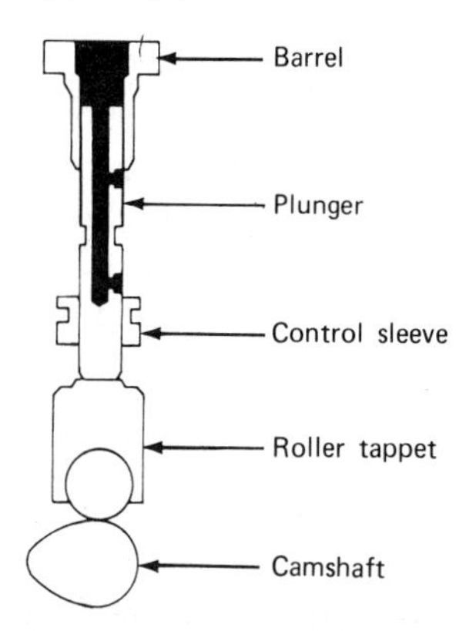

Figure 24.41 Shutoff position. Control sleeve nearest to BDC.

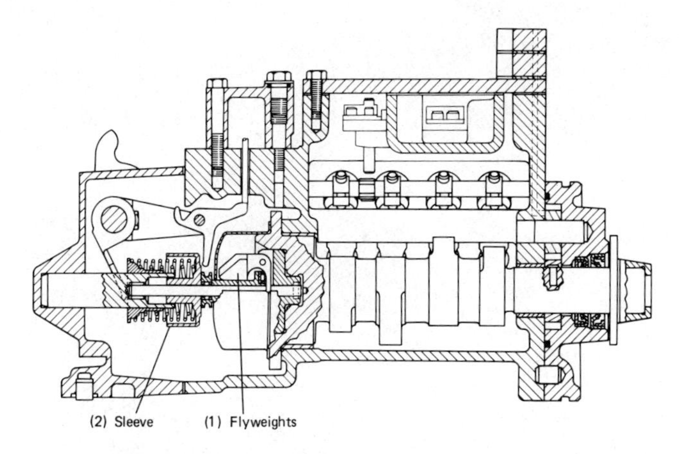
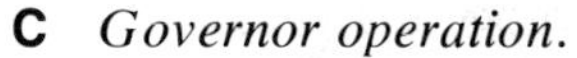

Figure 24.42 Governor assembly.

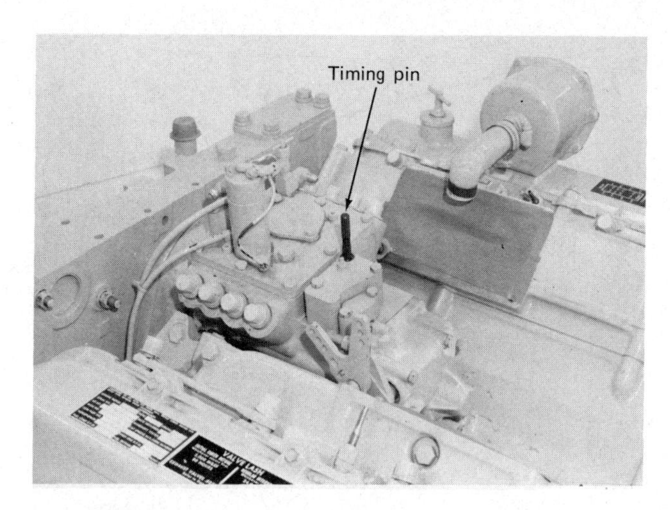

Figure 24.43 Installing timing pin on pump camshaft.

C *Governor operation.*

A mechanical governor is used on sleeve metering pumps. Some models also use a hydraulic dashpot to eliminate governor surge. The component parts of the governor are shown in Figure 24.42. As engine speed increases, the flyweights (Figure 24.42,1) fly outward and move the thrust sleeve (Figure 24.42,2). This linear movement is transferred to rotary movement to move the fuel pump control sleeves by a bellcrank and ball and socket joint. The sleeves are then moved to a position giving less fuel. The engine speed drops. The flyweights collapse slightly, causing the levers to rotate the control sleeves to a position giving more fuel (away from BDC). Engine speed then balances out.

D. Procedure for Removing the Pump from the Engine

Clean the entire engine before beginning.

1 Install the timing pin in injection pump, making certain it drops in the slot of the pump camshaft (Figure 24.43).

NOTE The engine may have to be rotated.

2 Remove the tachometer drive housing and tachometer drive shaft.

3 Remove the plug from the timing hole in the timing cover and install a $\frac{5}{16}$ in. NC bolt. Turn the engine crankshaft until the bolt can be screwed into the timing gear (Figure 24.44).

NOTE The screw must be able to be turned in by hand. Do *not* use a wrench.

4 Using tool 5P2371, remove drive gear from the pump camshaft.

5 Remove injection lines, fuel supply lines, and pump mounting bolts.

6 Slide the pump back and lift off from the engine.

E. Procedure for Disassembly

When a complete overhaul is to be done on the pump, first wash it completely in solvent and blow dry. Work

Figure 24.44 Installing timing bolt in timing gear.

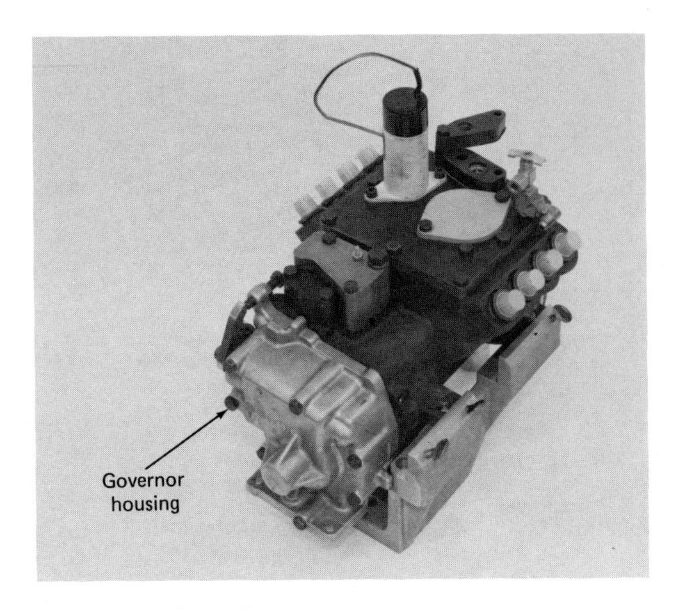

Figure 24.45 Removing governor.

Figure 24.46 Unscrewing pump element hold-down bushings with special tool.

on a clean bench! If the correct disassembly fixture isn't available, fabricate one to hold the pump safely in the vise.

1 Remove the pump housing cover and drain off the diesel fuel.

2 Remove the speed limiter from the governor.

NOTE This is no longer used on later pumps.

3 Unscrew the governor housing bolts and remove the governor (Figure 24.45).

4 Lift out the governor spring.

5 Remove the governor shaft and dowel from housing.

6 A light metal cover surrounds the governor weights. Remove it at this time.

7 Hold the pump camshaft stationary and remove the screws holding the flyweights to the camshaft.

8 Disassemble the throttle shaft and related parts from the governor housing.

9 Using a special serrated socket no. 8S2243, loosen and remove the bushings that hold the pumping elements in the housing (Figure 24.46).

10 Lift out the pump assembly (Figure 24.47), being careful not to drop or lose any parts.

11 Set the removed assemblies in a divided tray or pan to avoid any mix-up of parts.

NOTE Make certain to keep the control sleeves with their respective plungers. The plunger, barrel, and sleeve are one matched set.

12 If the pump units are to be disassembled, first remove the crescent retaining clip that holds the bonnet and barrel together. Remove the plunger spring and spring seat. Pull the plunger from the barrel (Figure 24.48). Keep all these parts in clean fuel oil.

13 Remove the transfer pump by first locking camshaft in position with tool no. 2P1544 as shown in Figure 24.49.

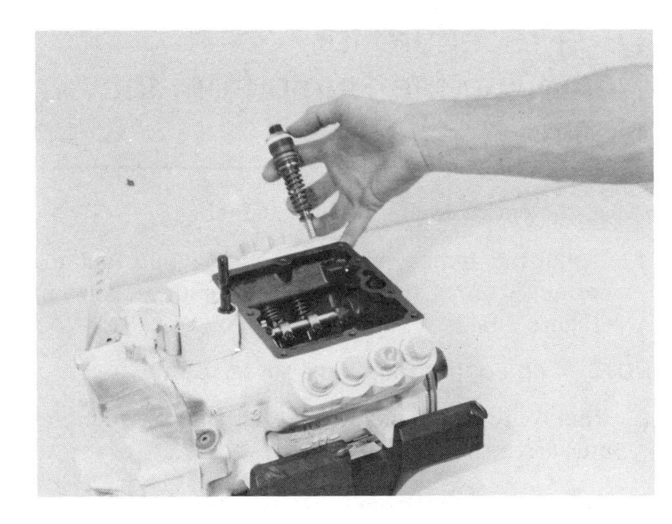

Figure 24.47 Removing pump element assembly.

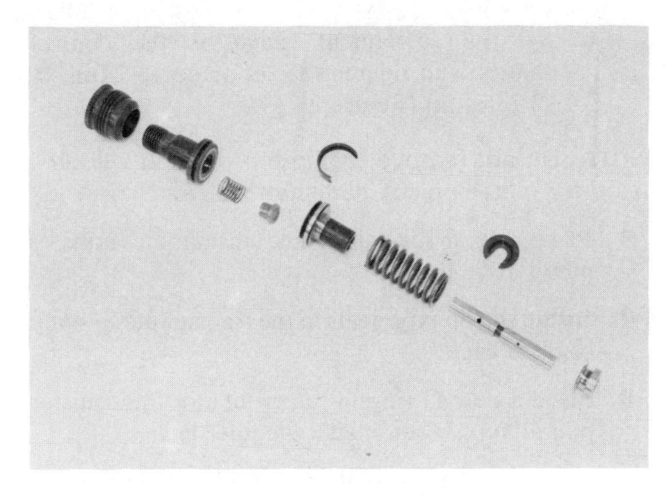

Figure 24.48 Pumping element disassembled.

14 Screw tool no. 2H3740 into the tapered drive sleeve at the front of the pump. As this tool is tightened the sleeve will be removed.

15 Remove the transfer pump screws and transfer pump.

16 Loosen the control levers on the governor control shaft(s) from the pump housing.

17 Lift the control levers out of the housing.

18 Remove the roller tappets from the pump housing and keep them in order for each cylinder.

19 While slowly rotating the camshaft, pull it out of the housing.

Figure 24.49 Locking camshaft in position prior to removing transfer pump.

F. Procedure for Inspection

Soak housings in parts cleaner, rinse in solvent, and blow dry. Lay the pump parts on a clean bench for inspection. Inspect the parts as listed below and repair or replace as necessary.

1 All gaskets, seals, and O rings must be replaced.

2 Inspect the pumping plungers for wear and scoring.

NOTE Remember that if the plungers need replacing, the barrels and control sleeves must be replaced also.

3 Closely examine the transfer pump housing for scoring and wear. Check the gears for flaking and chipping. Maximum clearance of gears to body as measured with Plastigage should not exceed 0.002 in. (0.05 mm).

4 Check all moving governor parts for wear. These would include the tips of the governor weights where they contact the thrust sleeve, the thrust sleeve itself, and connecting levers. The shield that covers the governor weights must fit snug.

5 Check the tips of the control levers for wear.

6 Check the lobes of the camshaft for flaking or galling.

7 Inspect the housing and repair any thread damage found.

a Check for correct camshaft housing clearance by measuring the rear journal of the camshaft and the internal bore of the housing. The difference (camshaft running clearance) should not exceed 0.006 in. (0.15 mm).

b Measure the front in a similar manner. Here the clearance should not exceed 0.002 in. (0.05 mm).

G. Procedure for Reassembly

1 Mount the pump housing in the holding fixture.

2 Lubricate the camshaft and gently install it in the housing.

3 Install the roller tappets in their correct bore, making certain to align the groove in the tappet with the pin in the housing.

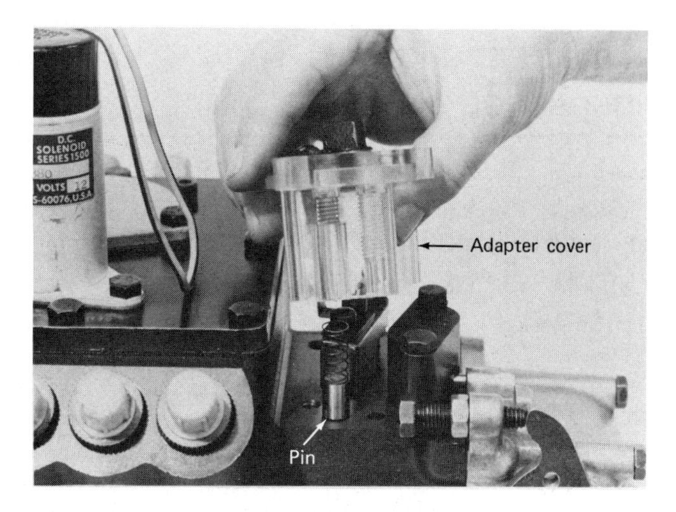

Figure 24.50 Torque control cover adapter.

4 Slide the governor control shaft(s) into the housing while slipping the control levers over the shaft(s).

NOTE Do not forget the lever for the crossover shaft on V pumps.

5 If working on a V pump the crossover levers must be adjusted at this time.

  a Install the timing pin (15 mm diameter) and correct the torque control cover adapter (Figure 24.50). Tighten the screw over the timing pin.

  b Set the adjustment gauge on the control shaft(s) and tighten the set screw to 24 in. lb (2.71 N·m) (Figure 24.51).

NOTE Do not remove timing pin yet as it will be used for installing the camshaft sleeve.

6 Place key in camshaft and install the transfer pump drive gear.

7 Install the lip type seals in the transfer pump body back to back.

8 Place a new O ring in the groove of the transfer pump body. Then install the idler gear.

9 Lubricate the camshaft seals and install the body to the pump (Figure 24.52).

10 Install the tapered drive sleeve (Figure 24.53) by pulling it on with washers and the correct size bolt to fit the camshaft.

11 Using a dial indicator, check the camshaft endplay. It must be 0.023 ± 0.018 in. (0.58 ± 0.46 mm).

NOTE If endplay is not as specified, the camshaft must be replaced.

12 After installing a new seal, slide the governor shaft into housing.

13 Install the lever on the shaft.

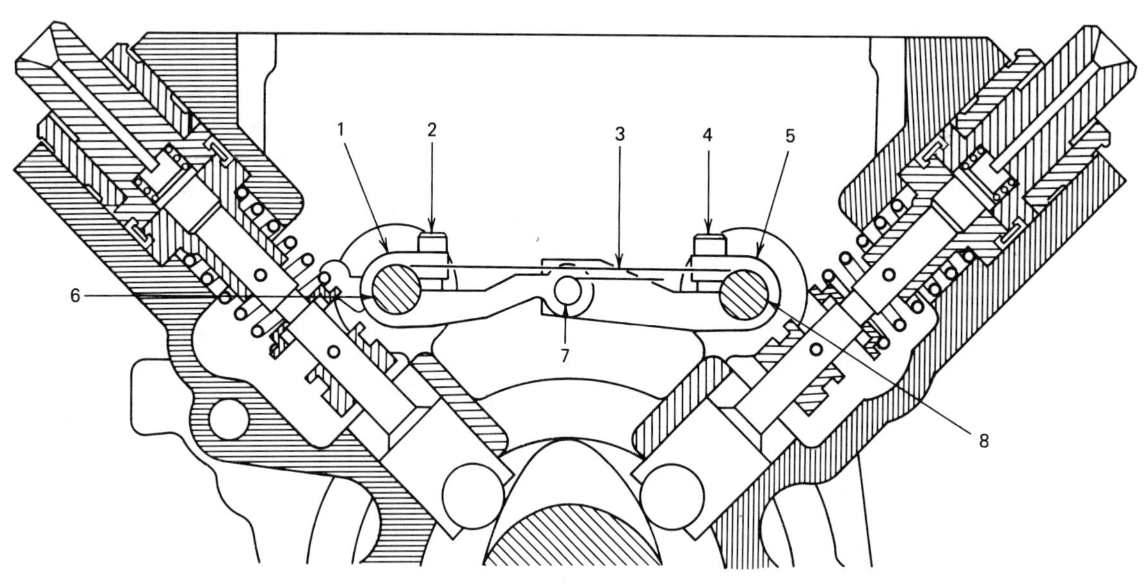

Crossover levers

1. Crossover lever.
2. Screw.
3. Dimension for adjustment .0828 in. (2.130 mm).
4. Screw.
5. Crossover lever.
6. Shaft.
7. Pin.
8. Shaft.

Figure 24.51 Setting the crossover levers on V pumps.

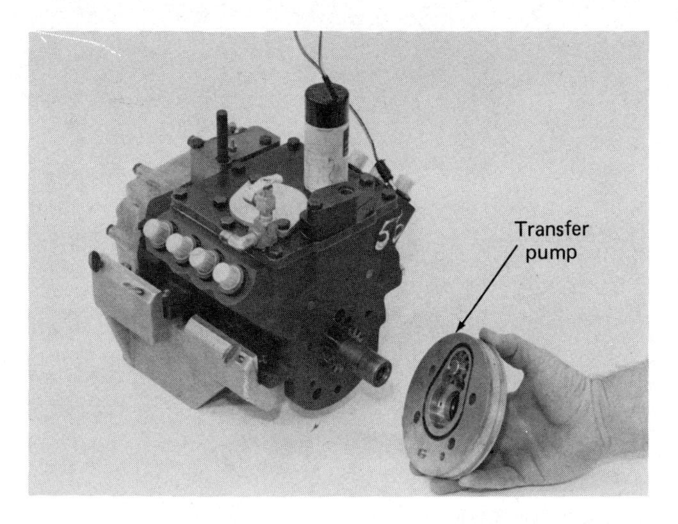

Figure 24.52 Installing transfer pump body.

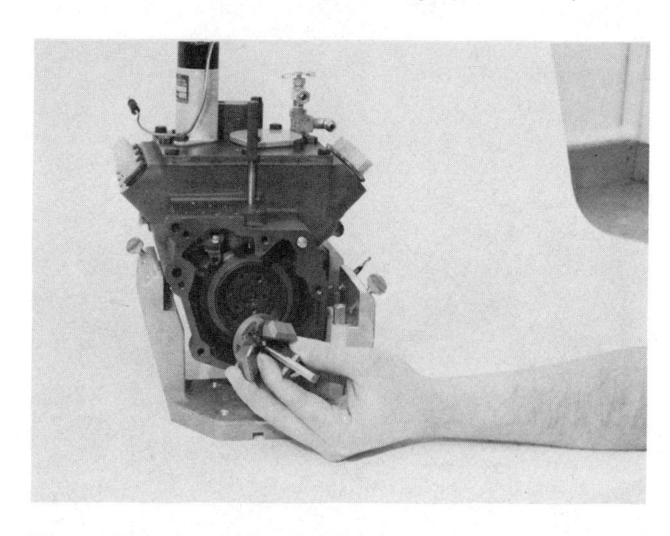

Figure 24.54 Assembling flyweight assembly to camshaft.

14 Bolt the flyweight assembly to the camshaft. Torque to specifications (Figure 24.54).

15 Tap the light metal shield in position over the flyweights.

16 Install the full load stop lever and dowel.

17 Install the pin and torque spring assembly.

18 Place the thrust sleeve and spring on the flyweight shaft.

19 Install the governor housing.

NOTE The following procedure for assembling and calibrating the injection pump differs greatly from other brands of pumps. Each individual pumping assembly will be assembled, and its control lever calibrated to give correct lift and fuel delivery. Then the actual pumping element will be installed to the pump. No further adjustments will be made to the pumping elements after assembly to the housing.

20 Install the calibration pin in the calibration hole (Figure 24.55).

21 Place the correct adapter over the pin.

22 Install the calibration pump assembly into no. 1 pump bore (Figure 24.56) with the flat of the plunger facing the control lever. Hold the plunger down and rotate it 180°.

23 Install the special bushing from the Caterpillar tool kit to hold the barrel tight in the housing.

24 Zero the calibration dial indicator as shown in Figure 24.57, using the microgauge.

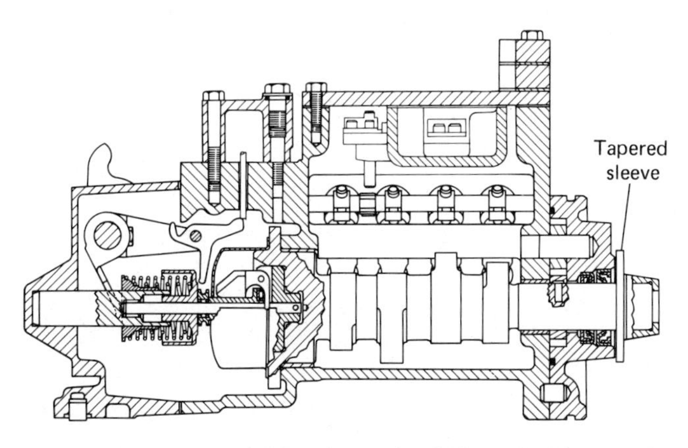

Figure 24.53 Tapered drive sleeve should be pulled into position with a bolt and washers.

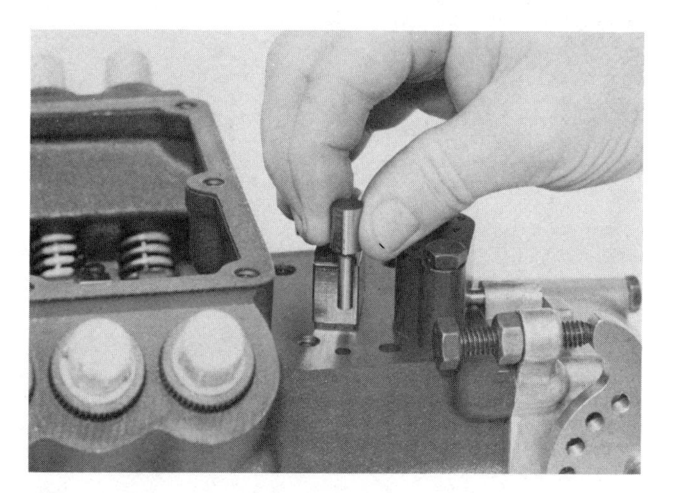

Figure 24.55 Install calibration pin.

Figure 24.56 Installing calibration pump.

25 Push the plunger of the calibration pump down below flush.

26 Move the throttle lever to the wide open position.

27 Insert an allen wrench into the adjusting screw of the control lever and loosen it. Push down on the allen wrench to raise the plunger to just below flush.

28 Set the calibration dial indicator on top of the calibration pump and hold it down tightly.

29 Push down again on the allen wrench to raise the plunger. Stop when the dial indicator reads zero (Figure 24.58) or slightly above. Tighten the adjusting screw to 24 in. lb (2.71 N·m).

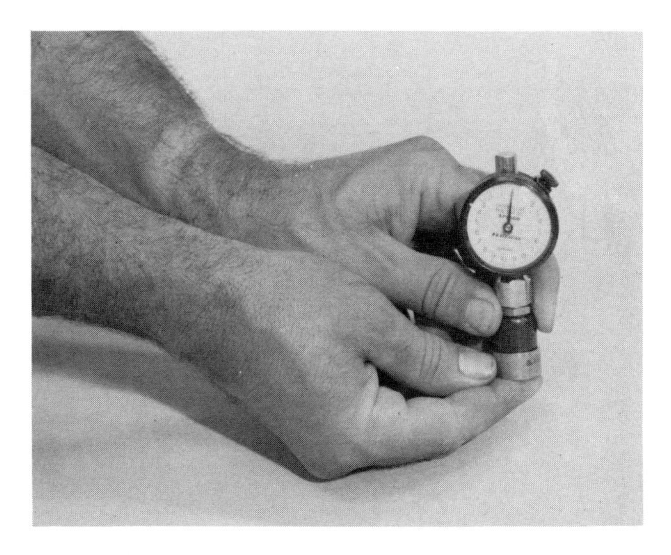

Figure 24.57 Zeroing the calibration indicator.

Figure 24.58 Correct adjustment of plunger.

30 Move the lever back and forth and then recheck the setting.

31 Remove the calibration pump and reinstall it in the no. 2 bore. Calibrate all remaining cylinders in the same manner.

32 Remove the calibration pin and the pin that locks the camshaft stationary. Plug the holes.

33 Assemble the pumping assemblies in the order shown in Figure 24.59. Lubricate each part with fuel oil when assembling.

NOTE When installing the pumping plunger, make certain the large hole is closer to the top. Also, the thin edge of the control sleeve must face up.

34 Rotate the camshaft so that the lobe of the no. 1 cylinder is at BDC.

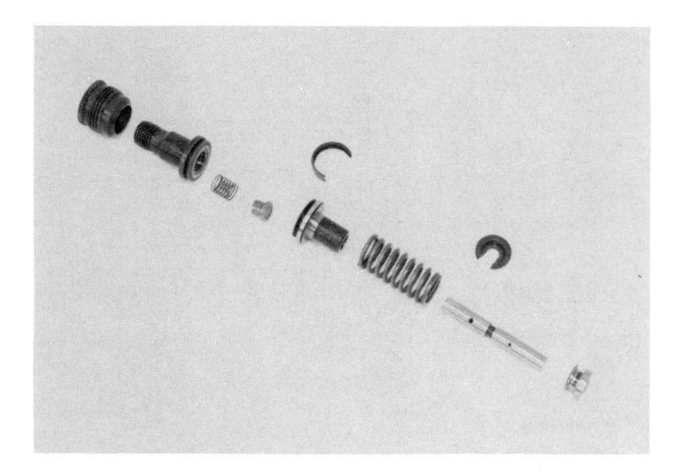

Figure 24.59 Order of assembly for pumping elements.

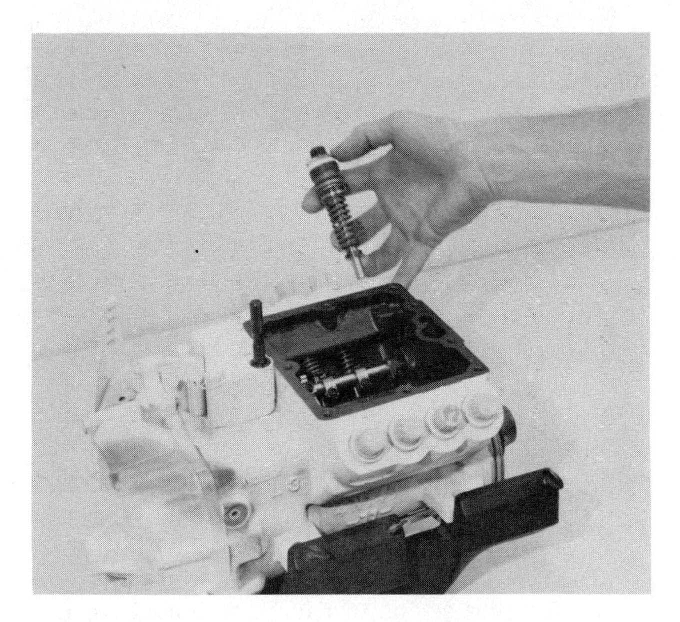

Figure 24.60 Installation of pumping element.

35 Install the pumping assembly in the bore, making sure the control sleeve engages the control lever (Figure 24.60).

36 Slide the bushing over the bonnet and torque.

37 Install the remaining elements in a similar manner, always rotating the camshaft so that the specific cam lobe is at BDC.

38 Install the bypass valve and spring and then the housing cover with a new gasket.

H. Procedure for Installation to Engine

1 With the timing pin installed in the injection pump and engine timing gear, slide the pump into position. Install mount bolts, injection lines, and supply lines.

2 Screw the tachometer drive into the pump camshaft and torque.

3 Remove the two timing pins.

4 Rotate the engine two revolutions and try to insert the pins again. They must still line up. If not, readjust the timing.

5 Remove the timing pins and reinstall the cover bolt and plugs.

6 Mount the tachometer drive housing.

I. Procedure for Calibration (Pump Shown)

After the pump is installed to the engine, the engine should be run up to operating temperature and the following adjustments made:

1 *Low idle.*
Check low idle with an accurate tachometer. If it is incorrect, adjust the screw (Figure 24.61) until it is within limits.

2 *High idle.*
Again using the tachometer, hold the throttle in the wide open position and check high idle speed. If incorrect, adjust the high idle stop screw (Figure 24.62) that is located under a small cap on top of the governor.

3 *Pump housing pressure.*

a Remove the small plug from the top cover of the pump.

b Install a 0 to 60 psi (0 to 4 kg/cm^2) pressure gauge in this port with a snubber to reduce pressure pulsations.

c Run the engine at full load rpm and check the pressure.

d Compare the reading to specifications.

4 *Balance point of the fuel system.*

a Connect a continuity light or circuit tester to the screw on top of the full load stop cover

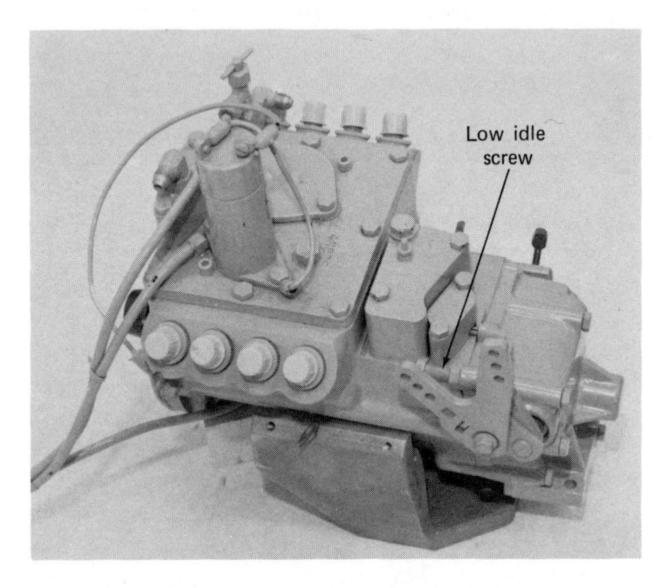

Figure 24.61 Low idle adjustment screw.

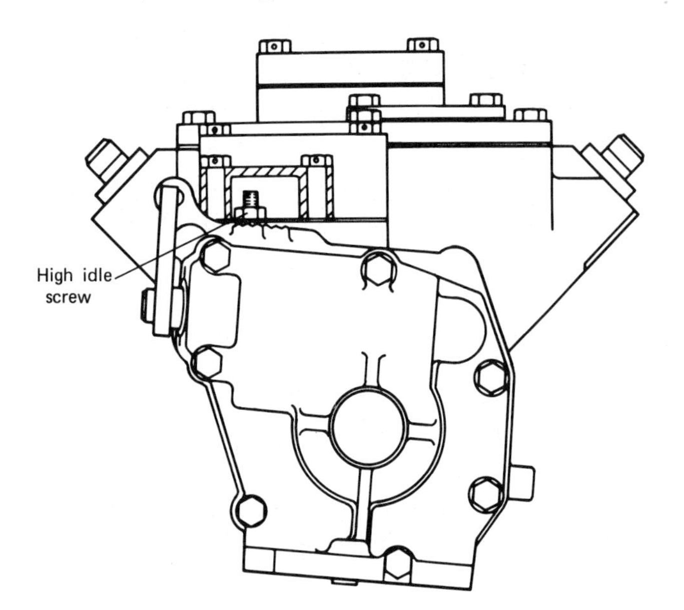

Figure 24.62 High idle adjustment screw.

(Figure 24.63). Connect the other end of the tester to a paint-free ground on the injection pump.

b Run engine at high idle and load the engine until the tester light comes on. *Record this speed.*

CAUTION Be sure to use all safety precautions when loading the engine, especially if using a chassis dynamometer.

c Repeat step b a couple of times to be sure the correct speed is recorded.

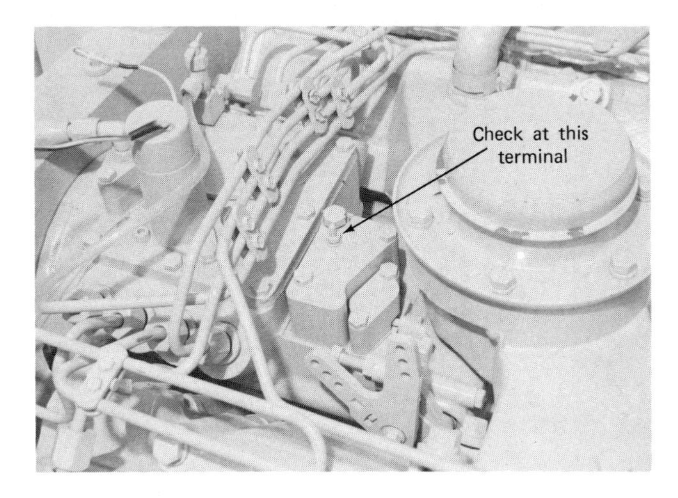

Figure 24.63 Adjusting the balance point.

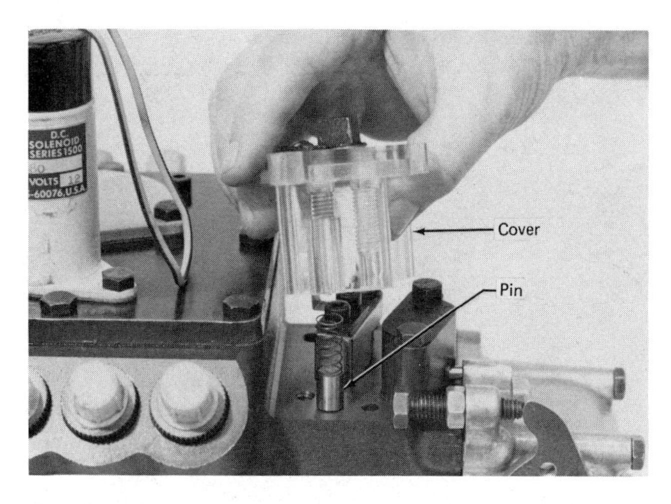

Figure 24.64 Install zero set pin and cover.

d Idle engine for two to three minutes before shutting it down.

e Check the Caterpillar specification book for the *balance point* speed. The speed obtained should correspond to the speed listed in the book.

f If it does not, readjust high idle speed until it does.

NOTE Always keep high idle within specified limits if adjustment is required.

5 *Maximum fuel setting.*

a Do not run engine for this setting.

b Remove the fuel shutoff solenoid.

c Install the *correct* zero set pin as listed in the service manual and adapter cover without gasket (Figure 24.64).

d Install a setscrew in the adapter cover and adjust it to hold the zero set pin down on the housing.

e Install dial indicator, extension, and clamp (Figure 24.65). Zero the dial.

f Turn the adjusting screw counterclockwise six to seven revolutions. The dial must follow this movement.

g Place the throttle lever at low idle.

h Using the circuit tester, connect one end to a good ground and the other to the torque spring.

i Advance the throttle lever to the *high idle* position until the light flickers. *Record readings on the dial indicator.*

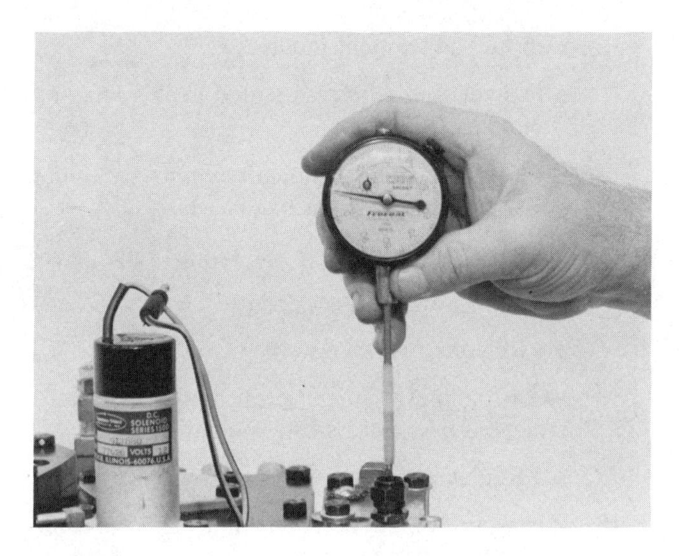

Figure 24.65 Install dial indicator with correct extension.

NOTE This procedure applies to V pumps only. On in-line pumps simply turn the screw in the adapter housing until the light comes on.

j Compare this reading to the one listed in the specification for the pump being worked on.

k If adjustment is required, proceed as follows:

(1) On pumps with power screw (Figure 24.66), adjust the power screw until the correct reading is obtained on the dial indicator.

(2) On pumps with shimmed torque spring assembly (Figure 24.67):

(a) Record dimensional difference between dial indicator reading taken in step i and the specified dimension in step j.

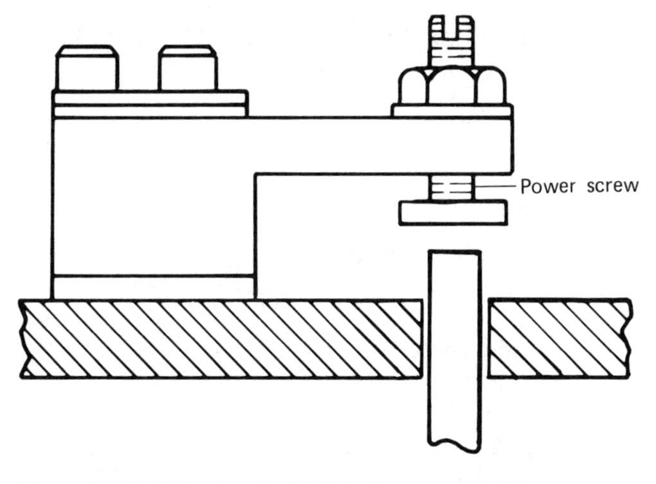

Figure 24.66 Adjustment of full fuel on pumps with power screw.

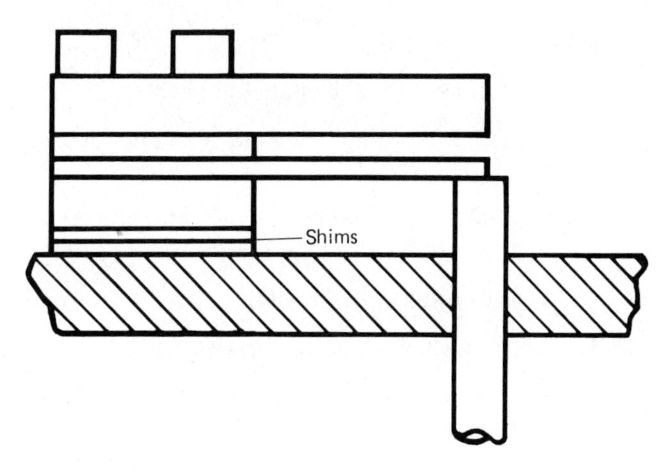

Figure 24.67 Adjustment of full fuel on pumps with torque spring.

(b) Remove adjusting cover and dial indicator from pump. Remove the torque spring assembly.

(c) Add or remove the exact thickness of shims necessary to correct the dimensional difference in step a.

(d) Reinstall the torque spring assembly, adjusting cover and dial indicator.

(e) Repeat the checking procedure, steps f through j. Repeat until the dial indicator reading is correct.

VII. Troubleshooting Caterpillar Fuel Systems

Troubleshooting Caterpillar fuel systems is somewhat different than for other types because most of it will be done with the pump on the engine. Various engine running faults must first be isolated to the fuel injection system. Do not attempt to remedy engine mechanical problems by changing the adjustments or replacing parts in the fuel system.

Following are some of the more common problems that occur on compact housing, forged body, and sleeve metering pumps.

A *Engine will not start.*

1 Dirty or plugged filters.

a Change fuel filters.

2 Defective transfer pump.

a Repair or replace transfer pump.

3 Engine out of time.

a Retime engine as stated in chapter.

4 Defective shutoff solenoid (except forged body).

a Test the solenoid and connections leading to it.

B *Uneven or rough running engine.*

1 Worn or stuck injection nozzles.

a Repair or replace as explained in nozzle chapter.

2 Transfer pump pressure too low.

a Change fuel filters.

b Check transfer pump and repair or replace.

3 System improperly bled.

a Tighten all fuel supply lines and note condition.

b Bleed air at nozzles.

4 Uneven fuel delivery.

a Replace plunger and barrel if worn (compact housing and forged body).

b Adjust fuel rate levers or crossover lever (sleeve metering pumps).

5 Advance device not working (if used).

a Repair or replace advance unit.

C *Loss of governor control or unexpected changes in engine speed.*

1 Worn governor components.

a Inspect and replace parts as necessary.

2 Broken crescent connecting link (hydramechanical governor).

a Replace connecting link.

3 Loss of engine oil pressure. (Engine will not overspeed, but full speeds cannot be reached.)

a Restore pressure to governor.

4 Speed limiter not working correctly (on some models).

a Check and repair as required.

D *Low horsepower.*

1 Some items in A, B, and C above will affect engine horsepower. Check as necessary.

2 Full fuel adjustment incorrect.

a Check and adjust as stated in this chapter under *testing*.

3 Air-fuel ratio control defective or not receiving boost pressure from turbocharger.

a Check turbocharger and connecting lines.

b Adjust air-fuel ratio control.

c Repair or replace unit.

4 Worn pumping plungers (all models), usually indicated by a hard starting condition.

a Replace as required.

5 Pump not timed correctly internally (except sleeve metering pumps).

a Check for correct timing with depth gauge and adjust lifter or change spacer to obtain correct dimension.

E *White smoke.*

1 Incorrect timing.

a Check and set to specification.

2 Improper fuel.

a Make certain to use a good quality 1 or 2 diesel fuel.

3 Faulty injection nozzle.

a Test all nozzles and repair or replace as needed.

F *Slow acceleration.*

1 Air-fuel ratio control adjusted incorrectly or defective

a Adjust or repair.

G *Excessive black smoke.*

1 Fuel rate too high.

a Reset to correct specification.

2 Turbocharger defective.

a Repair or replace as needed.

3 Air-fuel ratio control out of adjustment or defective.

a Adjust to specification or repair unit.

SUMMARY

This chapter has covered the three basic Caterpillar injection pumps in common use. With this information and the Caterpillar rack adjustment specification, the mechanic should be able to correctly overhaul and calibrate these injection pumps. Since many Caterpillar pumps are not placed on a test bench following overhaul, extreme caution and careful workmanship should be used during assembly, followed by a careful and thorough calibration on the pump on the engine. Information such as rack setting specifications and full fuel settings is available only to Caterpillar service stations and dealers using Caterpillar engines as original equipment. If you are unable to obtain this information at your school or shop, contact your nearest Caterpillar dealer or service station.

REVIEW QUESTIONS

1 What type fuel control do Caterpillar compact housing pumps use?

- a Inlet metering.
- b Sleeve metering.
- c Port and helix.
- d None of the above.

2 If port closure (internal timing) is incorrect on compact housing pumps:

- a The roller tappets must be replaced.
- b The plunger wear washer may be worn.
- c The timing spacer may be incorrect.
- d Both b and c.

3 Why is the control rack centered before installing the pump elements on compact housing pumps?

- a To give correct low idle.
- b To prevent possible engine overspeed.
- c To correctly align control gear teeth.
- d Both b and c.

4 What purpose does the air-fuel ratio control serve?

- a Limits maximum fuel delivery.
- b Aids acceleration.
- c Prevents engine surge.
- d Limits smoke during acceleration.

5 Why is engine oil routed to the governor on compact housing pumps?

- a It must drain through the governor to return to the engine.
- b To eliminate engine surge.
- c It assists the governor weights to regulate engine speed.
- d All of the above.

6 How is port closure changed on sleeve metering pumps?

- a Changing the roller tappets.
- b Adjusting levers on control shaft.
- c Changing control sleeves.
- d Port closure is not adjusted.

7 What is the correct test sequence for a sleeve metering pump?

- a Low idle, high idle, housing pressure, balance point, full fuel set.
- b Balance point, low idle, high idle, full fuel set, housing pressure.
- c Housing pressure, full fuel set, balance point, high idle, low idle.
- d Low idle, high idle, housing pressure, full fuel, balance point.

8 How would maximum fuel be set on a sleeve metering pump with a torque spring?

- a Full load screw.
- b Balance point screw.
- c Lever on control shaft.
- d Shims under torque spring.

9 Low idle on a sleeve metering pump:

- a Will occur when the control sleeve is at BDC.
- b Will occur when the control sleeve is at TDC.

c Will occur when the plunger is rotated to give the shortest helix length.

d Will occur when the control sleeve is near BDC.

10 The hydraulic dashpot serves to:

a Limit smoke during acceleration.

b Give excess fuel for overload.

c Improve fuel economy.

d Eliminate governor surge.

11 Transfer pumps used on all Caterpillar injection pumps are of what type?

a Internal gear.

b Vane type.

c Single acting piston type.

d External gear.

12 The crossover adjustment on sleeve metering pumps:

a Provides for even fuel delivery to all cylinders.

b Limits maximum fuel delivery.

c Coordinates control shaft movement on V pumps.

d Adjusts the balance point of the system.

13 If camshaft end play of sleeve metering pumps is out of limits:

a Adjust with shims behind the tapered drive sleeve.

b Adjust with shims under the governor weights.

c Replace the camshaft.

d None of the above.

14 How are the pump and governor lubricated on the forged body design?

a Diesel fuel.

b High temperature grease.

c Engine oil.

d Hydraulic oil.

15 Caterpillar advance units

a Are mechanical.

b Are hydraulically assisted.

c Advance the pump camshaft.

d May be all of the above.

25

Injection Pump Test Bench

The injection pump test bench or stand is an invaluable tool used in servicing diesel engine fuel systems. It enables the mechanic to test correctly and adjust all types of injection pumps and injectors before they are placed in service on the engine. The following tests can be made with the test bench:

1. Transfer or lift pump pressure.
2. Maximum fuel delivery.
3. Advance operation.
4. High idle delivery.
5. Low idle delivery.
6. Torque (overload) control.
7. Back leakage.
8. Internal pump timing.
9. Aneroid adjustment.
10. Governor operation.
11. Fuel or oil leakage.
12. Break-in of new or overhauled pumps.

A pump adjusted in this manner will provide the correct fuel delivery for the engine under all load and speed conditions; overfueling, incorrect power, and other problems of this nature are eliminated.

CAUTION These tests must be performed by qualified personnel using the proper tools in conjunction with a good test bench. Unskilled personnel and/or incorrect practices can cause damage to the injection pump.

Although several different types of test benches are available, they are all designed to do the previously mentioned tests with speed and accuracy.

Modern test stands are more capable of doing these tests than early ones and will be the only ones considered here. If an older bench is to be used, make certain you consult the operator's manual on correct operation and maintenance.

OBJECTIVES

Upon completion of this chapter the student will be able to:

1 Explain why an injection pump test stand is used.
2 List two manufacturers of test stands.
3 List five tests that it must be possible to make on a test stand.
4 List two cautions that should be observed when mounting the injection pump.
5 Describe and explain the use of a comparator.

Figure 25.2 Bacharach test stand.

GENERAL INFORMATION

The three major brands of test benches in common use today are:

Robert Bosch (Figure 25.1)
Bacharach (Figure 25.2)
Hartridge (Figure 25.3)

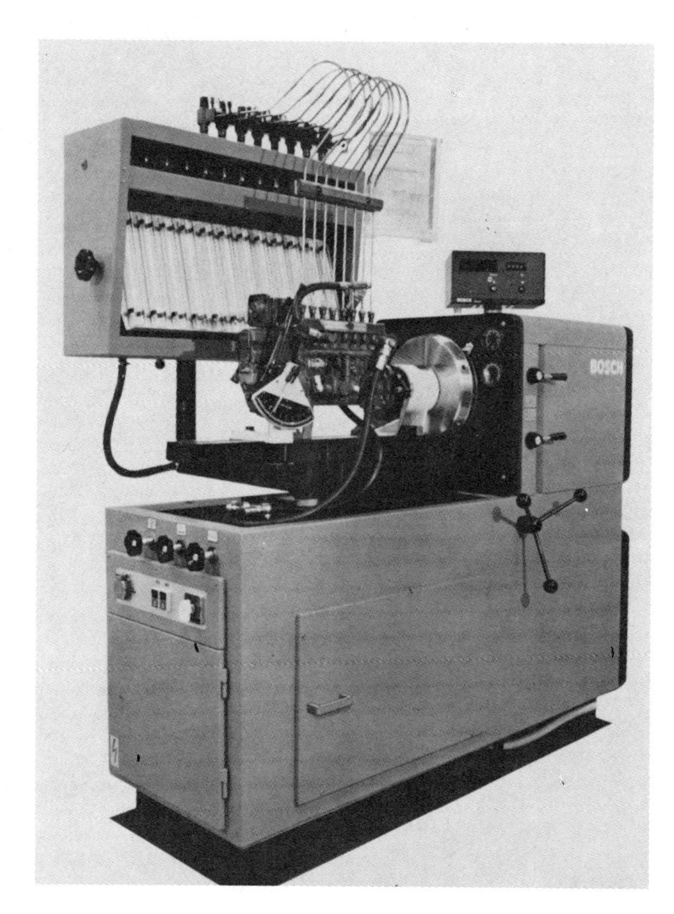

Figure 25.1 Robert Bosch test stand (courtesy Robert Bosch Corporation).

Figure 25.3 Hartridge test stand (courtesy Hartridge Equipment Co.)

These companies supply test benches to calibrate any injection pump on the market. The test stands they produce have all the features that are required of any bench. These features are:

1 Provision for mounting various pump types.

2 Motor of sufficient power to turn the pump at a steady speed under all conditions of fuel delivery and speed.

3 A drive system that provides easily variable speeds either by hydrostatic drive or a variable speed drive belt system.

4 An accurate counting system to count out the delivery strokes for comparison with the test figures.

5 A system of graduates that are divided into cubic centimeters (cc) to measure accurately fuel delivery.

6 Storage tanks for calibrating oil and waste oil.

7 An accurate tachometer (either digital or conventional) to register pump speed.

8 Various other controls to aid speed and accuracy of testing.

9 Calibrating oil. The oil used should meet the SAE requirements for a testing fluid. *Do not use diesel fuel.*

I. Pump Mounting

Most modern test benches have a heavy channel frame (Figure 25.4) on which to mount the injection pumps. This type of bed frame tends to minimize vibrations and twisting. In addition, a backlash-free coupling (Figure 25.5) should be used on the drive line. This coupling will compensate for some misalignment of drive line components and greatly reduces the noise level as compared to other types of couplings.

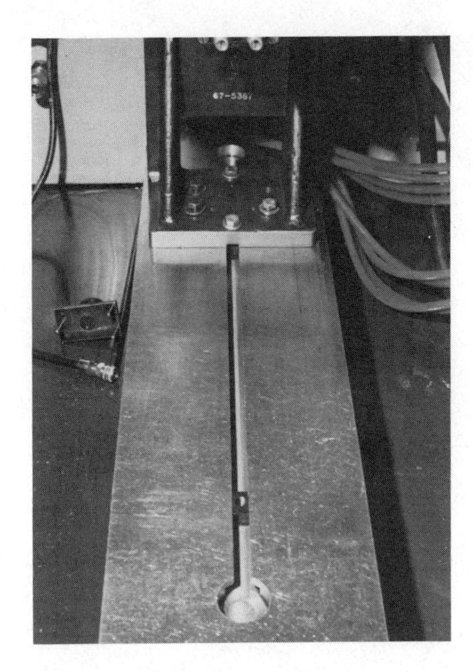

Figure 25.4 Channel type bed frame.

A. Procedure for Mounting the Pump

1 Make certain the correct mounting flange is used.

2 After the pump is correctly aligned, tighten all mount screws securely.

CAUTION If using an Oldham type coupling (without backlash-free drive), make sure there is at least 0.010 in. (0.254 mm) clearance between the coupling and the drive (Figure 25.6). If this is not done, seizure of distributor type pumps or bearing failure of in-line pumps could result.

3 Using the correct fittings, attach the fuel inlet and return fuel lines to the pump.

NOTE Many injection pumps use metric fasteners. *Do not* mix metric and standard SAE fittings!

Figure 25.5 Backlash-free coupling.

Figure 25.6 Measuring clearance of Oldham coupling.

4 Supply lubricating oil to the pump (except on Roosa Master, CAV and Robert Bosch distributor pumps). Lube oil must be added (approximately one-half pint) (0.47 liter) to all other pumps except American Bosch PSM, PSJ and 100 series. On those pumps pressure oil is supplied continuously through a separate line from the test stand.

5 Attach any gauges that may be required, such as transfer pressure gauges, rack travel gauges, and plunger lift.

NOTE More specific instructions may be obtained in the service manual for the particular pump being tested.

6 Attach high pressure injection lines from pump to test nozzles. These lines have a specific length and inside diameter that must be used for a particular pump if correct fuel delivery is to be obtained.

CAUTION Do not bend the injection lines to fit different pumps. Any bending done on the line will soon cause flaking of metal on the inside diameter. This flaking will change the inside diameter and also damage the injection pump as particles could become lodged in the injection pump.

7 Make sure the correct test nozzle or orifice plate is being used in the accumulator. Any deviance here will result in incorrect fuel delivery and a totally inaccurate test. *Also make sure the nozzles are set at the correct opening pressure.*

B *Terms.*

To do a proper job of calibrating an injection pump, the mechanic must know and understand the terms used. Listed below are a few of the most common:

1 *WOT.* Wide open or full throttle.

2 *Governor cutoff.* A speed somewhat above high idle at which the injection pump must cease delivery of fuel.

3 *Torque back-up.* The correct delivery of fuel at speeds simulating engine overload (lower speeds).

4 *Full load.* The maximum fuel delivery for which the pump has been set.

5 *Breakaway.* The start of governor action. High idle is slightly higher and cutoff after that.

6 *Transfer pump pressure.* Pressure of the low pressure pump. (This pump is usually an integral part of the injection pump.) Same as lift pump or supply pump pressure.

7 *Accumualtor rack.* Used on some test benches to hold the nozzles and collect the fuel after injection.

II. Procedure for Running the Pump

Initial mechanical testing should be done as follows:

1 Start the low pressure fuel pump and supply fuel to the pump inlet at recommended pressure (usually about 2 psi) (0.140 kg/cm^2).

2 Always start the test stand in low range. Run the pump at 100 rpm until fuel flows from the loosened connections at the nozzles. Tighten the lines.

3 Run the pump at 500 rpm for 10 minutes. During this time adjust the heaters to raise the temperature of the test oil to the recommended level. Check the pump for any leakage and repair.

4 Test the pump as per instructions.

III. Injector Comparator or Calibrator (Cummins and Detroit Diesel)

A test stand of somewhat different design is used to test the injectors from Cummins and Detroit Diesel engines (Figure 25.7). This type of stand, available in several brands, has a cambox and a hydraulic ram. The

Figure 25.7 Bacharach comparator.

injector is installed in the inverted position and tensioned by downward force of the ram. It uses a count mechanism and a graduate to flow test individual injectors.

A. Procedure for Installation of Injector

1 Set the cam shaft lever (Figure 25.8) to either the CUM or GM position depending on which type of injector is being tested.

2 Set the knurled wheel of the cam to the TDC position.

CAUTION If this is not done, the hydraulic ram will bottom out the injector and damage it.

3 Using the correct adapter and push rod extension, invert the injector and install it on the support springs. Install the correct discharge head.

NOTE For a complete listing of adapters and push rod extensions, refer to operator's manual supplied with the comparator.

4 Switch on the ram and press down, aligning the injector with the discharge head. Press down until the ram stops.

5 Lock the ram and stop the ram motor.

6 Rotate the comparator to make certain it is free.

Figure 25.8 Setting cam shift lever of comparator.

7 Attach the inlet and outlet fuel lines.

8 Run the stand and check injector for correct operation.

9 Take at least three flow readings before recording the flow.

10 Refer to the Cummins and/or Detroit Diesel shop manual to obtain the correct flow readings for the particular injector being worked on.

11 On Detroit Diesel injectors, move the control rack to the shutoff position and check for complete fuel shutoff.

B. Procedure for Removal of Injector

1 Turn off the comparator.

2 Disconnect the inlet and outlet lines.

3 Move the ram lever to the release position.

4 As the ram moves upward, catch the injector and lift it off the push rod.

5 Remove all adapters from the injector.

For additional information on Cummins and Detroit Diesel injectors, refer to Chapters 22 and 23.

IV. Future Requirements of the Design and Operation of Fuel Pump Test Equipment

With the diesel engine being used more and more in automobiles and light-duty trucks, the diesel pump itself must be very carefully calibrated to meet emission and economy standards. Shop methods presently in use are accurate and repeatable if sufficient care is exercised; however, a major increase in diesel-powered vehicles will put a strain on the shop's ability to handle the work *accurately* and *quickly*.

One method that may help alleviate this problem is a system that measures fuel flow and displays this flow instantly on a control panel (Figure 25.9). This eliminates running a complete 1000 stroke cycle, thus greatly reducing the time required to calibrate an injection pump. Also, since the system is closed, there would be no spilled fuel or fumes. It would also eliminate any variation in graduate readings by different test bench operators. With this type of unit, the output from each individual cylinder could be displayed (necessary for in-line pumps) or all cylinders collected together and one average reading displayed. If required, a printout of fuel delivery at various speeds could be made and recorded for each pump worked on.

This and other developments will enable the authorized diesel repair stations to give accurate *and* fast testing of injection pumps.

Figure 25.9 Bacharach fuel flow meter.

SUMMARY

The information in this chapter should help you understand the somewhat confusing test stand operation. If you use this material in connection with the information supplied with the injection pump, you should have very little trouble checking an injection pump on the test stand.

REVIEW QUESTIONS

1 Why are injection pump test stands used?

2 What are the three major brands of test stands in use today?

3 List five parts that should be incorporated in a diesel pump test stand.

4 What does the term WOT mean when referring to the calibration of an injection pump?

5 What is the purpose of an injector comparator?

6 What changes can be expected in diesel test stands in the future?

26

Troubleshooting and Tune-Up of Diesel Engines

One of the areas in which the diesel mechanic's job requires the most knowledge and skill is tune-up and troubleshooting. In troubleshooting you need to understand thoroughly the component or engine on which you are working. Mechanics who become proficient at troubleshooting can consider themselves masters of the trade and a very valuable asset to their employers.

OBJECTIVES

After completion of this chapter the student will be able to:

1. Demonstrate the correct procedures for a diesel engine tune-up (engine of student's choice).
2. Explain and troubleshoot a diesel engine for a low power complaint.
3. Explain and diagnose an engine noise.
4. Explain and locate a misfiring injector or injection nozzle in a diesel engine.
5. Explain, locate, and repair a diesel engine oil leak.

GENERAL INFORMATION

Troubleshooting is, without a doubt, a science that is learned by those mechanics who have diligently applied themselves over a span of several years. It requires a higher level of skill and knowledge than ordinary everyday repair work. Troubleshooting requires that you give conscientious thought to the problem and then apply your knowledge and experience to reach a decision. Any engine malfunction must be approached with common sense by checking the simplest things first. What looks like a major problem can often be easily solved in some simple way.

Before you begin troubleshooting an engine problem, it is advisable to follow these or similar procedures:

1. Discuss the problem with the operator. In many cases the operator holds the key to the problem. Ask him if the problem has occurred before and what the operating conditions were that preceded it.
2. Find out if the engine has been worked on recently; if so, what was done and by whom.
3. If possible, operate the engine and check for correct oil pressure, water temperature, and any unusual engine noise.

4 Make a decision based on what is known about the problem and check it out.

5 If your first choice was not correct, eliminate it and check the next possible problem.

I. Engine Will Not Start or Starts Hard

A *Insufficient cranking speed.*

Cranking speed is a very important factor in the starting of a diesel engine as proper speed must be obtained to generate heat within the cylinder to ignite the fuel and air mixture. This can be caused by:

1 *Low battery.* Check the battery for charge and conditions, as outlined in Chapter 27.

2 *Defective starting motor.* Check and repair the starter, as outlined in Chapter 28.

3 *Corroded battery cable.* Clean or replace the cable.

4 *Lube oil too heavy for climate conditions.* Change the lube oil and refill the engine with the correct weight oil for condition.

5 *Excessive resistance in the insulated or ground circuit.* Excessive resistance in either of the starter circuits does not allow enough current to flow through the circuit. The result will be a slow running starter. Check and eliminate excessive circuit resistance, as outlined in Chapter 28.

B *Low compression.*

Compression, like cranking speed, is a very important factor in creating enough heat in the cylinder to ignite the fuel and air mixture. Low compression can be caused by:

1 *Worn rings.* Run a compression or cylinder leakage test on the engine (Figure 26.1) to determine if the compression meets specifications.

2 *Leaking intake or exhaust valves.* This condition can be checked by a compression check as indicated above or in the following manner:

a If an intake valve is suspected, remove the intake manifold inlet pipe and crank the engine; a snapping or hissing sound indicates that an intake valve is not seating correctly.

NOTE Do not confuse the normal intake sound of valve opening and closing with a leaking valve.

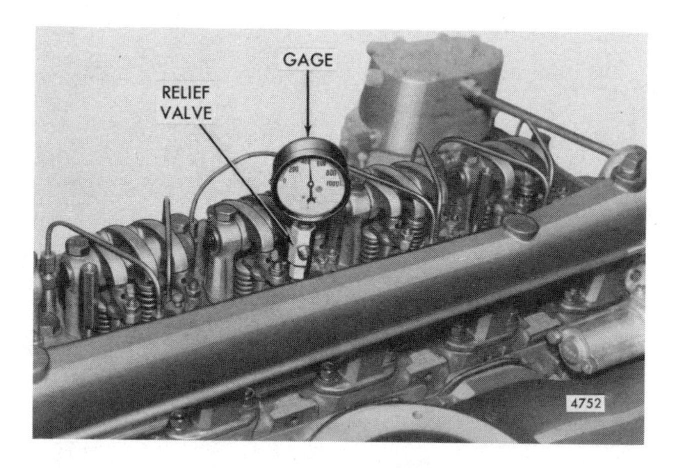

Figure 26.1 Compression testing a diesel engine (courtesy Detroit Diesel Allison, Division of General Motors Corporation).

This test will require some assistance from your instructor or experience.

b To check the exhaust valves, listen at the exhaust pipe while cranking the engine; again, a hissing or popping sound indicates a leaking valve. If a snapping noise is heard in either the intake or exhaust manifold, determine which cylinder is leaking by removing the intake or exhaust manifold and cranking the engine; the hissing sound will now be coming from a port or ports on the engine.

3 *Washed cylinder walls.* The lubricating oil along with the piston rings provides the cylinder seal. This seal can be destroyed by continued cranking, as when you are trying to start a cold engine, since no oil is being thrown off the connecting rod bearings to relubricate the cylinder wall. To relubricate the cylinder walls, remove the injectors and squirt some lube oil into the cylinders.

C *Insufficient fuel being injected into the engine cylinders.*

Diesel engines, like gas engines, require an overrich mixture to start properly. Many injection pumps have devices built into them to deliver excess fuel during engine starting. Hard starting due to low fuel delivery can be caused by:

1 *Clogged fuel filters.* Whenever checking a fuel system for any type of malfunctions, *always* change the fuel filter if the slightest doubt exists about its condition.

Figure 26.2 Checking suction leaks in diesel fuel pump fuel line (courtesy Cummins Engine Company).

2 *Air leak on suction side of pump.* The fuel lines leading to the fuel filter and injection pump should be checked closely for tight connections and worn or cracked spots. A leaking fuel line on the suction side of the transfer pump will be hard to find because it generally will not leak fuel. A device to check for suction leaks is shown in Figure 26.2.

3 *Injection pump not delivering the correct amount of fuel for starting.* As the injection pump wears, clearances increase, and at slow starting speeds the leakage within the pump may be great enough so that insufficient fuel is supplied to open the injection nozzle. Also many injection pumps use excess fuel devices that may not be operating correctly. On unit injector type fuel systems the wear would have to be in the injector since the pump only supplies fuel at low pressure to the injector. To remedy this situation, the injection pump or the injector must be repaired as outlined in fuel injection pump chapters.

4 *Injection pump not turning.* The injection pump may have seized, causing the drive to break. Check the pump to make sure it is rotating. In most cases this check can be made very easily. For example, on a Roosa Master pump remove the timing cover and view the weight retainer, and on a Cummins PT pump remove the tachometer cable and view the cable drive while cranking the engine.

D *Injection pump timing not correct.*

The timing of fuel injection is very critical in a diesel engine. Most diesel engines start best when timed to inject fuel at or very near top dead center because the air in the cylinder is hottest at this point. Incorrect injection timing may be caused by:

1 *Injection pump timed to the wrong cylinder.* In many cases it is possible to time the injection pump to the no. 6 cylinder rather than the no. 1 cylinder. Make sure the engine and the pump are on the no. 1 cylinder.

2 *Injection pump static timing too early or too late.* Static timing is engine stopped timing. This timing should be checked and reset to specifications if incorrect. This information is available in every engine service manual. On engines that use a unit injector, the timing is checked differently from a pump and nozzle setup. Check your service manual or consult your instructor.

II. Engine Uses Excessive Amounts of Oil

A *External oil leaks.*

Very often oil consumption is caused by external oil leaks. This should be checked closely before condemning the engine condition. Check the following for external oil leakage:

1 *Rear main bearing seal.* The rear main seal is supposed to seal the engine oil into the engine behind the main bearing. Most modern engines use a lip type spring-loaded seal. This seal can become hard after extended use and leak. It is often installed incorrectly, which causes it to leak. Clean area around seal thoroughly, run engine under load if possible, and note the seal area. No leaks should be visible.

2 *Front crankshaft seal.* Leakage at the front crankshaft seal can easily be detected visually since the area surrounding the seal will be wet with oil if it leaks. To replace the seal, the vibration damper must be removed.

CAUTION Remove the vibration damper, as outlined in Chapter 9, to prevent damage to damper.

3 *Engine gaskets.* Any gaskets on the engine, such as the pan gasket, rocker arm cover gasket, or any gasket that is supposed to keep oil in the engine, can allow oil to leak from the engine. To locate such a leak, clean the engine thoroughly and run it under load if possible. Replace any gaskets that leak. Always try to determine what caused the gasket to leak: improper installation, wrong gasket, etc.

B *Oil being lost into the engine air brake system (if used).*

Many times oil is being lost from the engine and ends up in the air system reservoir. To test for leakage into the air system, drain the air tanks and check for oil in the escaping air. Oil in the air tank is generally caused by:

1 *Worn air compressor rings.* Engine air compressors have pistons and rings much like an engine and eventually wear out. At this time the air compressor should be overhauled or replaced.

C *Oil lost into the combustion chamber.*

When all other possibilities for engine oil loss have been thoroughly checked, the engine must be burning the oil in the combustion chamber. Oil can get into the combustion chamber in the following ways:

1 *Oil leakage through the valve guides.* Oil leakage through the valve guides can be caused by worn or missing valve guide seals, worn valves, or guides. In most cases valve seals can be put on the valves without removing the cylinder head. If valve guides are worn, the cylinder head must be removed and the guides and possibly valves must be replaced.

2 *Leakage through the turbocharger or supercharger.* Turbochargers and superchargers are lubricated by engine oil that can be forced into the engine intake manifold if the seals in the turbocharger or supercharger leak. Turbochargers and superchargers can be overhauled as outlined in Chapter 13 if you have the correct tools and parts available; otherwise, they should be replaced with a rebuilt unit.

3 *Worn rings or new rings not sealed properly.* Dust and dirt entering the engine can wear out the rings and the cylinder walls. Worn rings allow oil to work into the combustion chambers, causing oil consumption. Run a cylinder compression check to determine ring condition. The only cure for worn rings is the installation of new ones. Most diesel engines have replaceable sleeves, rings, and piston kits that will repair the engine to like-new condition. There are times after an engine overhaul that new rings do not seat properly, causing engine oil consumption. An overhauled engine must be "broken in" properly to insure that the rings are seated to the cylinder wall as they should be. Normal break-in requires that the engine be loaded intermittently during the first forty hours of operation using the same oil that would normally be used in the engine. If any question exists about break-in procedures, ask your instructor or refer to your engine service manual.

III. Engine Misfires on One or More Cylinders

A *Low compression.*

Compression is one of the key elements in the proper operation of a diesel engine cylinder, since without it not enough heat is developed to ignite the fuel. Low cylinder compression could result from the following causes:

1 *Valve tappet clearance too tight.* If the tappet clearance is too tight, the valves are not allowed to seat properly, causing them to leak. Check clearance carefully as outlined in Chapter 7.

2 *Broken or bent push rod.* The push rods must operate the rocker arms to open the intake valves, which allow air into the cylinder to be compressed. If a push rod is bent or broken the valve may not open at all, causing a loss of compression. Check all push rods visually for bends or broken ends.

3 *Burned valves.* Valves that have burned spots on their face area will not seat properly and may cause a loss of compression. Check for valve leakage as outlined above in Section I, B, 2.

4 *Worn or broken rings.* Check for worn or broken rings as outlined above in Section I, B, 1.

B *Fuel incorrectly atomized or not injected into cylinder.*

Atomized fuel in the correct amount is very important to the operation of a diesel engine cylinder. Incorrect atomization or delivery can be attributed to any of the following:

1 *Injector or nozzle adjustment.* Proper injector adjustment on a unit injector type engine like a Cummins is necessary for good atomization and fuel delivery. Check injector adjustment. If cylinder still does not operate properly, remove the injector for service. Injection nozzles, used on many other engines, always should be removed to be tested for opening pressure and adjusted. To test for a misfiring injector or nozzle, bar the injector down in some manner or loosen the fuel line leading to the nozzle.

2 *Plugged nozzle tip holes.* Nozzle or injector tip holes can be plugged by carbon or other foreign material. If the nozzle is suspected of having plugged holes, remove and test.

3 *Stuck injection nozzle or injector.* An injection nozzle valve can become varnished and corroded to the point where the valve can no longer operate freely. Remove the nozzle or injector from the engine and test. Replace the tip or injector if needed.

C *Injection timing incorrect.*

1 *Injection pump static timing.* Check as outlined in Section I, D, 2.

2 *Injection pump automatic advance mechanism not operating properly.* Engine cylinder misfire can be caused by the advance mechanism not advancing the injection timing as the engine speeds up. The timing must be advanced to compensate for injection lag and ignition lag. In some cases the advance can be adjusted. In other situations the advance unit must be overhauled or repaired.

IV. Engine Smokes Excessively Under Normal Load (Black Smoke)

A *Insufficient air reaching the engine.*

A sufficient amount of air must be supplied to the engine if it is to run without excessive smoke. An insufficient air supply can be caused by any of the following:

1 *Plugged air cleaner cartridge (dry type air cleaner).* Air cleaner cartridges must be changed or cleaned periodically to prevent them from causing a restriction to the flow of air into the engine. Dry type air cleaner cartridges can be cleaned with compressed air or washed in soap and water. If any doubt exists about the air cleaner element, replace it.

2 *Collapsed air inlet hoses.* Check hoses closely for signs of collapse. Replace hoses as needed.

3 *Low turbo pressure.* Turbo pressure is directly related to the amount of air delivered to the engine. A turbo that does not run at the proper speed or has bad bearings, allowing it to drag, will not deliver the proper amount of air to allow the engine to run without excessive smoke. Repair or replace turbo as needed.

4 *Leaking or broken hose between turbo and intake manifold.* If the hose between the turbo and intake manifold is leaking air, the pressure developed in the manifold will be lowered and less air will be delivered to the engine.

B *Fuel injection system problems.*

Very often a diesel engine smokes black under load because of the diesel fuel system. Check it for the following:

1 *Plugged nozzle or injector orifice.* See III, B, 2 above.

2 *Incorrect injection timing.* Check as outlined above in III, C, 1.

3 *Incorrect injection pump or fuel system calibration.* All fuel injection pumps must be precisely adjusted (calibrated) to deliver the correct amount of fuel under different load and speed conditions.

V. Engine Uses Too Much Fuel (Poor Efficiency)

A *Incorrect fuel system settings.*

In many cases the fuel injection pump settings are changed to gain more power than the engine is designed to deliver. This practice, called "overfueling," can cause excessive fuel consumption and engine damage. If this condition is found, the fuel system should be recalibrated.

B *Restricted air intake.*

1 *Clogged air cleaner.* See Section IV, A, 1.

2 *Collapsed intake air pipe.* See Section IV, A, 2.

C *Incorrect engine operation.*

All engines have specified rpm's at which they operate most efficiently. An engine operated differently from factory recommendations will use excessive amounts of fuel.

1 *Operator lugging engine.* Most diesel engine fuel systems will deliver more fuel as the engine is lugged down. This characteristic is built into the fuel system so that the engine will recover from an overloaded condition quickly.

If the engine is operated in the lugged condition, it will burn more fuel than it should.

NOTE An engine is being "lugged" when the engine cannot pick up speed under load with the throttle in the full fuel position.

VI. Low Engine Oil Pressure

Engine oil pressure is one of the vital signs of engine condition. Whenever there is a change in oil pressure, an attempt to determine the cause must be made. Low engine oil pressure may result from the following causes:

A *Inoperative or incorrect oil pressure gauge.*
Oil pressure gauges are a common cause of low oil pressure complaint. Be sure about the gauge. Always use a master gauge when the system oil pressure is in question. Any of the following can cause gauge problems:

1. *Bad sending unit.* An electric type gauge uses a sending unit that is screwed into the engine oil galley at some point on the engine. It must be checked as outlined in Chapter 11.
2. *Broken wire between sending unit and dash gauge.* Check this wire closely for opens or grounds and repair as needed.
3. *No power supplied to dash gauge unit.* Check with a voltmeter or test light to determine if power is being delivered to the gauge. If not, trace the circuit back to locate the trouble.
4. *Incorrect reading dash unit (gauge).* In many cases a constant pressure gauge will not read correctly after it has been used for a long period of time. Compare the reading to a master gauge and replace if needed.

B *Restriction in system.*
Like any pump, the oil pump must receive a constant and adequate supply of oil if it is to maintain engine oil pressure. Check the following:

1. *Oil pump inlet screen or filter.* The oil pump pickup must be free from any restriction or clogging.
2. *Clogged filter. Always* change the filter when working on a lubrication system problem.

NOTE Systems that use partial flow filters will not be affected by a clogged oil filter.

C *Faulty oil pressure relief valve.*
Relief valves or bypass valves throughout the system control the system oil pressure. Check the system valves for the following:

1. *Broken relief valve spring.* A broken spring does not hold the valve on its seat properly, which causes low oil pressure. Replace the spring.
2. *Sticking relief valve piston.* A relief valve piston that does not return to its seat after opening will cause low oil pressure. Remove the valve and clean it with crocus cloth or replace it with a new one.
3. *Carbon or other foreign object under relief valve.* This is most likely to occur after an engine overhaul since loose pieces of carbon can be floating around in the system or a small piece of gasket material may find its way into the system. This is easily corrected by removing the valve and cleaning the foreign material out.
4. *Relief valve not seating properly.* After repeating opening and closing, the relief valve seat may become worn to the point where it allows oil to leak by lowering the oil pressure. Replace the valve with a new one.

D *Defective oil pump.*
The oil pump should be the *last* thing to check when troubleshooting a low oil pressure complaint since it usually is the most dependable. If the oil pump is suspected, it should be checked for:

1. *Worn gears.* Oil pump gears should be measured and visually checked for wear. If worn beyond specifications, they should be replaced.
2. *Worn pump body.* The oil pump body and gear cover should be checked for wear and replaced if needed. For complete oil pump service and checks, see Chapter 11.

VII. Crankcase Oil Dilution

To do its job correctly, lubricating oil must be free from contamination. Lube oil can be contaminated by:

A *Fuel oil.*
Fuel oil will thin the lube oil, causing a loss in oil pressure and oil lubricating quality. Fuel oil can get into the lube oil in the following manner:

1. *Injection nozzle stuck open.* If the injection nozzle is stuck open, raw fuel will be injected into the cylinder. Since no atomization is taking place, a small amount will burn while the

rest will find its way by the rings into the crankcase.

2 *Leaking fuel injection pump seal.* Injection pumps that use fuel oil to lubricate the pump have the injection pump housing full of fuel. If this fuel leaks from the housing, it may end up in the engine crankcase, diluting the engine oil.

B *Antifreeze solution (coolant) leaking into the crankcase.*

Antifreeze and engine oil do not mix well. If antifreeze is allowed to leak into the engine oil, the oil becomes a thick, heavy, tarlike substance that has no lubricating qualities. Antifreeze can leak into the engine oil in any of the following ways:

1 *Leaking cylinder liner O rings.* Any leakage by the cylinder liner O rings will leak directly into the oil pan. The O rings must be replaced to eliminate this problem.

2 *Head gasket blown or leaking.* A leaking head gasket can allow coolant to leak into the oil pan. When replacing the head gasket, check all mating surfaces carefully for warpage and erosion.

3 *Oil cooler O rings leaking.* If the O rings in the oil cooler are suspected of leaking, remove the oil cooler and test it for leakage.

4 *Oil cooler unit or element leaking.* The oil cooler element should be checked under pressure for leaks if it is suspected that it is leaking.

5 *Cracked block.* A cracked block can leak internally, causing crankcase oil dilution.

6 *Cracked cylinder head.* Most cracks in cylinder heads are found in the combustion chamber area, but occasionally a cylinder head will crack on the outside and allow antifreeze to leak into the crankcase.

VIII. Low Engine Power

A *Insufficient fuel delivery.*

Engine power is derived from the fuel burned in the cylinders. Low engine power can be caused by any of the following:

1 *Clogged fuel filter.* If there is doubt about the fuel filter, replace it.

2 *Suction leaks in fuel supply system.* Check as outlined in Section I, C, 2.

3 *Worn injector plunger.* If the injector plungers are worn, insufficient fuel will be supplied to the cylinder. Remove and check injectors for leakage.

4 *Worn or incorrectly adjusted injection pump governor.* Correctly operating governors are the key to a good operating engine. If governor operation is not correct, it should be removed for repair and calibration.

5 *Incorrect throttle travel.* Check to make sure that the throttle is moving to the full open position. If not, adjust leakage as needed to open throttle all the way.

B *Insufficient air reaching engine.*

The engine cannot develop full power unless sufficient air is reaching the engine. This may be prevented by the following:

1 *Clogged air filter.* Remove and clean or replace.

2 *Worn or defective turbocharger.* See Chapter 13.

C *Low cylinder compression.*

To burn the air and fuel mixture correctly, the cylinder must have good compression. Any of the following may cause low cylinder compression:

1 *Worn rings.* See Section I, B, 1.

2 *Burned valves.* See Section I, B, 2.

3 *Damaged piston.* A piston with a hole burned in it (Figure 26.3) will not develop any compression. As a result, the cylinder will not have any power. This condition is easily noticed because engine blow-by will be excessive. The damaged piston must be removed and replaced with a new one.

Figure 26.3 Hole burned in piston.

4 *Stuck rings.* Stuck rings, like a hole in the piston, are easily diagnosed because large amounts of blow-by are produced. The rings and piston must be replaced to remedy this problem.

D *Incorrect injector or injection nozzle adjustment.*

A malfunctioning nozzle or injector will not correctly atomize the fuel, causing a loss in engine power. Readjust the nozzle or injector to remedy this problem.

IX. Engine Uses Excessive Oil After an Overhaul

A *Oil consumption by piston rings.*

Oil consumption by piston rings after an overhaul can be caused by:

1 *Incorrectly installed piston rings.* Piston rings can easily be installed wrong if the mechanic does not pay close attention. After all other possible causes are checked, the piston must be removed and rings checked.

2 *Rings not seating properly.* Rings must seat to cylinder wall after installation. If they do not, excessive oil consumption can result. To insure that the rings seat correctly after an overhaul, break the engine in as indicated by the manufacturer. If the engine has been properly assembled and run-in according to the manufacturer's recommendations, the engine must be disassembled and new rings installed.

3 *Valve seals not installed or incorrectly installed.* In most modern engines, valve seals are a must. Check them closely to make sure they are correctly installed.

X. Head Gasket Leaks Compression

A *Damaged or blown head gasket.*

Most head gasket leakage problems are caused by one of the following:

1 *Loose head bolts.* Recheck all head bolts for correct torque.

2 *Broken head bolts.* Head bolts have to flex many times per minute during engine operation. After many hours of operation, the bolts may break due to fatigue. If one or two head bolts break in an engine, it is advisable to replace all of them. This should prevent a recurrence of head bolt breakage.

3 *Excessive fuel delivery.* Excessive fuel delivery usually results in the engine developing more power than it is designed for. This excessive power applies far more pressure on the head gasket than it can withstand, which usually results in a blown head gasket.

4 *Block top surface or cylinder head not straight (warped).* To seal the head gasket effectively, it must be installed between two straight surfaces. If head gasket failure occurs repeatedly, the block and head should be closely checked.

5 *Cylinder sleeve protrusion incorrect.* Cylinder sleeve protrusion must be correct if the head gasket is expected to seal the cylinder pressure. A general sleeve protrusion that is common among most engines is 0.001 to 0.005 in. (0.025 to 0.127 mm) above the top surface of the block.

NOTE Some engines, such as Detroit Diesel, don't have sleeve protrusion since they use a different type head gasket from many other engines. Instead they are designed with sleeve recession.

XI. Engine Starts Hard When Hot

A *Cranking speed low.*

The engine generally turns harder when it is hot since the rings provide a better seal; therefore, compression pressure is greater. If any problems exist in a cranking system, they are most likely to show up when the engine is hot or very cold. Hard hot starting can be caused by any of the following:

1 *Low battery.* Check the battery for state of charge as outlined in Chapter 27.

2 *High resistance in the cranking circuit, either ground or insulated circuit.* A loose cable or connection can cause high resistance, causing a drop in available amperage. Check the circuit as outlined in Chapter 28.

3 *Faulty starter.* A starter with dirty brushes, worn bearings, or other problems cannot deliver full cranking power. Check and repair the starter as outlined in Chapter 28.

B *Fuel not delivered to engine combustion chamber during cranking.*

1 *Worn injection pump.* This is a common occurrence with diesel fuel injection pumps. As the pump wears, because of dirt and other impurities within the fuel, the injection or delivery of fuel decreases since some of the fuel that

would normally be injected leaks by the pumping plungers or distributor rotor.

XII. Abnormal Engine Vibration

A *Broken crankshaft.*

In many cases engine crankshafts will break in such a way that the engine will continue to run but vibrate severely. The following items can cause crankshaft breakage:

1 *A damaged or faulty vibration damper.* Check damper as outlined in Chapter 9.

2 *Improper grinding of the crankshaft during overhaul.* Find out if the crankshaft was reground; if so, were the fillets reground?

B *Incorrect or faulty vibration damper.*

Vibration dampers are sometimes replaced with little concern about the correct application. "If it fits, bolt it on" is often heard among some mechanics. Vibration dampers must be for the engine they are installed on. Additional vibration dampers are sometimes damaged during engine overhaul by a mechanic who utilizes incorrect pulling (removal) and installation procedures. Check the damper as outlined in Chapter 9.

C *Loose flywheel.*

The flywheel must be held solid to the rear end of the crankshaft if the engine is to be free of vibration. A loose flywheel can be caused by:

1 *Broken bolts.* Flywheel bolts may break because of fatigue and allow the engine to vibrate.

2 *Loose bolts.* A bolt that has been left loose during engine assembly can cause engine vibration. Retighten the bolts to the proper torque.

D *Engine clutch pressure plate.*

1 *Out of balance.* Since the pressure plate bolts onto the engine flywheel, it must be balanced.

2 *Incorrectly installed.* When installing pressure plates, pay close attention to balance or match marks if they are used. Also make sure no dirt or other foreign material is trapped between the pressure plate and the flywheel during assembly.

XIII. Engine Knocks (Mechanical)

A *Scored pistons.*

A scored piston is one that has become so hot that it started to melt and fused itself to the cylinder liner. After the piston cools down (after scoring), it will have excessive clearance, causing engine knocks. Scored pistons can best be located by causing the cylinder to misfire by loosening an injection line or barring down on an injector plunger. The knock generally decreases or disappears when the cylinder does not fire. Scored pistons can be caused by:

1 *Excessive engine temperature.* The engine pistons and other parts must be cooled correctly since they are subjected to the heat of combustion; otherwise, they may be damaged. Check the engine cooling system for malfunctions as outlined in Chapter 12.

2 *Overfueling of the engine.* If the engine is overfueled and more combustion occurs, generally the cooling system does not have sufficient capability to cool the engine. This overheated condition can cause piston scoring. If this condition is suspected, have the fuel injection equipment calibrated.

3 *Malfunctioning injector or injection nozzle.* An injection nozzle that is stuck open delivers unatomized fuel into the cylinder. This raw fuel can wash away the lubricating oil and cause piston scoring. Injector or injection nozzle should be removed and repaired.

B *Loose connecting rod bearings (excessive clearance).*

1 *Worn bearing.* A bearing that has lost some metal due to wear can cause excessive clearance and will cause a knock.

2 *Low oil pressure.* If insufficient oil pressure is supplied to the engine, bearing noises may result since the oil provides a film separating the bearing and bearing journal. This film absorbs the shock and prevents engine noise. Check oil pressure as outlined in Chapter 11.

3 *A tapered or out-of-round crankshaft.* Bearings cannot be properly fitted to a crankshaft that is out-of-round or tapered. In most cases this condition can only be remedied by replacement or grinding of the crankshaft.

C *Worn timing gears.*

Timing gears, if worn, can cause a knock much like connecting rods. Unlike a connecting rod, this noise cannot be changed by causing a cylinder to misfire. If timing gears are noisy, replace them as outlined in Chapter 10.

XIV. Engine Does Not Cool Properly (Coolant Too Hot)

NOTE Make sure an accurate gauge is being used to check the coolant temperature.

A *Insufficient air flow across radiator.*

1. *Loose fan drive belt.* V-belts tend to loosen after they are used for a while. Check the adjustment to prevent slippage.
2. *Radiator fins clogged with dirt, grease, or other foreign material.* The radiator fins must be kept clean and free from obstructions if the radiator is to function properly. If the radiator fins are clogged, blow them out with compressed air.
3. *Shutters not operating properly.* If the shutters are sticking or do not open at the correct temperature, engine overheating can result. Check shutter operation as outlined in Chapter 12 on cooling systems.
4. *Fan clutch not operating properly.* If an air-operated fan clutch is used, check to see if it is engaging properly. If a viscous fan clutch is being used, check the fan at rated engine rpm.

NOTE Low fan speed is generally associated with low fan noise. If the fan noise is low at high engine speed, it might be that the fan is slipping. Fan speed can also be checked with a stroboscope.

B *Insufficient water circulation through the radiator.*

Engine cooling is directly related to water flow through the radiator. Restricted water flow can be caused by the following:

1. *Radiator core plugged with rust or scale.* If this condition is suspected, the radiator must be removed and checked for flow on a flow machine.
2. *Thermostat stuck.* After repeated use over a long period of time, the thermostat can stick closed, stopping all water flow through the radiator. If the thermostat is suspected, remove it and check with a tester or replace it.
3. *Collapsed water hose.* A collapsed hose on the inlet side of the water pump will restrict the flow of water to the engine cylinder block, creating a hot running engine. If this condition exists, replace the hose.
4. *Defective water pump.* A water pump with a loose or worn impeller will not circulate water through the engine, causing an overheated engine. If this condition is suspected, remove and check the water pump. If it is defective, follow the overhaul procedures outlined in the cooling systems in Chapter 12.

XV. Engine Does Not Warm Up (Coolant Temperature Too Low)

A *Too much water flow through the radiator.*

Cooling systems and radiators in particular are designed to have sufficient capacity for cooling the engine under heavy load during high ambient temperatures. If this condition is met, it becomes a problem then to warm up the engine sufficiently under no load during cold weather operation. Too much water flow through the engine can be caused by:

1. *Thermostat stuck open.* The thermostat must be closed to make the engine warm up properly. Check the thermostat as outlined in cooling systems in Chapter 12.
2. *Thermostat seal broken or leaking.* Thermostats used on diesel engines are usually of the bypass type that incorporate a lip type seal to prevent water flow through the radiator during engine warm-up. Check and replace the seal if necessary.

B *Engine operating under no load or at idle.*

During cold weather most diesel engines will not warm up unless they are being worked. It is recommended that diesel engines never be allowed to idle for long periods of time during cold weather.

C *Too much air moving across radiator fins.*

Air movement across the radiator must be controlled by shutters or clutch type fans.

1. *Shutters stuck open.* Check shutter operation. Adjust shutterstat if opening and closing temperatures are not correct.
2. *Fan clutch not operating.* If the fan clutch does not disengage the fan, too much air is passing across the radiator, causing the engine to run very cool. Check fan clutch operation as outlined in Chapter 12.

XVI. Engine Tune-Up

Diesel engine tune-ups are performed whenever the engine has lost power, is not running properly, or when the engine has a number of hours on it that constitute a normal tune-up interval.

Tune-ups fall into one of two categories, major or minor.

A *A minor tune-up generally includes the following:*

1. Retorque cylinder head. (This is optional; it depends, for example, on the type of head gasket and length of service, among other factors.)
2. Adjust tappet clearance.
3. Adjust injector timing or setting on engines using unit injectors.
4. Check pump static timing on engines using a pump-nozzle combination.
5. Change fuel filters and clean all water traps.
6. Check air filter. Change oil if oil bath type.
7. Check high idle speed.
8. Check low idle speed.
9. Dyno check engine for correct horsepower.
10. Visually check engine for leaks.

In addition to these 10 items, some engines may have an additional item that requires adjustment or checking before considering the tune-up complete.

B *A major tune-up should include the following items:*

1. Retorque cylinder head.
2. Adjust tappet clearance.
3. Clean and adjust injectors and/or injection nozzles.
4. Check pump static timing.
5. Change fuel filters and clean or drain all water traps.
6. Service air cleaner.
7. Check and overhaul injection pump if needed.
8. Check high idle speed.
9. Check low idle speed.
10. Dyno check engine for correct horsepower.
11. Visually check engine for leaks.

As you perform the procedures in a tune-up, always watch closely for any loose bolts or hose clamps that may be a potential trouble spot. Also replace all gaskets, such as tappet cover gaskets, pump timing cover gaskets, and any other gaskets that have been disturbed during the tune-up.

SUMMARY

This chapter has covered many of the common problems that you will encounter while troubleshooting a diesel engine. Troubleshooting will become a habit after you practice the procedures listed. Always add to your knowledge of troubleshooting as you become more experienced, remembering the problems you encounter.

Engine tune-up, like troubleshooting, follows a certain pattern and requires considerable skill if you are to do a quality job. These two subjects were covered in this chapter, since many times they go together when you are working on an engine.

REVIEW QUESTIONS

1. What are two important steps that you should follow when troubleshooting a diesel engine?
2. Give two reasons why a diesel engine might not start.
3. List two reasons for low compression in a diesel engine and explain them.
4. What are two causes of insufficient fuel being injected into the diesel engine?
5. List and discuss four reasons for a diesel engine using excessive amounts of oil.
6. How can a quick check be made to determine if engine oil is being pumped through the air brake system?

7 Give two reasons that would cause a diesel engine cylinder to misfire.

8 What can cause excessive black smoke from a diesel engine exhaust?

9 List and explain two problems that you would check for if a diesel engine is using an excessive amount of fuel.

10 Crankcase oil dilution can cause low oil pressure and engine damage. List two reasons for crankcase oil dilution.

11 Low engine power is a common complaint by engine operators. Give three possible causes of low engine power.

12 List two reasons for head gasket leakage.

13 Abnormal engine vibration can be difficult to diagnose. List three possible causes.

14 Engine mechanical noises can be very misleading. Explain and demonstrate how you would check for a scored piston.

15 At what times should a diesel engine be tuned up?

16 Explain the difference between a major and minor tune-up.

27

Basic Electrical Fundamentals and Batteries

The study of basic electricity, its application and how it works, is a very important part of becoming a proficient diesel engine mechanic since many electrical components are used on diesel engines. Electricity as applied to diesel engines is interesting and enjoyable to deal with. After you gain experience with the proper equipment, testing, repairing, and troubleshooting diesel engine electrical systems is relatively simple, and many mechanics prefer electrical repair over engine or fuel system repair. This chapter will cover the basics of electricity, magnetism, and batteries.

OBJECTIVES

Upon completion of this chapter the student will be able to:

1. Explain the "current" theory of electricity.
2. List two good conductors of electricity.
3. List two factors that affect the flow of current in a conductor.
4. Define amperage.
5. Be able to discuss and define voltage.
6. Be able to discuss and define resistance.
7. List two sources of electricity available for use on a diesel engine.
8. Explain Ohm's law and draw the triangle expression.
9. List and explain the three parts of a basic circuit.
10. List the two types of electricity.
11. List two functions of a battery in an electrical circuit.
12. Explain what happens within the battery during charging.
13. List two things that may happen to a battery, causing it to function incorrectly.

14 List two ways in which a battery should be checked.

15 List two precautions that should be observed when charging a battery.

GENERAL INFORMATION

To understand the electrical components that are used on a diesel engine, you must understand the principles of electricity and magnetism. In addition to understanding the electrical components, a thorough knowledge of basic electricity will help you to be more effective in troubleshooting and repairing any of the electrical equipment or circuits used on the engine. Very often mechanics are called upon to repair electrical circuits on the vehicle in which the diesel engine is installed, such as headlights, clearance lights, heaters, windshield wipers, turn signals, and engine shutdown solenoids. If thoroughly understood, electrical troubleshooting and repair can be easy.

I. Electricity

Electricity can be defined as the flow of electrons in a conductor. Electron movement through the conductor is created by an electrical unbalance.

Contained in all conductors are smaller parts called atoms. An atom is made up of protons and electrons. The protons are contained in the core of the atom while the electrons orbit around this core.

Different types of material contain different numbers of protons and electrons. A common material or element that is used extensively as a conductor in electrical equipment and electrical circuits is copper. Copper contains 29 protons and 29 electrons.

As stated earlier, the protons are contained in the core of the atom, while the electrons that encircle the core are contained in four separate rings. Each electron has a path around the protons much like a planet circling the sun (Figure 27.1).

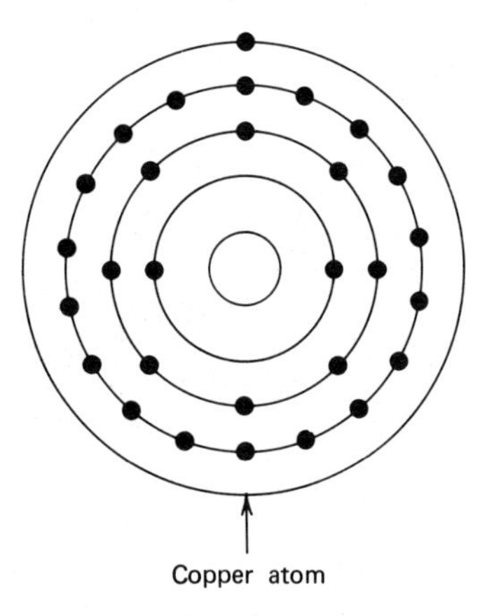

Figure 27.1 Electrons encircling a proton (courtesy Delco Remy).

In copper the first ring around the protons contains 2 electrons, the second ring 8 electrons, the third ring 18 electrons, and the fourth ring 1 electron. Copper is a good conductor of electricity, since any element that contains less than four electrons in its outer ring is considered a good conductor of electricity. Such electrons, which are distant from the proton core, are loosely bound to it and are considered free electrons. Elements that contain four or more electrons in the outer ring are called semiconductors. Silver is one of the best conductors of electricity since it has one electron in its outer ring; however copper, because its low cost, is the popular metal in use today.

To further clarify how electricity flows in the element copper, visualize a force such as a battery connected to the ends of a copper wire. The imbalance from the positive terminal to the negative terminal of the battery exerts a force on the free electrons, causing them to move from atom to atom through the copper wire, much like a row of falling dominoes. This is known as current flow. Current flow in a wire is created by electromagnetic induction or by a battery.

NOTE Current flow in a diesel engine electrical system is in one direction only. This current is direct current (DC). AC current signifies that the current flows first in one direction and then reverses, flowing in the other direction; therefore, it is called alternating current.

II. Ohm's Law

Ohm's law shows the relationship between the three parts of every electrical circuit:

1. *Voltage or electromotive force* (*E*). Voltage is known as the system push or pressure.
2. *Current or inductance* (*I*). Current is the flow of electricity in a circuit. The amount of flow (measured in amps) is dependent on voltage and resistance.
3. *Ohms or resistance* (*R*). Resistance to flow is measured in ohms. The amount of resistance in a circuit is determined by the type of conductors and the load.

Ohm's law states that there is a definite relationship between voltage, current, and resistance. Ohm's

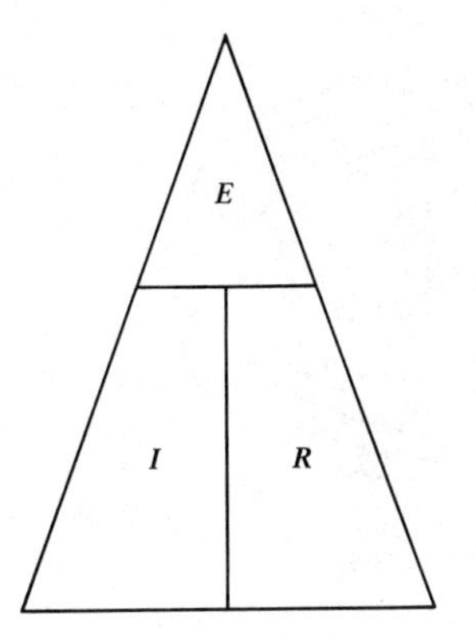

Figure 27.2 Ohm's law triangle.

law can be expressed in graphic form, as shown in Figure 27.2. This illustrates that if two parts of the total equation or circuit are known, the other can be calculated. The formulas for finding the missing part of the circuit are as follows:

1. $I \times R$ (amps $\times$ ohms) = volts.
2. $E \div I$ (volts $\div$ amps) = ohms.
3. $E \div R$ (volts $\div$ ohms) = amps.

One way of using the simple formula expressed in triangle form is shown in Figure 27.3. A circuit has a 12-volt power supply, and 3 amps of current are flowing in the circuit.

This relationship between voltage, amperage, and resistance is true for all electrical circuits. Once this relationship is understood by the mechanic, electrical circuit and component troubleshooting becomes easy.

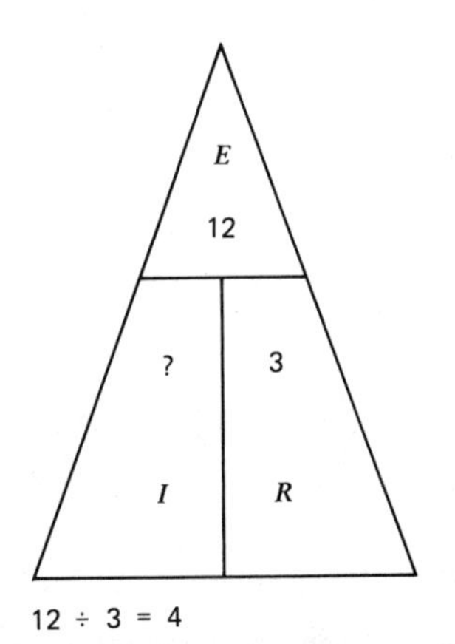

Figure 27.3 A typical Ohm's law problem.

III. Permanent Magnets and Electromagnets

A *Permanent magnets.*

Almost everyone has had some experience with magnetism, especially permanent magnets. The magnetic base on a dial indicator is a good example of a permanent magnet. Permanent magnetism was first discovered in iron called lodestone. It was discovered that this iron ore has polarity. This means that the one end is attracted to the earth's north pole and the other end is attracted to the south pole. All magnets, permanent and electro, have polarity. Further discoveries indicated that a permanent magnet (Figure 27.4) had a force field surrounding it that tended to attract other iron, hence the name magnet.

This force field is utilized in all electrical components of diesel engines. Since the control of this force field is an essential part of component operation, permanent magnets are not used in diesel engine components because they cannot be controlled. Electromagnets are used in most diesel engine electrical components because they can be controlled by regulating the amount of electrical power applied to them.

B *Electromagnets.*

Electromagnets, much like permanent magnets, are pieces of iron with wire wrapped around them. To make the magnet work, current must be flowing through the wire. To better understand how an electromagnet works, we must first look at a straight conductor or wire. If current is passed through this wire, a magnetic field is created around the wire (Figure 27.5). This magnetic field has direction that can be determined using the *right-hand rule for a current carrying conductor.*

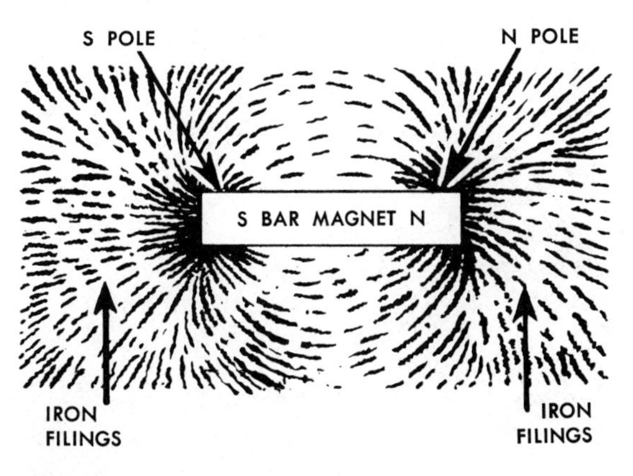

Figure 27.4 Permanent magnet with force field (courtesy Delco Remy).

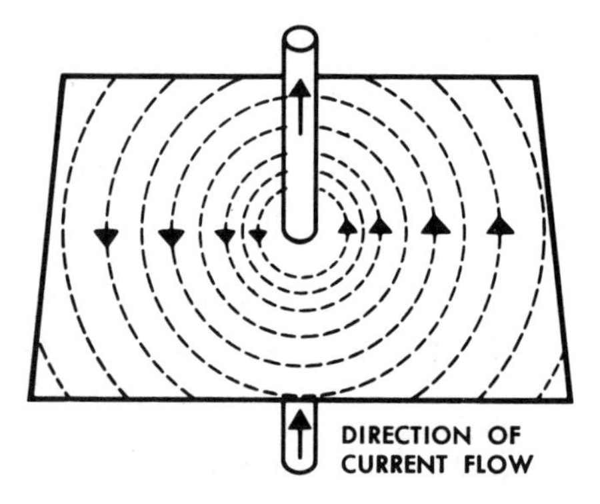

Figure 27.5 Force field encircling a straight conductor (courtesy Delco Remy).

To determine the force field direction around a straight conductor by using this rule, grasp the conductor with your right hand, letting the thumb point in the direction of current flow through the conductor. The other four fingers of your right hand will then point in the direction that the magnetic lines of force encircle the conductor (Figure 27.6). These magnetic lines of force will encircle the conductor as long as current is flowing through the wire. The lines of force are stationary and do not move. The size of the force field is dependent on the amount of current being passed through the conductor.

If a straight current carrying conductor is wound to create a coil of wire, the magnetic field around each loop of wire is added together to create an electromagnet (Figure 27.7). This electromagnet has polarity that can be determined by using the *right-hand rule for an electromagnet* (Figure 27.8). To identify the polarity of an electromagnet by using the right-hand rule, let your fingers encircle the coil in the direction the wire is wrapped and current is flowing. The thumb then will point to the north pole of the electromagent.

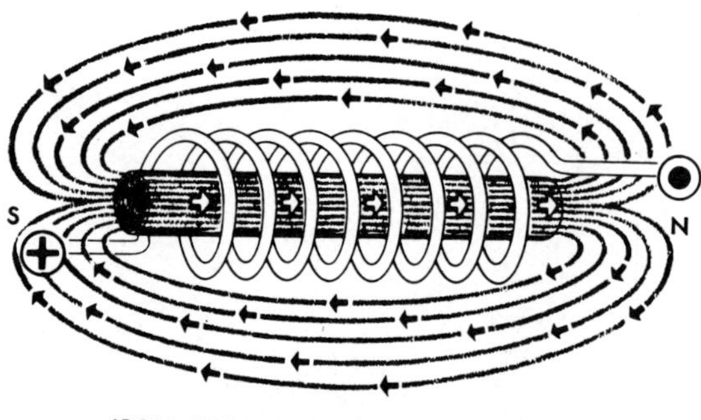

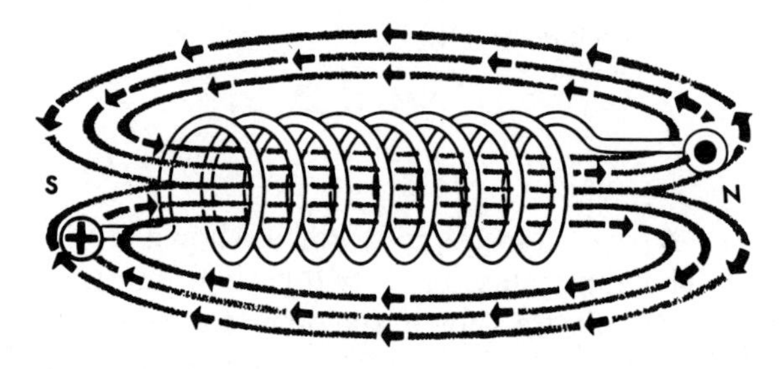

Figure 27.7 Typical electromagnet (courtesy Delco Remy).

Since a simple coil of wire would not be a very strong magnet, many coils are used, and in most cases a soft iron bar is placed in the center of the coil to increase its strength. This increase in strength occurs because the force field can pass through the soft iron core much easier than it could pass through the air.

Once the two right-hand rules as applied to a straight conductor and a electromagnet are understood, the electrical components used on a diesel engine will be easier understood.

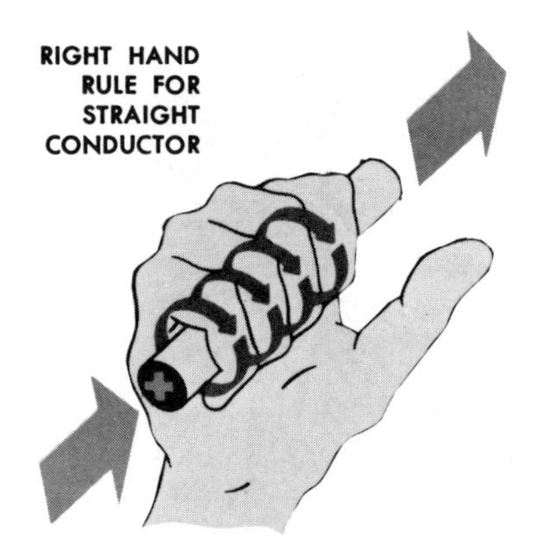

Figure 27.6 Right-hand rule applied to a current carrying conductor (courtesy Delco Remy).

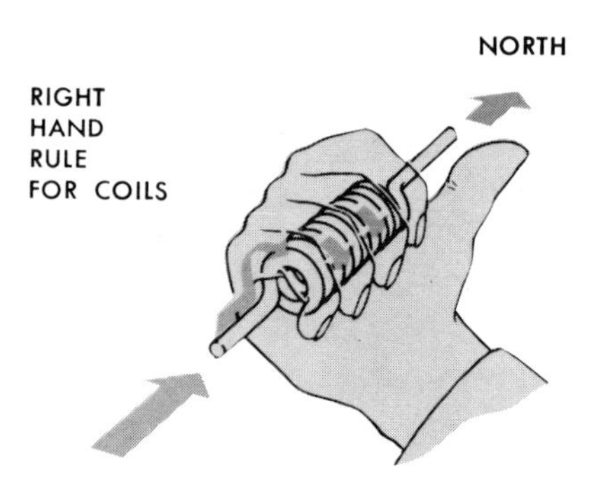

Figure 27.8 Right-hand rule for an electromagnet (courtesy Delco Remy).

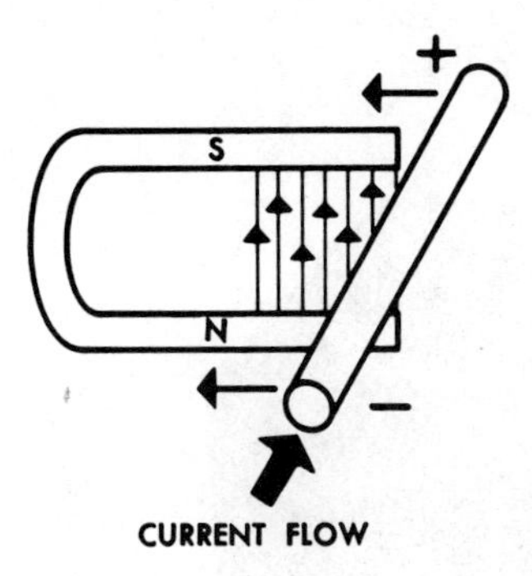

Figure 27.9 Illustration of electromagnetic induction (courtesy Delco Remy).

Figure 27.11 Typical battery.

IV. Electromagnetic Induction

As stated earlier, electron flow in a conductor is current flow. To get this current flowing, voltage is needed. The most common method used in diesel engine components is electromagnetic induction. This principle is widely used to produce current flow in a conductor. Figure 27.9 shows the three items needed to produce voltage by electromagnetic induction:

1 Conductor

2 Magnetism

3 Movement

A voltage is induced in the conductor as the conductor is moved through the force field. This principle is used in dc generators to produce voltage and current flow.

A simple DC generator could be constructed by placing a loop of wire, which serves as a conductor, between two electromagnets. The loop of wire is then rotated and voltage is induced into the wire. This principle is illustrated in Figure 27.10. AC generators (alternators) use the same basic principle but apply it differently. In this application the conductors are stationary and the magnet rotates, moving the magnetic field across the conductor, creating current flow. These two principles will be further explained in Chapter 29, "Charging Systems."

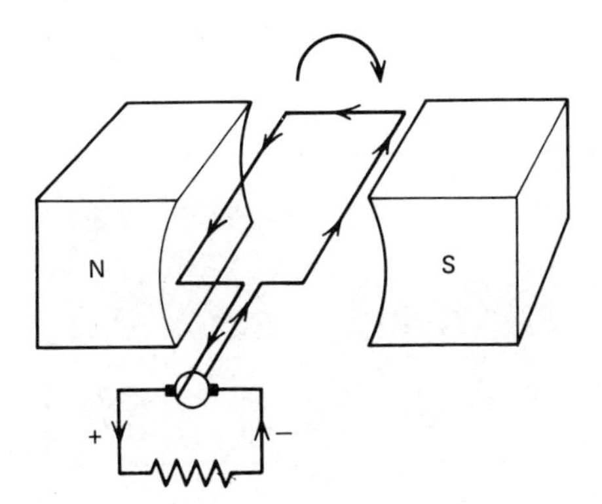

Figure 27.10 Generator principle (courtesy Delco Remy).

V. Batteries

The electricity produced in the charging system must be stored for use during starting. This job is handled by the battery (Figure 27.11). Batteries used in diesel-powered equipment are lead acid batteries.

A *Battery construction.*

1 Lead acid batteries are made up of cells that are separated from each other by compartments (Figure 27.12). Each individual cell contains negative and positive plates separated by porous separators and electrolyte (a mixture of water and acid).

2 The negative and positive plates are connected together by a molded strap across the top of the cells. This strap connects the cells in series, meaning that they are connected negative–positive through the entire battery (Figure 27.13).

3 Every battery contains a number of cells; 6-volt batteries contain three cells and 12-volt batteries contain six cells. Each cell has a voltage potential of 2 volts. Since the cells are connected in series, three 2-volt cells will have a total voltage of 6 volts, while six 2-volt cells will have a total voltage of 12 volts.

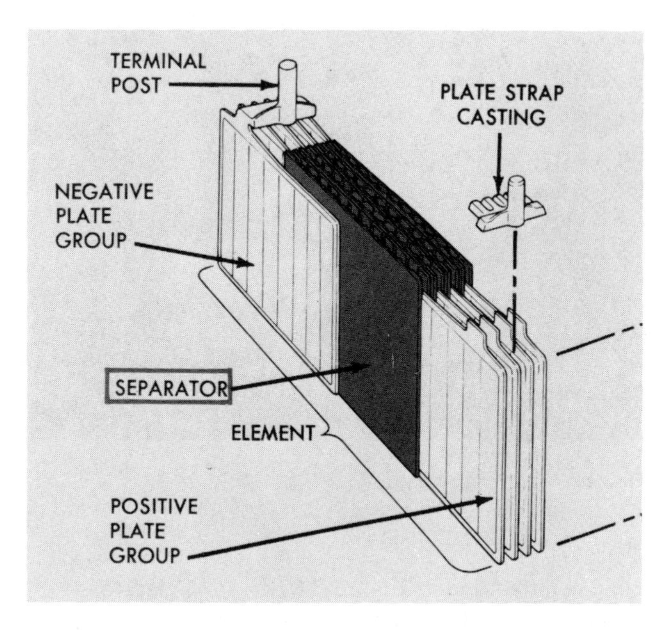

Figure 27.12 Battery cell construction (courtesy Delco Remy).

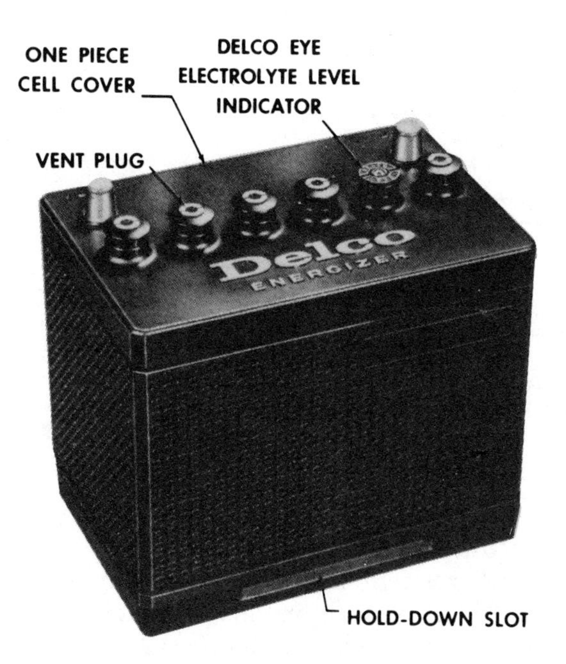

Figure 27.14 Typical hard top battery (courtesy Delco Remy).

4 All the cells are contained in a hard rubber case. Early battery cells were placed in the case connected together and the spaces between the cells filled with tar to seal the battery. These batteries were called soft top batteries. The soft top battery created some problems since the tar was soluble in gasoline or diesel fuel. Current batteries are constructed with a hard rubber top and are called hard top batteries (Figure 27.14). Each cell has an opening through which the battery is filled. This opening is covered with a cell cover or cap. One style of battery manufactured today is completely maintenance-free and does not have any cell openings or cell covers.

Figure 27.13 Battery cross section.

B *Battery operation.*

As stated previously, the battery cells are filled with water and sulfuric acid. Chemical action between this electrolyte solution and the cell plates produces electricity. This chemical action will eventually discharge the battery, since the active materials in the cells are reacting as the battery discharges.

The electrolyte in a fully charged battery consists of a solution of sulfuric acid in water that has a specific gravity of approximately 1.270 at 80°F (27°C). The solution is approximately 36 percent sulfuric acid (H_2SO_4) and 64 percent water (H_2O) (Figure 27.15).

When the battery is connected into a completed electrical circuit, current begins to flow from the battery. This current is produced by chemical reactions between the active materials in the two kinds of plates and the sulfuric acid in the

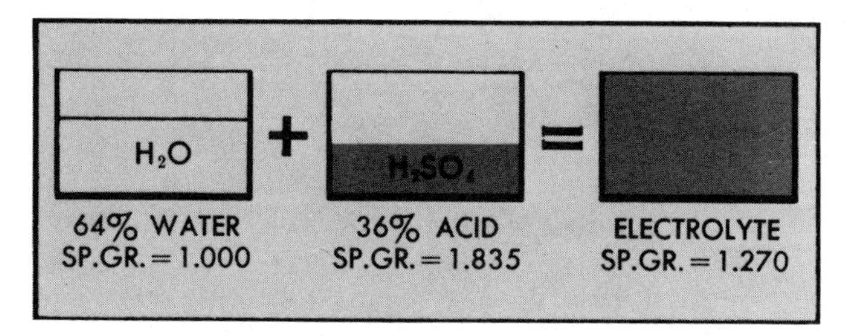

Figure 27.15 Battery solution contents (courtesy Delco Remy).

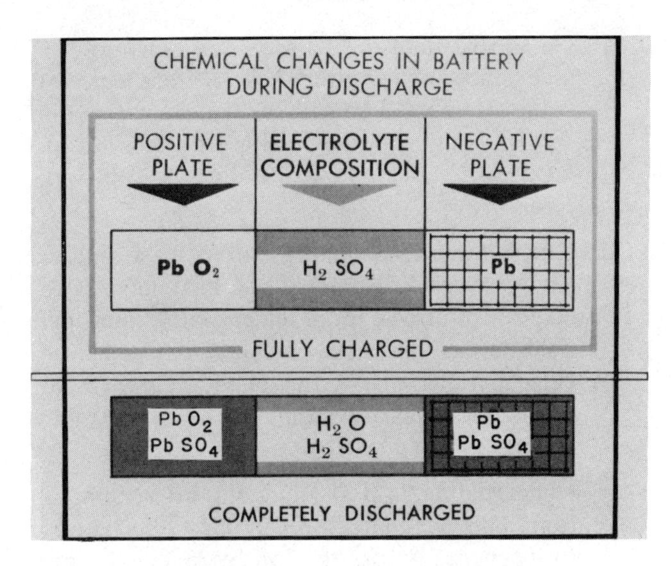

Figure 27.16 Chemical action during battery discharge (courtesy Delco Remy).

electrolyte. The chemical reactions during discharge are shown in Figure 27.16.

In a battery electrical energy is produced by chemical reaction between the active materials of the dissimilar plates and the sulfuric acid of the electrolyte. The availability and amount of electrical energy that can be produced by the battery in this manner is limited by the active area and weight of the materials in the plates and by the quantity of sulfuric acid in the electrolyte. After most of the available active materials have reacted, the battery can produce little or no additional energy. The battery then is said to be *discharged.* Before the battery can be further discharged, it must be recharged by being supplied with a flow of direct current from some external source. The charging current must flow through the battery in a direction *opposite* to the current flow from the battery during discharge. Therefore, the positive terminal of the charging unit should be connected to the positive terminal of the battery. This causes a reversal of the discharge reactions and the battery becomes recharged.

When the battery is in a charged condition, the active material in the positive plate is essentially lead peroxide (PbO_2). It is chocolate brown in color. The active material in the negative plate is spongy lead (Pb), which is gray in color.

The lead peroxide (PbO_2) in the positive plate is a compound of lead (Pb) and oxygen (O_2). Sulfuric acid is a compound of hydrogen (H_2) and the sulfate radical (SO_4). During discharge, oxygen in the positive active material combines with hydrogen in the electrolyte to form water (H_2O). At the same time, lead in the positive active material combines with the sulfate radical, forming lead sulfate ($PbSO_4$).

A similar reaction takes place at the negative plate where lead (Pb) of the negative active material combines with the sulfate radical to form lead sulfate ($PbSO_4$). Thus lead sulfate is formed at both plates as the battery is discharged.

The material in the positive plates and negative plates becomes similar chemically during discharge as the lead sulfate accumulates. This condition accounts for the loss of cell voltage, since voltage depends upon the difference between the two materials.

As the discharge continues, dilution of the electrolyte and accumulation of lead sulfate in both plates eventually cause the reactions to stop. However, the active materials never are completely used up during a discharge. This is because the lead sulfate formed on the plates acts as a natural barrier to diffusion of electrolyte into the plates. When the cell can no longer produce the desired voltage, it is said to be discharged.

The chemical reactions that occur in the cell during charging are essentially the reverse of those occurring during discharge, as shown in Figure 27.17. The lead sulfate on both plates is separated into lead (Pb) and sulfate (SO_4). At the same time,

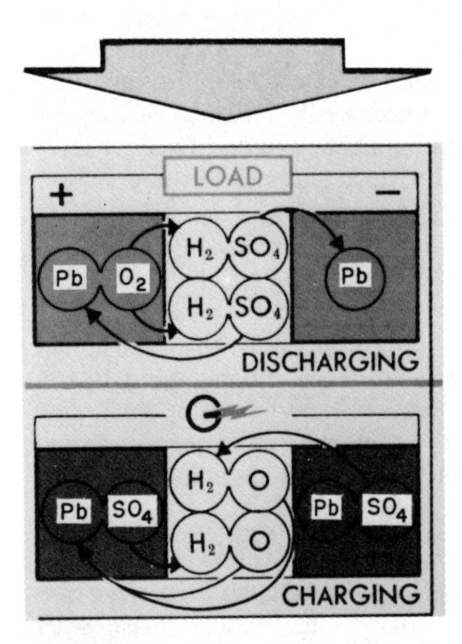

Figure 27.17 Chemical action during battery charge cycle (courtesy Delco Remy).

the oxygen (O_2) in the electrolyte combines with the lead (Pb) at the positive plate to form lead dioxide (PbO_2), and the negative plate returns to the original form of lead (Pb).

In normal operation the battery gradually loses water from the cells due to disassociation of the water into hydrogen and oxygen gases, which escape to the atmosphere through the vent caps. If this water is not replaced, the level of the electrolyte falls below the tops of the plates. This results in overconcentration of the electrolyte and also allows the exposed active material to dry and harden. Thus, if the water level is not properly maintained, premature failure of the battery is certain to occur. Since water loss is more rapid during high temperature operation than at low temperatures, the electrolyte level should be checked more frequently during the summer months.

It has been noted that water plays an important part in the chemical action of a storage battery. Colorless and odorless drinking water is satisfactory, but when obtained from a faucet it should be allowed to run for a few minutes before using.

C. Procedure for Battery Testing

Testing batteries requires a certain degree of know-how and equipment. Work at becoming a proficient battery tester since this is one place where many mechanics have problems. It has been estimated that approximately 50 percent of the batteries that are replaced are replaced needlessly. This unnecessary replacement of batteries could be avoided if battery tests were taken with more accuracy and less guesswork. Unfortunately, batteries sometimes have to be tested in a short period of time because many customers will not or cannot allow their vehicle to stand idle while the battery is being charged at a slow charge rate to determine if it is usable or not. In situations like this a quick, accurate test applied in a systematic approach is needed. Battery testing becomes confusing, however, because of the many different recommended procedures. The following information attempts to sort out and clarify some of them.

1 *Visual inspection.* Battery visual inspection plays an important part in making the decision of the battery's condition. Inspect the battery visually for the following:

a Check the date the battery was put into service, since an old battery has a better chance of being worn out. The date the battery was put in service is usually stamped on the battery or indicated on a tag fastened to the top of the battery.

b Check for cracks in the battery case and/or cover. A cracked battery case may have been caused by freezing of electrolyte, improper hold-down clamp or brackets, plugged vent caps that prevent venting of the hydrogen gas given off during charging, battery explosion, and excessive charging.

c Check battery top for acid and dirt accumulation. This accumulation can allow the battery to discharge across the top by making a connection through the dirt from the positive to negative cell of the battery. Clean this accumulation from the battery by washing it with a mixture of baking soda and water.

d Remove the vent caps and inspect the color of the electrolyte. Discolored electrolyte indicates cell problems. Note also the odor of the electrolyte. A very toxic odor indicates the cell is sulfated and will not take a charge.

e Check electrolyte level. Electrolyte level is important if the battery is going to function normally since cell capacity is reduced greatly when it is low on water.

f Check battery posts for looseness and signs of abuse, such as partially melted posts caused by arcing the battery from terminal to terminal.

NOTE If the battery is installed in the vehicle when making the inspection, check the cables and cable clamps for corrosion and correct size. Most 12-volt applications will require a 4 or 6 gauge cable (Figure 27.18). Also check the cables for corrosion or fraying.

2 *Testing battery state of charge with a hydrometer.* Specific gravity of the battery electrolyte is the main test in determining the state of charge of the battery or battery cells. It is checked by using a hydrometer (Figure 27.19). The hydrometer has a reading range of 1.100 to 1.300. The scale is based on pure water, which has a reading of 1.000. Test the battery electrolyte by:

NOTE The following instructions may not apply to the type of hydrometer you are using. Follow the instructions provided by the manufacturer of

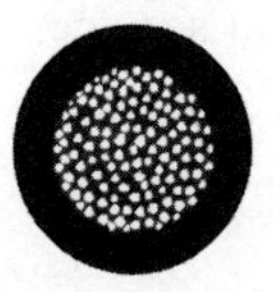

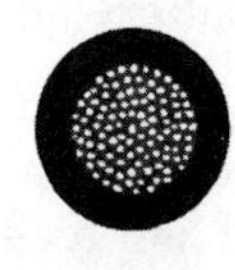
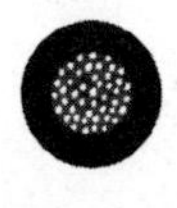
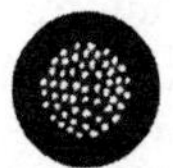
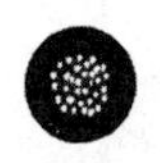

Figure 27.18 Battery cable size.

your tester if a question exists about the correct procedure. Also maintenance-free batteries cannot be checked with a hydrometer, since all cells are sealed. One type of maintenance-free battery, the Delco freedom battery, has a built-in hydrometer that can be used to check the state of charge. Instructions for reading this hydrometer are written on top of the battery.

a Removing the cell cap and squeezing the bulb of the hydrometer, expelling the air in it.

b Insert the hydrometer pick-up tube in the cell of electrolyte and release slowly, drawing electrolyte into the float bulb chamber.

c Draw in only enough electrolyte to cause the float to float.

d Read the number or letter directly at the electrolyte level.

Figure 27.19 Typical hydrometer (courtesy Snap On Tools).

e This reading must be corrected for temperature depending on the type hydrometer you have.

f Squeeze the bulb to force the electrolyte back into the cell and then flush the hydrometer with water.

CAUTION Be careful when handling the hydrometer filled with acid. Avoid splashing acid on your clothing or getting it into your eyes.

---

g After determining what the cell's specific gravity is, make your decision about the cell, based on the following:

If the specific gravity reading is 1.215 or more, the state of charge is satisfactory.

NOTE To doublecheck the state of charge, make a capacity test, outlined later in this chapter.

If the reading is 1.215 or less, recharge the battery.

NOTE The difference in specific gravity between cells should not exceed 0.050. If it does, one or more cells are probably defective. Replace the battery.

3 *Light load test (hard top batteries)*. This test is recommended by most battery manufacturers for use in determining if the battery is usable. The battery to be tested should not be charged before making the light load test.

NOTE A voltmeter with a dial divided into 0.01 increments is required for this test. Also required are cadmium cell probes that can be connected to the voltmeter leads.

a To make the test, remove the battery surface charge by one of the following methods.

(1) Run the starter motor for approximately three seconds.

NOTE When running the starter keep the fuel shutoff in the off position to prevent the engine from starting. If the battery is not in the vehicle, use a carbon pile resistance unit to place a 150 amp load on the battery for three seconds.

(2) Place a 10 to 15 amp load on the battery by turning on the vehicle headlights.

b With the lights on, check the voltage of each cell with the voltmeter. Record the voltage of each cell and remember the highest and lowest reading.

c Determine the condition of the battery by using the following information:

(1) The battery is good and needs no charging if you obtained a reading of 1.95 volts or higher from all the cells. The voltage difference between the cells should not exceed 0.05 volt.

(2) The battery is good but needs charging if all voltages are above and below 1.95 volts and the difference does not exceed 0.05 volt.

(3) The battery is probably bad if any cell has a reading higher than 1.95 volts with a difference of 0.05 volt between highest and lowest cell.

(4) If all cells read less than 1.95 volts recharge and retest the battery.

The above listed test is very effective in determining battery condition if used as outlined. This can only be used on batteries where individual cells can be tested.

4 *Battery load tests (high rate discharge test).* If equipment is not available to test individual battery cells, the battery load test can be made in addition to the specific gravity test.

A battery load tester, which is needed to make this test, is a device that enables you to put a variable current load on the battery. It consists of a carbon pile with which to load the battery, a voltmeter, and an ammeter. Test the battery as follows:

NOTE Make sure the battery temperature is between 60 and 100°F (16 and 38°C) when testing.

NOTE If battery has just been charged, remove surface charge by loading the battery 200 to 300 amp load for 15 seconds. Then wait 15 seconds for battery to recover before testing.

a Connect the tester to the battery with respect to polarity.

b Operate the tester to obtain an ampere draw of approximately 180 amps for a 12-volt battery and 200 amps for a 6-volt battery.

c Maintain this load for approximately 15 to 20 seconds.

d If the battery is in good condition, the battery voltage should stay above 4.5 volts for a 6-volt battery and 9.6 volts for a 12-volt battery.

e If the battery voltage meets the recommended voltage after the discharge test, the battery is generally good and will perform satisfactorily. Charge the battery and put it back in service.

f If the battery fails the test on the basis of voltage, do not condemn the battery until it has been charged and rechecked.

NOTE The above load test information is general in nature since there are many different ways recommended by battery manufacturers. Test the battery according to the specific instructions supplied by the manufacturer.

The above procedure is a common test in many shops. The danger in this test is that the mechanic may make a hasty decision and replace a perfectly good battery. Always recharge the battery and make a second test to insure that you are getting an accurate test.

5. Procedure for Three-minute Charge Test

All the tests we have performed so far have been made on batteries that were in some stage of charge. The three-minute charge test is made on batteries that are discharged to determine if cells are sulfated to the point where they will not accept a charge and the battery must be replaced.

NOTE A sulfated battery means that the sulfate compound on the battery plates, which is normally returned to the electrolyte during charging, is not doing so and the battery will not accept a charge.

To make the three-minute charge test, proceed as follows:

a Connect the battery charger to the battery with respect to polarity.

b Set the charger for a 3-minute charge.

NOTE Charge a 12-volt battery at 40 amps and a 6-volt battery at 75 amps.

c After charging for three minutes, connect a voltmeter from positive to negative posts on the battery with the charger still operating.

d The battery is acceptable if voltage is less than 15.5 volts on a 12-volt battery or less than 7.75

NOTE Voltages given are with battery at 70°F (21°C).

volts on a 6-volt battery. The battery can be recharged and put back in service.

e The battery is not acceptable if voltage is more than 15.5 for a 12-volt battery and more than 7.75 for a 6-volt battery.

f Depending upon how much time is available, you may place the battery on a slow charge (1 amp) for a 24-hour period; in many cases the battery will respond to this slow charge and be acceptable after charging. If this time is not available, replace the battery.

The three-minute charge test, like all other battery tests, is not 100 percent failsafe, but after you have gained some experience in making these tests, you will be able to test a battery and make sound recommendations on its continued use or replacement.

D *Battery General Maintenance*

The diesel mechanic will be called upon to perform general maintenance on batteries. Some of these maintenance procedures are as follows:

1. Procedure for Battery Charging

a *Slow charge.* To properly slow charge a battery, charge battery at 1 amp for approximately 12 to 16 hours. Slow charging is recommended if you think the battery is sulfated.

b *Quick charge.* To properly quick charge a battery, charge a 12-volt battery at 40 amps and a 6-volt battery at 75 amps for approximately one hour. This will not completely charge the battery but it should be sufficiently charged so it can be put back in service. To completely charge the battery, the fast charge must be followed with a slow charge.

NOTE During fast charging do not charge the battery at a rate that will cause the battery cell temperature to rise beyond 125°F (52°C).

c *Charging more than one battery at a time.* A number of batteries of the same voltage can be charged at the same time by connecting them in parallel (Figure 27.20), which is positive to positive and negative to negative. Two batteries may be charged hooked in series if the charger has the capability. Two 6-volt batteries hooked in series can be charged the same way as one 12-volt, since two 6-volt batteries in series equal one 12-volt.

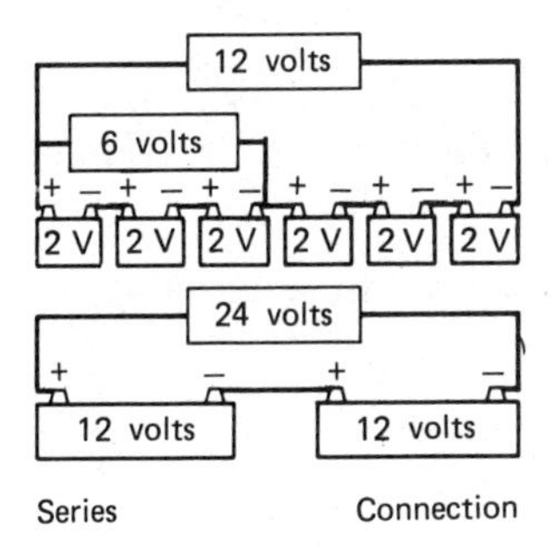

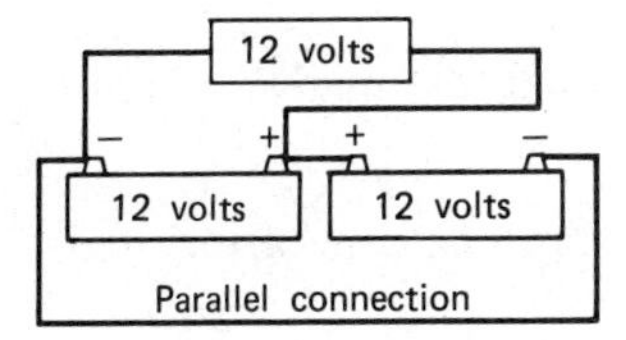

Figure 27.20 Batteries connected in parallel.

2. Procedure for Filling Dry Charged Batteries

Most batteries, with the exception of maintenance-free batteries (sealed), will need to be filled with acid before putting them in service. These batteries are dry charged at the factory and are shipped without electrolyte.

a Carefully fill each cell with electrolyte.

CAUTION Make sure protective goggles and gloves are worn to prevent injury when filling the battery with electrolyte.

b After filling, charge battery at a rate of approximately 30 to 40 amps until the electrolyte has a specific gravity reading of 1.240 or higher with an 80°F (27°C) temperature.

3. Procedure for Installation of Battery into Vehicle

Special care should be taken when installing a battery, if we expect it to provide trouble-free power for a long period of time.

a Check battery box for rocks, corrosion, and foreign objects. Also make sure the battery box or compartment is solid, since a loose battery compartment can ruin a battery in a short time.

b Check all battery cables to make sure they are free

from corrosion. Replace any bolts that show signs of deterioration.

c If you have two or more batteries in one vehicle, place the batteries in the compartment in a manner that will enable you to connect the cables. Install hold-down brackets or clamps.

d Install and tighten cables on the batteries. Coat cables with a special battery cable preservative or a spray paint.

After installing the battery or batteries, check the starter operation to make sure all cables are connected correctly.

4. Procedure for Battery Troubleshooting

It is not enough to be able to determine what is wrong with a battery and whether it should be replaced or not; an effort must be made to determine why the battery failed. Some of the common reasons for battery failure are:

a *Overcharging.* In many cases the voltage regulator in the charging system is not functioning correctly and the battery is continually being overcharged. The first symptom of this condition is the excessive use of water in the battery.

b *Undercharging.* The voltage regulator may be set to cause the battery to be in a low undercharged condition at all times. Undercharging the battery can cause the battery to become sulfated.

c *Battery too small for application.* A battery that does not have sufficient capacity for the vehicle load will quickly fail since the battery will be discharged in large amounts and may not have time to charge adequately before it is again called on to deliver large amounts of current, such as during engine starting.

d *Improper or lack of maintenance.* If the battery is not properly maintained as outlined in the battery maintenance section, the battery will age prematurely and fail much sooner than normal.

If you have any further questions concerning battery service and testing, consult your instructor or the information supplied with your battery tester.

SUMMARY

This chapter has explained the theory of electricity used in automotive electrical components and has discussed theory, testing, and maintenance of batteries. If you master the material in this chapter, you are well on the way toward becoming a proficient electrical mechanic.

REVIEW QUESTIONS

1. Name two good conductors of electricity and explain why they are good conductors.
2. Explain electricity as you understand it.
3. Explain Ohm's law and discuss how you would use it in troubleshooting an electrical system problem.
4. Describe a permanent magnet and explain how it might be used.
5. Describe an electromagnet and explain where it would be used in a diesel engine electrical system.
6. How can the right-hand rule be applied to a current-carrying conductor?
7. How can the right-hand rule help you determine the polarity of an electromagnet?
8. What three factors are required to generate electricity?
9. In what way do batteries produce electricity?
10. How many cells will a 12-volt battery have?
11. What two battery checks do you consider most important?
12. How is the battery light load test made? Explain in detail.
13. What are four reasons for premature battery failure?

14 How can a high charge rate affect battery life?

15 How would you make a decision regarding the condition of a battery that has been brought to you for testing? Explain in detail.

16 What can cause a battery explosion? Explain.

28

Starting Systems

Two common types of starting systems found on diesel engines are electric and air. As the name implies, the starting system provides the power to turn the engine to start it. Starting system repair and maintenance become a major part of the diesel mechanic's job since the engine must be started under all types of conditions, occasionally many times a day.

OBJECTIVES

Upon completion of this chapter the student will be able to:

1 Explain and draw a simple electric starting circuit.
2 Name and explain the function of five starter components.
3 Explain the basic electric starter principle.
4 Demonstrate the ability to overhaul and test a starter (electric).
5 Demonstrate the ability to test an electric starter circuit for excessive ground circuit resistance.
6 Demonstrate the ability to test an electric starter circuit for excessive insulated circuit resistance.
7 Explain how an air type starter circuit works.
8 Explain the function of two component parts in an air starter circuit.
9 Demonstrate the ability to turn the commutator on a starter armature.

GENERAL INFORMATION

Starting systems for diesel engines have been greatly improved in the last few years. Early starting systems had marginal cranking power. Because of the diesel engine's higher compression, 24 volts were needed to get enough cranking power. This voltage presented a problem since the accessory circuit of the vehicle along with the generator were 12 volts. Diesel engines required a series parallel switch, which would switch the circuit to 24 volts for starting and 12 volts for

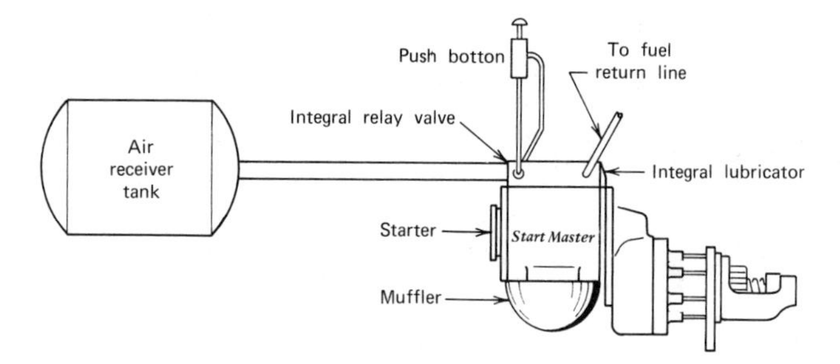

Figure 28.1 Air starting motor (courtesy Stanadyne).

charging and accessory operation. The development of high torque 12 volts starters eliminated the need for 24 volts and series parallel switches. Some early diesel engines used an additional small gas engine to crank the diesel engine. This engine proved to be effective with respect to turning power but presented many additional maintenance problems in addition to the high initial cost.

Another type of starting system used on modern diesel engines is a starting motor powered by air supplied from a reservoir. The air supply is replenished by the vehicle air compressor during operation (Figure 28.1). The advantage of this system is that it required no larger batteries or heavy electrical cables. Nevertheless the electrical starting system is the most widely used system on engines today.

I. Electric System Components

All electric starting systems are made up of the following components (Figure 28.2):

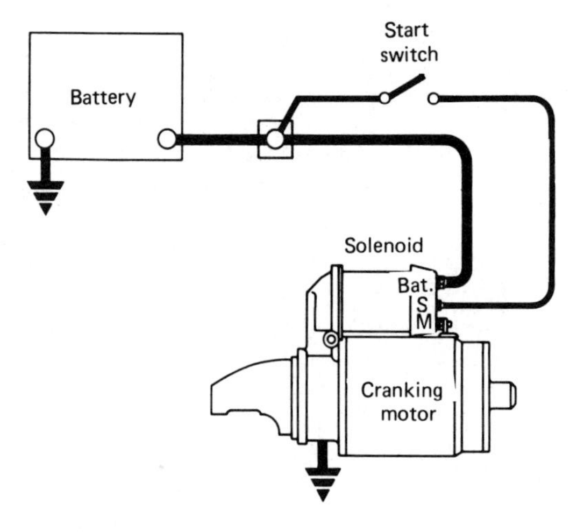

Figure 28.2 Typical electric starting system (courtesy Delco Remy).

A *Starting motor.* A dc electric motor that converts electrical energy into cranking power to rotate the engine for starting.

B *Solenoid switches.* An electrical magnetic switch that makes and breaks the circuit between the starter and battery. It also shifts the starter drive in and out of the flywheel ring gear.

C *Cables.* Large cables are required to transmit the huge amount of current needed by the starter motor to crank the engine.

D *Battery.* The battery provides the source of power to operate the starter motor. In many systems more than one battery is required, since one battery does not contain sufficient amperage to turn the starter.

II. Starter Component Parts and Operating Principle

A *Starter component parts* (Figure 28.3).

All electrical starters must have the following component parts:

1 *Field frame.* The field frame provides a place to mount the fields and also the front and rear bearing housing.

2 *Brush end bearing housing.* This housing provides a place for the commutator end bushing or bearing.

3 *Armature assembly.* This assembly is composed of many conductors (heavy copper ribbons) mounted between iron laminations on an iron shaft. On one end of the armature is the commutator and on the other end is the starter drive.

4 *Starter drive.* It is mounted onto the armature shaft and transmits the power of the starting motor to the flywheel. On all drives is a pinion that engages the flywheel ring gear when the starter motor is operating. To allow the starter motor to turn faster than the engine it is cranking, a gear reduction of approximately 15 to 1 is utilized. The pinion mechanism must be designed to disengage from the flywheel or overrun after the engine starts; otherwise, the starting motor would be rotated by the engine at too fast a speed and cause damage to the starter. Many different types of starter drives are used today; some of the most common ones are listed here:

a *Inertia type (Bendix)* (Figure 28.4). The

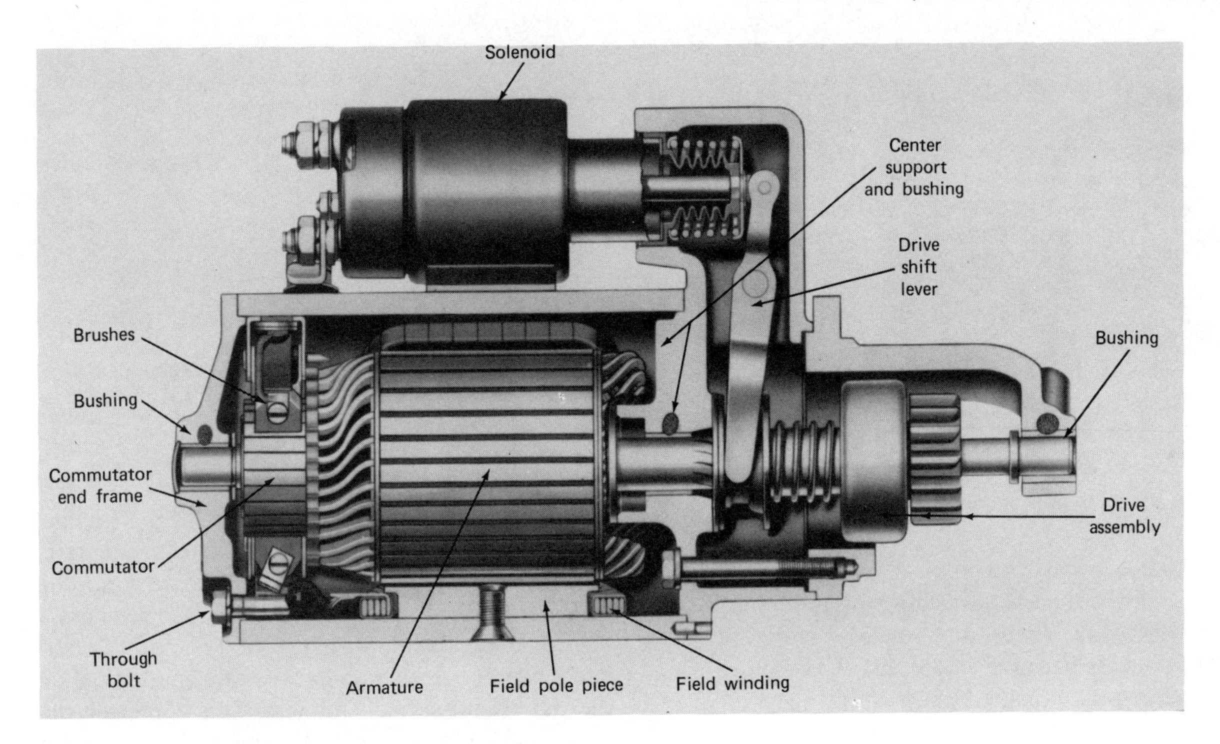

Figure 28.3 Component parts of an electric starter (courtesy Deere and Company).

inertia starter drive is designed so that the pinion gear has a counterweight on one side. This unbalance causes the gear to move along the shaft of the armature, which has screw threads cut on it. The pinion gear is forced into the flywheel by the threads.

b *Dyer drive* (Figure 28.5). The dyer drive is a standard type drive used on diesel engines for many years. The dyer drive pinion is meshed into the flywheel by a shift lever that connects to the solenoid. This shift is completed before the motor circuit is closed to avoid grinding of the pinion gear and flywheel during engagement.

c *Sprag clutch drive* (Figure 28.6). The sprag clutch is an overrunning clutch that locks the pinion to the armature shaft in one direction and allows it to rotate freely in the other direction. It is composed of an inner and outer shell that are locked together by sprags. It is engaged by the starter solenoid through a shift lever.

5 *Brushes*. They are made from a carbon and graphite mixture, are square or oblong in

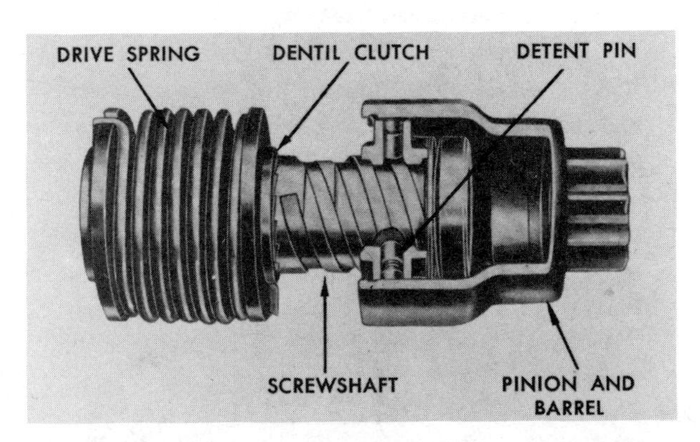

Figure 28.4 Inertia type starter drive (courtesy Delco Remy).

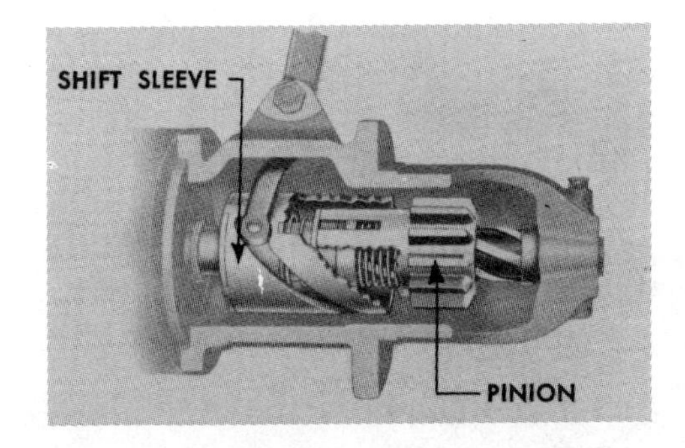

Figure 28.5 Dyer drive (courtesy Delco Remy).

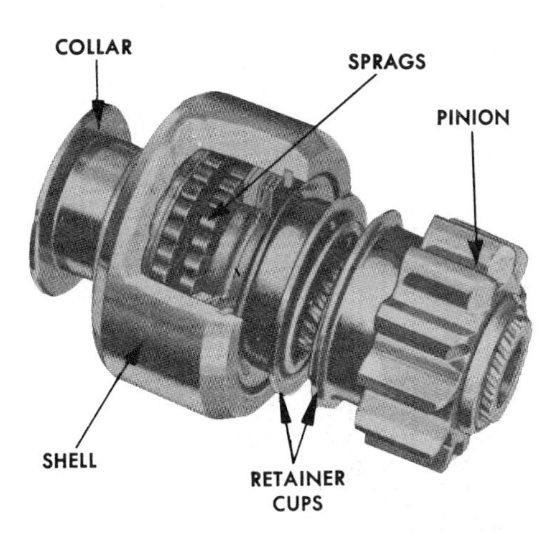

Figure 28.6 Sprag clutch drive (courtesy Delco Remy).

shape, and connect the starter commutator segments to the generator terminals. They are called brushes because they brush the commutator segments to make contact.

6 *Drive end housing.* The starter housing that provides a means of mounting the starter onto the engine.

7 *Bearings and bushings.* The starter armature is supported in the field frame by bushings or bearings.

B *Starter operating principle.*

To understand the electrical principle on which a cranking motor operates, consider a straight wire conductor located in the magnetic field of a horseshoe-shaped magnet with current flowing through the wire as shown by the red arrow in Figure 28.7. With this arrangement there will be two separate magnetic fields: one produced by the horseshoe magnet and one produced by the current flow through the conductor.

Since magnetic lines leave the N pole and enter the S pole, the direction of the magnetic lines between the two poles of the horseshoe magnet

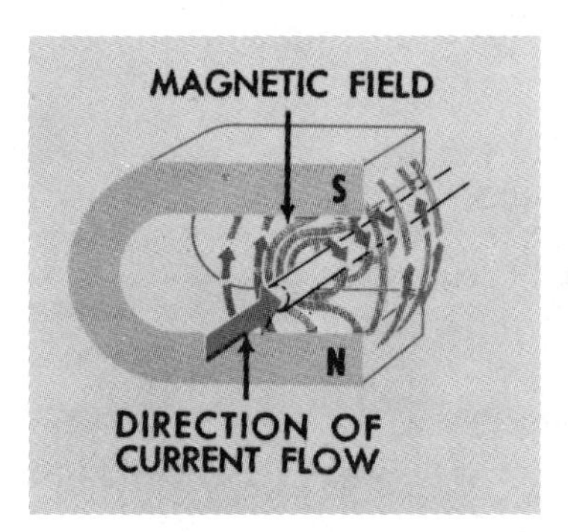

Figure 28.7 Magnetic field and conductor (courtesy Delco Remy).

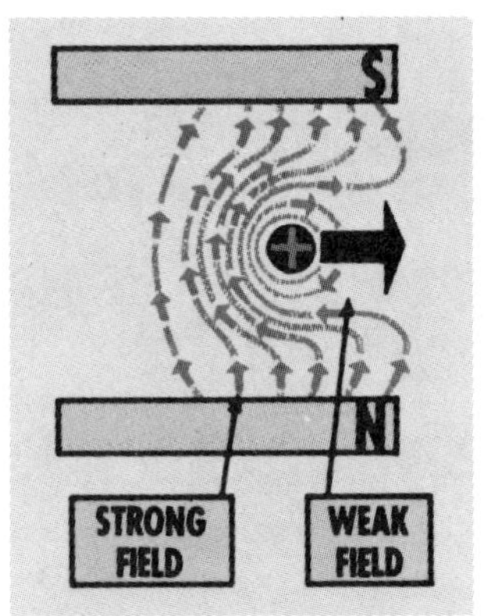

Figure 28.8 Magnetic field between two magnets (courtesy Delco Remy).

will be upward as shown in Figure 28.8. The current-carrying conductor will produce a magnetic field consisting of concentric circles around the wire in the direction illustrated. The net result is a heavy concentration of magnetic lines on the left-hand side of the wire and a weak magnetic field on the right-hand side of the wire. This condition occurs on the left side where the magnetic lines are in the same direction and add together; it occurs on the right-hand side where the magnetic lines are in the opposite direction and tend to cancel each other out.

With a strong field on one side of the conductor and a weak field on the other side, the conductor will tend to move from the strong to the weak field, or from left to right, as shown in Figure 28.8. The stronger the magnetic field produced by the horseshoe magnet and the higher the current flow in the conductor, the greater will be the force tending to move the conductor from left to right. The resultant force illustrates the electrical principle on which a cranking motor operates.

A basic motor is shown in Figure 28.9. A loop of wire is located between two iron pieces and is

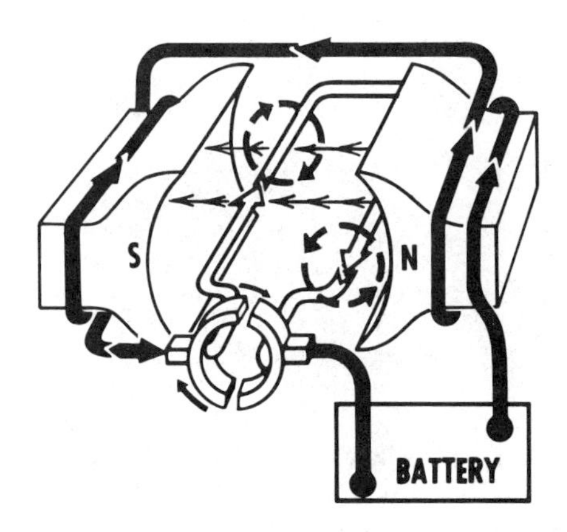

Figure 28.9 A simple starter motor (courtesy Delco Remy).

Figure 28.10 Force exerted on starter conductors by force field (courtesy Delco Remy).

connected to two separate commutator segments or bars. Riding on the commutator bars are two brushes that are connected to the battery and to the windings located over the pole pieces.

With this arrangement, current flow can be traced from the battery through the pole piece windings to a brush and commutator bar, through the loop of wire to the other commutator bar and brush, and then back to the battery. The resulting magnetic fields impart a turning or rotational force on the loop of wire as illustrated in Figure 28.10.

When the wire loop has turned one-half turn, the commutator bars will have interchanged positions with the two brushes, so that the current through the wire loop will move in the opposite direction. But since the wire loop has exchanged positions with the pole pieces, the rotational effect will still be in the same clockwise direction as previously shown in Figure 28.10.

III. Solenoid Switch Component Parts and Operating Principles

A *Component parts.*

A starter solenoid (Figure 28.11) is made up of the following component parts:

1. *Terminal bolts.* Bolts to which the battery cable and motor terminal are connected.

2. *Contact plate.* The plate that makes the contact between the terminal bolts.

3. *Pull-in coil.* A coil within the solenoid that helps engage the solenoid shift lever.

4. *Hold-in coil.* A coil within the solenoid that holds the solenoid in the engaged position.

5. *Plunger.* The iron core of the solenoid, which is connected to the starter shift lever.

B *Operation.*

The solenoid switch is used to engage the starter pinion and close the circuit between the starter and the battery. When the starter switch on the vehicle instrument panel is closed, the solenoid operates as follows:

1. The hold-in and pull-in coils work together to pull the solenoid plunger into the solenoid.

2. As the plunger is pulled into the solenoid housing, the contact plate shorts the pull-in coil and the hold-in coil holds the switch engaged (Figure 28.12).

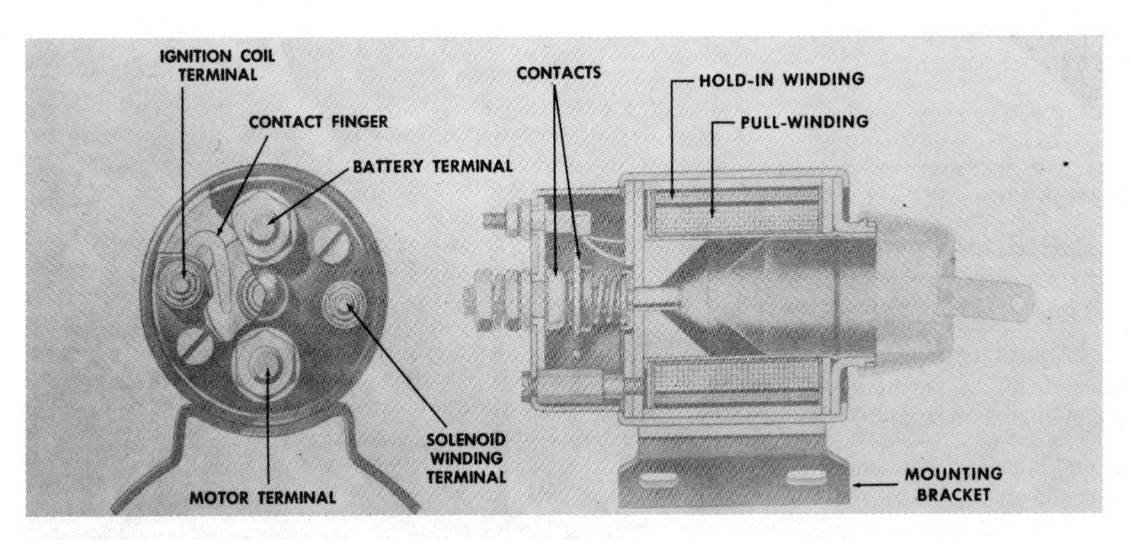

Figure 28.11 A typical starter solenoid (courtesy Delco Remy).

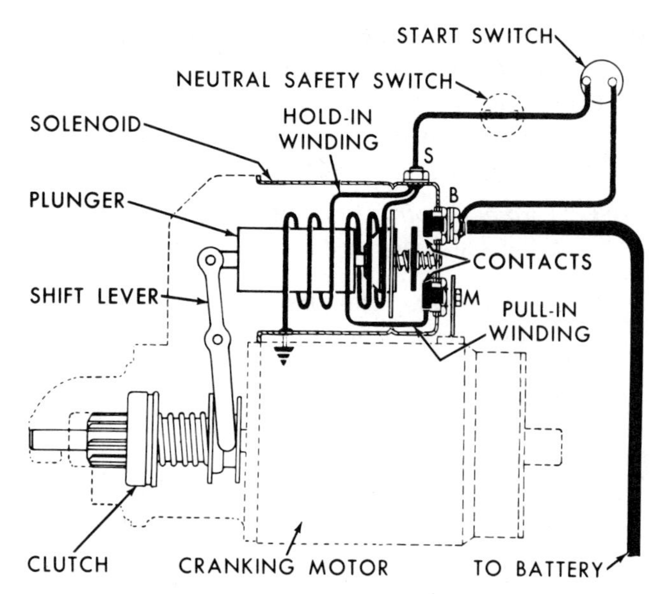

Figure 28.12 Operation of hold-in and pull-in coil (courtesy Delco Remy).

3 In addition to closing the circuit between the battery and starter with the contact plate, the solenoid operates the drive shift lever, which moves the drive into the flywheel.

4 The switch remains in this position until the starter switch on the instrument panel is released, causing the solenoid to disengage the shift lever and break the contact between the battery and starter. When this happens, the starter stops turning and the pinion is disengaged from the flywheel.

IV. Disassembly, inspection, and Overhaul of Electric System Components

A. Procedure for Starter Disassembly

It is assumed that the starter has been removed from the engine. If not, disconnect the battery ground cable and remove the starter from the engine. Disassemble the starter in the following manner:

1 Match mark the end housing to the end frame with a center punch or chisel (Figure 28.13).

2 Remove the bolts that hold the commutator end frame and the field frame together.

NOTE Depending on which type starter you have, the bolts may be threaded into the main field frame or may be through bolts that reach through the field frame housing and thread into the shift lever housing.

3 At this point the front bearing end frame can be removed from some starters without further disassembly, while other starters must have the brushes disconnected before the end frame can be removed.

4 Remove the bolts that hold the lever and nose housing to the main field frame.

5 Withdraw the nose and lever housing along with the armature from the field frame assembly.

NOTE The solenoid plunger is still connected to the shift lever and will slide out of the solenoid when the lever housing is removed.

6 To remove the armature from the lever housing and drive assembly of a starter that does not use through bolts, simply lift the armature from the lever housing and drive.

7 On starters that use through bolts to attach the lever housing to the main frame, the bolts that hold the nose housing to the lever housing must be removed. This allows the nose housing to be removed, providing access to a snap ring and collar that must be removed before the armature can be withdrawn from the lever housing.

8 Remove the solenoid housing from the starter field frame.

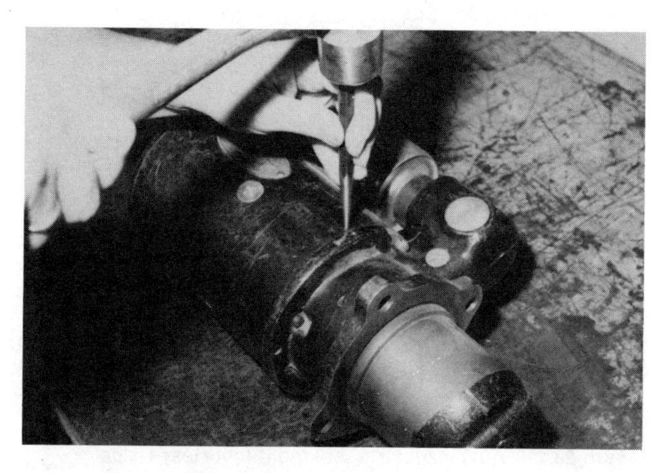

Figure 28.13 Match marking starter housings.

B. Procedure for Starter Component Testing and Inspection

1 Visually inspect the starter armature commutator for roughness, burned spots, and ridging. If these problems are noticed, the commutator should be turned in a commutator lathe.

NOTE Do not turn the commutator until you have made the electrical checks on the armature in case it is bad electrically and will have to be replaced.

2 Place the armature in an armature growler and check for:

a *Shorts.* Perform this check by turning the growler switch on and placing the hacksaw blade crossways on top of the armature center (Figure 28.14). The hacksaw blade should not vibrate. If it does, the armature is shorted and must be replaced.

NOTE Before condemning an armature that tests shorted, check the commutator carefully for pieces of copper between the segments. A piece of copper wedged between the segments will short the armature. Clean the grooves between the segments and recheck the armature.

b *Grounds.* The armature can be checked for grounds by using the growler test light. Touch one probe of the test light to a commutator bar and the other probe to the armature shaft (Figure 28.15). The test light should not light. If it does, the armature must be replaced.

c *Opens.* The armature can be checked for opens by using the meter on the growler. Place the armature on the growler and turn on the power switch. Place the meter probe on two adjacent commutator bars.

NOTE On some armatures it may be necessary to place the test probe on one bar, then skip one bar and place the other probe on the next bar to get a reading.

Turn the armature to obtain the highest reading. Generally the probes will have to be placed on the armature below the center line to obtain a reading.

In addition to the growler check, open circuits may be detected by visually inspecting the commutator bars. The bar that has an open circuit is usually burned and easily spotted.

3 Testing field coils (fields not removed from the starter).

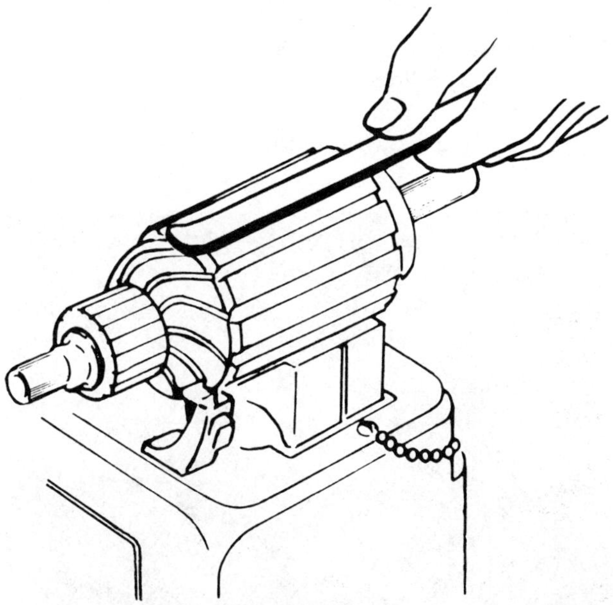

Figure 28.14 Checking armature for shorts using a hacksaw blade and growler (courtesy Delco Remy).

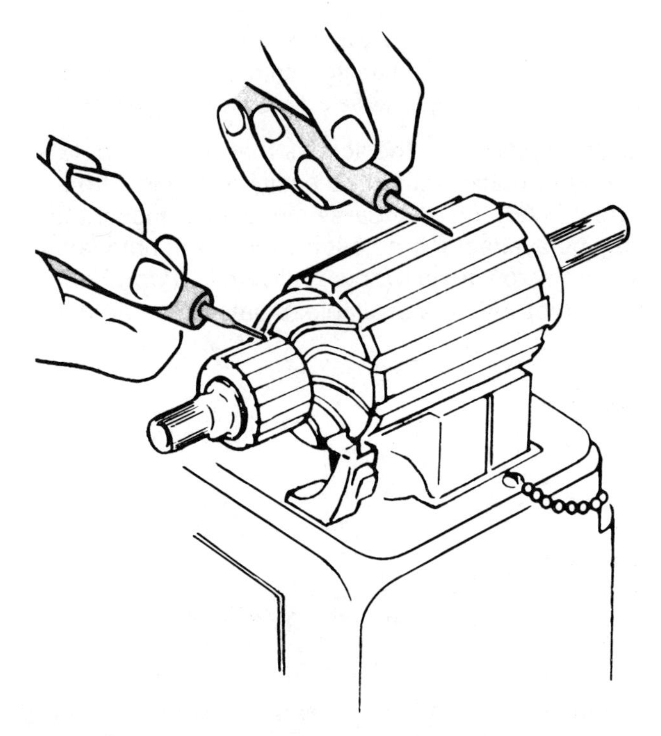

Figure 28.15 Testing armature for grounds using a test light (courtesy Delco Remy).

NOTE Before testing the field coils, make sure that they are disconnected from ground or the starter solenoid, since such a connection will cause an incorrect reading during the test.

The field coils should be tested for:

a *Open circuits.* Using a test light or ohmmeter, test the field circuit for opens by placing one probe on one end of the field brush connection and the other probe on the starter terminal connection (Figure 28.16). The light should light or the ohmmeter should show continuity if the fields are good. If not, the fields are open.

NOTE To insure complete testing of the field circuit, touch all field brush connections with the one test probe while holding the other test probe on the starter connector terminal.

b *Field ground circuit test.* Test the field circuit for grounds with a test light or ohmmeter. Touch one probe of the test light to a field connection and the other probe to the ground. The test light or ohmmeter should not show any ground. If it does, the field circuit is grounded to the starter case and must be repaired or replaced. If the test light does not light, the field is not grounded and can be considered good. If the fields are grounded and require replacement, follow this procedure:

(1) With a large screwdriver or square drive handle, remove the screws that hold the field pole pieces in place.

NOTE Sometimes the screws that hold the pole pieces in place will not be easily removed with a screwdriver. In some cases a special removal tool may be available. If no such tool is available, an impact screwdriver works really well. If none of these methods are available, a small pin punch ($\frac{1}{4}$ in. or 6 to 7 mm) and hammer can be used to jar them loose (Figure 28.17).

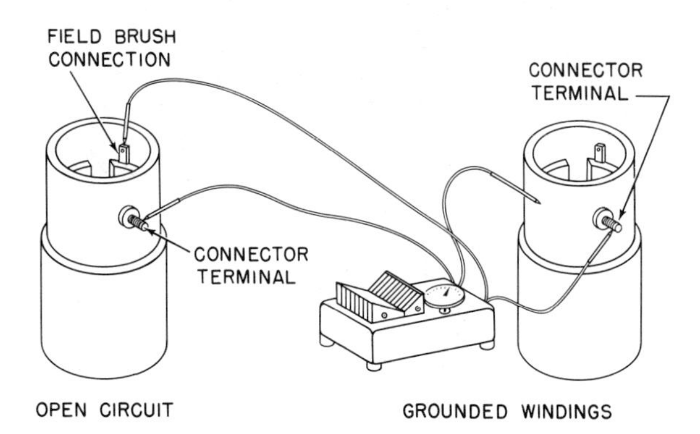

Figure 28.16 Testing field coils for opens using test light or ohmmeter (courtesy Deere and Company).

Figure 28.17 Using small punch to loosen field pole screws.

(2) After the screws have been loosened, remove them with a screwdriver.

NOTE Before removing the fields and pole pieces, match mark them to the housing with a marking pencil for reference during reassembly.

(3) Remove the fields and field pole pieces.

(4) Install the field pole pieces into the new fields if new fields are to be installed and install the assembly into the starter main frame with reference to match marks. Tighten the pole screws securely.

4 The insulated brush holder must be tested for grounds as follows:

Using a test light or ohmmeter, touch one probe of the test light to the insulated brush holder

Figure 28.18 Testing insulated brush holder with test light.

while touching the other probe to the brush end frame (Figure 28.18). The light or ohmmeter should not show continuity. If it does, the brush holder is grounded and must be repaired.

5 All bearings and bushings should be checked closely for wear and roughness. If any questions or doubt remains about a bearing or bushing, replace it.

NOTE Special grease has been developed for use in starter bushings or bearings and should be used instead of wheel or chassis lubricant. This grease is available from starter parts suppliers.

6 Insulated terminal testing.

The starter insulated terminal should be checked for grounds with a test light or ohmmeter. If the terminal is grounded, remove it and check the insulation; replace if visually damaged or cracked. Replace bolt and check for grounds.

7 Solenoid disassembly and inspection.

The starter solenoid on a diesel engine starter performs two important jobs: (1) engaging the starter drive pinion and (2) closing the circuit between the starting motor and battery. To disassemble and check the starter solenoid, proceed as follows:

a Remove the screws that hold the solenoid cover onto the solenoid.

b Lift the cover from the solenoid.

c Inspect the contact plate and contact bolts.

d Replace the contact bolts and contact plate if they show wear, pitting, or burning.

e Replace the solenoid cover and install and tighten the screws.

C. Procedure for Reconditioning or Turning of the Starter Commutator

Visually inspect the starter commutator for roughness or scoring. It should provide a clean, smooth surface for the brushes to ride on. If the commutator is rough or scored, it must be reconditioned on an armature lathe. Recondition the commutator as follows:

1 Mount the armature in an armature lathe as shown in Figure 28.19.

2 Adjust the armature end stops so that the armature does not move endways in the lathe while it is being turned.

Figure 28.19 Mounting armature in armature lathe.

3 After the armature has been positioned so that it will run correctly, adjust the cutting tool so that it just contacts the commutator surface on the right-hand side (Figure 28.20).

CAUTION Make all preliminary adjustments with the lathe motor stopped; if this is not done, serious damage to the armature may result.

4 After adjustment, move the commutator cutting tool sideways (to the right, away from the commutator) with the advance handle and start the lathe motor.

5 With the advance handle, slowly move the cutting tool sideways toward the commutator until it starts to cut the commutator.

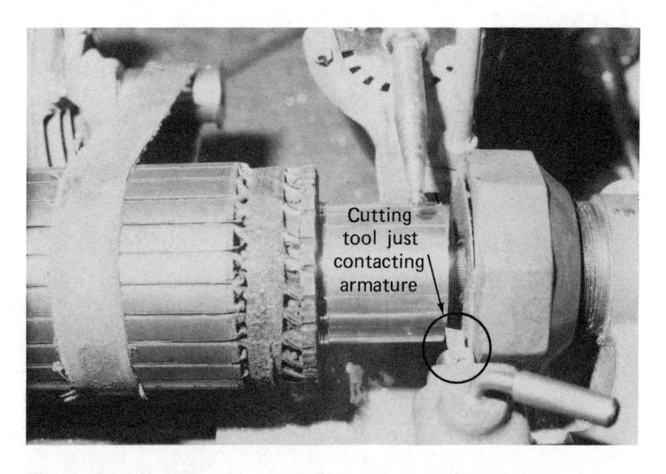

Figure 28.20 Adjusting cutting tool on armature lathe.

CAUTION Make sure you wear safety glasses during this operation to prevent getting copper chips in your eyes.

6 Continue moving the cutting tool sideways across the commutator with a slow, even speed.

CAUTION Once you have started the cut, do not stop the sideways movement of the tool. To do so will cause ridging and grooving of the commutator surface.

7 After completing one cut all the way across the commutator from right to left, back out the cutting tool feed screw slightly and move the cutting tool sideways (left to right) with the advance handle back to the starting position.

8 Continue cutting the commutator as outlined until all worn or ridged spots have been removed.

9 After the commutator has been turned until all ridging or scoring has been removed, check for mica between the commutator bars.

NOTE In most cases starter armatures are not undercut since they do not run continuously or at high speeds like the generator. High mica in a starter armature can be removed by using a strip of sandpaper held against the commutator while the commutator is rotating.

10 Recheck the commutator after reconditioning since any small piece of brass wedged between two commutator segments can cause a short circuit in the armature.

D. Procedure for Starter Reassembly and Testing

After all starter parts have been inspected and repaired or replaced, the starter can be reassembled.

1 Place the armature in a vise and install the lever housing and starter drive on the armature.

2 Install the nose housing on the lever housing.

3 Reclamp the armature and lever housing in the vise in the upright position.

4 Place the field frame on the lever housing, guiding the solenoid plunger into the solenoid.

5 Align the match marks and install the bolts in the field frame and lever housing.

6 Install the brushes in the brush holders.

7 Install the commutator end housing with brushes on the field frame and armature. Align the match marks and install the bolts and tighten.

NOTE The brushes must be wedged in the brush holder so that the brushes can be slipped onto the commutator.

8 Connect the field connections to the brush holders and install the access plug or cover plate.

9 Test the starter operation with a battery or battery charger.

10 With all leads disconnected from the solenoid, connect a jumper wire from the solenoid motor terminal to ground and connect the battery leads to the solenoid switch terminal and to ground.

11 Check the pinion adjustment with a feeler gauge as shown in Figure 28.21.

12 Run a starter no-load test on the starter before installing it on the engine as follows:

a Connect a fully charged battery of the correct voltage and an ammeter in series to the starter solenoid battery terminal and starter frame (Figure 28.22).

NOTE Most starters will draw a starting amperage of 100 to 150 amps; make sure the ammeter being used is of sufficient capacity.

b Connect a voltmeter from the starter motor terminal and starter frame.

c Connect a remote starter switch to the battery terminal of the solenoid and to the S terminal of the solenoid.

d Close the remote starter switch and with the motor running check the rpm's with a handheld tachometer on the end of the armature.

Figure 28.21 Checking pinion adjustment.

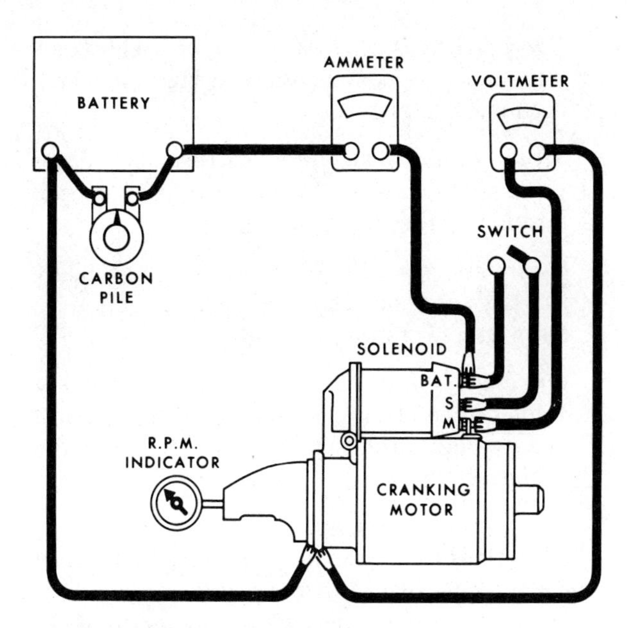

Figure 28.22 Running starter no load test.

e Continue running the motor and observe the ammeter (if used) and voltmeter.

f Compare voltage, amperage, and rpm's to starter specifications.

g If the free speed is low, with high current draw, the following problems may be present:

(1) The armature is dragging on the field pole shoes; this condition is caused by worn bearings or a bent armature.

(2) The bearings are fitted too tightly.

(3) The fields or armature are grounded.

(4) The armature is shorted.

h High current draw, no starter operation may be caused by:

(1) Field or terminal ground.

(2) A stuck or bound armature.

i No current draw, no operation may occur because of:

(1) No brush contact, caused by broken springs, or an oily or dirty commutator.

(2) An open armature; starter must be disassembled to check.

(3) An open field circuit; starter must be disassembled to check.

13 Install starter onto the engine and connect all cables.

V. Electric Starting System Maintenance

The starting system must be properly maintained if it is to deliver many hours of trouble-free operation. Maintenance of the system should include:

1 Checking the battery water level and state of charge.

2 Checking all battery connections for corrosion and tightness. Battery cables should be cleaned often and coated with a corrosion resistor.

3 Checking all connections to the starter terminal for tightness and corrosion.

4 Lubricating the starter bushings and/or bearings at frequent intervals.

5 Checking the starter mounting bolts for tightness.

VI. Air System Components

Air starting systems have been used for a long time but only recently has their dependability level been perfected to where they meet or exceed that of electric starting systems. Air systems have many advantages. They are virtually unaffected by cold weather, and the motor part of the starter is practically trouble-free and will operate for many years without service. One of the major problems with air starting systems in the past has been air leaks, but new types of sealers have been developed that now make them virtually leak-proof. The following components are common to all air starting systems (Figure 28.23):

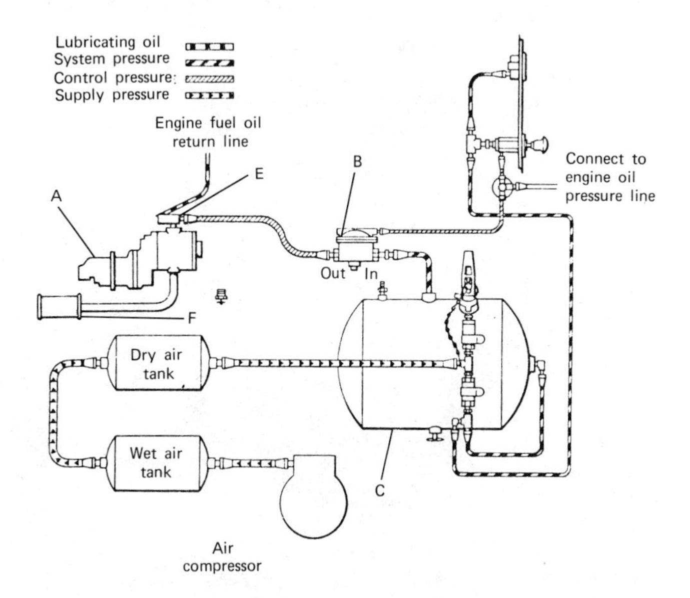

Figure 28.23 Air system components (courtesy Stanadyne).

A *Starter motor.* An air motor of the vane type that converts the air pressure and volume in the air tank to mechanical energy.

B *Relay valve.* A valve that is operated by the starter push button on the dash. This valve supplies full air flow to the starting motor.

C *Air receiver.* A large tank that holds the air supply to operate the starter motor.

D *Starter drive* (not shown). The gear or pinion that connects and disconnects the starter to the flywheel for starting. The starter drive is somewhat similar to the sprag type overrunning clutch in an electric starter.

E *Lubricator.* A device fitted to the air motor that supplies lubrication to the air motor each time the motor is run.

F *Muffler.* Connected to the air motor outlet, the muffler decreases the exhaust noise of the air starter, which can be quite loud without a muffler.

VII. Starter Component Parts and Operating Principles

A *Starter component parts* (Figure 28.24).

Almost all air starting motors have the following component parts:

1. *Motor housing.* This housing provides a place to mount the end bearing housings.
2. *Bearing housings (front and rear).* The bearings housings contain the bearings on which the motor rotor is mounted.
3. *Motor rotor.* The rotor carries the motor vanes, which trap the air and provide the torque to turn the engine.
4. *Gear reduction.* Some air starters, especially for large engines, have a gear reduction to allow greater torque buildup.
5. *Drive assembly.* The drive assembly is very similar to an electric starter drive. It is used to engage and disengage the starter from the engine flywheel.

B *Starter operating principles.*

The air starter system operation is really very simple and much easier to understand than an electric starter system. Air system operation is as follows:

1. Air supplied by the vehicle air compressor is stored in a reservoir (tank).
2. This air is supplied to the air starting motor through piping when the relay valve is opened.
3. Air flows to the vanes in the air motor, forcing the motor rotor to turn.
4. After turning the rotor, the air is exhausted through the muffler.

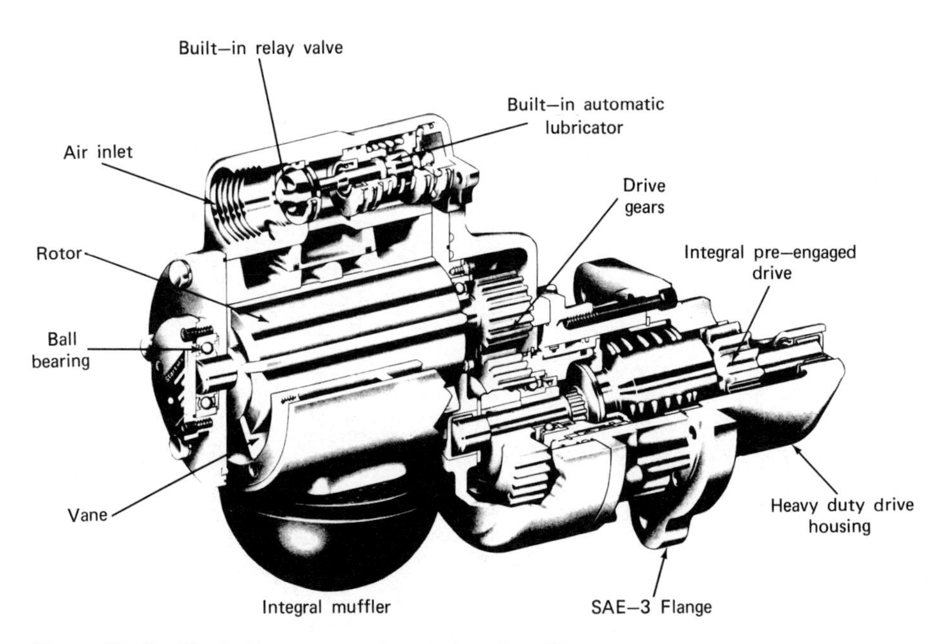

Figure 28.24 Air starter component parts (courtesy Stanadyne).

5 As the motor starts to turn, the drive is moved toward the engine flywheel for engagement.

NOTE Some air starters use the inertia engagement principle while others use a pre-engagement type. The inertia type starter uses a drive much like the electric starter Bendix drive that relies on drive pinion inertia for engagement. The pre-engagement type starter is similar to an electric starter that uses a shift lever to engage the pinion. The pinion is engaged into the flywheel with air pressure *before* the starting motor starts to turn. This system is designed to prevent the starter pinion from damaging the flywheel ring gear during engagement.

VIII. Air Starter System Maintenance

To have the air starter motor operate at its best efficiency, the following maintenance procedures should be followed:

A *Eliminate any leaks that occur in the system.*
A system that reduces air pressure by leaking overnight is of very little use to the operator. Check all fittings closely for leaks. When applying sealer to the fittings within the system, apply a paste type sealer to all fittings. Remember that one of the major problems of an air system is leaks.

B *Lubricate the system.*
Because of the high speed of the air motor, some form of lubrication is required. Most air starters employ a device that supplies a lubricant (generally diesel fuel) to the air that is applied to the vane motor of the starter.

C *Keep moisture from entering the system.*
The air system must be kept free from moisture to prevent the air starter from freezing up. Many modern air systems have automatic moisture ejection valves on the air storage tanks to insure that the air supplied to the system is free of moisture.

D *Make sure the engine is in good condition and all starting aids are working.*
It is almost a necessity to have some type of starting aid available for diesel engines that are used in cold climates. Types of starting aids commonly used are:

1 *Block water heater,* powered by electricity, gasoline, or propane. This is by far the most acceptable type of cold weather starting aid.

2 *Intake manifold air heaters.* These heaters heat the air that is supplied to the engine cylinder.

3 *Starting fluid.* This is the least acceptable from the standpoint of engine longevity, since its use puts a severe strain on the engine parts such as piston rings and head gaskets. Starting fluid should be used very sparingly to prevent engine damage. Starting fluid can be supplied to the engine through the intake air system in one of two ways: spraying directly from a hand-held can or from a measured shot starting device that will allow only a certain amount of starting fluid into the intake manifold when it is activated.

CAUTION Starting fluid should not be used in conjunction with electric starting aids such as glow plugs or intake air heaters since an explosion can occur when the starting fluid comes in contact with the heating element. Check your engine operator's manual before using starting fluid.

IX. System Testing, Troubleshooting, and Repair

As a diesel mechanic you will be asked many times to troubleshoot and repair a starting system that does not operate or does not operate correctly. Both electric and air systems will be covered here.

A. Procedure for Electric System Troubleshooting

1 The starter does not operate when the switch is closed.

a Check the battery for charge with a voltmeter or battery load tester. Recharge or replace the battery.

NOTE If the vehicle is equipped with lights, a quick check can be utilized. Turn on the vehicle lights; if lights die out when the starter switch is closed, check the battery and cable connections. If lights remain bright when starter switch is closed, check the solenoid plate and starting motor.

b Check all battery cable connections for corrosion and tightness. Clean and retighten if needed.

c After all connections and battery conditions have been checked, check the starter switch.

NOTE The starter switch operates the starter solenoid. It may be a push button type switch or incorporated into a key switch. To test it for operation, supply power (from the battery) with a jumper wire to the switch terminal of the solenoid. If the solenoid and starter operate at this time, the starter switch or wiring from the starter switch to the solenoid must be bad. Check and replace the wiring and/or switch.

- **d** If the solenoid operates and the starting motor does not operate, the contact plate in the solenoid could be bad or the starter motor is inoperable.
- **e** Check the solenoid contact plate by removing the solenoid cover, if removable. If not removable, replace the solenoid.
- **f** If the solenoid checks out all right, the starting motor must be bad and should be removed for repair. Repair the starting motor as outlined in Section III.

2 The starter operates, but the drive does not turn engine.

- **a** The drive clutch is slipping and does not turn the drive gear.
- **b** The drive not engaging into the flywheel is usually caused by an inoperative solenoid.

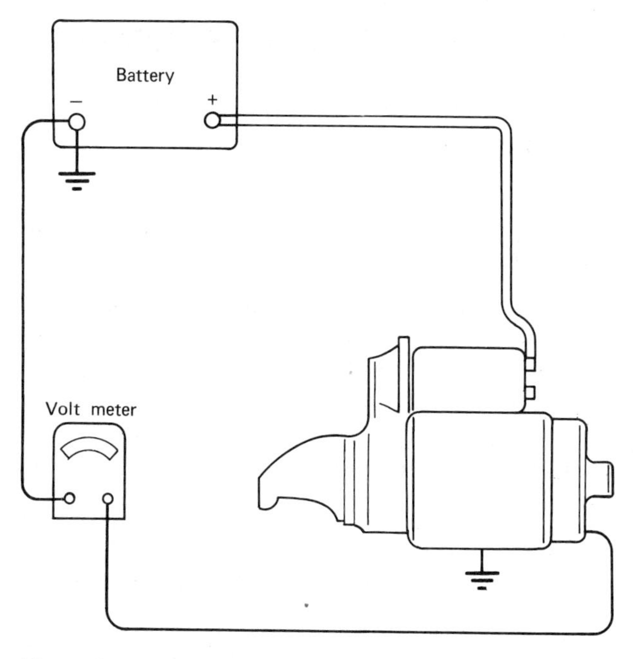

Figure 28.25 Voltmeter connection for ground circuit resistance test.

- **c** Replace the drive clutch or adjust the shift levers.

3 The starting system "clicks" when the switch is operated, but the starter does not operate.

- **a** Check all battery connections and cables.
- **b** Check the solenoid contact plate as outlined above in step A, 1, e.

4 The starter operates and the drive engages, but the engine turns slowly.

- **a** Check the battery's state of charge and condition.
- **b** Check the battery cables.
- **c** Check the ground and insulated resistance as outlined in Section B.

B. Procedure for Electric System Resistance Testing

One of the least understood and rarely used tests in starter circuit troubleshooting is the circuit resistance test. This test can be used on either the ground or insulated side of the starter circuit. It has been my experience that many starter problems can be traced to excessive ground circuit resistance. Make starter circuit resistance tests as outlined in the following information:

1 *Ground circuit resistance check.* This test is made while the starter, installed on the vehicle, is cranking the vehicle. The test should be made in this sequence:

- **a** Connect one lead of a voltmeter to the ground terminal of the battery and the other lead to the starter brush end frame (Figure 28.25).

NOTE The voltmeter will have two leads on it, one positive and one negative. When you operate the starter in this test, if the voltmeter needle goes in the wrong direction, reverse the leads.

- **b** Set the voltmeter on the 4 volts scale so that you can read volts accurately down to 0.1 volt.
- **c** Operate the starting motor. The voltmeter reading should not exceed 0.2 volts.
- **d** A voltage reading greater than 0.2 indicates a resistance in the ground circuit. Excessive resistance can be found in the following places:

(1) Ground cable at the battery.

(2) Ground cable at the engine connection.

(3) Where the starter bolts onto the engine.

(4) Where the starter main frame contacts the end frames.

(5) In any connection between the battery ground post and the starter brush end frame.

Excessive resistance at any of these points can be caused by dirt, acid corrosion, loose connections, or any other foreign material that prevents good metal-to-metal contact.

2 *Insulated circuit resistance check.* This test, like the ground circuit resistance test, is made with the starter installed on the vehicle. A voltmeter is required. The test should be made in the following sequence:

a Connect one lead of a voltmeter to the positive post of the battery.

b Set the voltmeter on the 4 volts scale or use a voltmeter on which voltage can be read in tenths.

c Operate the starter by pushing the starter switch.

NOTE Make sure that the engine does not start by putting the fuel shutoff in the shutoff position.

d While the engine is being cranked, touch the other voltmeter lead to the starter input terminal (Figure 28.26).

NOTE Make this connection *only* while the starter is running or the voltage reading will be 12 volts. Failure to do so could damage your voltmeter.

e The voltmeter reading should be 0.4 to 0.6 volts while the starter is cranking the engine. A higher voltage reading indicates there is a high resistance somewhere in the insulated side of the circuit. This high resistance could be in any of the following places:

(1) Cable connection onto the battery.

(2) Cable connection onto the solenoid.

(3) Starter solenoid.

(4) Battery cable with high resistance.

Clean and check all points of resistance listed and then recheck the insulated circuit resistance.

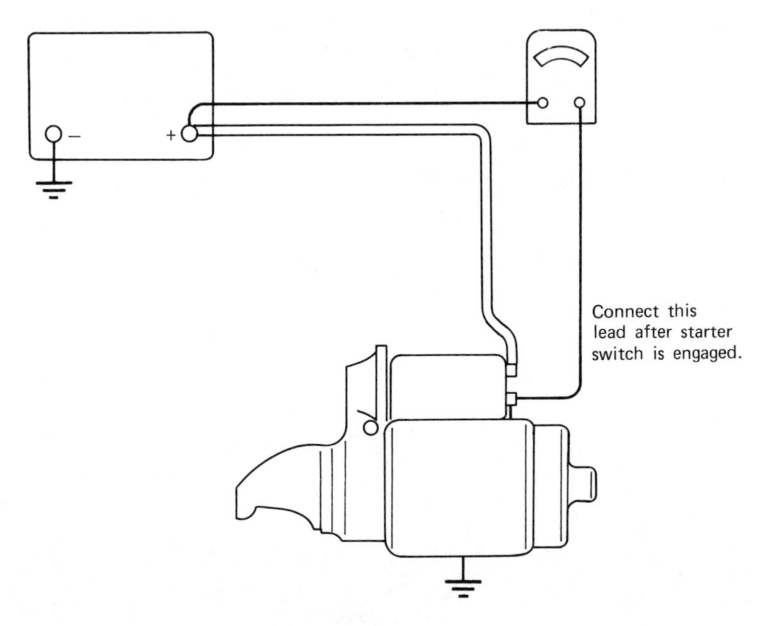

Figure 28.26 Voltmeter connection for insulated circuit resistance check.

C. Procedure for Air System Troubleshooting

1 If the cranking motor does not operate when the control is pushed, one of the following conditions may be the cause:

  a The motor blades are frozen from moisture in the air. Remove the starter from engine and clean the air motor.

  b No air pressure. Check the air supply in the pressure tank and pump the system up to operating pressure.

  c Relay valve is inoperable. Replace the relay valve.

2 The cranking motor may crank the engine very slowly because:

  a The air pressure is low. Pump up the air system.

  b The air motor is dry. Lubricate the motor with diesel fuel or other air motor lubricant.

  c The muffler is restricted or plugged. Remove and clean the muffler or replace it.

  d The air motor is worn. Remove and overhaul the starter.

  e There is incorrect oil in the engine for operating conditions. Change the engine oil.

SUMMARY

As stated earlier in this chapter, the electric and air starting systems are widely used on diesel engines. Make an effort to understand how all the systems work and how they should be tested and repaired. A diesel mechanic will be expected to repair effectively a starting system that does not operate correctly. Try to become familiar with all new types of starting systems as they become available.

REVIEW QUESTIONS

1 What two types of starting systems are commonly found on modern diesel engines?

2 What are two disadvantages of an electric starting system?

3 What are two advantages of an electric starting system?

4 What are two advantages of an air starting system?

5 List and explain the function of three electric starting system components.

6 What are two common types of starter drives used in electric starters?

7 List two checks that should be made on an electric starter during overhaul.

8 List two possible causes of an electric starter not turning the engine.

9 List two possible causes of an air starter not turning the engine.

10 What is one of the major problems associated with an air starting system?

29

Charging Circuits

Figure 29.1 shows a typical charging circuit used on a diesel engine. The primary function of this charging circuit is to provide electricity (current) for recharging the vehicle battery and operating any electrical circuits used, such as lights, heaters, and air conditioners. As a diesel mechanic you will be called upon to repair and/or troubleshoot the charging system on the diesel engine. To do this you must have a good knowledge of the system.

OBJECTIVES

Upon completion of this chapter the student will be able to:

1. Explain the generator operating principle.
2. Discuss and list the three things required for the generation of electricity.
3. List and discuss the two different types of generator field circuits used in diesel engine generators.
4. List and explain the differences between generators and alternators.
5. Explain the procedure for making a simple generator output test.
6. Demonstrate the ability to disassemble and test a generator.
7. Demonstrate the ability to properly turn a generator armature commutator or the slip rings on an alternator rotor.
8. Demonstrate the ability to test a complete generator regulator circuit using a voltmeter and ammeter.
9. Explain the alternator principle.
10. Explain the procedure for making a quick test on an alternator.
11. Demonstrate the ability to disassemble and test an alternator.
12. Demonstrate the ability to test a complete alternator and regulator circuit.

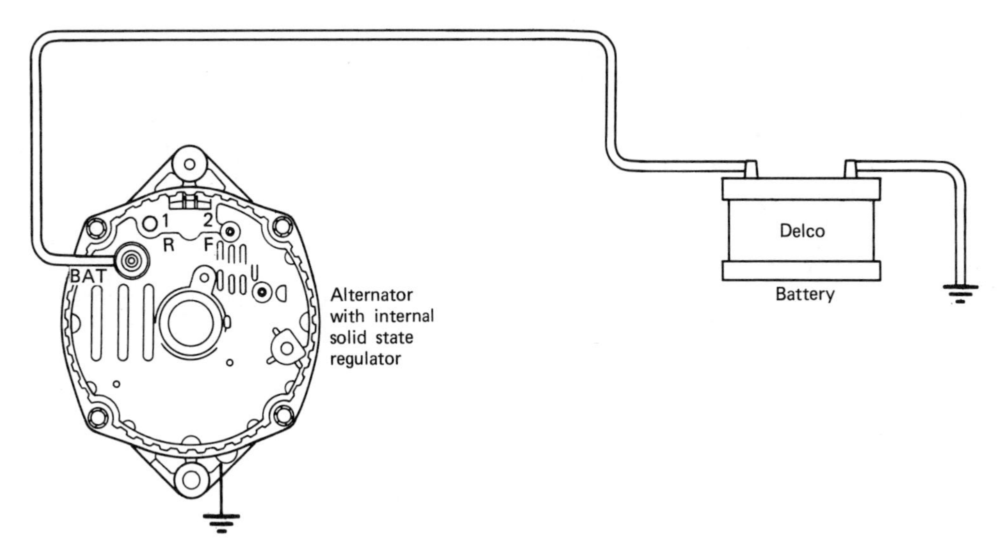

Figure 29.1 Alternator charging circuit (courtesy Delco Remy).

GENERAL INFORMATION

Charging systems on early diesel engines were always direct current (DC) generators. Most modern diesel engines use alternators. The modern charging system using an alternator senses the vehicle and battery requirements and supplies power as needed. Every vehicle has different current requirements, so different size alternators or generators are required. The amperage rating of the generator may be from 35 to 100 A (amps). If the generator had an amperage rating of 45 A and the system requirements were 55 A, the battery would quickly be discharged, since the additional 10 A would be supplied from the battery. When replacing a generator or alternator always make sure you have the correct amperage.

I. System Components

Most charging systems are made up of the following components:

A *Generator or alternator.* The generator or alternator is a device that converts mechanical energy to electrical energy.

NOTE The terms generator and alternator can and will be used synonymously throughout this chapter. In reality an alternator generates electricity and can be called a generator. Alternators differ from generators in the way voltage is produced. They are called alternators because the original voltage developed within the alternator is alternating current (ac). It then must be changed to direct current (dc) by the diode rectifier in the electrical system.

B *Regulator.* The regulator is considered the brain of the system and controls the generator output (Figure 29.2). It must sense the state of charge in the battery along with the current load demand of the components. Many different types of regulators are used. Some are separate units while others are an integral part of the generator.

C *Battery.* Within the system the battery is a source of electricity. It provides the electricity to start the alternator charging after the engine is started.

Figure 29.2 Vibrating-type relay regulator and solid state regulator.

Figure 29.3 Typical voltmeter.

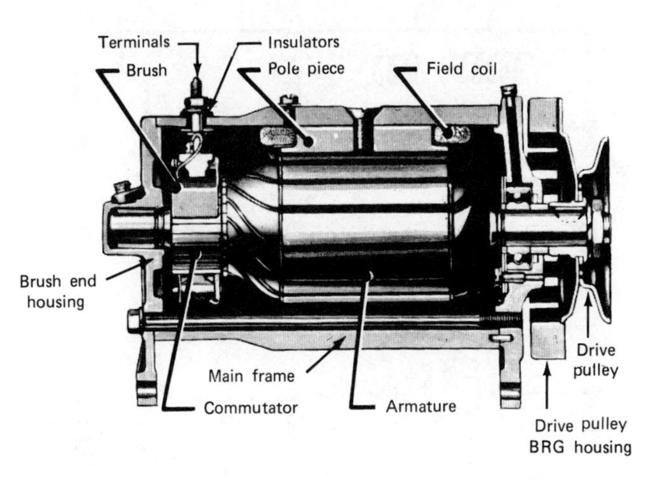

Figure 29.4 Component parts of DC generator (courtesy Delco Remy).

D *Wiring.* Wires or cables are used to connect the generator, regulator, and battery together.

E *Ammeter.* A meter which, when connected in series with the circuit, indicates the rate of current flow in the circuit. On many older vehicles an ammeter was installed in the dash or instrument panel at the factory.

F *Voltmeter.* A meter (Figure 29.3) that shows voltage in the circuit. This meter will be found as original equipment in many late model vehicle instrument panels.

G *Indicator light.* A panel mounted light indicating if the generator or alternator is charging.

II. Generator Component Parts and Operating Principles

A *Generator component parts.*

Common component parts in a dc generator (Figure 29.4) are:

1. *Brush end housing (commutator end housing).* This housing provides a place for the commutator end bushing or bearing. In addition, most generators have the brushes mounted on this housing.

2. *Main frame.* The main frame provides a place to mount the fields along with the front and rear bearing housings.

3. *Armature.* Made up of many conductors and fitted into the main frame between the fields. It collects the voltage.

4. *Drive pulley bearing housing.* This housing supports the front armature bearing.

5. *Drive pulley and fan.* Used to drive the armature and cool the generator.

6. *Terminals and insulators.* Bolts to which the output wire and field wire are connected. Insulators are used to prevent the bolts from grounding to the generator frame.

7. *Brushes.* Made from a carbon and graphite mixture and generally square or oblong in shape. They connect the generator commutator segments to the generator terminals. They are called brushes because they brush the commutator segments to make contact.

8. *Fields.* Electromagnets used to create the magnetism within the generator.

B *Generator operating principles.*

A generator is a device that converts mechanical energy into electrical power. This power is necessary to charge the vehicle battery and operate the vehicle electrical components. The generator is mounted somewhere on the engine and may be driven by a gear or belt.

The generator employs three things in the generation of electricity:

1. *Magnetism (force field).* In a generator magnetism is provided by two electromagnets called fields. These fields are made of windings (many turns of wire) and soft iron pole pieces (pieces of iron around which the windings are wrapped).

   Current is supplied to the field windings from the generator main brush. The field windings then become electromagnets, the strength of which is regulated by the current supplied to them. Since this strength determines generator

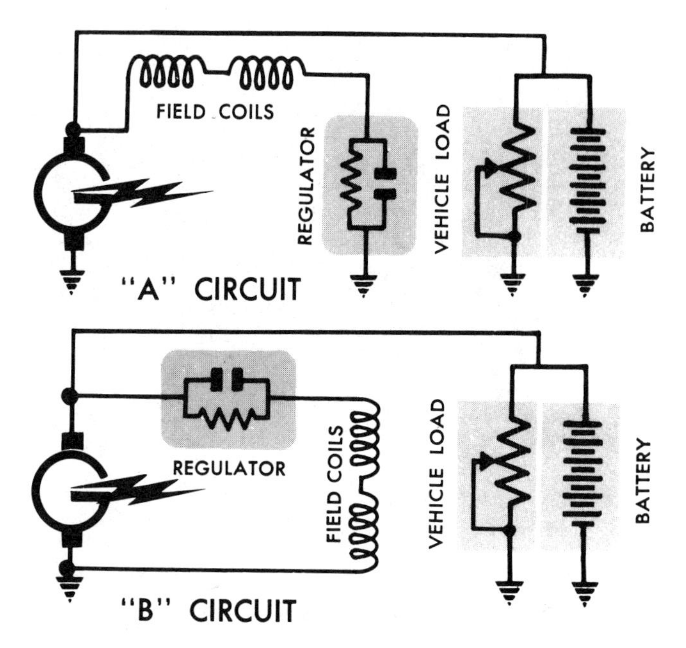

Figure 29.5 Generator A and B circuit (courtesy Delco Remy).

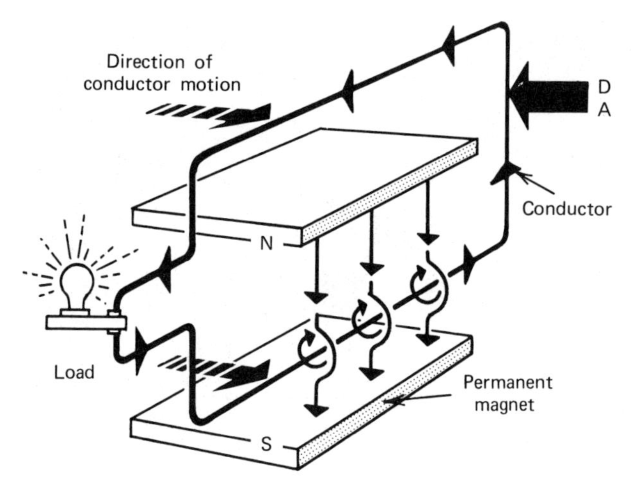

Figure 29.6 Conductor moving through force field (courtesy Delco Remy).

output, the fields can be used to regulate the generator output.

The fields, therefore, have a dual role in the generator: They provide (1) a magnetic field and (2) a means of controlling generator output.

Field circuits in a generator can be of two types, *A* or *B* (Figure 29.5). *A* circuit fields have the current supplied to them from the main brush with the generator field terminal grounded through the regulator to complete the circuit. *B* circuit fields are grounded internally at the generator ground brush with current supplied to the generator field terminal from the regulator. Most generators that you will work on will have *A* circuit fields.

2 *Conductors (armature)* (Figure 29.6). The conductors in the generator make up the armature. These conductors transmit voltage to the brushes via the commutator segments connected to the conductors. The entire assembly is designed so that it can be supported in the generator with bearings.

3 *Movement.* Movement of the conductor (armature) through the magnetic field is essential if current is going to be generated. This movement is provided by the engine, which is connected to the armature with a pulley and belt arrangement. The armature rotates on its two mounting bearings.

The principle known as electromagnetic induction is utilized in the operation of a generator. Electromagnetic induction produces voltage when a conductor is moved through the force field between two magnets. As the conductor moves through the force field, the lines of force are distorted around it until they break. This action creates a voltage in the conductor that causes current to flow.

NOTE To determine the direction of current flow in the conductor, use the right-hand rule for a current carrying conductor as outlined in Chapter 27.

Figure 29.7 shows a simple generator making use of the principle of electromagnetic induction. This basic generator has an armature represented by one loop of wire between two pole pieces. Because of the residual magnetism (magnetism that is retained in the field pole pieces after the generator stops charging), a magnetic field is set up between them. This magnetic field flows from the north pole of one magnet to the south pole of the other magnet. If the conductor is rotated within the magnetic field, each side of the loop is cutting across the force field. By cutting across the force field, voltage is induced in both sides of the conductor. This induced voltage causes current to flow in the conductor if the circuit is complete. Completing the circuit on either end of the conductor is a commutator segment that is connected by

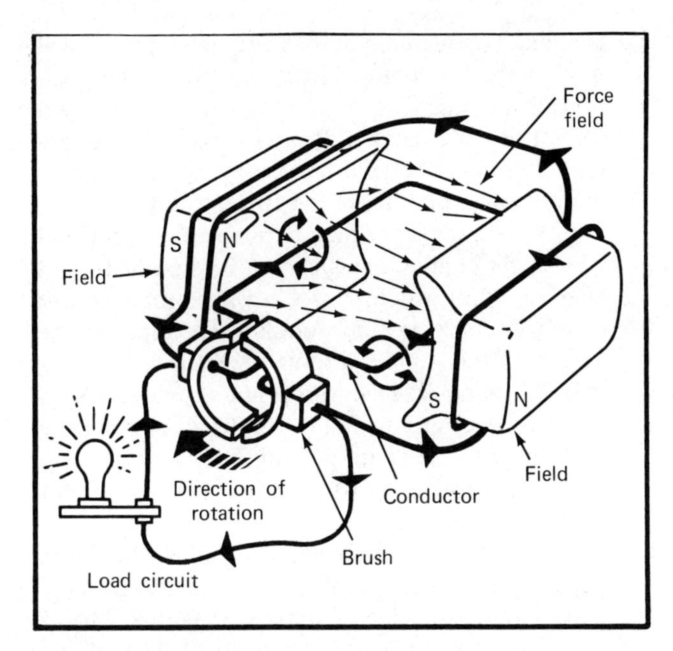

Figure 29.7 Simple illustration of generator principle (courtesy Delco Remy).

brush to the circuit load. This load completes the circuit and current flows through the conductor and the load.

In Figure 29.7 the electromagnetic fields are also connected to the brushes. As current flows in the armature loop, some of it will flow in the field windings that are wrapped around the field poles. As current flows in the field windings, the pole pieces become stronger electromagnets; as a result, the force field that exists between the two pole pieces is stronger and more voltage is produced when the conductor moves through the force field. The result is increased generator output. It can be seen that generator output can be controlled by the strength of the fields.

An actual generator will have an armature that contains many conductors connected to commutator segments, which make up the commutator. Riding on the commutator are the brushes, which provide a sliding contact between the commutator bars and the load and field circuits. In actual operation all conductors in the armature work together to provide a flow of current at the armature terminal of the generator.

The generator output is controlled by controlling the amount of current flowing in the fields. This flow is controlled by a regulator connected to the field terminal of the generator. Since the field terminal in an *A* circuit needs to be grounded back to the generator frame to complete the field circuit, the regulator senses the system's need and completes the circuit directly to ground for high output or through a resistance to ground for low output.

III. Alternator Component Parts and Operating Principles

A *Alternator component parts.*

Common parts found in an ac generator (Figure 29.8) are:

1 *Bearing and pulley end frame.* This end frame supports the front bearing and drive pulley and provides a place where the rear end frame can be mounted.

2 *Magnetic rotor.* An electromagnet mounted on a shaft that rotates inside the stator. It provides the magnetism with which the alternator generates voltage.

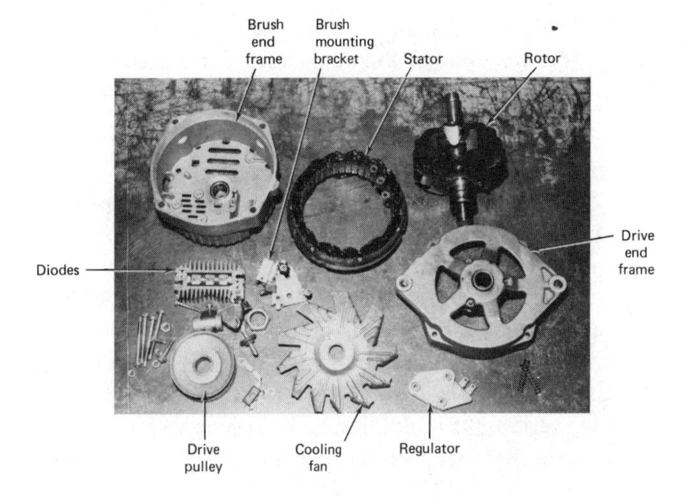

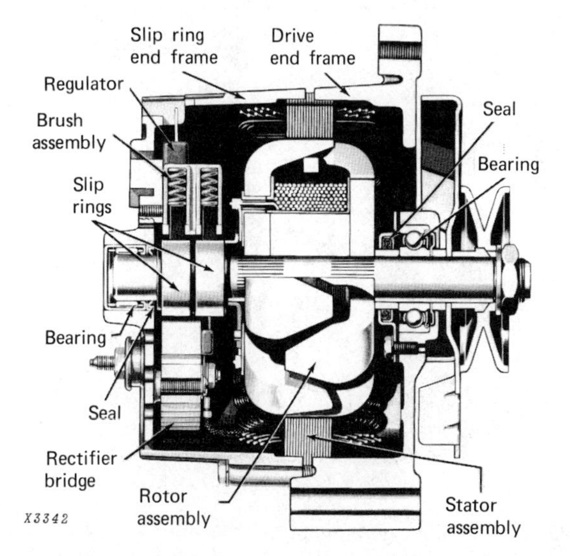

Figure 29.8 Alternator component parts.

3. *Stator*. A number of conductors mounted on a laminated iron frame. The conductors collect the voltage and provide a path for current flow, while the iron frame acts as a spacer between the front and rear bearing housings.
4. *Slip ring end frame*. The slip ring end frame is used to support the back rotor bearing, heat sink (a bracket that contains the insulated side of the circuit diodes usually positive) and the ground or return side diodes.
5. *Drive pulley and cooling fan*. Used to drive the rotor and cool the alternator.
6. *Brushes*. Made from carbon and graphite and oblong in shape. They carry the field current to the magnetic rotor through the slip rings.
7. *Slip rings*. Copper rings mounted on one end of the magnetic rotor that provide a place for the brushes.
8. *Diodes*. Considered electric check valves, the diodes change the ac current generated by the alternator to dc current. Most alternators contain three negative and three positive diodes.

B *Alternator operating principles*.

The basic alternator operating principle is shown in Figure 29.9. A permanent magnet is mounted on a shaft so that it can be rotated. Surrounding the bar magnet is a magnetic field. As the bar magnet rotates, this field is rotated across the conductor, inducing voltage into the conductor. Figure 29.10 shows a close-up view of what the magnetic field looks like as it cuts across the conductor. By using the right-hand rule for a current carrying conductor, you can see how current will flow in the conductor. Note also that as the position of the magnetic rotator changes with respect to polarity, the current flow will change. In an actual alternator this will happen many times a second, creating an alternating current output.

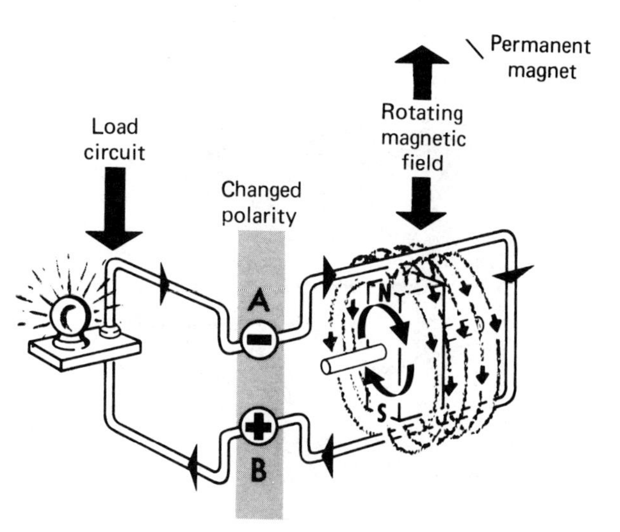

Figure 29.9 Alternator operating principle (courtesy Delco Remy).

An alternator with a permanent magnet and one single conductor as shown in the illustration (Figure 29.9) would not be very practical nor would it produce much output. To improve the output of this simple alternator, an iron frame is placed around it. This iron frame provides a path for the force field to follow as it moves from the north pole to the south pole of the magnet. It also becomes a frame on which the conductor (stator winding) can be mounted.

The voltage produced by the alternator is dependent on the strength of the magnetic field cutting across the conductor or conductors. Therefore, the strength of the magnetic field and the speed of the rotating magnet can control alternator output.

The simple alternator discussed so far would not meet the needs of today's vehicles; therefore, the actual alternator has many conductors instead of one. These conductors are made into what is called the stator. The rotating magnet is an electromagnet instead of a permanent magnet. In addition, diodes are used to change the ac output to dc as it leaves the stator windings.

NOTE Stator windings may be of two different designs, the Y or delta. The Y design stator is connected so as to represent a Y (Figure 29.11) while the delta stator is connected in the shape of the delta symbol, Δ (Figure 29.12).

A circuit schematic (Figure 29.13) shows the complete alternator internal circuit. In operation voltage is produced in the three stator windings as the magnetic rotor turns. As explained earlier, the voltage produced within the stator windings will be ac. This ac current must be rectified (changed to dc) before it leaves the alternator. The diodes perform this function. To understand better how the diode rectifier works, we must remember that with six diodes, three negative and three positive, there are six possible ways current can be conducted through the rectifier, changing it from ac to dc.

Figure 29.14 shows phase 1 of stator winding. In this illustration voltage is produced in the *B* and *A* windings. Current is now flowing through *B* and *A* windings from *B* to *A*. It then passes through two diodes, one negative and one positive, to complete the circuit to the battery.

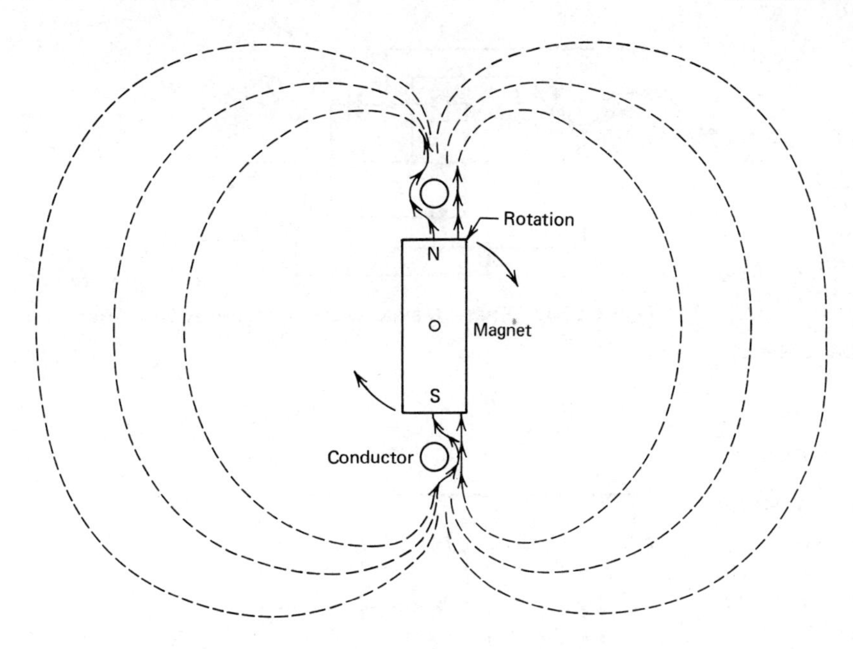

Figure 29.10 Close up view of magnetic field around rotor.

NOTE It should be observed that diodes 1 and 5 are conducting, while 6, 2, and 3 are blocking. Note also that the current leaving the battery terminal of the alternator and going to the battery will always enter the positive battery terminal.

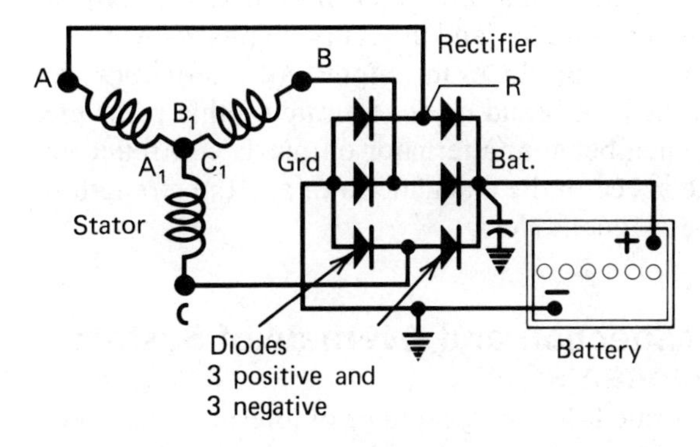

Figure 29.13 Circuit schematic (courtesy Delco Remy).

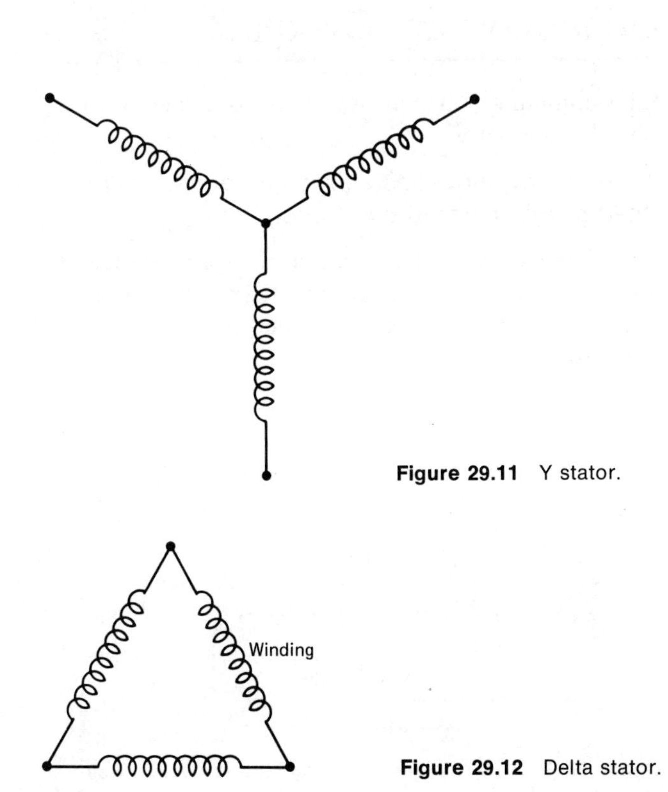

Figure 29.11 Y stator.

Figure 29.12 Delta stator.

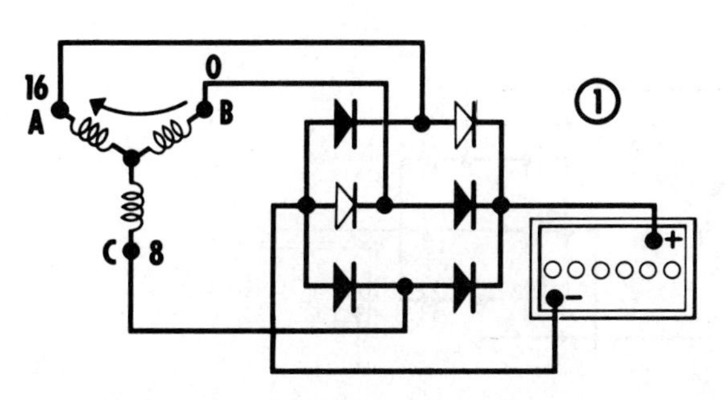

Figure 29.14 Phase 1, stator winding (courtesy Delco Remy).

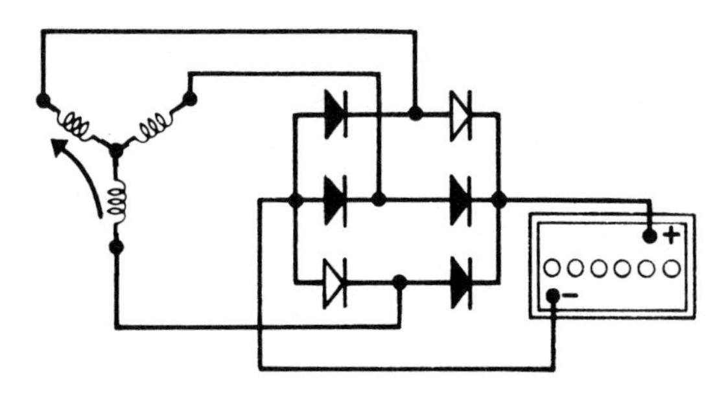

Figure 29.15 Phase 2, stator winding (courtesy Delco Remy).

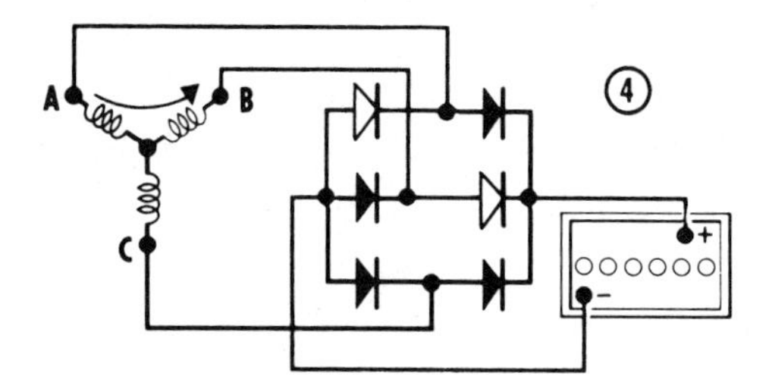

Figure 29.17 Phase 4, stator winding (courtesy Delco Remy).

As the magnetic rotor turns, two new stator windings, *C* and *A*, produce current as shown in Figure 29.15, phase 2. Diodes 1 and 4 are conducting, while diodes 2, 3, and 6 are blocking. Continued turning of the rotor will produce the following phases: phase 3 (*C*,*B*) in Figure 29.16, phase 4 (*A*,*B*) in Figure 29.17, phase 5 (*A*,*C*) in Figure 29.18, phase 6 (*B*,*C*) in Figure 29.19.

NOTE Regardless of the direction current is flowing in the alternator, it always leaves the battery terminal of the alternator in the same direction.

In the actual alternator this phasing is happening very rapidly, and in reality all phases work together to produce an output. All alternators, regardless of brand name, operate on this principle. Remember that alternator output is controlled by the speed of the magnetic rotor and the strength of the magnetic field.

IV. Inspection and Overhaul of System Components

It is assumed that the generator or alternator has been removed from the engine. If it has not been removed, refer to Chapter 5 on engine disassembly.

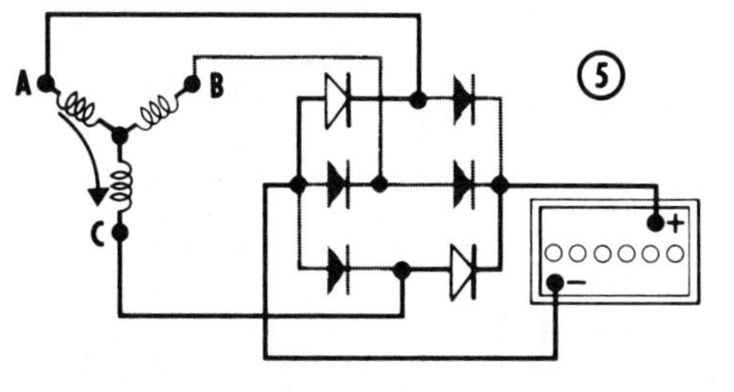

Figure 29.18 Phase 5, stator winding (courtesy Delco Remy).

A. Procedure for Generator (dc) Overhaul

It is common practice during engine overhaul to overhaul the generator.

NOTE Many shops install rebuilt generators instead of rebuilding the old one.

As the engine is put back into service, an overhauled or rebuilt generator can be expected to provide many hours of trouble-free operation and prevent needless down time.

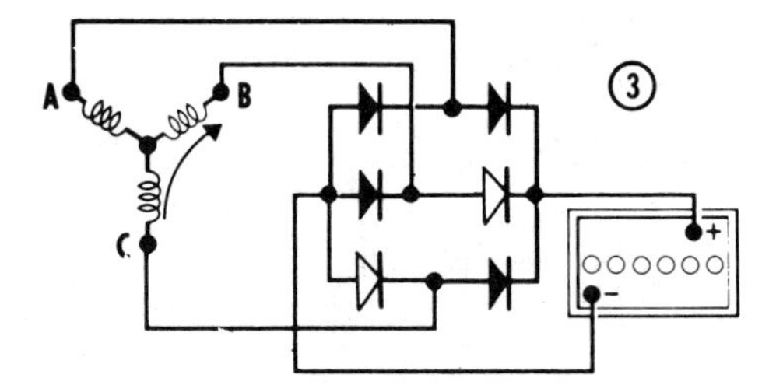

Figure 29.16 Phase 3, stator winding (courtesy Delco Remy).

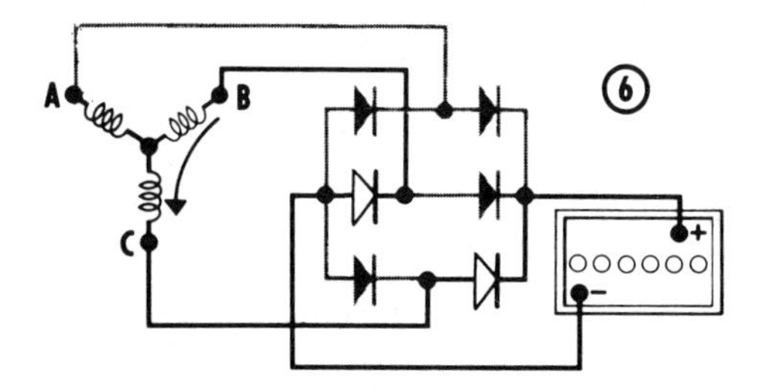

Figure 29.19 Phase 6, stator winding (courtesy Delco Remy).

1 "Match mark" the generator end housing for aid in correct assembly.

NOTE Some generators use dowel pins for alignment and will not require match marks.

2 Remove the three bolts that hold the generator together.

3 Remove the brush end housing by tapping on it with a rubber hammer.

NOTE If the end housing cannot be tapped off with a rubber hammer, use a screwdriver and wedge it between the end housing and the main generator frame.

CAUTION Do not wedge on only one side because you may break the end housing.

4 Remove the main frame from the armature bearing end housing.

5 Clamp the armature in a vise and remove the nut that holds the pulley and fan assemblies on.

6 Hold the pulley and fan assembly with one hand and tap on the armature shaft with a rubber or plastic hammer. The armature should drive out of the pulley. If it does not, a puller may have to be used.

NOTE Select a puller which will fit around the pulley to prevent damage to the pulley.

B. Procedure for Generator Component Testing

1 *Armature*. After removing the pulley and fan assembly from the armature, test the armature as follows:

a *Shorts*. Place armature on a growler and use a hacksaw blade to check it for shorts (Figure 29.20). Turn the growler switch to on and place the hacksaw blade crossways on top of the armature. The hacksaw blade should not vibrate. It it does, this indicates that the armature is shorted and must be replaced.

b *Grounds*. Using the growler test light (Figure 29.21), touch one probe of the test light to a commutator bar and the other probe to the armature iron shaft. The test light should not light. If it does, the armature is grounded and must be replaced.

c *Opens* (open circuits). Using the meter on the growler (Figure 29.22), place the armature on the growler and turn on the power switch. Place meter probes on two adjacent commutator bars. Then turn the armature to obtain the highest reading. Generally the probes will have

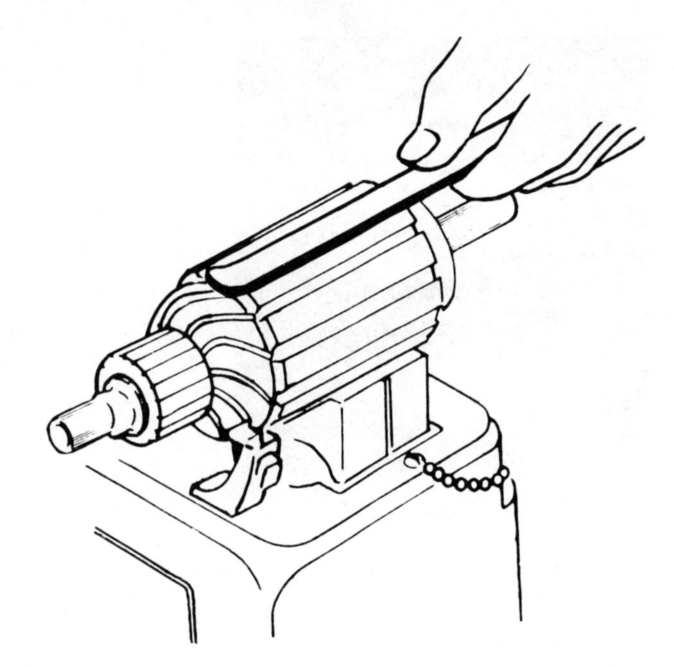

Figure 29.20 Checking armature for shorts (courtesy Delco Remy).

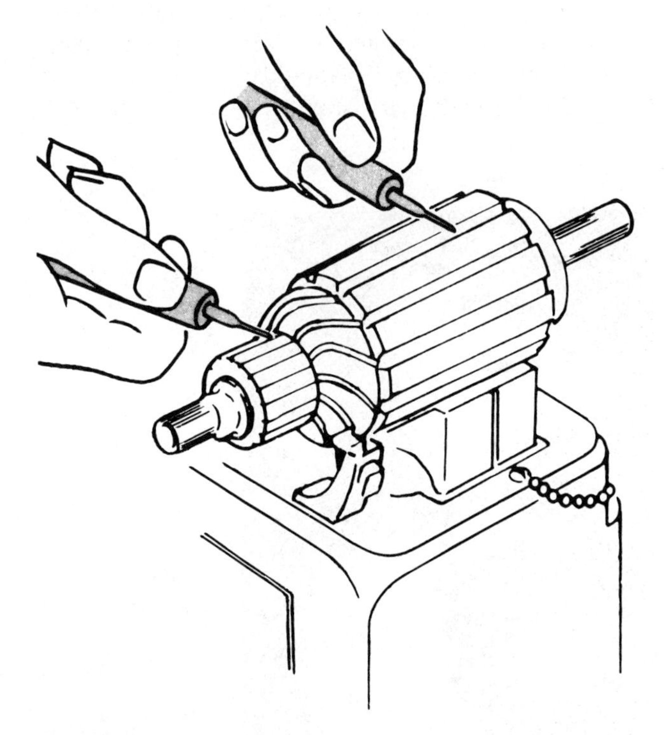

Figure 29.21 Checking armature for grounds (courtesy Delco Remy).

Figure 29.22 Checking armature for opens.

to be placed on the armature below the center line to obtain a reading.

If an armature growler with a meter is not available, opens in a generator armature can sometimes be found by using an ordinary hacksaw blade. To make this test, place the armature in the growler, turn power switch to on and use a hacksaw blade or thin strip of metal to short the two commutator bars together (Figure 29.23). As the blade is brought into contact with two commutator bars, a small spark should be seen. This indicates that current is flowing in the bars, proving there is no open circuit. If a spark is not produced, this indicates the armature has an open circuit.

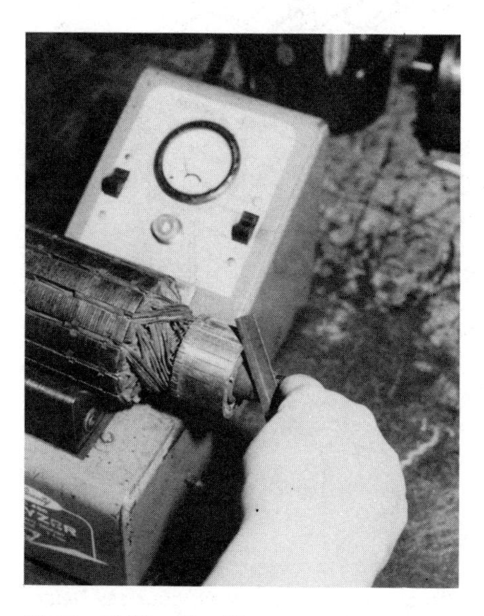

Figure 29.23 Checking generator armature for opens using a hacksaw blade and growler.

NOTE When making this test, the blade must be held below the center line of the armature to obtain an accurate reading.

Open circuits may also be detected by visually inspecting the commutator bars. The bar with an open circuit is usually burned and easily spotted.

NOTE Not all burned commutator bars indicate an open circuit. Test the armature when in doubt.

2 *Field coils* (fields not removed from the generator). The field coils in the generator should be checked during generator overhaul as follows:

a *Open circuit test.* Using a test light, test the field circuit for opens by placing one probe on the field terminal and the other probe on the wire that supplies power to the field from the main brush (Figure 29.24).

NOTE On *B* circuit generators touch one probe to the field terminal and one to the field wire that is connected to the ground brush.

The light should light or the ohmmeter show continuity; if not, the fields are open.

b *Field ground circuit test.* Test the field circuit for grounds with the test light.

NOTE On *B* circuit generators the field wire that is connected to the ground brush must be removed before making this test.

Touch one probe of the test light to the generator main frame (case) and one probe to the generator field terminal. If the test light lights, the field is grounded and must be replaced. If the test light does not light, the field is all right. If the fields are grounded and require replacement, the following procedure should be followed:

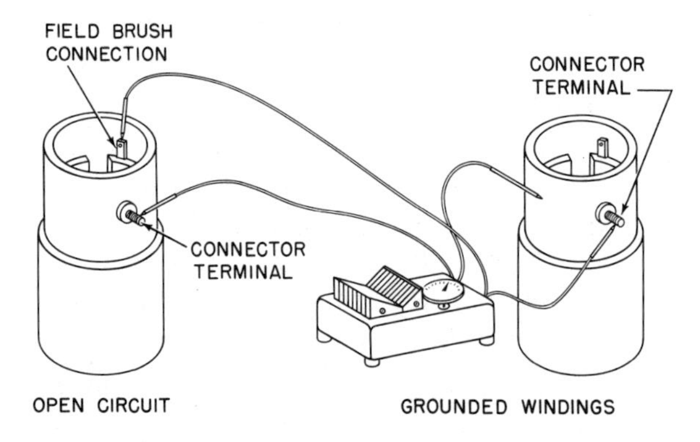

Figure 29.24 Testing generator fields using a test light.

Figure 29.25 Loosening field pole screws with a hammer and punch.

(1) With a large screwdriver, remove the screws that hold the field pole pieces in place.

NOTE Sometimes the screws that hold the pole pieces in place will not be easily removed with a screwdriver. If so, some shops will have a special removal tool to remove them. If no tool is available, an impact screwdriver will work well. If none of these methods is available, a small pin punch (1/4 in. or 6 to 7 mm) and hammer can be used to jar them loose (Figure 29.25).

(2) After the screws have been loosened, remove them with a screwdriver.

(3) Remove fields and field pole pieces.

NOTE It is good practice to keep the field pole pieces separate so that they can be returned to the same position they came from.

(4) Place new fields and field pole pieces into the main frame and install field pole screws. Tighten securely.

3 *Insulated brush holder.* The insulated brush holder must be tested for grounds as follows:

Using the test light on the growler or any battery-operated test light, touch one probe of the test light to the insulated brush holder while touching the other probe to the brush end frame (Figure 29.26). The light should not light. If it does, the brush holder is grounded and must be repaired.

4 *Bearings and bushings.* All bearings and bushings should be checked closely for wear and roughness.

Figure 29.26 Checking brush holders with test light.

If any question exists, replace them with new ones.

NOTE Special grease has been developed for use in generator bushings or bearings and should be used instead of wheel or chassis lubricant. This grease is available from parts supply stores.

C. Procedure for Reconditioning or Turning of Generator Commutator

Very often during generator overhaul the generator commutator is found to be worn out-of-round or scored. Before the generator is reassembled with new brushes, the commutator must be reconditioned. Recondition the commutator as follows:

1 Mount the armature in an armature lathe as shown in Figure 29.27.

Figure 29.27 Armature in armature lathe.

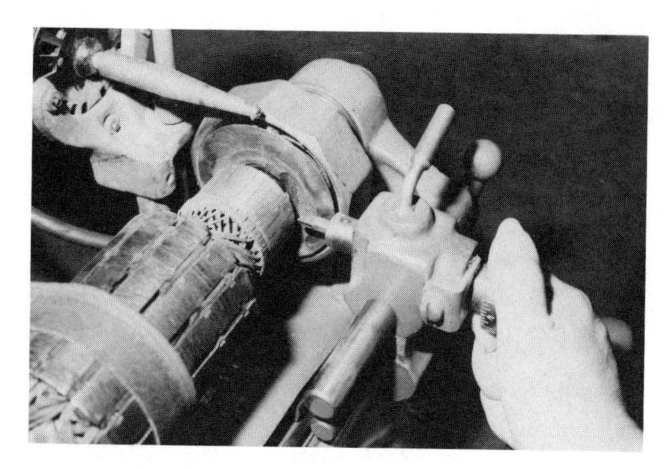

Figure 29.28 Adjusting cutting tool to just contact commutator.

Figure 29.29 Undercutting the mica on a generator armature.

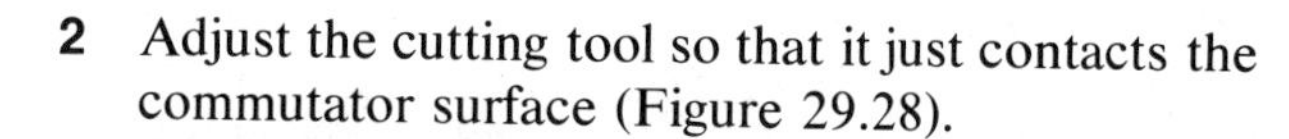

2 Adjust the cutting tool so that it just contacts the commutator surface (Figure 29.28).

CAUTION Do this with the armature lathe motor stopped or serious damage to the armature may result.

---

3 Back the commutator cutting tool sideways from the commutator with the advance handle and start the lathe motor.

4 Move the cutting tool sideways slowly toward the commutator with the advance handle until it starts to cut.

5 Continue moving the tool sideways across the commutator with a slow, even speed.

CAUTION Once you have started the cut, do not stop the sideways movement of the tool. To do so will cause ridging and grooving of the commutator.

---

6 After completing one cut all the way across the commutator from right to left, back out the cutting tool feed screw slightly and move the cutting tool sideways (left to right) with the advance handle.

7 Continue cutting the commutator as outlined above until all worn and ridged spots have been removed from the commutator.

8 After the commutator has been turned until it is clean all the way around, it may need to be undercut.

NOTE Undercutting the armature involves cutting the insulating mica (insulating material) from between the commutator bars. (High mica causes brush jumping and arcing that will eventually ruin the commutator.) Many armature lathes have a material undercutting attachment to undercut the mica (Figure 29.29).

9 To undercut the mica, leave the armature in the lathe and adjust the cutting tool so that it undercuts approximately $\frac{1}{32}$ inch (0.80 mm).

10 After adjusting the undercutting tool, move it from right to left using the feed handle.

CAUTION Guide the cutting tool by turning the armature slightly to line the mica groove. If the cutting tool slips off to the side, damage to the commutator will result.

---

11 After the commutator has been undercut, back the undercutter away from the armature and start the armature drive motor. Using commutator paper (fine sandpaper), polish the commutator by holding the sandpaper against the rotating commutator.

12 Remove the armature from the lathe and recheck it on the growler for shorts.

NOTE A small piece of brass wedged between two commutator segments can cause a short-circuited armature. Rechecking after commutator reconditioning is very important.

D *Generator reassembly (dc) and testing.*
After all the generator components have been

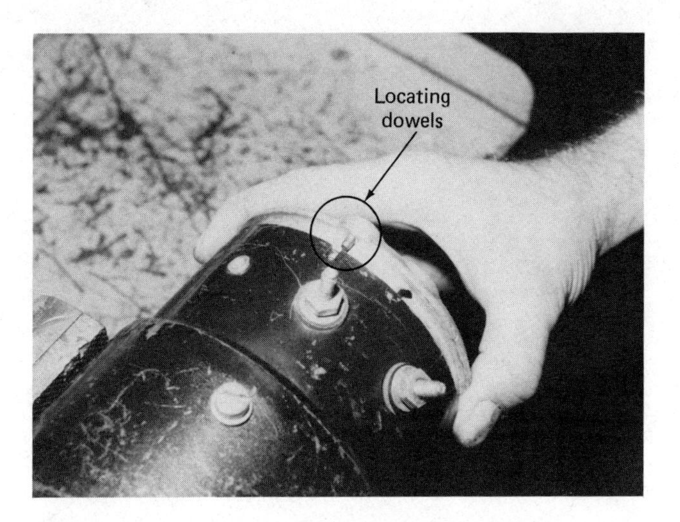

Figure 29.30 Aligning end frame alignment dowels.

inspected and repaired or replaced, the generator can be reassembled.

1. Place the drive end bearing and housing on the armature shaft.
2. Install the pulley drive key, if used.
3. Install the cooling fan and drive pulley.
4. Install the lock washer and nut that secure the pulley to the armature and tighten securely.
5. Install the generator main frame and field assembly onto the armature and drive end frame.
6. Install the brush end frame and align the alignment dowels (Figure 29.30).
7. Install the two through bolts that hold the generator together and tighten securely.
8. Turn the armature to insure that no binding or dragging exists. If the armature does bind, recheck the alignment of the brush end frame and retighten the field pole screws.
9. If test equipment is available, test the generator by operating it. (Specific test equipment instructions must be followed.)
10. If no test equipment is available, install the generator back on the engine and check it for output.

NOTE A quick check for generator output can be made by grounding the field terminal of the generator to the generator main frame with a jumper wire. Run the generator; the output should be at a maximum rated output.

E *Alternator disassembly and overhaul.*

Remove the alternator from the engine by disconnecting all wires and attaching bolts.

1. Match mark the alternator end frames and center section for reference during reassembly.

NOTE The following disassembly procedure is general in nature and may have to be supplemented by using the service manual for the specific alternator being worked on.

2. Remove the through bolts that hold the alternator together.
3. Remove the drive end frame drive pulley and rotor as an assembly. It may be necessary to pry the drive end frame drive pulley and rotor away from the stator and rear end frame using a screwdriver.

CAUTION Caution must be used when prying on the end frame with a screwdriver as damage to the end frame could result.

NOTE When the rotor is removed, the brushes and brush springs will fall out of place. The brushes are attached to the alternator with small wires and will not fall completely out. The springs are not attached to anything and may be lost if you are not careful.

4. Remove the stator from the slip ring end frame by disconnecting the three stator leads that are connected to the diodes.
5. Place the rotor, which has the drive pulley and drive end frame on it, into the vise and clamp it securely.

CAUTION Tighten the vise on the rotor just tight enough so that you can remove the pulley retaining nut. Tightening the vise tighter than this may damage the rotor.

6. Using the correct size socket, remove the pulley retaining nut, the washer, pulley, fan, and spacer.
7. Next remove the drive end frame from the rotor shaft.

NOTE You may have to tap on the rotor shaft slightly with a plastic hammer in order to remove it from the drive end bearing.

8. The alternator is now completely disassembled with the exception of the diode heat sink that does not have to be removed unless a bad diode is found during testing.

F *Procedure for Alternator Component Testing.*

1 Check all bearings and bushings for roughness or scoring.

NOTE Ball bearings are used in the front end frame of an alternator and should be spun by hand to check for roughness. If any roughness is felt, replace the bearing. Needle bearings are used in the slip ring end frame and are usually replaced as a matter of practice if the alternator has many hours of use on it.

2 Check brushes for wear. If they are worn over half of the original length, they should be replaced with new ones.

3 Check pulley groove for wear. If worn excessively, replace the pulley.

4 Check the stator windings with a test light or ohmmeter for:

a *Grounds.* Touch one lead of the test light to the slip ring and the other to the rotor shaft.

b *Opens.* Use an ohmmeter connected between each pair of windings at the lead connections.

NOTE Checking the stator for shorts requires special equipment found in a repair shop. If all other checks on the rotor prove it to be all right and the alternator still does not work, replace the stator. Many times a shorted stator can be identified by the color of the windings. Burned or blackened windings usually indicate a shorted winding.

5 Check the rotor with a test light or ohmmeter for:

a *Grounds.* Touch one lead of the tester to the shaft and the other lead to the slip ring. There should be no continuity between the rotor shaft and slip ring.

b *Shorts.* Using a battery and ammeter in series with the two slip rings, touch one lead from the battery to one slip ring. Connect one ammeter lead to the other battery post and then touch the other ammeter lead to the other slip ring (Figure 29.31). Current flow should be as specified in specification sheet.

CAUTION Do not touch the ammeter lead from the ammeter to the face of the slip ring. As the wire is disconnected from the slip ring, a spark will be created and the slip ring surface will be damaged.

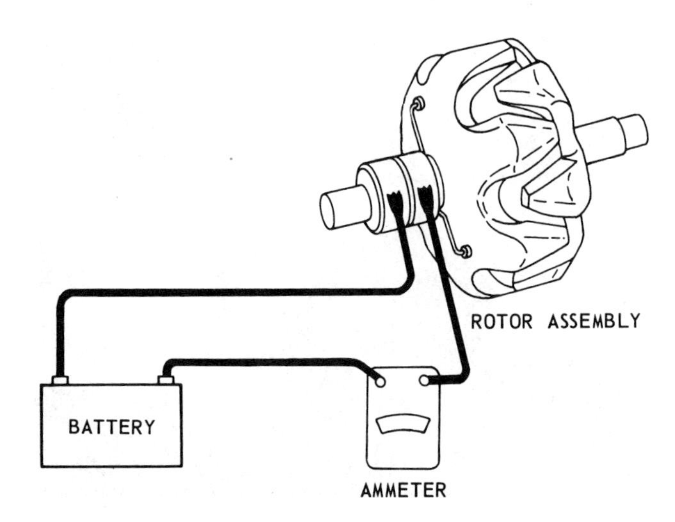

Figure 29.31 Testing rotor with ammeter and battery (courtesy Deere and Company).

c *Opens.* The rotor is easily checked with a test light for opens by touching the two test leads to the two rotor slip rings. (The light should light.) After testing the rotor electrically, the slip rings should be checked for wear. The brushes will wear grooves in the slip rings after a long period of use. These grooves must be removed before the alternator is reassembled to prevent premature brush wear. The rotor should be mounted in an armature lathe to recondition the slip rings (Figure 29.32).

NOTE Slip rings should be reconditioned to 0.002 in. (0.05 mm) out-of-round.

Figure 29.32 Alternator rotor mounted in an armature lathe.

CAUTION Remove only enough metal from the slip rings to clean them up because they are very thin and may be ruined.

6 Check the diodes by using a test light or ohmmeter.

NOTE Diodes are mounted in different alternators in many different ways. It can be expected that removal and checking of diodes will differ from alternator to alternator and the specific repair manual should be followed in this area. The following checking procedure is general in nature.

a Isolate the diode you are checking from the other diodes. Touch one lead from the test light to the alternator lead and the other test lead to the diode case. If the light glows, reverse the leads; the glow should cease. If the light glows both ways, the diode is shorted and must be replaced.

b If the diode passes the electrical tests, check the diode lead for looseness in the diode case. If the lead is loose, the diode should be replaced.

G *Regulators.*

As stated previously, regulators are the brain of the system. They are designed to serve the demands on the system and adjust the system output to the demand.

1 Regulators for dc generators (Figure 29.33) contain the following:

a The *cutout relay* is an electromagnetic relay that automatically opens and closes the circuit between the generator and battery during generator operation. If the generator was not disconnected from the battery after the engine was shut down, the battery would discharge back through the generator.

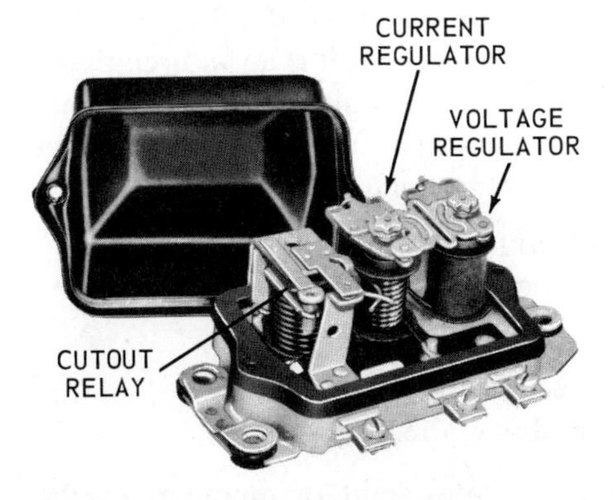

Figure 29.33 DC generator regulator (courtesy Deere and Company).

Figure 29.34 Voltage adjustment of dc regulator (courtesy Deere and Company).

b The *current regulator* is an electromagnetic relay that controls the strength of the field circuit within the generator. Without the current regulator the generator would exceed its rated output and cause generator damage.

c The main function of the *voltage regulator* is to protect the vehicle electrical system from excessive voltage that would damage it. It also controls the generator output by regulating the strength of the field circuit.

The regulator is a rugged, dependable device and requires little maintenance. After many hours of operation, a voltage adjustment (Figure 29.34) may be required.

2 Regulators for alternators are usually solid state (Figure 29.35), although some early alternators did use the vibrating-relay type regulator similar to the one used with DC generators. In most cases solid-state regulators are not adjustable and, if the generator voltage is not correct, the regulator must be replaced. Exceptions to this would be some Delco alternators

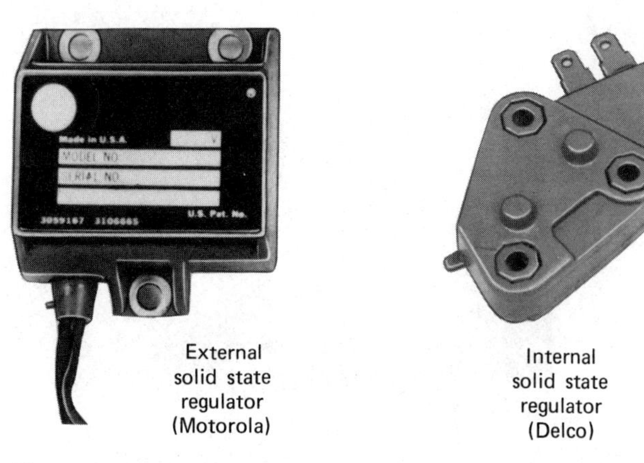

Figure 29.35 Solid state regulator (courtesy Deere and Company).

that have a high-low voltage adjustment (Figure 29.36).

H. Procedure for Alternator Assembly and Testing

1. Install the front bearing into the front bearing housing and install the bearing retainer.
2. Install the rotor into bearing and then install the cooling fan and drive pulley.
3. Install and tighten the drive pulley retaining nut and lock washer.
4. Clamp the rotor into a vise and torque the drive pulley nut.
5. Install the brushes or brush assembly into the slip ring end frame and insert the brush-holding device.

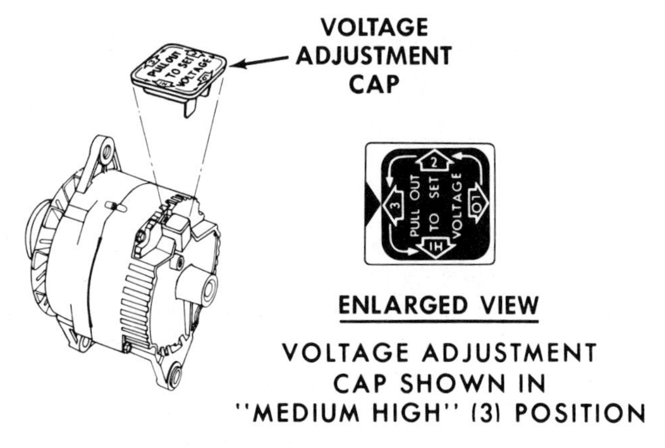

Figure 29.36 High-low voltage adjustment on regulator (courtesy Delco Remy).

NOTE Some alternators are assembled before the brushes are installed. Follow the manufacturer's recommendation closely in this area.

6. Install stator assembly onto the slip ring end frame and connect the leads.
7. Install the rotor and drive end frame into the stator and slip ring end frame.
8. Install the bolts that hold alternator end frames together.
9. Testing:
   - **a** Mount the alternator on a test stand or engine.
   - **b** Test output by running the machine or engine.
   - **c** If alternator was tested on test stand, mount the alternator on the engine and connect all wires.
   - **d** Run and test alternator.

V. Testing and Troubleshooting

Testing and troubleshooting the charging circuit or system is a skill that must be mastered since you must diagnose the problem before you can repair it. While troubleshooting the charging system you will need the following equipment: hand tools, voltmeter and ammeter, jumper wire with alligator clips and a test light.

Since this chapter covers both dc and ac generators, each system will be covered separately.

A. Procedure for Generator Testing

Listed below are system troubles and the tests that you should follow in determining the problem:

1. System not showing a charge.
   - **a** Connect a jumper wire from the F terminal of the generator to ground if it is an *A* circuit generator. If testing a *B* circuit generator a jumper wire must be connected from a source of power to the generator terminal.

NOTE Most *A* circuit generators have two terminals while many *B* circuit generators have three terminals.

   - **b** After connecting the field jumper wire, run the engine just above idle.
   - **c** Watch the indicator light or ammeter of the vehicle. The generator should charge during this test if it is all right.

CAUTION Do not run the engine faster than 700 to 800 rpm since the generator may produce excessive voltage, which could damage an electrical component.

2 The generator charges with the field jumper wire attached and shows no charge when the field jumper is removed.

This indicates the problem is elsewhere in the system.

- **a** Disconnect the jumper wire from the generator field terminal and reconnect it to the field terminal of the regulator.
- **b** Run the engine. If system now shows a charge, this indicates that the regulator is faulty.
- **c** Check the regulator ground strap.
- **d** Remove the regulator cover and check the voltage regulator points. Clean them with fine sandpaper or a piece of ordinary paper.
- **e** Remove the jumper wire from the field terminal of the regulator and check the system for charge by running the engine. If system still does not charge, it may be necessary to replace the regulator.

3 Generator does not show charge with jumper wire connected as in step 2a above.

- **a** This condition indicates a broken field wire between the generator and the regulator, or a loose or corroded connection.
- **b** Replace the field wire and recheck the system.

4 Generator charges but does not keep the battery fully charged.

- **a** This condition may be the result of low generator output caused by a loose generator drive belt, dirty or worn brushes, and incorrect voltage regulator adjustment.
- **b** Check system charging voltage by connecting a wire lead of a voltmeter to armature terminal of the generator and the other lead to ground. Run the engine. Voltage reading should be 14.4 to 14.8. If the voltage reading is lower, the regulator may have to be adjusted.

NOTE Do not attempt to make this check without first charging the battery since a low battery may give you a lower voltage reading than the system is set at. To obtain a 100 percent accurate adjustment of the voltage regulator, it is recommended that $\frac{1}{4}$ ohm resistance be connected in series with the battery. For complete details on regulator adjustment, see regulator service manual or consult your instructor.

5 Excessive generator output.

This condition is usually indicated by excessive use of water by the battery or burned-out light bulbs and/or accessories.

- **a** Disconnect the field wire at the field terminal of the generator.
- **b** Connect a voltmeter between the output terminal of the generator and the ground.
- **c** Run the engine and observe the charge rate. If generator still charges with the field wire disconnected, the problem is probably within the generator. Remove and disassemble the generator; repair as needed.
- **d** If generator does not charge with the field disconnected, the problem must be in the field wire or regulator.
- **e** Connect the field to the generator and disconnect it at the regulator. If the output drops to zero, the problem is in the regulator. If the output remains high with the field wire disconnected from the regulator, the problem is probably a grounded field wire. Check the wire closely and replace if necessary.
- **f** If the generator does not charge with the field wire disconnected at the regulator, the regulator must be defective or out of adjustment; replace or adjust.

6 Excessive generator noise.

Generator noise is usually caused by worn bearings or bushings. Worn bearings or bushings allow the armature to come in contact with the field pole shoes and make a noise.

- **a** Remove the generator and disassemble.
- **b** Replace the bearings and/or bushings.

B. Procedure for Alternator Testing

Listed below are some common system problems and the correct procedure with which to diagnose the problem:

1 System does not show charge.

- **a** Connect a voltmeter to the alternator battery terminal and to ground.

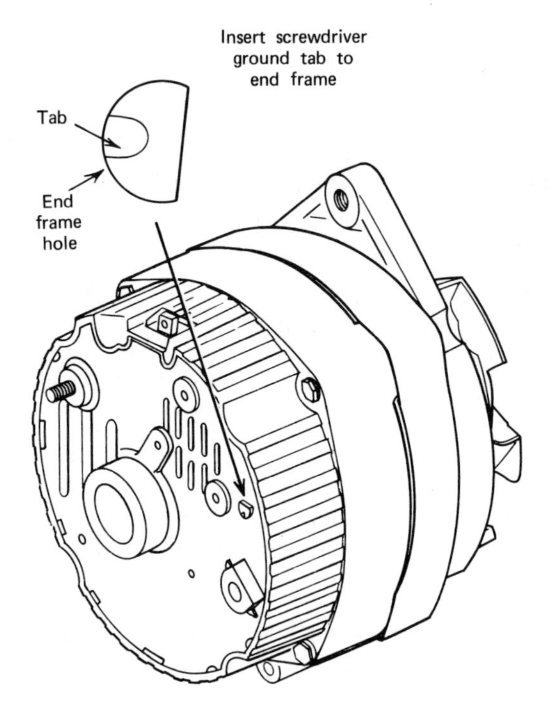

Figure 29.37 Test hole for testing alternator output (courtesy Delco Remy).

b Disconnect the alternator field wire at the alternator if an external type regulator is being used.

c With engine running at idle, *momentarily* connect a wire from the field terminal of the alternator to the battery terminal; alternator should show charge. If it does not, it is probably defective. Remove it and repair it.

NOTE Since there are many different types of alternators in use, different procedures exist for checking them for maximum output. This last step (c) is general and cannot be used on all alternators. For example, current production Delco Remy alternators in many cases will have an internal regulator. To test this system insert a screwdriver into a crescent-shaped hole in the back of the alternator; this grounds a tab that protrudes from the regulator. With this tab grounded the alternator should show full output. (Figure 29.37). Check with your instructor or service manual.

d If you find the alternator is working properly after making the test outlined in step c, proceed to step 2.

2 Alternator charges when the field is connected directly to the battery terminal. (The regulator is bypassed.)

a Check wiring between the alternator and the remote-mounted regulator, if used. If internal regulation is used, the alternator must be disassembled and a new regulator installed.

b Remove the regulator and replace.

NOTE Most regulators used in current charging systems are solid state and cannot be repaired. If the regulator checks out bad, replacement is the only alternative.

3 Low charging system output.

a Connect the voltmeter from the alternator battery terminal to the ground.

b Bypass the regulator.

c Run the engine and note the output.

NOTE Many times alternators will have sufficient voltage when tested, as outlined above, but will not keep the battery charged. This condition can be caused by an open or shorted diode and cannot be detected accurately with a voltmeter. The most accurate way to test for maximum alternator current output on the vehicle is to use an ammeter connected in series with the battery terminal of the alternator and the positive terminal of the battery. A current load must be put on the battery, preferably by a carbon pile resistor across the battery. Maximum current output of the alternator can be tested in this manner.

d If current output of the alternator is low when checked as indicated in note above, remove and repair the alternator.

e If voltage output of the alternator is sufficient with regulator bypassed, replace the regulator.

4 Excessive system output.

Excessive output is generally indicated when the battery requires unusual amounts of water.

a Connect the voltmeter to the alternator, as in all other tests.

b Run the engine and note the reading. If it is excessive (greater than 14.2 to 14.6), adjust the regulator. If regulator is not adjustable, it must be replaced.

5 Excessive alternator noise.

Alternator noise generally is caused by worn bearings or bushings and shorted diodes. If alternator makes excessive noise, remove and repair it, checking all bearings and electrical components as outlined under alternator repair.

SUMMARY

This chapter has covered the operating principles, overhaul, and troubleshooting of generators and alternators. You will find that charging system repair is easy once you master the basics. As you gain experience on charging circuits, it is troubleshooting and repair that you look forward to.

REVIEW QUESTIONS

1 What is the function of the charging system on a diesel engine?

2 What is the difference between an *A* circuit generator and a *B* generator?

3 Name four component parts of a generator and their function.

4 Explain the basic generator principle.

5 Explain the basic alternator principle.

6 How does an alternator change ac current to dc?

7 How should the generator armature be checked during generator overhaul?

8 Explain how a quick check on a generator charging circuit can be made.

9 How should the stator windings in an alternator be correctly checked?

10 What functions do the diodes perform in an alternator?

11 What check would you make on a generator charging system with a no output complaint? Explain what tools and equipment you would need.

12 How would you test a generator system that is charging excessively?

13 How can alternator systems be quickly checked for output?

14 On a charging system, what is the first sign of an excessive charging rate?

Glossary

ACCUMULATOR A reservoir in which hydraulic fluid is stored.

ADAPTER POT An adapter in which a cylindrical injector must be inserted to be tested on the test stand.

ALLOY A metal composed of many metals mixed together.

ALTERNATING CURRENT (AC) Current flow that alternates in direction of flow, first flowing in one direction and then in another.

ALTERNATOR A device used to generate electricity.

AMBIENT Temperature of surrounding air.

AMMETER A meter used to measure current flow (amperes) in an electrical circuit.

AMPERE Current flow is measured in amperes (abbreviated A or amps).

ANGULARITY Denotes an angle, such as between the connecting rod and cylinder wall, or between the connecting rod and the crankshaft throw.

ANNULAR Pertaining to a ring-shaped cavity.

ANNULUS A ringed structure or space.

ARMATURE The rotating part of the generator that contains the conductors.

ARMATURE GROWLER A tester used to test starter armature.

ATMOSPHERIC PRESSURE The pressure that exists all around us, created by the weight of air within the atmosphere. At sea level, atmospheric pressure is about 14.7 psi (1.03 kg/cm^2).

ATOMIZATION The breaking up of fuel into many small particles, similar to the action of a spray nozzle on the end of a garden hose.

AUTOMATIC ADVANCE DEVICE A device that senses engine speed and automatically changes the pump-to-engine timing.

BABBITT A mixture of lead and tin, used to line the inside of a main or rod bearing.

BAFFLE SCREWS High hardness screws used to divert high pressure fuel away from the soft main housing of the pump.

BALANCE POINT SPEED In Caterpillar sleeve metering pumps the point where full fuel delivery and rated speed are correct.

BANJO A fuel line end that is shaped like a banjo to accept a hollow line bolt.

BELLCRANK An angled arm that connects the governor weights with the face of the thrust sleeve.

BELLOWS A diaphragm-like device made of neoprene.

BELLOWS TYPE THERMOSTAT A thermostat that uses bellows to operate the thermostat.

BIG END BORE The bore of the connecting rod into which the connecting rod fits.

BLOCK MATING SURFACE The surface of the block that mates to a corresponding surface on the cylinder head.

BONNET A metal cover used to cover the check valve in the area above the barrel on caterpillar compact housing and sleeve metering pumps.

BORED TYPE BLOCK A block containing no sleeves, with pistons and rings inserted directly into the cylinder block bores.

BOURDON GAUGE A temperature gauge that is operated by a tube filled with bourdon gas.

BOURDON TUBE TYPE OIL PRESSURE GAUGE A gauge that is operated by expansion and contraction of a flat, circular-shaped tube.

BUFFER SCREW A screw used to eliminate engine roll or surge at idle.

BURRED A burred surface is a rough surface caused by dirt, insufficient lubrication, or extreme surface loading.

CALIBRATION Precise adjustment of an injection pump.

CALIBRATOR A machine used to test injectors.

CALIPER A device used to gauge sizes.

CAM BUSHINGS A bearing surface used to support the camshaft.

CAM FOLLOWERS A tube-shaped device that rides on the camshaft lobes and provides a place to install the push rods.

CAM LOBES Semicircular lobes on the camshaft that operate the valves.

CAMSHAFT JOURNALS A round machined part on the camshaft used to support it in the block.

CAMSHAFT RUN-OUT A bent camshaft will not be in a straight line from one end to the other. This will cause the dial indicator reading to vary as the camshaft is rotated when supported on V blocks.

CATALYST Something that accelerates the occurrence of an event or reaction.

CAVITATION EROSION Erosion caused by tiny air bubbles in the coolant.

CENTRALITY A part that is centered relative to another part.

CENTRIFUGAL A force developed outward.

CENTRIFUGAL FORCE Force exerted on a body in motion that causes it to move away from a pivot or center point.

CENTRIFUGAL PUMP A pump that uses the centrifugal principle to pump water or other fluid.

CHAMFERED Tapered or beveled on the edge.

CHARGING PORT A port in the hydraulic head of an injection pump through which fuel passes to fill the pumping chamber.

CIRCLIPS Keepers or snap rings.

CIRCUIT A connection from the battery to load and back to the battery.

CLEVIS A U-shaped clamp connected with a hole for a connecting pin.

COKING Plugging with solid carbon. This carbon results from the incomplete burning, which sometimes occurs in the diesel engine cylinder.

COMBUSTION The burning of fuel and air.

COMBUSTION CHAMBER The area above the piston within the engine where the combustion occurs.

COMMUTATOR Part of the armature on which the starter brushes ride.

COMMUTATOR SEGMENTS Sections of the commutator made of copper that are connected to the conductors within the armature.

COMPARATOR A machine used to test injectors.

COMPRESSION RATIO The ratio between the cylinder volume at bottom dead center and the cylinder volume at top dead center.

COMPRESSOR A device used to compress air to high pressure.

CONCAVE Sloping inward and downward from all sides to the center.

CONCENTRICITY Used when referring to a valve seat with reference to valve guide center. *Example:* The valve seat must be concentric to the valve guide center. Explained another way, it means the radius from the valve guide center to the valve seat surface is the same all the way around the seat.

CONCENTRICITY INDICATOR A dial indicator and fixture used to check concentricity.

CONDUCTOR A material such as copper or lead used to conduct electricity.

CONFIGURATION A special pattern produced by this design.

CONICAL Shaped like a cone.

CORROSION The eating away of metal caused by chemicals or acids.

COUNTERBORE A lip cut in the top of the block in which the cylinder sleeve is fitted.

COUNTERWEIGHT A weight added to the crankshaft that counteracts the inertia forces developed by the piston and the connecting rod.

CRANKPIN JOURNALS A round part of the crankshaft into which the connecting rod is fastened.

CRANKSHAFT JOURNAL A round machined part of the crankshaft throw on which the connecting rod is connected.

CRESCENT PUMP A pump shaped like a half moon or crescent that can be used either in lubrication or hydraulic systems.

CROCUS CLOTH A cloth used to polish metal.

CROSSHEAD GUIDE PIN A pin in the cylinder head that guides the crosshead.

CYLINDER A bored hole in the block in which pistons and rings are inserted.

CYLINDRICAL IMPELLER A round impeller that may be used to pump a fluid, not considered a positive type pump.

DELETE To omit or drop from use.

DIAL INDICATOR An indicator usually calibrated in 0.001 inch or 0.01 millimeter that shows the reading on a dial.

DIAPHRAGM A thin membrane made of neoprene, steel, or cloth and used to separate a vacuum chamber, similar to those used on pneumatic governors.

DIRECT COMBUSTION Combustion that occurs in the main chamber of the engine.

DIRECT CURRENT (DC) Current that flows in one direction only.

DOWEL PINS Pins used to align two objects together.

DROOP SCREW A screw that is used to control the amount of fuel delivered to the engine under overload.

DRY SLEEVE A cylinder wearing surface inserted into a block that makes no contact with the engine coolant.

DYNAMOMETER A machine used to measure engine horsepower.

ECCENTRIC STUD A stud that has an eccentric on it, used as an adjusting screw. When turned, it provides a camlike action.

ELECTRIC GAUGE A gauge that shows psi or kg/cm^2; it is connected to the engine via a sending unit rather than an oil line.

ELECTROLYTE The solution contained in a battery cell made up of sulfuric acid and water.

ELECTRONS Particles of an atom that encircle the nucleus.

EMERY CLOTH A metal particle coated paper used to clean or polish metal.

EPOXIED Cemented together with an adhesive that bonds by chemical interaction.

EQUILIBRIUM A balance between two opposing forces.

ERMETO A special type of fitting on a nozzle fuel line.

EROSION Metal wear or degeneration usually caused by coolant water.

EXPANDABLE PILOTS A valve seat grinding pilot that is expanded into the valve guide to hold it in place.

EXPANSION GAUGE A gauge that is connected directly to the engine via an oil line and registers oil pressure.

EXPANSION PLUGS A soft metal plug used to plug openings in water jackets. A safety device in case of freezing; the opening also allows removal of core sand when the block is molded at the factory.

FILLET AREA An angled ground area between the crank journal and the crankshaft.

FIRING ORDER Order in which engine cylinders fire on power stroke.

FLUTTER Small, rapid vibrations.

FLYWEIGHT An integral part of a mechanical governor that senses speed changes.

FLYWHEEL A large, heavy wheel connected to the crankshaft that absorbs the power impulses of the engine.

FORGED BODY An early Caterpillar pump design that uses the barrel and plunger and the port and helix fuel control. The barrel is one forged body that can be removed separately.

FRICTION TYPE BEARINGS Bearings that do not have rollers or balls; bushing type bearings.

FUEL INJECTION SYSTEM The system that injects or supplies the fuel to the engine cylinders.

FULCRUM LEVER A support lever that connects the control rack to the guide lever.

GALLING Wearing away of metal by friction or chafing.

GEAR BACKLASH The clearance between gear teeth.

GEAR SEGMENT The part of a gear that is segmented to provide a reference for timing, such as combining two teeth together.

GLOW PLUG An electrically heated coil or plug that aids in starting a diesel engine.

"GLYD" RING A teflon ring used on the Cummins AFC plunger.

GOVERNOR A speed control device, driven by the engine, that senses changes in engine speed and makes corrections in fuel delivery to correct the speed.

GRADUATE (TEST BENCH) A thin glass receptacle used to hold and measure fuel.

GROWLER An electrical test instrument used to check generator armatures for opens, shorts, and grounded circuits.

HAND PRIMING To fill fuel lines with a small hand pump before starting the engine.

HELICOIL A threaded insert made of steel.

HELIX A tapered groove, cut on a pumping plunger, which is used to control fuel delivery.

HONE To enlarge a cylinder or sleeve using a special grinding device.

HUNTING Rapid oscillation of the governed engine speed.

HYDRAULIC Operated or effected by oil, conveying power by oil or other fluid.

HYDRAULIC DASHPOT A device used with some governors to prevent low idle underrun and give a more stable idle speed.

HYDRAULIC RAM A hydraulic cylinder with a bracket for removal of the cylinder sleeves.

HYDROMETER A device that measures the specific gravity of a fluid such as battery electrolyte or antifreeze.

HYDROSTATIC DRIVE A test stand drive that is completely hydraulic with no mechanical connection. This system allows an easily variable method of speed control.

IGNITION Process of igniting the air-fuel mixture within the cylinder so that it will burn.

IGNITION LAG The time after fuel has been injected before it starts to burn.

IMMERSE To dip or cover with fuel oil or lubricating oil.

IMPELLER The part of a water pump that moves the water and is connected to the water pump drive pulley.

INDIRECT COMBUSTION Combustion that takes place in a small prechamber.

INERTIA The tendency of air, fluid, or metal to stay in motion if it is in motion.

INHIBITOR An additive added to the radiator coolant that will prevent rust.

INJECTION LAG After the injection pump moves fuel into the injection line, no injection will occur until the elasticity of the line and the compressibility of the fuel are taken up. This is known as injection lag.

INTAKE MANIFOLD A series of pipes or pipe fittings that connect one pipe with several others. Used on a diesel engine to connect the air cleaner outlet air pipe to each cylinder inlet.

INTEGRAL A component part of an assembly that makes a complete unit.

INTEGRAL GUIDES Guides that are part of the cylinder head and not removable.

INTERCOOLER A device mounted between the turbocharger and intake manifold that is used to cool the intake air.

INTERFERENCE ANGLE Denotes the difference in angle between the valve seat and the valve (*Example:* 45° seat, 44° valve) that aids in seating the valve.

INTERNAL COMBUSTION The combustion that occurs within the engine cylinder.

INTERNAL COMBUSTION ENGINE An engine that burns the fuel within the engine cylinder.

INTRAVANCE An automatic timing device that advances the injection timing while the engine is running.

INVERT To turn over.

JUMPER PIPE A small pipe used to connect the inlet and return fuel manifold to the injectors.

KG/CM^2 Kilograms per square centimeter. The metric equivalent of psi. 1 kg/cm^2 = 14.2 psi.

KNURLING A ridging process applied to the inside of a valve guide to decrease the inside diameter of the guide, compensating for the normal wear.

LAMP PETROLEUM A product used in early days that is similar to kerosene.

LAPPED To cut or polish metal with a fine compound

and metal plate.

LINE BORE To cut or ream all bores or holes in a line.

LINE RETRACTION Removing a certain amount of fuel from a line between fuel injection.

LINEAR FORCE Force exerted in a straight line.

LOBE An oblong ground section on the camshaft used to lift the cam followers.

MAGNA-FLUX MACHINE A machine that uses magnetic lines of force to check a part, such as a cylinder head or block, for cracks.

MAIN BEARING BORE ALIGNMENT Having all main bearing bores in an exact alignment from one end of block to the other.

MALFUNCTION To function in some way other than the intention of the original design.

MANDREL PILOT A machined rod that is the same size as the valve, onto which the valve seat grinding stone mandrel is placed.

"MATCH MARKS" Marks on the connecting rods or main bearings used to insure that the caps are placed in the correct position during reassembly.

MICROMETER A precision measuring device used to measure diameters and thickness.

MUTTON TALLOW A special tallow used to lap and clean nozzle parts.

N^2 PRESSURE Pressure that operates the torque piston and advance piston servo on the CAV DP-15 pump.

NATURALLY ASPIRATED An engine that is naturally aspirated does not have a turbocharger.

NC BOLT National Coarse, a coarse thread bolt.

NEGATIVE The negative side of a battery that tends to gain electrons and then becomes negative.

NEOPRENE A rubberlike material that resists oil, grease, and antifreeze. Used in making sleeve seals, injection pump seals, and other engine seals.

NO AIR BLEED SCREW An adjusting screw used in the Cummins AFC pump to adjust the amount of fuel that is supplied to the engine when the AFC plunger is in the no air position.

NONFERROUS METAL Metals that contain little or no iron. Aluminum, brass, and copper are nonferrous metals.

NONPOSITIVE DISPLACEMENT Used to describe a pump that does not displace a given amount of fluid each time it makes a revolution.

OAKITE An acid used to clean grease and dirt from metal parts.

OAKITE SOLUTION A solution of oakite and water.

OFFSET WOODRUFF KEY A woodruff key that is offset so that the gear that it times to the shaft allows the shaft to be retarded or advanced.

OHMMETER A meter used to measure resistance.

OIL GALLEY A drilled or cored passageway used to pipe oil to various engine parts.

OIL SLINGER A metal disc that fits over the end of a camshaft or crankshaft and prevents oil from overloading a seal.

OPTIMUM The most favorable conditions in any given situation.

ORIFICES Small holes in the nozzle tip.

PELLET TYPE THERMOSTAT A pellet type thermostat uses a heat sensitive pellet to operate the thermostat.

PHASE Two items in correct relation to each other.

PILOT BEARING A bearing that fits in the center of the crankshaft and supports the transmission input shaft.

PINION A small gear used to drive a larger gear.

PLASTIGAGE A thin plastic thread used to check bearing clearance.

PNEUMATIC Relating to the use of air pressure or lack of it (vacuum).

PORT A small drilled hole.

PORT AND HELIX A specially designed pumping plunger and barrel assembly that is used to control fuel delivery.

POSITIVE Electrical potential of a battery terminal from which electrons flow, tending to make the terminal positive.

POSITIVE DISPLACEMENT Used when describing a pump that displaces a certain amount of water or oil each time it makes a revolution. A pump of this type is either an oil or hydraulic pump.

PRECHAMBER A small chamber next to the main combustion chamber of a diesel engine in which fuel and air burning begins.

PRECOMBUSTION Combustion that occurs before the main combustion takes place within the cylinder.

PRESS RAM The movable part of the cylinder used on a hydraulic press.

PRESSURE RELIEF VALVE A spring-loaded plunger or

ball valve used to regulate oil pressure.

PRIME MOVER An engine, gas or diesel, is called a prime mover.

PROTRUSION The distance the sleeve stands out or protrudes above the block top surface.

PSI Pounds per square inch. 1 psi = 0.07 kg/cm^2.

PULLER A mechanical device used to remove pulleys or gears.

PURGE (BLEED) To remove air from a fuel system.

RADIAL PRESSURE Pressure exerted radially, that is, around the entire cylinder wall.

REAMING Enlarging a valve guide or bushing by cutting with a round cutter.

REBORE To drill or cut a cylinder to an oversize.

RECIPROCATE To move back and forth, such as a piston and connecting rod assembly.

REFACE To grind the face of a valve or valve seat.

REFERENCE MARKS (MATCH MARKS) Marks on the connecting rods or main bearings used to insure that the caps are placed in the correct position during reassembly.

RESERVOIR A tank in which compressed air can be stored.

RESISTOR An electrical device that creates a resistance to the flow of electricity.

RETRACTION Relating to the fuel that is allowed to drain from the injection line after injection. This retraction prevents nozzle dribbling.

RETRACTION LAND A small piston that seals the injection line from the injection pump. Part of the delivery valve.

RETRACTION VALUE The amount of retraction travel of the delivery valve.

RIFLE DRILLING Denotes a connecting rod that has a drilling in order to supply oil under pressure to the connecting rod.

ROCKER ARMS Devices to open the valves that are operated by cam followers and push rods.

ROD JOURNAL A smooth ground surface on the crankshaft to which the connecting rod and bearing are connected.

ROOTS TYPE BLOWER A positive displacement type of blower used on two stroke cycle engines.

RPM Revolutions per minute.

SAE Society of Automotive Engineers.

SATURATION Thoroughly filled or penetrated to a high degree.

SCAVENGING The removal of burned gases from the cylinder after combustion occurs.

SCORING Scratching or marring of a surface that occurs when the metal is hot enough to melt.

SCRIBING To mark with a sharp object.

SEMITORIDAL A piston design used in direct injection type combustion chambers, shaped like a Mexican hat.

SENSING UNIT A device that senses the coolant temperature and changes the resistance in the temperature gauge circuit.

SERRATED SOCKET A socket wrench with many small serrations (lands) instead of the usual six or twelve.

SERVO PISTON A hydraulic piston that moves another part on command from a servo valve.

SHIM A thin washer used to change nozzle opening pressure.

SHIM PACK A pack of thin washers used in conjunction with a spring.

SHUTTERS Air-operated louver-like doors that fit in front of the radiator and regulate air flow.

SHUTTERSTAT A temperature-operated device that controls the radiator shutters.

SHUTTLE To move back and forth in a bore and provide fuel movement.

SLIDE HAMMER A mechanical puller that uses a sliding weight to apply force to a pulley or gear.

SLOT A vertical groove machined into the pumping plunger that connects the top of the plunger to the area below the helix.

SNUBBER A restrictor valve used in conjunction with a gauge to reduce gauge flutter and give a more stable reading.

SPECIFIC GRAVITY The weight of electrolyte as compared to the weight of water.

SPLINED DRIVE SHAFT A shaft with equally spaced splines rather than a tapered shaft.

SPRAY PENETRANT METHOD A method used to check metal parts for cracks.

STABILIZED A balanced condition between the governor weights and governor spring.

STELLITE SEATS Valve seats made of a special steel alloy called stellite.

SUPERCHARGER A gear-driven air blower that supercharges the intake manifold of diesel engines.

TANG A piece of metal that looks like a screwdriver fixed to the end of the governor plunger, which causes the tang to turn as the governor weights turn.

TAPER OR CHAMFER TOOL Tool or fixture used with the valve machine to chamfer valves.

TAPERED PILOTS A valve seat grinding pilot that is tapered; it is tightened when it is forced into the valve guide.

TAPPET CLEARANCE Clearance between the rocker arm (tappet) and the valve stem.

TDC Top dead center.

TEMPILSTICK A stick of crayonlike material used to measure the temperature of an object; melts at a given temperature.

TEMPLATE TORQUE METHOD Used by some engine manufacturers to tighten the main bearing and rod caps in place of a conventional torque method.

TENSION Pressure exerted on the governor spring by the throttle control lever.

TERMINAL An eyelet attached to a wire used to connect the wire to a generator or regulator.

THERMOSTAT A temperature-controlled device that regulates the flow of water through the radiator and engine block.

THERMOSTAT HOUSING The housing in which the thermostat is placed.

THRUST FACE OR SURFACE A bearing surface machined on the crankshaft that accepts crankshaft end thrust.

TOP DEAD CENTER (TDC) VOLUME Volume of the cylinder when the piston is at top dead center.

TORQUE Twisting effort supplied by an engine or hand-held wrench.

TORQUE RISE An increase in engine torque as the engine is put under an increasingly heavier load.

TORQUE TURN METHOD A method similar to the template method used to tighten main or connecting rod bolts.

TORSIONAL VIBRATION Twisting vibration that develops in the crankshaft as the engine runs.

TRIMMER SCREW An adjusting screw that provides only a small amount of adjustment.

TRUNK TYPE PISTON A one-piece piston used in most modern diesel engines.

TURBOCHARGER An exhaust-driven air pump used to pump air into the engine intake manifold.

TURBULENCE Fast, violently moving air.

VALVE CHUCK HEAD The part of the valve machine into which the valve is inserted to hold it while grinding.

VALVE CROSSHEADS A bridgelike device that sits on top of the valves and allows the rocker arm to open both valves at once.

VALVE OVERLAP Time in the cylinder cycle operation when both the intake and exhaust valves are open. Specifically designed into the engine to aid in removing exhaust gases from the cylinder and in bringing fresh air in.

VALVE ROTATORS A device installed on the valve that either allows it to rotate or makes it rotate during engine operation.

VANE TYPE TRANSFER PUMP A transfer pump that uses vanes to pump the liquid.

VARIABLE DROOP Percent of regulation from high idle no load speed to full load speed.

VELOCITY Related to speed; we use it to refer to the speed of air entering the intake manifold.

VERNIER CALIPER A caliper with a sliding rule for extremely fine measurement.

VIA By way of.

VIABLE Something that is possible or probable in a given situation.

VIBRATION DAMPER A large rubber-mounted wheel fitted onto the front of the crankshaft to absorb torsional vibration.

VISCOSITY The resistance of a fluid to flow.

VISCOUS Thickness of a fluid, pertaining to viscosity.

VISCOUS DAMPER A vibration damper filled with a fluid.

VOLATILE Easily ignited, as, a volatile fuel.

VOLTMETER A meter used to measure voltage in an electrical system.

VOLUMETRIC EFFICIENCY In relation to an engine, it is the ability to breathe or fill its cylinders with air during operation. The efficiency figure is a comparison between the actual air volume the cylinder can hold and the air volume with which it is filled during engine operation.

WATER PLATE A metal plate that is bolted to the cylinder head to plug all water outlets so that the head can be pressure checked.

WET SLEEVE A cylinder-wearing surface inserted into a block that contacts the engine coolant.

WOODRUFF KEY A half moon shaped key.

WOT Wide open throttle.

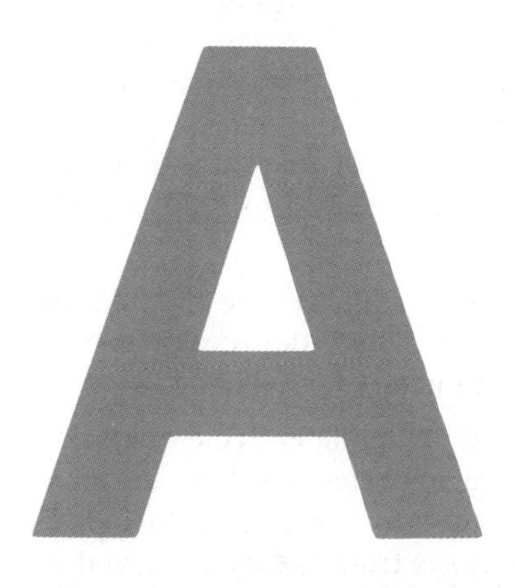

Conversion Factors

METRIC TO ENGLISH

LENGTH	
Millimeters to inches	multiply by 0.03937
Centimeters to inches	0.3937
Meters to feet	3.2808
Meters to yards	1.0936
Kilometers to miles	0.6214
WEIGHT	
Grams to ounces	multiply by 0.03527
Kilograms to pounds	2.2046
Tons (metric) to tons	0.9842
LIQUID	
Liters to U.S. gallons	multiply by 0.2642
Liters to U.S. quarts	1.057
Cubic centimeters to cubic inches	0.06102
PRESSURE	
Kg/cm² to lb/in.²	multiply by 14.223
TEMPERATURE	
°Centigrade to °Fahrenheit	multiply by $\frac{9}{5}$ and add 32
TORQUE	
Meter kilograms to foot pounds	multiply by 7.23
Meter kilograms to inch pounds	86.8
Centimeter kilograms to foot pounds	0.0723
Centimeter kilograms to inch pounds	0.868
Newton meters to foot pounds	0.736
Newton meters to inch pounds	8.851

ENGLISH TO METRIC

LENGTH	
Inches to millimeters	multiply by 25.4
Inches to centimeters	2.54
Feet to meters	0.3048
Yards to meters	0.9144
Miles to kilometers	1.609
WEIGHT	
Ounces to grams	multiply by 28.35
Lbs to kilograms	0.4536
Tons to tons (metric)	1.016
LIQUID	
U.S. gallons to liters	multiply by 3.785
U.S. quarts to liters	0.946
Cubic inches to cubic centimeters	16.387
Fluid ounces to cubic centimeters	29.57
PRESSURE	
$Lb/in.^2$ to kg/cm^2	multiply by 0.07031
TEMPERATURE	
°Fahrenheit to °Centigrade	subtract 32 and multiply by $\frac{5}{9}$
TORQUE	
Foot pounds to meter kilograms	multiply by 0.138
Foot pounds to centimeter kilograms	13.83
Foot pounds to inch pounds	12
Inch pounds to meter kilograms	0.0115
Inch pounds to centimeter kilograms	1.153
Foot pounds to Newton meters	1.3558
Inch pounds to Newton meters	.11298

Conversion Tables

CONVERSION FORMULAS

For use in obtaining **approximate** length equivalents, English to metric and metric to English.

inches = millimeters × 0.04	millimeters = inches × 25
inches = centimeters × 0.4	centimeters = feet × 30
yards = meters × 1.1	meters = yards × 0.9
miles = kilometers × 0.6	kilometers = miles × 1.6

FRACTIONS TO DECIMALS TO MILLIMETERS

Inches		
Fraction	Decimal	mm
1/64	0.0156	0.3969
1/32	0.0312	0.7938
3/64	0.0469	1.1906
1/16	0.0625	1.5875
5/64	0.0781	1.9844
3/32	0.0938	2.3812
7/64	0.1094	2.7781
1/8	0.1250	3.1750
9/64	0.1406	3.5719
5/32	0.1562	3.9688
11/64	0.1719	4.3656
3/16	0.1875	4.7625
13/64	0.2031	5.1594
7/32	0.2188	5.5562
15/64	0.2344	5.9531
1/4	0.2500	6.3500
17/64	0.2656	6.7469
9/32	0.2812	7.1438
19/64	0.2969	7.5406
5/16	0.3125	7.9375
21/64	0.3281	8.3344
11/32	0.3438	8.7312
23/64	0.3594	9.1281
3/8	0.3750	9.5250
25/64	0.3906	9.9219
13/32	0.4062	10.3188
27/64	0.4219	10.7156
7/16	0.4375	11.1125
29/64	0.4531	11.5094
15/32	0.4688	11.9062
31/64	0.4844	12.3031
1/2	0.5000	12.7000
33/64	0.5156	13.0969
17/32	0.5312	13.4938
35/64	0.5469	13.8906
9/16	0.5625	14.2875
37/64	0.5781	14.6844
19/32	0.5938	15.0812
39/64	0.6094	15.4781
5/8	0.6250	15.8750
41/64	0.6406	16.2719
21/32	0.6562	16.6688
43/64	0.6719	17.0656
11/16	0.6875	17.4625
45/64	0.7031	17.8594
23/32	0.7188	18.2562
47/64	0.7344	18.6531
3/4	0.7500	19.0500
49/64	0.7656	19.4469
25/32	0.7812	19.8437
51/64	0.7969	20.2406
13/16	0.8125	20.6375
53/64	0.8281	21.0344
27/32	0.8438	21.4312
55/64	0.8594	21.8281
7/8	0.8750	22.2250
57/64	0.8906	22.6219
29/32	0.9062	23.0188
59/64	0.9219	23.4156
15/16	0.9375	23.8125
61/64	0.9531	24.2094
31/32	0.9688	24.6062
63/64	0.9844	25.0031
1	1.0000	25.4000

METRIC TO ENGLISH CONVERSION/0.01 TO 2.50 mm

mm	Inch	mm	Inch	mm	Inch	mm	Inch	mm	Inch
0.01	.00039	0.51	.02008	1.01	.03976	1.51	.05945	2.01	.07913
0.02	.00079	0.52	.02047	1.02	.04016	1.52	.05984	2.02	.07953
0.03	.00118	0.53	.02087	1.03	.04055	1.53	.06024	2.03	.07992
0.04	.00157	0.54	.02126	1.04	.04094	1.54	.06063	2.04	.08032
0.05	.00197	0.55	.02165	1.05	.04134	1.55	.06102	2.05	.08071
0.06	.00236	0.56	.02205	1.06	.04173	1.56	.06142	2.06	.08110
0.07	.00276	0.57	.02244	1.07	.04213	1.57	.06181	2.07	.08150
0.08	.00315	0.58	.02283	1.08	.04252	1.58	.06220	2.08	.08189
0.09	.00354	0.59	.02323	1.09	.04291	1.59	.06260	2.09	.08228
0.10	.00394	0.60	.02362	1.10	.04331	1.60	.06299	2.10	.08268
0.11	.00433	0.61	.02402	1.11	0.4370	1.61	.06339	2.11	.08307
0.12	.00472	0.62	.02441	1.12	.04409	1.62	.06378	2.12	.08346
0.13	.00512	0.63	.02480	1.13	.04449	1.63	.06417	2.13	.08386
0.14	.00551	0.64	.02520	1.14	.04488	1.64	.06457	2.14	.08425
0.15	.00591	0.65	.02559	1.15	.04528	1.65	.06496	2.15	.08465
0.16	.00630	0.66	.02598	1.16	.04567	1.66	.06535	2.16	.08504
0.17	.00669	0.67	.02638	1.17	.04606	1.67	.06575	2.17	.08543
0.18	.00709	0.68	.02677	1.18	.04646	1.68	.06614	2.18	.08583
0.19	.00748	0.69	.02717	1.19	.04685	1.69	.06654	2.19	.08622
0.20	.00787	0.70	.02756	1.20	.04724	1.70	.06693	2.20	.08661
0.21	.00827	0.71	.02795	1.21	.04764	1.71	.06732	2.21	.08701
0.22	.00866	0.72	.02835	1.22	.04803	1.72	.06772	2.22	.08740
0.23	.00906	0.73	.02874	1.23	.04843	1.73	.06811	2.23	.08780
0.24	.00945	0.74	.02913	1.24	.04882	1.74	.06850	2.24	.08819
0.25	.00984	0.75	.02953	1.25	.04921	1.75	.06890	2.25	.08858
0.26	.01024	0.76	.02992	1.26	.04961	1.76	.06929	2.26	.08898
0.27	.01063	0.77	.03032	1.27	.05000	1.77	.06969	2.27	.08937
0.28	.01102	0.78	.03071	1.28	.05039	1.78	.07008	2.28	.08976
0.29	.01142	0.79	.03110	1.29	.05079	1.79	.07047	2.29	.09016
0.30	.01181	0.80	.03150	1.30	.05118	1.80	.07087	2.30	.09055
0.31	.01220	0.81	.03189	1.31	.05157	1.81	.07126	2.31	.09094
0.32	.01260	0.82	.03228	1.32	.05197	1.82	.07165	2.32	.09134
0.33	.01299	0.83	.03268	1.33	.05236	1.83	.07205	2.33	.09173
0.34	.01339	0.84	.03307	1.34	.05276	1.84	.07244	2.34	.09213
0.35	.01378	0.85	.03346	1.35	.05315	1.85	.07283	2.35	.09252
0.36	.01417	0.86	.03386	1.36	.05354	1.86	.07323	2.36	.09291
0.37	.01457	0.87	.03425	1.37	.05394	1.87	.07362	2.37	.09331
0.38	.01496	0.88	.03465	1.38	.05433	1.88	.07402	2.38	.09370
0.39	.01535	0.89	.03504	1.39	.05472	1.89	.07441	2.39	.09409
0.40	.01575	0.90	.03543	1.40	.05512	1.90	.07480	2.40	.09449
0.41	.01614	0.91	.03583	1.41	.05551	1.91	.07520	2.41	.09488
0.42	.01654	0.92	.03622	1.42	.05591	1.92	.07559	2.42	.09528
0.43	.01693	0.93	.03661	1.43	.05630	1.93	.07598	2.43	.09567
0.44	.01732	0.94	.03701	1.44	.05669	1.94	.07638	2.44	.09606
0.45	.01772	0.95	.03740	1.45	.05709	1.95	.07677	2.45	.09646
0.46	.01811	0.96	.03780	1.46	.05748	1.96	.07717	2.46	.09685
0.47	.01850	0.97	.03819	1.47	.05787	1.97	.07756	2.47	.09724
0.48	.01890	0.98	.03858	1.48	.05827	1.98	.07795	2.48	.09764
0.49	.01929	0.99	.03898	1.49	.05866	1.99	.07835	2.49	.09803
0.50	.01969	1.00	.03937	1.50	.05906	2.00	.07874	2.50	.09843

METRIC TO ENGLISH CONVERSION/2.51 TO 5.00 mm

mm	Inch	mm	Inch	mm	Inch	mm	Inch	mm	Inch
2.51	.09882	3.01	.11850	3.51	.13819	4.01	.15787	4.51	.17756
2.52	.09921	3.02	.11890	3.52	.13858	4.02	.15827	4.52	.17795
2.53	.09961	3.03	.11929	3.53	.13898	4.03	.15866	4.53	.17835
2.54	.10000	3.04	.11969	3.54	.13937	4.04	.15906	4.54	.17874
2.55	.10039	3.05	.12008	3.55	.13976	4.05	.15945	4.55	.17913
2.56	.10079	3.06	.12047	3.56	.14016	4.06	.15984	4.56	.17953
2.57	.10118	3.07	.12087	3.57	.14055	4.07	.16024	4.57	.17992
2.58	.10158	3.08	.12126	3.58	.14095	4.08	.16063	4.58	.18032
2.59	.10197	3.09	.12165	3.59	.14134	4.09	.16102	4.59	.18071
2.60	.10236	3.10	.12205	3.60	.14173	4.10	.16142	4.60	.18110
2.61	.10276	3.11	.12244	3.61	.14213	4.11	.16181	4.61	.18150
2.62	.10315	3.12	.12284	3.62	.14252	4.12	.16221	4.62	.18189
2.63	.10354	3.13	.12323	3.63	.14291	4.13	.16260	4.63	.18228
2.64	.10394	3.14	.12362	3.64	.14331	4.14	.16299	4.64	.18268
2.65	.10433	3.15	.12402	3.65	.14370	4.15	.16339	4.65	.18307
2.66	.10472	3.16	.12441	3.66	.14409	4.16	.16378	4.66	.18347
2.67	.10512	3.17	.12480	3.67	.14449	4.17	.16417	4.67	.18386
2.68	.10551	3.18	.12520	3.68	.14488	4.18	.16457	4.68	.18425
2.69	.10591	3.19	.12559	3.69	.14528	4.19	.16496	4.69	.18465
2.70	.10630	3.20	.12598	3.70	.14567	4.20	.16535	4.70	.18504
2.71	.10669	3.21	.12638	3.71	.14606	4.21	.16575	4.71	.18543
2.72	.10709	3.22	.12677	3.72	.14646	4.22	.16614	4.72	.18583
2.73	.10748	3.23	.12717	3.73	.14685	4.23	.16654	4.73	.18622
2.74	.10787	3.24	.12756	3.74	.14724	4.24	.16693	4.74	.18661
2.75	.10827	3.25	.12795	3.75	.14764	4.25	.16732	4.75	.18701
2.76	.10866	3.26	.12835	3.76	.14803	4.26	.16772	4.76	.18740
2.77	.10906	3.27	.12874	3.77	.14843	4.27	.16811	4.77	.18780
2.78	.10945	3.28	.12913	3.78	.14882	4.28	.16850	4.78	.18819
2.79	.10984	3.29	.12953	3.79	.14921	4.29	.16890	4.79	.18858
2.80	.11024	3.30	.12992	3.80	.14961	4.30	.16929	4.80	.18898
2.81	.11063	3.31	.13032	3.81	.15000	4.31	.16969	4.81	.18937
2.82	.11102	3.32	.13071	3.82	.15039	4.32	.17008	4.82	.18976
2.83	.11142	3.33	.13110	3.83	.15079	4.33	.17047	4.83	.19016
2.84	.11181	3.34	.13150	3.84	.15118	4.34	.17087	4.84	.19055
2.85	.11221	3.35	.13189	3.85	.15158	4.35	.17126	4.85	.19095
2.86	.11260	3.36	.13228	3.86	.15197	4.36	.17165	4.86	.19134
2.87	.11299	3.37	.13268	3.87	.15236	4.37	.17205	4.87	.19173
2.88	.11339	3.38	.13307	3.88	.15276	4.38	.17244	4.88	.19213
2.89	.11378	3.39	.13347	3.89	.15315	4.39	.17284	4.89	.19252
2.90	.11417	3.40	.13386	3.90	.15354	4.40	.17323	4.90	.19291
2.91	.11457	3.41	.13425	3.91	.15394	4.41	.17362	4.91	.19331
2.92	.11496	3.42	.13465	3.92	.15433	4.42	.17402	4.92	.19370
2.93	.11535	3.43	.13504	3.93	.15472	4.43	.17441	4.93	.19409
2.94	.11575	3.44	.13543	3.94	.15512	4.44	.17480	4.94	.19449
2.95	.11614	3.45	.13583	3.95	.15551	4.45	.17520	4.95	.19488
2.96	.11654	3.46	.13622	3.96	.15591	4.46	.17559	4.96	.19528
2.97	.11693	3.47	.13661	3.97	.15630	4.47	.17598	4.97	.19567
2.98	.11732	3.48	.13701	3.98	.15669	4.48	.17638	4.98	.19606
2.99	.11772	3.49	.13740	3.99	.15709	4.49	.17677	4.99	.19646
3.00	.11811	3.50	.13780	4.00	.15748	4.50	.17717	5.00	.19685

METRIC TO ENGLISH CONVERSION/5.01 TO 7.50 mm

mm	Inch	mm	Inch	mm	Inch	mm	Inch	mm	Inch
5.01	.19724	5.51	.21693	6.01	.23661	6.51	.25630	7.01	.27598
5.02	.19762	5.52	.21732	6.02	.23701	6.52	.25669	7.02	.27638
5.03	.19803	5.53	.21772	6.03	.23740	6.53	.25709	7.03	.27677
5.04	.19843	5.54	.21811	6.04	.23780	6.54	.25748	7.04	.27717
5.05	.19882	5.55	.21850	6.05	.23819	6.55	.25787	7.05	.27756
5.06	.19921	5.56	.21890	6.06	.23858	6.56	.25827	7.06	.27795
5.07	.19961	5.57	.21929	6.07	.23898	6.57	.25866	7.07	.27835
5.08	.20000	5.58	.21969	6.08	.23937	6.58	.25906	7.08	.27874
5.09	.20039	5.59	.22008	6.09	.23976	6.59	.25945	7.09	.27913
5.10	.20079	5.60	.22047	6.10	.24016	6.60	.25984	7.10	.27953
5.11	.20118	5.61	.22087	6.11	.24055	6.61	.26024	7.11	.27992
5.12	.20158	5.62	.22126	6.12	.24095	6.62	.26063	7.12	.28032
5.13	.20197	5.63	.22165	6.13	.24134	6.63	.26102	7.13	.28071
5.14	.20236	5.64	.22205	6.14	.24173	6.64	.26142	7.14	.28110
5.15	.20276	5.65	.22244	6.15	.24213	6.65	.26181	7.15	.28150
5.16	.20315	5.66	.22284	6.16	.24252	6.66	.26221	7.16	.28189
5.17	.20354	5.67	.22323	6.17	.24291	6.67	.26260	7.17	.28228
5.18	.20394	5.68	.22362	6.18	.24331	6.68	.26299	7.18	.28268
5.19	.20433	5.69	.22402	6.19	.24370	6.69	.26339	7.19	.28307
5.20	.20472	5.70	.22441	6.20	.24409	6.70	.26378	7.20	.28347
5.21	.20512	5.71	.22480	6.21	.24449	6.71	.26417	7.21	.28386
5.22	.20551	5.72	.22520	6.22	.24488	6.72	.26457	7.22	.28425
5.23	.20591	5.73	.22559	6.23	.24528	6.73	.26496	7.23	.28465
5.24	.20630	5.74	.22598	6.24	.24567	6.74	.26535	7.24	.28504
5.25	.20669	5.75	.22638	6.25	.24606	6.75	.26575	7.25	.28543
5.26	.20709	5.76	.22677	6.26	.24646	6.76	.26614	7.26	.28583
5.27	.20748	5.77	.22717	6.27	.24685	6.77	.26654	7.27	.28622
5.28	.20787	5.78	.22756	6.28	.24724	6.78	.26693	7.28	.28661
5.29	.20827	5.79	.22795	6.29	.24764	6.79	.26732	7.29	.28701
5.30	.20866	5.80	.22835	6.30	.24803	6.80	.26772	7.30	.28740
5.31	.20906	5.81	.22874	6.31	.24843	6.81	.26811	7.31	.28780
5.32	.20945	5.82	.22913	6.32	.24882	6.82	.26850	7.32	.28819
5.33	.20984	5.83	.22953	6.33	.24921	6.83	.26890	7.33	.28858
5.34	.21024	5.84	.22992	6.34	.24961	6.84	.26929	7.34	.28898
5.35	.21063	5.85	.23032	6.35	.25000	6.85	.26969	7.35	.28937
5.36	.21102	5.86	.23071	6.36	.25039	6.86	.27008	7.36	.28976
5.37	.21142	5.87	.23110	6.37	.25079	6.87	.27047	7.37	.29016
5.38	.21181	5.88	.23150	6.38	.25118	6.88	.27087	7.38	.29055
5.39	.21221	5.89	.23189	6.39	.25158	6.89	.27126	7.39	.29095
5.40	.21260	5.90	.23228	6.40	.25197	6.90	.27165	7.40	.29134
5.41	.21299	5.91	.23268	6.41	.25236	6.91	.27205	7.41	.29173
5.42	.21339	5.92	.23307	6.42	.25276	6.92	.27244	7.42	.29213
5.43	.21378	5.93	.23347	6.43	.25315	6.93	.27284	7.43	.29252
5.44	.21417	5.94	.23386	6.44	.25354	6.94	.27323	7.44	.29291
5.45	.21457	5.95	.23425	6.45	.25394	6.95	.27362	7.45	.29331
5.46	.21496	5.96	.23465	6.46	.25433	6.96	.27402	7.46	.29370
5.47	.21535	5.97	.23504	6.47	.25472	6.97	.27441	7.47	.29409
5.48	.21575	5.98	.23543	6.48	.25512	6.98	.27480	7.48	.29449
5.49	.21614	5.99	.23583	6.49	.25551	6.99	.27520	7.49	.29488
5.50	.21654	6.00	.23622	6.50	.25591	7.00	.27559	7.50	.29528

METRIC TO ENGLISH CONVERSION/7.51 TO 10.00 mm

mm	Inch	mm	Inch	mm	Inch	mm	Inch	mm	Inch
7.51	.29567	8.01	.31535	8.51	.33504	9.01	.35472	9.51	.37441
7.52	.29606	8.02	.31575	8.52	.33543	9.02	.35512	9.52	.37480
7.53	.29646	8.03	.31614	8.53	.33583	9.03	.35551	9.53	.37520
7.54	.29685	8.04	.31654	8.54	.33622	9.04	.35591	9.54	.37559
7.55	.29724	8.05	.31693	8.55	.33661	9.05	.35630	9.55	.37598
7.56	.29764	8.06	.31732	8.56	.33701	9.06	.35669	9.56	.37638
7.57	.29803	8.07	.31772	8.57	.33740	9.07	.35709	9.57	.37677
7.58	.29843	8.08	.31811	8.58	.33780	9.08	.35748	9.58	.37717
7.59	.29882	8.09	.31850	8.59	.33819	9.09	.35787	9.59	.37756
7.60	.29921	8.10	.31890	8.60	.33858	9.10	.35827	9.60	.37795
7.61	.29961	8.11	.31929	8.61	.33898	9.11	.35866	9.61	.37835
7.62	.30000	8.12	.31969	8.62	.33937	9.12	.35906	9.62	.37874
7.63	.30039	8.13	.32008	8.63	.33976	9.13	.35945	9.63	.37913
7.64	.30079	8.14	.32047	8.64	.34016	9.14	.35984	9.64	.37953
7.65	.30118	8.15	.32087	8.65	.34055	9.15	.36024	9.65	.37992
7.66	.30158	8.16	.32126	8.66	.34095	9.16	.36063	9.66	.38032
7.67	.30197	8.17	.32165	8.67	.34134	9.17	.36102	9.67	.38071
7.68	.30236	8.18	.32205	8.68	.34173	9.18	.36142	9.68	.38110
7.69	.30276	8.19	.32244	8.69	.34213	9.19	.36181	9.69	.38150
7.70	.30315	8.20	.32284	8.70	.34252	9.20	.36221	9.70	.38189
7.71	.30354	8.21	.32323	8.71	.34291	9.21	.36260	9.71	.38228
7.72	.30394	8.22	.32362	8.72	.34331	9.22	.36299	9.72	.38268
7.73	.30433	8.23	.32402	8.73	.34370	9.23	.36339	9.73	.38307
7.74	.30472	8.24	.32441	8.74	.34409	9.24	.36378	9.74	.38347
7.75	.30512	8.25	.32480	8.75	.34449	9.25	.36417	9.75	.38386
7.76	.30551	8.26	.32520	8.76	.34488	9.26	.26457	9.76	.38425
7.77	.30591	8.27	.32559	8.77	.34528	9.27	.36496	9.77	.38465
7.78	.30630	8.28	.32598	8,78	.34567	9.28	.36535	9.78	.38504
7.79	.30669	8.29	.32638	8.79	.34606	9.29	.36575	9.79	.38543
7.80	.30709	8.30	.32677	8.80	.34646	9.30	.36614	9.80	.38583
7.81	.30748	8.31	.32717	8.81	.34685	9.31	.36654	9.81	.38622
7.82	.30787	8.32	.32756	8.82	.34724	9.32	.36693	9.82	.38661
7.83	.30827	8.33	.32795	8.83	.34764	9.33	.36732	9.83	.38701
7.84	.30866	8.34	.32835	8.84	.34803	9.34	.36772	9.84	.38740
7.85	.30906	8.35	.32874	8.85	.34843	9.35	.36811	9.85	.38780
7.86	.30945	8.36	.32913	8.86	.34882	9.36	.36850	9.86	.38819
7.87	.30984	8.37	.32953	8.87	.34921	9.37	.36890	9.87	.38858
7.88	.31024	8.38	.32992	8.88	.34961	9.38	.36929	9.88	.38898
7.89	.31063	8.39	.33032	8.89	.35000	9.39	.36969	9.89	.38937
7.90	.31102	8.40	.33071	8.90	.35039	9.40	.37008	9.90	.38976
7.91	.31142	8.41	.33110	8.91	.35079	9.41	.37047	9.91	.39016
7.92	.31181	8.42	.33150	8.92	.35118	9.42	.37087	9.92	.39055
7.93	.31221	8.43	.33189	8.93	.35158	9.43	.37126	9.93	.39095
7.94	.31260	8.44	.33228	8.94	.35197	9.44	.37165	9.94	.39134
7.95	.31299	8.45	.33268	8.95	.35236	9.45	.37205	9.95	.39173
7.96	.31339	8.46	.33307	8.96	.35276	9.46	.37244	9.96	.39213
7.97	.31378	8.47	.33347	8.97	.35315	9.47	.37284	9.97	.39252
7.98	.31417	8.48	.33386	8.98	.35354	9.48	.37323	9.98	.39291
7.99	.31457	8.49	.33425	8.99	.35394	9.49	.37362	9.99	.39331
8.00	.31496	8.50	.33465	9.00	.35433	9.50	.37402	10.00*	.39370

* 10.00 millimeters = 1.0 centimeter.

METRIC TO ENGLISH CONVERSION/11 TO 1000 mm

mm	Inch	mm	Inch	mm	Inch	mm	Inch
11	.43307	36	1.41732	61	2.40157	86	3.38583
12	.47244	37	1.45669	62	2.44094	87	3.42520
13	.51181	38	1.49606	63	2.48031	88	3.46457
14	.55118	39	1.53543	64	2.51969	89	3.50394
15	.59055	40	1.57480	65	2.55906	90	3.54331
16	.62992	41	1.61417	66	2.59843	91	3.58268
17	.66929	42	1.65354	67	2.63780	92	3.62205
18	.70866	43	1.69291	68	2.67717	93	3.66142
19	.74803	44	1.73228	69	2.71654	94	3.70079
20	.78740	45	1.77165	70	2.75591	95	3.74016
21	.82677	46	1.81102	71	2.79528	96	3.77953
22	.86614	47	1.85039	72	2.83465	97	3.81890
23	.90551	48	1.88976	73	2.87402	98	3.85827
24	.94488	49	1.92913	74	2.91339	99	3.89764
25	.98425	50	1.96850	75	2.95276	100	3.93701
26	1.02362	51	2.00787	76	2.99213	200	7.87402
27	1.06299	52	2.04724	77	3.03150	300	11.81102
28	1.10236	53	2.08661	78	3.07087	400	15.74803
29	1.14173	54	2.12598	79	3.11024	500	19.68504
30	1.18110	55	2.16535	80	3.14961	600	23.62205
31	1.22047	56	2.20472	81	3.18898	700	27.55906
32	1.25984	57	2.24409	82	3.22835	800	31.49606
33	1.29921	58	2.28346	83	3.26772	900	35.43307
34	1.33858	59	2.32283	84	3.30709	1000*	39.37008
35	1.37795	60	2.36220	85	3.34646		

* 1000 millimeters = 1 meter

METRIC TO ENGLISH CONVERSION/.001 TO .249 INCH

Inch	mm	Inch	mm	Inch	mm	Inch	mm	Inch	mm
.001	0.0254	.044	1.1176	0.96	2.4384	.148	3.7592	.199	5.0546
.0015	9.0381	.045	1.1430	0.97	2.4638	.149	3.7846	.200	5.0800
.002	0.0508	.046	1.1684	.098	2.4892	.150	3.8100	.201	5.1054
.0025	0.0635	.047	1.1938	.099	2.5146	.151	3.8354	.202	5.1308
.003	0.0762	.048	1.2192	.100	2.5400	.152	3.8608	.203	5.1562
.0035	0.0889	.049	1.2446	.101	2.5654	.153	3.8862	.204	5.1816
.004	0.1016	.050	1.2700	.102	2.5908	.154	3.9116	.205	5.2070
.0045	0.1143	.051	1.2954	.103	2.6162	.155	3.9370	.206	5.2324
.005	0.1270	.052	1.3208	.104	2.6416	.156	3.9624	.207	5.2578
.0055	0.1397	.053	1.3462	.105	2.6670	.157	3.9878	.208	5.2832
.006	0.1524	.054	1.3716	.106	2.6924	.158	4.0132	.209	5.3086
.0065	0.1651	.055	1.3970	.107	2.7178	.159	4.0386	.210	5.3340
.007	0.1778	.056	1.4224	.108	2.7432	.160	4.0640	.211	5.3594
.0075	0.1905	.057	1.4478	.109	2.7686	.161	4.0894	.212	5.3848
.008	0.2032	.058	1.4732	.110	2.7940	.162	4.1148	.213	5.4102
.0085	0.2159	.059	1.4986	.111	2.8194	.163	4.1402	.214	5.4356
.009	0.2286	.060	1.5240	.112	2.8448	.164	4.1656	.215	5.4610
.0095	0.2413	.061	1.5494	.113	2.8702	.165	4.1910	.216	5.4864
.010	0.2540	.062	1.5748	.114	2.8956	.166	4.2164	.217	5.5118
.011	0.2794	.063	1.6002	.115	2.9210	.167	4.2418	.218	5.5372
.012	0.3048	.064	1.6256	.116	2.9464	.168	4.2672	.219	5.5626
.013	0.3302	.065	1.6510	.117	2.9718	.169	4.2926	.220	5.5880
.014	0.3556	.066	1.6764	.118	2.9972	.170	4.3180	.221	5.6134
.015	0.3810	.067	1.7018	.119	3.0226	.171	4.3434	.222	5.6388
.016	0.4064	.068	1.7272	.120	3.0480	.172	4.3688	.223	5.6642
.017	0.4318	.069	1.7526	.121	3.0734	.173	4.3942	.224	5.6896
.018	0.4572	.070	1.7780	.122	3.0988	.174	4.4196	.225	5.7150
.019	0.4826	.071	1.8034	.123	3.1242	.175	4.4450	.226	5.7404
.020	0.5080	.072	1.8288	.124	3.1496	.176	4.4704	.227	5.7658
.021	0.5334	.073	1.8542	.125	3.1750	.177	4.4958	.228	5.7912
.022	0.5588	.074	1.8796	.126	3.2004	.178	4.5212	.229	5.8166
.023	0.5842	.075	1.9050	.127	3.2258	.179	4.5466	.230	5.8420
.024	0.6096	.076	1.9304	.128	3.2512	.180	4.5720	.231	5.8674
.025	0.6350	.077	1.9558	.129	3.2766	.181	4.5974	.232	5.8928
.026	0.6604	.078	1.9812	.130	3.3020	.182	4.6228	.233	5.9182
.027	0.6858	.079	2.0066	.131	3.3274	.183	4.6482	.234	5.9436
.028	0.7112	.080	2.0320	.132	3.3528	.184	4.6736	.235	5.9690
.029	0.7366	.081	2.0574	.133	3.3782	.185	4.6990	.236	5.9944
.030	0.7620	.082	2.0828	.134	3.4036	.186	4.7244	.237	6.0198
.031	0.7874	.083	2.1082	.135	3.4290	.187	4.7498	.238	6.0452
.032	0.8128	.084	2.1336	.136	3.4544	.188	4.7752	.239	6.0706
.033	0.8382	.085	2.1590	.137	3.4798	.189	4.8006	.240	6.0960
.034	0.8636	.086	2.1844	.138	3.5052	.190	4.8260	.241	6.1214
.035	0.8890	.087	2.2098	.139	3.5306	.191	4.8514	.242	6.1468
.036	0.9144	.088	2.2352	.140	3.5560	.192	4.8768	.243	6.1722
.037	0.9398	.089	2.2606	.141	3.5814	.193	4.9022	.244	6.1976
.038	0.9652	.090	2.2860	.142	3.6068	.194	4.9276	.245	6.2230
.039	0.9906	.091	2.3114	.143	3.6322	.195	4.9530	.246	6.2484
.040	1.0160	.092	2.3368	.144	3.6576	.196	4.9784	.247	6.2738
.041	1.0414	.093	2.3622	.145	3.6830	.197	5.0038	.248	6.2992
.042	1.0668	.094	2.3876	.146	3.7084	.198	5.0292	.249	6.3246
.043	1.0922	.095	2.4130	.147	3.7338				

ENGLISH TO METRIC CONVERSION/.250 TO .504 INCH

Inch	mm	Inch	mm	Inch	mm	Inch	mm	Inch	mm
.250	6.3500	.301	7.6454	.352	8.9408	.403	10.2362	.454	11.5316
.251	6.3754	.302	7.6708	.353	8.9662	.404	10.2616	.455	11.5570
.252	6.4008	.303	7.6962	.354	8.9916	.405	10.2870	.456	11.5824
.253	6.4262	.304	7.7216	.355	9.0170	.406	10.3124	.457	11.6078
.254	6.4516	.305	7.7470	.356	9.0424	.407	10.3378	.458	11.6332
.255	6.4770	.306	7.7724	.357	9.0678	.408	10.3632	.459	11.6586
.256	6.5024	.307	7.7978	.358	9.0932	.409	10.3886	.460	11.6840
.257	6.5278	.308	7.8232	.359	9.1186	.410	10.4140	.461	11.7094
.258	6.5532	.309	7.8486	.360	9.1440	.411	10.4394	.462	11.7348
.259	6.5786	.310	7.8740	.361	9.1694	.412	10.4648	.463	11.7602
.260	6.6040	.311	7.8994	.362	9.1948	.413	10.4902	.464	11.7856
.261	6.6294	.312	7.9248	.363	9.2202	.414	10.5156	.465	11.8110
.262	6.6548	.313	7.9502	.364	9.2456	.415	10.5410	.466	11.8364
.263	6.6802	.314	7.9756	.365	9.2710	.416	10.5664	.467	11.8618
.264	6.7056	.315	8.0010	.366	9.2964	.417	10.5918	.468	11.8872
.265	6.7310	.316	8.0264	.367	9.3218	.418	10.6172	.469	11.9126
.266	6.7564	.317	8.0518	.368	9.3472	.419	10.6426	.470	11.9380
.267	6.7818	.318	8.0772	.369	9.3726	.420	10.6680	.471	11.9634
.268	6.8072	.319	8.1026	.370	9.3980	.421	10.6934	.472	11.9888
.269	6.8326	.320	8.1280	.371	9.4234	.422	10.7188	.473	12.0142
.270	6.8580	.321	8.1534	.372	9.4488	.423	10.7442	.474	12.0396
.271	6.8834	.322	8.1788	.373	9.4742	.424	10.7696	.475	12.0650
.272	6.9088	.323	8.2042	.374	9.4996	.425	10.7950	.476	12.0904
.273	6.9342	.324	8.2296	.375	9.5250	.426	10.8204	.477	12.1158
.274	6.9596	.325	8.2550	.376	9.5504	.427	10.8458	.478	12.1412
.275	6.9850	.326	8.2804	.377	9.5758	.428	10.8712	.479	12.1666
.276	7.0104	.327	8.3058	.378	9.6012	.429	10.8966	.480	12.1920
.277	7.0358	.328	8.3312	.379	9.6266	.430	10.9220	.481	12.2174
.278	7.0612	.329	8.3566	.380	9.6520	.431	10.9474	.482	12.2428
.279	7.0866	.330	8.3820	.381	9.6774	.432	10.9728	.483	12.2682
.280	7.1120	.331	8.4074	.382	9.7028	.433	10.9982	.484	12.2936
.281	7.1374	.332	8.4328	.383	9.7282	.434	11.0236	.485	12.3190
.282	7.1628	.333	8.4582	.384	9.7536	.435	11.0490	.486	12.3444
.283	7.1882	.334	8.4836	.385	9.7790	.436	11.0744	.487	12.3698
.284	7.2136	.335	8.5090	.386	9.8044	.437	11.0998	.488	12.3952
.285	7.2390	.336	8.5344	.387	9.8298	.438	11.1252	.489	12.4206
.286	7.2644	.337	8.5598	.388	9.8552	.439	11.1506	.490	12.4460
.287	7.2898	.338	8.5852	.389	9.8806	.440	11.1760	.491	12.4714
.288	7.3152	.339	8.6106	.390	9.9060	.441	11.2014	.492	12.4968
.289	7.3406	.340	8.6360	.391	9.9314	.442	11.2268	.493	12.5222
.290	7.3660	.341	8.6614	.392	9.9568	.443	11.2522	.494	12.5476
.291	7.3914	.342	8.6868	.393	9.9822	.444	11.2776	.495	12.5730
.292	7.4168	.343	8.7122	.394	10.0076	.445	11.3030	.496	12.5984
.293	7.4422	.344	8.7376	.395	10.0330	.446	11.3284	.497	12.6238
.294	7.4676	.345	8.7630	.396	10.0584	.447	11.3538	.498	12.6492
.295	7.4930	.346	8.7884	.397	10.0838	.448	11.3792	.499	12.6746
.296	7.5184	.347	8.8138	.398	10.1092	.449	11.4046	.500	12.7000
.297	7.5438	.348	8.8392	.399	10.1346	.450	11.4300	.501	12.7254
.298	7.5692	.349	8.8646	.400	10.1600	.451	11.4554	.502	12.7508
.299	7.5946	.350	8.8900	.401	10.1854	.452	11.4808	.503	12.7762
.300	7.6200	.351	8.9154	.402	10.2108	.453	11.5062	.504	12.8016

ENGLISH TO METRIC CONVERSION/.505 TO .758 INCH

Inch	mm	Inch	mm	Inch	mm	Inch	mm	Inch	mm
.505	12.8270	.555	14.0970	.606	15.3924	.657	16.6878	.708	17.9832
.506	12.8524	.556	14.1224	.607	15.4178	.658	16.7132	.709	18.0086
.507	12.8778	.557	14.1478	.608	15.4432	.659	16.7386	.710	18.0340
.508	12.9032	.558	14.1732	.609	15.4686	.660	16.7640	.711	18.0594
.509	12.9286	.559	14.1986	.610	15.4940	.661	16.7894	.712	18.0848
.510	12.9540	.560	14.2240	.611	15.5194	.662	16.8148	.713	18.1102
.511	12.9794	.561	14.2494	.612	15.5448	.663	16.8402	.714	18.1356
.512	13.0048	.562	14.2748	.613	15.5702	.664	16.8656	.715	18.1610
.513	13.0302	.563	14.3002	.614	15.5956	.665	16.8910	.716	18.1864
.514	13.0556	.564	14.3256	.615	15.6210	.666	16.9164	.717	18.2118
.515	13.0810	.565	14.3510	.616	15.6464	.667	16.9418	.718	18.2372
.516	13.1064	.566	14.3764	.617	15.6718	.668	16.9672	.719	18.2626
.517	13.1318	.567	14.4018	.618	15.6972	.669	16.9926	.720	18.2880
.518	13.1572	.568	14.4272	.619	15.7226	.670	17.0180	.721	18.3134
.519	13.1826	.569	14.4526	.620	15.7480	.671	17.0434	.722	18.3388
.520	13.2080	.570	14.4780	.621	15.7734	.672	17.0688	.723	18.3642
.521	13.2334	.571	14.5034	.622	15.7988	.673	17.0942	.724	18.3896
.522	13.2588	.572	14.5288	.623	15.8242	.674	17.1196	.725	18.4150
.523	13.2842	.573	14.5542	.624	15.8496	.675	17.1450	.726	18.4404
.524	13.3096	.574	14.5796	.625	15.8750	.676	17.1704	.727	18.4658
.525	13.3350	.575	14.6050	.626	15.9004	.677	17.1958	.728	18.4912
.526	13.3604	.576	14.6304	.627	15.9258	.678	17.2212	.729	18.5166
.527	13.3858	.577	14.6558	.628	15.9512	.679	17.2466	.730	18.5420
.528	13.4112	.578	14.6812	.629	15.9766	.680	17.2720	.731	18.5674
.529	13.4366	.579	14.7066	.630	16.0020	.681	17.2974	.732	18.5928
.530	13.4620	.580	14.7320	.631	16.0274	.682	17.3228	.733	18.6182
.531	13.4874	.581	14.7574	.632	16.0528	.683	17.3482	.734	18.6436
.532	13.5128	.582	14.7828	.633	16.0782	.684	17.3736	.735	18.6690
.533	13.5382	.583	14.8082	.634	16.1036	.685	17.3990	.736	18.6944
.534	13.5636	.584	14.8336	.635	16.1290	.686	17.4244	.737	18.7198
.535	13.5890	.585	14.8590	.636	16.1544	.687	17.4498	.738	18.7452
.536	13.6144	.586	14.8844	.637	16.1798	.688	17.4752	.739	18.7706
.537	13.6398	.587	14.9098	.638	16.2052	.689	17.5006	.740	18.7960
.538	13.6652	.588	14.9352	.639	16.2306	.690	17.5260	.741	18.8214
.539	13.6906	.589	14.9606	.640	16.2560	.691	17.5514	.742	18.8468
.540	13.7160	.590	14.9860	.641	16.2814	.692	17.5768	.743	18.8722
.541	13.7414	.591	15.0114	.642	16.3068	.693	17.6022	.744	18.8976
.542	13.7668	.592	15.0368	.643	16.3322	.694	17.6276	.745	18.9230
.543	13.7922	.593	15.0622	.644	16.3576	.695	17.6530	.746	18.9484
.544	13.8176	.594	15.0876	.645	16.3830	.696	17.6784	.747	18.9738
.545	13.8430	.595	15.1130	.646	16.4084	.697	17.7038	.748	18.9992
.546	13.8684	.596	15.1384	.647	16.4338	.698	17.7292	.749	19.0246
.547	13.8938	.597	15.1638	.648	16.4592	.699	17.7546	.750	19.0500
.548	13.9192	.598	15.1892	.649	16.4846	.700	17.7800	.751	19.0754
.549	13.9446	.599	15.2146	.650	16.5100	.701	17.8054	.752	19.1008
.550	13.9700	.600	15.2400	.651	16.5354	.702	17.8308	.753	19.1262
.551	13.9954	.601	15.2654	.652	16.5608	.703	17.8562	.754	19.1516
.552	14.0208	.602	15.2908	.653	16.5862	.704	17.8816	.755	19.1770
.553	14.0462	.603	15.3162	.654	16.6116	.705	17.9070	.756	19.2024
.554	14.0716	.604	15.3416	.655	16.6370	.706	17.9324	.757	19.2278
		.605	15.3670	.656	16.6624	.707	17.9578	.758	19.2532

ENGLISH TO METRIC CONVERSION/.759 TO 12.0 INCHES

Inch	mm
.759	19.2786
.760	19.3040
.761	19.3294
.762	19.3548
.763	19.3802
.764	19.4056
.765	19.4310
.766	19.4564
.767	19.4818
.768	19.5072
.769	19.5326
.770	19.5580
.771	19.5834
.772	19.6088
.773	19.6342
.774	19.6596
.775	19.6850
.776	19.7104
.777	19.7358
.778	19.7612
.779	19.7866
.780	19.8120
.781	19.8374
.782	19.8628
.783	19.8882
.784	19.9136
.785	19.9390
.786	19.9644
.787	19.9898
.788	20.0152
.789	20.0406
.790	20.0660
.791	20.0914
.792	20.1168
.793	20.1422
.794	20.1676
.795	20.1930
.796	20.2184
.797	20.2438
.798	20.2692
.799	20.2946
.800	20.3200
.801	20.3454
.802	20.3708
.803	20.3962
.804	20.4216
.805	20.4470
.806	20.4724
.807	20.4978
.808	20.5232
.809	20.5486

Inch	mm
.810	20.5740
.811	20.5994
.812	20.6248
.813	20.6502
.814	20.6756
.815	20.7010
.816	20.7264
.817	20.7518
.818	20.7772
.819	20.8026
.820	20.8280
.821	20.8534
.822	20.8788
.823	20.9042
.824	20.9296
.825	20.9550
.826	20.9804
.827	21.0058
.828	21.0312
.829	21.0566
.830	21.0820
.831	21.1074
.832	21.1328
.833	21.1582
.834	21.1836
.835	21.2090
.836	21.2344
.837	21.2598
.838	21.2852
.839	21.3106
.840	21.3360
.841	21.3614
.842	21.3868
.843	21.4122
.844	21.4376
.845	21.4630
.846	21.4884
.847	21.5138
.848	21.5392
.849	21.5646
.850	21.5900
.851	21.6154
.852	21.6408
.853	21.6662
.854	21.6916
.855	21.7170
.856	21.7424
.857	21.7678
.858	21.7932
.859	21.8186
.860	21.8440

Inch	mm
.861	21.8694
.862	21.8948
.863	21.9202
.864	21.9456
.865	21.9710
.866	21.9964
.867	22.0218
.868	22.0472
.869	22.0726
.870	22.0980
.871	22.1234
.872	22.1488
.873	22.1742
.874	22.1996
.875	22.2250
.876	22.2504
.877	22.2758
.878	22.3012
.879	22.3266
.880	22.3520
.881	22.3774
.882	22.4028
.883	22.4282
.884	22.4536
.885	22.4790
.886	22.5044
.887	22.5298
.888	22.5552
.889	22.5806
.890	22.6060
.891	22.6314
.892	22.6568
.893	22.6822
.894	22.7076
.895	22.7330
.896	22.7584
.897	22.7838
.898	22.8092
.899	22.8346
.900	22.8600
.901	22.8854
.902	22.9108
.903	22.9362
.904	22.9616
.905	22.9870
.906	23.0124
.907	23.0378
.908	23.0632
.909	23.0886
.910	23.1140
.911	23.1394

Inch	mm
.912	23.1648
.913	23.1902
.914	23.2156
.915	23.2410
.916	23.2664
.917	23.2918
.918	23.3172
.919	23.3426
.920	23.3680
.921	23.3934
.922	23.4188
.923	23.4442
.924	23.4696
.925	23.4950
.926	23.5204
.927	23.5458
.928	23.5712
.929	23.5966
.930	23.6220
.931	23.6474
.932	23.6728
.933	23.6982
.934	23.7236
.935	23.7490
.936	23.7744
.937	23.7998
.938	23.8252
.939	23.8506
.940	23.8760
.941	23.9014
.942	23.9268
.943	23.9522
.944	23.9776
.945	24.0030
.946	24.0284
.947	24.0538
.948	24.0792
.949	24.1046
.950	24.1300
.951	24.1554
.952	24.1808
.953	24.2062
.954	24.2316
.955	24.2570
.956	24.2824
.957	24.3078
.958	24.3332
.959	24.3586
.960	24.3840
.961	24.4094
.962	24.4348

Inch	mm
.963	24.4602
.964	24.4856
.965	24.5110
.966	24.5364
.967	24.5618
.968	24.5872
.969	24.6126
.970	24.6380
.971	24.6634
.972	24.6888
.973	24.7142
.974	24.7396
.975	24.7650
.976	24.7904
.977	24.8158
.978	24.8412
.979	24.8666
.980	24.8920
.981	24.9174
.982	24.9428
.983	24.9682
.984	24.9936
.985	25.0190
.986	25.0444
.987	25.0698
.988	25.0952
.989	25.1206
.990	25.1460
.991	25.1714
.992	25.1968
.993	25.2222
.994	25.2476
.995	25.2730
.996	25.2984
.997	25.3238
.998	25.3492
.999	25.3746
1.0	25.4000
2.0	50.8000
3.0	76.2000
4.0	101.6000
5.0	127.0000
6.0	152.4000
7.0	177.8000
8.0	203.2000
9.0	228.6000
10.0	254.0000
11.0	279.4000
12.0	304.8000

TEMPERATURE CONVERSION

Use this table to convert Fahrenheit degrees (F°) directly to Centigrade degrees (C°) and vice versa. It covers the range of temperatures used in most hardening, tempering and annealing operations.

$$F° = \frac{C° \times 9}{5} + 32$$

Lower, higher and intermediate coversions can be made by substituting a known Fahrenheit (F°) or Centigrade (C°) temperature figure in either of the following formulas.

$$C° = \frac{F° - 32}{9} \times 5$$

F°	C°	F°	C°	F°	C°	F°	C°	F°	C°
−160	−107	340	171	840	449	1340	727	1840	1004
−140	− 96	360	182	860	460	1360	738	1860	1016
−120	− 84	380	193	880	471	1380	749	1880	1027
−100	− 73	400	204	900	482	1400	760	1900	1038
− 80	− 62	420	216	920	493	1420	771	1920	1049
− 60	− 51	440	227	940	504	1440	782	1940	1060
− 40	− 40	460	238	960	516	1460	793	1960	1071
− 20	− 29	480	249	980	527	1480	804	1980	1082
0	− 18	500	260	1000	538	1500	816	2000	1093
20	− 7	520	271	1020	549	1520	827	2020	1104
40	4	540	282	1040	560	1540	838	2040	1116
60	16	560	293	1060	571	1560	849	2060	1127
80	27	580	304	1080	582	1580	860	2080	1138
100	38	600	316	1100	593	1600	871	2100	1149
120	49	620	327	1120	604	1620	882	2120	1160
140	60	640	338	1140	616	1640	893	2140	1171
160	71	660	349	1160	627	1660	904	2160	1182
180	82	680	360	1180	638	1680	916	2180	1193
200	93	700	371	1200	649	1700	927	2200	1204
220	104	720	382	1220	660	1720	938	2220	1216
240	116	740	393	1240	671	1740	949	2240	1227
260	127	760	404	1260	682	1760	960	2260	1238
280	138	780	416	1280	693	1780	971	2280	1249
300	149	800	427	1300	704	1800	982	2300	1260
320	160	820	438	1320	716	1820	993	2320	1271

Index